About Island Press

Island Press is the only nonprofit organization in the United States whose principal purpose is the publication of books on environmental issues and natural resource management. We provide solutions-oriented information to professionals, public officials, business and community leaders, and concerned citizens who are shaping responses to environmental problems.

In 1999, Island Press celebrates its fifteenth anniversary as the leading provider of timely and practical books that take a multidisciplinary approach to critical environmental concerns. Our growing list of titles reflects our commitment to bringing the best of an expanding body of literature to the environmental community throughout North America and the world.

Support for Island Press is provided by The Jenifer Altman Foundation, The Bullitt Foundation, The Mary Flagler Cary Charitable Trust, The Nathan Cummings Foundation, The Geraldine R. Dodge Foundation, The Charles Engelhard Foundation, The Ford Foundation, The Vira I. Heinz Endowment, The W. Alton Jones Foundation, The John D. and Catherine T. MacArthur Foundation, The Andrew W. Mellon Foundation, The Charles Stewart Mott Foundation, The Curtis and Edith Munson Foundation, The National Fish and Wildlife Foundation, The National Science Foundation, The New-Land Foundation, The David and Lucile Packard Foundation, The Pew Charitable Trusts, The Surdna Foundation, The Winslow Foundation, and individual donors.

About World Wildlife Fund

Known worldwide by its panda logo, World Wildlife Fund is dedicated to protecting the world's wildlife and the rich biological diversity that we all need to survive. The leading privately supported international conservation organization in the world, WWF has sponsored more than 2,000 projects in 116 countries and has more than 1 million members in the United States.

Through its Living Planet Campaign, WWF aims to make the remaining days of this century a turning point in the worldwide struggle to preserve species and habitats. The campaign calls on governments, corporations, and others to take significant actions to help preserve the world's endangered spaces—places we call the Global 200; to protect endangered species; and to address the global threats that put all living things in harm's way. To learn more about the Living Planet Campaign, visit the WWF Web site at www.worldwildlife.org or write to us.

Terrestrial Ecoregions
of North America

Terrestrial Ecoregions
of North America

A CONSERVATION ASSESSMENT

Taylor H. Ricketts
Eric Dinerstein
David M. Olson
Colby J. Loucks
William Eichbaum
Dominick DellaSala
Kevin Kavanagh
Prashant Hedao
Patrick T. Hurley
Karen M. Carney
Robin Abell
and Steven Walters

WORLD WILDLIFE FUND—
UNITED STATES AND CANADA

ISLAND PRESS
WASHINGTON, D.C. • COVELO, CALIFORNIA

Library of Congress Cataloging-in-Publication Data
Terrestrial ecoregions of North America : a conservation assessment /
    Taylor H. Ricketts . . . [et al.].
        p.   cm.
    ISBN 1-55963-722-6 (pbk.)
    1. Biological diversity conservation—North America.
    I. Ricketts, Taylor H.
    QH77.N56T47   1999
    333.95′16′097—dc21   99-18912
                                                      CIP

Printed on recycled, acid-free paper ✪ ∞
Using soy-based inks
Manufactured in the United States of America

10 9 8 7 6 5 4 3 2 1

Contents

List of Special Essays

List of Figures

List of Figures

List of Tables

Abbreviations and Acronyms

AFB   Air Force Base
ASL   Above Sea Level
AVHRR   Advanced Very High Resolution Radiometer
BDI   Biological Distinctiveness Index
BLM   Bureau of Land Management
CONABIO   Comisión Nacional para el Conocimiento y Uso de la Biodiversidad
CRP   Conservation Reserves Program
DoD   Department of Defense
EPA   Environmental Protection Agency
EROS   Earth Resources Observation Systems
ESWG   Ecological Stratification Working Group
FSC   Forest Stewardship Council
GAP   Gap Analysis Program
INEGI   Instituto Nacional de Estadística, Geografía, e Informática
IUCN   International Union for the Conservation of Nature (now called World Conservation Union)
MNR   Ministry of Natural Resources
NGO   Nongovernmental Organization
NPS   National Park Service
NWR   National Wildlife Refuge
TEC   Terrestrial Ecoregions of Canada
TNC   The Nature Conservancy
USDA   United States Department of Agriculture
USFS   United States Forest Service
USFWS   United States Fish and Wildlife Service
WRI   World Resources Institute
WSR   Wild and Scenic River
WWF   World Wildlife Fund
WWF-US   World Wildlife Fund–United States

Foreword

With this report, WWF-US and WWF Canada provide a frame of reference for action to conserve biodiversity in our two countries. Through this and other analyses supporting our Latin American program, WWF is also examining conservation needs in Mexico.

Across North America, the demands we all make for land and resources crowd out species, degrade their habitat, and stress the underlying ecological processes that are the lifeblood of natural systems. Conservationists are responding to these threats in many ways. But, by conducting an ecoregion-based assessment of biodiversity, WWF aims to speed up conservation planning and action across our continent. At a minimum, this will help to focus our own efforts as an organization. We hope it will also help knit together the work of others working to safeguard North America, regardless of how they are approaching this challenge. For example, our principal funding partner for this project, the Commission for Environmental Cooperation, has an important role to play in this regard as it exercises its mandate to report on the state of the North American environment.

The key message we hope readers will take away is one of urgency. In some ecoregions this urgency is due to the losses already incurred and the need to hang on to what's left or carry out major efforts at restoration. In other ecoregions, the situation is urgent because the opportunity remains to prevent similar losses from occurring in the first place.

Given this urgency, what is WWF doing? Increasingly, all our conservation programs are designed to achieve ecoregion-based conservation, employing a two-pronged strategy of establishing protected areas and achieving sustainable management of the lands and waters outside protected areas. In the United States, our efforts are now concentrated on urgent conservation issues in five globally outstanding and endangered ecoregions: the Klamath-Siskiyou; southern Florida, including the Everglades; the Chihuahuan Desert; the Bering Sea; and the rivers and streams of the southeastern United States. In Canada, our priority is to complete an ecologically representative system of terrestrial protected areas nationwide by the year 2000, and of marine areas by 2010. In both countries these efforts are reinforced by WWF's campaign to protect the globe's oceans and forests and to reverse global warming.

By preparing this assessment of North America's biodiversity, we now have a road map for conservation action and a yardstick for judging the success or failure of WWF's own conservation work, as well as that of others. We look forward to cooperating with all those who share our mission so that, together, we do succeed.

James Leape   Monte Hummel
Senior Vice President   President
WWF–US   WWF Canada

Preface

This book is part of a long-term effort undertaken by the Conservation Science Program of World Wildlife Fund–United States to conduct conservation assessments of terrestrial, freshwater, and marine ecoregions around the world. This effort began with a biodiversity assessment of the Russian Federation (Krever et al. 1994), followed by Latin America and the Caribbean (Dinerstein et al. 1995), mangrove ecoregions of Latin America (Olson et al. 1996), and freshwater ecoregions of Latin America and the Caribbean (Olson et al. 1997). Forthcoming in the series are conservation assessments of freshwater ecoregions in North America and of terrestrial ecoregions in Asia and Africa. Ecoregions from these and other analyses that were identified as globally outstanding in their biodiversity values recently were combined in a map of the Global 200 Ecoregions (Olson and Dinerstein 1998). As the new millennium begins, the Global 200 will serve as a framework for WWF's activities to conserve our planet's full range of biological diversity.

Eric Dinerstein
Chief Scientist
WWF-US

Acknowledgments

We thank first and foremost the participants in our expert workshop, who contributed their expertise, their years of accrued experience, and their valuable time to the project. Several of these participants, in addition to several other experts, were invited to write the nineteen special essays included throughout this volume. Their contributions have expanded the breadth of the report enormously. All of these contributors are listed on pages 467–470.

Within WWF, many people helped in innumerable ways at every stage of the project. David Schorr has lent his continuous support since the beginning. Belaine Lehman organized with humor and great competence the considerable logistics of a large workshop. The staff of the Conservation Science Program—Emma Underwood, Wes Wettengel, Steven Walters, and Steve Primm—despite the many ongoing projects in the lab, continually found time to help in vital ways when needed.

Gavin McGhie at UC Santa Barbara and Jesslyn Brown at the EROS data center were very helpful and forthcoming in supplying us with advanced versions of their protected areas and AVHRR land-cover coverages, respectively. John Kartesz, director of the Biota of North America Program, greatly enriched our study by sharing extensive databases on the flora of North America.

Reed Noss (Conservation Biology Institute), David Wilcove (Environmental Defense Fund), Larry Master (The Nature Conservancy), John Robinson (Wildlife Conservation Society), and Whitney Tilt (National Fish and Wildlife Foundation) peer-reviewed the final draft. Bill Eichbaum, Arlin Hackman, Dominick DellaSala, and Randall Snodgrass supplied helpful comments on earlier drafts, which have resulted in many improvements.

A number of outstanding regional biologists from The Nature Conservancy have contributed in major ways to the completion of this assessment. In particular, we would like to acknowledge Alan Weakley, Steve Chaplin, Steve Buttrick, Randy Hagenstein, and Sam Gon. Their spirit of cooperation sets a fine example for the rest of us in the conservation community.

The Commission for Environmental Cooperation, and in particular Victor Lichtinger, Janine Ferretti, and Irene Pisanty, provided core support for this project at an early stage, as well as invaluable guidance throughout. We owe them a debt of gratitude. We thank Jim Omernik, Robert Bailey, and their colleagues for review of our mapping effort and for their constructive comments on early drafts of our maps. Tom Born and his colleagues from the U.S. EPA supported this effort, and we are most grateful.

We thank the many authors, cartographers, biogeographers, and conservation biologists upon whose work this assessment is based. They are cited within but deserve thanks here. Additionally, we would like to thank Carla Langeveld, who assisted in finding many of the literature sources used.

Finally, we would particularly like to thank and acknowledge those who have provided support for this conservation effort: The Commission for Environmental Cooperation, Ford Motor Company, WMX, Inc., Environmental Protection Agency, the Center for Conservation Biology at Stanford University, Environmental Systems Research Institute, Hewlett-Packard Company, Melvin B. Lane, and Hope Stevens. Eric Dinerstein was supported in part by the Armand G. Erpf Conservation Fellowship.

Terrestrial Ecoregions
of North America

CHAPTER 1

Introduction

The biodiversity of North America offers many superlatives. Among terrestrial ecosystems, the Chihuahuan Desert supports one of the world's most diverse desert plant communities. The California coastal sage scrub, an ecosystem characterized by extraordinary plant richness and endemism, is one of only five Mediterranean shrublands worldwide. The temperate broadleaf forests of the Appalachian and Blue Ridge Mountains form one of the richest temperate forest regions in the world for plants, land snails, and salamanders. The longleaf pine forests of the southeastern United States support one of the richest herbaceous floras on earth. Mexico harbors the world's most diverse tropical dry forests and subtropical pine-oak forests. The vast tundra, taiga, and boreal forests of Canada and Alaska feature the largest caribou migrations in the world, as well as relatively intact predator-prey systems, increasingly a global rarity (Olson and Dinerstein 1998).

In freshwater and marine ecosystems, the list continues. Among temperate freshwater ecosystems, the streams and rivers of central Appalachia and the southeastern United States contain the largest variety of temperate fish, mussel, and crayfish faunas globally. In the marine realm, the Bering and Beaufort Seas are among the richest polar seas, and the Monterey and Chesapeake Bays support two of the world's largest and most biodiverse coastal marine systems. Globally important migratory fish populations and extraordinary concentrations of seabirds, marine mammals, waterfowl, and shorebirds add to the rich tapestry of the continent's biological wealth (Olson and Dinerstein 1998).

Despite this wealth, most North Americans equate biodiversity with exotic, faraway places, such as the tropical rain forests of Brazil or Indonesia. In truth, tropical rain forests contain the highest concentration of species on earth, harboring perhaps as many as half of all known species (Wilson 1988). However, the other half lives outside tropical rain forests in unique ecosystems and communities reaching to the poles. Losing representative examples of these ecosystems would be a significant loss of biodiversity at the global scale. To help conserve the North American component of this biodiversity, World Wildlife Fund–United States and World Wildlife Fund Canada have undertaken a conservation assessment of terrestrial ecoregions of the United States and Canada as a first step in prioritizing efforts to save them.

1

This effort could not come at a more critical time. As we approach the end of the millennium, many natural habitats in North America are severely threatened by degradation, fragmentation, and outright conversion. Across much of North America, only remnants of natural habitat support the rich flora and fauna of our continent. While conservationists fight to halt the loss of primary rain forests in the tropics, the temperate rain forests of North America are under equally heavy pressure from logging and conversion to tree plantations. Less than 5 percent of the original forests remain unlogged in the continental United States. In the Sierra Madre Occidental of Mexico—part of the richest pine-oak forests in the world—there remain only a few blocks of forest large enough to maintain the rich biota of the area. The Great Plains of the United States and Canada was the most extensive grassland region in the world. Within the last century, tallgrass prairies have been reduced to 1 percent of their historical range (Noss and Peters 1995). If the United States, Canada, and Mexico are to assume leadership roles in global conservation, we must confront an enormous set of challenges and obligations in our own backyard.

Assessment Overview

This book is designed to help the conservation community in North America to be as strategic as possible in confronting these challenges and in conserving the extraordinary biodiversity of our continent. In setting conservation priorities on broad scales, it is important to measure not only the biodiversity value of an area but also the conservation threats it faces. In this assessment, we capture these two main factors with an index of "biological distinctiveness" (incorporating species richness, endemism, and other factors) and of "conservation status" (including habitat loss, degree of existing protection, etc.). By integrating these two indices, we are able to evaluate the combination of biological and threat factors in each area and to recommend appropriate conservation activities accordingly.

Ecoregions, the geographic units of our analysis, are relatively coarse biogeographic divisions of a landscape; they delineate areas that share broadly similar environmental conditions and natural communities (chapter 2). We choose to base our analyses on ecoregions, rather than political jurisdictions such as states or provinces, because ecoregions provide a biologically meaningful geographic framework for biodiversity conservation and management at a broad scale (Bailey 1996). The distributions of communities and habitat types rarely follow political boundaries. For example, the state of California contains habitats ranging from wet temperate rain forests in the northwest, to oak woodlands and grasslands in the central part, Mediterranean shrub in the south, and deserts in the southeast. Dividing California into twelve ecoregions, as we do in this book, offers a framework for representing all of its unique habitats and species assemblages in conservation programs, recognizing the extensions of these habitats into neighboring states and

Mexico, capturing the geographic area over which ecological processes operate, and defining the arena for future restoration programs. In contrast, focusing on state boundaries runs the risk of not only overlooking important features and conservation needs specific to each ecoregion but also investing in redundant or poorly coordinated efforts in ecoregions that span political borders.

Broad-scale conservation is based on the idea that the maximum level of biodiversity will be conserved if the maximum diversity of habitats are represented in protected area networks or other biodiversity management schemes (Scott et al. 1993; Noss and Peters 1995; The Nature Conservancy 1997; Olson and Dinerstein 1998). Representation of the full variety of North American habitat and ecosystem types is therefore essential to the conservation of the continent's biodiversity, and it is thus one of the foundations of our analysis. Biomes such as tundra, deserts, temperate broadleaf forests, and grasslands each harbor distinctive species, ecological processes, and evolutionary phenomena. Although they may not support communities as rich as those seen in tropical rain forests or coral reefs, they contain species assemblages that have adapted to distinct environmental conditions and that reflect unique evolutionary histories. In our assessment, we address representation by dividing North America into ten major habitat types (MHTs, roughly equivalent to biomes) and analyzing each one separately, thus ensuring that all MHTs are represented among the final priority ecoregions.

Some of the information used in our assessment is available from published sources (e.g., species distribution data, which we used to produce species richness and endemism indices for each ecoregion). However, for the most part, information on critical variables at the scale of ecoregions is unavailable in published form. For example, quantitative data on habitat loss over entire ecoregions, available in standardized form for the whole continent, simply do not exist. To gather this kind of important information efficiently, we relied heavily on expert assessment. We convened an expert workshop, at which thirty-five ecologists and conservationists collectively assessed the ecoregions for a variety of biological distinctiveness and conservation status criteria. We asked the experts first to critique our ecoregion boundaries and to adjust them if necessary. Regional sub-groups then assessed the ecoregions and, based on their collective experience and knowledge, placed them in broad categories for each criterion. The entire process was guided by explicit decision rules and category thresholds to standardize the treatment of all ecoregions. In this manner, we were able to gather an enormous amount of subjective, but quantified, information that is unavailable from published sources. Perhaps most importantly, the decision rules and methods for calculating all criteria are contained in appendices in this book, so that the process is as transparent and repeatable as possible.

Studies of this geographic extent are necessarily coarse. Ecoregion boundaries are approximations of what in reality are gradual shifts in ecological communities. The information we have gathered, from published data and the expert workshop, are rough categorical rankings. Nevertheless, this level of data precision is appropriate for the type of analysis we conduct: a broad-scale assessment of biodiversity and the threats facing it. This book is intended to help conservation planners take the first, broad-scale step toward strategically setting priorities in North America; it is not an end in itself. Appropriate conservation policies and activities at finer scales (i.e., within ecoregions or small groups of them) must follow if the value of this study is to be realized (see essay 15 by Reed Noss, pages 89–92).

During the course of the expert workshop, we gathered a wealth of information that should help those interested in taking the more fine-scale approach of within-ecoregion conservation. For each ecoregion, we asked experts to identify the important remaining blocks of original habitat, a set of priority actions that would best advance biodiversity conservation, and major conservation organizations operating in the region. We have compiled these details in the ecoregion descriptions of appendix F. Therefore, this book combines a broad-scale framework and inter-ecoregional assessment with more detailed information on each ecoregion to help direct more local initiatives. We hope it will lead to better choices and implementation of local-scale conservation activities by placing them in the context of continental and global priorities.

Structure of the Book

This project involved an enormous amount of information and a large number of separate decisions and analyses, which we synthesized into the main assessment. Details of the analyses are contained in this book, but we have organized it into three levels so the reader can choose the degree of detail most appropriate to his or her interests. First, the six chapters of the main text describe our assessment approach and summarize the key results and main conclusions of the analyses. Scattered throughout the chapters are nineteen essays, written by workshop participants and other regional experts, that focus on a variety of important conservation issues in North America and worldwide. Second, appendices A through E contain the specific methods used to assess the ecoregions and the details of the results summarized in the main text. Third, appendix F provides a detailed description of each ecoregion, emphasizing its important biological features and conservation threats, in addition to listing the remaining habitat blocks, priority activities, and local conservation organizations mentioned above. Throughout the chapters and appendices, numbers referring to WWF ecoregions are enclosed in square brackets to distinguish them clearly from frequent references to the mapping units of other studies, which are enclosed in parentheses.

The Challenge

As citizens of North America, we have a global responsibility to conserve the biodiversity within and on our shores. Unfortunately, the results of our assessments illustrate that the United States often is doing a worse job of protecting its biological wealth than many countries with fewer resources for biodiversity conservation. The Florida Everglades has received a great deal of attention in both the international and American press as a place where the U.S. government has made a major commitment to conserve and restore a globally outstanding ecoregion. The Everglades, however, is not an isolated case. This study identifies thirteen additional ecoregions that match the Everglades in global biodiversity importance and that face even greater threats (chapter 6). Clearly, conserving and restoring the landscapes and biota of these ecoregions must be given top priority. If the United States and Canada are to earn their places as leaders in international conservation, conserving and restoring the landscapes and biota of these thirteen ecoregions must begin immediately with the same level of commitment as that demonstrated in the Everglades.

The combination of broad- and fine-scale analyses being conducted recently by a host of conservation groups and agencies, such as The Nature Conservancy, Defenders of Wildlife, the WWF Canada Endangered Spaces Campaign, State Gap Analysis programs, and CONABIO in Mexico, has begun to provide a valuable framework and strategic planning tool to help ensure the conservation of North America's biological diversity. Ecoregion-level analyses have much to contribute to this growing collective framework, and ecoregion-based conservation has been adopted by some of the leading conservation groups in North America. This project contributes to that effort by providing an efficient and transparent method of recommending areas for restoration, increased protection, and policy reform to further biodiversity conservation.

The conservation movement in North America is at an important crossroads. We can continue with a business-as-usual approach—engaging in a series of piecemeal, small-scale, and largely reactionary efforts—and bear witness to the rapid erosion of our natural heritage. Or we can choose a more coordinated, integrative path, where local projects are informed by global and regional perspectives and priorities are established more proactively. Restoring the "River of Grass" in the Everglades is a noble beginning. Conserving the extraordinary biodiversity of North America now must become a continent-wide imperative.

CHAPTER 2   **Approach**

Delineation of Ecoregions and Geographic Scope of This Study

An ecoregion is defined as a relatively large area of land or water that contains a geographically distinct assemblage of natural communities. These communities (1) share a large majority of their species, dynamics, and environmental conditions, and (2) function together effectively as a conservation unit at global and continental scales (Dinerstein et al. 1995). This initial study focuses exclusively on terrestrial ecoregions of the United States and Canada; conservation assessments of the terrestrial ecoregions of Mexico (see essay 1), North American freshwater ecoregions (including Mexico), and Pacific Coast marine ecoregions are forthcoming (Abell

---

ESSAY 1

A Conservation Assessment of the Terrestrial Ecoregions of Mexico: A Status Report

David M. Olson

A conservation assessment of North American biodiversity is incomplete without a comprehensive evaluation of the ecoregions of Mexico. A map of Mexican ecoregions was presented in an earlier study (Dinerstein et al. 1995) and was based largely on a habitat classification by Flores et al. (1971). Recent efforts to update ecoregion boundaries—spearheaded by the Comisión Nacional para el Conocimiento y Uso de la Biodiversidad (CONABIO), the Instituto Nacional de Estadística, Geografía e Informática (INEGI), WWF-US, and regional experts—are nearly complete (figure 2.2). The revised work is based on new habitat and ecosystem data and a wealth of information on patterns of biodiversity acquired since 1972 but only recently incorporated into a new ecoregion map.

A detailed description of the biodiversity attributes of the revised Mexican ecoregions is part of a forthcoming effort by CONABIO and its colleagues (J. Soberon, Executive Secretary, CONABIO, pers. comm., 1998). CONABIO, INEGI, and others have already developed habitat and taxa databases for the entire country, and this information will be analyzed within the context of the new ecoregion boundaries. Overall, the portion of North America contained within Mexico harbors a substantial portion of the world's most globally outstanding biodiversity. For a glimpse of how Mexican ecoregions compare to others around the world, see chapter 6, essay 16: "Globally Outstanding Biodiversity in Our Own Backyard."

et al. 1999; Ford, in prep.). Descriptions of how freshwater and marine ecoregions were delineated are presented as short essays at the end of this chapter.

The 116 ecoregions that form the conservation targets for this study (figure 2.1) are largely based on three established ecoregion mapping projects: Omernik (1995b) for the contiguous United States, Ecological Stratification Working Group (ESWG 1995) for Canada, and Gallant et al. (1995) for Alaska. We chose these as foundations because their published maps approximate well-documented patterns of biodiversity in North America north of Mexico.

Figure 2.1   Terrestrial ecoregions of the United States and Canada.

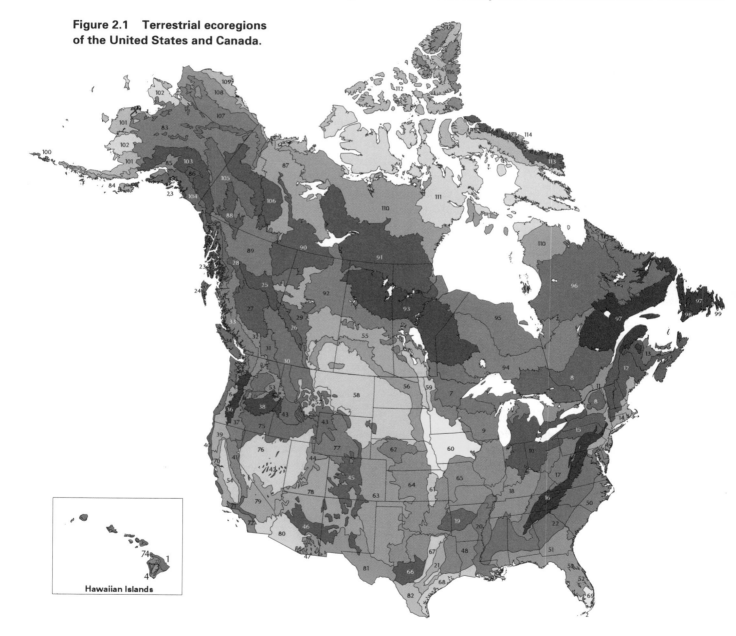

Puerto Rico

We joined ecoregion maps of Canada, the United States, and Mexico (figure 2.2) into a single coherent map that spans international boundaries and then modified certain areas to better reflect patterns of biodiversity. Our modifications are of two general types. First, where biogeographic and biodiversity patterns do not shift substantially across the boundaries of two or more areas mapped as distinct units by the original authors, we combined the areas into a single ecoregion. An example is our Northwestern Mixed Grasslands ecoregion [58], which combines Omernik ecoregions

LEGEND

1.1 Tropical Moist Broadleaf Forests MHT

1. Hawaiian Moist Forests
2. South Florida Rocklands
3. Puerto Rican Moist Forests

1.2 Tropical Dry Broadleaf Forests MHT

4. Hawaiian Dry Forests
5. Puerto Rican Dry Forests

2.1 Temperate Broadleaf and Mixed Forests MHT

6. Willamette Valley Forests
7. Western Great Lakes Forests
8. Eastern Forest/Boreal Transition
9. Upper Midwest Forest/Savanna Transition Zone
10. Southern Great Lakes Forests
11. Eastern Great Lakes Lowland Forests
12. New England/Acadian Forests
13. Gulf of St. Lawrence Lowland Forests
14. Northeastern Coastal Forests
15. Allegheny Highlands Forests
16. Appalachian/Blue Ridge Forests
17. Appalachian Mixed Mesophytic Forests
18. Central U.S. Hardwood Forests
19. Ozark Mountain Forests
20. Mississippi Lowland Forests
21. East Central Texas Forests
22. Southeastern Mixed Forests

2.2 Temperate Coniferous Forests MHT

23. Northern Pacific Coastal Forests
24. Queen Charlotte Islands
25. Central British Columbia Mountain Forests
26. Alberta Mountain Forests
27. Fraser Plateau and Basin Complex
28. Northern Transitional Alpine Forests
29. Alberta/British Columbia Foothills Forests
30. North Central Rockies Forests
31. Okanagan Dry Forests
32. Cascade Mountains Leeward Forests
33. British Columbia Mainland Coastal Forests
34. Central Pacific Coastal Forests
35. Puget Lowland Forests
36. Central and Southern Cascades Forests
37. Eastern Cascades Forests
38. Blue Mountains Forests

39. Klamath-Siskiyou Forests
40. Northern California Coastal Forests
41. Sierra Nevada Forests
42. Great Basin Montane Forests
43. South Central Rockies Forests
44. Wasatch and Uinta Montane Forests
45. Colorado Rockies Forests
46. Arizona Mountains Forests
47. Madrean Sky Islands Montane Forests
48. Piney Woods Forests
49. Atlantic Coastal Pine Barrens
50. Middle Atlantic Coastal Forests
51. Southeastern Conifer Forests
52. Florida Sand Pine Scrub

3.1 Temperate Grasslands/Savanna/Shrub MHT

53. Palouse Grasslands
54. California Central Valley Grasslands
55. Canadian Aspen Forest and Parklands
56. Northern Mixed Grasslands
57. Montana Valley and Foothill Grasslands
58. Northwestern Mixed Grasslands
59. Northern Tall Grasslands
60. Central Tall Grasslands
61. Flint Hills Tall Grasslands
62. Nebraska Sand Hills Mixed Grasslands
63. Western Short Grasslands
64. Central and Southern Mixed Grasslands
65. Central Forest/Grassland Transition Zone
66. Edwards Plateau Savannas
67. Texas Blackland Prairies
68. Western Gulf Coastal Grasslands

3.2 Flooded Grasslands MHT

69. Everglades

4.1 Mediterranean Scrub and Savanna MHT

70. California Interior Chaparral and Woodlands
71. California Montane Chaparral and Woodlands
72. California Coastal Sage and Chaparral

4.2 Xeric Shrublands/Deserts MHT

73. Hawaiian High Shrublands
74. Hawaiian Low Shrublands
75. Snake/Columbia Shrub Steppe

76. Great Basin Shrub Steppe
77. Wyoming Basin Shrub Steppe
78. Colorado Plateau Shrublands
79. Mojave Desert
80. Sonoran Desert
81. Chihuahuan Desert
82. Tamaulipan Mezquital

6.1 Boreal Forest/Taiga MHT

83. Interior Alaska/Yukon Lowland Taiga
84. Alaska Peninsula Montane Taiga
85. Cook Inlet Taiga
86. Copper Plateau Taiga
87. Northwest Territories Taiga
88. Yukon Interior Dry Forests
89. Northern Cordillera Forests
90. Muskwa/Slave Lake Forests
91. Northern Canadian Shield Taiga
92. Mid-Continental Canadian Forests
93. Midwestern Canadian Shield Forests
94. Central Canadian Shield Forests
95. Southern Hudson Bay Taiga
96. Eastern Canadian Shield Taiga
97. Eastern Canadian Forests
98. Newfoundland Highland Forests
99. South Avalon–Burin Oceanic Barrens

6.2 Tundra MHT

100. Aleutian Islands Tundra
101. Beringia Lowland Tundra
102. Beringia Upland Tundra
103. Alaska/St. Elias Range Tundra
104. Pacific Coastal Mountain Tundra & Ice Fields
105. Interior Yukon/Alaska Alpine Tundra
106. Ogilvie/MacKenzie Alpine Tundra
107. Brooks/British Range Tundra
108. Arctic Foothills Tundra
109. Arctic Coastal Tundra
110. Low Arctic Tundra
111. Middle Arctic Tundra
112. High Arctic Tundra
113. Davis Highlands Tundra
114. Baffin Coastal Tundra
115. Torngat Mountain Tundra
116. Permanent Ice

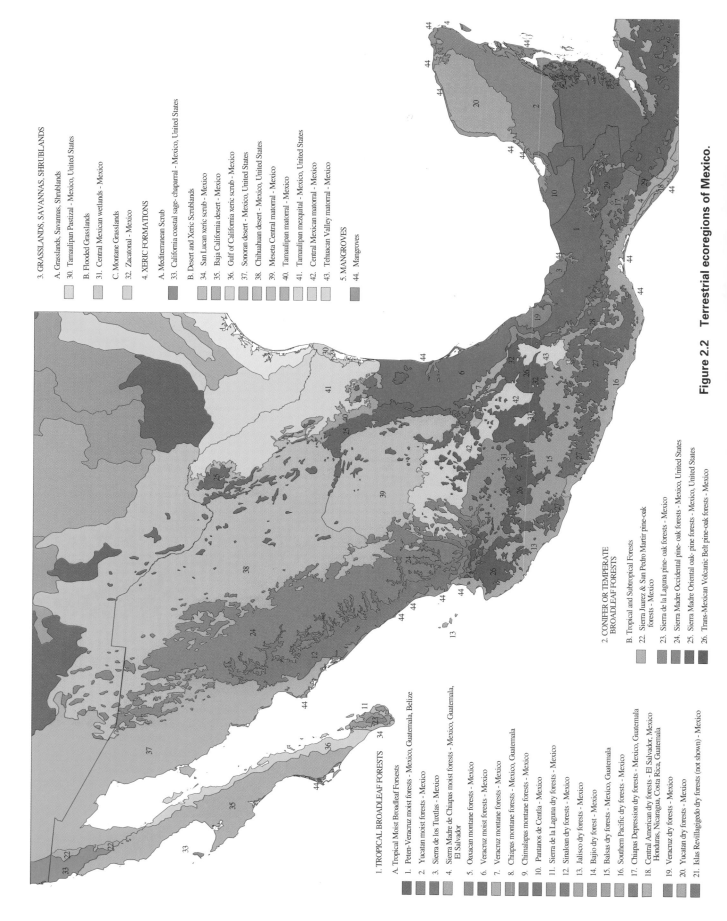

1. TROPICAL BROADLEAF FORESTS

A. Tropical Moist Broadleaf Forests

1. Peten-Veracruz moist forests - Mexico, Guatemala, Belize
2. Yucatan moist forests - Mexico
3. Sierra de los Tuxtlas - Mexico
4. Sierra Madre de Chiapas moist forests - Mexico, Guatemala, El Salvador
5. Oaxacan montane forests - Mexico
6. Veracruz moist forests - Mexico
7. Veracruz montane forests - Mexico
8. Chiapas montane forests - Mexico, Guatemala
9. Chimalapas montane forests - Mexico
10. Pantanos de Centla - Mexico

B. Tropical and Subtropical Forests

11. Sierra de la Laguna dry forests - Mexico
12. Sinaloan dry forests - Mexico
13. Jalisco dry forests - Mexico
14. Bajío dry forest - Mexico
15. Balsas dry forests - Mexico, Guatemala
16. Southern Pacific dry forests - Mexico
17. Chiapas Depression dry forests - Mexico, Guatemala
18. Central American dry forests - El Salvador, Mexico Honduras, Nicaragua, Costa Rica, Guatemala
19. Veracruz dry forests - Mexico
20. Yucatan dry forests - Mexico
21. Islas Revillagigedo dry forests (not shown) - Mexico

2. CONIFER OR TEMPERATE BROADLEAF FORESTS

22. Sierra Juarez & San Pedro Martir pine-oak forests - Mexico
23. Sierra de la Laguna pine- oak forests - Mexico
24. Sierra Madre Occidental pine- oak forests - Mexico, United States
25. Sierra Madre Oriental oak- pine forests - Mexico, United States
26. Trans-Mexican Volcanic Belt pine-oak forests - Mexico

3. GRASSLANDS, SAVANNAS, SHRUBLANDS

A. Grasslands, Savannas, Shrublands

30. Tamaulipan Pastizal - Mexico, United States

B. Flooded Grasslands

31. Central Mexican wetlands - Mexico

C. Montane Grasslands

32. Zacatonal - Mexico

4. XERIC FORMATIONS

A. Mediterranean Scrub

33. California coastal sage- chaparral - Mexico, United States

B. Desert and Xeric Scrublands

34. San Lucan xeric scrub - Mexico
35. Baja California desert - Mexico
36. Gulf of California xeric scrub - Mexico
37. Sonoran desert - Mexico, United States
38. Chihuahuan desert - Mexico, United States
39. Meseta Central matorral - Mexico
40. Tamaulipan matorral - Mexico
41. Tamaulipan mezquital - Mexico, United States
42. Central Mexican matorral - Mexico
43. Tehuacan Valley matorral - Mexico

5. MANGROVES

44. Mangroves

Figure 2.2   Terrestrial ecoregions of Mexico.

41 (Northern Montana Glaciated Plains), 42 (Northwestern Glaciated Plains), 43 (Northwestern Great Plains), 45 (Northeastern Great Plains), and 46 (Northwestern Glaciated Plains) and the prairies ecozone (ESWG 1995) in Canada. The broad species and community ranges in this region and the relatively subtle differences in community dominance and structure among the original ecoregions justify the use of Northwestern Mixed Grasslands ecoregion as a single unit for purposes of biodiversity conservation at this scale.

In the second type of modification, we split into separate ecoregions those areas of an original ecoregion that contained exceptionally distinct assemblages of species or unique habitats. Examples include separating the coastal redwood forests of California and Oregon in the Northern California Coastal Forests [40] from the coastal coniferous forests of the Central Pacific Coastal Forests [34], and separating the Florida Sand Pine Scrub [52] from the surrounding pine forests of Florida, the Southeastern Conifer Forests [51]. Specific modifications (if any) for each ecoregion, along with reasons and references for them, are found within the ecoregion descriptions in appendix F.

Our ecoregion map was further modified as regional and local experts critiqued the drafts and as new data were gathered. Ecoregions for Puerto Rico were taken from a prior conservation assessment for Latin America (Dinerstein et al. 1995), and Hawaii's ecoregions were delineated based on potential natural vegetation as mapped by the Hawaii Chapter of The Nature Conservancy (Gon 1996). The ecoregion boundaries were reviewed and modified by each of the workshop's regional groups. The result is a map that reflects patterns of biodiversity across the United States, Canada, and Mexico that can be used for effective large-scale conservation planning, reporting, and monitoring.

A concern expressed by some conservationists and management agencies is that there is not a single ecoregion map for North America, but rather several maps: Bailey et al.'s (1994) map of the United States, Omernik's map of the United States, The Nature Conservancy's ecoregion map derived from Bailey, and the WWF map derived from Omernik and ESWG. Several points are in order:

1. All of these mapping efforts show a great deal of overlap. In Alaska, for example, the ecoregion boundaries of Gallant (1995) and Bailey et al. (1994) are nearly identical. The map of Mexican ecoregions, being developed by CONABIO in collaboration with WWF-US and the Instituto Nacional de Estadística, Geografía e Informática (INEGI), will be used as the standard map by a variety of groups in Mexico and the United States for conservation planning, including CONABIO, WWF-US, WWF-Mexico, and The Nature Conservancy.

2. When this mapping effort began, the Bailey ecoregions for Canada and Mexico were too coarse to be suitable for this study, so we chose to adopt ESWG's map for further modification.

3. Most important, the centroids (center points) of ecoregions from each map show a strong degree of overlap. While boundaries may differ, they should be looked on as transition lines or,

in the parlance of statisticians, confidence intervals. Ecologists know that an ecoregion boundary delineated on a map is often difficult to identify precisely on the ground. Unfortunately, where lines from Omernik and Bailey do differ, the original vegetation and communities are often no longer there to determine the exact placing of the boundary.

4. Finally, all of these mapping efforts, at least in the United States, are strongly derivative of Küchler's (1975) potential vegetation map of the United States.

Thus, all of these worthwhile mapping efforts show more similarity than differences when scrutinized carefully.

Assignment of Major Habitat Types

Ecological processes, general patterns of biodiversity, and responses to disturbance vary widely in their scale and importance among different habitat types (deciduous forest, grasslands, etc.). To address this variation, we grouped ecoregions into ten major habitat types (MHTs), following a global framework applied in other regional analyses (Dinerstein et al. 1995; Olson and Dinerstein 1998) (see figure 2.3). The U.S. and Canada MHTs are: tropical moist broadleaf forests, tropical dry broadleaf forests, temperate

Figure 2.3   Hierarchy of spatial units used in conservation assessment framework.

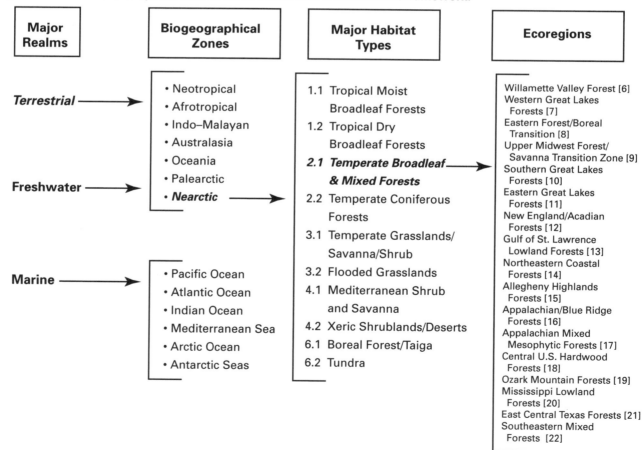

broadleaf and mixed forests, temperate coniferous forests, temperate grasslands/savannas/shrub, flooded grasslands, Mediterranean scrub and savanna, xeric shrublands/deserts, boreal forest/taiga, and tundra (figure 2.4). MHTs are not geographically defined units; rather, they refer to the dynamics of ecological systems and to the broad vegetative structures and patterns of species diversity

Figure 2.4 Major habitat types of the United States and Canada.

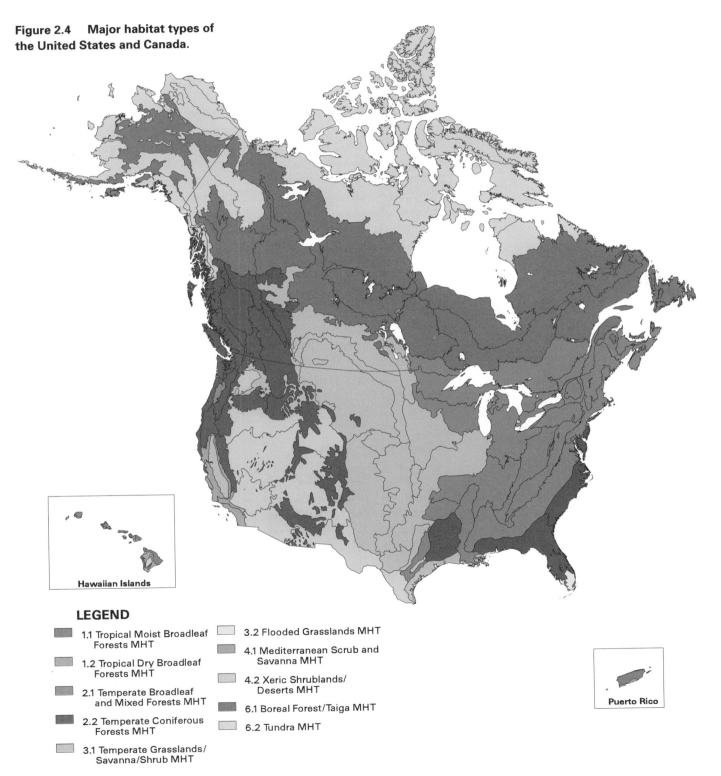

Hawaiian Islands

Puerto Rico

LEGEND

1.1 Tropical Moist Broadleaf Forests MHT

1.2 Tropical Dry Broadleaf Forests MHT

2.1 Temperate Broadleaf and Mixed Forests MHT

2.2 Temperate Coniferous Forests MHT

3.1 Temperate Grasslands/ Savanna/Shrub MHT

3.2 Flooded Grasslands MHT

4.1 Mediterranean Scrub and Savanna MHT

4.2 Xeric Shrublands/ Deserts MHT

6.1 Boreal Forest/Taiga MHT

6.2 Tundra MHT

that define them. In this way they are roughly equivalent to biomes. We have kept MHT coding in figure 2.4 consistent with the global WWF framework (Olson and Dinerstein 1998). The Mangroves MHT (#5), represented in Florida and Puerto Rico, was treated previously as part of a Latin America/Caribbean assessment (Dinerstein et al. 1995).

In all of the analyses described in the following sections, we treat each MHT separately for three main reasons. First, comparing species richness and endemism among ecoregions is much more relevant and powerful if the MHTs are analyzed separately. For example, a comparison of tree species richness between the Southeastern Mixed Forests and the Sonoran Desert does not yield much useful information. Second, the conservation status measures (described below) are designed to discriminate among ecoregions and can be tailored to reflect the different patterns of biodiversity, response to disturbance, and ecological dynamics of the different MHTs. For example, natural disturbance regimes (e.g., wildfire) operate on vastly larger scales in the boreal forest/taiga MHT than in the xeric shrublands/deserts MHT. Therefore, remaining intact habitat blocks need to be larger in the boreal forest/taiga so that entire habitat blocks are not decimated by a single fire event and some unburned areas remain to support the late-successional species assemblages and processes characteristic of this MHT. Our minimum habitat block criteria is adjusted to reflect varying requirements among MHTs. Third, there is consensus within the conservation biology community that representation of all of the earth's habitats and major ecosystem types is critical in conserving the full complement of the planet's biodiversity. By organizing ecoregions within MHTs, and by including all MHTs in North America, we can facilitate the goal of ensuring representation in regional conservation strategies (Noss and Cooperrider 1994; Olson and Dinerstein 1998).

Discriminators

To better assess the relative conservation status of ecoregions within each MHT, we use two major discriminators: biological distinctiveness and conservation status (Dinerstein et al. 1995). Each of these indices is calculated using several criteria (detailed descriptions are found in appendices A and B):

Biological Distinctiveness Index

1. Species richness
2. Species endemism
3. Rare ecological or evolutionary phenomena
4. Rare habitat type

Conservation Status Index

1. Habitat loss
2. Remaining habitat blocks
3. Degree of fragmentation
4. Degree of protection
5. Future threat

Biological Distinctiveness Index (BDI): Overview

For this study, we follow Dinerstein et al. (1995) in interpreting the biological importance of an ecoregion as the degree to which its biodiversity is distinctive at different biogeographic scales. We use this scale-dependent assessment to assign ecoregions to one of four categories: globally outstanding, regionally (e.g., Nearctic, Neotropical) outstanding, bioregionally (e.g., eastern North America) outstanding, or nationally important.

Biodiversity assessments typically focus at the level of the species. Our use of the term *biological distinctiveness* invokes a broader definition of biodiversity—besides species, we incorporate ecosystem diversity and ecological processes that sustain biodiversity. Specifically, biological distinctiveness is based on broad measures of species richness, endemism, unusual ecological and evolutionary phenomena, and the global rarity of MHTs (figure 2.5).

Criteria and Evaluation

Species richness and endemism were assessed by systematically comparing our ecoregion map with the ranges of over twenty

Figure 2.5   Component criteria and weighting used in the biological distinctiveness index.

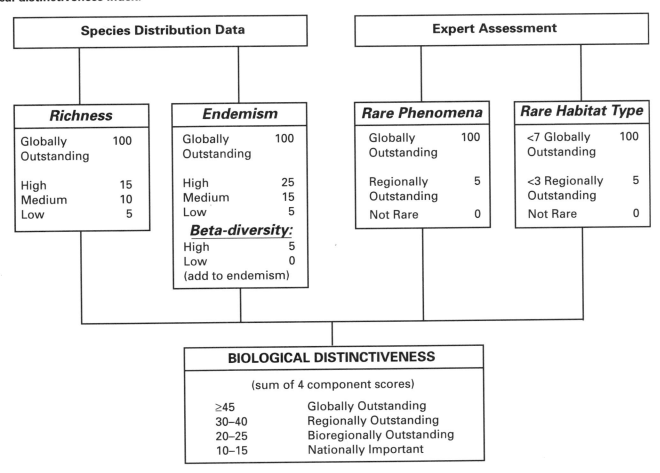

thousand North American species representing seven taxonomic groups: native vascular plants, birds, butterflies, mammals, reptiles, amphibians, and terrestrial mollusks.

These groups were chosen because continent-wide range data were available for them. They also represent a diverse subset of the North American biota. We feel that, in aggregate, they can be used with caution as an effective proxy for more numerous and less well-known groups such as insects, and thus as indicators of overall bio-diversity patterns. From the resulting database we constructed richness and endemism indices and assigned each ecoregion into one of four categories for each index: globally outstanding, high, medium, or low.

Besides species distributions, we include two ecoregion-scale criteria in the BDI: rare ecological or evolutionary phenomena and the global rarity of habitat types. Rare evolutionary phenomena include globally outstanding centers of evolutionary radiation, higher-level taxonomic diversity, and unique species assemblages. Examples of rare ecological phenomena are large-scale migrations of larger vertebrates, extraordinary seasonal concentrations of wildlife, or distinctive processes such as the world's most extensive sheet-flow grasslands (i.e., the Everglades). Rarity can exist in two ways: original rarity and human-induced rarity. Because this crite-rion is an attempt to measure the remaining opportunities for con-servation of a certain phenomenon, both kinds of rarity are taken into account. Thus, an ecological process that was once widespread but has been disrupted in the majority of its original locations is rated, in its remaining locations, the same as a process that was originally rare (for example, intact large predator-prey assemblages that were once common across North America but are much reduced today). Overall, we emphasize phenomena that are truly globally outstanding in their magnitude, expression, or devel-opment, particularly those that involve a broad range of higher taxa or species, or those that profoundly influence the structure of whole communities. We also highlight phenomena whose patterns or dynamics are largely confined within a single ecoregion. Conti-nental or hemispheric-scale migrations of vertebrates and inverte-brates each involve many different ecoregions. The geographic breadth of these important and distinctive ecological phenomena preclude using them as effective discriminators among ecoregions. Clearly, particular sites and habitat types all along migration routes are critical stopover, resting, or feeding areas for migrating species, but their broad geographic distribution and often localized nature are difficult to characterize within the context of an ecoregion-scale analysis. However, in some cases, we do note ecoregions that are migration endpoints with extraordinary concentrations of breeding birds. Migrations of larger terrestrial vertebrates are also distin-guished because of their current rarity and their general limitation to single ecoregions.

Although it is difficult to define and measure ecoregion-scale phenomena, and the criteria developed here are a preliminary characterization, we stress that biodiversity assessments must increasingly move beyond a narrow focus on species richness and endemism to better conserve the full expression of life on earth.

The rarity of ecological and evolutionary phenomena in ecoregions was assessed by experts, and globally outstanding and regionally outstanding ecoregions were scored accordingly.

The global rarity of habitat types also was scored by expert assessment and uses the same two categories as above. An ecoregion is considered globally outstanding if fewer than seven ecoregions worldwide contain its habitat type, and regionally outstanding if fewer than three occur in the Nearctic or Neotropics. This measure represents the number of opportunities to conserve this habitat type worldwide and the corresponding importance of a North American ecoregion that contains it.

We give additional recognition to ecoregions that contain species assemblages with high levels of beta-diversity. Beta-diversity is a measure of turnover, or replacement of species, with distance or along environmental gradients. Beta-diversity reflects the biological complexity of an ecoregion. It also provides an important indicator of the level of effort needed to conserve an ecoregion. Typically, ecoregions with very high levels of beta-diversity will require multiple protected areas distributed across the landscape to conserve scattered, highly distinct assemblages of plants and animals. Ecoregions in some MHTs, such as boreal forests and tundra, show little turnover in species, while desert ecoregions often exhibit high beta-diversity.

Combining BDI Criteria

Ecoregions were awarded points corresponding to their ranks in the four criteria described above. The scores from the four criteria were then summed to place each ecoregion in one of four overall biological distinctiveness categories: globally outstanding, regionally outstanding, bioregionally outstanding, and nationally important (see appendix A for full methods and discussion).

Ecoregions categorized as globally outstanding are exceptional among ecoregions of their MHT worldwide, in terms of the level and the rarity of biodiversity they contain. Regionally outstanding refers to those ecoregions that are most distinctive at the level of biogeographic provinces such as the Nearctic or Neotropics (North America has ecoregions in both the Nearctic and the Neotropics). Bioregionally outstanding ecoregions are those that may not harbor distinctive biodiversity at a global or regional scale, but are noteworthy among ecoregions within the same MHT and subregional areas of North America (bioregions). Bioregions are geographic clusters of ecoregions that may span several habitat types but have strong biogeographic affinities, particularly at taxonomic levels higher than the species level (genus, family). The Northern Mexico bioregion and Eastern North America bioregion are examples (figure 2.6). The biogeographic resolution of bioregions used in this analysis may appear broad to North American biodiversity specialists. However, they are intended to approximate the scale of bioregion classifications used in assessments on other continents (e.g., Latin America and the Caribbean—Dinerstein et al. 1995; Indo-Pacific Region—Wikramanayake et al. 1997).

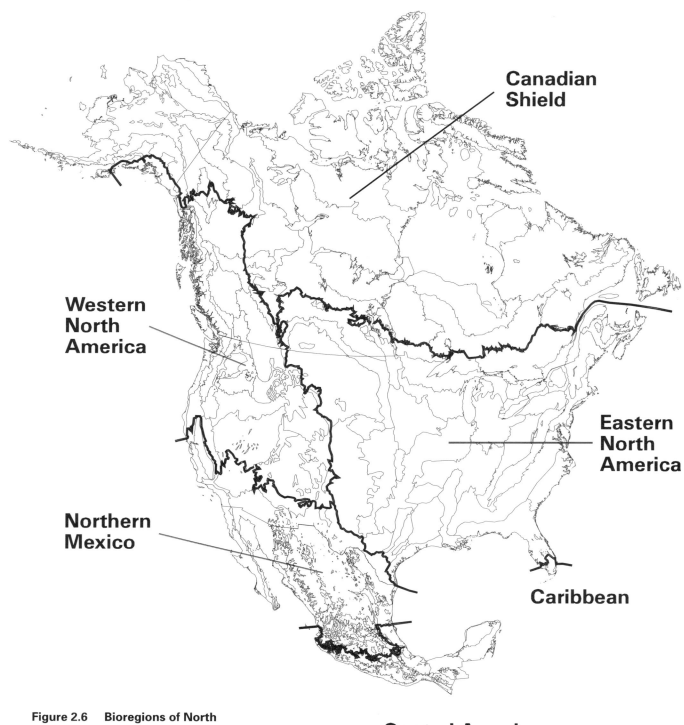

Canadian Shield

Western North America

Eastern North America

Northern Mexico

Caribbean

Figure 2.6   Bioregions of North America.

Central America

The biodiversity of every ecoregion is at a minimum nationally important. Areas within ecoregions categorized here as nationally important frequently contain important habitat for rare or endangered species. Every ecoregion also provides essential ecosystem services, natural resources, and natural recreation opportunities for local human communities (Daily 1997). However, some ecoregions within each MHT in North America are so extraordinarily rich or unusual in their biodiversity that conservation efforts within that ecoregion take on global or regional (continental) importance.

Conservation Status Index: Overview

The second major discriminator, conservation status, is designed to estimate the current and future ability of an ecoregion to meet three fundamental goals of biodiversity conservation: maintaining viable species populations and communities, sustaining ecological processes, and responding effectively to short- and long-term environmental change. The snapshot conservation status is based on four essential landscape-level criteria: habitat loss and degradation, presence of large blocks of remaining habitat, degree of habitat fragmentation, and degree of existing protection. This index is then modified by an assessment of future threats over the next twenty years to arrive at the threat-modified conservation status, or final conservation status. A complete description of the conservation status index, its components and design, is found in appendix B.

Criteria

We rely on landscape-level measures for this analysis because the loss of biodiversity and the degradation of ecosystem function, especially for a region as large and diverse as North America, are difficult to measure directly. Even in countries as relatively well studied as the United States and Canada, comprehensive taxonomic surveys, population monitoring, and ecosystem monitoring are not sufficient to allow consistent and rigorous analyses. Therefore, we must rely on landscape-level features and assume that they can be used to predict the ability of an ecoregion to meet the fundamental goals of biodiversity conservation (Noss 1992; Primack 1993; Meffe and Carroll 1994; Noss and Cooperrider 1994; Kareiva and Wennergren 1995).

Habitat loss and degradation are probably the two most important factors contributing to the reduction of species populations, extinctions, and the disruption of ecosystem function (essay 2). Mechanisms include the elimination of geographically restricted species through the destruction of their specific habitat, and the reduction of the area of intact habitat below a certain minimum critical size or distribution needed to maintain stable population numbers or to sustain critical ecological processes.

Besides the simple reduction in habitat area, the spatial pattern of habitat loss is critically important to maintaining native species, communities, and ecological processes across large landscapes. Large blocks of habitat generally contain larger and more stable species populations, and are uniquely able to support species with naturally low population densities or large home ranges (Noss and Cooperrider 1994). At least some habitat blocks must be large enough to absorb a typical natural disturbance event without being completely decimated so that they may serve as refugia or source pools for populations of wide-ranging species. Large blocks of

Habitat Loss in North America Fuels Species Extinctions

David M. Olson

The tremendous loss of habitat in many of North America's ecoregions suggests an increasingly high rate of extinction among a wide range of invertebrates, fungi, bacteria, and other less charismatic taxa (Bean 1993; Flowers 1993). Many species in these taxa have not made it onto the official list of endangered, threatened, and extinct species and subspecies maintained by national authorities for two reasons. First, incomplete scientific knowledge, agency biases, politics, and limited resources keep the list from ever being complete. Second, many of these taxa are poorly researched, and "unencountered" species may go extinct before they become known to science (what E.O. Wilson has named centinelan extinctions). We must recognize as a scientific community and at the national level that species extinctions are happening today, a concern not for the next generation, but for us.

A majority of these less-known or unencountered species are highly specialized inhabitants of endangered habitats, such as the old-growth temperate rain forests of the Pacific Northwest, the sand pine scrubs of central Florida, and the dune communities of central California. The canopies of old-growth forests on Vancouver Island are known to harbor at least seventy-one species of oribatid mite, most of which occur only in this habitat and have been only recently discovered (Winchester and Ring 1997). Similarly high diversity and specificity in forest-floor and canopy mites in Oregon's old-growth forests has been recorded (Moldenke 1990; Lattin and Moldenke 1992). Such extreme specialization on increasingly rare habitats and the propensity of some species to have very limited geographic distributions make them highly prone to extinction from habitat loss. Indeed, the status of several species of millipedes and harvestmen known only from a few localities within the coastal rain forests of Oregon and California is questionable given the wholesale clear-cutting of their habitat after their discovery (Olson 1992). Many other coastal rain forest taxa—beetles, land snails, fungi, lichens, mosses, and bacteria—are believed to have very limited geographic distributions and be highly specialized on old-growth habitats.

Unfortunately, less than 4 percent of lowland temperate rain forests remain along the Pacific Northwest, and the last patches are very small, fragmented, and isolated. Severe habitat loss and the propensity of specialized species to have restricted ranges means that invertebrates and other important taxa—not as charismatic as the California condor or spotted owl but extremely important for maintaining ecological integrity—are fast going extinct. Federal agencies are slowly adding not just the endangered but also the extinct invertebrates to their lists: from California alone there is the Xerces blue butterfly, Antioch sphecid wasp, yellow-banded andrenid bee, San Joaquin Valley tiger beetle, Pasadena freshwater shrimp, Valley flower-loving fly, Antioch shield-back katydid, sooty crayfish, Mono Lake hygrotus diving beetle, Sthenele satyr butterfly, El Segundo flower-loving fly, Atossa fritillary butterfly, Antioch robber fly . . . the list goes on. In some parts of North America, Wilson's "little things that run the world" (1992) are running out of habitat and time.

habitat also maintain ecological processes more effectively than do smaller blocks. However, small blocks are important as source pools for geographically restricted species (e.g., rare plants) and for maintaining species distributed as metapopulations.

Habitat fragmentation often results in many small blocks of habitat lacking the critical functions described above. Populations are less able to interbreed, to forage over large areas for scattered resources, and to disperse in response to habitat disruption or long-term environmental change (Berger 1990; Wilcove, McLellan, and Dobson 1986). As fragmentation continues, the amount of core habitat area, unaffected by surrounding human activity and exotic species, declines sharply. Key ecological processes such as pollination and seed dispersal may also cease functioning in highly fragmented landscapes.

The degree-of-protection criterion assesses how well the existing network of national parks, wilderness areas, and other reserves protects sufficiently large blocks of habitat, in sufficient number, within the ecoregion. Protected areas equivalent to GAP Categories I and II (Scott et al. 1993) were considered in this analysis. Categories I–II refer to areas actively managed to remain in natural states; they include national and provincial parks, congressionally designated wildernesses, and national monuments. In certain cases, state parks in the United States that offer good protection of large blocks of habitat (e.g., Adirondack State Parks) were also included. Generally, we assume that all these designated areas are managed adequately to meet their designations, an assumption that likely is wrong in some areas. Although extremely important, incorporating detailed considerations of the quality of management for biodiversity conservation is beyond the scope of this assessment.

Although it may be ideal to include imperiled species as a criterion for the conservation status index, we did not do so for four main reasons. First, lists of imperiled species are often biased toward vertebrates as opposed to invertebrates. Second, political rather than biological factors also preclude the inclusion of certain taxa that otherwise should be listed. Third, there are likely to be many species that should be listed for which there are only limited data. Finally, a comprehensive database for imperiled species for the entire continent, or even the United States, was unavailable when we initiated this analysis.

Assessment of Criteria

Several national and continental data sets and maps exist that are potentially useful for quantitatively assessing some of the four criteria (e.g., EROS 1996; McGhie 1996; Küchler 1975; Scott et al. 1993). However, we found none of them to be of sufficient spatial resolution or consistency to rely upon completely for the entire analysis at this scale. To gather high-quality information in a timely fashion, we relied on the ecologists and conservationists at the expert workshop to assess each ecoregion in terms of the four landscape-level features. The workshop participants used the various data sets and maps as useful guides in their evaluation, but they were able to modify these information sources with their expert knowledge of regions to arrive at a more accurate assessment. For example, remotely sensed land-cover maps, such as those generated from AVHRR data (EROS 1996), typically do not distinguish between old-growth forest and recently regenerated clear-cuts. Different land cover types may have similar satellite spectral signatures, but from the perspective of biodiversity conservation they are vastly different. Also, some management areas not designed to conserve biodiversity, such as military artillery ranges, sometimes do so incidentally and represent some of the last remaining intact habitat in an ecoregion, even though they are not usually included in lists of protected areas.

Methods and decision rules employed at the workshop are described in appendix B. Each group also was asked to list up to ten priority sites or activities for conservation within each ecoregion and to provide contact information for important groups working toward biodiversity conservation in the ecoregion. This information is included in the ecoregion descriptions in appendix F.

Snapshot Conservation Status

The four criteria were weighted and combined into a single index, from which five categories of conservation status were derived: critical, endangered, vulnerable, relatively stable, and relatively intact. These categories follow those of Dinerstein et al. (1995) and were developed in the tradition of the IUCN Red Data Book series (IUCN 1994; Collar et al. 1992). The IUCN Red Data Books list species under various levels of threat in order to call attention to species and populations on trajectories toward extinction. WWF has adopted the use of similar categorical schemes to describe ecoregions and to assess their conservation status (Dinerstein et al. 1995). The rationale for using these criteria is simple: Almost 90 percent of all species found in the Red Data Books are listed as endangered because of loss of habitat (Wilcove et al. 1996). As many more species that share those same ecoregions are either undescribed or unlikely ever to be officially listed, it makes sense to apply Red Data Book criteria directly to ecoregions to determine where overall species loss or declines are most likely to occur. Moreover, landscape-level integrity also will reflect the persistence of distinctive and important ecological processes.

Methods, weighting, and thresholds used in deriving the conservation status index categories are detailed in appendix B. Below we provide a generalized and qualitative description of each category in terms of the ecoregion's landscape integrity and likely predicted ecological impacts. The descriptions reflect how, with increasing habitat loss, degradation, and fragmentation, ecological processes cease to function naturally, populations no longer occur within the natural range of variation, and major components of biodiversity are eroded (Dinerstein et al. 1995).

Critical. The remaining intact habitat is restricted to isolated small fragments with low probabilities of persistence over the next five or ten years without immediate or continuing protection and restoration. Many species are already extirpated or extinct due to the loss of viable habitat. Remaining habitat fragments do not meet the minimum area requirements for maintaining viable populations of many species and ecological processes. Land use in areas between remaining fragments is often incompatible with maintaining most native species and communities. Spread of alien species may be a serious ecological problem, particularly on islands. Top predators have, or have almost, been exterminated.

Endangered. The remaining intact habitat is restricted to isolated fragments of varying size (a few larger blocks may be present) with medium to low probabilities of persistence over the next ten to fifteen years without immediate or continuing protection or restoration. Some species are already extirpated because of loss of viable habitat. Remaining habitat fragments do not meet minimum area requirements for most species populations and large-scale ecological processes. Land use in areas between remaining fragments is largely incompatible with maintaining most native species and communities. Top predators are almost exterminated.

Vulnerable. The remaining intact habitat occurs in habitat blocks ranging from large to small; many intact clusters will likely persist over the next fifteen to twenty years, especially if given adequate protection and moderate restoration. In many areas, some sensitive or exploited species have been extirpated or are declining, particularly top predators and game species. Land use in areas between remaining fragments is sometimes compatible with maintaining most native species and communities.

Relatively stable. Natural communities have been altered in certain areas, causing local declines in exploited populations and disruption of ecosystem processes. These distributed areas can be extensive but are still patchily distributed relative to the area of intact habitats. Ecological linkages among intact habitat blocks are still largely functional. Guilds of species that are sensitive to human activities, such as top predators and ground-dwelling birds, are present but at densities below the natural range of variation.

Relatively intact. Natural communities within an ecoregion are largely intact with species, populations, and ecosystem processes occurring within their natural ranges of variation. Guilds of species that are sensitive to human activities, such as top predators and ground-dwelling birds, occur at densities within the natural range of variation. Biota move and disperse naturally within the ecoregion. Ecological processes fluctuate naturally throughout largely contiguous natural habitats.

Final (Threat-Modified) Conservation Status

The snapshot conservation status of each ecoregion is modified according to the degree of expected future threat, as assessed at the expert workshop. This measure looks beyond the ecological threats implicit in existing habitat loss and fragmentation to evaluate the future trajectories of these phenomena. Experts estimated the cumulative impact of all threats on habitat conversion, habitat degradation (note: this category includes the impacts of alien species), and wildlife exploitation over the next twenty years to categorize ecoregions into three levels of threat: high, medium, and low. An ecoregion with high threat is promoted to the next highest conservation status category to arrive at its modified conservation status. (For example, an endangered ecoregion with high threat is promoted to critical.) Conservation status for ecoregions with moderate or low threat is unchanged. For further details, see appendix B.

Integrating Biological Distinctiveness and Conservation Status

The biological distinctiveness and conservation status indices are two essential discriminators for biodiversity conservation planning at large scales. They combine an evaluation of the relative biological importance of ecoregions with a measure of anthropogenic impacts, both current and projected, facing each ecoregion. Considered together, the two indices provide a powerful tool for indicating appropriate conservation activities within ecoregions and for setting regional priorities when limited resources require careful and strategic planning.

To integrate the two discriminators, we modified a matrix developed by Dinerstein et al. (1995). The biological distinctiveness categories lie along the vertical axis, the conservation status categories along the horizontal axis (figure 2.7). Ecoregions, based on their categories for both indices, fall into one of the twenty cells of the matrix. The entire assessment process, including this integration step, is carried out independently for each MHT to ensure representation of MHTs in the final analysis. We have organized the twenty cells into five classes (figure 2.7), that reflect the nature and extent of the management activities likely to be required for effective biodiversity conservation:

Figure 2.7   Integration matrix for conservation status and biological distinctiveness, with recommended conservation action categories.

	CRITICAL	ENDANGERED	VULNERABLE	RELATIVELY STABLE	RELATIVELY INTACT
GLOBALLY OUTSTANDING	I	I	I	III	III
REGIONALLY OUTSTANDING	II	II	II	III	III
BIOREGIONALLY OUTSTANDING	IV	IV	V	V	V
NATIONALLY IMPORTANT	IV	IV	V	V	V

I. Globally outstanding ecoregions requiring immediate protection of remaining habitat and extensive restoration.

II. Regionally outstanding ecoregions requiring immediate protection of remaining habitat and extensive restoration.

III. Globally or regionally outstanding ecoregions that present rare opportunities to conserve large blocks of intact habitat.

IV. Bioregionally and nationally important ecoregions requiring protection of remaining habitat and extensive restoration.

V. Bioregionally and nationally important ecoregions requiring protection of representative habitat blocks and proper management elsewhere for biodiversity conservation.

Class I. Globally outstanding ecoregions requiring immediate protection of remaining habitat and extensive restoration. These ecoregions contain elements of biodiversity that are of extraordinary global value or rarity and are under extreme threat. Conservation actions in these ecoregions must be swift and immediate to protect the remaining source pools of native species and communities for restoration efforts. Restoration efforts at high priority sites are likely to be extensive and costly.

Class II. Regionally outstanding ecoregions requiring immediate protection of remaining habitat and extensive restoration. These ecoregions have high regional biodiversity and are under serious threat. Conservation actions should be swift and may include extensive and costly habitat restoration.

Class III. Globally or regionally outstanding ecoregions that present rare opportunities to conserve large blocks of intact habitat. Ecoregions contain globally or regionally high levels of biodiversity or rare ecological processes. Conservation action in these ecoregions is not immediately needed, but these ecoregions represent some of the last remaining areas where it is possible to conserve large patches of intact, globally or regionally outstanding habitat.

Class IV. Bioregionally and nationally important ecoregions requiring protection of remaining habitat and extensive restoration. Ecoregions contain bioregionally or nationally important elements of biodiversity that are under extreme threat. Conservation actions include protection of remaining habitat and extensive restoration of degraded habitat. Proper stewardship or expansion of protected areas, conservation management on native lands, and vigilant monitoring of ecological integrity are needed.

Class V. Bioregionally and nationally important ecoregions requiring protection of representative habitat blocks and proper management elsewhere for biodiversity conservation. Conservation actions include proper stewardship or expansion of protected areas, conservation management on public and private lands, and vigilant monitoring of ecological integrity.

Biodiversity conservation is important in every ecoregion because naturally functioning ecosystems provide many services to natural and human communities. Conservation of natural areas in all ecoregions ensures preservation of many distinct species and communities as well as the genetic and functional diversity of populations across species ranges. Flood control, groundwater recharge, freshwater purification, pest control, and innumerable recreational opportunities are all examples of local ecosystem services that must be maintained within each ecoregion (Daily 1997).

However, besides providing ecological services, there are ecoregions around the world that warrant more immediate attention from conservationists because they are of global importance biologically, containing unique or extraordinarily diverse floras, faunas, or unusual ecological phenomena (Olson and Dinerstein

1998). They may also be at extreme risk from anthropogenic forces such as extensive habitat loss or fragmentation. With limited resources and time available for conservation, it is important to strategically allocate and time conservation effort and funds. This decision matrix is designed to assist in this process and to highlight the extraordinary nature of North American biodiversity.

We recognize that others might choose to reorganize the cells in a different manner than we have done in this report. For example, we chose to lump globally outstanding ecoregions that are considered vulnerable with Class I. We make this choice to be conservative in cases where ecoregions evaluated as vulnerable are actually endangered. The same logic applies to ecoregions classified as regionally outstanding and vulnerable, which we lumped in Class II rather than Class III. In contrast, we feel that ecoregions classified as vulnerable but with less distinct levels of biodiversity fit best the description for Class V. Because the raw data for each of the ecoregions are provided in the appendices, readers are welcome to revisit the organization of our classes with these data or by adding data of their own.

Biodiversity in the Freshwater and Marine Realms

A complete conservation assessment of North America requires addressing the needs of freshwater and marine species and habitats. Thus, a single ecoregion framework to map terrestrial, freshwater, and marine biodiversity has been widely sought. However, differences in the physical features of these environments, biogeography, species life histories, and levels of endemism preclude the use of one system for defining ecoregions across the three realms. WWF-US recently undertook a conservation assessment of freshwater ecoregions of North America and, in collaboration with R. Glenn Ford, an assessment of marine ecoregions of western North America. Short summaries of these studies, to be published in 1999, are presented in essays 3 and 4.

ESSAY 3

Do Freshwater Ecoregion Boundaries Correlate with Terrestrial?

Robin Abell

The natural distributional limits of species confined to freshwater systems correspond with larger drainage areas, lake margins, and spring locations. This pattern contrasts with terrestrial species distributions, which are closely linked to vegetation, topography, and soil type. An assessment of freshwater biodiversity thus can benefit from a separate set of ecoregions based on the zoogeographic patterns of freshwater biota.

For the freshwater portion of WWF's North American Conservation Assessment, we began our ecoregion delineation using the subregions described by Maxwell et al. (1995). These subregions,

---

based on fish distributions, were then modified following the recommendations of experts from the United States, Canada, and Mexico familiar both with specific regions and with other freshwater taxa such as crayfish, freshwater mussels, and freshwater herpetofauna. The resulting map contains seventy-six ecoregions (figure 2.8a). Twenty-eight of these ecoregions are partly or entirely located in Mexico. The freshwater ecoregions for the most part represent aggregations of catchments, also known as watersheds or drainage basins. A catchment includes all of the land draining into a particular river (or lake, in the case of closed-basin systems without exterior drainage).

The species found in an individual catchment are not always widely distributed in it, but instead are associated with distinct habitats, such as small headwater streams or large downstream river reaches. A full complement of species defines an ecoregion, explaining why ecoregions rarely divide drainage basins, except in the case of large rivers such as the Mississippi. As with terrestrial systems, endemic species often provide the best indication of how catchments should be aggregated.

An overlay of terrestrial and freshwater ecoregions reveals that there is little concordance between their boundaries (figure 2.8b). The only region with obvious alignment is the terrestrial Great Basin Shrub Steppe [76], covered by the Bonneville and Lahontan freshwater ecoregions (8 and 9). This agreement occurs because the Bonneville and Lahontan, both closed-basin systems, lie entirely within the terrestrial ecoregion. Elsewhere, and particularly from the Rockies westward, portions of several other terrestrial and freshwater ecoregions display the same general shapes but without strict agreement of boundaries. Often such concordance occurs in areas of abrupt topographic change, such as at the tops of mountain ranges; in these places, topography defines both catchment boundaries and the limits of terrestrial species. What is perhaps most obvious from figure 2.8b, though, is that terrestrial ecoregions often run in bands across freshwater ecoregions. This again is largely a function of topography, with these bands corresponding roughly to headwater, middle, and downstream reaches of river systems.

The different corridors and linkages within terrestrial and aquatic realms largely account for the lack of congruity among freshwater and terrestrial ecoregions. While this incongruity may complicate priority setting for decision makers, forcing one classification to fit the other could weaken a comprehensive biodiversity conservation effort. Ecoregion boundaries may not overlap perfectly, but conservation planners can nonetheless use the complementary information provided by the two schemes to choose intervention activities that will have the greatest benefit across all taxa. As many threats to freshwater biodiversity are tied to catchment-scale land uses, real opportunities may exist for protecting both terrestrial and freshwater species simultaneously.

As an example, the freshwater Tennessee-Cumberland ecoregion (35) overlaps with three terrestrial ecoregions [16], [17], and [18], which correspond to different forest types. The Tennessee-Cumberland is of globally outstanding biodiversity value, with extraordinarily high species richness and endemism across all faunal groups measured; for instance, 65 of 227 (29 percent) fish species and 40 of 65 (61 percent) crayfish species are endemic. If terrestrial ecoregions were

The Colorado squawfish (*Ptychocheilus lucius*) is adapted to live in the swift currents of the Colorado River. Growing up to five feet in length, it is the largest minnow in North America. Dams, non-native fish, and degradation of water quality have pushed this extraordinary fish to the brink of extinction. Photo credit: USFWS/H. Stewart.

used to quantify freshwater biodiversity, aquatic species inhabiting the same drainage would be split among several ecoregions and patterns of biodiversity would be lost.

WWF-US's freshwater biodiversity assessment, to be published separately from this report, will be available to conservation planners in late 1999.

Figure 2.8a   Freshwater ecoregions of North America.

Pacific Bioregion
Coastal Complex
1. North Pacific Coastal
2. Columbia Glaciated
3. Columbia Unglaciated
4. Upper Snake
5. Pacific Mid-Coastal
6. Pacific Central Valley
7. South Pacific Coastal

Great Basin Complex
8. Bonneville
9. Lahontan
10. Oregon Lakes
11. Death Valley

Colorado Complex
12. Colorado
13. Vegas-Virgin
14. Gila

Arctic-Atlantic Bioregion
Rio Grande Complex
15. Upper Rio Grande
    (Rio Bravo Del Norte)
16. Guzmán
17. Rio Conchos
18. Pecos
19. Mapimí
20. Lower Rio Grande
    (Rio Bravo Del Norte)
21. Rio Salado
22. Cuatro Ciénegas
23. Rio San Juan

Mississippi Complex
24. Mississippi
25. Mississippi Embayment
26. Upper Missouri
27. Middle Missouri
28. Central Prairie
29. Ozark Highlands
30. Ouachita Highlands
31. Southern Plains
32. East Texas Gulf
33. West Texas Gulf
34. Teays-Old Ohio
35. Tennessee-Cumberland
36. Mobile Bay
37. Apalachicola
38. Florida Gulf

Atlantic Complex
39. Florida
40. South Atlantic
41. Chesapeake Bay
42. North Atlantic

St. Lawrence Complex
43. Superior
44. Michigan-Huron
45. Erie
46. Ontario
47. Lower St. Lawrence
48. North Atlantic-Ungava

Hudson Bay Complex
49. Canadian Rockies
50. Upper Saskatchewan
51. Lower Saskatchewan
52. English-Winnipeg Lakes
53. South Hudson
54. East Hudson

Arctic Complex
55. Yukon
56. Lower Mackenzie
57. Upper Mackenzie
58. North Arctic
59. East Arctic
60. Arctic Islands

Mexican Transition Bioregion
61. Sonoran
62. Sinaloan Coastal
63. Santiago
64. Manantlan-Ameca
65. Chapala
66. Llanos El Salado
67. Rio Verde Headwaters
68. Tamaulipas-Veracruz
69. Lerma
70. Balsas
71. Papaloapan
72. Catemaco
73. Coatzacoalcos
74. Tehuantepec
75. Grijalva-Usumacinta
76. Yucatán

Ecoregion boundaries and names derived from:

Maxwell et. al.

Ecoregion boundaries subsequently modified during WWF-sponsored North America Freshwater Workshop, April 11-12, 1997.

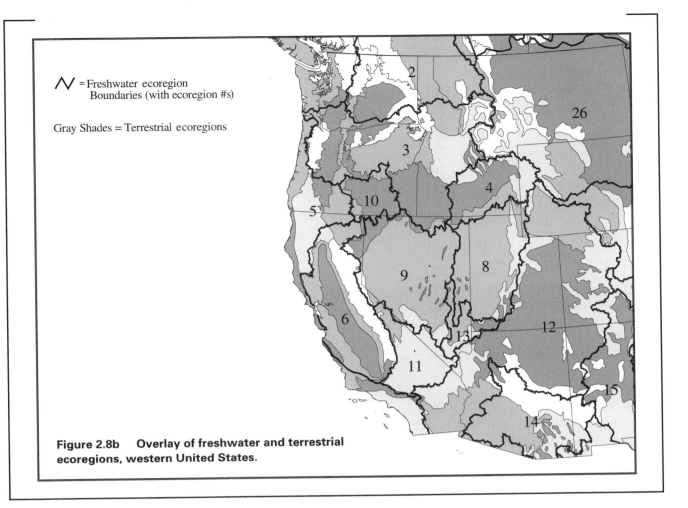

= Freshwater ecoregion
Boundaries (with ecoregion #s)

Gray Shades = Terrestrial ecoregions

Figure 2.8b   Overlay of freshwater and terrestrial ecoregions, western United States.

ESSAY 4

Defining Marine Ecoregions of the Pacific Continental United States

R. Glenn Ford

Marine ecoregions in temperate and arctic zones differ from terrestrial ecoregions in that there are usually no fixed geologic or vegetative features that can be used to define ecoregion boundaries. Rather, marine ecoregions are best defined as assemblages of species with an affinity for a water mass having particular characteristics at a particular time of year. Many marine animals migrate over vast distances during their annual cycles, moving from temperate or tropical to arctic regions, or even from hemisphere to hemisphere over the course of the year. Local species assemblages are usually transitory and opportunistic, often based on short-lived patches of plankton that vary enormously in their richness. Within a relatively small area, it is common for temperate and arctic marine species to vary in density by a factor of hundreds or even tens of thousands.

Over the years, several worldwide classification schemes for marine biogeographic regions have been proposed. The system originally developed by Ekman (1953) was based on global-scale patterns of circulation and species distributions and subdivided the world's oceans into large "provinces." This system was refined by Briggs (1974) and Hayden, Ray, and Dolan (1984) and in various forms is widely accepted as a coarse-scale system of classification. These provinces, however, are defined on the basis of physical factors, and within them there is great variation in both ecosystem structure and the threats to those ecosystems. For the purposes of monitoring the health of and assessing the risk to those ecosystems, it is necessary to further subdivide these vast provinces into smaller components.

The best approach to defining temperate and arctic marine ecoregions is to define them by using the distributions of the organisms themselves. However, species assemblages alone do not entirely define marine units. Within a large region where the mix of species is reasonably uniform, there are areas where certain species periodically congregate in vast numbers due to the local richness of the food supply. These concentrations do not represent different ecoregions, since the same species present at a very high density in one place can typically be found at much lower densities in a much larger surrounding area. Ecoregion delineations are therefore based not only on taxonomic composition, but also on areas of concentration that are of special importance to the local and regional ecosystem.

To define ecoregions and areas of concentration, data from a number of Pacific Coast studies were used to describe major groups of marine animals:

- seabirds—alcids (murres, auklets, etc.), tubenoses (shearwaters, petrels, albatrosses), loons, grebes, scoters
- marine mammals—northern fur seal; otters, pinnipeds; toothed whales; baleen whales
- fish—salmon, herring, flatfish, bass

Distributional data for these groups along the U.S. Pacific Coast were summarized as relative abundance per 15-foot block (an area about 21 km X 28 km) averaged over the entire year. A technique similar to that, developed by Ray and Hayden (1992) for Alaskan waters was used to construct ecoregions for the Pacific Coast. First, the densities of the species groups were reduced to three principal component axes. Distributions of the principal component values were then used as input to a cluster analysis to generate contiguous non-overlapping regions of biological similarity (Eastman 1992). Clusters were then assembled to form ecoregions, and within the ecoregions, distributional data were used to define areas of especially high animal densities.

Based on this analysis, there are five differentiable marine ecoregions in addition to the major estuarine systems along the Pacific Coast (figure 2.9).

- Columbian—The oceanography of the U.S. north Pacific Coast is dominated by the cool, southerly flowing California current offshore and the northerly flowing Davidson current inshore. Prevailing northwesterly winds generate massive upwelling throughout the area. Seabird densities are especially high along the Washington coast and between Cape Blanco and Cape Mendocino, whereas marine mammals are present at lower densities than farther south.
- Montereyan—This ecoregion represents a transition zone between the temperate Columbian region to the north and the subtropical Southern California Bight to the south, and tends to represent the northern extent of the ranges of many subtropical species and the southern extent of the ranges of many temperate species. Both seabirds and marine mammals are often found in high densities in the Gulf of the Farallones and in Monterey Bay.
- Southern California Bight (Outer)—The cool California current moves offshore in the vicinity of Point Conception, where the continental shelf widens to form the Southern California Bight. This ecoregion represents the northernmost extent of subtropical waters and is typically warmer and (with the exception of the area around Point Conception) contains less upwelling than the Montereyan and Columbian ecoregions to the north. The area north of Point Conception and around the northern Channel Islands contains especially high densities of marine mammals. Although the northern Channel Islands contain some seabird colonies, seabirds are generally less abundant and marine mammals more abundant in this ecoregion than farther north.

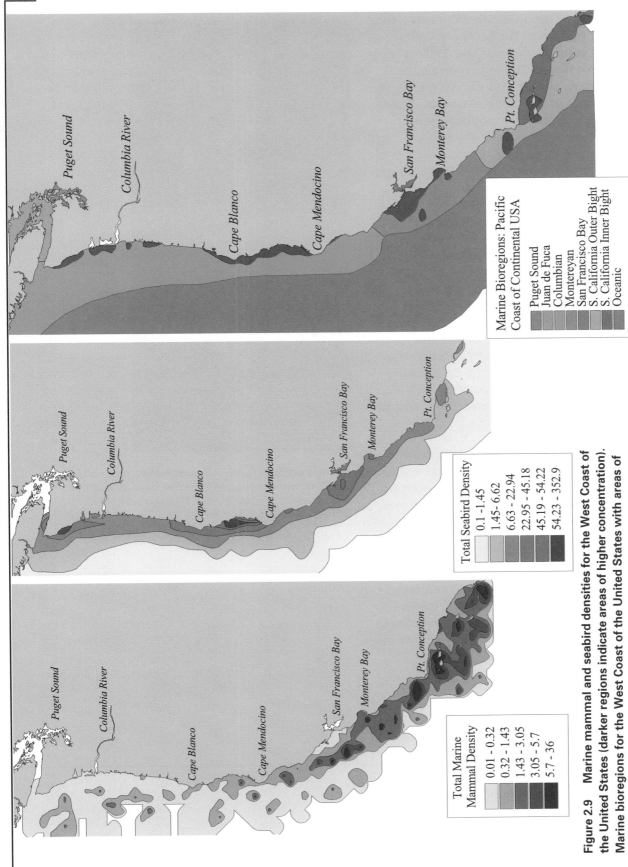

Figure 2.9   Marine mammal and seabird densities for the West Coast of the United States (darker regions indicate areas of higher concentration). Marine bioregions for the West Coast of the United States with areas of high marine fauna concentrations.

Marine Bioregions: Pacific Coast of Continental USA

Puget Sound
Juan de Fuca
Columbian
Montereyan
San Francisco Bay
S. California Outer Bight
S. California Inner Bight
Oceanic

Concentration Area

Total Seabird Density

0.1 - 1.45
1.45 - 6.62
6.63 - 22.94
22.95 - 45.18
45.19 - 54.22
54.23 - 352.9

Total Marine Mammal Density

0.01 - 0.32
0.32 - 1.43
1.43 - 3.05
3.05 - 5.7
5.7 - 36

Puget Sound
Columbia River
Cape Blanco
Cape Mendocino
San Francisco Bay
Monterey Bay
Pt. Conception

- Southern California Bight (Inner)—The inner portion of the Southern California Bight is insulated from the cooling influence of the California current by the abrupt curvature of the coast and by the presence of the Channel Islands. These warm inshore waters contain few seabirds but are especially rich in marine mammals.
- Oceanic—Species composition changes abruptly at the point where the continental shelf drops off. In deeper waters, both density and diversity decrease. This area shows much less variation than the shallower waters over the shelf.
- Estuarine—There are several major estuarine systems along the Pacific Coast that are distinct from the pelagic regions and from each other. The largest and most important estuarine systems along the Pacific Coast of the United States are the Juan de Fuca Straits, Puget Sound, and San Francisco Bay and San Pablo Bay.

WWF is delineating other U.S. marine ecoregions as part of an ongoing conservation assessment of U.S. marine ecoregions. For purposes of planning a representative, marine protected-areas system, WWF Canada is developing a marine natural region and nested seascape framework for Canadian territory in the Pacific, Arctic, and Atlantic Oceans as well as a related framework for the Canadian portion of the Great Lakes. A report detailing the methodology, including case studies, will be available from WWF Canada early in 1999.

Atlantic puffins (*Fratercula arctica*) are marine birds found along the Atlantic Coast of Canada and the northeastern United States. Photo credit: USFWS/Susan Steinacher.

CHAPTER 3

Biological Distinctiveness of North American Ecoregions

This chapter offers an overview of the most biologically distinct ecoregions in North America and the reasons for their recognition as such. We then follow with a brief analysis of patterns of species richness and endemism that contribute to calculating part of the biological distinctiveness index.

Globally Outstanding Ecoregions

The BDI elevates thirty-two North American ecoregions (28 percent), excluding Mexican ecoregions, to a globally outstanding ranking (table 3.1). These ecoregions are most prominent along the West Coast of North America, along most of the border with Mexico, across much of the southeastern United States, and in a belt across the northern boreal forests and the low arctic tundra (figure 3.1).

All ten MHTs are well represented in the globally outstanding category. Temperate coniferous forests (ten ecoregions) have the highest total, followed by boreal forests/taiga (five ecoregions). All three ecoregions in the Mediterranean scrub and savanna MHT, as well as the single flooded grassland ecoregion (the Everglades), are

TABLE 3.1. Distribution of Ecoregions by Major Habitat Type for the Biological Distinctiveness Index.

	Biological Distinctiveness Index					
Major Habitat Type	Globally Outstanding	Regionally Outstanding	Bioregionally Outstanding	Nationally Important	Not Classified	Total
1.1 Tropical Moist Broadleaf Forests	1	1	0	0	1	3
1.2 Tropical Dry Broadleaf Forests	1	0	0	0	1	2
2.1 Temperate Broadleaf and Mixed Forests	3	2	5	7	0	17
2.2 Temperate Coniferous Forests	10	3	8	9	0	30
3.1 Temperate Grasslands/ Savanna/Shrub	2	7	1	6	0	16
3.2 Flooded Grasslands	1	0	0	0	0	1
4.1 Mediterranean Scrub and Savanna	3	0	0	0	0	3
4.2 Xeric Shrublands/Deserts	2	3	4	1	0	10
6.1 Boreal Forest/Taiga	5	2	5	5	0	17
6.2 Tundra	4	4	4	5	0	17
Total	32	22	27	33	2	116

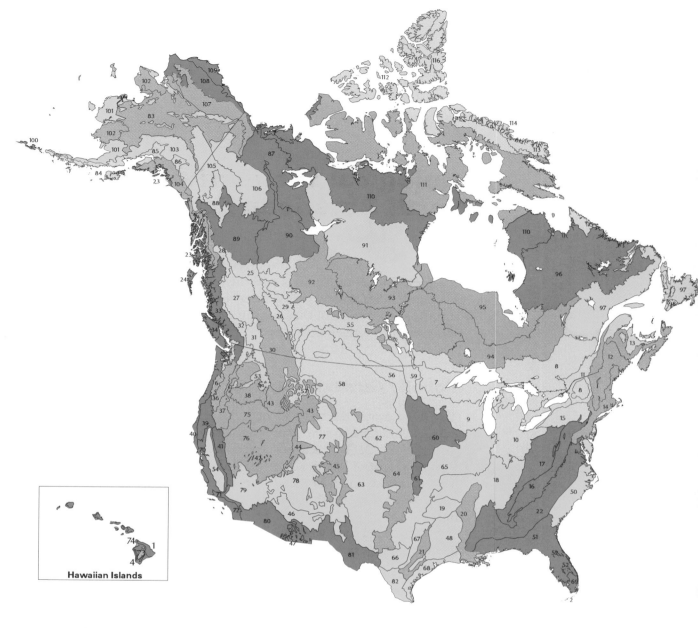

Hawaiian Islands

LEGEND

Globally Outstanding

Regionally Outstanding

Bioregionally Outstanding

Nationally Important

Figure 3.1   Biological distinctiveness categories by ecoregion.

ranked globally outstanding. Among the six MHTs containing at least ten ecoregions, temperate coniferous forests (33 percent) and boreal forest/taiga (29 percent) have the highest percentage in this category, followed closely by tundra (24 percent) (table 3.2). With all MHTs considered together, ecoregions are distributed evenly among the different biological distinctiveness categories. Besides the thirty-two globally outstanding ecoregions, there are twenty-two regionally outstanding, twenty-seven bioregionally outstanding, and thirty-three nationally important ecoregions. The two Puerto Rican ecoregions [numbers 3 and 5] are not classified because information comparable to that for the rest of the ecoregions was unavailable. They are included in a previous assessment of Latin American and Caribbean ecoregions (Dinerstein et al. 1995).

The BDI ranks an ecoregion as globally outstanding if it scores as such for any of the component criteria (see appendix A). Nine ecoregions received globally outstanding scores for more than one criterion (e.g., the Southeastern Conifer Forests [51] for richness, endemism, and rare phenomena). A further twenty ecoregions received globally outstanding scores in only one criterion (table 3.3).

Three ecoregions (Appalachian Mixed Mesophytic Forests [17], Klamath-Siskiyou Forests [39], Sonoran Desert [80]) did not receive a globally outstanding score in any single criterion but accrued enough points from all the criteria to achieve that rank.

The list of globally outstanding ecoregions includes twenty-five ecoregions that would receive that rank regardless of the total number of species or endemics found within them. Nine are elevated for the rare habitat type criterion, and sixteen more are designated for the rare ecological or evolutionary phenomena criterion (table 3.3). Mapping this subset shows a concentration along the west coast of North America and in the boreal/low arctic tundra belt (figure 3.2). Six of the twenty-five elevated by these two criteria would also earn the rank of globally outstanding because they are also rich in species or endemics (Hawaiian Moist Forests [1], Hawaiian Dry Forests [4], Southeastern Conifer Forests [51], Cal-

TABLE 3.2. Percentage of Ecoregions by Major Habitat Type for the Biological Distinctiveness Index.

	Biological Distinctiveness Index				
Major Habitat Type	*Globally Outstanding*	*Regionally Outstanding*	*Bioregionally Outstanding*	*Nationally Important*	*Not Classified*
1.1 Tropical Moist Broadleaf Forests	33%	33%	0%	0%	33%
1.2 Tropical Dry Broadleaf Forests	50%	0%	0%	0%	50%
2.1 Temperate Broadleaf and Mixed Forests	18%	12%	29%	41%	0%
2.2 Temperate Coniferous Forests	33%	10%	27%	30%	0%
3.1 Temperate Grasslands/ Savanna/Shrub	13%	44%	6%	38%	0%
3.2 Flooded Grasslands	100%	0%	0%	0%	0%
4.1 Mediterranean Scrub and Savanna	100%	0%	0%	0%	0%
4.2 Xeric Shrublands/Deserts	20%	30%	40%	10%	0%
6.1 Boreal Forest/Taiga	29%	12%	29%	29%	0%
6.2 Tundra	24%	24%	24%	29%	0%

TABLE 3.3. Ecoregions in the United States and Canada Designated as Having Globally Outstanding (GO) Biodiversity.

Ecoregion Number	Ecoregion Name	Biological Distinctiveness Index Criteria				
		Richness	Endemism	Rare Habitat Type	Rare Phenomena	Point Accumulation
1	Hawaiian Moist Forests		X		X	
4	Hawaiian Dry Forests		X		X	
16	Appalachian/Blue Ridge Forests	X	X			
17	Appalachian Mixed Mesophytic Forests					X
22	Southeastern Mixed Forests	X	X			
23	Northern Pacific Coastal Forests			X		
24	Queen Charlotte Islands			X		
33	British Columbia Mainland Coastal Forests			X		
34	Central Pacific Coastal Forests			X		
39	Klamath-Siskiyou Forests					X
40	Northern California Coastal Forests			X		
41	Sierra Nevada Forests				X	
47	Madrean Sky Islands Montane Forests				X	
51	Southeastern Conifer Forests	X	X		X	
52	Florida Sand Pine Scrub				X	
60	Central Tall Grasslands				X	
61	Flint Hills Tall Grasslands				X	
69	Everglades				X	
70	California Interior Chaparral and Woodlands	X	X	X		
71	California Montane Chaparral and Woodlands	X		X		
72	California Coastal Sage and Chaparral	X		X		
80	Sonoran Desert					X
81	Chihuahuan Desert	X	X			
87	Northwest Territories Taiga				X	
89	Northern Cordillera Forests				X	
90	Muskwa/Slave Lake Forests				X	
96	Eastern Canadian Shield Taiga				X	
99	South Avalon–Burin Oceanic Barrens			X		
100	Aleutian Islands Tundra		X			
108	Arctic Foothills Tundra				X	
109	Arctic Coastal Tundra				X	
110	Low Arctic Tundra				X	

NOTE: "X" designates a score of globally outstanding for that criterion.

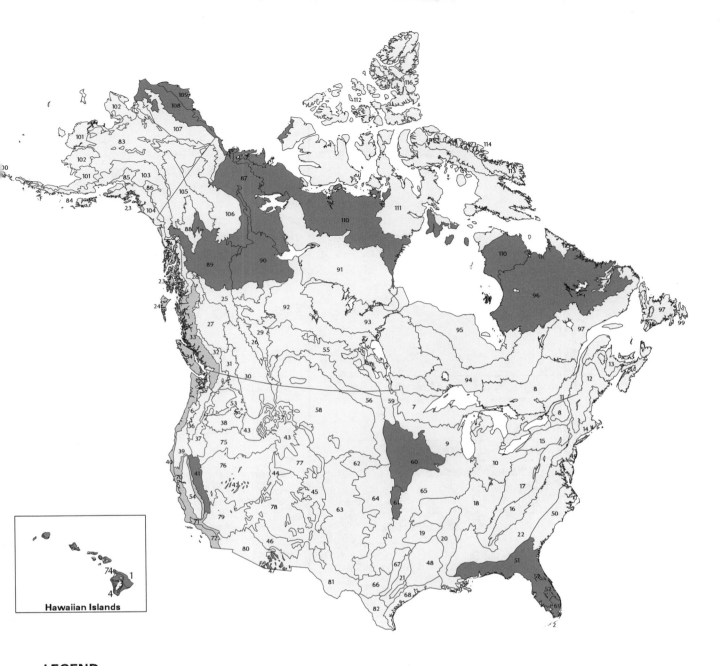

LEGEND

Globally rare habitat types

Globally outstanding phenomena

Figure 3.2   Ecoregions with globally rare habitat types or evolutionary and ecological phenomena.

Hawaiian Islands

ifornia Interior Chaparral and Woodlands [70], California Montane Chaparral and Woodlands [71], and California Coastal Sage and Chaparral [72]).

Table 3.4 briefly lists justifications for elevating these twenty-five ecoregions, and full explanations are detailed in appendix F. Two such justifications are the presence of large-scale migrations and calving areas of terrestrial herbivores, and the presence of intact large predator–prey assemblages, both of which are biological phenomena disappearing around the world (essays 5 and 6). In addition to those listed in table 3.3, we also rank twenty-two ecoregions in intermediate categories for these two criteria. These ecoregions can be located in appendix D, and the reasons for their designation are included in the ecoregion descriptions in appendix F.

TABLE 3.4. Ecoregions Designated Globally Outstanding for Rare Habitat Type or Phenomena.

Ecoregion Number	Ecoregion Name	Reason for Designation
Rare Habitat Type		
23	Northern Pacific Coastal Forests	Global rarity of temperate rain forests.
24	Queen Charlotte Islands	Global rarity of temperate rain forests.
33	British Columbia Mainland Coastal Forests	Global rarity of temperate rain forests.
34	Central Pacific Coastal Forests	Global rarity of temperate rain forests.
40	Northern California Coastal Forests	Global rarity of temperate rain forests.
70	California Interior Chaparral and Woodlands	Global rarity of Mediterranean scrub and savanna habitat type.
71	California Montane Chaparral and Woodlands	Global rarity of Mediterranean scrub and savanna habitat type.
72	California Coastal Sage and Chaparral	Global rarity of Mediterranean scrub and savanna habitat type.
99	South Avalon–Burin Oceanic Barrens	Rare headland tundra.
Rare Ecological or Evolutionary Phenomena		
1	Hawaiian Moist Forests	High beta-diversity and species radiations.
4	Hawaiian Dry Forests	High beta-diversity and species radiations.
41	Sierra Nevada Forests	Forest structure dominated by world's most massive trees.
47	Madrean Sky Islands Montane Forests	Pronounced local endemism.
51	Southeastern Conifer Forests	High beta-diversity and species radiations.
52	Florida Sand Pine Scrub	High beta-diversity (endemism/unit area) and relictual taxa.
60	Central Tall Grasslands	Tallest temperate grasslands, ecological phenomena.
61	Flint Hills Tall Grasslands	Tallest temperate grasslands, ecological phenomena.
69	Everglades	Sheetflow grasslands, globally rare ecological process.
87	Northwest Territories Taiga	Intact predator-prey assemblages.
89	Northern Cordillera Forests	Intact predator-prey assemblages.
90	Muskwa/Slave Lake Forests	Intact predator-prey assemblages.
96	Eastern Canadian Shield Taiga	Large-scale caribou migrations.
108	Arctic Foothills Tundra	Large-scale caribou migrations, bird nesting sites.
109	Arctic Coastal Tundra	Large-scale caribou migrations, bird nesting sites.
110	Low Arctic Tundra	Large-scale caribou migrations.

Caribou Migrations and Calving Grounds: Globally Outstanding Ecological Phenomena

Anne Gunn

Large-scale migrations of large terrestrial mammals are disappearing around the world. Grass-lands and forested habitats that once supported extraordinary movements and seasonal concentrations of large herbivores are becoming increasingly threatened by development activities. Ecoregions of the North American boreal forest and tundra MHTs of North America represent some of the last strongholds of this important ecological phenomenon.

The annual return of barren-ground caribou (*Rangifer arcticus*) to their traditional calving grounds is an unforgettable spectacle that serves as one of the best examples of the migratory phenomenon. Few places in the world can equal the scale of this migration, unfettered by fences, roads, agriculture, or people.

Throughout northern Canada and Alaska, free-roaming caribou are an important subsistence food for aboriginal peoples. In Russia, another important site for caribou, and reindeer, many populations are semidomesticated and managed by local people.

In the larger herds, tens of thousands of cows stream from the tree-covered taiga across the snow-covered barrens to the areas that they have used for generations. The cows unerringly return to their own calving grounds, where they would have been born and where some have calved before. Some barren-ground caribou herds spend the whole year on the barrens but exhibit the same pattern of migrating to a traditional calving ground from their winter ranges.

The cows follow traditional routes to the calving grounds, but the chosen route varies between years, being dictated by where the caribou wintered and snow conditions during the spring migration. In contrast, the location of the traditional calving ground is highly predictable over decades.

Usually, the calving grounds are sufficiently north of the winter range that snow melt is just beginning when the cows reach the area. The melting snow uncovers plants, which begin their annual cycle of growth and flowering. Flower and leaf buds are highly nutritious and avidly sought by the caribou. When the cows reach the calving grounds, they are thin. Once they calve, they need high-quality food to produce milk for their rapidly growing calves. Caribou prefer variable terrain, which offers more choice in possible feeding sites. At this time, wolves are less abundant, because they are denning along the treeline when the caribou are calving. Additionally, there is safety in numbers. Caribou calves are almost all born within two or three days of each other, and their high numbers satiate predators such as wolves and grizzly bears.

There are fifteen major caribou herds in Canada, each with a different spring migration route. The Eastern Canadian Shield Taiga ecoregion [96] is home to the George River herd, the single largest caribou herd in the world. The Northwest Territories contain an additional 1.6 million caribou. The annual range of many of these herds, including the Bluenose, Bathurst, Beverly, Queen Maud Gulf, and Qamanirjuaq herds, fall within the Low Arctic Tundra ecoregion [110].

In Alaska, twenty-five caribou herds totaling approximately a million animals annually stream between wintering and calving grounds. Caribou range across virtually all of Alaska's ecoregions (except the Northern Pacific Coastal Forests [23] and Pacific Coastal Mountains [104]). Introduced herds even occupy islands in the Bering Sea and Aleutian Islands.

Caribou cows, especially when they have newborn calves, are highly sensitive to human activities. Caribou, especially in large groups, may gallop away if startled by people or machinery. Calves may be knocked over and injured or may flounder in wet snow and become exhausted. When cows run, walk, or even stand in response to human activity, they are expending energy without feeding. Reduced food intake in turn reduces milk production, which could limit calf growth.

Not only do cows and calves need protection while on calving grounds, but the calving ground habitat itself needs protection. Loss of sedge meadows through the alteration of surface drainage could lead to reduced forage availability. Mining and mine exploration pose serious threats to the ecological integrity of caribou calving grounds and migration routes. There are already two active mines on the spring migration route and calving grounds of the Bathurst herd. Oil exploration and development activities also pose threats to the integrity of migration routes. Impacts of oil development along Alaska's North Slope are a significant concern. Although caribou in the Central Arctic herd have become somewhat habituated over time to oil field activities, increased development, including habitat loss, animal disturbance, and disruption of migrations, remains a concern, particularly in calving grounds.

Recognizing these threats, the governments of Canada and the Northwest Territories are working with the aboriginal people through wildlife comanagement boards and other stakeholders to develop a protective strategy for the caribou calving grounds. Alaskan Native peoples are also increasingly playing a role in management decisions regarding caribou. The maintenance of one of the last great spectacles of nature depends on all stakeholders acting together.

Caribou (*Rangifer* spp.) from the Porcupine herd graze in the Arctic National Wildlife Refuge in Alaska. Photo credit: WWF/Randall Snodgrass.

ESSAY 6

Reconnecting Grizzly Bear Populations in Fragmented Landscapes

Steve Primm and Emma Underwood

Perhaps the next biological calamity in the current global extinction crisis is the loss of large terrestrial predatory mammals in most natural areas of the world (Wikramanayake et al., in press). The North American conservation assessment correctly focuses on ecoregions that still maintain the full complement of large terrestrial predators (wolves, bears, puma, lynx, etc.) and their prey, and where predator-prey populations still fluctuate within their natural range of variation. For example, the Northern Cordillera Forests [89] and the Muskwa/Slave Lake Forests [90] are two ecoregions with intact predator-prey assemblages—some Canadian biologists refer to these ecoregions as the Canadian Serengeti—and are elevated as globally outstanding in this analysis.

But farther down the Rocky Mountain chain, how can we preserve large carnivores in landscapes with more people, roads, and fragmented habitat blocks? If large carnivores are vital to the health of ecosystems, how do we design effective corridors to reconnect isolated populations?

An excellent example is the grizzly bear (*Ursus arctos*) population of the Greater Yellowstone ecosystem. Even by the most optimistic measures, this grizzly population is too small ($n < 400$) for long-term viability, especially since it likely has a negative growth rate. The population's chances of persistence would be markedly increased if it could expand into new habitat and ultimately reconnect with other grizzly populations in the Rockies. An evaluation of unoccupied habitat in southwestern Montana that could serve as "linkage zones" has been undertaken in a WWF study, illustrated in part on the adjacent maps. The initial focal area for analysis is a 70,000-hectare mountainous landscape on the western fringe of the Greater Yellowstone ecosystem (figure 3.3a). Grizzly recolonization of these mountains could form the first stepping stone in a series that would ultimately lead to functional connectivity with other grizzly populations. The GIS approach used here provides a framework to apply to grizzlies moving between blocks of natural habitat connected by corridors into other ecosystems, as well as to other carnivores, such as mountain lions or jaguars.

One of these ecosystems is grassland-shrub steppe. The southwest Montana region being evaluated offers a rare opportunity for grizzlies to re-inhabit grasslands, since much of the landscape consists of high, remote, grassland-sagebrush communities. Grizzly bears once ranged across the prairies to central Illinois. The presence of grizzly bears in grasslands is a globally outstanding ecological phenomenon that is now restricted to less than three sites in North America. Successful conservation efforts to maintain connectivity in southwest Montana would help preserve this phenomenon.

The analysis examines human-related disturbances across the landscape and evaluates them according to generalized responses of grizzlies to human activities. Linkage zones for grizzlies must meet two key requirements: First, there must be ample undisturbed habitat that is contiguous with source populations of grizzlies. Such habitat greatly increases the likelihood of movement and exchange of individual bears. Second, these linkage habitats must supply food resources sufficient to support some resident animals.

The first requirement, of undisturbed habitat, is crucial because human-caused mortality and disturbance led to the extirpation of grizzlies in most of southwest Montana. Our disturbance modeling approach assumes the behavior of a wary, or unhabituated bear, traveling across the landscape. The GIS analysis has five input data layers: roads, surface mines, towns, campgrounds, and cow camps. Each of these features has a distance of disturbance and also an intensity of disturbance (based on the Cumulative Effects Model, Weaver et al. 1985). For example, the noise from a road decreases in a linear fashion with distance away from it. Attention is also given to the biophysical landscape; for example, the effect of vegetation and topography serves to modify the disturbance of human-related features. Once the disturbance has been calculated for each feature, all are combined into what can be visualized as a three-dimensional surface; the highest peaks represent the most disturbed areas, while the flat areas can be considered relatively secure habitat for grizzlies (figure 3.3b).

Within this "disturbed landscape" the easiest route for a grizzly to cross the landscape can be determined, therefore highlighting where conservation attention should be focused. One area of particular interest is a potential crossing point over Route 287, from the source grizzly populations of Yellowstone and the Madison Range into the Beaverhead National Forest.

The second part of the analysis assesses the degree to which these blocks of relatively secure habitat can act as a "stepping stone" in the landscape to ensure safe passage to larger contiguous habitat blocks. Input layers here include calving areas and ungulate winter range; areas of appropriate nutritional vegetation, such as white bark pine (*Pinus albicaulis*); slopes greater than 45 percent, which are relatively secure from logging or grazing; and the undisturbed blocks elucidated from the prior analysis. The weighting and combining of these inputs highlight the southern part of the Beaverhead National Forest as having high-quality, multiseasonal food sources, with other parts of the national forest having two-season food sources (figure 3.3c).

A holistic evaluation of grizzly habitat needs to incorporate the issue of public and private lands in the region; based on these analyses, WWF-US is making some recommendations that are applicable to both public and private lands. Some of these include appropriate sanitation and food storage, public land area closures, removal of logging roads, and conservation of important riparian habitat as a corridor for movement along valley bottoms. On private land, measures such as conservation easements may be appropriate.

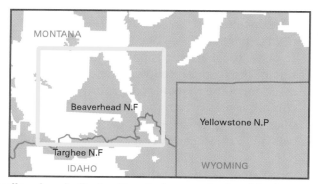

Figure 3.3 Maps of habitat disturbance and dispersal for grizzly bears.

(a) The yellow box indicates the study area depicted in the analyses below.

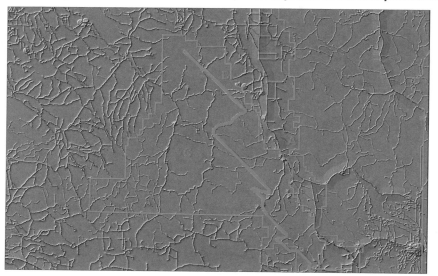

(b) Potential dispersal paths in a disturbed landscape.

Figure 3.3b represents a surface of disturbance. The smooth, flat areas can be considered relatively secure habitat for grizzlies, whereas the gray elevated lines and features in the image represent the disturbance associated with human-related features. For example,

A = linear disturbances associated with Route 287
B = radial disturbances associated with mine sites
C = area of minimal disturbance

The pink line represents the least disturbed, easiest route through the landscape. The path starts on the Targhee National Forest, near a potential source population of grizzlies in Yellowstone, and an ending point specified in the northern Gravelly Range. The path is disturbance driven, i.e., its trajectory is determined sequentially as each new disturbance in the landscape is encountered.

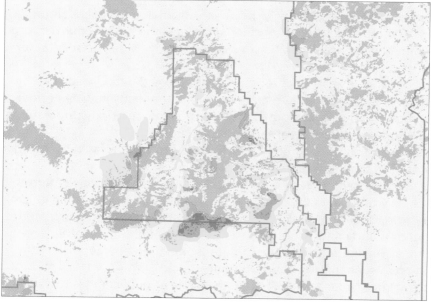

(c) Potential nutritional density of various habitats.

The analysis of potential nutritional density (figure 3.3c) displays the cumulative results of potential food sources and areas considered relatively safe owing to their limited economic value. From the analysis, the following classes can be identified:

CRITICAL FOOD RESOURCES FOR BEARS;

High-quality, multiseasonal food sources
■ Level I
■ Level II

Two-season food sources
■ Level I
■ Level II

Limited seasonal food availability/productivity
□ Level I
■ Level II

Dispersed Food Resources
□ General habitat
∧ National forest boundary

Geographic Patterns of Species Richness and Endemism

Although our BDI treats each MHT independently when comparing levels of richness and endemism, the broader patterns of diversity in these taxa across all MHTs are worth noting. Combined data for species richness (figure 3.4a) and endemism (figure 3.4b) reveal a strong pattern in distribution of species among ecoregions. From visual inspection, the well-documented latitudinal gradient of species richness is clear (Brown 1995; Currie 1991).

The ecoregions vary substantially in their area, possibly influencing the richness and endemism scores in an ecoregion. However, removing area statistically does not change results in a significant manner (Ricketts et al. 1999).

Geographic Patterns of Species Richness and Endemism within and among Higher Taxa

Geographic patterns of species richness and endemism within and among individual higher taxa vary considerably across the continent (figure 3.5a–t; see appendix C for raw scores of richness and endemism by ecoregion). Between groups there is sometimes high

Figure 3.4a Total richness of analyzed taxa (amphibians, birds, butterflies, mammals, vascular plants, reptiles, snails) for ecoregions in the United States and Canada.

LEGEND

- 3650 - 4295 species
- 3237 - 3649 species
- 2834 - 3236 species
- 2423 - 2833 species
- 2012 - 2422 species
- 1601 - 2011 species
- 1189 - 1600 species
- 778 - 1188 species
- 595 - 777 species
- 112 - 594 species

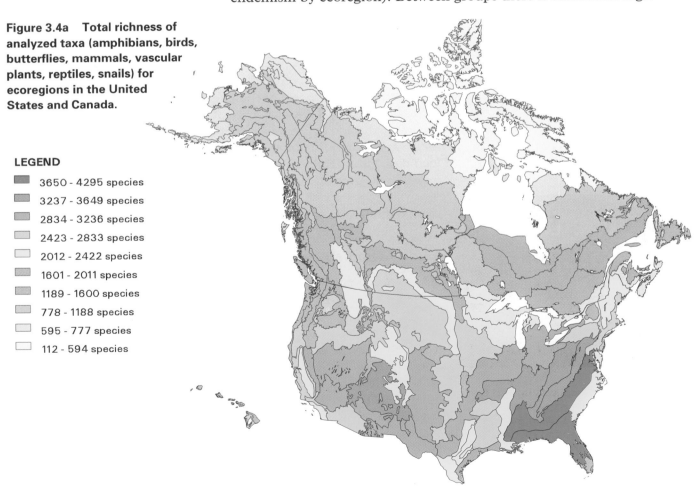

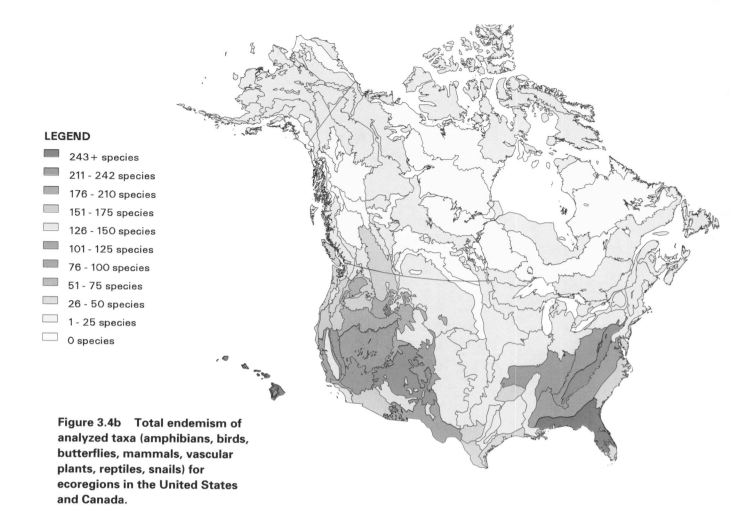

Figure 3.4b   Total endemism of analyzed taxa (amphibians, birds, butterflies, mammals, vascular plants, reptiles, snails) for ecoregions in the United States and Canada.

More butterfly species inhabit the Chihuahuan Desert than any other ecoregion in the United States and Canada. The Mormon metalmark (*Apodemia mormo*) occurs in arid habitats throughout the West, including the Chihuahuan. Photo credit: USFWS.

geographic concordance of these features, while others show diverging trends. This variability derives from many factors, but certainly the ecological and evolutionary opportunities for different taxa vary by major habitat type, and the role of history and chance can have considerable influence on the diversity and distinctiveness of different groups. We have organized the taxa included in our analysis by groups that show the greatest similarities in their continental-scale patterns.

Birds, Mammals, and Butterflies

Three taxa show similar continental-scale richness and endemism patterns. The birds (figure 3.5a,b), mammals (figure 3.5c,d), and butterflies (figure 3.5e,f) display similar east-west bands of richness that decrease northward to the edge of the boreal forests, with richness maxima centered in the southwestern United States. If these southwestern ecoregions were considered to their full extent into Mexico, richness values would be even higher. The Chihuahuan Desert [81] stands out as the ecoregion that supports the highest number of these three taxa; it probably is the most species-rich ecoregion in the entire Nearctic zone for these taxa and may be the richest desert ecosystem in the world.

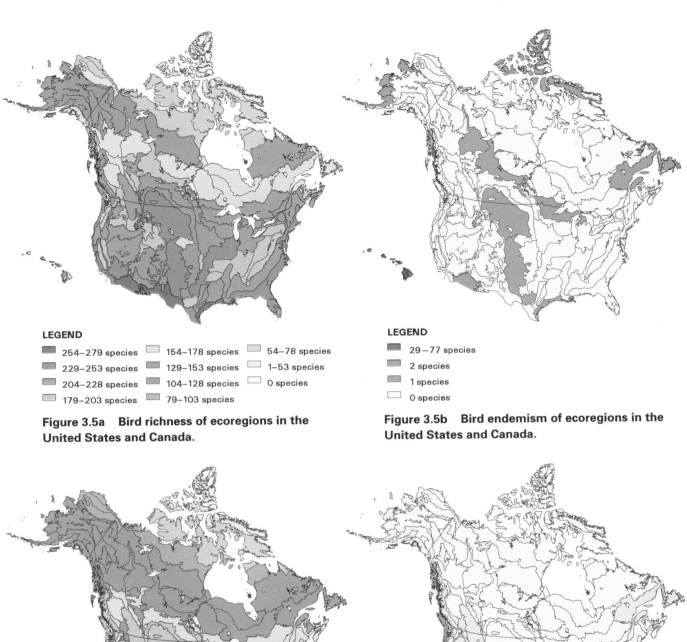

LEGEND

254–279 species	154–178 species	54–78 species
229–253 species	129–153 species	1–53 species
204–228 species	104–128 species	0 species
179–203 species	79–103 species	

Figure 3.5a Bird richness of ecoregions in the United States and Canada.

LEGEND

29–77 species
2 species
1 species
0 species

Figure 3.5b Bird endemism of ecoregions in the United States and Canada.

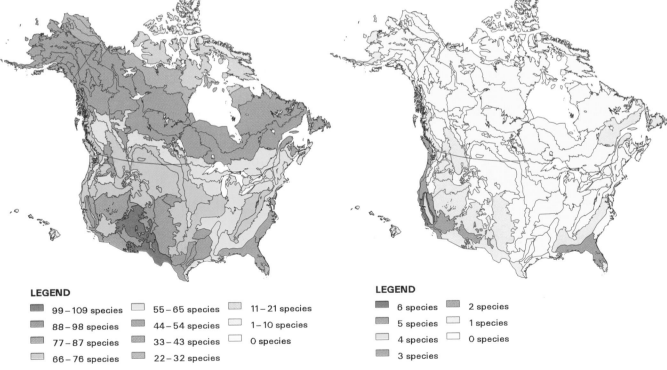

LEGEND

99–109 species	55–65 species	11–21 species
88–98 species	44–54 species	1–10 species
77–87 species	33–43 species	0 species
66–76 species	22–32 species	

Figure 3.5c Mammal richness of ecoregions in the United States and Canada.

LEGEND

6 species	2 species
5 species	1 species
4 species	0 species
3 species	

Figure 3.5d Mammal endemism of ecoregions in the United States and Canada.

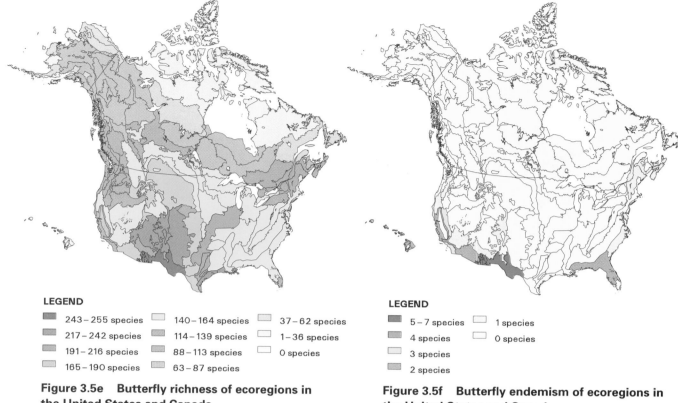

LEGEND

■ 243 – 255 species	140 – 164 species	37 – 62 species
■ 217 – 242 species	114 – 139 species	1 – 36 species
■ 191 – 216 species	88 – 113 species	0 species
■ 165 – 190 species	63 – 87 species	

Figure 3.5e   Butterfly richness of ecoregions in the United States and Canada.

LEGEND

■ 5 – 7 species	1 species
■ 4 species	0 species
3 species	
2 species	

Figure 3.5f   Butterfly endemism of ecoregions in the United States and Canada.

Within Mexico, some Neotropical moist forest ecoregions in the south likely support more total species for each of these taxa. Bird richness in higher latitudes is bolstered by the seasonal influx of breeding species.

Overall, few ecoregions support high numbers of endemic mammals, birds, or butterflies. Endemism is low in these taxa largely because the ranges of temperate, boreal, and polar species are large compared to tropical forms, and many species in these taxa are highly vagile. The few ecoregions where foci occur are in the California region and in the Sonoran [80] and Chihuahuan Deserts [81] (although <7 species of mammals occur there, these values will increase when the full extent of the Chihuahuan and Sonoran Deserts is considered). The Chihuahuan Desert has the highest butterfly endemism, with lower levels of endemism in the Sierra Nevada Forests [41], Northern California Coastal Forests [40], and California Coastal Sage and Chaparral [72]. Mammal endemism is low across the United States and Canada, with a small number of endemics found in the California Interior Chaparral and Woodlands [70], the Sierra Nevada Forests [41], and the California Central Valley Grasslands [54]. This pattern largely reflects localized ranges seen in certain groups of rodents (e.g., Kangaroo rats, *Dipodomys* spp.). The California Coastal Sage and Chaparral [72], the British Columbia Mainland Coastal Forests [33], the Central Pacific Coastal Forests [34], and the Southeastern Conifer (longleaf pine, *Pinus palustris*) Forests [51] also harbor three endemic mammals each.

Reptiles

Reptiles also show a maximum for species richness in the Chihuahuan Desert [81] (103 species), although the Western Gulf Coastal Grasslands [68] (84 species), Southeastern Conifer Forests [51] (85 species), Southeastern Mixed Forests [22] (73 species), and Central Forest/Grassland Transition Zone [65] (79 species) are also particularly diverse (figure 3.5g). Only the Great Sandy Desert of Australia supports a richer desert reptile fauna than the Chihuahuan Desert (Cogger 1992; Flannery 1994). Reptile richness drops off quickly in more northerly ecoregions and in the California region, perhaps reflecting the importance of warm temperatures as a precondition for rich assemblages of these organisms. The low reptile diversity of the California region is puzzling but may be related to the cool, wet winters characteristic of Mediterranean climates. Reptile endemism is highest in the Southeastern Conifer Forests, where 16 species occur; this level is likely a global maximum for endemic reptile species in temperate forests (figure 3.5h). The Southeastern Mixed Forests and the Chihuahuan Desert also harbor 5 endemic species; consideration of the Mexican portion of the Chihuahuan Desert ecoregion would add several more endemics.

Amphibians and Land Snails

Amphibians (figure 3.5i,j) and land snails (figure 3.5k,l) display similar patterns in both richness and endemism. The forests of the Appalachia region, composed of the Appalachian Mixed

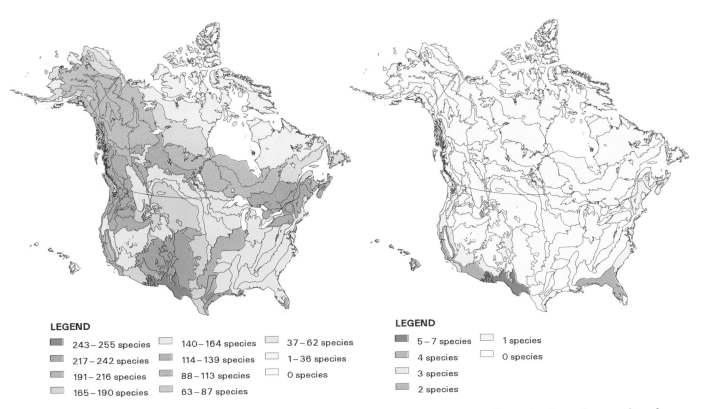

LEGEND

- 243 – 255 species
- 217 – 242 species
- 191 – 216 species
- 165 – 190 species
- 140 – 164 species
- 114 – 139 species
- 88 – 113 species
- 63 – 87 species
- 37 – 62 species
- 1 – 36 species
- 0 species

Figure 3.5g   Reptile richness of ecoregions in the United States and Canada.

LEGEND

- 5 – 7 species
- 4 species
- 3 species
- 2 species
- 1 species
- 0 species

Figure 3.5h   Reptile endemism of ecoregions in the United States and Canada.

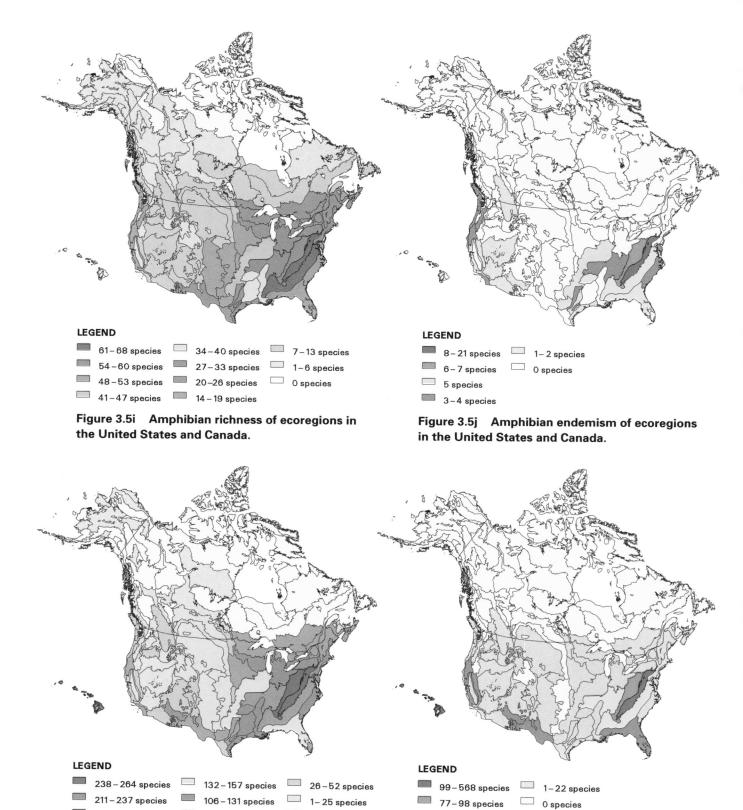

LEGEND

▪ 61–68 species	▫ 34–40 species	▪ 7–13 species	
▪ 54–60 species	▪ 27–33 species	▫ 1–6 species	
▪ 48–53 species	▪ 20–26 species	▫ 0 species	
▫ 41–47 species	▪ 14–19 species		

Figure 3.5i Amphibian richness of ecoregions in the United States and Canada.

LEGEND

▪ 8–21 species	▫ 1–2 species
▪ 6–7 species	▫ 0 species
▫ 5 species	
▪ 3–4 species	

Figure 3.5j Amphibian endemism of ecoregions in the United States and Canada.

LEGEND

▪ 238–264 species	▫ 132–157 species	▪ 26–52 species	
▪ 211–237 species	▪ 106–131 species	▫ 1–25 species	
▪ 185–210 species	▪ 79–105 species	▫ 0 species	
▫ 158–184 species	▪ 53–78 species		

Figure 3.5k Snail richness of ecoregions in the United States and Canada.

LEGEND

▪ 99–568 species	▫ 1–22 species
▪ 77–98 species	▫ 0 species
▫ 44–76 species	
▪ 23–43 species	

Figure 3.5l Snail endemism of ecoregions in the United States and Canada.

Many invertebrates of the Pacific Northwest rain forests cannot survive outside old-growth forest habitat, such as that found in the Hoh Rain Forest of Olympic National Park. Photo credit: WWF/C. Loucks.

Mesophytic Forests [17], the Appalachian/Blue Ridge Forests [16], and the Southeastern Mixed Forests [22], support the most species of these groups in the Nearctic region. Amphibian richness is highest in the latter two ecoregions, with up to 68 species in the Appalachian/Blue Ridge Forests. These ecoregions harbor remnants of a formerly widespread temperate forest that now occurs only in central China and the eastern part of North America. This region is the global center of diversity for Plethodontid (wood) salamanders, an ancient group that has thrived in these old, warm, and moist temperate broadleaf forests. The Appalachian/Blue Ridge Forests show the highest amphibian endemism (21 species) because wood salamanders display considerable local endemism, with some found only on single small mountain tops. Frogs and toads are also very diverse in this region. Indeed, the freshwater ecoregions of the Southeast support the richest temperate aquatic reptile and amphibian faunas in the world, with many endemic species (Abell et al. 1999).

Land snails show a similar pattern, with a center of diversity in the Appalachian Mixed Mesophytic Forests and the Appalachian/Blue Ridge Forests. Again, this group has survived and diversified in these relictual forests. The great age of these forests has allowed considerable diversification within the land snails, as well as the amphibians. Endemism is also highest for land snails in the Appalachian/Blue Ridge Forests (122 species), many species occurring only within small areas, particularly on localized limestone and shale formations.

Other Invertebrates

For the vast majority of invertebrate taxa, little or no distribution information is available. However, many of the taxa, particularly those with strong affinities and specializations for moist temperate broadleaf forests, may follow the richness and endemism patterns seen in the land snails. Such groups would include flatworms, collembola, rove beetles, weevils, mold beetles, and leaf beetles (as well as many nonvascular plants such as mosses, lichens, and liverworts). Tiger beetles follow richness and endemism patterns similar to those seen in butterfly and mammal distributions (figure 3.5m,n). Some bees appear to be most diverse in the Mediterranean-climate shrublands and deserts of the Southwest, while mites and spiders may be most diverse in the few remaining old-growth temperate rain forests of the Pacific Northwest. Indeed, some studies indicate that mite diversity in those ecosystems surpasses that of even the richest tropical forests (Marcot 1997).

Native Vascular Plants

Native vascular plants are the most diverse taxa of those analyzed, their numbers in any one ecoregion probably surpassed only by certain groups of unanalyzed invertebrates, such as beetles. The Southeastern Mixed Forests [22] support the richest assemblage of vascular plants (3,363 species), followed by the Southeastern Conifer Forests [51] (figure 3.5o). The Southeastern Mixed

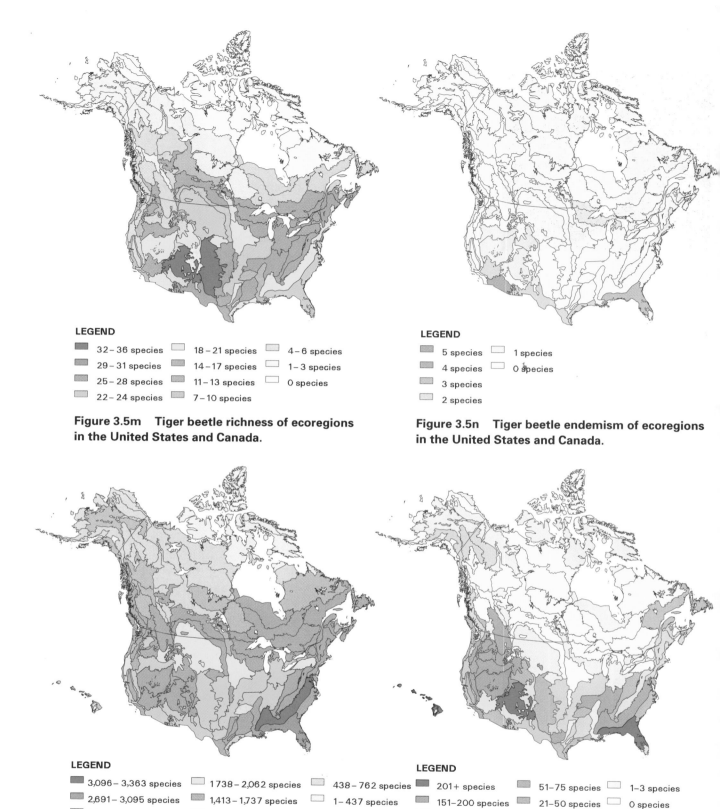

LEGEND

32 – 36 species	18 – 21 species	4 – 6 species
29 – 31 species	14 – 17 species	1 – 3 species
25 – 28 species	11 – 13 species	0 species
22 – 24 species	7 – 10 species	

Figure 3.5m   Tiger beetle richness of ecoregions in the United States and Canada.

LEGEND

5 species	1 species
4 species	0 species
3 species	
2 species	

Figure 3.5n   Tiger beetle endemism of ecoregions in the United States and Canada.

LEGEND

3,096 – 3,363 species	1738 – 2,062 species	438 – 762 species
2,691 – 3,095 species	1,413 – 1,737 species	1 – 437 species
2,388 – 2,690 species	1,088 – 1,412 species	
2,063 – 2,387 species	763 – 1,087 species	

Figure 3.5o   Vascular plant richness of ecoregions in the United States and Canada.

LEGEND

201+ species	51 – 75 species	1 – 3 species
151 – 200 species	21 – 50 species	0 species
111 – 150 species	11 – 20 species	
76 – 110 species	4 – 10 species	

Figure 3.5p   Vascular plant endemism of ecoregions in the United States and Canada.

Many endemic plants such as this pitcher plant (*Darlingtonia californica*) occur in the bog and serpentine communities of the Klamath-Siskiyou region. Photo credit: WWF/D. Olson.

Forests are particularly high in species because they share rich floras with the adjacent Appalachian ecoregions and the Southeastern Conifer Forests. The longleaf pine forests of the Southeast are remarkable for having the most diverse herbaceous understory floras on earth. These floras exist in the relatively high-light environments of the open longleaf pine forests—a phenomenon maintained by periodic ground fires—whose great age and stability have likely contributed to diversification. The Appalachian/Blue Ridge Forests [16] and Mixed Mesophytic Forests [17] are also particularly rich. Another rich plant region is found within the ecoregions of the Colorado Plateau Shrublands [78], the Mojave Desert [79], and the Great Basin Shrub Steppe [76]. The complex terrain, soils, and diverse environmental conditions found within these ecoregions have contributed to pronounced diversification and local endemism within several plant groups. Indeed, the Great Basin and the Colorado Plateau show some of the highest levels of plant endemism in the United States and Canada (figure 3.5p) (Mexico contains many ecoregions with many endemic species and higher taxa). Only the Southeastern Conifer Forests rival the Colorado Plateau in endemism. The Chihuahuan Desert and the California Coastal Sage and Chaparral may also be considerably higher in plant richness and endemism when their full extent into Mexico is evaluated. The Aleutian Islands Tundra [100] and the Channel Islands (part of California Coastal Sage and Chaparral [72]) are notable for their numbers of endemic plants relative to their area and habitat type. Other interesting patterns are described in essay 7.

Essay 7

Multiscale Analysis of Endemism of Vascular Plant Species

John Kartesz and Amy Farstad

Since endemic taxa often include members of our flora that are considered to be rare, or at risk of becoming so, the topic of endemism generates a disproportionately high level of interest among biologists and conservationists. Ever since A. C. de Candole coined the term *endemism,* numerous theories have been proposed to explain why certain taxa are restricted geographically. Over the decades, the term has been subdivided by various workers into a myriad of concepts and definitions to explain the types of endemism. The two most widely accepted terms used for this purpose are *paleoendemism* and *neoendemism. Paleoendemism,* as the word implies, describes relictual or ancient taxa that have lost much of their original range or taxa that may never have had broad-based distributions over the landscape. In contrast, *neoendemism* is used to describe newly evolved or incipient taxa that have become morphologically and genetically distinct from their ancestral lines and may in fact be expanding their geographic ranges. For many endemic taxa it is difficult to ascertain a complete picture of historical distributions; thus, we must understand endemism from what we can glean from their current distributions, fossil remains, pollen cores, etc.

Any serious discussion of endemism must be predicated on the concept of geographic scale—i.e., the physical size of the landscape occupied by the endemics. In conjunction with topography, resource availability, and historical patterns of disturbance such as glaciation and volcanism, knowledge of geographic scale helps to explain the types and potential numbers of endemics that may occur within a given area. Generally speaking, the larger the geographic scale, the higher the number of endemics found there, due to the increased number of isolating factors such as deserts, mountains, rivers, and lakes. However, a large geographic area in itself—i.e., a vast, undifferentiated landscape—appears to be insufficient alone as a factor promoting endemism. Consider the enormous area of the Great Plains, which includes portions of thirteen U.S. states and six ecoregions, including: Northern Mixed Grasslands [56], Northwestern Mixed Grasslands [58], Flint Hills Tall Grasslands [61], Nebraska Sand Hills Mixed Grasslands [62], Western Short Grasslands [63], and Central and Southern Mixed Grasslands [64]. Despite its size, this entire area contains a relatively small number of endemic species: 23. Other large areas, such as the vast interior of the state of Alaska, which encompasses portions of more than thirteen ecoregions, have equally few endemics (16), whereas a disproportionately high number of endemics occurs within the Aleutian archipelago (approximately 9). Conversely, on a much smaller geographic scale, the Hawaiian Islands (covering about 16, 950 km²) have historically been known for their high concentration of endemic species: 850 in total. Another small ecoregion on the continental mainland, the Madrean Sky Islands Montane Forests [47], with a much hotter, drier, and more stressful environment than that of Hawaii, has produced a relatively high number of endemic species: 31. The most likely explanation for these smaller areas producing such high numbers of endemic species is the influence of isolating mechanisms acting over prolonged time periods.

Assessing endemism on a continental scale, especially for an area the size of the North American continent north of Mexico, presents a formidable challenge. The difficulties inherent to such an assessment are numerous, not the least of which must include resolving taxonomic problems relating to classification, i.e., the splitting versus the lumping of taxa. Since taxonomic views vary widely, they can alter dramatically the number of endemics thought to occur within any biogeographic area. Inasmuch as these views have direct bearing on the final tally for a region, it is nearly as important to consider the political consequences of the exercise, as the biological ones, to ensure that a uniform and accurate assessment of endemism is determined. Thus, for the purpose of this study the number of taxa considered to be endemic within any particular ecoregion is based on the single taxonomic standard of Kartesz (1994). It should also be noted that during the course of this study, only fully recognized species were considered and that any plant found outside of a particular ecoregional boundary was eliminated from consideration.

The results from our study suggest clearly that the ecoregion representing the highest rate of endemism within continental North America north of Mexico, at least in terms of actual numbers of species, is the Colorado Plateau Shrublands [78]. This ecoregion, which spans more than 326,390 km² of broad desert plains, interspersed by volcanic mountains and transversed by the Colorado River, contains more than 290 endemic species. It is a region dominated by a harsh, dry environment, one that has historically placed intense environmental stress on its flora. Due to its large physical size and varied topography, it is reasonable to expect that this ecoregion would be dotted by many small pockets of unusual habitat with numerous isolated plant populations, representing both paleoendemic and neoendemic species.

Adjacent to this region is the Great Basin Shrub Steppe [76], an area of the Intermountain West that includes the Lahontan, Bonneville, and Central Great Basin sections, as well as much of the Tonopah and Reno sections of Nevada and Utah. It, like the region discussed previously, is dominated by dry, hot, desert valleys, interspersed by mountains. Here, 151 endemic species occur, including numerous members of the Fabaceae, Apiaceae, Brassicaceae, Polygonaceae, and Asteraceae families, examples of which are *Astragalus porrectus* (Lahontan milk-vetch), *Cymopterus davisii* (Davis' spring-parsley), *Arabis falcatoria* (Grouse Creek rockcress), *Eriogonum soredium* (Frisco wild buckwheat), and *Iva nevadensis* (Nevada marsh-elder). The Great Basin Montane

Forests ecoregion [42], which includes desert ranges such as the Toiyabe, Toquima, Monitor, and Ruby Mountains of central and eastern Nevada, along with other ranges found within this area, supports 23 endemic species, an unusually high number for such a small area. Similarly, farther south is the Madrean Sky Island Montane Forests ecoregion, which includes both the Chiricahua and Huachuca Mountains of Arizona, along with other ranges, which have been known historically to support relatively large numbers of endemics. This ecoregion, which has produced 31 endemic species (as discussed previously), is embedded within the much larger Chihuahuan Desert ecoregion [81]. Due in part to its larger size, the Chihuahuan Desert, which lies adjacent to and east of the Sonoran Desert, has a much higher number of endemics (138) than does the Madrean Sky Island Montane Forests ecoregion.

In contrast to these areas, the Sonoran Desert [80], supports a rather low number of endemics, only 13 species. This region, which covers 116,770 km^2, has fewer endemic species than one might expect; however, many of its regional endemics extend well into Mexico, thus eliminating them from consideration in this study.

North of the Great Basin Shrub Steppe [76], the numbers of endemic species decrease dramatically. For example, in the areas of southern Idaho through the northern Rockies and Pacific Northwest, the number of endemics varies from a high of 75 endemic species in the Snake/Columbia Shrub Steppe [75], to a low of 1 endemic species in each of two ecoregions, the Puget Lowland Forests [35] and the Okanagan Dry Forests [31]. In part, this is because of the large number of individual ecoregions (fifteen) found within this area, which divide the landscape into much smaller units than occur farther south. The same is true of the enormously diverse flora of California, which is divided into no less than thirteen ecoregions. It should be noted that if all thirteen were to be treated as one, its 1,169 endemic species would far exceed the number of endemics of any North American ecoregion. Among the ecoregions of California the greatest concentration of endemics appears to occur within the California Interior Chaparral and Woodlands [70], with nearly 150 endemic species.

Endemism in eastern North America differs dramatically from that of the West, both by way of actual numbers and by way of taxonomic representation. There appear to be two highly diverse ecoregions in eastern North America that support numerous endemics. The first, the Southeastern Conifer Forests [51], occurs along the southeastern coastal plain, while the second, the Appalachian/ Blue Ridge Forests [16], is centered within the Appalachian Mountains. Although the Southeastern Conifer Forests lack the obvious types of biological isolating mechanisms often associated with endemic species of western North America, endemism totals are high nonetheless, the second highest of those tallied for North America. Certainly, within this region there are isolated habitats, as evidenced by the Apalachicola area of northern Florida, where well-known endemics occur, such as *Torreya taxifolia* (Florida nutmeg), *Taxus floridana* (Florida yew), *Magnolia ashei* (Ashe's magnolia), and a few others, but these are exceptional. The Southeastern Conifer Forest is an ecoregion characterized by baking heat, intense light, and an abundance of moisture, which together are responsible for producing its extremely rich and varied flora as well as its high number of endemics. However, unlike the relatively few large plant families that dominate endemic counts in the West, here plant families representing the bulk of the endemic species appear to be more numerous.

Extending northeastward along the entire stretch of the Appalachian range are the Appalachian/ Blue Ridge Forests. This region, which arguably supports the most primitive flora of the globe, also supports a relatively high number of endemics. Along its rugged high peaks, mountain balds, and dense forests are found some 87 endemics, including the 16 shale barren endemics of Pennsylvania, Virginia, and Maryland.

Proceeding still farther north, in both easterly and westerly directions, the number of endemics declines dramatically. Even with notable increases in geographic scale, northern ecoregions have quite predictably produced the fewest number of endemics, due to a variety of factors, including the influence of glaciation, shorter growing seasons, colder temperatures, more homogenous habitats, etc.

Although much can be said about the causes and occurrences of endemism within North America, additional biological data are needed to help explain peculiar situations. This is especially true of the Southeastern Conifer Forests ecoregion, with its high concentration of endemics yet lack of obvious isolating mechanisms that are generally associated with their evolution. For this ecoregion, and others, additional insights must be offered to provide a more comprehensive understanding of endemic members of their floras.

Tree richness and endemism (figure 3.5q,r), trees being a subset of vascular plants, also peak in the Southeastern Conifer Forests. These ancient forests are a complex mosaic of conifer forests, broadleaf forests, and diverse wetland and swamp forest habitats, whose distinctive floras all contribute to the ecoregion's tree richness. In general, tree richness is greatest in the southeastern United States. Tree endemism is also pronounced in southwest deserts and in California. An even smaller subset, that of the conifers, is richest in the Klamath-Siskiyou ecoregion [39], with up to 16 conifer species found at a single site (figure 3.5s,t). Only the unique assemblage of ancient and primitive gymnosperms in distant New Caledonia surpasses this richness, and only a few subtropical conifer forests in Mexico can rival this ecoregion's conifer diversity. The Sierra Nevada, North Central Rockies, California Montane Woodland (Transverse Range), and Cascade Forests also support rich conifer assemblages. Conifer endemism is highest in the temperate rain forests of California and the California Montane Woodlands and Chaparral, with secondary endemism maxima in the Klamath-Siskiyou, Sierra Nevada, California Interior Chaparral and Woodlands, and Coastal Sage ecoregions. The Southeastern Conifer Forests also have four endemic species of conifer.

The longleaf pine communities of the Southeastern Conifer Forest ecoregion harbor one of the richest herbaceous floras in the world. Photo credit: USFWS/ K. Hollingsworth.

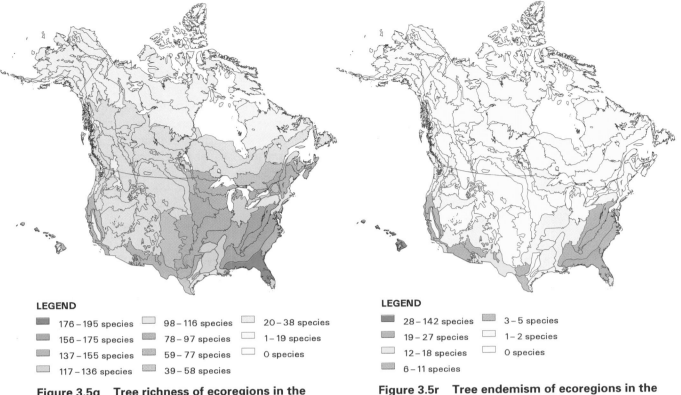

LEGEND

▇ 176 – 195 species	▨ 98 – 116 species	▨ 20 – 38 species	
▇ 156 – 175 species	▨ 78 – 97 species	☐ 1 – 19 species	
▨ 137 – 155 species	▨ 59 – 77 species	☐ 0 species	
▨ 117 – 136 species	▨ 39 – 58 species		

Figure 3.5q   Tree richness of ecoregions in the United States and Canada.

LEGEND

▇ 28 – 142 species	▨ 3 – 5 species
▇ 19 – 27 species	☐ 1 – 2 species
☐ 12 – 18 species	☐ 0 species
▨ 6 – 11 species	

Figure 3.5r   Tree endemism of ecoregions in the United States and Canada.

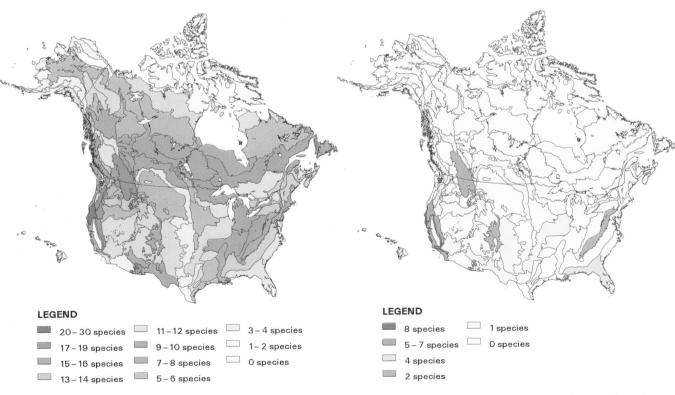

LEGEND

▇ 20 – 30 species	▨ 11 – 12 species	▨ 3 – 4 species	
▇ 17 – 19 species	▨ 9 – 10 species	☐ 1 – 2 species	
▨ 15 – 16 species	▨ 7 – 8 species	☐ 0 species	
▨ 13 – 14 species	▨ 5 – 6 species		

Figure 3.5s   Conifer richness of ecoregions in the United States and Canada.

LEGEND

▇ 8 species	☐ 1 species
▇ 5 – 7 species	☐ 0 species
☐ 4 species	
▨ 2 species	

Figure 3.5t   Conifer endemism of ecoregions in the United States and Canada.

Despite the differences mentioned above, patterns of richness in the seven taxa considered in this assessment are generally well correlated (Ricketts et al. 1999). These correlations are due largely to the strong latitudinal and resultant climatic gradients in North America. All taxa respond to these gradients with higher richness in the temperate latitudes and lower richness in the polar latitudes, resulting in a higher degree of correlation among groups. The concordance in diversity patterns must be examined at a variety of scales or by controlling for the effects of latitude in order to evaluate the utility of indicator taxa in predicting overall biodiversity patterns.

Analyses at Finer Geographical Scales

The distribution of several of the taxa described above is now being mapped at much finer scales than that of the ecoregion. For example, the distribution of over twenty thousand vascular plant species has been mapped at the county level in the United States (see essay 7). These patterns are vital data layers for within-ecoregion analyses. Range maps of vascular plant species also help to validate and correct the boundaries of ecoregions.

Except for bats, subterranean organisms often escape mention in biodiversity reviews, yet their contribution to local levels of endemism is high. The mapping of patterns of richness and endemism for subterranean species has often been restricted to individual sites, because many species have very restricted ranges, yet there are continent-wide patterns as well. These are discussed in essay 8, which includes a map of subterranean hot spots developed by a group of North American experts (U.S. data only).

ESSAY 8

Ecosystem and Species Diversity beneath Our Feet

David Culver

Some of the most unusual species occur only underground in caves and voids in perpetual darkness. Hidden from view, these "ecosystems beneath our feet" are prominent in karst landscapes. Springs, sinkholes, blind valleys, and cave entrances signal the presence of karst, a unique landscape typically occurring in carbonate rock, especially limestone. Karst results from the dissolution of the rock by chemical action rather than erosion. Lava can also develop karst-like features (especially caves) when cooling lava flows crust over.

Caves and associated subsurface habitats harbor over 1,300 described species and several times that number of undescribed species of stygobites and troglobites (Culver and Holsinger 1992; Peck 1997). (*Troglobite* is the name given to an obligate terrestrial cave-dwelling species, and *stygobite* is the name given to an obligate aquatic cave-dwelling species.) Troglobites and stygobites usually have a long evolutionary history in caves, resulting in reduced eyes and pigment, elongated appendages and antennae, and the elaboration of other sensory structures such as those involved in taste. In North America, groups with troglobite or stygobite species include: flatworms, oligochaetes, snails, amphipods, isopods, crayfish, shrimps, harvestmen, spiders, pseudoscorpions, mites, millipedes, centipedes, springtails, bristletails, crickets, flies, beetles, fish, and salamanders. Nine genera of invertebrates have over 25 troglobitic or stygobitic species, and two genera—*Stygobromus* (amphipods) and *Pseudanophthalmus* (beetles)—have over 100 species.

There are six areas of the continental United States with at least 25 troglobite and stygobite species (figure 3.6). They include the following regions:

- the Appalachians (including the eastern margin of the Allegheny Plateau) in Pennsylvania, Maryland, West Virginia, Virginia, Tennessee, Georgia, and Alabama
- the Interior Low Plateaus (including the western margin of the Cumberland Plateau) in Alabama, Tennessee, Kentucky, and Indiana
- the Ozark Plateaus in Arkansas, Missouri, Illinois, and Oklahoma
- the Lime Sink region in Florida (part of the Floridian Aquifer)
- the Edwards Plateau and Balcones Escarpment in Texas
- the Mother Lode region in California

Diversity ranges from 25 species in the Lime Sink region to over 200 in the Appalachians. The lava tubes of Hawaii, not shown on this map, are another area of high regional diversity.

Endemism is a hallmark of cave species. For example, in the Appalachians, over one third of the species are known from a single cave, and over 90 percent of the species are endemic to the Appalachians.

In spite of high regional diversity and high endemism, diversity in a single cave rarely reaches twenty troglobites and stygobites. This is largely because of very limited opportunities for dispersal (cave animals must stay underground) and food scarcity. There are four sites with twenty or more stygobites and troglobites (figure 3.6):

- Mammoth Cave system in central Kentucky (Central U.S. Hardwood Forests [18]), largest known cave in the world, with over 500 km of passages
- Parker Cave in central Kentucky (Central U.S. Hardwood Forests)
- Shelta Cave in northern Alabama (Central U.S. Hardwood Forests)
- San Marcos Springs artesian well and springs in south-central Texas (Edwards Plateau Savannas [66])

Two of these sites, Shelta Cave and San Marcos artesian well and springs, provide access to permanent groundwater. Permanent groundwater (phreatic water) in karst may often harbor high-diversity communities, but it has not been extensively studied. At two of the sites, San Marcos Springs artesian wells and Parker Cave, there is chemoautotrophic production, that is, energy production in which chemical bonds rather than sunlight are the energy source. This results in higher than normal productivity.

Besides troglobites and stygobites, karst and pseudo-karst (lava) areas are of great biological value for other reasons. Caves and lava tubes are habitat for many bats, including the Mexican free-tailed bat (*Tadarida brasiliensis*) and the endangered gray bat (*Myotis grisescens*). Springs in karst areas, especially in deserts, often have species with highly restricted ranges and ecological requirements. The classic example of this is the Devil's Hole pupfish (*Cyprinodon diabolis*), the lone endemic of Devil's Hole, Nevada. Its range of 23 square yards is the smallest of any vertebrate species (Sigler and Sigler 1994). On the surface, karst landscapes also harbor unique plant communities. The best known of these are the cedar glades of Kentucky, Tennessee, Alabama, and Virginia. A variety of federally endangered species, such as the Tennessee purple coneflower (*Echinacea tennesseensis*), are limited to this habitat.

Figure 3.6 displays karst regions only in the United States, highlighting those that contain globally outstanding biotas. There are extensive karst regions in Canada, especially Ontario and Northwest Territories, yet these regions lack the high degree of biological diversity found in those in the United States. This is primarily because of glaciation, which forced many species to migrate south

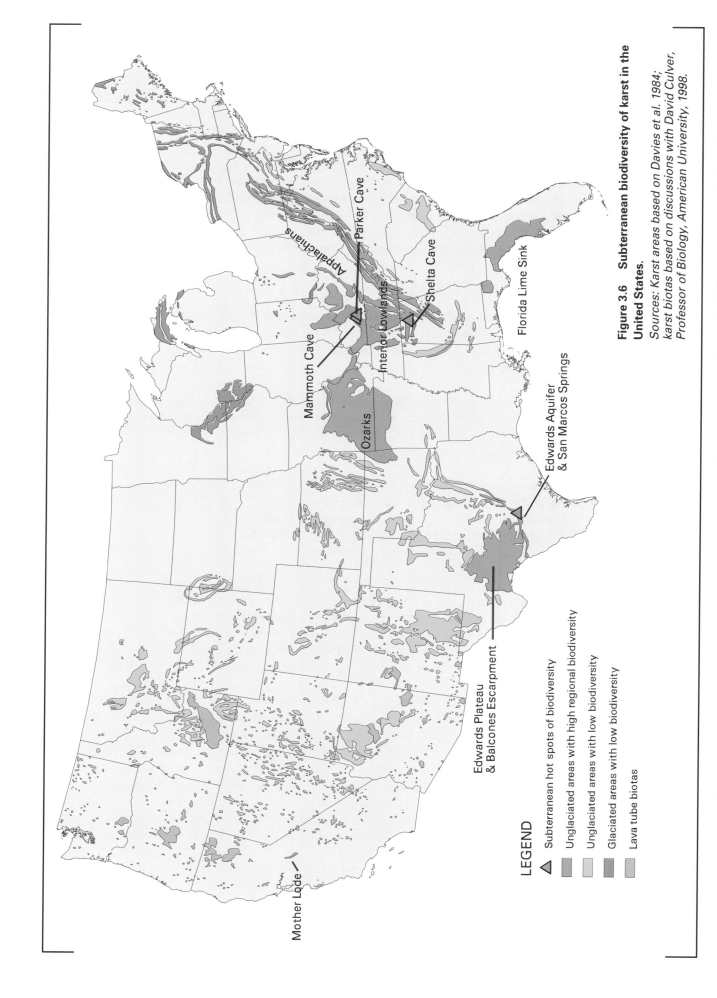

Figure 3.6 Subterranean biodiversity of karst in the United States.

Sources: Karst areas based on Davies et al. 1984; karst biotas based on discussions with David Culver, Professor of Biology, American University, 1998.

LEGEND

△ Subterranean hot spots of biodiversity

▨ Unglaciated areas with high regional biodiversity

▨ Unglaciated areas with low biodiversity

▨ Glaciated areas with low biodiversity

▨ Lava tube biotas

Mother Lode

Edwards Plateau & Balcones Escarpment

Edwards Aquifer & San Marcos Springs

Ozarks

Mammoth Cave

Parker Cave

Appalachians

Interior Lowlands

Shelta Cave

Florida Lime Sink

with the ice flows. However, the species that inhabit the karst regions of Canada are unique because they survived glaciation by moving downward in the earth's strata. A notable example in Banff National Park is an isopod that survived in streams in Castleguard Cave below the Columbia ice fields of Alberta.

Karst areas in general, and caves in particular, face a number of threats. Since caves are connected to the surface by numerous large and small cavities, any environmental degradation on the surface has subsurface effects. We need to support efforts for responsible environmental use of landfills and containment of the refuse within them. Illegal dumping is a major problem, which often goes unpunished or unnoticed and can have disastrous consequences on subterranean biota. An added problem with illegal dumping is that there is no clear agency to which to report offenders. Additionally, caves are fragile ecosystems and can be seriously harmed by minor intrusions, much the same as coral reefs and arctic tundra. Human activities in caves can be particularly harmful to bats, whose hibernacula and maternity colonies are especially sensitive to disturbance. To alleviate threats to karst areas, and to caves specifically, access should be restricted in areas that support large bat colonies. Lastly, responsible water use should be promoted because increased groundwater use has resulted in the drying up of many subterranean habitats. Growing awareness of the biological value of these areas will greatly help preserve sensitive karst ecosystems.

CHAPTER 4

Conservation Status of North American Ecoregions

In this chapter, we present broad trends detected in the two components of the conservation status analysis. The first part presents the snapshot conservation status—the situation as of 1996—and the second part presents the final assessment modified by the threat analysis (projected threats over the next ten or twenty years). Interspersed are six essays that highlight some of the threats to biodiversity that typically transcend the boundaries of ecoregions and MHTs. More detailed information on the threats to specific ecoregions is available in appendix F.

Snapshot Conservation Status

The most striking observation from the snapshot conservation status map (figure 4.1) is the large number of critical or endangered ecoregions covering most of the central and eastern part of the United States and Canada. This region is mainly composed of two MHTs, the temperate grasslands/savanna/shrub and the temperate broadleaf and mixed forests. Thirteen ecoregions (82 percent) of the temperate grasslands/savanna/shrub MHT and fourteen ecoregions (81 percent) of the temperate broadleaf and mixed forests are classified as critical or endangered (table 4.1). Together these

Military bases in the Southeast contain the largest blocks of intact longleaf pine forests. These fire-dependent communities harbor many endemic species and support one of the world's richest herbaceous floras. Photo credit: WWF/D. DellaSala.

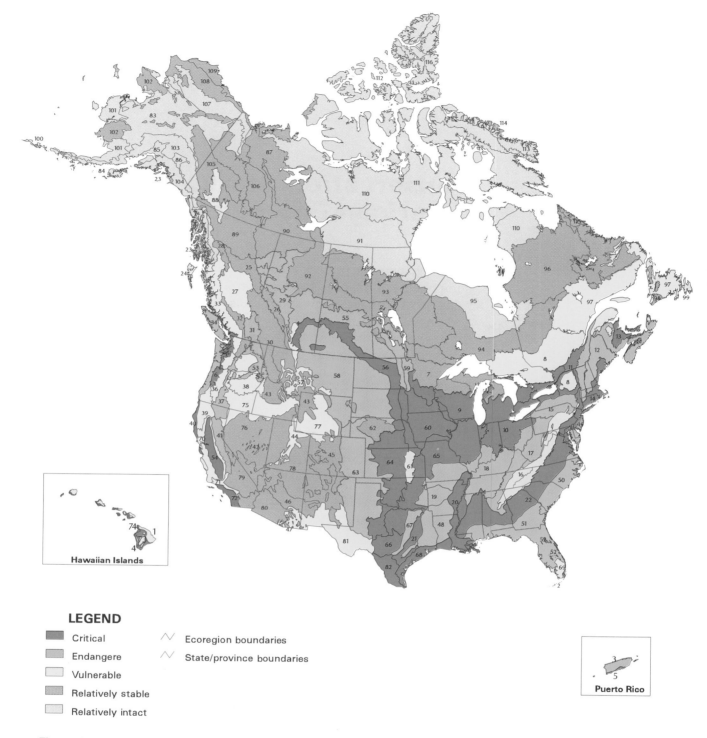

LEGEND

▇ Critical	∧	Ecoregion boundaries
▇ Endangere	∧	State/province boundaries
▢ Vulnerable		
▨ Relatively stable		
▢ Relatively intact		

Hawaiian Islands

Puerto Rico

**Figure 4.1   1996 snapshot
conservation status.**

Many areas of the Sonoran Desert support "forests" of large columnar cacti, such as the saguaro (*Carnegiea gigantea*), an ecological phenomenon found in only three other regions in the world. Photo credit: WWF/G. Huey.

TABLE 4.1. Distribution of Ecoregions by MHT for the Snapshot (1996) Conservation Status Index.

	Snapshot Conservation Status Index					
Major Habitat Type	*Critical*	*Endangered*	*Vulnerable*	*Relatively Stable*	*Relatively Intact*	*Total*
1.1 Tropical Moist Broadleaf Forests	1	1	1	0	0	3
1.2 Tropical Dry Broadleaf Forests	2	0	0	0	0	2
2.1 Temperate Broadleaf and Mixed Forests	9	5	2	1	0	17
2.2 Temperate Coniferous Forests	3	8	8	11	0	30
3.1 Temperate Grasslands/Savanna/Shrub	7	6	2	1	0	16
3.2 Flooded Grasslands	0	0	1	0	0	1
4.1 Mediterranean Scrub and Savanna	1	0	2	0	0	3
4.2 Xeric Shrublands/Deserts	2	0	4	4	0	10
6.1 Boreal Forest/Taiga	0	0	2	9	6	17
6.2 Tundra	0	0	0	7	10	17
Total	25	20	22	33	16	116

two MHTs account for 60 percent of all the critical or endangered ecoregions in the United States and Canada.

The West Coast, which contains a mix of MHTs, has also been subjected to large-scale habitat destruction and degradation pressures (e.g., conversion of native forests to plantations). The loss of this region's biodiversity is pervasive, irrespective of MHT. Widespread, non-habitat-specific degradation is also occurring in the Hawaiian and Puerto Rican Islands. These regions, with quite different habitats and species, all contain human pressures for land, timber, and development that continue to alter the natural habitats.

In contrast, the conservation status of the boreal forest/taiga and tundra MHTs remains relatively intact. With a few exceptions, these areas have been free from large-scale human disturbance, and their ecological integrity has remained intact. The majority of the xeric shrublands/deserts are in fair condition but are subject to intensive livestock grazing, which together with exotic species invasions is a major factor in the ecological degradation of the West as an entire bioregion (essay 9).

Keeping the Cows Off: Conserving Riparian Areas in the American West

Thomas L. Fleischner

Western riparian zones are one of the most productive habitats in North America (Johnson, Haight, and Simpson 1977), providing essential wildlife habitat for breeding, wintering, and migration (Brode and Bury 1984; Gaines 1977; Laymon 1984; Lowe 1985; Stevens et al. 1977). Riparian habitats in the Southwest are home to the North American continent's highest breeding bird density (Carothers, Johnson, and Aitchison 1974; Carothers and Johnson 1975), one of the rarest forest types, and more than one hundred state and federally listed threatened and endangered species (Johnson 1989). Approximately three quarters of the vertebrate species in Arizona and New Mexico depend on riparian habitat for at least a portion of their life cycles (Johnson, Haight, and Simpson 1977; Johnson 1989). Even xeroriparian habitats—normally dry corridors that intermittently carry floodwaters through low deserts—support five to ten times the bird density and species diversity of surrounding desert uplands (Johnson and Haight 1985).

Riparian habitats are widely distributed throughout the West, but they are in extreme danger, a fact that cannot be gleaned from the conservation status map, which generalizes threats across entire ecoregions (see figure 4.1). The primary threat to riparian habitats is livestock grazing. The threat is so serious that it has led the American Fisheries Society, the Society for Conservation Biology, and The Wildlife Society to issue position statements calling for a drastic overhaul of riparian zone and rangeland management (Armour, Duff, and Elmore 1991; Fleischner et al. 1994; The Wildlife Society 1996).

Livestock grazing (primarily by beef cattle) is the most pervasive influence on native ecosystems of western North America (Wagner 1978; Crumpacker 1984). Approximately 70 percent of eleven western states (Montana, Wyoming, Colorado, New Mexico, and westward) is grazed by livestock (Council for Agricultural Science and Technology 1974; Crumpacker 1984; Longhurst, Hafenfeld, and Connolly 1982). Grazing occurs on the majority of federal lands in the West, including most of the domains of the U.S. Bureau of Land Management and the U.S. Forest Service, as well as many national wildlife refuges, federal wilderness areas, and even in some national parks. In the sixteen western states, approximately 165 million acres of BLM land and 103 million acres of Forest Service land are grazed by 7 million head of livestock, primarily cattle (U.S. General Accounting Office 1988). Ninety-four percent of the BLM lands in these states is grazed. Thirty-five percent of federal wilderness areas in the United States has active livestock grazing allotments (Reed et al. 1989—this figure is from a nationwide survey; the percentage for the West is probably higher).

Cattle are not considered terribly intelligent, but they are not as dumb as we sometimes think—they prefer riparian areas for the same reasons we humans do: shade, cooler temperatures, and water, not to mention more abundant food. While public lands grazing allotments may stretch over thousands of acres, livestock spend a disproportionate amount of their time in riparian zones (Ames 1977; Gillen, Krueger, and Miller 1984; Kennedy 1977; Roath and Krueger 1982; Thomas, Maser, and Rodick 1979; Van Vuren 1982).

Riparian habitats are not only biologically rich, but also easily damaged. The U.S. Environmental Protection Agency concluded that riparian conditions throughout the West are now the worst in American history (Chaney, Elmore, and Plaits 1990). Over 90 percent of Arizona's original riparian habitat is gone (Johnson 1989). Less than 5 percent of the riparian habitat in California's Central Valley Grasslands [54] remains, and 85 percent of that is in disturbed or degraded condition (Franzreb 1987). The Oregon-Washington Interagency Wildlife Committee (1979), composed of biologists from several government agencies, concluded that grazing is the most important factor in degrading wildlife and fisheries habitat throughout the eleven western states.

A great deal of research concurs (Carothers 1977; Mosconi and Hutto 1982; Szaro 1989; and Chaney, Elmore, and Plaits 1990).

Livestock alter riparian vegetation in several ways: (1) they compact soil, which increases runoff and decreases water availability to plants; (2) they remove herbage, which causes soil temperatures to rise, thereby increasing evaporation; (3) they physically damage vegetation by rubbing, trampling, and browsing; and (4) they alter the growth form of plants by removing terminal buds and stimulating lateral branching (Kauffman and Krueger 1984; Szaro 1989). Livestock grazing is one of the principal factors contributing to the decline of native trout in the West; cattle activities especially deleterious to fishes are the removal of vegetative cover and the trampling of overhanging streambanks (Behnke and Zarn 1976). Livestock have been shown to decrease water quality of streams (Buckhouse and Gifford 1976; Diesch 1970). Changes in water chemistry (Jeffries and Klopatek 1987) and temperature (Van Velson 1979), in effect, create an entirely new aquatic ecosystem (Kauffman and Krueger 1984; Kennedy 1977). Livestock disturb: (1) streamside vegetation, (2) stream channel morphology, (3) shape and quality of the water column, and (4) structure of streambank soil (Kauffman and Krueger 1984; Ohmart 1996; Platts 1979, 1981, 1983; Platts and Nelson 1989).

Are we willing to trade rich riparian communities, bursting with birdsong, for trampled mud and befouled water? Public policy has yet to catch up with science. If the lacy network of green that harbors so much of western biodiversity is to be saved, government action must be bold and change must be immediate.

What must be done:

1. Reverse the long-standing U.S. government policy that assumes that livestock grazing is appropriate on federal lands in the West. Instead, evaluate the ecological costs and appropriateness of livestock grazing on a site-by-site basis. A site should be considered appropriate for grazing only if grazing contributes to long-term productivity and maintenance of native biodiversity.
2. Remove livestock immediately from all damaged riparian areas, except in the rare cases where grazing can be shown to provide a specific tool for ecological restoration.
3. Establish a network of significant livestock exclosures to provide landscape-scale benchmark areas for scientific monitoring of human impacts (see Fleischner 1994).
4. Eliminate grazing on U.S. public lands, especially wilderness areas and wildlife refuges, where it leads to eradication of native predators and native vascular plants.

Conservation Snapshot Criteria

The snapshot conservation status assessment includes four criteria: habitat loss, size and number of habitat blocks, habitat fragmentation, and habitat protection. We summarize the broad trends illuminated by analysis of each of these criteria below. Detailed methods and discussion can be found in appendix B.

Habitat Loss

We estimate the percentage of remaining intact habitat using mapped habitat information, satellite data, and expert opinion as guides (figure 4.2). This map mirrors the snapshot conservation status (figure 4.1) to a certain extent but also underscores the amount of intact habitat that remains.

Sixteen ecoregions (14 percent) have dangerously high levels of habitat loss, and five of these are globally outstanding: the Hawaiian Dry Forests [4], Central Pacific Coastal Forests [34], Northern

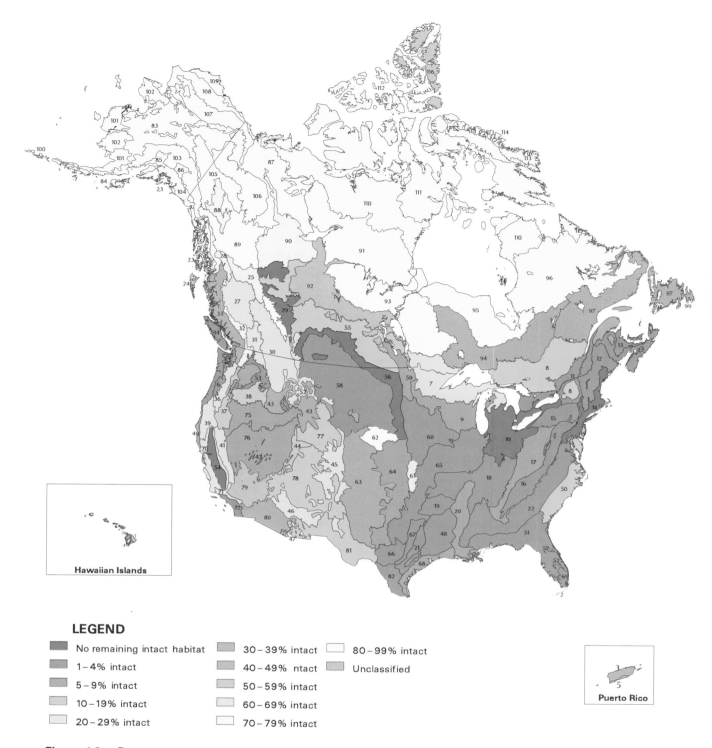

- No remaining intact habitat
- 1–4% intact
- 5–9% intact
- 10–19% intact
- 20–29% intact
- 30–39% intact
- 40–49% ntact
- 50–59% intact
- 60–69% intact
- 70–79% intact
- 80–99% intact
- Unclassified

Hawaiian Islands

Puerto Rico

Figure 4.2   Percentage remaining habitat by ecoregion.

California Coastal Forests [40], Central Tall Grasslands [60], and the California Coastal Sage and Chaparral [72] (see appendix D). Regionally, the most altered areas are the central grasslands of the United States and southern Canada, the southeastern United States, and the entire West Coast, specifically most of California. The northern sections of Canada, primarily the boreal forest/taiga and tundra, were the only MHTs to receive low scores for habitat loss.

Large blocks of unfragmented forest habitat distributed over the landscape are necessary for the successful breeding of many neotropical migrants, such as this black and white warbler (*Mniotilta varia*). Photo credit: B. Dyer.

Size and Number of Habitat Blocks

The widespread loss of large representative examples of intact habitat in ecoregions in the contiguous United States and Hawaii is truly staggering. Thirty-seven ecoregions (32 percent) have few or no large blocks left, and eight of these are globally outstanding: the Hawaiian Dry Forests [4], Appalachian Mixed Mesophytic Forests [17], Southeastern Mixed Forests [22], Northern California Coastal Forests [40], Sierra Nevada Forests [41], Florida Sand Pine Scrub [52], Central Tall Grasslands [60], and the Everglades [69] (appendix D). Habitat block size criteria for these evaluations are described in appendix B.

Regionally, most of the temperate broadleaf and mixed forests east of the Mississippi River lack large blocks of intact habitat, as do most of the grasslands/savanna/shrub ecoregions in the United States and Canada. The lack of a sufficient number of large blocks of habitat for migratory songbirds is a conservation challenge (essay 10).

ESSAY 10

Saving Migratory Songbirds

Scott Robinson

The majority of birds that breed in the forests of North America are neotropical migrants: they spend the winter in the tropics of Latin America. More than 100 species of forest birds are neotropical migrants, and as consumers of defoliating insects, they may play a crucial role in maintaining forest health. They are also a major part of the bird communities of tropical forests, especially in the northern part of their wintering range in Central America and the West Indies.

Most neotropical migrants have large populations and vast geographical ranges, making their conservation needs very different from less mobile taxa that show more endemism. Regardless, neotropical migrants, neither endangered nor of direct economic importance, have become the target of one of the largest conservation and research efforts ever aimed at wildlife. Neotropical migrants face severe problems in North America. First, many species are area-sensitive, meaning they are absent from small habitat patches even when suitable habitat is present and there is enough space for multiple territories. Second, some species are showing steep population declines in many parts of their ranges, as in the Great Smoky Mountains and the Adirondacks. Third, the development of beach property has led to extensive loss of migratory stopover habitat. Fourth, habitats such as southern floodplain forests, midwestern savannas, and Pacific rain forests have been severely degraded by human activities. Fifth, nesting success of many species appears to be much lower in small habitat patches than in the larger forest tracts, which suggests that forest-dwelling neotropical migrants are vulnerable to fragmentation. And sixth, neotropical migrants are also threatened by a host of other factors that affect most North American birds, such as pesticides and other agrochemicals, collisions with windows and radio towers, and domestic cats.

Only if these species are abundant can they continue to play their roles in ecosystem function. Conservation plans thus aim to maintain or restore the large populations of these species rather than to prevent the imminent extinction of a whole suite of species. One approach to the conservation of neotropical migrants in North America is to emphasize the preservation and restoration of large blocks of forest, for the following reasons:

- Large tracts are likely to contain significant populations of area-sensitive species.
- Large tracts are more likely to contain reproductively viable populations of many forestdwelling species. Large tracts minimize losses to nest predators and brood parasitic brown-headed cowbirds (*Molothrus ater*) that thrive in fragmented landscapes composed of small forest tracts with many openings. Populations of many species in small forest tracts may act as "sinks," producing too few young to compensate for adult mortality but being maintained by immigration from large tracts that act as "sources" by producing a surplus of young.
- Larger tracts will provide more buffering from pesticides and agrochemicals and from the effects of heavy deer browsing that could alter forest structure unfavorably for many species.
- Larger tracts are better candidates for managing and restoring natural ecosystem processes such as fire and flooding, on which many species depend for the creation of habitats. Even some species characteristic of the forest interior may depend on periodic disturbance to create suitable microhabitats (e.g., cane stands for Swainson's warblers [*Limnothlypis swainsonii*] in the Southeast).

Given a general approach emphasizing large tracts, the first step in a conservation strategy should be to create and maintain a network of large forest tracts both within and among ecoregions. Operationally, we do not know how large these tracts have to be to escape from the pervasive effects of fragmentation. Almost certainly, the necessary size will vary regionally. In the Central U.S. Hardwoods Forests [18], for example, tracts of 2,500 hectares still have high levels of cowbird parasitism and nest predation. Nesting success is much higher in the extensive forest tracts (10,000–30,000 ha) of the Missouri Ozarks and the Hoosier National Forest area within this ecoregion. Within ecoregions that contain a few large and many small forest tracts (e.g., Central U.S. Hardwoods Forests [18], Mississippi Lowland Forests [20]), the most effective way to conserve forest bird populations may be to maintain contiguous forest cover within the largest remaining forest tracts.

Maintaining public ownership of the cores of these areas may be crucial to the long-term prospects of preventing forest fragmentation. The Shawnee National Forest within the Central U.S. Hardwoods Forests [18] contains so many agricultural inholdings that nesting success of many species is low even in the largest tracts available.

In some ecoregions that are mostly forested (e.g., Western Great Lakes Forests [7], New England/Acadian Forests [12], Allegheny Highlands Forests [15], and Appalachian/Blue Ridge Forests [16]), even small forest tracts have relatively low levels of parasitism and nest predation. In these areas, the major issues are the management of forests to maintain high nesting success and to contain significant breeding populations of all of the native forest species. Even disturbance-dependent species appear to nest much more successfully within forested areas than they do in highly fragmented habitat patches. Indeed, some heavily forested ecoregions help to maintain populations in nearby ecoregions where there are no large forest patches.

Incorporating migratory songbirds into ecoregional planning therefore requires different approaches than those required for most other taxa. Further research on the dispersal of forest birds may show that the heavily forested ecoregions of North America are sufficient for maintaining continental populations. Until we have developed methods to test this hypothesis, however, the most prudent conservation strategy may be to preserve and restore large forest tracts whenever possible to avoid the most extreme problems associated with fragmentation. If this approach is not combined with conservation of stopover and wintering habitats, though, many species may continue to decline no matter how well we manage breeding habitat.

Habitat Fragmentation

The remaining intact habitat in twenty ecoregions (17 percent) is severely affected by fragmentation. Two of these are globally outstanding: the Southeastern Mixed Forests [22] and the Central Tall Grasslands [60] (appendix D). Continued fragmentation and degradation of the grasslands of the Midwest jeopardize grassland bird communities (essay 11).

Regionally, most of the temperate broadleaf and mixed forests of the eastern United States and Canada, excepting the Appalachian/ Blue Ridge Forests [16], are extremely fragmented.

ESSAY 11

The Most Threatened Birds of Continental North America

Scott Robinson

One of the most rapidly declining groups of birds in North America and around the world are the birds that nest in temperate zone grasslands. Grassland habitats are ideal for conversion to row crops, hayfields, and pastures and are therefore one of the first habitats to be permanently altered following human settlement. Unlike many other species of wildlife, however, grassland birds adapted well to the human agricultural landscapes that dominated the continent through the 1950s, at least in the Midwest and in the East. Grassland birds tolerate the introduced cool-season grasses that replaced the native warm-season grasses, and many species tolerate and even require disturbance in the form of grazing and fires to create suitable vegetation structure. Grass- land species were probably much more abundant in the Northeast following European settlement than they were before it.

Over the last three decades, however, changing agricultural practices have led to some catastrophic declines in many of the species that were formerly abundant in agricultural landscapes. Some, such as Henslow's sparrow (*Ammodramus henslowii*) and the mountain plover (*Charadrius montanus*), are now candidates for the federal list of threatened and endangered species. Formerly abundant species such as the bobolink (*Polichonyx oryzivorus*) and the lark bunting (*Calamospiza melanocorys*) are becoming increasingly patchy in their distributions. For these reasons, grassland bird conservation has become a very high conservation priority for Partners in Flight and other bird conservation organizations. Some of the problems faced by grassland birds are: (1) loss of winter habitat in both North and South America, especially in the Western Gulf Coastal Grasslands [68] and in the pampas of Argentina; (2) earlier cutting of hayfields, which destroys many nests; (3) a decrease in the area of hayfields and pastures available; (4) fragmentation of grasslands, especially the many small patches in conservation reserves, which are often too small to contain populations of area-sensitive species; (5) invasion of woody vegetation in grasslands, which provides cowbirds and nest predators with perches from which to search for nests; (6) changing fire regimes and grazing pressures, which have altered the variety of vegetation structure; and (7) increased exposure to pesticides and other agrochemicals.

Ecoregions that contain a high proportion of grassland habitat, such as the Nebraska Sand Hills Mixed Grasslands [62] and the Flint Hills Tall Grasslands [61], should be of special priority because they are presumably large enough both to contain all area-sensitive species characteristic of the region and to escape the most negative consequences of habitat fragmentation (increased nest predation and brood parasitism by brown-headed cowbirds). We do not know yet whether nesting success is consistently higher in landscapes dominated by grasslands than it is in heavily forested landscapes. It is possible, however, that these large grassland areas are crucial to the maintenance of regional grassland bird populations. For this reason, a prudent conservation strategy would be to focus conservation efforts on both private and public land in these ecoregions and to develop management practices that increase populations and nesting success of all of the species characteristic of these ecoregions.

Possible management practices that would benefit grassland birds include: (1) increasing the size of prairie restoration sites, such as the 10,000 hectares of the Midewin National Tallgrass Prairie in the Central Forest/Grassland Transition Zone [65]; (2) maintaining a network of grassland reserves that can act as refugia for grassland birds during periods when agricultural policy or high grain prices reduce the amount of private land available to grassland birds, especially in the East, where there are few large public grasslands; (3) removing encroaching woody vegetation from large tracts (unless it is along a riparian corridor); (4) maintaining some areas that are not grazed or burned for at least three years to provide habitat for species that require taller, denser vegetation; (5) removing drainage tiles from selected watersheds to restore wet grasslands, in which many species nest (including waterfowl); (6) developing razing rotations that provide habitat for species that need both tall and short grass; (7) minimizing early-season mowing of hayfields; (8) minimizing early-season cutting of fields in the Conservation Reserve Program; (9) aggregating fields in the Conservation Reserve Program to create a few large rather than many small grasslands; and (10) giving special conservation attention to grasslands in the Western Gulf Coastal Grasslands [68].

Habitat Protection

Habitat protection was given the lowest weight in the snapshot conservation status index, although this criterion had the greatest number of ecoregions receiving the highest point score. Forty-six ecoregions (40 percent) scored the worst in the category of habitat protection, and nine of these are globally outstanding: the Hawaiian Dry Forests [4], Appalachian Mixed Mesophytic Forests [17], Northern California Coastal Forests [40], Florida Sand Pine Scrub [52], Central Tall Grasslands [60], Flint Hills Tall Grasslands [61], California Coastal Sage and Chaparral [72], Eastern Canadian Shield Taiga [96], and the Arctic Coastal Tundra [109] (see appendix D). Protected-areas size criteria on which these evaluations are based are listed in appendix B.

Most of the central and eastern United States and much of Canada lack adequate protection for remaining native habitat. Likewise, the central grasslands and western coast have been severely degraded and contain little protected habitat. Of the forty-six ecoregions considered to have poor protection of their habitats, twenty-nine (63 percent) are designated as critical. The majority of these ecoregions belong to the temperate broadleaf and mixed forests, temperate coniferous forests, and temperate grasslands/savanna/shrub MHTs.

Final Conservation Status: Threat Analysis

We derive the final conservation status by applying a threat analysis to the snapshot conservation status assessment (see appendix B for a description of methods for the threat analysis). Twenty-three (20

Giant sequoias (*Sequoiadendron gigantea*), the most massive trees on earth, form unique groves in the Sierra Nevada Forests ecoregion [41]. Photo credit: WWF/D. Olson.

percent) ecoregions are considered more threatened after future threats are applied to their snapshot assessment (figure 4.3, appendix D). For seven other ecoregions, there is no change because the snapshot assessment was already determined to be critical, even though threat is considered high. Seven ecoregions "rise" from endangered to critical, two of which are globally outstanding: Appalachian Mixed Mesophytic Forests [17] and Southeastern Conifer Forests [51]. Nine ecoregions jump from vulnerable to endangered, two of which are globally outstanding: Klamath-Siskiyou Forests [39] and Hawaiian Moist Forests [1].

The final conservation status assessment reveals that nearly half of all ecoregions are considered either critical (thirty-two, 28 percent) or endangered (twenty-two, 18 percent). Additionally, twenty-one (18 percent) are vulnerable, twenty-five (21 percent) relatively stable, and sixteen (14 percent) relatively intact (table 4.2, figure 4.3). A comparison of the snapshot and final conservation status maps (figures 4.1 and 4.3) demonstrates a large increase in vulnerable, endangered, and critical areas in Canada, as well as on the West Coast of the United States. Because of the severely degraded nature of the central and eastern United States, the final conservation status of ecoregions in this area changed little. Only the boreal forest/taiga and tundra MHTs still contain relatively intact ecoregions.

Threat Analysis Criteria

Habitat Conversion

Thirty-three ecoregions (28 percent) face serious threats from habitat conversion. Among MHTs, nearly half of the temperate coniferous forests ecoregions have major, proximal threats to their ecological integrity. These threats are primarily due to population pressures and agriculture in Florida, mining and timber extraction

TABLE 4.2. Distribution of Ecoregions by MHT for the Final Conservation Status Index (Snapshot Modified by Threat).

	Final Conservation Status Index					
Major Habitat Type	Critical	Endangered	Vulnerable	Relatively Stable	Relatively Intact	Total
1.1 Tropical Moist Broadleaf Forests	1	2	0	0	0	3
1.2 Tropical Dry Broadleaf Forests	2	0	0	0	0	2
2.1 Temperate Broadleaf and Mixed Forests	10	4	2	1	0	17
2.2 Temperate Coniferous Forests	7	9	7	7	0	30
3.1 Temperate Grasslands/Savanna/Shrub	9	5	1	1	0	16
3.2 Flooded Grasslands	0	0	1	0	0	1
4.1 Mediterranean Scrub and Savanna	1	0	2	0	0	3
4.2 Xeric Shrublands/Deserts	2	1	3	4	0	10
6.1 Boreal Forest/Taiga	0	1	5	5	6	17
6.2 Tundra	0	0	0	7	10	17
Total	32	22	21	25	16	116

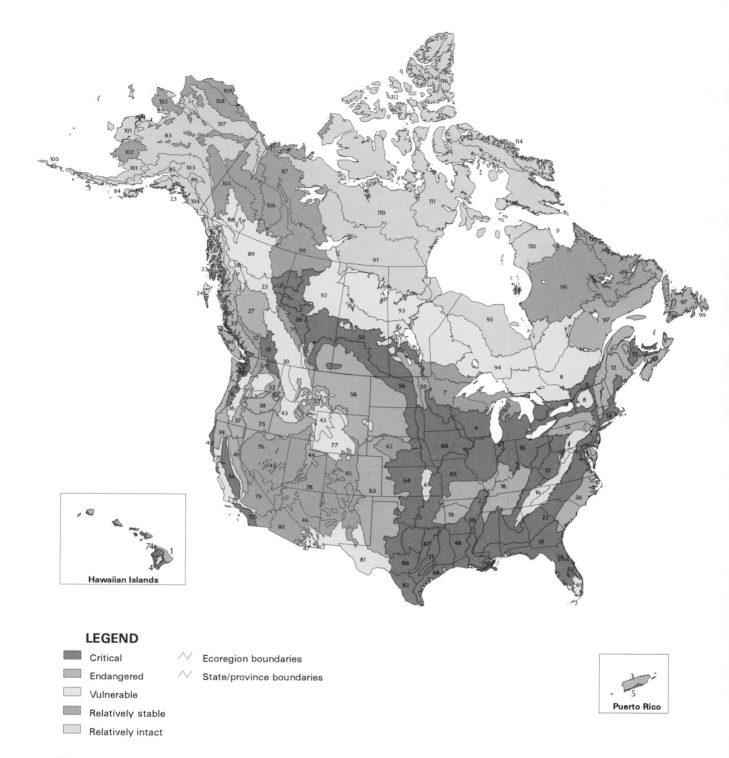

LEGEND

	Critical	$\wedge$	Ecoregion boundaries
	Endangered	$\wedge$	State/province boundaries
	Vulnerable		
	Relatively stable		
	Relatively intact		

Hawaiian Islands

Puerto Rico

Figure 4.3   Final conservation status (snapshot modified by threats).

in Utah and Wyoming, exploitation of remaining forests in the Piney Woods Forests [48], and timber extraction in the lowland valleys of the Sierra Nevada Forests [41], Klamath-Siskiyou Forests [39], Central and Southern Cascades Forests [36], Central Pacific Coastal Forests [34], and British Columbia Mainland Coastal Forests [33]. The temperate grasslands/savanna/shrub ecoregions have been largely converted to grain crop production and pasture for grazing stock. Furthermore, the threat analysis indicates that this trend of land conversion will continue in the future, mainly in the last remaining Canadian grassland ecoregions. In addition, one third of the boreal/taiga ecoregions faces high conversion threat due to logging. Together, these ecoregions span a broad swath of central Canada.

Habitat Degradation

Twenty-two ecoregions (19 percent) have the highest level of threats by habitat degradation. In particular, all three (100 percent) of the Mediterranean scrub and savanna ecoregions receive this score. Globally, there are only five regions containing Mediterranean scrub, and all are endangered (see chapter 6, "North America's Global Responsibilities for Biodiversity Conservation"). The Mediterranean scrub and savanna ecoregions have been severely altered due to vast human development and sprawl, particularly in Southern California, and will be further susceptible to this type of degradation unless lands are set aside for protection. Geographically, the West Coast and central to eastern parts of the United States and Canada contain the most severely degraded habitats (see figure 4.3). Much of the degradation is a result of fire exclusion, particularly in eastern habitats, which require regular fire events to maintain the full suite of plants and animals typical of that bioregion (essay 12).

ESSAY 12

Fire Exclusion in the Eastern Ecoregions of the United States and Canada

Alan Weakley

Urban and suburban development, agricultural conversion, pollution, drainage of wetlands, other hydrologic alterations, destructive silvicultural practices, destructive grazing practices, habitat fragmentation, direct exploitation of species, impacts of exotic species, removal of top predators—the litany of forces that degrade the environment are familiar to conservationists. But the alteration of fire regimes—less obvious and less dramatic—is sometimes overlooked, especially in the eastern United States and Canada. Even in the generally temperate, moist, and forested East, most terrestrial ecosystems were dramatically or subtly shaped by fire prior to European settlement, and many of the plants and animals still depend on the habitats and landscape diversity created and maintained by fire.

What was the source of fire in eastern North America? Especially in the Southeast, humid air combined with hot summer temperatures to create conditions suitable for thunderstorms, and lightning-set fires were common in many landscapes. Depending on fuel conditions and the extent of natural fire compartments (determined by natural fire breaks), such fires could be local, or could spread and burn hundreds of thousands of acres. The intensity and ecological effects were also variable; fires could be "catastrophic" (removing nearly all vegetation), or could burn as low-intensity surface fires.

Increasing evidence suggests that Native Americans were also important setters of fire in eastern North America. In some ecoregions (e.g., Southeastern Mixed Forests [22]), Native Americans were likely the primary source of landscape-affecting fire, while in others (such as the Middle Atlantic Coastal Forests [50]), they apparently augmented the frequency of lightning-set fires. The relative importance of "natural" fire and anthropogenic fire in various ecoregions is an interesting, and poorly understood, issue; we do know that Native Americans and their fire practices were a part of eastern North American ecoregions since shortly after the most recent glaciation. As ecosystems changed with the climate, fire helped shape the transition to the post-Pleistocene ecoregional landscape.

In the southeastern coastal plain, fire dominated the ecoregion and shaped the structure and composition of most communities. Many species of vascular plants and vertebrate animals are endemic to the southeastern coastal plain (one of the most important concentrations of biodiversity in the temperate world), and the great majority of these depend on fire-maintained ecosystems. Most celebrated among these ecosystems are the pine savannas and woodlands, which occupied most upland and wet flats in the coastal plain and were dominated by longleaf pine (*Pinus palustris*), but also contained slash pine (*P. elliottii*), pond pine (*P. serotina*), sand pine (*P. clausa*), and (northward) pitch pine (*P. rigida*). Fire-adapted and fire-promoting grasses, pines, and shrubs also encouraged fire frequency and extent in the overall landscape.

These fire-maintained pinelands formed a matrix for other pyrogenic communities of lesser extent, including seepage bogs, glades, barrens, scrub, pocosins, Atlantic white cedar (*Chamaecyparis thyoides*) swamps, bayheads, baygalls, canebrakes, wet prairies, dry prairies, upland depression marshes, and pond-cypress (Taxodium ascendens) savannas. Most of these communities burned frequently with ground fires, though some, such as sand pine scrub, Atlantic white cedar swamps, and pocosins, burned less frequently and more catastrophically.

In the interior, non-montane portions of the East (the piedmont, plateaus, and hilly areas), fire was less omnipresent but still important. We have the impression that these areas were uniformly and densely forested, but open habitats were not uncommon. Oak woodlands, pine woodlands, glades, barrens, and even prairies provided a diversity of habitats. These more open communities occurred primarily where unfavorable soil conditions (such as rock outcrops, heavy clays, soil toxicity, or droughty sands or karst) impeded tree growth, but the droughty conditions also promoted fire, which maintained and expanded these communities at the expense of forests. The accounts of early explorers and settlers describe expansive grasslands and forests with open grassy understories; following European settlement, later writers observed that the forests had become, for example, "gloomy and intricate," choked with shrubs. In these portions of the eastern United States, many of the rarest and most threatened species are endemics dependent on open habitats.

In the montane parts of the eastern United States and Canada, pine communities on ridges, variously dominated by shortleaf pine (*P. echinata*), Table Mountain pine (*P. pungens*), pitch pine (*P. rigida*), Virginia pine (*P. virginiana*), red pine (*P. resinosa*), white pine (*P. strobus*), and jack pine (*P. banksiana*), depend on fire for their regeneration. South of the boreal zone, eastern North America is largely a land of oak forests; yet it now appears that many oak-dominated communities also depend on periodic fire events for regeneration. In an era of fire exclusion, many oak forests are undergoing succession toward fire-intolerant trees, such as red maple (*Acer rubrum*), sugar maple (*A. saccharum*), and beech (*Fagus grandifolia*).

With European settlement of eastern North America, fire regimes were soon altered. Northern European settlers saw fire as a hostile and destructive force, and settlement fragmented the natural landscape and increased the need for the protection of resources and infrastructure. Still, natural and anthropogenic fires remained common in rural parts of the region well into the twentieth century, as difficulties in access and communications hampered the effectiveness of the desired fire suppression, and rural residents learned to use fire to improve timber growth, hunting, berry picking, and forage for free-range livestock. During the latter part of the twentieth century, increased landscape fragmentation decreased the extent of fires, and fire suppression (sometimes necessary in a settled landscape) became nearly universally effective. Fire was nearly eliminated from the remaining natural lands of the eastern United States, except for limited areas of federal, state, and private lands (especially in the southeastern Coastal Plain), where natural resource managers continued to use prescribed fire as an important tool in multiple resource management.

Now, even the use of fire as an ecological tool for management of our remnant natural areas is threatened. Urbanization has expanded, complicating prescribed burning. Ironically, environmental concerns about resulting air pollution have led some federal and state agencies to regulate controlled burning, despite the great ecological benefits of prescribed fire.

The conservation of thousands of species depends on maintaining fire as an ecological force in their habitats. Fire as a management tool is efficient and effective, because it maintains the ecological system in which the species evolved. Improved public education about the important role of fire in the natural landscape is needed. The responsible use of fire as a resource management tool must remain legal and practical. In many areas, fire frequency on natural lands should be increased, and the season of burn should be varied. While the impacts of fire exclusion are not dramatic or direct, they are as serious—plants and animals dependent on the ecosystem-shaping force of fire will slowly and silently disappear.

Among MHTs, broad trends indicate that the temperate broadleaf and mixed forests have undergone the most significant degradation (59 percent of ecoregions classified as critical). The temperate broadleaf and mixed forests have been subjected to intensive logging to make way for agriculture, mining, and development over the past several centuries. In addition, the threat analysis identifies the temperate coniferous forests MHT as the habitat type likely to be most exploited over the next twenty years.

Wildlife Exploitation

Only three ecoregions face severe threat to their wildlife via direct exploitation. They are the Appalachian/Blue Ridge Forests [16], the Southern Great Lakes Forests [10], and the Southeastern Conifer Forests [51]. These ecoregions retain a large portion of their native flora and fauna, yet populations of medicinal herbs and black bear (*Ursus americanus*) are under increasing threats.

The rich Klamath-Siskiyou Forests are threatened by logging. Photo credit: WWF/D. DellaSala.

Exotic Species

We consider the threats posed by exotic species to be part of habitat degradation rather than a separate category. The introduction and spread of exotic mammals has decimated Hawaiian plant and animal communities. The spread and establishment of cheat grass (*Bromus tectorum*) in the inland West and the Great Basin Shrub Steppe [76] may be permanent unless new techniques for eradication are developed. The potential impact of exotic plant species by ecoregion is presented in essay 13.

ESSAY 13

Exotic Vascular Plant Species: Where Do They Occur?

John Kartesz

Imagine planting a vegetable or herb garden in continental North America and not having to worry about problematic exotic weeds competing with your crops. Moreover, consider the enormous economic benefits that commercial North American agriculture might reap if mechanical and chemical controls were unnecessary in managing the invasive and often toxic exotic species that continue to plague our farmlands and rangelands. In early America, long before European immigrants settled the coastal perimeter of this continent, aboriginal people farmed the rich glacial till of the northern territories, as well as the fertile valleys and lowlands to the south, without worry of exotic weeds competing with their domestic crops. In fact, Asian, Australian, European, South American, and other foreign vascular plant introductions into this country are relatively recent phenomena, dating back to the time when the earliest settlers landed upon our coastal shores. Over the last several centuries the vast spread of virtually all of the pernicious exotics that now invade our beaches, rivers, fields, forests, and roadsides began with either the deliberate or the unwitting introduction of these species onto the North American continent, via early travelers and their seafaring vessels.

Distributional patterns of exotic plant species across the North American landscape are anything but random (figure 4.4a,b). Quite predictably, the largest concentrations occur in and around our large coastal cities, which have historically served as centers of commerce and settlement. Port cities such as Montreal, Quebec; Boston, Massachusetts; New York, New York; Baltimore, Maryland; Miami, Florida; Mobile, Alabama; San Francisco, California; and Portland, Oregon represent some of the most notable entry points for exotic plants. It is within these cities that international trade vessels commonly brought grain and livestock shipments contaminated with foreign seeds, and where large oceanic freighters dumped huge ore deposits and ship ballast laden with foreign plant material onto the harbor shores and around their docks. From here the exotic problem spread inland and ultimately to virtually every corner of the continent.

Once upon North American soil, cargos contaminated with foreign seeds were transported via rail, road, and other corridors throughout America's heartland, dispersing viable foreign seeds liberally along the way. Vast quantities of contaminated foreign grain, most notably wheat, were transported and planted in the fertile agricultural belts of the Great Plains and Midwest. Here, exotic species such as *Buglossoides arvensis* (corn-gromwell), *Convolvulus arvensis* (field bindweed), *Abutilon theophrasti* (velvetleaf), *Malva neglecta* (dwarf mallow), *Salsola kali* (Russian-thistle), *Carpuus acanthoides* (spiny plumeless-thistle), and *Verbascum thapsus* (great mullein) escaped commonly, causing significant crop infestation and rangeland modification. Other exotics, such as *Halogeton glomeratus* (saltlover, a toxic species responsible for the deaths of

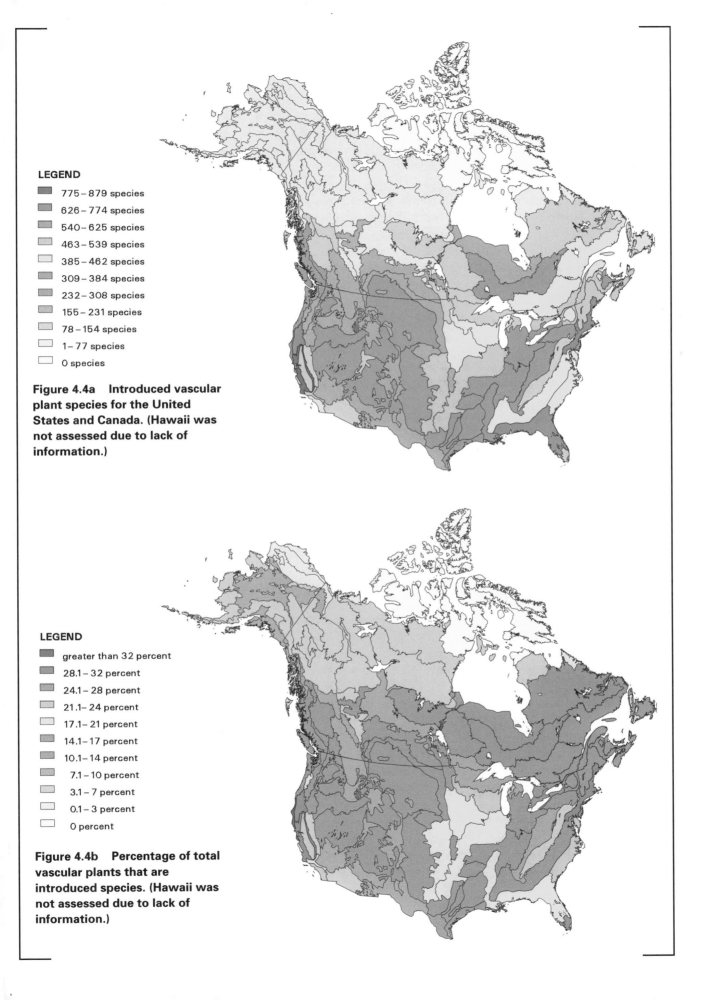

LEGEND

- 775 – 879 species
- 626 – 774 species
- 540 – 625 species
- 463 – 539 species
- 385 – 462 species
- 309 – 384 species
- 232 – 308 species
- 155 – 231 species
- 78 – 154 species
- 1 – 77 species
- 0 species

Figure 4.4a   Introduced vascular plant species for the United States and Canada. (Hawaii was not assessed due to lack of information.)

LEGEND

- greater than 32 percent
- 28.1 – 32 percent
- 24.1 – 28 percent
- 21.1 – 24 percent
- 17.1 – 21 percent
- 14.1 – 17 percent
- 10.1 – 14 percent
- 7.1 – 10 percent
- 3.1 – 7 percent
- 0.1 – 3 percent
- 0 percent

Figure 4.4b   Percentage of total vascular plants that are introduced species. (Hawaii was not assessed due to lack of information.)

thousands of livestock animals), were carried farther westward, into the southwest deserts, where hundreds of square miles of rangeland were infected.

The number of exotics deliberately brought into North America for food, forage, fiber, horticultural and aquarium trade, erosion control, and medicinal purposes has been equally devastating. Some of these introductions include the most noxious plants of all, such as *Hydrilla verticillata* (water-thyme), *Eichhornia crassipes* (common water-hyacinth), *Schinus terebinthifolius* (Brazilian peppertree), *Sorghum halepense* (Johnson grass), *Hyoscyamus niger* (black henbane), *Datura stramonium* (jimsonweed), *Cynodon dactylon* (European Bermuda grass), *Isatis tinctoria* (Dyer's-woad), *Rosa multiflora* (rambler rose), *Linaria vulgaris* (greater butter-and-eggs), and *Linaria dalmatica* (Dalmatian toadflax). Regrettably, federal laws governing the introduction of these plants into North America have been outrageously lax, allowing thousands of new introductions to be imported annually by the nursery trade alone.

We can no longer be smug about the significance of the exotic plant problem impacting and threatening our native flora. The annual cost to North American agriculture alone, regarding the control of weeds and exotic plants, is estimated to be in the billions of dollars. Regrettably, over the past several centuries we have witnessed the homogenization of our native flora with exotic species and at the same time the direct and indirect increases of toxic pests and other associated problems.

Currently, a full 17 percent of the 22,031 North American species occurring north of Mexico, or some 3,758 species, are considered to be exotic, including all but 4 species discussed in *The World's Worst Weeds* (Holm 1997) (*Digitaria scalarum, Leptochloa chinensis, Mikania cordata,* and *Monochoria hastata*).

Interestingly, not all North American plant families are equally represented by the exotic members. The percentage of exotics varies between and among plant families, as it does between and among plant duration types. For example, among the three largest North American plant families, Asteraceae, Fabaceae, and Poaceae, clearly the latter family leads in the percentage of its species being exotics, 29.6 percent, followed by the Fabaceae with 22.4 percent, and the Asteraceae, with 11.7 percent. Among the three major plant duration types, annual, biennial, and perennial, clearly the percentage of perennial introductions, 69 percent, far surpasses those of annual and biennial introductions, 29.7 percent and 5.5 percent, respectively (a total greater than 100 percent because some species exhibit more than one duration type).

Just as plant families and duration types vary regarding the number of exotics, so do North American ecoregions. As discussed earlier, coastal ecoregions along the entire eastern seaboard from eastern Canada to Florida, as well as those along the Pacific Coast, from British Columbia to California, suffer most from plant introductions. The number of introductions in these areas is staggering. Consider, for example, that of the 3,479 species known from New York state, 1,110, or 32 percent, are introduced. Similarly, the introductions for the states of Massachusetts, Pennsylvania, and Florida are nearly as high, with 32 percent, 30 percent, and 23 percent of the floras, respectively. Introductions along the West Coast, although not as dramatic, are equally troubling. Fortunately, however, ecoregions elsewhere in North America, such as tundra ecoregions, densely forested ecoregions, and those of the intensely hot deserts of the Southwest, e.g., the Sonoran Desert ecoregion [80], appear to be more resilient to introductions. To a similar degree, the savanna communities of the Middle Atlantic Coastal Forests [50] and the Southeastern Conifer Forests [51] of the southeastern United States appear to be equally resilient. Possibly due to high species diversity within at least some of these areas, or perhaps in combination with other stabilizing factors, these ecoregions appear to display a remarkable resistance to plant invasion. However, with increasing environmental impacts even these areas are now showing signs of infestation, as evidenced by the increased presence of species such as *Ligustrum vulgare* (European privet), *Lonicera japonica* (Japanese honeysuckle), *Microstegium vimineum* (Nepalese browntop), *Pueraria montana* (kudzu), *Amaranthus retroflexus* (red-root amaranth), *Imperata cylindrica* (cogon grass), and *Rosa bracteata* (Chickasaw rose).

To curb the continued infiltration of exotics into North America, the scientific and conservation communities must join efforts with governmental agencies to develop and implement stronger guidelines to protect our wildlands from this scourge. Moreover, nurserymen, plant collectors, and others involved in the horticultural trade need to be more responsible, and more aware of the potential problems created by introducing new species and new genotypes into North America on experimental bases. We can also reduce the spread of noxious species by convincing the nurseries that propagate and sell them to discontinue offering these species. Finally, we need to increase public awareness of the devastation brought about by exotic plants and to encourage people to assist with efforts to eliminate further spread of these plants.

Conservation Threats Moving Northward

Most of the boreal forests/taiga ecoregions are currently listed as relatively stable or relatively intact, but this MHT has the second-highest increase in conservation status when threat is applied. Close to 30 percent of the ecoregions were elevated in conservation status once threats were factored into the analysis (essay 14). This is an indication of the potential for large-scale exploitation of these ecoregions. The leading sources for future degradation are mineral exploration and timber extraction, opening the way for development.

The Arctic National Wildlife Refuge, located within the Arctic Coastal Tundra ecoregion, is one of the last places in the United States where intact large mammal assemblages and migrations still occur. Photo credit: USFWS.

Conservation Threats Moving Northward into Canada

Kevin Kavanagh and Arlin Hackman

Globally, Canada contains one third of the world's boreal forests and a large component of arctic ecosystems. Also within Canada's jurisdiction are 25 percent of the world's temperate rain forests and extensive temperate conifer forests. Northern reaches of deciduous forests and prairies further add to the ecological diversity of Canada. However, results presented in this book show that many of the ecoregions in southern Canada have little intact habitat remaining, while in others, natural habitats are under increasing threat. Of the twenty-three North American ecoregions where the conservation status was predicted to significantly worsen over the next twenty years, fourteen (61 percent) were Canadian (figure 4.5). As resource development continues to expand northward, Canada's options to protect natural areas and wildlife species are diminishing at a rapid rate.

Logging allocations on Canadian public lands are among the largest in the world, spanning millions of hectares of the boreal forest regions of Quebec, Ontario, Manitoba, Saskatchewan, and Alberta. In British Columbia, where approximately 85–90 percent of forests are under the control of the Ministry of Forests or other agencies, major new road access and timber harvesting will result in high levels of fragmentation and conversion of forested valleylands in coastal and interior parts of the province within the next two decades (Harding and McCullum 1994). Even as of 1990, coastal Douglas fir (*Pseudotsuga menziesii*), interior Douglas fir, and ponderosa pine (*Pinus ponderosa*) forests were so fragmented by roads that only 10 percent of these forest types existed in blocks greater than 50 km^2 (Harding and McCullum 1994). Forest harvesting has now begun to enter the southern Yukon and Northwest Territories and is predicted to extend into the few remaining patches of primary forest in New Brunswick.

Figure 4.5   Ecoregions elevated in conservation status following the threat analysis.

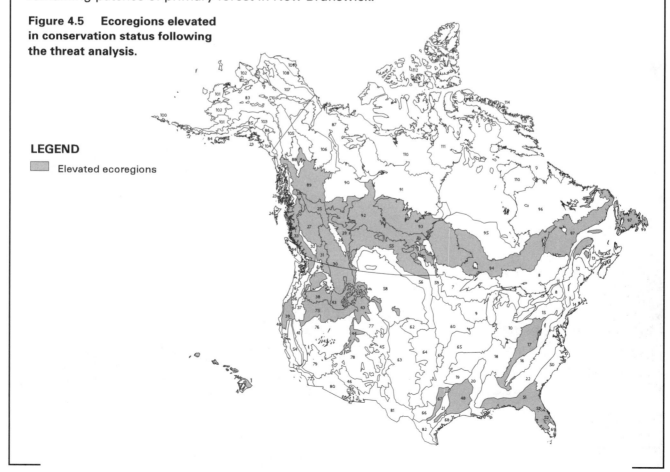

LEGEND

Elevated ecoregions

The extent of some of these timber concessions is staggering. Between September 1987 and December 1988, the Alberta government leased public timberlands almost the size of Great Britain to a dozen firms. Among these, two Japanese firms obtained leased rights to tracts covering 15 percent of the province (nearly 100,000 km²). As of 1989, Alberta was the only jurisdiction in North America where 221,000 km² of public forests, maintained by public dollars, could be signed away without public hearings (Nikiforuk and Struzik 1989).

Even where forests have been previously logged and converted to young successional stages, levels of harvest continue to increase by targeting "pioneer" tree species such as poplar (*Populus* spp.) and white birch (*Betula papyrifera*). The Ontario Ministry of Natural Resources (MNR) has actively encouraged the forest industry to retool its mills to utilize large amounts of these species. Thus, rather than strengthening protection of forest resources or enforcing more sustainable logging practices, provincial government and industry have shifted their attention to the potential of the province's newly identified "surplus of hardwood." In the fall of 1994, the MNR committed the government to allocating a 50 percent increase in the harvest level of poplar and birch over existing harvest levels in the next ten-year period, which could result in close to an additional 1,000 km² of Ontario's boreal forest being clear-cut. By early 1997, a 12 percent harvest increase had already been achieved (Canadian Parks and Wilderness Society 1997). Another recent study determined that there remain only seven roadless areas larger than 1,000 km² within the approximately 500,000 km² of productive boreal and mixed-wood forests in Ontario (Federation of Ontario Naturalists 1997).

Clearly, if Canada is going to continue to pride itself as a country wealthy in its natural resources and wildlife, *urgent* action needs to be taken to complete a nationwide system of protected areas by the year 2000 that represent the full breadth of Canada's terrestrial ecological diversity. This is the goal of WWF Canada's Endangered Spaces Campaign and one that has been committed to by all of Canada's provinces, territories, and federal governments. Despite the fact that the total area protected in Canada has nearly doubled since the launch of the campaign in 1989, with only three years remaining, the pace of site designations must increase significantly if governments are going to meet their commitments.

CHAPTER 5

Setting the Conservation Agenda: Integrating Biological Distinctiveness and Conservation Status

Efforts to conserve biodiversity must be undertaken in every ecoregion, but some ecoregions support such outstanding biological diversity and face such severe threats that they deserve immediate and proportionally greater attention from conservationists. Some ecoregions are so degraded that their globally outstanding biodiversity may erode even further without intensive restoration efforts and protection of remaining source pools containing the native biota. Finally, another subset of ecoregions offers rare opportunities to maintain the ecological integrity of very large landscapes over the long term.

Integrating the results from the biological distinctiveness and conservation status indices helps to categorize ecoregions facing these different conservation trajectories. Based on the categories they receive for the two component indices, ecoregions are placed in one of twenty cells in the integration matrix (figure 2.7). Using this matrix, conservation planners and organizations can decide which scenarios best describe the situation in their ecoregions of concern and develop appropriate conservation strategies. We have grouped the twenty cells into five conservation classes, each of which is associated with a set of recommended conservation actions (figure 2.7). Below are examples of ecoregions in each conservation class. For a complete definition of the conservation classes, see chapter 2.

Class I. Globally outstanding ecoregions requiring immediate protection of remaining habitat and extensive restoration. An example is the California Coastal Sage and Chaparral [72], where human population growth and urban development continue to severely and rapidly alter one of the richest and most unusual habitat types on earth.

Class II. Regionally outstanding ecoregions requiring immediate protection of remaining habitat and extensive restoration. An example is the Middle Atlantic Coastal Forests [50], where agricultural practices, logging, and fire suppression have severely altered and degraded many natural processes.

Class III. Globally or regionally outstanding ecoregions that present rare opportunities to conserve large blocks of intact habitat. An example is the Low Arctic Tundra [110], where large-scale caribou migrations and intact predator-prey species assemblages still exist over large undisturbed portions of the landscape. It is important to preserve these large areas before habitat threats such as fragmentation and degradation permanently alter the landscape and natural disturbance regimes.

Class IV. Bioregionally and nationally important ecoregions requiring protection of remaining habitat and extensive restoration. An example is the Northwestern Mixed Grasslands [58], where dryland farming and livestock grazing have displaced a majority of the natural grasses and ecological processes found in the ecoregion.

Class V. Bioregionally and nationally important ecoregions requiring protection of representative habitat blocks and proper management elsewhere for biodiversity conservation. An example is the Southern Hudson Bay Taiga [95], where most of the ecoregion remains as an intact block, but expansion of protected areas to encompass polar bear denning areas is needed.

The specific nature of conservation efforts will vary among ecoregions in the same class because of differences in habitat type, levels of beta-diversity, and resiliency.

These brief descriptions of the five conservation classes reflect our belief that biodiversity conservation is important in every ecoregion. Functioning ecosystems provide numerous services to natural and human communities. Also, conservation of natural areas in all ecoregions ensures the preservation of distinct communities and the functional and genetic diversity of populations throughout a species' range.

However, timing, sequence, types of activities, levels of urgency, and amounts of effort required differ among ecoregions. Some contain globally unique species assemblages, others support some of the last examples of important ecological processes, and still others face extreme and rapid anthropogenic degradation. In the face of severely limited resources for conservation, knowledge of the differing factors facing different ecoregions will help to allocate resources more effectively and strategically. These results, and this entire assessment, are an attempt to assist in this process.

The results of the integration are summarized in table 5.1. The identity of the ecoregions in each cell can be found in appendix E, in which the matrix is reproduced for each MHT separately and the individual ecoregions are listed in the cells they occupy.

Ecoregions show a relatively even distribution among the five classes, excepting Class V, which contains six cells as opposed to three or four in the other classes (tables 5.2 and 5.3). The integration results are mapped onto the ecoregion map, showing the geographic distribution of the ecoregions in each class (figure 5.1).

Class I and III ecoregions contain the highest biodiversity values. Class I ecoregions are located along the Pacific Coast and in the southeastern United States, with additional ecoregions in the desert Southwest and central prairies. The Pacific Coast aggrega-

TABLE 5.1. Final Integration Matrix between Biological Distinctiveness and Conservation Status.

Biological Distinctiveness \ Conservation Status	Critical	Endangered	Vulnerable	Relatively Stable	Relatively Intact	Total
Globally Outstanding	8	5	8	9	2	32
Regionally Outstanding	8	6	0	5	3	22
Bioregionally Outstanding	6	5	8	3	5	27
Nationally Important	9	5	5	8	6	33
Unclassified	1	1	0	0	0	2
Total	32	22	21	25	16	116

TABLE 5.2. Distribution of Ecoregions by Conservation Classes.

	Class I	Class II	Class III	Class IV	Class V	Total
Number of Ecoregions	21	14	19	25	35	114
Percentage of Total	18%	12%	17%	21%	32%	100%

tion includes the Northern Cordillera Forests [89] of the boreal forest/taiga MHT in the north, a majority of the coniferous and coastal temperate rain forests of the temperate coniferous forests MHT moving south along the coast, and all of the Mediterranean scrub and savanna ecoregions in the Mediterranean scrub MHT of California [70–72]. The southeast aggregation includes a majority of the Appalachian temperate broadleaf and mixed forests [16, 17] as well as the Piedmont and Southeastern Mixed Forests [22]. The Southeastern Conifer Forests [51], the Florida Everglades [69], and the Sand Pine Scrub [52] fill out this grouping of Class I ecoregions. Class I also includes the Central [60] and Flint Hills Tall Grasslands [61] in the central prairies, as well as the Chihuahuan Desert [81] in the Southwest. The Chihuahuan Desert also extends far beyond the U.S. border into Mexico.

Class III ecoregions, those that present rare opportunities for conservation of globally or regionally outstanding levels of biodiversity, are found primarily in polar areas and in the southwestern deserts and mountains. The polar group includes both tundra and boreal forest/taiga ecoregions. Tundra ecoregions extend east from the Aleutian Islands [100] and the Alaskan north slope [108–109] through the St. Elias Range and Yukon [103] into the Low Arctic Tundra [110] of the Northwest Territories and Quebec. The boreal forests include a majority of the Northwest Territories, including the Muskwa/Slave Lake Forests [90] and Northern Canadian Shield Taiga [91], and extend east to the North Atlantic Ocean in the Eastern Canadian Shield Taiga [96]. The southwestern aggregation includes Mojave [79] and Sonoran Deserts [80] and Colorado Plateau Shrublands [78], as well as the coniferous forest–dominated mountain ranges of Arizona [46] and the Madrean Sky Islands [47].

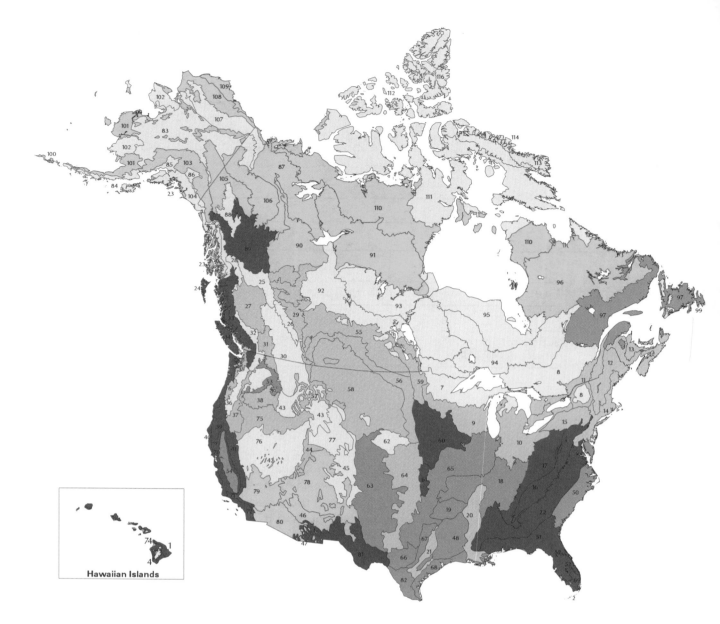

LEGEND

Class I. Globally outstanding ecoregions requiring immediate protection of remaining habitat and extensive restoration.

Class II. Regionally outstanding ecoregions requiring immediate protection of remaining habitat and extensive restoration.

Class III. Globally or regionally outstanding ecoregions that present rare opportunities to conserve large blocks of intact habitat.

Class IV. Bioregionally and nationally important ecoregions requiring protection of remaining habitat and extensive restoration

Class V. Bioregionally and nationally important ecoregions requiring protection of representative habitat blocks and proper management elsewhere for biodiversity conservation.

Figure 5.1 Conservation classes and recommended conservation actions.

TABLE 5.3. Distribution of Ecoregions by Integration Class.

Integration Class

Class I	Class II	Class III	Class IV	Class V
Hawaiian Moist Forests [1]	South Florida Rocklands [2]	Northern Pacific Coastal Forests [23]	Willamette Valley Forests [6]	Western Great Lakes Forests [7]
Hawaiian Dry Forests [4]	Central U.S. Hardwood Forests [18]	Arizona Mountains Forests [46]	Upper Midwest Forest/Savanna Transition Zone [9]	Eastern Forest/Boreal Transition [8]
Appalachian/Blue Ridge Forests [16]	Ozark Mountain Forests [19]	Madrean Sky Islands Montane Forests [47]	Southern Great Lakes Forests [10]	Central British Columbia Mountain Forests [25]
Appalachian Mixed Mesophytic Forests [17]	Piney Woods Forests [48]	Colorado Plateau Shrublands [78]	Eastern Great Lakes Lowland Forests [11]	Alberta Mountain Forests [26]
Southeastern Mixed Forests [22]	Middle Atlantic Coastal Forests [50]	Mojave Desert [79]	New England/Acadian Forests [12]	Northern Transitional Alpine Forests [28]
Queen Charlotte Islands [24]	Palouse Grasslands [53]	Sonoran Desert [80]	Gulf of St. Lawrence Lowland Forests [13]	North Central Rockies Forests [30]
British Columbia Mainland Coastal Forests [33]	California Central Valley Grasslands [54]	Northwest Territories Taiga [87]	Northeastern Coastal Forests [14]	Cascade Mountains Leeward Forests [32]
Central Pacific Coastal Forests [34]	Western Short Grasslands [63]	Muskwa/Slave Lake Forests [90]	Allegheny Highlands Forests [15]	Central and Southern Cascades Forests [36]
Klamath-Siskiyou Forests [39]	Central Forest/Grassland Transition Zone [65]	Northern Canadian Shield Taiga [91]	Mississippi Lowland Forests [20]	Great Basin Montane Forests [42]
Northern California Coastal Forests [40]	Edwards Plateau Savannas [66]	Eastern Canadian Shield Taiga [96]	East Central Texas Forests [21]	South Central Rockies Forests [43]
Sierra Nevada Forests [41]	Texas Blackland Prairies [67]	South Avalon–Burin Oceanic Barrens [99]	Fraser Plateau and Basin Complex [27]	Colorado Rockies Forests [45]
Southeastern Conifer Forests [51]	Western Gulf Coastal Grasslands [68]	Aleutian Islands Tundra [100]	Alberta/British Columbia Foothills Forests [29]	Atlantic Coastal Pine Barrens [49]
Florida Sand Pine Scrub [52]	Tamaulipan Mezquital [82]	Beringia Lowland Tundra [101]	Okanagan Dry Forests [31]	Nebraska Sand Hills Mixed Grasslands [62]
Central Tall Grasslands [60]	Eastern Canadian Forests [97]	Alaska/St. Elias Range Tundra [103]	Puget Lowland Forests [35]	Hawaiian High Shrublands [73]
Flint Hills Tall Grasslands [61]		Interior Yukon/Alaska Alpine Tundra [105]	Eastern Cascades Forests [37]	Great Basin Shrub Steppe [76]
Everglades [69]		Ogilvie/MacKenzie Alpine Tundra [106]	Blue Mountains Forests [38]	Wyoming Basin Shrub Steppe [77]
California Interior Chaparral and Woodlands [70]		Arctic Foothills Tundra [108]	Wasatch and Uinta Montane Forests [44]	Interior Alaska/Yukon Lowland Taiga [83]
California Montane Chaparral and Woodlands [71]		Arctic Coastal Tundra [109]	Canadian Aspen Forest and Parklands [55]	Alaska Peninsula Montane Taiga [84]
California Coastal Sage and Chaparral [72]		Low Arctic Tundra [110]	Northern Mixed Grasslands [56]	Cook Inlet Taiga [85]
Chihuahuan Desert [81]			Montana Valley and Foothill Grasslands [57]	Copper Plateau Taiga [86]
Northern Cordillera Forests [89]			Northern Short Grasslands [58]	Yukon Interior Dry Forests [88]
			Northern Tall Grasslands [59]	Mid-Continental Canadian Forests [92]
			Central and Southern Mixed Grasslands [64]	Midwestern Canadian Shield Forests [93]
			Hawaiian Low Shrublands [74]	Central Canadian Shield Forests [94]
			Snake/Columbia Shrub Steppe [75]	Southern Hudson Bay Taiga [95]
				Newfoundland Highland Forests [98]
				Beringia Upland Tundra [102]
				Pacific Coastal Mountain Tundra and Ice Fields [104]
				Brooks/British Range Tundra [107]
				Middle Arctic Tundra [111]
				High Arctic Tundra [112]
				Davis Highlands Tundra [113]
				Baffin Coastal Tundra [114]
				Torngat Mountain Tundra [115]
				Permanent Ice [116]

The Class II ecoregions represent some of the last opportunities to conserve regionally (continentally) important areas for biodiversity in ecoregions facing high levels of threat. They are clustered in the southern half of the lower forty-eight states. Three of these Class II ecoregions, the Middle Atlantic Coastal Forests [50], California Central Valley Grasslands [54], and Edwards Plateau Savannas [66], are noted for harboring many endemic species.

Class IV and Class V ecoregions are generally located in the central and far northern portions of the continent. Although not highly distinctive at global or continental scales, the biodiversity they contain is important and includes notable habitats such as the forests of the Cascades, Colorado Rockies, Yellowstone, central and eastern Canada, and north central Rockies, as well as the grasslands of the northern plains. Some of these ecoregions are also notable for endemic plants, such as the Great Basin Shrub Steppe [76] (>2,400 spp.).

When MHTs are examined individually, different patterns become apparent (table 5.4). In temperate broadleaf and mixed forests (MHT 2.1) a majority of the ecoregions are in Class IV, bioregionally and nationally important ecoregions requiring protection of remaining habitat and extensive restoration, indicating that while not globally or regionally outstanding in biodiversity, their conservation status is still under serious threat. Temperate coniferous forest (MHT 2.2) and xeric shrublands/desert (MHT 4.2) ecoregions are relatively evenly distributed among all five classes. The relatively even distribution among classes allows all levels of conservation activities to be employed within these MHTs. In temperate grasslands/savanna/shrub (MHT 3.1) and temperate broadleaf and mixed forests (MHT 2.2), there are no Class III ecoregions, underscoring the absence of large stable or intact habitat blocks with high biodiversity. In addition, 82 percent of the ecoregions in temperate grasslands/savanna/shrub (MHT 3.1) fall into Classes II and IV. This calls attention to the poor conservation status of the temperate grasslands and the extensive need for restoration. In boreal forest/taiga and tundra, a majority of the ecoregions are in only two classes, Classes III and V. This reflects the relative stability of these ecoregions in terms of conservation status but the varying degree of biological distinctiveness within the major habitat type.

TABLE 5.4. Distribution of Conservation Classes by Major Habitat Type (by percentage).

Major Habitat Type	Conservation Class				
	Class I	Class II	Class III	Class IV	Class V
2.1 Temperate Broadleaf and Mixed Forests	18%	12%	0%	59%	12%
2.2 Temperate Coniferous Forests	27%	7%	10%	23%	33%
3.1 Temperate Grasslands/Savanna/Shrub	13%	44%	0%	38%	6%
4.2 Xeric Shrublands/Deserts	10%	10%	30%	10%	40%
6.1 Boreal Forest/Taiga	6%	6%	29%	0%	59%
6.2 Tundra	0%	0%	50%	0%	50%
All Major Habitat Types	18%	12%	17%	21%	32%

Overall, the integration exercise underscores that the conservation outlook is rather grim in a number of very important ecoregions. In addition to the preservation of future lands for conservation, a significant effort to restore degraded lands is needed in many areas of the United States and parts of Canada. It is our hope that this assessment complements other efforts in identifying conservation and restoration activities and time scales with which to focus these efforts (essay 15).

ESSAY 15

Conservation Assessments: A Synthesis

Reed F. Noss

In a world with relentless human population growth and habitat destruction, it is not possible for conservationists to devote intensive effort everywhere. They are forced to focus much of their attention, at any given time, on a subset of regions, landscapes, and sites that need the most urgent help. Choosing which areas to protect is not an easy task. Every region deserves a rigorous conservation plan, and local people often need assistance from national and international conservation organizations to put such plans in motion. Yet conservation resources are always limited, and the global consequences of habitat destruction in some regions are more severe than in others. Priorities must be established. Hence, conservationists have developed ways to assess the conservation value of regions and sites—and the imminence of the threats they face—so that resources can be directed to those places that have the most to lose if not protected quickly.

A conservation assessment is essentially a way of considering regions and sites within a broader, ultimately global context. From a broad perspective it can readily be established that some areas are of greater biological value—for example, as hot spots of species richness or endemism—than others, and that some areas face greater risks to their ecological integrity than others. In this short essay I review some conservation assessments, ongoing or completed, in North America that complement World Wildlife Fund's assessment of terrestrial ecoregions, and speculate about how they might all work together.

As many authors have noted, most parks, wilderness areas, and other protected areas in North America were not established because of their biological or ecological values, but because they were scenic, provided recreational opportunities, or conflicted little with resource (e.g., timber) production objectives (see Noss and Cooperrider 1994). Although scientists in North America have recognized the need to use biological criteria in selecting protected areas at least since the 1920s and 1930s (e.g., Shelford 1926, 1933), their recommendations had relatively little influence on reserve establishment. It was not until the mid-1970s, when The Nature Conservancy implemented its first natural heritage programs, that an objective basis for ranking sites of differing conservation value became available and began to inform reserve selection decisions.

The heritage program methodology established by The Nature Conservancy allows sites to be ranked in terms of the elements of diversity—chiefly species and natural communities—they contain (Jenkins 1985, 1988). Elements are ranked at state (or provincial) and global scales in terms of their relative rarity and endangerment. Sites with higher-ranked elements (e.g., those classified as critically imperiled globally) and higher-quality occurrences of those elements are generally considered of higher conservation value. The elements themselves can also be mapped and tracked throughout their ranges. The global context and objective methodology provided by the heritage programs revolutionized the way many conservation agencies thought about conservation priorities. Although there is still some resistance among agency bureaucrats to looking beyond political boundaries, the heritage approach made parochial and unscientific decisions about conservation much more difficult to defend.

One limitation of the heritage program methodology is that it tends to focus attention on the rarest elements, for example, an endemic lily occurring on just a few acres. Although the heritage approach was designed to protect "the last of the least and the best of the rest" (Jenkins 1985), in practice the last of the least have received almost all the attention. Furthermore, although the objective of the heritage approach is to apply a "coarse filter" of community- or ecosystem-level protection, supplemented by a "fine filter" to protect the rarest species (Noss 1987), the coarse filter approach has been seldom used. When it is used, the focus has been on narrowly defined plant associations rather than on large, heterogeneous landscapes or "functional mosaics" of communities that have a greater chance of long-term viability (Noss 1987).

The idea of coarse-filter conservation is essentially equivalent to the long-established conservation strategy of representation. The goal behind this strategy is to represent every kind of habitat, community, or ecosystem in protected areas, thus assuring conservation of most species. Because it protects common as well as rare species, this strategy has greater potential to be proactive and to sustain entire assemblages before individual species become so rare as to warrant protection under endangered species laws. In the United States the major representation assessment is the gap analysis project (GAP) of the Department of Interior in collaboration with state agencies, university researchers, and others. GAP is a complex, geographic information system–based program with many potential applications (see Scott et al. 1993; Scott, Tear, and Davis 1996). Among the useful information produced by GAP is an analysis of the level of representation in protected areas of the existing vegetation types of a region. Beginning with LANDSAT Thematic Mapper scenes and supplemented by ancillary information such as soil maps and aerial photographs, a map of current vegetation is produced for a state or other region. The boundaries of protected areas and other lands managed for their natural values are overlaid on the vegetation map, producing an assessment of how well the various vegetation types are currently protected (for example, see the assessment for Idaho by Caicco et al. 1995). Through simple habitat association models, vertebrate species (and often selected groups of invertebrates) are linked to vegetation types and mapped to correspond to polygons of suitable habitat within their known geographic ranges. Detailed maps of predicted distribution can therefore be produced for each species and, by overlaying the distributions of many species, hot spots of species richness for any group of interest can be delineated (Scott et al. 1993). The coverage of species groups in protected areas is another analytic product. Thus, GAP produces an up-to-date assessment of where the gaps in protection are for a geographic area of interest and helps determine priorities for new protected areas and changes in land management. In Canada, a gap analysis undertaken by World Wildlife Fund Canada has similar objectives but is based on a framework of physically defined habitats called "enduring features" (Kavanagh and Iacobelli 1995).

GAP, as usually conducted, is a snapshot of current conditions only. It does not consider history. A useful supplement to GAP is an assessment of vegetation change through time (Strittholt and Boerner 1995). A study sponsored by the U.S. National Biological Service (Noss, LaRoe, and Scott 1995) reviewed the literature on the decline and degradation of ecosystems (vegetation types, habitats, etc.) in the United States since European settlement. The results showed that many of the country's native ecosystems have declined to a fraction—often less than 5 percent—of their former extent. Among the ecosystems showing the greatest overall losses are grasslands of many types, savannas and woodlands, barrens, undammed streams, many kinds of coastal habitats, longleaf pine forests, the old-growth stage of all eastern deciduous forests, Fraser fir forest in the southern Appalachians, red pine and white pine forests in the Great Lakes region, ponderosa pine forests with natural structure throughout the West, shrublands with a native herbaceous component, and many kinds of riparian forests and wetlands.

A follow-up study sponsored by Defenders of Wildlife (Noss and Peters 1995) summarized the data on endangered ecosystems from the National Biological Service study, identified the "top 21" most endangered ecosystems in the United States, and, because many decisions of importance to conservation take place at a state level, analyzed threats to ecosystems and their associated species by state boundaries. An overall risk index was calculated by combining three factors: ecosystems at risk, species at risk, and risk from development. The ecosystem risk index was based on the number of the most highly endangered ecosystems nationally that occur in each state. The species risk index is a measure of the percentage of a state's species that are imperiled, as expressed by the proportion of vascular plants, vertebrates, and aquatic invertebrates (mussels and crayfish) that are ranked as imperiled or critically imperiled globally by The Nature Conservancy. The development risk index is composed of two subindices: development status and development trend. A state's development status was assessed by four equally weighted measures: human population density, percentage of state developed as of 1992, percentage of state in agricultural land, and rural road density. The development trend subindex measured the rate of development in recent years, as determined by change in development status during the decade 1982–92. The results showed that the southeastern states (especially Florida), Texas, California, and Hawaii have the most imperiled species and ecosystems and are losing them to development most rapidly (Noss and Peters 1995). Similarly, a study using county-level distributional data for endangered species to determine hot spots of threatened biodiversity found the highest concentrations in Hawaii, Southern California, the southeastern coastal states, and southern Appalachia (Dobson et al. 1997).

How can these various studies be interpreted in light of WWF's North American Terrestrial Ecoregion Assessment, and how can their results be applied to real-life conservation work? To begin, I suggest a hierarchical framework, with both top-down and bottom-up approaches nested within it. Conservation planning and development of sustainable life styles and economies must proceed in each region from the ground up, starting with grassroots knowledge and enthusiasm. But conservation groups operating on national and international scales need to see the "big picture" in order to allocate resources effectively, and, as noted earlier, some regions have more urgent conservation needs than others. A major challenge is reconciling the top-down approach of the national and international groups with the bottom-up approach of the grassroots groups. A case study in the Klamath-Siskiyou region, cosponsored by a grassroots group (the Siskiyou Project) and an international group (WWF), is attempting to show how reconciliation of global and local goals might be accomplished.

The following steps suggest a way to apply existing data and tools to conduct globally informed but locally relevant conservation planning:

1. Conduct a continental-scale analysis of conservation value and actions needed, with results interpreted from a global perspective. Identify regions of greatest global conservation significance that require immediate protection or restoration, as well as other regions of lesser conservation value but nevertheless in need of protection, restoration, and management. This is precisely what has been done by WWF for North America (this book) and several other regions.

2. Initiate regional conservation planning efforts for every region, informed by global understanding and priorities; conducted by conservation biologists in consultation with national, regional, and local groups; and implemented by local and regional groups. National and international organizations should devote most resources to those regions considered globally outstanding but should also support case studies and planning in regions requiring different sorts of actions.

3. For each region, use more detailed information to identify the individual sites and landscapes of greatest conservation value. I recommend a three-pronged analytic approach focusing on: (1) special elements (e.g., occurrences of imperiled species and communities, critical watersheds, and roadless areas); (2) unrepresented or poorly represented habitats, communities, or ecosystems (i.e., from gap analysis); and (3) specific habitat requirements of wide-ranging or other fragmentation-sensitive or potential umbrella species (Noss 1996). From this analysis, identify several alternative reserve networks consisting of core areas, buffer or transition zones, and corridors or other connectivity zones.

4. For each region, conduct additional socioeconomic and ecological studies to determine the effects of implementing alternative reserve designs. Identify a preferred alternative and appropriate management actions for each zone and site.

5. Implement the preferred alternative in each region by acquiring or otherwise protecting key areas, establishing an ecological monitoring and adaptive management program, and carrying out management actions.

CHAPTER 6

Recommendations

North America's Global Responsibilities for Biodiversity Conservation

This study helps put the biological distinctiveness of North American ecoregions in a global context. We rank thirty-two North American ecoregions as globally outstanding—their biodiversity attributes equal or exceed levels found in the most distinct ecoregions sharing the same MHT on other continents. Another way of illustrating the extraordinary biodiversity value of North American ecoregions is to display them along with other outstanding examples of the world's terrestrial MHTs (essay 16). The Global 200 ecoregions map portrays these units, coded in colors by their MHT (figure 6.1). The map shows that North American ecoregions are well represented among the MHTs that lie within the subtropical, temperate, boreal, and tundra regions.

ESSAY 16

Globally Outstanding Biodiversity in Our Own Backyard

Eric Dinerstein and David M. Olson

WWF-US recently released the Global 200 ecoregions, a four-year study to identify and map the ecoregions of the world that support globally outstanding and representative biodiversity (Olson and Dinerstein 1998). The Global 200 promotes representation of biodiversity across nineteen major habitat types distributed among the three realms—terrestrial, freshwater, and marine—and on each continent or ocean basin (biogeographic zones) (figure 6.1). Terrestrial ecoregions are color coded by their major habitat types (MHTs) to illustrate how they are distributed across continents.

The map highlights as globally outstanding a number of North American terrestrial ecoregions that lie within the subtropical, temperate, boreal, and tundra MHTs. As examples, the Chihuahuan Desert [81] and the Namib-Karoo of Southern Africa are the most diverse warm deserts in the world, with the Chihuahuan Desert ranking globally outstanding in reptile, bird, mammal, and cactus richness (Olson and Dinerstein 1998). The Appalachian/Blue Ridge Forests [16] and the Southwest China Temperate Forests are the two most diverse temperate broadleaf and mixed forests on earth. The Klamath-Siskiyou Forests [39] are among the richest ecoregions of the world for

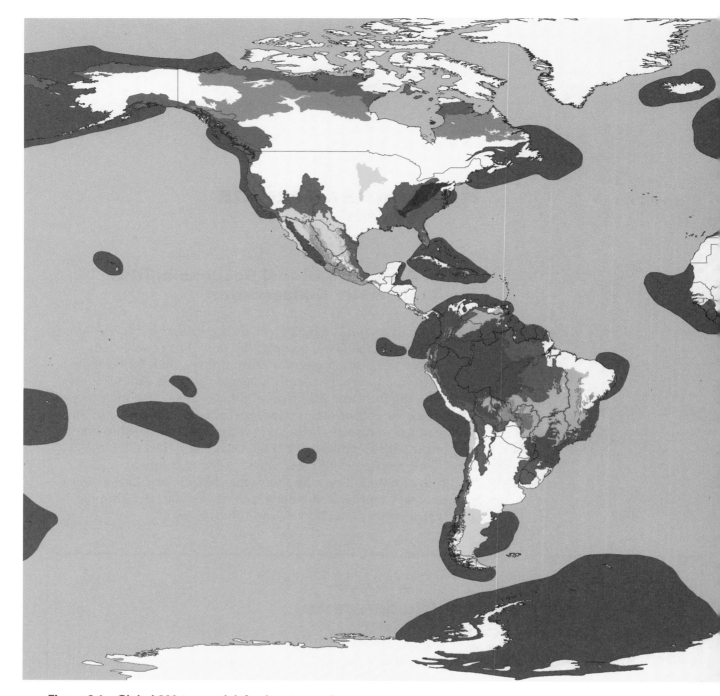

Figure 6.1   Global 200 terrestrial, freshwater, and
marine ecoregions. The Global 200 ecoregions are
recognized as global conservation priorities because of
their representative and globally outstanding
biodiversity. To achieve representation goals at a
global scale, we selected ecoregions of each major
habitat type (e.g., temperate broadleaf forests) from
each biogeographic realm (e.g., Nearctic) to represent
distinctive ecosystems and species assemblages from
around the world. Ecoregions with outstanding
biodiversity features are also highlighted in the Global
200, including those with extraordinary richness,
pronounced endemism, unique higher taxa, unusual
ecological or evolutionary phenomena, or global rarity
of major habitat types.

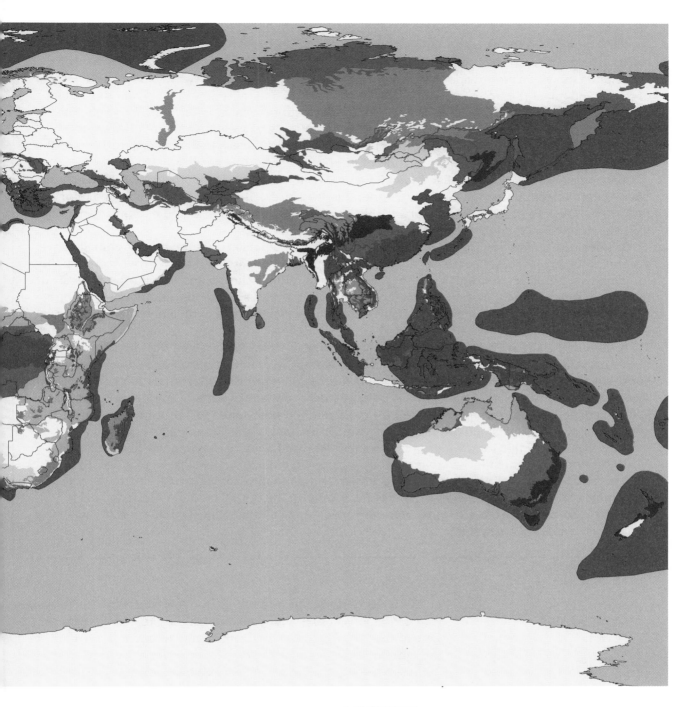

LEGEND

Tropical and Subtropical Moist Broadleaf Forests
Tropical Dry Forests
Temperate Broadleaf and Mixed Forests
Temperate Coniferous Forests
Tropical and Subtropical Coniferous Forests
Boreal Forests and Taiga
Tropical and Subtropical Grasslands, Savannas, and Shrublands
Temperate Grasslands, Savannas, and Shrublands

Flooded Grasslands
Montane Grasslands
Tundra
Mediterranean Scrub
Deserts and Xeric Shrublands
Mangroves
Freshwater ecoregions
Marine ecoregions

conifers, along with the Mexican pine-oak forests and the tropical moist forests of New Caledonia. North America also contains ecoregions that represent rare habitat types, such as Mediterranean shrublands (only five worldwide) and temperate rain forests (only eight worldwide and 50 percent of remaining intact rain forest habitat). The Southeastern Conifer Forests [51] contain the richest assemblages of herbaceous plants of any temperate or subtropical conifer forest on earth. The Florida Sand Pine Scrub [52] contains remarkably high levels of endemism in relatively small patches of habitat along the central spine of Florida. And the Eastern Canadian Shield Taiga [96] may contain the most extensive population of migratory caribou on earth.

Also extraordinary are Hawaiian Moist Forests [1] and Hawaiian Dry Forests [4], which contain nearly complete endemism in plants, invertebrates, and birds, as well as spectacular evolutionary radiations. The Puerto Rican Moist Forests [3] contain some interesting island endemics and relictual taxa characteristic of the distinct Greater Antillean fauna.

Besides its Chihuahuan Desert [81], Mexico's Jalisco and Balsas dry forests rank globally as do several of its conifer forest ecoregions. The Sierra Madre Occidental and the Sierra Madre del Sur, to name two, support communities with globally outstanding richness and extraordinary levels of endemism in a wide range of taxa. Finally, another of its desert ecoregions, southern and central Baja, contains some of the world's most structurally unusual habitats, with convergent systems occurring only in southern Madagascar.

Together, the Great Plains ecoregions of North America once constituted the largest area of native grasslands in the world and would have figured prominently in the Global 200 if they still supported their native biota and the once extensive bison migrations that linked them. Most North Americans are unaware that the biomass of ungulates reported for parts of the Great Plains historically exceeded levels currently recorded for the Serengeti grasslands of Tanzania and Kenya. The absence of any ecological integrity for even a small remnant of the Great Plains is a global tragedy. To remedy this loss, we urge that conservationists unite to demand the restoration of a North American Serengeti—the creation of at least one giant conservation area of at least 25,000 km², perhaps in the Northwestern Mixed Grasslands [58], that restores populations of the full complement of herbivores (bison, elk, pronghorn, and prairie dogs), predators (grizzly bears, wolves, black-footed ferrets [*Mustela nigripes*]), and native plants to levels that are self-sustaining in a native grassland ecosystem.

With the designation of these globally outstanding North American ecoregions comes a global responsibility to conserve them. One of these ecoregions, the Everglades, has become a symbol in the United States for the type of effort needed to restore a Global 200 ecoregion. Restoring a critical ecological process—sheet flow of nutrient-poor water over limestone bedrock—to re-create a functioning ecosystem of about 6,500 km² is estimated to cost $3–5 billion over the next five to ten years (Tipton 1996).

The Everglades ecoregion has received a large amount of attention from the press and the U.S. government, but it is not an isolated case in North America. This assessment highlights thirteen other Class I ecoregions whose biodiversity values are equal to those of the Everglades and are even more threatened (figure 6.2). These ecoregions also are ranked as globally outstanding in our biological distinctiveness index and as either critical or endangered in the conservation status index. Like the Everglades, they will require immediate protection and restoration on a massive scale.

1. The Hawaiian Moist Forests [1], supporting an extraordinary diversity of endemic plants, birds, and invertebrates, are highly threatened by invasive species and changes in land use.

British Columbia mainland coastal forests

Northern California coastal forests

Klamath-Siskiyou forests

California coastal sage and chaparral

Central tall grasslands

Sierra Nevada forests

Appalachian mixed mesophytic forests

Southeastern mixed forests

Florida sand pine scrub

Southeastern conifer forests

Hawaiian moist forests

Hawaiian dry forests

Figure 6.2 Ecoregions with biological diversity equal to that of the Everglades and more threatened.

2. The Hawaiian Dry Forests [4], located in the lee of volcanoes, contain numerous endemic plants and invertebrates. Protection of remnant patches and restoration offer the only hope for the biota of this ecoregion.

3. Appalachian Mixed Mesophytic Forests [17] harbor the most diverse temperate forests in North America. Although pockets of undisturbed native forests are quite rare, restoration is tractable.

4. Southeastern Mixed Forests [22], also known as the Appalachian Piedmont, contain some of the richest plant communities on the continent, but agriculture and development have left only small fragments of native habitat.

5. Northern California Coastal Forests [40], featuring the magnificent groves of coastal redwoods, have been so decimated by unbridled logging that only a few, relatively small examples remain.

6. Southeastern Conifer Forests [51], featuring the fire-dependent longleaf pine communities, support some of the richest herbaceous floras on earth. Unfortunately, intact native forests are largely limited to a few military bases.

7. Florida Sand Pine Scrub [52], an archipelago of ancient ice-age sand communities, supports a wealth of endemic plants and invertebrates, but rampant development threatens to extinguish these hotbeds of biodiversity.

8. British Columbia Mainland Coastal Forests [33], once comprising some of the earth's largest expanses of temperate rain

forests, are threatened by widespread logging that is rapidly creating barren landscapes and conversion to tree plantations over much of this ecoregion.

9. Central Pacific Coastal Forests [34], a temperate rain forest ecoregion of globally outstanding richness, have been so intensively logged that only small pockets remain, especially in the lowlands.

10. Klamath-Siskiyou Forests [39], one of the most endemism-rich and complex temperate conifer forests on earth, are also threatened by intensive logging poised to eliminate all lowland forests in the near future, and mining threatens serpentine communities rich in endemics.

11. Sierra Nevada Forests [41] are the only home to the sequoia, the world's largest tree species. Logging and development pressures continue to fell these unique forests.

12. Central Tall Grasslands [60], the tallest temperate grasslands on earth, once constituted a vast ecosystem in which millions of bison roamed but are now restricted to a few, very small fragments in a sea of agriculture.

13. California Coastal Sage and Chaparral [72], a representative of a globally rare habitat, is rapidly succumbing to housing developments and golf courses that imperil many plants, invertebrates, reptiles, and birds.

The critical ecoregions listed above mirror patterns of human settlement and demography, and the poor conservation status of these regions is equivalent to the conservation status of ecoregion in many poor developing countries. In many of these critical ecoregions, little native flora and fauna remains, and conservation activities in these areas will have to focus on saving the last source pools of native biodiversity and restoring natural ecological processes.

However, if all conservation efforts are little more than "emergency room" measures, we face long odds in conserving representative examples of North America's ecoregions. On the more positive side, our assessment highlights eleven ecoregions that offer rare opportunities to conserve globally outstanding biodiversity in relatively intact landscapes. Most of these ecoregions are found in Canada and Alaska:

1. Northern Pacific Coastal Forests [23] (temperate coniferous forests MHT)

2. Madrean Sky Islands [47] (temperate coniferous forests MHT)

3. Sonoran Desert [80] (xeric shrublands/desert MHT)

4. Northwest Territories Taiga [87] (boreal forest/taiga MHT)

5. Muskwa/Slave Lake Forests [90] (boreal forest/taiga MHT)

6. Eastern Canadian Shield Taiga [96] (boreal forest/taiga MHT)

7. South Avalon–Burin Oceanic Barrens [99] (boreal forest/taiga MHT)

8. Aleutian Islands Tundra [100] (tundra MHT)

9. Arctic Foothills Tundra [108] (tundra MHT)

10. Arctic Coastal Tundra [109] (tundra MHT)

11. Low Arctic Tundra [110] (tundra MHT)

Specific Targets Requiring Urgent Action

At a finer scale, we invited experts from across North America to recommend priority sites and activities to support conservation of biodiversity in each ecoregion, and to identify nongovernmental conservation organizations active in each unit. The results of these assessments appear in the ecoregion descriptions included in appendix F.

The expert assessment workshop, an integral part of this study, identified several important measures that must be taken to conserve—and in many cases restore—the biodiversity in the MHTs of the lower forty-eight states and southern Canada. While the measures are relatively simple, implementation will be challenging and will require the united efforts of many conservation groups, resource agencies, and local communities. Most of all, the answers call for a demonstration of political will by leaders in government and the public and private sectors. Specifically:

- Complete a network of ecologically representative protected areas in Canada and the United States that conserves the last remnants of intact original habitat. For several habitat types (e.g., old-growth forests, native grasslands, Mediterranean scrub habitats), only 2 to 5 percent of the original intact habitat in the lower forty-eight states remains. These remnants are the only source pools for much of the diversity of these ecoregions, and they are being eliminated rapidly, precluding future opportunities for restoration.
- Save representative examples of the large blocks of undisturbed forests in North America. There are great opportunities to achieve this goal in central and northern Canada and parts of Alaska. Few blocks of intact (old-growth) forest habitat occur in all of the contiguous United States and southern Canada, but many forests, especially in the eastern United States, are coming back. Over the course of the next one hundred years, these forests will begin to take on some of the characteristics of original old-growth habitats. It is essential to protect large blocks of these forests and restrict logging in them if we are to save and recover old-growth associated species. Land managers must think ahead of the short (12–60-year) harvesting cycles that are implemented in many forested habitats in the United States and Canada. (In Canada harvesting cycles usually range between 50 and 100 years.) This is essential for the conservation of songbirds, the return of large predators to the natural landscape, and the survival of numerous invertebrates and plants (essay 17).
- Greatly increase the number of certified forests where timber is being harvested sustainably (essay 18). This is essential for maintaining the integrity of ecosystems outside protected areas.
- Allow fire to play its crucial role in maintaining the biodiversity and habitat complexity of many eastern, inland west, and boreal forests and most of the Great Plains grasslands. Decades of fire exclusion, even on protected federal, state, and provincial lands, have dramatically and in some cases permanently altered the composition of plant and animal communities. Wherever possible,

Getting Ahead of the Cutting Cycles: What Happens When the Trees Grow Back?

Gordon Orians

Most modern forestry practices are designed to maximize wood production, which typically involves clear-cutting and removal of a large proportion of the structural legacy of a forest. In addition, advances in wood technology have increased the array of products that can be manufactured from small trees. Because the rate at which new wood is added declines with a forest's age, maximizing wood production in most cases is accomplished by planting even-aged, single-species stands that are harvested on short rotation cycles. These forest plantations lack many of the structural features of natural forests, and they are poor providers of some of the other valuable goods and services.

Not surprisingly, as forests have been increasingly replaced by managed tree plantations, concerns about maintaining natural forests composed of many species of trees, ranging in age from saplings to large canopy dominants, have been heightened. Recently, most attention has been directed to those places on earth where large tracts of old-age forests and roadless areas remain. Dominating concerns are the virgin temperate rain forests of the north Pacific Coast of North America; southern Chile and Tasmania as well as the tropical moist forests of the Amazon Basin; central Africa; and parts of Malaysia, Indonesia, and Papua New Guinea. Because no large stands of old-age temperate deciduous forests remain, this important component of the world's forests has been relatively neglected.

Sustained efforts to preserve significant tracts of virgin forests, both temperate and tropical, are vital and should not be diminished, but more attention needs to be paid to regions in which forests are regenerating after they have been cut. This is particularly true of the deciduous forests of eastern North America, where nearly all of them have been logged one or more times since the arrival of Europeans in America. Initially, these forests were cut primarily to clear land for agriculture, but as the more productive prairie soils of the Midwest and Great Plains were plowed, large areas of agricultural land in the East were abandoned. The young forests that have grown on these former fields have been increasing steadily. However, the market value of the wood accumulating in these forests is also increasing; extensive cutting will soon become economically attractive.

The current status and trends in the deciduous forests of eastern North America create an unusual opportunity for conservation biologists and land managers to develop a management plan for a habitat type whose conservation status is improving. This contrasts with the usual situation, which is to react in a crisis mode to save fragments of a dwindling resource. A comprehensive deciduous forest plan should include large unharvested tracts that are maintained as ecological preserves, with proper attention paid to the size and configuration of regrowing forest blocks. We have an opportunity to restore whole landscapes of forested habitat that enhance a region's ecological integrity and promote the conservation of its characteristic biodiversity over the long term. Other areas should be designated for testing of experimental harvesting techniques that can combine extraction of valuable market products with maintenance or even enhancement of other values (e.g., through independent forest certification). Still other areas should be allocated to intensive harvest of wood while maintaining some important structural components of biodiversity as source pools for forest regeneration. Specific plans should be developed for forest communities that are maintained by fire so that controlled burning can be carried out in ways that minimize potential damage to other ecological communities and to human property.

Unharvested and lightly harvested forest tracts should be distributed so that organisms with poor dispersal abilities can move among them and so that ecological communities can shift geographically in response to the climatic changes expected to occur during the coming century. Forest management should provide forest continuity at boundaries between deciduous forest ecoregions, because those areas are likely to be especially important for shifts in ranges of species. Because urban and suburban sprawl is prevalent in the region, methods to control and direct such sprawl need to be incorporated into planning. Control of suburban sprawl has the potential to assist efforts to reduce disease, such as Lyme disease, that result from increased contact of people with wildlife.

Because land ownership patterns in eastern deciduous forests are complex, development of a comprehensive plan will require cooperation of federal, state, and local agencies as well as private landowners. Although such a planning process would not be easy, all landowners and managers stand to benefit from the existence of a plan that maintains and enhances a broad array of forest values and offers more long-term stability of land-use options than would be available without a plan. Much is at stake, and the potential benefits are great!

ESSAY 18

Diverse Forest Ecoregions of North America: Their Protected Status and Importance in Forest Certification

Dominick DellaSala

The forests of the United States and Canada are a resource of incalculable value and global significance (figure 6.3a). Collectively these two countries make up 13–18 percent of the world's forests, one third of the total boreal forest cover, and one half of the world's temperate rain forest (FAO 1990; World Conservation Monitoring Center, London, pers. comm., 1997; Kuusela 1990; Alaback 1991). More than one third of the fifty-two forested ecoregions are ranked globally outstanding (figure 6.3b), and three fourths are endangered or critically endangered (figure 6.3c). As a group, temperate coniferous forests include the greatest number (seven) of critical or endangered ecoregions of any major forest type.

In the United States and Canada, keystone habitats such as old-growth (Norse 1990; Noss, LaRoe, and Scott 1995; Davis 1996) and riparian forests (Brinson et al. 1981; Swift 1984; Kauffman 1988) have experienced widespread conversion to biologically simplistic plantations—all but 2–5 percent of old-growth forests in the continental United States have been logged at least once (Noss, LaRoe, and Scott 1995), and comparable declines have occurred in old-growth white pine (*Pinus strobus*) forests of southern Ontario (Buchert et al. 1997). Only 5 percent of all forest ecoregions are legally protected from commercial logging and mining (e.g., national and provincial parks, congressionally designated wilderness, and national monuments, IUCN 1994 categories I–III) (based on data from McGhie 1996). Most ecoregions are poorly represented in protected areas, with particularly large gaps in the eastern United States and Canada (figure 6.3a). In general, most protected areas (69 percent, n = 1083) are too small to conserve large-scale ecological processes and species with large home-range requirements (e.g., grizzly bear [*Ursus arctos horribilis*] and anadromous salmon [*Oncorhynchus* spp.]).

This assessment identifies important conservation measures in the United States and Canada that can contribute to global conservation targets established by WWF's International Forests for Life campaign. The campaign has two primary, short-term goals: (1) establish an ecologically representative network of protected areas covering a minimum of 10 percent of the world's forests by the year 2000 (only 6 percent of the world's forests are protected and only 5 percent in the United States and Canada); and (2) ensure independent certification of 25 million acres of forests outside protected areas by 1998 (7 million acres have been certified globally, with 20 percent of these forests in the United States and none in Canada).

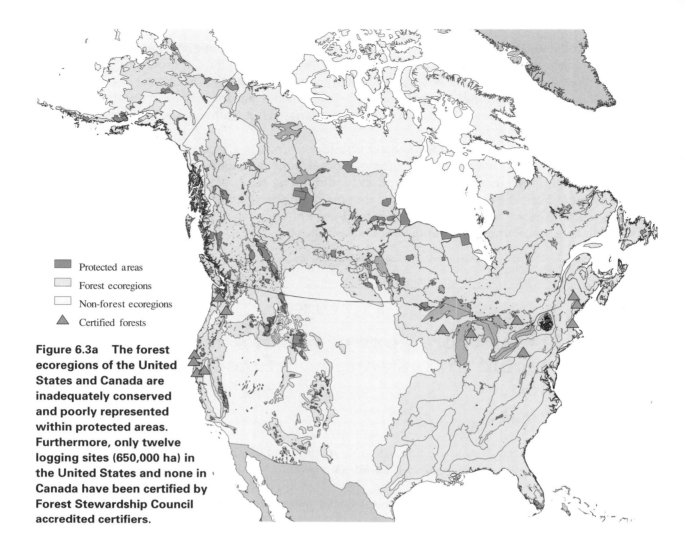

Protected areas

Forest ecoregions

Non-forest ecoregions

▲ Certified forests

Figure 6.3a   The forest ecoregions of the United States and Canada are inadequately conserved and poorly represented within protected areas. Furthermore, only twelve logging sites (650,000 ha) in the United States and none in Canada have been certified by Forest Stewardship Council accredited certifiers.

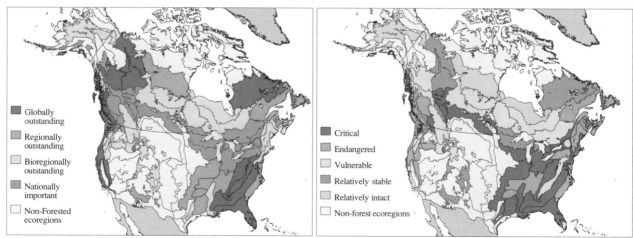

Globally outstanding

Regionally outstanding

Bioregionally outstanding

Nationally important

Non-Forested ecoregions

Figure 6.3b   Nearly one third of all forest ecoregions in the United States and Canada have extraordinary biodiversity features when compared to similar forests around the world.

Critical

Endangered

Vulnerable

Relatively stable

Relatively intact

Non-forest ecoregions

Figure 6.3c   Nearly three fourths of all forest ecoregions in the United States and Canada are considered critical or endangered, in terms of biodiversity loss and ecological integrity.

Independent forest certification is a new initiative developed to improve forest management outside protected areas. The Forest Stewardship Council (FSC) is an independent, nonprofit organization founded in 1993 by conservationists, indigenous peoples, forest product certifiers, and timber trade groups from twenty-five countries. The FSC supports environmentally appropriate, socially beneficial, and economically viable management of the world's forests based on agreed-upon regional and national standards.

WWF believes that in order for forests to be sustainably managed for both people and wildlife, the following measures need to enacted by the year 2000:

- Double the amount of forests protected in the United States and Canada and triple the amount of certified forests to better achieve ecological representation.
- Protect large representative blocks of intact forests and focus certification of commercial logging outside of the biologically richest areas of ecoregions with significant primary forests (e.g., some Canadian boreal/taiga, Northern Pacific Coastal Forests [23]).
- Protect remaining patches of primary forests and focus certification on restoration of degraded forests in critical and endangered ecoregions.
- Include ecologically representative protected areas as part of landscape-scale certification efforts for large holdings.

development on lands prone to large, intense forest fires must be limited. Controlled burning programs that are compatible with biodiversity conservation must be implemented in many of these ecoregions.

- Restrict livestock, sheep, and horse grazing in a number of the ecoregions identified by this study, and make this a continental priority, particularly where grazing occurs on U.S. public lands. Some grassland ecoregions of the Great Plains evolved with large native herbivores, making grazing a prominent form of natural disturbance. In other ecoregions, where grazing was much less common, domestic livestock have had severe impacts on natural communities. Federal and provincial agencies must recognize the extensive degradation caused by grazing in wilderness areas, riparian zones, and other areas managed for biodiversity conservation and begin restoration efforts in such areas.
- Restore at least one large grassland site within the ecoregions of the western Great Plains with a full complement of large herbivores (bison, elk), small keystone herbivores (prairie dogs), predators (grizzly bears, wolves, black footed ferrets), and native plants.
- Integrate landscape conservation with zoning laws to reduce development in critical habitats within ecoregions that are subject to major natural disturbance events, including fire-prone areas, floodplains, and barrier islands.
- Make efforts to incorporate existing military bases and bases slated for closing into biodiversity reserves. For several ecoregions, some of the best examples of native biodiversity occur on military installations, where access is restricted and natural processes, such as fire regimes, are allowed to play out (essay 19).
- Accelerate efforts to control invasive species in many natural systems. These steps will be essential to maintain ecological integrity. In some ecoregions, this will require financing research to develop new control techniques. Invasive species are a particular problem in the American West, where exotic

The Conservation Potential of Military Bases and Related Installations

Steve Gatewood and Rod Mondt

Lands withdrawn from the public domain for military purposes are far more than mere training grounds for the Army, Marines, or Air Force. Some of the finest examples of intact ecosystems occur on military reservations. Because of the restricted nature of these bases, relatively undisturbed lands have been left to their own devices. Military bases also provide needed habitat for threatened or endangered plants and animals. In some cases, a species' entire known distribution is restricted to a military reservation, such as the Okaloosa darter (*Etheostoma okaloosae*), an endangered fish found only on Eglin Air Force Base (AFB) in the Florida panhandle (Southeastern Conifer Forests [51]). In other cases, bases offer the best opportunity to protect more widely distributed elements of biodiversity, like the California Coastal Sage and Chaparral [72] and its associated species at Camp Pendleton in Southern California or a swath of intact boreal forest (Mid-Continental Canadian Forests [92]) straddling the Alberta-Saskatchewan border on the Cold Lake Air Weapons Range.

Many land managers, conservationists, and military staff are starting to view the conservation value of these pristine lands in a new light and requiring a new set of management guidelines. Several of the larger installations have natural resource management staff focused on remaining natural areas. In the 1980s, the U.S. Department of Defense created and funded a Legacy Program to inventory resources and develop management plans that address protection of critical natural values. Working with civilian partners like The Nature Conservancy, many bases moved toward ecosystem management of the entire land base.

Other partnerships include the Cabeza Prieta National Wildlife Refuge and the Barry M. Goldwater Bombing Range in the southwestern corner of Arizona (Sonoran Desert [80]), providing vital habitat for Sonoran pronghorn (*Antilocarpa americana sonorensis*) and the endangered lesser long-nosed bat (*Leptonycteris sanbornii*). These lands that border Mexico to the south and Organ Pipe Cactus National Monument to the east are critical to the ecological health and integrity of this desert region. The Cabeza-Goldwater region may be one of the least disturbed and most botanically significant deserts in the world, partly because of the absence of cattle grazing, mining, and off-road vehicle use (R. Felger, Director, Drylands Institute, pers. comm., 1998). Outside of these military lands, this ecosystem type is becoming increasingly fragmented, and what remains is in dire need of protection, restoration, and management.

Even in the East, large tracts of intact natural habitats exist on military land. Florida has several large bases (>100,000 acres), including Avon Park Air Force Range, which supports the greatest concentration in the world of cutthroat seeps, a seepage wetland ecosystem dominated by a single grass found only in four counties of central Florida. At Eglin AFB, in addition to the Okaloosa darter mentioned above, federally listed species include the red-cockaded woodpecker (*Picoides borealis*), pine barrens treefrog (*Hyla andersonii*), gopher tortoise (*Gopherus polyphemus*), Choctawhatchee beach mouse (*Peromyscus polionotus allophrys*), and the carnivorous white-topped pitcher plant (*Sarracenia leucophylla*).

Enlightened fire management is a hallmark of some installations, while nearby, federally protected national wildlife refuges continue to deteriorate ecologically from unhealthy fire exclusion programs. Fires set by explosives have secondary benefits of maintaining conditions essential for populations of plants and animals dependent on frequent fire events. At Eglin AFB, historic conversion of mature longleaf pine forest to commercial sand pine plantation by base forestry staff has been halted, and longleaf pine communities are being restored by replanting of converted sites, selective thinning, and prescribed burning that mimics natural fire. Because they are intact and frequently burned, some military landscapes also provide much needed benchmarks in our efforts to restore degraded habitats occupying the same ecoregion.

In these days of base closures and downsizing, large tracts of land that could be released from the public domain are significant assets to local communities, and, if they are to be preserved, critical decisions will need to be made soon. The restoration of North America's biological landscape must include the intact ecosystems found on military installations. As military holdings are consolidated, we urge that size, biological value, and degree of intactness be considered in the decision-making process regarding base closures. We also strongly recommend that any biologically rich base that remains open after review be required to partner with a local conservation organization to protect and manage the resources on the installation. Finally, we suggest that a nationally recognized conservation biologist be asked to join any review panel for base closings to provide technical expertise on the contribution the installation makes to a national biodiversity conservation plan. In some cases, military bases provide the last opportunities for conservation of rare elements of biodiversity both on the U.S. and Canadian mainland and in Hawaii, Okinawa, and the Vieques Islands.

grasses such as cheat grass (*Bromus tectorum*) have replaced native species to the extent that whole ecosystems have been converted. Much of Hawaii faces ecological disaster because of the invasion of exotic species.

- Promote the development of ecoregion conservation programs by federal, state, and provincial agencies that span state, provincial, and national boundaries. Biodiversity patterns and ecological dynamics do not correspond to political units.

Ecoregion-based planning for conservation has now been adopted by the leading conservation groups in North America. This project hopes to contribute to that effort by offering an efficient and scientifically credible method for recommending areas for restoration, increased protection, and policy reform in ways that support biodiversity conservation. The combination of broad- and fine-scale analyses being conducted by a host of conservation groups and agencies will provide a valuable framework and strategic planning tool to help ensure the conservation of North America's diverse natural heritage. The alternative is to watch the extinction crisis engulf each ecoregion. We hope that the results of this study contribute to the successful implementation of ecoregion-based conservation across North America.

Methods for Assessing the Biological Distinctiveness of Terrestrial Ecoregions

The biological distinctiveness index is designed to provide an objective measure of the degree to which the biodiversity of an ecoregion is distinctive over a range of biogeographic scales (i.e., from that of the entire earth, to those of biogeographic regions or provinces such as Nearctic or Neotropical, to those of bioregions within provinces such as the northern Mexico bioregion or eastern North America bioregion, to those of the local vicinity of single ecoregions, referred to as nationally important). We use four main criteria to evaluate biological distinctiveness:

- species richness
- species endemism
- rare ecological and evolutionary phenomena
- global rarity of habitat type

The points awarded for these criteria are summed to yield a biological distinctiveness score for each ecoregion. The ecoregions are then assigned to one of four categories, which reflect the distinctiveness of the ecoregion's biodiversity at different biogeographic scales: globally outstanding, regionally (e.g., continentally) outstanding, bioregionally outstanding, or nationally important. General explanations and the rationale for the four criteria are presented in chapter 2. The design of the index is summarized in figure 2.5, and below we describe the evaluation methods in detail.

Species Presence or Absence and Endemism Data

We collected published and unpublished range and distribution data for North American species in seven taxonomic groups, comprising over twenty thousand species. The taxonomic groups chosen, the total number of North American species considered in each group, and the information sources used are listed in table A.1.

For all groups except vascular plants and western land snails, the data were in the form of range maps. We compared the range map of each species to the ecoregion map and recorded the species as present or absent in each ecoregion, as well as whether the species was endemic to the ecoregion (see below for endemism decision rules). This method is much more efficient than exact GIS analyses, which would involve

TABLE A.1. Taxonomic Groups Used in the Analysis, with Number of Species per Group and the Sources of Information.

Taxonomic Group	Number of Species	Source(s)
Birds	622	(Scott 1995)
Mammals	313	(Banfield 1974)
		(Burt and Grossenheider 1980)
		(Dobbyn 1994)
		(Forsyth 1985)
Butterflies	570	(Scott 1986)
Amphibians	205	(Conant and Collins 1991)
		(Cook 1984)
		(Stebbins 1985)
Reptiles	267	(Conant and Collins 1991)
		(Cook 1984)
		(Stebbins 1985)
Land snails	495	(Hubricht 1985, analysis by B. Roth)
Vascular plants	> 20,000	(analysis by J. Kartesz)
		(Farrar 1995)
		(Lauriault 1989)
		(Little 1976a,b, 1977)
		(Petrides 1988, 1992)

digitizing each range map into a GIS and overlaying all of them. When using thousands of nonstandard range maps, a GIS approach is cumbersome, requiring extensive digitizing, scanning, and rescaling of maps to correspond to the ecoregion base map.

The maps included in many field guides represent only coarse outlines of species occurrence, not exact boundaries. This level of resolution is useful for the scale of an ecoregional assessment, and any recording error in the analysis used here is likely to be of a lower magnitude than the error inherent in the maps themselves. Moreover, significant portions of a species' mapped range do not in reality support populations of the species. By using not only the range maps but also the text of each species account, we were able to use habitat preference information to modify the species ranges shown on the maps. For example, the range map for the desert horned lizard (*Phrynosoma platirhinos*) places it not only in the Great Basin Shrub Steppe [76] but also in the Great Basin Montane Forests [42] embedded within ecoregion [76]. This lizard, however, is associated mainly with sagebrush (*Artemisia* spp.), saltbush (*Atriplex* spp.), and greasewood (*Sarcobatus* spp.), which are found at the lower elevations. We used this information to restrict the lizard's presence in this region to ecoregion [76]. This type of modification was frequently necessary, suggesting that strict overlays of these coarse range maps would result in many misplaced species.

For birds (Scott 1995), we included winter, summer (breeding), and year-round ranges for comparison with the ecoregion map, so a species occupying an ecoregion in any season was recorded as present within it. Similarly, for butterflies (Scott 1986), we included not only the areas in the range maps designated as year-round range but also the areas shown (with hatching) to be occupied occasionally by migrants.

Species were recorded as present, endemic, or absent in each ecoregion. For native vascular plants and western land snails, no published range data were available, and unpublished data were provided by regional specialists J. Kartesz and B. Roth, respectively. Distribution data were aggregated into an ecoregion format and reported as a richness and endemism total for each ecoregion.

For a species to be considered endemic to an ecoregion, in the strictest sense, it would occur

there and nowhere else on earth. We have relaxed this definition of endemism to better achieve conservation goals in two important ways. First, we treat a near endemic as endemic, so that the importance of conserving the species in the ecoregion containing most of its range is captured. Therefore, if ≥75 percent of a species' global range is found within only one ecoregion, it is counted as endemic in that ecoregion. Second, species with highly restricted ranges occasionally display distributions that straddle ecoregion boundaries, with no single ecoregion containing ≥75 percent of their ranges. In order not to lose restricted-range species that are at high risk of extinction from habitat loss, we scored species with such restricted global ranges ($<50,000$ km^2) as endemic to all ecoregions containing them. Based on these modifications, the assignment of endemism was made following the set of decision rules below:

1. Total species range $>50,000$ km^2
   a. A single ecoregion contains 75–100 percent of the species' range: species recorded as endemic to that ecoregion, present in all others containing it.
   b. No single ecoregion contains >75 percent of the species' total range: species recorded as present in all ecoregions containing it.
2. Total species range $<50,000$ km^2
   a. Species occurs in five or fewer ecoregions: species recorded as endemic in all ecoregions containing it.
   b. Species occurs in more than five ecoregions (many small, disjunct areas): species recorded as present in all ecoregions containing it.

We chose the 50,000 km^2 threshold for narrow-range endemics following Birdlife International's classification for endemic species (Bibby 1992). The 75 percent threshold for the proportion of a range within a single ecoregion is an arbitrary threshold but represents most of the total range. The point here is to highlight ecoregions that represent the only practical opportunity to conserve a certain species in the wild. An ecoregion containing over 75 percent of a species range fits that description much more than one containing less than 25 percent.

A species that fits the criteria for endemism in the mapped area but whose range extends outside North America was not treated as endemic. Where ecoregions extend into Mexico, endemism was considered for the entire ecoregion. However, species restricted entirely to Mexico were not counted in this study for the United States and Canada. Therefore, richness and endemism values for ecoregions along the Mexico–United States border were reduced relative to their full ecoregion levels.

Richness

Species richness comparisons were made only among ecoregions sharing the same MHT. A simple approach in comparing species richness among ecoregions sharing the same MHT is to add up the totals for each of the seven taxa groups analyzed and use the combined value as a proxy for overall richness. However, the values of different higher taxa often differ by several orders of magnitude in certain ecoregions, up to a factor of 30 in some cases. For instance, the Southeastern Mixed Forests ecoregion [22] is home to 62 mammal species and 3,196 native vascular plant species. Therefore, a sum of taxon values for a particular ecoregion was often biased toward those taxa with the highest num-

ber of species, often native vascular plants. Is an additional plant species worth the same in ecosystem function, ecological or evolutionary distinctiveness, or biodiversity value as an additional mammal species? We made the assumption that each of the higher taxa (genera, families) analyzed were equivalent in their biodiversity value and of greater conservation value relative to individual species. To reduce the effect of speciose taxa and, perhaps, better recognize the underlying ecological patterns and processes that support communities, several transformations of the data were performed to reduce the influence of highly species-rich taxa. We transformed the raw data using a ranked and log transform method (see below for details). Below we compare the results of each of these transformed data sets for a forested and a nonforested major habitat type.

The results of all three methods (i.e., sum, rank, log transform) for MHT 2.1, temperate broadleaf and mixed forests, are shown in tables A.2, A.3, and A.4. The sum data method simply adds species richness from all taxa to arrive at a sum total for the ecoregion (table A.2). The ecoregion with the highest total species richness was given a rank of 1, and the ecoregion with the

TABLE A.2. Raw Values for Species Richness for MHT 2.1: Temperate Broadleaf and Mixed Forests.

Ecoregion Number	Ecoregion Name	Amphibians	Birds	Butterflies	Vascular Plants	Mammals	Reptiles	Snails	Total Richness	Rank
22	Southeastern Mixed Forests	63	212	156	3176	62	73	185	4094	1
17	Appalachian Mixed Mesophytic Forests	56	200	155	2308	67	52	248	3252	2
16	Appalachian/Blue Ridge Forests	66	194	157	2224	67	48	264	3178	3
18	Central U.S. Hardwood Forests	54	203	161	2205	60	65	210	3103	4
10	Southern Great Lakes Forests	32	224	150	2117	62	33	120	2859	5
15	Allegheny Highlands Forests	27	198	132	1768	59	28	85	2398	6
14	Northeastern Coastal Forests	28	251	159	1580	61	34	101	2320	7
19	Ozark Mountain Forests	37	191	151	1621	50	58	89	2316	8
21	East Central Texas Forests	28	210	203	1435	41	60	62	2154	9
20	Mississippi Lowland Forests	35	222	159	1334	58	52	89	2076	10
12	New England/Acadian Forests	20	222	136	1396	61	21	76	2004	11
11	Eastern Great Lakes Lowland Forests	22	222	136	1378	57	26	72	1998	12
7	Western Great Lakes Forests	20	216	144	1385	62	19	63	1972	13
9	Upper Midwest Forest/ Savanna Transition Zone	20	216	149	1331	58	33	80	1964	14
8	Eastern Forest/Boreal Transition	20	206	127	1157	56	18	74	1714	15
6	Willamette Valley Forests	6	177	94	1047	33	8	27	1409	16
13	Gulf of St. Lawrence Lowland Forests	15	195	77	975	48	6	2	1361	17

lowest, a rank of 17. If there were any ties in the numbers, both ecoregions received the score of the higher ranking.

In the ranked method, ecoregions were ranked first for each taxon separately; then the taxa ranks were summed for each ecoregion. The final position of an ecoregion is based on the summed total of the ranks (table A.3). Thus, the ecoregion with the lowest value had the highest total species diversity and was given a rank of 1.

In the log transform method we first took the natural log of the raw species richness values for each taxon (Krebs 1989). These log values were then summed for each ecoregion, giving a total logarithmic score for the ecoregion (table A.4). The ecoregion with the highest logarithmic value was assumed to have the highest species richness across all taxa and was given a rank of 1.

The log transform method was chosen for the analysis of species richness, because it condensed the range of the data yet preserved differences among taxa. The ranking method was not chosen because it does not preserve the relative scale among ecoregion values for all taxa. For instance, the butterfly richness of the East Central Texas Forests [21] is 203, for the Central U.S. Hardwood Forests [18] is 161, and for the Northeastern Coastal Forests [14] is 159. Those

ecoregions would receive ranks of 1, 2, and 3, respectively, but the large gap between 1 and 2 would be lost in the rankings.

Comparison of Methods

We conducted a sensitivity analysis on the outcomes of the different methods. Table A.5 gives the ranked outcomes of the three methods for MHT 2.1, temperate broadleaf and mixed forests.

We compared the log transform method to both the raw data method and the rank transformation method using the Pearson product moment correlation coefficient, r, as a test of similarity among the methods (Tukey 1977). The correlation coefficient was significant at $\alpha = 0.05$ for each comparison. The Pearson correlation coefficient for the logarithmic–raw data comparison was $r = 0.938$ ($p < 0.00005$), and the logarithmic-ranking method correlation coefficient was $r = 0.968$ ($p < 0.00005$). The high correlation coefficients among the three methods indicate a strong similarity in each method's interpretation of the data.

While the Pearson correlation coefficient was significant for each comparison, individual ecoregions, notably the Allegheny Highlands

TABLE A.3. Ranked Values for Species Richness for MHT 2.1: Temperate Broadleaf and Mixed Forests.

Ecoregion Number	Ecoregion Name	Amphibians	Birds	Butterflies	Vascular Plants	Mammals	Reptiles	Snails	Summed Ranks	Overall Rank
					Taxonomic Categories					
22	Southeastern Mixed Forests	2	8	6	1	3	1	4	26	1
17	Appalachian Mixed Mesophytic Forests	3	12	7	2	1	5	2	34	2
16	Appalachian/Blue Ridge Forests	1	15	5	3	1	7	1	36	3
18	Central U.S. Hardwood Forests	4	11	2	4	8	2	3	38	4
10	Southern Great Lakes Forests	7	2	9	5	3	9	5	46	5
15	Allegheny Highlands Forests	10	13	14	6	9	11	9	82	11
14	Northeastern Coastal Forests	8	1	3	8	6	8	6	49	6
19	Ozark Mountain Forests	5	16	8	7	14	4	7	68	8
21	East Central Texas Forests	9	9	1	9	16	3	15	70	9
20	Mississippi Lowland Forests	6	3	3	13	10	5	7	52	7
12	New England/Acadian Forests	12	3	12	10	6	13	11	80	10
11	Eastern Great Lakes Lowland Forests	11	3	12	12	12	12	13	86	14
7	Western Great Lakes Forests	12	6	11	11	3	14	14	85	13
9	Upper Midwest Forest/ Savanna Transition Zone	12	6	10	14	10	9	10	83	12
8	Eastern Forest/Boreal Transition	12	10	15	15	13	15	12	107	15
6	Willamette Valley Forests	17	17	16	16	17	16	16	132	17
13	Gulf of St. Lawrence Lowland Forests	16	14	17	17	15	17	17	129	16

Forests [15] and the Mississippi Lowland Forests [20], did differ in their final ranking. These differences can be attributed to the large influence that native vascular plant richness has on the totals. The Allegheny Highlands Forests are rich in native vascular plants but less rich in the remaining taxa. This results in the ecoregion having a high raw data score. The Mississippi Lowlands Forests display the opposite phenomena. They are low in richness of native

TABLE A.4. Logarithmic Values for Species Richness for MHT 2.1: Temperate Broadleaf and Mixed Forests.

Ecoregion Number	Ecoregion Name	Amphibians	Birds	Butterflies	Vascular Plants	Mammals	Reptiles	Snails	Total Richness	Rank
22	Southeastern Mixed Forests	1.799	2.326	2.193	3.502	1.792	1.863	2.267	17.966	1
17	Appalachian Mixed Mesophytic Forests	1.748	2.301	2.190	3.363	1.826	1.716	2.394	17.759	3
16	Appalachian/Blue Ridge Forests	1.820	2.288	2.196	3.347	1.826	1.681	2.422	17.778	2
18	Central U.S. Hardwood Forests	1.732	2.307	2.207	3.343	1.778	1.813	2.322	17.665	4
10	Southern Great Lakes Forests	1.505	2.350	2.176	3.326	1.792	1.519	2.079	16.830	5
15	Allegheny Highlands Forests	1.431	2.297	2.121	3.247	1.771	1.447	1.929	16.248	10
14	Northeastern Coastal Forests	1.447	2.400	2.201	3.199	1.785	1.531	2.004	16.593	8
19	Ozark Mountain Forests	1.568	2.281	2.179	3.210	1.699	1.763	1.949	16.725	7
21	East Central Texas Forests	1.447	2.322	2.307	3.157	1.613	1.778	1.792	16.478	9
20	Mississippi Lowland Forests	1.544	2.346	2.201	3.125	1.763	1.716	1.949	16.750	6
12	New England/Acadian Forests	1.301	2.346	2.134	3.145	1.785	1.322	1.881	15.772	13
11	Eastern Great Lakes Lowland Forests	1.342	2.346	2.134	3.139	1.756	1.415	1.857	15.919	12
7	Western Great Lakes Forests	1.301	2.334	2.158	3.141	1.792	1.279	1.799	15.605	14
9	Upper Midwest Forest/ Savanna Transition Zone	1.301	2.334	2.173	3.124	1.763	1.519	1.903	16.004	11
8	Eastern Forest/Boreal Transition	1.301	2.314	2.104	3.063	1.748	1.255	1.869	15.403	15
6	Willamette Valley Forests	0.778	2.248	1.973	3.020	1.519	0.903	1.431	13.103	16
13	Gulf of St. Lawrence Lowland Forests	1.176	2.290	1.886	2.989	1.681	0.778	0.301	12.736	17

TABLE A.5. Comparison of Analysis Methods in Determining Species Richness in MHT 2.1: Temperate Broadleaf and Mixed Forests.

Ecoregion Number	Ecoregion Name	Log Method	Raw Data	Rank Method
22	Southeastern Mixed Forests	1	1	1
17	Appalachian Mixed Mesophytic Forests	3	2	2
16	Appalachian/Blue Ridge Forests	2	3	3
18	Central U.S. Hardwood Forests	4	4	4
10	Southern Great Lakes Forests	5	5	5
15	Allegheny Highlands Forests	10	6	11
14	Northeastern Coastal Forests	8	7	6
19	Ozark Mountain Forests	7	8	8
21	East Central Texas Forests	9	9	9
20	Mississippi Lowland Forests	6	10	7
12	New England/Acadian Forests	13	11	10
11	Eastern Great Lakes Lowland Forests	12	12	14
7	Western Great Lakes Forests	14	13	13
9	Upper Midwest Forest/Savanna Transition Zone	11	14	12
8	Eastern Forest/Boreal Transition	15	15	15
6	Willamette Valley Forests	16	16	17
13	Gulf of St. Lawrence Lowland Forests	17	17	16

vascular plants but relatively rich in number for other taxa. Both the log and the rank methods capture this subtle difference.

As an additional test, we applied the different methods to a second major habitat type, MHT 3.1, temperate grasslands/savanna/shrub. The correlation coefficient was significant at $\alpha = 0.05$ for each comparison. The Pearson correlation coefficient for the log-sum comparison was $r = 0.702$ ($p < 0.0024$), and the log-rank method correlation coefficient was $r = 0.930$ ($p < 0.00005$). Therefore, while the correlation was not quite as strong, it did suggest a robustness in the methodology across MHTs and supports the use of the log method in assessing biodiversity.

Assigning Ranks

Using the log transform method of analysis, a sum of logs for all taxa was calculated for each ecoregion (see table A.4 for an example). These values were used to define the biodiversity richness categories of high, medium, and low. Once the category threshold points were obtained (see below), the ecoregions that received a high ranking were then reassessed to determine if they were globally outstanding in the richness of their species assemblages. This determination was made by comparing flora and fauna lists for selected taxa with ecoregions of the same MHT but on different continents (i.e., biogeographic realms) (Olson and Dinerstein 1998). Within each MHT the total log values for each ecoregion were plotted in increasing order. In this way, natural ecoregion groupings could be identified by looking at the difference in slope between ecoregions. Threshold values were set where there was a sharp increase in slope between points (figure A.1). In the rare cases where no sharp increases existed but rather a steady increasing slope occurred, we used set threshold values at the quartiles of the distribution.

After each ecoregion was assigned a richness level, we assigned point values to each level: high = 15 points; medium = 10 points; and low = 5 points. Seven ecoregions were designated globally outstanding for species richness, including two ecoregions in MHT 2.1, temperate broadleaf and mixed forests. Ecoregions designated globally outstanding were awarded 100 points to assure that they would obtain a globally outstanding designation in the synthesis of the final biological distinctiveness index. The richness scores for all the ecoregions can be found in appendix C.

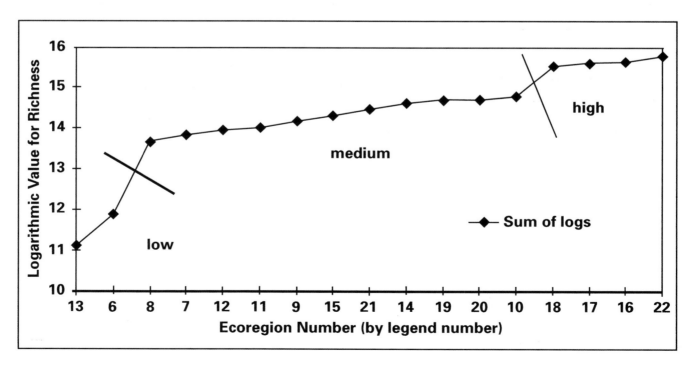

Figure A.1 Logarithmic values for ecoregions in MHT 2.1, Temperate Broadleaf and Mixed Forests, showing breaks where thresholds were placed.

Endemism

We collected endemism data in the same manner and from the same sources as richness data. As with richness, there is a wide disparity of endemism levels in ecoregions both within and among taxa. The values for native vascular plant endemism far exceed those for the other taxa in most ecoregions. For instance, in the Klamath-Siskiyou Forests [39], the total number of endemic native vascular plant species is 131, while the total number of endemics from the other six taxa combined is 37. While conserving areas with the highest absolute levels of endemism is desirable, we feel that the conservation of regions with assemblages of endemic species across many taxa may help to conserve endemic species in other taxa not analyzed in this study but influenced by similar evolutionary and biogeographic processes. To address the taxon bias in the endemism data, we considered two transformations of the data: an endemism-richness ratio method and a logarithmic method. The endemism-richness ratio method computes, for each taxon, the proportion of an ecoregion's richness that is endemic to that ecoregion, by using the simple equation:

$$E\ (\%) = \frac{\text{number of endemic species by taxon}}{\text{number of species by taxon}} \times 100$$

The percentage of endemism for all taxa are analyzed, and the ecoregions are ranked based on a measure of average total endemism. Table A.6 illustrates this analysis, using MHT 2.1, temperate broadleaf and mixed forests, as an example. As with richness, we compared the two transformations to a simple sum of the raw endemism values in order to assess the features of the different measures.

The endemism information for native vascular plants was provided in predefined ranges (Kartesz 1997). Therefore, we did not have precise endemic values by ecoregion for this taxon. To be able to incorporate this important data set into the analysis, we chose the mean value of the reported ranges as a surrogate for the actual number of endemic vascular plants (table A.7).

Comparison of Methods

As with the richness analysis, we chose the logarithmic transformation method for analysis because it gave greater weight to the speciose taxa, while dampening their influence on the

TABLE A.6. Endemism-Richness Ratio Method for Species Endemism for MHT 2.1: Temperate Broadleaf and Mixed Forests (as percentages).

		Taxonomic Categories								
Ecoregion Number	Ecoregion Name	Amphibians	Birds	Butterflies	Vascular Plants	Mammals	Reptiles	Snails	Average Percentage	Total Richness
16	Appalachian/Blue Ridge Forests	31.82%	0.00%	0.00%	3.88%	0.00%	2.08%	46.21%	12.00%	1
17	Appalachian Mixed Mesophytic Forests	12.50%	0.00%	0.00%	0.64%	0.00%	3.85%	39.52%	8.07%	2
22	Southeastern Mixed Forests	7.94%	0.00%	0.00%	1.07%	1.61%	6.85%	28.11%	6.51%	3
18	Central U.S. Hardwood Forests	5.56%	0.00%	0.00%	1.54%	0.00%	0.00%	33.81%	5.84%	4
19	Ozark Mountain Forests	13.51%	0.00%	0.00%	0.92%	0.00%	0.00%	24.72%	5.59%	5
6	Willamette Valley Forests	0.00%	0.00%	0.00%	0.66%	3.03%	0.00%	18.52%	3.17%	6
21	East Central Texas Forests	3.57%	0.00%	0.00%	0.45%	0.00%	1.67%	11.29%	2.43%	7
12	New England/Acadian Forests	0.00%	0.00%	0.74%	0.47%	0.00%	4.76%	3.95%	1.42%	8
10	Southern Great Lakes Forests	0.00%	0.00%	0.00%	0.09%	0.00%	3.03%	5.00%	1.16%	9
14	Northeastern Coastal Forests	0.00%	0.00%	0.00%	0.41%	0.00%	0.00%	5.94%	0.91%	10
15	Allegheny Highlands Forests	0.00%	0.00%	0.00%	0.00%	0.00%	3.57%	2.35%	0.85%	11
20	Mississippi Lowland Forests	0.00%	0.00%	0.00%	0.14%	0.00%	0.00%	5.62%	0.82%	12
9	Upper Midwest Forest/ Savanna Transition Zone	0.00%	0.00%	0.00%	0.14%	0.00%	0.00%	3.75%	0.56%	13
8	Eastern Forest/Boreal Transition	0.00%	0.00%	0.00%	0.00%	0.00%	0.00%	2.70%	0.39%	14
7	Western Great Lakes Forests	0.00%	0.46%	0.00%	0.00%	0.00%	0.00%	1.59%	0.29%	15
11	Eastern Great Lakes Lowland Forests	0.00%	0.00%	0.00%	0.14%	0.00%	0.00%	1.39%	0.22%	16
13	Gulf of St. Lawrence Lowland Forests	0.00%	0.00%	0.00%	0.00%	0.00%	0.00%	0.00%	0.00%	17

TABLE A.7. Native Vascular Plant Endemism Ranges and Corresponding Mean Value Used in the Endemism Analysis.

Reported Endemism Range (Kartesz 1997)	Endemism Value Used in the Analysis
0	0
1–3	2
4–10	7
11–20	16
21–50	36
51–75	63
76–110	93
111–150	131
151–200	175
201+	225

entire analysis. In this way the less speciose taxa have a relatively greater effect on the outcome, without sacrificing the conservation of high endemic areas. For instance, an ecoregion with twenty-five total endemics, five each of land snails, native vascular plants, birds, mammals, and reptiles, would score higher than an ecoregion with fifty native vascular plants and no other endemics. This is the primary reason for not using the untransformed endemism totals to analyze endemism—it places too great an emphasis on the speciose taxa, predominantly native vascular plants and amphibians. The ratio method was rejected because it is overly sensitive to the low richness value of less speciose taxa. For example, in MHT 3.1, temperate grass-

lands/savanna/shrub, the ratio method identifies the Nebraska Sand Hills Mixed Grasslands [62] as the ecoregion with the highest endemism-richness ratio. On closer inspection of the data, this ranking is based entirely on the land snail ratio, which has 50 percent endemism. However, the raw data show a richness of two land snails, one of which is endemic. The total endemism for this ecoregion is three, whereas the total endemism for the Edwards Plateau Savannas [66] is fifty.

The ranked outcomes of the three methods for MHT 2.1, temperate broadleaf and mixed forests, are given in table A.8. While ecoregion order is slightly different, the top five and top ten ecoregions are the same with a single exception.

We compared the logarithmic method to both the raw data method and the endemism-richness ratio method using the Pearson product moment correlation coefficient, r, as a test of similarity among methods. The correlation coefficient was significant at $\alpha = 0.05$ for each comparison. The correlation coefficients were $r = 0.925$ ($p < 0.00005$) and $r = 0.877$ ($p < 0.00005$), respectively.

We also tested the similarity among methods, using the Pearson correlation coefficient, for MHT 3.1, temperate grasslands/savanna/shrub, to analyze the similarities in a nonforested MHT, as well as to assess the effects of the ratio method's bias discussed earlier. The Pearson cor-

TABLE A.8. Comparison of Analysis Methods for Endemism for MHT 2.1: Temperate Broadleaf and Mixed Forests.

Ecoregion Number	Ecoregion Name	Analysis Methods		
		Log Method	Raw Data	Rank Method
16	Appalachian/Blue Ridge Forests	2	1	1
17	Appalachian Mixed Mesophytic Forests	3	2	2
18	Central U.S. Hardwood Forests	7	3	4
22	Southeastern Mixed Forests	1	4	3
19	Ozark Mountain Forests	4	5	5
21	East Central Texas Forests	5	6	7
6	Willamette Valley Forests	10	7	6
14	Northeastern Coastal Forests	6	7	10
12	New England/Acadian Forests	8	9	8
10	Southern Great Lakes Forests	11	10	9
20	Mississippi Lowland Forests	9	11	12
9	Upper Midwest Forest/Savanna Transition Zone	11	12	13
11	Eastern Great Lakes Lowland Forests	11	13	16
15	Allegheny Highlands Forests	11	13	11
7	Western Great Lakes Forests	15	15	15
8	Eastern Forest/Boreal Transition	15	15	14
13	Gulf of St. Lawrence Lowland Forests	15	17	17

relation coefficient was significant at $\alpha = 0.05$. The correlation coefficient for the log-sum comparison was 0.9458 ($p < 0.00005$), and the log-ratio comparison was 0.6232 ($p < 0.0099$). These correlation coefficients suggest that the results of the log method are similar to those of the raw data method, and that the endemism analysis is not overly sensitive to application of one analytical method over another.

Assigning Ranks

Endemism ranks were assigned in the same manner as richness ranks. Using graphical representation of the distribution of total endemism by ecoregion, we assigned threshold values and gave them high, medium, or low rankings. Ecoregions receiving a high ranking were compared to similar MHT ecoregions around the world to determine if they were globally outstanding in terms of species endemism. Our assessment drew upon WWF regional conservation assessments, thorough review of available literature, and consultation with regional experts (Olson and Dinerstein 1998). We felt that endemism was a highly significant factor in determining an ecoregion's distinctive biodiversity value and accordingly gave it more weight in the rankings than richness. The point values for each level were: high = 25, medium = 15, low = 5. Globally outstanding ecoregions received 100 points to assure that they would obtain a globally outstanding designation in the synthesis of the final biological distinctiveness index. (Endemism scores can be found in appendix C.)

Rare Ecological or Evolutionary Phenomena

Ecological and evolutionary phenomena also contribute to the distinctiveness of an ecoregion's biodiversity. These criteria are included to capture important biodiversity features that are otherwise difficult to measure using quantitative methods. Distinctive ecological phenomena include the presence of intact vertebrate faunas, including intact predator assemblages (an increasingly rare phenomenon around the world); extraordinary aggregations of breeding vertebrates, such as occurs in the tundra in summer; large-scale migrations of large vertebrates (e.g., large caribou migrations, also a disappearing phenomenon); and habitat types whose

structures are highly distinctive because of the dominance of globally large trees or cacti. For this analysis, the presence of continental-scale migration of birds, bats, or invertebrates was not highlighted because such phenomena are common to many ecoregions (making it a poor discriminator) and the highly localized (e.g., critical stopover sites) or, conversely, highly diffuse (i.e., across broad migration fronts) nature of migrations makes them difficult to characterize at ecoregion scales. We did note large terrestrial vertebrate migrations that are generally confined to single ecoregions. Migrations of large terrestrial vertebrates that are extinct, such as the past migrations of American bison throughout the short- and mixed-grass prairies, were not considered. Outstanding evolutionary phenomena include spectacular adaptive radiations within multiple taxa, such as that seen in honeycreepers, tarweeds, and other plants and invertebrates in Hawaii. A preponderance of unusual or unique higher taxa (i.e., genera, families), or primitive or relict groups, was also considered a distinctive evolutionary phenomenon. High levels of beta-diversity (turnover of species over distance or along gradients) reflect unusual ecological and evolutionary phenomena (e.g., high beta-diversity is often associated with pronounced local endemism). Because beta-diversity directly links phenomena with spatial patterns, we measured it separately from the phenomena criteria. Where beta-diversity was considered high for a given major habitat type, we added 5 points to the ecoregion's BDI total (thus, the ecoregion could still receive additional points for unusual ecological or evolutionary phenomena). Examples of ecoregions that display high beta-diversity include the complex Klamath-Siskiyou Forests [39], Central Pacific Coastal Forests [34], and Southeastern Conifer Forests [51]; the archipelagic Florida Sand Pine Scrub [52]; and the isolated basins and ranges of the Chihuahuan Desert [81].

At the expert workshop, regional working groups were asked to list any appropriate phenomena occurring within their assigned regions of expertise. They also were asked to describe the level of rarity of each phenomenon (e.g., four examples worldwide, unique in North America) and the phenomenon's distribution within the ecoregion. We reviewed all the phenomena described by the experts and compared them to information gathered for similar phenomena around the world (Olson and Dinerstein 1998).

We placed ecoregions into one of three categories: globally outstanding phenomena, regionally outstanding phenomena, or not applicable. This criterion emphasizes only those phenomena that are truly outstanding at global or continental (i.e., biogeographic realm) scales.

Ecoregions that were judged to contain globally outstanding ecological or evolutionary phenomena were automatically categorized as globally outstanding in the overall BDI. Such ecoregions were awarded 100 points for this criterion, to ensure a final rank of globally outstanding no matter their scores for other criteria. Therefore, ecoregions without extraordinarily high richness or endemism that nevertheless contained globally important and rare phenomena could be categorized as globally outstanding. Ecoregions that were categorized as regionally outstanding received 5 points toward the BDI. For each ecoregion that was categorized as globally outstanding or regionally outstanding, we have included in its ecoregion description a summary of the phenomena involved. The ecoregions that were awarded points for rare ecological and evolutionary phenomena can be found by looking in the scoring summaries in appendix D, and their ecoregion descriptions are included in appendix F.

Rare Habitat Type

The global rarity of habitat types was evaluated to identify those habitat types that offer few opportunities for conservation. This criterion encompasses ecological and evolutionary phenomena, but it addresses those characteristics at the scale of whole ecosystems and biotas, as well as structural features of ecosystems and habitats.

The rarity of an ecoregion's general habitat type was evaluated based on expert assessment and a recently completed global analysis (Olson and Dinerstein 1998). Ecoregions were placed in one of three categories: globally rare, regionally rare, and not applicable, based on the following decision rules:

1. Globally rare: Fewer than eight ecoregions occur globally that contain the habitat type in question.

2. Regionally rare: Fewer than three ecoregions occur in North America that contain the habitat type in question.
3. Not applicable: Otherwise.

Like the rare ecological and evolutionary phenomena criterion, an ecoregion that was judged to contain globally rare habitat type(s) was automatically categorized as globally outstanding in the overall BDI. Such ecoregions were awarded 100 points for this criterion, to ensure a final rank of globally outstanding, no matter their scores for the other criteria. Ecoregions with regionally rare habitat types were awarded 5 points.

An example of a globally rare habitat type is the Mediterranean-climate scrub habitats, found in North America in ecoregions [70, 71, and 72]. This habitat type is found elsewhere only in small portions of Chile, South Africa, Australia, and the Mediterranean itself. Opportunities for conservation of Mediterranean scrubs are so few that wherever they occur around the world, they should be made a global conservation priority. Very high threats and extraordinary richness and endemism in plants (together they support 20 percent of the world's plant species) contribute to the global priority status of this MHT. Temperate rain forests, a habitat type subset of the temperate conifer forest MHT, occur only in seven very limited areas around the world, and all are highly threatened or extinct (e.g., Black Sea temperate rain forests).

For each ecoregion that was categorized as globally outstanding or regionally outstanding, we have included in its ecoregion description a summary of the habitat(s) involved and any other reasons for the decision. The ecoregions that were awarded points for rare habitat type can be found by looking in the scoring summaries in appendix D, and their ecoregion descriptions are included in appendix F.

Final Biological Distinctiveness Categorization

The point totals for the four component criteria were added to yield a grand total for the biological distinctiveness index. Based on these totals, ecoregions were placed into one of four categories:

Globally outstanding	45 points and over
Regionally outstanding	30 or 40 points
Bioregionally outstanding	20 or 25 points
Nationally important	10 or 15 points

Therefore, an ecoregion can earn the designation of globally outstanding by accruing 45 or more points, or by being designated globally outstanding in any of the four component criteria.

Methods for Assessing the Conservation Status of Terrestrial Ecoregions

The conservation status index is designed to measure different degrees of habitat alteration and spatial patterns of remaining natural habitats across landscapes. Using landscape features as indicators for the ecological integrity of ecosystems, the index reflects how, with increasing habitat loss, degradation, and fragmentation, ecological processes cease to function naturally, or at all, and major components of biodiversity are steadily eroded. Here we assess the conservation status of ecoregions in the tradition of the IUCN Red Data Books categories for threatened and endangered species (critical, endangered, and vulnerable), except that we estimate the state of whole biotas, ecological processes, and ecosystems. The conservation status categories we created for the analysis are: critical, endangered, vulnerable, relatively stable, and relatively intact.

We relied on the assessments of regional experts to evaluate the component criteria outlined below. The following databases were consulted and proved very helpful as guides: an AVHRR satellite image developed by EROS Data Center (EROS 1996), a GIS database of managed areas in the United States (McGhie 1996), and a GIS database of managed areas in Canada.

We found existing continental-scale databases of current habitat and land use to be of inadequate spectral resolution, spatial coverage, or appropriate classification and unhelpful used by themselves in evaluating all of the criteria of the CSI. A useful and efficient approach at the continental scale is to employ the knowledge and experience of experts to evaluate the criteria, using the databases as guides and tools. In the majority of cases, the heightened accuracy available from a strictly GIS-based landscape analysis would be unlikely to change the conservation status we assigned to

ecoregions because we made the categories within the criteria sufficiently broad. The pooled knowledge of each of the eight regional working groups formed an authoritative source from which to evaluate the criteria of the CSI. In designing the CSI, we have tried to keep the evaluation process objective and transparent to facilitate future efforts to reexamine results and to analyze relationships among variables.

Determination of Snapshot Conservation Status: Categories and Weighting

The criteria outlined below are used to assess the snapshot conservation status for each ecoregion. The snapshot assessment, as of 1996, characterizes current patterns associated with loss of diversity. A threat analysis subsequently modifies the results, if necessary, to produce the final conservation status for the ecoregions. Some of the parameters of the CSI criteria described below are quite detailed in order to tailor them to different MHTs and ecoregion size classes. MHTs vary in terms of: (1) minimum threshold sizes of habitat blocks for maintenance of ecological processes, (2) degree of resiliency and response to disturbance, and (3) types of ecological processes that occur. For example, habitat block size requirements, one of our four CSI criteria, differ between boreal forests, which experience large forest fires, and xeric shrublands, where disturbances are smaller in spatial extent.

The snapshot conservation status index, the total of all indicators, has a point range from 0 to 100, with higher values denoting a higher level of endangerment. The point thresholds for different categories of conservation status are listed below:

Critical   89–100 points
Endangered   65–88 points
Vulnerable   37–64 points
Relatively stable   7–36 points
Relatively intact   0–6 points

We feel that some landscape parameters should be given greater weight in the determination of the snapshot conservation status of an ecoregion. The relative contributions of the different parameters are given below:

40%   Habitat loss
25%   Size and number of larger habitat blocks
20%   Habitat fragmentation
15%   Degree of protection

We gave greatest weight to habitat loss because it is widely recognized as a primary factor in the reduction and loss of terrestrial populations, species, and ecosystems. This criterion underscores the rapid loss of species predicted to occur in ecosystems when the total area of remaining habitat falls below minimum critical levels. We gave second highest weight to the number and extent of blocks of contiguous intact habitat. Large blocks of habitat sustain larger species populations that are more likely to be viable over the long term. Large blocks also permit a broader range of species and ecosystem dynamics to persist.

What Is "Intact" Habitat?

The analysis requires a definition of what constitutes intact habitat. We propose a three-class system in which land is categorized as: intact, altered, or heavily altered. Intact habitat represents relatively undisturbed areas that are characterized by the maintenance of most original ecological processes and by communities with most of their original suite of native species. Altered habitat represents areas more substantially affected by human disturbance but retaining the potential to sustain species and processes. Heavily altered habitat represents areas that have been degraded to the point of retaining little or no potential value for biodiversity conservation without long-term and extensive restoration.

The definition of intactness varied by MHT to better address each habitat type's distinctive

patterns of biodiversity and ecological characteristics. We created two broad categories that encompass the nine most common MHTs:

Broadleaf and Conifer Forests (MHTs 1.1, 1.2, 2.1, 2.2, 6.1)

*   *Intact.* Canopy disturbance through human activities such as logging is restricted to less than 10 percent of defined habitat block. Understory largely undisturbed by timber extraction, intensive management, or grazing. Natural fire regimes still present. Although large mammals and birds may presently be absent from some blocks of habitat due to exploitation, insufficient area, or diminished resources, such blocks sustain many native communities and populations of plant, invertebrate, and vertebrate species and their associated ecological processes.
*   *Altered.* Canopy and understory have been significantly disturbed by human activities, but habitat remains suitable for some native species. Species composition and community structure are altered, and a significant proportion of native species are absent but likely to return given sufficient time for recovery and appropriately dispersed source pools. Examples are large expanses of selectively logged forests, forests in which natural fires have been suppressed, areas where clear-cutting covers between 10 percent and 25 percent of the landscape and is patterned to facilitate natural ecological processes and recolonization, and clear-cut landscapes that have been allowed to regenerate with adequate source pools for more than one hundred years.
*   *Heavily altered.* Habitat is almost completely altered. Substrate alteration, exotic species introduction, and distance from source pools make recovery of original habitat unlikely without large and expensive restoration efforts. Examples are urban and suburban development, forests converted to pasture or cropland, extensive clear-cuts, and intensively managed plantation forests.

Grasslands/Savanna/Shrub, Xeric Shrublands/Deserts, Tundra (MHTs 3.1, 4.1, 4.2, 6.2)

*   *Intact.* Habitat has not been plowed or altered by major changes in hydrologic patterns. The

full suite of native plant species is still present, each in abundance within its natural range of variation, and successional patterns follow natural cycles (e.g., grazing by domestic livestock has no significant impact on species composition or seral stages). Natural fire regimes are still present. Although large mammals and birds may presently be absent from some blocks of habitat due to exploitation, insufficient area, or diminished resources, such blocks still sustain many native communities and populations of plant, invertebrate, and vertebrate species, as well as their associated ecological processes.

- *Altered.* Heavy grazing has altered dominance patterns of plant species. Some exotic species are present and surface water patterns may be altered, but the substrate has not been disturbed or plowed. Natural fire regimes have been significantly altered. Original habitat is likely to return with time, moderate restoration, and adequate source pools.
- *Heavily altered.* Habitat is almost entirely altered, such as by human development, plowing, or crop cultivation. Native species have been almost entirely replaced by exotics and crops. Surface water patterns have been extensively altered. Natural fire regimes have been significantly altered.

Total Habitat Loss

Habitat loss has been widely recognized as one of the primary factors contributing to the reduction and loss of terrestrial populations, species, and ecosystems. This criterion underscores the rapid loss of species predicted to occur in ecosystems when the total area of remaining habitat falls below minimum critical levels. Loss of habitat reduces biodiversity by: (1) eliminating species or communities limited to particular geographic localities, and (2) decreasing the area of available original habitat below the minimum size needed to maintain viable populations or important ecosystem dynamics.

Table B.1 is used to derive an ecoregion's score for total habitat loss. An ecoregion receives both an heavily altered score and an altered score, which represent the amount of original habitat in each category. For example, consider an ecoregion with 35 percent heavily altered habitat (10 points), 55 percent altered habitat (10 points), and, therefore, 10 percent intact

habitat. By summing the scores for heavily altered and altered, the ecoregion would receive a total score of 20 points.

TABLE B.1. Points Associated with Percentage of Lost Habitat in Determination of the Total Habitat Loss Index.

Percentage of Original Habitat	Point Values	
	Heavily Altered	Altered
90–100%	40	20
75–89%	30	15
50–74%	20	10
10–49%	10	5
0–9%	0	0

Habitat Blocks

Two important criteria for assessing conservation status are the number and extent of blocks of contiguous intact habitat large enough for populations and ecosystem dynamics to function naturally. Large blocks of habitat have the potential to sustain viable species populations, and they permit a broader range of species and ecosystem dynamics to persist. The geographic coverage of multiple large blocks also has the potential to conserve a wider range of habitats, environmental gradients, and species ranges.

The number of large habitat blocks in different size categories is an important component of this criterion. Conservation biology theory suggests that the presence of three or more examples of an ecosystem significantly increases its probability of long-term persistence. Factors such as fire, disease, pollution, invasion of exotics, deforestation, and degradation can eliminate species or natural habitats within blocks. The presence of several habitat blocks with similar communities allows for recolonization and persistence of particular habitat types and species over time. Multiple blocks that are well distributed across the landscape are particularly important for conserving species and habitats in ecoregions that are characterized by a high degree of beta-diversity (species turnover with distance or along environmental gradients).

The threshold size requirements for blocks of habitat are tailored to each MHT to reflect the different scales of ecological processes and disturbances that operate within different MHTs. Two examples illustrate the need for attention to these details. First, as latitude increases, species have wider geographic ranges on average (Brown 1995). They also depend on a variety of habitats that may be spatially separate. Therefore, we require larger intact blocks of habitat in the boreal forest/taiga and tundra MHTs than in the more temperate MHTs. Larger blocks of boreal forest help ensure that a given block of habitat includes sufficient area for viable species populations, particularly top predators living at low densities, and sufficient heterogeneity to provide the broad range of habitats needed. Second, natural disturbance events (such as forest fires) occur on much larger scales in the boreal forests than, for instance, in xeric shrublands. Blocks of intact habitat should be large enough to absorb natural disturbance events without permanently losing components or processes.

Because the ecoregions vary greatly in area, we have divided them into two or three size categories and used a different set of habitat block thresholds for each category. This avoids drawing potentially misleading conclusions by applying continental size thresholds to very small, disjunct, or island ecoregions. For example, the Great Basin Montane Forests [42] (5,784 km^2)

should not be held to the same habitat block standards as the Colorado Rockies Forests [45] (132,740 km^2). If they were, approximately 70 percent of their total area would have to be contained within a single block of intact habitat. The smaller size category applies to the fourteen ecoregions that are smaller than 10,000 km^2 and to the two additional polar ecoregions (MHT 6.1 and 6.2) that are smaller than 20,000 km^2. Scale issues are similarly addressed in the protected-area criterion.

In the following tables for habitat block analysis, the information in the rows, unless stated otherwise, refers to the minimal requirement for at least one block of intact habitat. For an ecoregion within its appropriate size column, the table should be read from top to bottom until a statement true of the ecoregion is reached. The ecoregion was then assigned the corresponding point value. The text ">500" should be interpreted as: "the ecoregion contains at least one block of intact habitat greater than 500 km^2." MHT 3.2 contains only one ecoregion, Everglades [69]. Since standardization within the MHT was not needed, it was not included in the following analyses.

Habitat Block Analysis for MHTs 1.1 and 1.2

	Ecoregion Size		
Point Value	>10,000 km^2	<10,000 km^2 (contiguous)	<10,000 km^2 (disjunct)
2	>5,000 or ≥3 blocks >2,000	>3,000 or ≥3 blocks >1,000	≥2 blocks >500
6	>3,000 or ≥3 blocks >1,000	>1,000	≥2 blocks >300
12	>2,000	>500	≥2 blocks >200
18	>1,000	>250	>200
25	None >1000	None >250	None >100

Habitat Block Analysis for MHTs 2.1 and 2.2

	Ecoregion Size	
Point Value	>10,000 km^2	<10,000 km^2
2	>4,000 or ≥3 blocks >1,500	>2,500 or ≥3 blocks >800
6	>3,000 or ≥3 blocks >1,000	>800
12	>2,000	≥3 blocks >250
18	>1,000	>250
25	None >1,000	None >250

Habitat Block Analysis for MHT 3.1

	Ecoregion Size	
Point Value	>10,000 km^2	<10,000 km^2
2	>2,000 or ≥3 blocks >800	>1,000 or ≥3 blocks >500
6	>1,000	>500
12	>500	>250
18	>250	>100
25	None >250	None >100

Habitat Block Analysis for MHTs 4.1 and 4.2

Point Value	Ecoregion Size	
	>10,000 km²	<10,000 km²
2	≥2 blocks >750 or ≥3 blocks >500	≥2 blocks >500 or ≥3 blocks >300
6	>750	>500
12	>500	>250
18	>250	>100
25	None >250	None >100

Habitat Block Analysis for MHTs 6.1 and 6.2

Point Value	Ecoregion Size	
	>20,000 km²	<20,000 km²
2	>10,000 or ≥3 blocks >5,000	>4,000 or ≥3 blocks >1,500
6	>6,000 or ≥3 blocks > 2,500	>1,500
12	>4,000	>1,000
18	>2,000	>500
25	None >2,000	None >500

Habitat Fragmentation

Habitat fragmentation is widely perceived as a major threat to the conservation of terrestrial species for two major reasons. First, the resulting diminishment and dissection of species populations places many low-density species in demographic jeopardy (Berger 1990; Laurance 1991; Newmark 1991; Wilcove, McLellan, and Dobson 1986). Second, as fragmentation increases, the amount of unaltered "core" habitat area decreases, and ecosystems increasingly experience "edge effect" degradation from hunting pressure, fires from surrounding human activity, changes in microclimates, high levels of predation or parasitism, and invasion of exotic species over a large percentage of their area (Lovejoy 1980; Saunders, Hobbs, and Margules 1991; Skole and Tucker 1993).

Fragments under 100 km² are generally inadequate for maintaining viable populations of most large vertebrates. However, small fragments can be particularly valuable for conserving populations of other species with very localized habitat requirements and small ranges, particularly in regions characterized by high levels of beta-diversity. Many invertebrates, plants, fungi, and small vertebrates can be effectively conserved within small blocks of original habitat. More vagile species such as birds (although not breeding populations of Neotropical migrants) and

butterflies also may be able to persist as metapopulations in sets of small remaining habitat fragments (Hanski and Gilpin 1991; Shafer 1995).

The point values associated with different categories of this criterion reflect the relatively greater severity of ecosystem disruption in landscapes where habitat fragmentation is more advanced (Groom and Shumaker 1993).

Description	Points
High: very low connectivity, little core habitat due to edge effects (i.e., most fragments small and/or noncircular), most individual fragments and clusters of fragments highly isolated, and intervening landscapes preclude dispersal for most taxa.	20
Advanced: low connectivity, more larger fragments than in "High" category, fragments highly isolated, and intervening landscape precludes dispersal for most taxa.	16
Medium: intermediate connectivity, fragments somewhat clustered, intervening landscape allows for dispersal of many taxa through some parts of ecoregion.	12
Low: higher connectivity, more than half of all fragments clustered to some degree (i.e., have some degree of interaction with other intact habitat blocks).	5
Relatively contiguous: high connectivity, fragmentation low, long-distance dispersal along elevational and climatic gradients still possible.	0

In ecoregions that are composed of several naturally disjunct areas, fragmentation was assessed within individual areas and not on the basis of connectivity between them. The state of the largest areas contribute the most to the assessment.

Degree of Protection

The degree-of-protection criterion assesses how well humans have conserved sufficiently large blocks of original intact habitat for biodiversity conservation. Protected areas managed for other purposes (e.g., national monuments, state parks) are not emphasized in this criterion, with some notable exceptions.

Several important aspects should be considered in a comprehensive analysis of protected areas:

- the degree to which large remaining blocks of habitat are adequately protected within a system of protected areas;

- the level of redundancy of protected areas required to help ensure the long-term persistence of habitat types, communities, endangered species, and critical habitats for species or ecological processes;
- the degree to which representative habitat types, communities, ecological gradients, endangered species, and critical habitats for resident or migratory species or ecological processes are contained within a system of protected areas;
- the degree of connectivity among reserves for the dispersal of species and contiguity of large-scale ecosystem processes; and
- the effectiveness of management of protected areas and the landscapes surrounding them. (Most protected areas currently are too few and too small to encompass complete ecosystems and will be effective only if the surrounding landscape is managed intelligently for biodiversity conservation.)

Of these five aspects, only the first two are feasibly addressed at the continental scale and are considered here. The last three should be treated in more detailed analyses at the intra-ecoregion level. For example, in Canada these analyses are being conducted as part of WWF Canada's Endangered Spaces Campaign. Because of the importance of these last three considerations and our inability to consider them fully, the degree-of-protection criterion is weighted relatively lightly in the conservation status analysis.

One could view the lack of formally protected areas as a threat factor for an ecoregion; however, some ecoregions may not be threatened due to their remoteness or rugged terrain. Assessing threats using negative criteria (i.e., absence of protected areas) increases the probability of making a poor decision compared to basing conclusions on existing parameters.

The thresholds below were set using a definition of protection that would include the Gap Analysis Project's Management Status 1 and 2 (Scott et al. 1993):

- *Management Status 1.* An area with an active management plan in operation that is maintained in its natural state and within which natural disturbance events are either allowed to proceed without interference or mimicked through management.
- *Management Status 2.* An area that is generally managed for its natural values but may

receive use that degrades the quality of natural communities that are present.

Experts at the workshop were provided with database summaries listing the managed areas in each ecoregion, their areas, and their GAP categories (McGhie 1996). The experts were able to use these summaries as guides in categorizing ecoregions, but they often found it necessary to augment the information in the database with their personal knowledge of the quality and management of some managed areas. The working groups also were asked to indicate on maps other areas not explicitly managed for biodiversity conservation that nevertheless contain important blocks of intact habitat (for example, military lands, artillery ranges, etc.).

In the following tables, which form the basis for the degree-of-protection analysis, the information in the rows, unless stated otherwise, refers to the minimal requirement for at least one block of protected habitat. For an ecoregion within its appropriate size column, the table should be read from top to bottom until a statement true of the ecoregion is reached. The text ">500" should be interpreted as: "the ecoregion contains at least one block of protected habitat greater than 500 km^2."

Degree of Protection Analysis for MHTs 1.1, 1.2, 2.1, and 2.2

Point Value	Ecoregion Size	
	>10,000 km^2	<10,000 km^2
1	≥2 blocks >2,000 or ≥3 blocks >1,500	≥2 blocks >1,000 or ≥3 blocks >800
4	>2,000	>1,000
8	>1,000	>500
12	>500	>250
15	None >500	None >250

Degree of Protection Analysis for MHTs 3.1, 4.1, and 4.2

Point Value	Ecoregion Size	
	>10,000 km^2	<10,000 km^2
1	≥2 blocks >750 or ≥3 blocks >500	≥2 blocks >500 or ≥3 blocks >300
4	>750	>500
8	>500	>250
12	>250	>100
15	None >250	None >100

	Ecoregion Size			
Point Value	>200,000 km^2	50,000–200,000 km^2	20,000–50,000 km^2	<20,000 km^2
1	>20,000	>10,000	>8,000 or ≥3 blocks >4,000	>3,000 or ≥3 blocks >1,500
4	≥3 areas >5,000	≥3 areas >4,000	>6,000 or ≥3 blocks > 2,500	>1,500
8	>5,000	>4,000	>4,000	>1,000
12	>2,000	>2,000	>2,000	>500
15	None >2,000	None >2,000	None >2,000	None >500

Modifying Snapshot Conservation Status with Degree of Threat

The point totals for each of the preceding landscape parameters are summed and a snapshot conservation status is assigned to each ecoregion based on the value ranges described at the beginning of this section. The snapshot conservation status assessments are subsequently modified by the predicted threat analyses to arrive at a final conservation status assessment. For example, an ecoregion may be largely unaffected by human disturbance and would receive a snapshot status of relatively stable, but immediate threats of extensive mining activity or large timber concessions might warrant shifting the ecoregion's final conservation status to vulnerable. The final conservation status assessments should reflect the urgency of conservation action as well as the presumed ecological integrity of ecoregions over the next two decades.

Threat analyses are inherently complex, as factors may affect ecosystems directly or indirectly and many feedback loops and interactions among factors are poorly understood. To evaluate as objectively as possible the range and severity of threats facing the ecoregions, we categorize threats into three major types: conversion threats, degradation threats, and wildlife exploitation threats. Each category is assigned points based on the anticipated severity of the threat and the time frame over which the threat is expected to occur. This analysis is necessarily a coarse assessment that treats only the aggregate effects of the threats in each class, not individual sources (e.g., individual timber sales, proposed mine sites, etc.). As in the degree-of-protection analysis, a more comprehensive and

detailed assessment of threats is appropriate at intra-ecoregional scales.

We used an index of 0–100 points to determine pending threats to an ecoregion. Points were attributed to the three major types of threat as follows: conversion threats (maximum 50 points), degradation threats (maximum 30 points), and wildlife exploitation threats (maximum 20 points). Conversion threats were weighted most heavily because the effects of habitat conversion are generally more far-reaching and difficult to reverse than those of either degradation or wildlife exploitation. The regional working groups at the expert workshop assigned the appropriate points for each of the threat types to each ecoregion based on the tables that follow. The general level of threat can be estimated from the point totals:

70–100	high threat
20–69	medium threat
0–20	low threat

Ecoregions with a threat rank of high were most often moved up one class in their conservation status ranking (e.g., from vulnerable to endangered). Medium threat estimates influenced the rankings depending upon our judgment. Generally, they did not alter the snapshot status, but instead impending threats are described in detail in the ecoregion descriptions. Low-threat estimates were not used to modify snapshot conservation status rankings.

The first list below gives the types of threats in each category. The second shows the points assigned to each category depending on the intensity and time frame of the anticipated threats.

Type of Threat

Conversion Threats
- intensive logging and associated road building
- intensive burning or grazing leading to habitat loss
- agricultural expansion and clearing for development
- permanent alteration from burning

Degradation Threats
- pollution (e.g., oil, pesticides, herbicides, mercury, heavy metals, defoliants)
- burning frequencies and intensities outside of the natural range of variation
- loss of habitat, resources, or individual organisms from introduced species
- firewood extraction
- unsustainable extraction of nontimber products
- grazing patterns, frequencies, and intensities outside the natural range of variation
- road building and associated erosion and landslide damage
- off-road vehicle damage
- selective logging
- excessive recreational impacts

Wildlife Exploitation
- hunting and poaching
- unsustainable extraction of wildlife and plants as commercial products
- harassment and displacement by commercial and recreational users

Intensity and Time Frame

Conversion Threats
- May significantly alter 25 percent or more of remaining habitat within 20 years   50 points
- May significantly alter between 10 percent and 24 percent or more of remaining habitat within 20 years   20 points
- May significantly alter between 5 percent and 9 percent or more of remaining habitat within 20 years   10 points
- No conversion threat(s) recognized for ecoregion   0 points

Degradation Threats
- High: Many populations of native plant species experience high mortality and low recruitment due to degradation factors. Succession and disturbance processes significantly altered. Low habitat quality for sensitive species. Abandonment and disruption of seasonal/migratory/breeding movements. Pollutants and/or linked effects widespread in ecosystem (e.g., recorded in several trophic levels).   30 points
- Medium: Populations of native plant species experience significant mortality and poor recruitment due to degradation factors. Succession and disturbance processes modified. Some abandonment and underuse of seasonal/migratory/breeding movements by species. Pollutants and/or linked effects commonly found in target species or assemblages.   15 points
- No degradation threats recognized for ecoregion.   0 points

Wildlife Exploitation
- High intensity of wildlife exploitation in region; elimination of local populations of most target species imminent or complete.   20 points
- Moderate levels of wildlife exploitation; populations of game/trade species persist but in reduced numbers.   10 points
- No wildlife exploitation recognized for ecoregion.   0 points

APPENDIX C

Species Richness and Endemism Data for Ecoregions

Ecoregion	Number	Amphibian Richness	Bird Richness	Butterfly Richness	Conifer Richness	Mammal Richness	Vascular Plant Richness	Reptile Richness	Snail Richness	Tree Richness	Total Richness	Amphibian Endemism	Bird Endemism	Butterfly Endemism	Conifer Endemism	Mammal Endemism	Vascular Plant Endemism	Vascular Plant Mean	Reptile Endemism	Snail Endemism	Tree Endemism	Total Endemism
1.1: Tropical Moist Broadleaf Forests																						
Hawaiian Moist Forests	1	0	60	2	0	1	935	0	570	145	1,568	0	0	3	0	0	710	225	0	0	0	228
South Florida Rocklands	2	15	176	136	6	28	1,034	52	64	162	1,505	0	0	0	1	1	11–20	16	3	26	5	46
Puerto Rican Moist Forests	3	1	0	0	0	0	1,705	0	0	0	1,706	0	0	0	0	0	113	131	0	0	0	131
1.2: Tropical Dry Broadleaf Forests																						
Hawaiian Dry Forests	4	0	79	2	0	1	659	0	380	114	1,121	0	0	3	0	0	490	225	0	0	0	228
Puerto Rican Dry Forests	5	0	0	0	0	0	819	0	0	0	819	0	0	0	0	0	37	36	0	0	0	36
2.1: Temperate Broadleaf and Mixed Forests																						
Willamette Valley Forests	6	6	177	94	3	33	1,067	8	27	17	1,412	0	0	0	0	1	4–10	7	0	5	0	13
Western Great Lakes Forests	7	20	216	144	10	62	1,459	19	63	63	1,983	0	0	0	0	0	0	0	0	1	1	2
Eastern Forest/Boreal Transition	8	20	206	127	12	56	1,228	18	74	56	1,729	0	0	0	0	0	0	0	0	2	0	2
Upper Midwest Forest/Savanna Transition Zone	9	20	216	149	11	58	1,420	33	80	77	1,976	0	0	0	0	0	1–3	2	0	3	0	5
Southern Great Lakes Forests	10	32	224	150	10	62	2,243	33	120	121	2,864	0	0	0	0	0	1–3	2	1	6	0	9
Eastern Great Lakes Lowland Forests	11	22	222	136	13	57	1,381	26	72	85	1,916	0	0	0	0	0	1–3	2	0	1	1	3
New England/Acadian Forests	12	20	222	136	14	61	1,496	21	76	72	2,032	0	0	1	0	0	4–10	7	1	3	0	12
Gulf of St. Lawrence Lowland Forests	13	15	195	77	11	48	1,033	6	2	43	1,376	0	0	0	0	0	0	0	0	0	0	0
Northeastern Coastal Forests	14	28	251	159	15	61	1,695	34	101	106	2,329	0	0	0	0	0	4–10	7	0	6	0	13
Allegheny Highlands Forests	15	27	198	132	13	59	1,883	28	85	101	2,412	0	0	0	0	0	0	0	1	2	1	3
Appalachian/Blue Ridge Forests	16	66	194	157	15	67	2,398	48	264	158	3,194	21	0	0	2	0	76–110	93	1	122	6	237
Appalachian Mixed Mesophytic Forests	17	56	200	155	9	67	2,487	52	248	166	3,265	7	0	0	0	0	11–20	16	2	98	8	123
Central U.S. Hardwood Forests	18	54	203	161	9	60	2,332	65	210	145	3,085	3	0	0	0	0	21–50	36	0	71	1	110
Ozark Mountain Forests	19	37	191	151	4	50	1,743	58	89	119	2,319	5	0	1	0	0	11–20	16	0	22	0	43
Mississippi Lowland Forests	20	35	222	159	7	58	1,468	52	89	127	2,083	0	0	0	0	0	1–3	2	0	5	2	7
East Central Texas Forests	21	28	210	203	3	41	1,553	60	62	115	2,157	1	0	0	0	1	4–10	7	1	7	0	16
Southeastern Mixed Forests	22	63	212	156	14	62	3,363	73	185	167	4,114	5	0	0	1	1	21–50	36	5	52	10	99
2.2: Temperate Coniferous Forests																						
North Pacific Coastal Forests	23	4	166	36	8	44	615	1	10	14	876	0	0	0	0	2	1–3	2	0	0	0	4
Queen Charlotte Islands	24	1	153	7	8	47	459	0	0	13	667	0	0	0	0	1	1–3	2	0	0	0	3
Northern British Columbia Mountain Forests	25	4	174	69	11	57	909	2	1	23	1,216	0	0	0	0	0	0	0	0	0	0	0
Alberta Mountain Forests	26	4	179	94	10	57	660	2	1	22	997	0	0	0	1	0	1–3	0	0	0	0	2
Fraser Plateau and Basin Complex	27	5	172	94	12	52	1,012	3	0	21	1,344	0	0	0	0	0	0	0	0	0	0	0
Northern Transitional Alpine Forests	28	5	144	64	18	52	876	1	0	19	1,142	0	0	0	1	0	0	0	0	0	0	0
Alberta/British Columbia Foothills Forests	29	5	182	95	8	50	740	2	0	20	1,074	1	0	0	2	1	1–3	2	0	0	0	2
North Central Rockies Forests	30	9	219	145	19	74	1,695	11	50	37	2,203	1	0	1	2	2	21–50	36	0	10	1	48
Okanogan Forests	31	8	199	137	14	54	1,355	15	24	17	1,792	0	0	0	0	1	1–3	2	0	4	0	7
Cascade Mountains Leeward Forests	32	4	187	97	16	55	1,328	4	0	24	1,675	0	0	0	0	1	11–20	16	0	0	0	16
British Columbia Mainland Coastal Forests	33	12	225	118	16	66	1,325	9	1	35	1,756	0	0	0	1	1	4–10	7	0	0	0	8
Central Pacific Coastal Forests	34	19	227	88	16	66	1,109	12	40	34	1,561	4	0	1	3	3	11–20	16	1	7	2	30
Puget Lowland Forests	35	11	200	78	11	53	1,100	9	34	33	1,485	0	0	0	0	0	1–3	2	0	6	0	9
Central and Southern Cascades Forests	36	19	202	130	18	64	1,296	14	10	36	1,735	5	0	1	1	1	21–50	36	0	0	2	44
Eastern Cascades Forests	37	4	204	135	17	67	1,224	13	15	31	1,662	0	0	0	1	0	21–50	36	0	0	1	36
Blue Mountains Forests	38	7	196	115	13	65	1,134	14	30	26	1,561	3	0	0	0	1	21–50	36	0	8	0	45
Klamath-Siskiyou Forests	39	14	222	141	30	69	1,859	18	54	60	2,377	4	0	3	7	2	111–150	131	0	31	4	168
Northern California Coastal Forests	40	17	241	123	16	63	1,212	19	35	57	1,710	5	0	0	8	2	11–20	16	0	18	4	43
Sierra Nevada Forests	41	13	197	162	20	77	2,373	17	47	50	2,886	5	0	3	5	5	51–75	63	1	32	3	108
Great Basin Montane Forests	42	4	160	123	12	55	1,043	8	12	21	1,405	0	0	1	1	2	21–50	36	1	5	0	44
South Central Rockies Forests	43	6	201	173	13	69	1,933	10	12	28	2,404	6	0	0	0	0	51–75	63	0	1	0	64

Ecoregion																						
Wasatch and Uinta Montane Forests	44	7	190	165	14	78	1,109	15	22	32	1,586	0	0	0	1	0	51–75	63	0	2	0	65
Colorado Rockies Forests	45	7	210	224	15	81	1,626	7	37	35	2,192	0	0	0	2	1	76–110	93	0	2	0	96
Arizona Mountains Forests	46	11	208	201	16	79	2,204	31	83	67	2,817	1	1	1	1	2	76–110	93	1	34	5	132
Madrean Sky Islands Montane Forests	47	10	207	186	13	62	1,139	34	79	62	1,717	1	1	1	2	2	21–50	36	4	53	13	100
Piney Woods Forests	48	34	205	158	6	53	1,729	57	82	135	2,318	0	0	0	1	0	4–10	7	0	7	1	14
Atlantic Coastal Pine Barrens	49	13	213	106	7	37	632	29	36	62	1,066	0	0	0	0	0	1–3	2	0	4	0	6
Middle Atlantic Coastal Forests	50	49	237	154	12	47	1,488	64	108	147	2,147	0	2	0	0	0	11–20	16	16	12	4	32
Southeastern Conifer Forests	51	53	236	157	12	47	3,095	85	138	193	3,811	3	5	0	4	3	>201	225	4	42	26	293
Florida Sand Pine Scrub	52	16	173	142	6	33	951	51	67	175	1,433	1	0	0	1	1	21–50	36	4	22	5	63
3.1: Temperate Grasslands/Savanna/Shrub																						
Palouse Grasslands	53	8	199	131	6	54	1,290	15	24	16	1,721	1	0	0	0	0	21–50	36	0	4	1	41
California Central Valley Grasslands	54	6	184	78	0	53	1,682	15	3	12	2,021	4	0	0	0	0	11–20	16	4	1	0	21
Canadian Aspen Forest and Parklands	55	9	206	140	7	57	1,464	9	2	36	1,887	0	0	0	0	0	0	0	0	0	0	0
Northern Mixed Grasslands	56	13	222	160	2	72	1,595	18	15	38	2,095	0	0	0	0	0	0	0	0	0	0	0
Montana Valley and Foothill Grasslands	57	8	214	157	6	57	1,197	11	13	23	1,657	0	0	1	0	0	1–3	2	0	1	0	1
Northern Short Grasslands	58	8	231	187	5	76	1,867	19	21	28	2,409	0	0	0	0	0	1–3	2	0	1	0	4
Northern Tall Grasslands	59	12	212	145	6	69	1,055	13	33	50	1,539	0	0	0	0	0	0	0	0	1	0	3
Central Tall Grasslands	60	19	228	160	6	68	1,779	39	69	80	2,362	0	0	0	0	0	0	0	0	2	0	2
Flint Hills Tall Grasslands	61	14	199	151	5	59	1,174	44	51	65	1,692	0	0	0	0	0	1–3	2	0	2	0	2
Nebraska Sand Hills Mixed Grasslands	62	8	160	110	2	49	1,185	21	2	23	1,535	1	0	0	1	0	11–20	16	1	3	0	3
Western Short Grasslands	63	17	245	230	3	86	2,359	61	13	40	3,011	0	0	0	0	1	4–10	7	0	0	0	22
Central and Southern Mixed Grasslands	64	17	228	166	2	71	2,081	69	5	64	2,637	0	1	0	2	1	4–10	7	0	16	2	9
Central Forest/Grassland Transition Zone	65	42	234	202	3	68	2,124	79	136	137	2,885	1	0	1	1	3	21–50	36	3	8	6	25
Edwards Plateau Savannas	66	17	191	171	2	48	2,361	64	59	69	2,911	2	1	2	2	1	1–3	2	3	7	6	50
Texas Blackland Prairies	67	32	216	197	2	53	1,531	69	64	62	2,162	4	2	4	0	0	1–3	2	1	0	0	15
Western Gulf Coastal Grasslands	68	31	269	222	5	57	2,165	84	80	102	2,908	1	4	1	1	0	11–20	16	1	11	2	30
3.2: Flooded Grasslands																						
Everglades	69	19	210	132	1	25	1,362	45	62	123	1,855	0	0	0	0	1	4–10	7	1	28	15	37
4.1: Mediterranean Scrub and Savanna																						
California Interior Chaparral and Woodlands	70	15	236	133	7	67	2,105	34	49	59	2,639	5	0	2	5	6	111–150	131	1	31	8	176
California Montane Chaparral and Woodlands	71	14	227	136	17	51	2,075	34	40	59	2,577	5	0	4	8	2	21–50	36	0	20	8	67
California Coastal Sage and Chaparral	72	12	241	137	11	51	1,491	40	74	70	2,046	4	0	3	5	3	51–75	63	3	56	16	132
4.2: Xeric Shrublands/Deserts																						
Hawaiian High Shrublands	73	0	73	1	0	1	124	0	63	4	262	0	0	0	0	0	110	93	0	0	0	93
Hawaiian Low Shrublands	74	0	45	1	0	1	384	0	254	4	685	0	0	0	0	0	270	225	0	0	1	225
Snake/Columbia Shrub Steppe	75	8	207	135	6	70	2,169	23	11	24	2,623	0	1	0	0	1	51–75	63	1	3	0	65
Great Basin Shrub Steppe	76	9	204	160	4	79	2,519	32	12	23	3,015	1	0	0	0	0	151–200	175	0	2	0	181
Wyoming Basin Shrub Steppe	77	6	189	159	3	72	1,557	7	8	20	1,998	0	0	0	0	0	11–20	16	0	2	0	16
Colorado Plateau Shrublands	78	12	222	225	4	107	2,556	61	41	38	3,224	2	0	0	0	1	>201	225	1	2	2	230
Mojave Desert	79	8	230	147	3	71	2490	46	23	31	3,015	2	2	0	2	2	76–110	93	1	19	3	120
Sonoran Desert	80	12	261	168	1	82	2068	58	35	36	2,684	2	2	0	7	0	11–20	16	2	21	6	47
Chihuahuan Desert	81	22	279	259	9	109	2263	103	70	76	3,105	2	7	0	7	1	111–150	131	1	37	13	181
Tamaulipan Mezquital	82	20	245	181	1	50	1487	60	36	43	2,079	0	0	0	0	1	11–20	16	0	10	3	29
6.1: Boreal Forest/Taiga																						
Interior Alaska/Yukon Lowland Taiga	83	1	137	63	5	36	810	0	3	16	1,050	0	0	0	0	0	1–3	2	0	0	0	2
Alaska Peninsula Montane Taiga	84	1	116	18	2	27	510	0	8	7	680	0	0	0	0	0	1–3	2	0	0	0	2
Cook Inlet Taiga	85	1	127	40	2	34	738	0	1	10	941	0	0	0	0	0	1–3	2	0	0	0	2
Copper Plateau Taiga	86	1	116	51	2	34	407	0	0	11	609	0	0	0	0	0	0	0	0	0	0	0
Northwest Territories Taiga	87	2	131	65	5	47	576	0	0	13	821	0	2	0	0	0	1–3	2	0	0	0	2
Yukon Interior Dry Forests	88	4	123	73	5	47	692	1	1	14	940	0	0	0	0	0	0	0	0	0	0	0
Northern Cordillera Forests	89	6	165	84	8	53	823	0	0	24	1,132	0	0	0	0	0	0	0	0	0	0	0
Muskwa/Slave Lake Forests	90	3	158	74	6	49	722	0	0	21	722	1	0	0	2	0	0	0	0	0	0	1
Western Canadian Shield Taiga	91	3	131	58	5	45	720	0	1	15	958	0	0	0	0	0	0	0	0	0	0	0
Western Canadian Forests	92	6	183	93	8	49	613	3	8	31	955	1	0	1	0	0	0	0	0	0	0	1

(continues)

Ecoregion	Number	Amphibian Richness	Bird Richness	Butterfly Richness	Conifer Richness	Mammal Richness	Vascular Plant Richness	Reptile Richness	Snail Richness	Tree Richness	Total Richness	Amphibian Endemism	Bird Endemism	Butterfly Endemism	Conifer Endemism	Mammal Endemism	Vascular Plant Endemism	Vascular Plant Mean	Reptile Endemism	Snail Endemism	Tree Endemism	Total Endemism
Midwestern Canadian Shield Forests	93	7	177	80	7	44	797	3	0	24	1,108	0	0	0	0	0	0	0	0	0	0	0
Central Canadian Shield Forests	94	12	160	78	9	48	1246	2	0	25	1,546	0	0	0	0	0	0	0	0	0	0	0
Southern Hudson Bay Taiga	95	7	160	64	7	48	1178	1	0	20	1,458	0	0	0	0	0	1–3	2	0	0	0	2
Eastern Canadian Shield Taiga	96	5	106	36	6	41	925	1	0	14	1,114	0	0	0	0	0	0	0	0	0	0	0
Eastern Canadian Forests	97	12	167	55	9	50	1140	1	0	29	1,425	0	1	0	0	1	11–20	16	0	0	0	18
Newfoundland Highland Forests	98	0	122	41	5	20	473	0	0	19	656	0	0	0	0	0	0	0	0	0	0	0
South Avalon–Burin Oceanic Barrens	99	0	120	37	5	18	258	0	0	11	433	0	0	0	0	0	0	0	0	0	0	0
6.2: Tundra																						
Aleutian Islands Tundra	100	0	100	3	0	5	388	0	10	1	506	0	2	0	0	2	4–10	7	0	0	0	11
Beringia Lowland Tundra	101	1	122	38	1	33	553	0	0	6	747	0	2	0	0	1	0	0	0	0	0	3
Beringia Upland Tundra	102	1	118	36	0	33	538	0	1	3	727	0	2	0	0	0	1–3	2	0	0	0	4
Alaska/St. Elias Range Tundra	103	2	153	67	2	44	747	0	0	9	1,013	0	0	0	0	0	4–10	7	0	0	0	7
Pacific Coastal Mountain Icefields and Tundra	104	5	162	65	12	50	792	1	0	12	1,075	0	0	0	0	0	0	0	0	0	0	0
Interior Yukon/Alaska Alpine Tundra	105	1	121	74	4	43	617	0	0	11	856	0	0	0	0	0	4–10	7	0	0	0	7
Ogilvie/MacKenzie Alpine Tundra	106	0	112	71	0	39	589	0	0	8	811	0	0	0	0	0	4–10	7	0	0	0	7
Brooks/British Range Tundra	107	0	52	48	0	28	593	0	0	5	721	0	0	0	0	0	1–3	2	0	0	0	2
Arctic Foothills Tundra	108	0	69	37	1	25	580	0	1	3	712	0	0	0	0	0	0	0	0	0	0	0
Arctic Coastal Tundra	109	0	76	21	0	26	539	0	2	2	664	0	0	0	0	0	1–3	2	0	0	0	2
Low Arctic Tundra	110	0	73	26	3	28	497	0	0	4	624	0	0	0	0	0	0	0	0	0	0	0
Middle Arctic Tundra	111	0	59	20	0	13	371	0	0	1	463	0	0	0	0	0	1–3	2	0	0	0	2
High Arctic Tundra	112	0	47	11	0	9	245	0	0	0	312	0	1	0	0	0	0	0	0	0	0	1
Davis Highlands Tundra	113	0	39	11	0	8	216	0	0	0	274	0	0	0	0	0	0	0	0	0	0	0
Baffin Coastal Tundra	114	0	30	11	0	7	135	0	0	0	183	0	0	0	0	0	0	0	0	0	0	0
Torngat Mountain Tundra	115	0	43	12	4	21	286	0	0	2	362	0	0	0	0	0	0	0	0	0	0	0
Permanent Ice	116	0	0	0	0	0	112	0	0	0	112	0	0	0	0	0	0	0	0	0	0	0

Sources of data are presented in table A.1 in appendix A on page 107.

Summary of Ecoregion Scores of the Biological Distinctiveness, Conservation Status, and Integration Matrix Analyses

Ecoregion	Ecoregion Number	Ecoregion Size (km²)	Richness Index[a]	Endemism Index[b]	Rarity of Habitat Type[c]	Rare Phenomena[d]	Beta-Diversity[e]	Biological Distinctiveness[f]	Biological Distinctiveness Index[g]	Percent Intact Habitat[h]	Habitat Loss[i]	Habitat Blocks[j]	Fragmentation[k]	Protection[l]	Conservation Status[m]	Conversion Threats[n]	Degradation Threats[o]	Wildlife Threats[p]	Total Threats[q]	Snapshot (1996) Conservation Status[r]	Final Conservation Status[s]	Conservation (Integration) Class[t]
1.1: Tropical Moist Broadleaf Forests																						
Hawaiian Moist Forests	1	6,944	15	100	0	100	0	215	1	45	25	12	5	8	50	30	30	10	70	3	2	1
South Florida Rocklands	2	2,157	10	25	0	5	0	40	2	2	40	25	20	15	100	10	15	0	25	1	1	2
Puerto Rican Moist Forests	3	7,544	NA	NA	0	0	0	NA	NA	35	35	18	16	12	81	10	0	0	10	2	2	NA
1.2: Tropical Dry Broadleaf Forests																						
Hawaiian Dry Forests	4	6,632	15	100	0	100	0	215	1	5	40	25	16	15	96	20	30	0	50	1	1	1
Puerto Rican Dry Forests	5	1,311	NA	NA	0	0	0	NA	NA	37	37	25	16	15	93	20	0	0	20	1	1	NA
2.1: Temperate Broadleaf and Mixed Forests																						
Willamette Valley Forests	6	14,850	5	15	0	0	0	20	3	1	40	25	20	15	100	10	15	10	35	1	1	4
Western Great Lakes Forests	7	273,623	10	5	0	0	0	15	4	20	20	2	12	1	35	20	30	0	50	4	4	5
Eastern Forest/Boreal Transition	8	346,724	10	5	0	0	0	15	4	10	25	6	5	4	40	10	8	5	23	3	3	5
Upper Midwest Forest/Savanna Transition Zone	9	166,122	10	5	0	0	0	15	4	5	35	25	20	15	95	50	0	0	50	1	1	4
Southern Great Lakes Forests	10	244,469	10	5	0	0	0	15	4	0	40	25	20	15	100	50	30	20	100	1	1	4
Eastern Great Lakes Lowland Forests	11	115,538	10	5	0	0	0	15	4	3	35	25	20	15	95	20	30	10	60	1	1	4
New England/Acadian Forests	12	236,894	10	15	0	0	0	25	3	3	30	18	12	8	68	20	15	0	35	2	2	4
Gulf of St. Lawrence Lowland Forests	13	39,426	5	5	0	0	0	10	4	3	35	25	16	15	91	50	15	10	75	1	1	4
Northeastern Coastal Forests	14	89,691	10	15	0	0	0	25	3	0	35	25	20	15	95	50	15	10	75	1	1	4
Allegheny Highlands Forests	15	83,881	10	5	0	0	0	15	4	1	25	18	12	15	70	10	15	0	25	2	2	4
Appalachian/Blue Ridge Forests	16	159,266	100	100	0	5	0	205	1	3	17	12	16	8	53	10	15	20	60	3	3	1
Appalachian Mixed Mesophytic Forests	17	192,152	15	25	0	5	0	45	2	1	25	25	20	15	81	50	30	15	80	2	3	1
Central U.S. Hardwood Forests	18	296,019	10	15	0	0	0	30	2	1	25	25	16	15	85	20	30	10	60	2	2	2
Ozark Mountain Forests	19	62,011	10	25	0	5	0	40	2	3	20	25	16	15	76	10	15	10	35	2	2	2
Mississippi Lowland Forests	20	112,284	10	15	0	0	0	25	3	5	40	25	20	15	100	10	30	5	45	2	2	2
East Central Texas Forests	21	52,637	10	15	0	5	0	25	3	1	40	25	12	15	92	10	15	10	35	1	1	4
Southeastern Mixed Forests	22	347,803	100	100	0	5	0	205	1	1	35	25	20	15	95	20	15	5	40	1	1	1
2.2: Temperate Coniferous Forests																						
Northern Pacific Coastal Forests	23	60,894	5	5	100	5	0	115	1	85	10	2	12	1	25	20	8	5	33	4	4	3
Queen Charlotte Islands	24	9,972	5	5	100	5	0	110	1	50	15	6	12	8	41	20	15	5	40	3	3	1
Northern British Columbia Mountain Forests	25	71,747	5	5	0	0	0	10	4	75	5	2	5	15	27	50	15	10	75	4	3	5
Alberta Mountain Forests	26	39,769	5	5	0	0	0	10	4	80	5	2	12	1	20	10	15	10	35	4	4	5
Fraser Plateau and Basin Complex	27	137,055	5	5	0	0	0	10	4	25	20	2	12	4	38	50	15	5	70	3	2	4
Northern Transitional Alpine Forests	28	25,662	5	5	0	0	0	10	4	75	5	2	5	15	27	50	15	5	70	3	3	5
Alberta/British Columbia Foothills Forests	29	120,460	5	5	100	0	0	10	4	0	25	25	16	12	78	50	15	5	70	2	3	4
North Central Rockies Forests	30	245,430	10	15	100	0	0	25	3	20	20	25	12	1	35	50	15	10	75	4	3	5
Okanogan Forests	31	53,210	10	5	0	0	0	15	4	20	5	2	16	15	20	50	15	5	70	4	4	4
Cascade Mountains Leeward Forests	32	46,626	5	5	0	5	0	10	4	70	20	2	12	1	76	50	15	5	40	4	4	1
British Columbia Mainland Coastal Forests	33	142,153	10	5	0	0	0	110	1	40	40	12	12	1	38	50	15	5	70	3	2	1
Central Pacific Coastal Forests	34	73,696	10	15	0	5	5	135	1	8	40	25	16	4	72	10	15	5	25	2	2	1
Puget Lowland Forests	35	22,488	10	5	100	0	5	15	4	5	40	25	16	1	100	0	15	0	25	3	3	4
Central and Southern Cascades Forests	36	44,848	10	15	0	0	0	25	3	20	25	18	12	8	53	10	15	0	25	2	3	5
Eastern Cascades Forests	37	55,200	10	15	0	0	0	25	3	15	15	18	16	12	71	20	15	5	40	3	2	4
Blue Mountains Forests	38	64,702	10	25	0	0	0	25	3	10	15	18	12	8	53	50	15	0	75	3	3	4
Klamath-Siskiyou Forests	39	50,299	10	15	0	5	5	45	1	25	40	25	12	1	53	50	30	10	80	2	2	4
Northern California Coastal Forests	40	13,274	10	25	100	0	0	130	1	5	25	25	16	15	96	0	0	5	0	3	1	1
Sierra Nevada Forests	41	52,832	10	15	0	100	0	135	1	25	5	25	16	4	70	20	15	5	35	2	2	2
Great Basin Montane Forests	42	5,784	5	15	0	0	0	20	3	37	15	2	0	15	45	10	15	5	30	3	3	5
South Central Rockies Forests	43	159,348	10	15	0	0	0	25	3	40	15	25	12	1	30	50	15	5	70	4	3	5

3.1: Temperate Grasslands/Savanna/Shrub
3.2: Flooded Grasslands
4.1: Mediterranean Scrub and Savanna
4.2: Xeric Shrublands/Deserts
6.1: Boreal Forest/Taiga

#	Ecoregion	Area	1	2	3	4	5	6	7	8	9	10	11	12	13	14	15	16	17	18	19
44	Wasatch and Uinta Montane Forests	41,465	10	15	0	0	25	3	25	10	12	16	8	61	50	30	10	90	3	2	4
45	Colorado Rockies Forests	132,740	10	15	0	0	25	3	24	20	2	5	1	28	20	15	5	40	4	4	5
46	Arizona Mountains Forests	109,080	15	15	0	0	30	3	25	10	2	12	4	28	20	8	5	28	4	4	3
47	Madrean Sky Islands Montane Forests	11,420	10	25	100	0	135	1	75	5	2	2	12	19	20	8	5	23	4	4	3
48	Piney Woods Forests	140,892	15	15	0	0	30	2	3	25	25	20	15	85	50	15	5	70	2	1	2
49	Atlantic Coastal Pine Barrens	8,975	10	5	0	0	15	4	10	15	6	5	8	34	10	15	0	25	4	4	5
50	Middle Atlantic Coastal Forests	133,855	15	15	0	5	35	2	12	25	25	16	15	81	20	30	10	60	2	2	2
51	Southeastern Conifer Forests	236,759	100	100	100	5	305	1	2	35	12	16	4	67	50	30	20	100	1	2	2
52	Florida Sand Pine Scrub	3,889	10	15	100	5	130	1	15	35	25	16	15	91	50	30	5	85	1	1	1
53	Palouse Grasslands	46,826	15	25	0	0	30	2	1	40	12	20	8	80	10	15	10	35	2	2	2
54	California Central Valley Grasslands	55,064	5	25	0	0	30	2	0	40	25	20	15	100	50	30	0	0	1	1	2
55	Canadian Aspen Forest and Parklands	394,376	5	5	0	0	10	4	15	35	18	20	12	85	50	30	0	90	1	1	4
56	Northern Mixed Grasslands	218,897	5	5	0	0	10	4	0	35	25	20	15	95	50	30	5	85	3	1	4
57	Montana Valley and Foothill Grasslands	81,666	5	5	0	0	10	4	25	20	2	12	15	49	50	15	5	70	2	3	4
58	Northern Short Grasslands	638,300	5	5	0	0	10	4	2	20	25	12	8	65	10	15	8	33	1	2	4
59	Northern Tall Grasslands	73,148	5	5	0	0	10	1	5	35	18	16	12	81	20	30	0	50	1	2	4
60	Central Tall Grasslands	248,435	10	5	100	100	115	1	3	40	25	0	15	100	20	15	0	50	2	2	4
61	Flint Hills Tall Grasslands	29,612	10	5	100	100	115	1	70	20	18	0	15	53	20	15	0	35	1	3	1
62	Nebraska Sand Hills Mixed Grasslands	61,117	5	5	0	0	10	4	85	10	2	0	12	24	10	0	0	25	3	4	5
63	Western Short Grasslands	435,154	25	25	0	0	35	2	40	35	25	12	15	72	10	0	0	10	4	4	2
64	Central and Southern Mixed Grasslands	282,139	10	15	0	0	25	3	5	35	25	20	15	95	50	0	0	50	2	2	4
65	Central Forest/Grassland Transition Zone	406,980	25	25	0	0	40	2	1	40	25	20	15	100	20	0	0	10	1	2	2
66	Edwards Plateau Savannas	61,800	25	25	0	0	40	2	2	40	25	20	15	100	50	15	10	75	1	1	2
67	Texas Blackland Prairies	50,255	15	25	0	0	40	2	1	30	25	12	15	82	50	15	5	70	2	1	2
68	Western Gulf Coastal Grasslands	77,425	15	25	0	0	40	2	5	35	25	16	15	91	20	30	10	60	1	1	2
69	Everglades	20,117	15	15	100	0	135	1	2	25	25	5	1	56	10	30	5	45	3	3	1
70	California Interior Chaparral and Woodlands	64,605	100	100	0	0	300	1	20	25	12	14	8	59	20	30	0	50	3	3	1
71	California Montane Chaparral and Woodlands	20,408	25	25	100	0	225	1	20	20	6	12	8	46	20	30	10	60	3	3	1
72	California Coastal Sage and Chaparral	24,356	25	25	100	0	225	1	2	45	18	16	15	94	10	30	10	50	1	1	1
73	Hawaiian High Shrublands	1,853	5	5	0	0	25	3	70	15	6	5	12	38	10	15	10	35	3	3	5
74	Hawaiian Low Shrublands	1,522	5	5	5	0	20	4	99	40	25	20	15	100	50	30	0	80	1	1	4
75	Snake/Columbia Shrub Steppe	218,111	10	5	0	0	25	4	5	35	2	12	1	50	50	30	5	85	4	2	4
76	Great Basin Shrub Steppe	335,868	5	5	0	0	25	3	5	15	2	5	1	23	10	8	5	23	3	4	5
77	Wyoming Basin Shrub Steppe	132,361	15	10	0	0	15	4	10	25	2	5	12	44	20	15	5	40	4	3	5
78	Colorado Plateau Shrublands	326,390	5	15	0	0	30	2	15	15	2	12	1	30	10	8	5	23	3	4	3
79	Mojave Desert	130,634	25	10	0	0	35	1	50	5	2	5	1	13	20	8	5	33	4	4	5
80	Sonoran Desert	116,770	15	15	5	0	45	1	40	15	2	12	1	30	20	15	10	35	4	4	3
81	Chihuahuan Desert	205,439	100	100	0	0	205	1	15	25	2	12	1	40	10	8	5	23	3	4	3
82	Tamaulipan Mezquital	57,776	15	15	0	0	30	2	2	35	25	16	15	91	50	15	10	75	1	3	1
83	Interior Alaska/Yukon Lowland Taiga	443,319	5	10	5	0	20	3	99	0	2	0	1	3	10	0	5	15	3	5	5
84	Alaska Peninsula Montane Taiga	47,885	5	0	5	0	15	4	99	0	2	0	1	3	0	8	5	13	1	5	5
85	Cook Inlet Taiga	27,963	5	20	0	0	10	4	90	0	2	0	4	6	20	8	10	38	5	5	5
86	Copper Plateau Taiga	17,170	5	10	5	0	15	4	90	0	2	0	4	6	10	8	5	23	5	5	5
87	Northwest Territories Taiga	345,833	5	5	100	0	110	4	90	5	6	5	8	10	20	15	5	38	4	4	3
88	Yukon Interior Dry Forests	62,370	15	15	0	0	15	1	75	25	6	12	15	38	20	15	15	50	3	3	5
89	Northern Cordillera Forests	262,780	10	5	100	0	120	1	85	5	2	5	8	20	50	8	8	73	3	3	3
90	Muskwa/Slave Lake Forests	262,295	15	5	100	0	115	1	75	5	2	5	8	20	20	15	15	45	4	4	1
91	Western Canadian Shield Taiga *	609,238	10	10	20*	0	35	2	92	25	2	0	1	3	10	8	10	28	5	4	3
92	Western Canadian Forests	361,759	15	15	0	0	20	3	50	35	12	12	15	40	50	15	5	35	4	5	5
93	Midwestern Canadian Shield Forests	538,770	15	15	0	0	20	3	80	10	2	5	4	16	10	8	10	23	3	3	3
94	Central Canadian Shield Forests	454,382	15	15	0	0	20	3	40	20	2	5	8	35	50	15	5	70	3	3	3
95	Southern Hudson Bay Taiga	373,502	15	5	0	0	20	3	98	0	2	0	1	3	10	10	5	15	5	1	5

(continues)

Ecoregion	Ecoregion Number	Ecoregion Size (km²)	Richness Index[a]	Endemism Index[b]	Rarity of Habitat Type[c]	Rare Phenomena[d]	Beta-Diversity[e]	Biological Distinctiveness[f]	Biological Distinctiveness Index[g]	Percent Intact Habitat[h]	Habitat Loss[i]	Habitat Blocks[j]	Fragmentation[k]	Protection[l]	Conservation Status[m]	Conversion Threats[n]	Degradation Threats[o]	Wildlife Threats[p]	Total Threats[q]	Snapshot (1996) Conservation Status[r]	Final Conservation Status[s]	Conservation (Integration) Class[t]
Eastern Canadian Shield Taiga	96	741,510	10	5	0	100	0	115	1	95	0	2	0	15	17	10	0	5	15	4	4	3
Eastern Canadian Forests	97	480,929	15	25	0	0	0	40	2	40	20	2	5	12	39	50	15	5	70	3	2	2
Newfoundland Highland Forests	98	16,338	5	5	0	0	0	10	4	80	5	2	0	8	15	20	8	5	33	4	4	5
South Avalon–Burin Oceanic Barrens	99	2,030	5	5	100	0	0	110	1	95	0	2	0	8	10	0	8	0	8	4	4	3
6.2: Tundra																						
Aleutian Islands Tundra	100	5,479	5	100	0	5	0	110	1	90	0	2	0	1	3	10	15	10	35	5	5	3
Beringia Lowland Tundra	101	151,053	10	15	0	5	0	30	2	99	0	2	0	1	3	0	0	5	5	5	5	3
Beringia Upland Tundra	102	97,364	10	15	0	0	0	25	3	90	5	2	5	1	13	0	8	5	13	4	4	5
Alaska/St. Elias Range Tundra	103	169,049	15	25	0	0	0	40	2	99	0	2	0	1	3	0	0	5	5	5	5	3
Pacific Coastal Mountain Icefields and Tundra	104	137,279	15	5	0	0	0	20	3	95	0	2	0	1	3	5	8	5	18	5	5	5
Interior Yukon/Alaska Alpine Tundra	105	232,600	15	25	0	0	0	40	2	85	5	2	5	15	27	5	8	10	23	4	4	3
Olgive/MacKenzie Alpine Tundra	106	208,421	15	25	0	0	0	40	2	95	0	2	0	15	17	10	8	10	28	4	4	3
Brooks/British Range Tundra	107	159,481	10	15	0	0	0	25	3	99	0	2	0	1	3	0	8	5	8	5	5	5
Arctic Foothills Tundra	108	123,514	10	5	0	100	0	115	1	99	0	2	0	8	10	10	8	5	23	4	5	3
Arctic Coastal Tundra	109	103,814	10	15	0	100	0	125	1	96	0	2	0	15	17	10	8	5	23	4	4	3
Low Arctic Tundra	110	796,674	10	5	0	100	0	115	1	95	0	2	0	1	3	5	15	5	25	5	5	5
Middle Arctic Tundra	111	1,033,010	5	15	0	5	0	25	3	95	0	2	0	1	3	0	8	5	13	5	5	5
High Arctic Tundra	112	463,688	5	5	0	0	0	10	4	95	0	2	0	4	6	0	8	5	13	5	5	5
Davis Highlands Tundra	113	87,882	5	5	0	0	0	10	4	95	0	2	0	1	3	0	8	5	13	5	5	5
Baffin Coastal Tundra	114	9,103	5	5	0	0	0	10	4	100	0	2	0	15	17	0	8	5	13	4	4	5
Torngat Mountain Tundra	115	32,288	5	5	0	5	0	15	4	100	0	2	0	15	17	0	15	10	25	4	4	5
Permanent Ice	116	112,906	5	5	0	0	0	10	4	100	0	2	0	1	3	0	0	0	0	5	5	5

Explanation of Numerical Codes:

[a] Richness index: 5 (low), 10 (medium), 15 (high), 100 (globally outstanding)
[b] Endemism index: 5 (low), 10 (medium), 15 (high), 100 (globally outstanding)
[c] Rare habitat type: 0 (not rare), 5 (regionally outstanding), 100 (globally rare)
[d] Rare ecological or evolutionary phenomena: 0 (not rare), 5 (regionally outstanding), 100 (globally rare)
[e] Beta-diversity: 0 (no globally outstanding beta-diversity), 5 (globally outstanding beta-diversity)
[f] Biological distinctiveness: Total score from components (a–e)
[g] Biological distinctiveness index: 1 (globally outstanding), 2 (regionally outstanding), 3 (bioregionally outstanding), 4 (nationally important)
[h] Percent intact habitat: Percentage intact
[i] Habitat loss: Index from 0 (no habitat loss) to 40 (habitat heavily altered)
[j] Habitat blocks: Index from 2 (large and numerous blocks) to 25 (small and few blocks)
[k] Fragmentation: Index from 0 (contiguous, high-connectivity habitat) to 20 (low-connectivity, highly fragmented habitat)
[l] Protection: Index from 1 (best protection) to 15 (least protection)
[m] Conservation status: Total score from components j–l
[n] Conversion threats: Index from 0 (low conversion of natural habitat) to 50 (high conversion pressures on natural habitat)
[o] Degradation threats: Index from 0 (least degraded) to 30 (most degraded)
[p] Wildlife exploitation: Index from 0 (no exploitation) to 20 (high intensity of exploitation)
[q] Total threats: Total score from components n–p
[r] Snapshot (1996) conservation status index: 1 (critical), 2 (endangered), 3 (vulnerable), 4 (relatively stable), 5 (relatively intact)
[s] Final conservation status index: 1 (critical), 2 (endangered), 3 (vulnerable), 4 (relatively stable), 5 (relatively intact)
[t] Conservation (integration) class: I–V
* Based on expert review, this ecoregion was given a score of 20 under "Rare Phenomena" because it has major caribou migrations but not globally outstanding caribou migrations.

Integration Matrices for the Ten Major Habitat Types

INTEGRATION MATRIX MHT 1.1: Tropical Moist Broadleaf Forests

	Critical	Endangered	Vulnerable	Relatively Stable	Relatively Intact
Globally Outstanding	I	I 1. Hawaiian Moist Forests—(HI)	I	III	III
Regionally Outstanding	II 2. South Florida Rocklands—(FL)	II	II	III	III
Bioregionally Outstanding	IV	IV	V	V	V
Nationally Important	IV	IV	V	V	V

I. Globally outstanding ecoregions requiring immediate protection of remaining habitat and extensive restoration.

II. Regionally outstanding ecoregions requiring immediate protection of remaining habitat and extensive restoration.

III. Globally or regionally outstanding ecoregions that present rare opportunities to conserve large blocks of intact habitat.

IV. Bioregionally and nationally important ecoregions requiring protection of remaining habitat and extensive restoration.

V. Bioregionally and nationally important ecoregions requiring protection of representative habitat blocks and proper management elsewhere for biodiversity conservation.

INTEGRATION MATRIX MHT 1.2: Tropical Dry Broadleaf Forests

	Critical	Endangered	Vulnerable	Relatively Stable	Relatively Intact
Globally Outstanding	I 4. Hawaiian Dry Forests—(HI)	I	I	III	III
Regionally Outstanding	II	II	II	III	III
Bioregionally Outstanding	IV	IV	V	V	V
Nationally Important	IV	IV	V	V	V

I. Globally outstanding ecoregions requiring immediate protection of remaining habitat and extensive restoration.

II. Regionally outstanding ecoregions requiring immediate protection of remaining habitat and extensive restoration.

III. Globally or regionally outstanding ecoregions that present rare opportunities to conserve large blocks of intact habitat.

IV. Bioregionally and nationally important ecoregions requiring protection of remaining habitat and extensive restoration.

V. Bioregionally and nationally important ecoregions requiring protection of representative habitat blocks and proper management elsewhere for biodiversity conservation.

INTEGRATION MATRIX MHT 2.1: Temperate Broadleaf and Mixed Forests

	Critical	Endangered	Vulnerable	Relatively Stable	Relatively Intact
Globally Outstanding	17. Appalachian Mixed Mesophytic Forests—(PA,WV,OH, KY,TN,VA,GA,AL), 22. Southeastern Mixed Forest— (PA,MD,WV,VA,NC,SC,GA,AL, MS,LA,TN) — I		16. Appalachian/Blue Ridge Forests—(NY,NJ,PA,WV,MD,VA, NC,SC,TN,GA,AL) — I	III	III
Regionally Outstanding		18. Central U.S. Hardwood Forests (OH,IN,IL,KY,TN,AL,MS,AR,MO,OK) 19. Ozark Mountain Forests—(OK, AR) — II	II	III	III
Bioregionally Outstanding	6. Willamette Valley Forests—(WA,OR) 14. Northeastern Coastal Forests— (ME,NH,VT,MA,RI,CO,NY,NJ, PA,DE,MD) 20. Mississippi Lowland Forests (IL,KY,TN,AR,MO,MS,LA) 21. East Central Texas Forests—(TX) — IV	12. New England/Acadian Forests— (NY,MA,VT,NH,ME,QC,NB,NS) — IV	V	V	V
Nationally Important	9. Upper Midwest Forest/Savanna Transition Zone—(MN,WI,MI,IL,IA) 10. Southern Great Lakes Forests— (IL,OH,PA,MI,ON) 11. Eastern Great Lakes Lowland Forests—(NY,VT,ON,QC) 13. Gulf of St. Lawrence Lowland Forests—(NB,NS,PE) — IV	15. Allegheny Highlands Forests— (NY,PA,OH,NJ) — IV	8. Eastern Forest/Boreal Transition—(NY,ON,QC) — V	7. Western Great Lakes Forests— (MI,WI,MN,ON, MB) — V	V

I. **Globally outstanding ecoregions requiring immediate protection of remaining habitat and extensive restoration.**

II. **Regionally outstanding ecoregions requiring immediate protection of remaining habitat and extensive restoration.**

III. **Globally or regionally outstanding ecoregions that present rare opportunities to conserve large blocks of intact habitat.**

IV. **Bioregionally and nationally important ecoregions requiring protection of remaining habitat and extensive restoration.**

V. **Bioregionally and nationally important ecoregions requiring protection of representative habitat blocks and proper management elsewhere for biodiversity conservation.**

INTEGRATION MATRIX MHT 2.2: Temperate Coniferous Forests

	Critical	Endangered	Vulnerable	Relatively Stable	Relatively Intact
Globally Outstanding	40. Northern California Coastal Forests—(OR,CA) / 51. Southeastern Conifer Forests—(GA,AL,MS,LA,FL) / 52. Florida Sand Pine Scrub—(FL) — I	33. British Columbia Mainland Coastal Forests—(BC,WA) / 34. Central Pacific Coastal Forests—(BC,WA,OR) / 39. Klamath-Siskiyou Forests—(OR,CA) / 41. Sierra Nevada Forests—(CA,NV) — I	24. Queen Charlotte Islands—(BC) — I	23. Northern Pacific Coastal Forests—(BC,AK) / 47. Madrean Sky Islands Montane Forests—(AZ,NM) — III	III
Regionally Outstanding	48. Piney Woods Forests—(OK,AR,TX,LA) — II	50. Middle Atlantic Coastal Forests—(NJ,DE,MD,VA,NC,SC,GA) — II	II	46. Arizona Mountains Forests—(AZ, NM,TX) — III	III
Bioregionally Outstanding	IV	37. Eastern Cascades Forests—(WA,OR,CA) / 38. Blue Mountains Forests—(WA,OR,ID) / 44. Wasatch and Uinta Montane Forests—(ID,WY,UT) — IV	30. North Central Rockies Forests—(BC,AB,WA,ID,MT) / 36. Central and Southern Cascades Forests—(WA,OR) / 42. Great Basin Mountain Forests—(NV,CA) / 43. South Central Rockies Forests—(MT,ID,WY,SD) — V	45. Colorado Rockies Forests—(WY,CO,NM) — V	V
Nationally Important	29. Alberta/British Columbia Foothills Forests—(BC,AB) / 31. Okanagan Dry Forests—(BC,WA) / 35. Puget Lowland Forests—(BC,WA) — IV	27. Fraser Plateau and Basin Complex—(BC) — IV	25. Central British Columbia Mountain Forests—(BC) / 28. Northern Transitional Alpine Forests—(BC) — V	26. Alberta Mountain Forests—(AB,BC) / 32. Cascade Mountains Leeward Forests—(BC,WA) / 49. Atlantic Coastal Pine Barrens—(MA,NY,NJ) — V	V

I. **Globally outstanding ecoregions requiring immediate protection of remaining habitat and extensive restoration.**

II. **Regionally outstanding ecoregions requiring immediate protection of remaining habitat and extensive restoration.**

III. **Globally or regionally outstanding ecoregions that present rare opportunities to conserve large blocks of intact habitat.**

IV. **Bioregionally and nationally important ecoregions requiring protection of remaining habitat and extensive restoration.**

V. **Bioregionally and nationally important ecoregions requiring protection of representative habitat blocks and proper management elsewhere for biodiversity conservation.**

INTEGRATION MATRIX MHT 3.1: Temperate Grasslands/Savanna/Scrub

	Critical	Endangered	Vulnerable	Relatively Stable	Relatively Intact
Globally Outstanding	60. Central Tall Grasslands—(ND,SD,MN,IA,NE,MO,KS) — I	I	61. Flint Hills Grasslands—(KS, OK) — I	III	III
Regionally Outstanding	54. California Central Valley Grasslands—(CA) 65. Central Forest/Grassland Transition Zone—(WI,MI,IN,IL,IA,MO,KS,OK,TX) 67. Texas Blackland Prairies—(TX) 68. Western Gulf Coastal Grasslands—(MS,LA,TX) — II	53. Palouse Grasslands—(WA,OR,ID) 63. Western Short Grasslands—(SD,NE,WY,CO,KS,OK,NM, TX) — II	II	III	III
Bioregionally Outstanding	64. Central and Southern Mixed Grasslands—(NE,KS,OK,TX) — IV	IV	V	V	V
Nationally Important	55. Canadian Aspen Forest and Parklands—(BC,AB,SK,MB,ND) 56. Northern Mixed Grasslands—(AB,SK,MB,ND,SD,NE) — IV	57. Montana Valley and Foothill Grasslands—(AB,MT,ID) 58. Northern Short Grasslands—(AB,SK,MT,WY,ND,SD,NE) 59. Northern Tall Grasslands—(MB,ND,SD,MN) — IV	V	62. Nebraska Sand Hills Mixed Grasslands—(SD,NE) — V	V

I. **Globally outstanding ecoregions requiring immediate protection of remaining habitat and extensive restoration.**

II. **Regionally outstanding ecoregions requiring immediate protection of remaining habitat and extensive restoration.**

III. **Globally or regionally outstanding ecoregions that present rare opportunities to conserve large blocks of intact habitat.**

IV. **Bioregionally and nationally important ecoregions requiring protection of remaining habitat and extensive restoration.**

V. **Bioregionally and nationally important ecoregions requiring protection of representative habitat blocks and proper management elsewhere for biodiversity conservation.**

	Critical	Endangered	Vulnerable	Relatively Stable	Relatively Intact
Globally Outstanding	I	I	I 69. Everglades—(FL)	III	III
Regionally Outstanding	II	II	II	III	III
Bioregionally Outstanding	IV	IV	V	V	V
Nationally Important	IV	IV	V	V	V

I. Globally outstanding ecoregions requiring immediate protection of remaining habitat and extensive restoration.

II. Regionally outstanding ecoregions requiring immediate protection of remaining habitat and extensive restoration.

III. Globally or regionally outstanding ecoregions that present rare opportunities to conserve large blocks of intact habitat.

IV. Bioregionally and nationally important ecoregions requiring protection of remaining habitat and extensive restoration.

V. Bioregionally and nationally important ecoregions requiring protection of representative habitat blocks and proper management elsewhere for biodiversity conservation.

INTEGRATION MATRIX MHT 4.1:   Mediterranean Scrub and Savanna

	Critical	Endangered	Vulnerable	Relatively Stable	Relatively Intact
Globally Outstanding	I 72. California Coastal Sage and Chaparral—(CA,Mex.)	I	I 70. California Interior Chaparral and Woodlands—(CA) 71. California Montane Chaparral and Woodlands—(CA)	III	III
Regionally Outstanding	II	II	II	III	III
Bioregionally Outstanding	IV	IV	V	V	V
Nationally Important	IV	IV	V	V	V

I. Globally outstanding ecoregions requiring immediate protection of remaining habitat and extensive restoration.

II. Regionally outstanding ecoregions requiring immediate protection of remaining habitat and extensive restoration.

III. Globally or regionally outstanding ecoregions that present rare opportunities to conserve large blocks of intact habitat.

IV. Bioregionally and nationally important ecoregions requiring protection of remaining habitat and extensive restoration.

V. Bioregionally and nationally important ecoregions requiring protection of representative habitat blocks and proper management elsewhere for biodiversity conservation.

INTEGRATION MATRIX MHT 4.2: Xeric Shrublands/Desert

	Critical	Endangered	Vulnerable	Relatively Stable	Relatively Intact
Globally Outstanding	I	I	81. Chihuahuan Desert—(AZ, TX, NM,Mex.) I	80. Sonoran Desert—(CA,AZ, Mex.) III	III
Regionally Outstanding	82. Tamaulipan Mezquital—(TX, Mex.) II	II	II	78. Colorado Plateau Shrublands—(CO,UT,NV,NM, AZ) 79. Mojave Desert—(UT,NV,CA,AZ) III	III
Bioregionally Outstanding	74. Hawaiian Low Shrublands—(HI) IV	75. Snake/Columbia Shrub Steppe—(WA,OR,CA,NV,ID,WY) IV	73. Hawaiian High Shrublands—(HI) V	76. Great Basin Shrub Steppe—(OR,CA,NV,ID,UT) V	V
Nationally Important	IV	IV	77. Wyoming Basin Shrub Steppe—(MT,ID,WY,CO,UT) V	V	V

I. **Globally outstanding ecoregions requiring immediate protection of remaining habitat and extensive restoration.**

II. **Regionally outstanding ecoregions requiring immediate protection of remaining habitat and extensive restoration.**

III. **Globally or regionally outstanding ecoregions that present rare opportunities to conserve large blocks of intact habitat.**

IV. **Bioregionally and nationally important ecoregions requiring protection of remaining habitat and extensive restoration.**

V. **Bioregionally and nationally important ecoregions requiring protection of representative habitat blocks and proper management elsewhere for biodiversity conservation.**

INTEGRATION MATRIX MHT 6.1: Boreal Forest/Taiga

	Critical	Endangered	Vulnerable	Relatively Stable	Relatively Intact
Globally Outstanding	I	I	89. Northern Cordillera Forests—(YT,NT,BC) **I**	87. Northwest Territories Taiga—(YT,NT) 90. Muskwa/Slave Lake Forests—(NT,YT,BC,AB) 96. Eastern Canadian Shield Taiga—(QC,NF) 99. South Avalon–Burin Oceanic Barrens—(NF) **III**	91. Northern Canadian Shield Taiga—(NT,AB,SK,MB) **III**
Regionally Outstanding	II	97. Eastern Canadian Forests—(QC,NB,NS,NF) **II**	II	III	III
Bioregionally Outstanding	IV	IV	92. Mid-Continental Canadian Forests—(NT,AB,SK,MB) 93. Midwestern Canadian Shield Forests—(AB,SK,MB,ON) 94. Central Canadian Shield Forests—(ON,QC) **V**	V	83. Interior Alaska/Yukon Lowland Taiga—(AK,YT) 95. Southern Hudson Bay Taiga—(ON,MB,QC) **V**
Nationally Important	IV	IV	88. Yukon Interior Dry Forests—(YT,BC) **V**	98. Newfoundland Highland Forests—(NF) **V**	84. Alaska Peninsula Montane Taiga—(AK) 85. Cook Inlet Taiga—(AK) 86. Copper Plateau Taiga—(AK) **V**

I. Globally outstanding ecoregions requiring immediate protection of remaining habitat and extensive restoration.

II. Regionally outstanding ecoregions requiring immediate protection of remaining habitat and extensive restoration.

III. Globally or regionally outstanding ecoregions that present rare opportunities to conserve large blocks of intact habitat.

IV. Bioregionally and nationally important ecoregions requiring protection of remaining habitat and extensive restoration.

V. Bioregionally and nationally important ecoregions requiring protection of representative habitat blocks and proper management elsewhere for biodiversity conservation.

INTEGRATION MATRIX MHT 6.2:   Tundra

	Critical	Endangered	Vulnerable	Relatively Stable	Relatively Intact
Globally Outstanding	I	I	I	III 108. Arctic Foothills Tundra—(AK) 109. Arctic Coastal Tundra—(AK,YT,NT)	III 100. Aleutian Islands Tundra—(AK) 110. Low Arctic Tundra—(YT,NT,QC)
Regionally Outstanding	II	II	II	III 105. Interior Yukon/Alaska Alpine Tundra—(AK,YT) 106. Ogilvie/MacKenzie Alpine Tundra—(AK,YT,NT)	III 101. Beringia Lowland Tundra—(AK) 103. Alaska/St. Elias Range Tundra—(AK,YT,BC)
Bioregionally Outstanding	IV	IV	V	V 102. Beringia Upland Tundra—(AK)	V 104. Pacific Coastal Mountain Tundra and Icefields—(AK,YT,BC) 107. Brooks/British Range Tundra—(AK,YT,NT) 111. Middle Arctic Tundra—(NT,QC)
Nationally Important	IV	IV	V	V 114. Baffin Coastal Tundra—(NT) 115. Torngat Mountain Tundra—(QC,NF)	V 112. High Arctic Tundra—(NT) 113. Davis Highlands Tundra—(NT) 116. Permanent Ice—(NT)

I. **Globally outstanding ecoregions requiring immediate protection of remaining habitat and extensive restoration.**

II. **Regionally outstanding ecoregions requiring immediate protection of remaining habitat and extensive restoration.**

III. **Globally or regionally outstanding ecoregions that present rare opportunities to conserve large blocks of intact habitat.**

IV. **Bioregionally and nationally important ecoregions requiring protection of remaining habitat and extensive restoration.**

V. **Bioregionally and nationally important ecoregions requiring protection of representative habitat blocks and proper management elsewhere for biodiversity conservation.**

Ecoregion Descriptions

This appendix contains individual descriptions of the 116 ecoregions in this study, written in collaboration with local experts who participated in the WWF North American workshop. Each description introduces the environmental and biological setting of the ecoregion and summarizes the important factors that determine its rankings in this broad-scale analysis. Also presented in each description is a wealth of more detailed information intended to assist those interested in contributing to the conservation efforts in that ecoregion. The authors identify important remaining blocks of original habitat, suggest a suite of priority conservation activities to enhance biodiversity conservation, and list some central conservation organizations active in the area. Full contact information for the organizations is listed alphabetically in appendix G.

Interspersed among the descriptions are three essays focusing on important clusters of similar ecoregions: those of the Hawaiian Islands (pages 147–149), the Great Plains grasslands (pages 273–275), and the Mediterranean habitats of California (pages 313–314). Each of these areas contains ecoregions that display similar (and extraordinary) biodiversity characteristics, face similar threats, or both. The essays illustrate that, since groups of ecoregions often present similar conservation situations, a synthetic approach, in addition to continental and intra-ecoregional efforts, is often useful.

Focus on Hawaiian Biodiversity and Ecoregions

Biological Distinctiveness

Vast stretches of the Pacific Ocean have isolated the Hawaiian Islands for 70 million years, producing one of the world's most spectacular evolutionary phenomena. A few successful colonists have undergone explosive evolution and radiated into over 1,870 species of flowering plants, 700 fungi, 800 lichens, 180 pteridophytes, 260 mosses, 130 bird species, 1,000 land snails, and 10,230 terrestrial arthropods, of which 509 species are fruit flies alone, one third of the fruit fly species on earth (Mueller-Dombois and Ellenberg 1974; Sohmer and Gustafson 1987; Wagner, Herbst, and Sohmer 1990; Howarth and Mull 1992; Kay 1994; Scott 1995). The degree of endemism is the world's highest, with 89 percent of the plants and 99 percent of the insects found only in Hawaii (Mueller-Dombois and Ellenberg 1974; Wagner, Herbst, and Sohmer 1990; Wagner and Funk 1995; Noss and Peters 1995). Fifteen percent of plant genera are endemic, one of the highest rates in the world (Sohmer and Gon 1995).

The diversity of Hawaiian habitats, ranging from the world's wettest rain forest (Mount Waialeale, rainfall average of 1,143 cm, i.e., 450 inches) to arid deserts and alpine habitats over 4,000 meters, has helped produce extremely complex distribution patterns of communities and species and a high degree of habitat special-ization and localized ranges. Several plants and beetles (e.g., *Rhynchogonus giffardi* weevil) are known only from habitats of less than an acre, making them particularly susceptible to extinction through habitat loss (see Wagner, Herbst, and Sohmer 1990; Howarth and Mull 1992). Island isolation has selected many native plant groups to evolve low dispersal characteristics (e.g., heavy seeds), reduced defenses against herbivores, and woody structure, even in those species of taxa that are typically herbaceous on continents (Sohmer and Gustafson 1987). Many invertebrate and bird taxa evolved flightlessness, as predators are few. No terrestrial amphibians or reptiles colonized the archipelago, and only a single mammal, a bat, is native.

Conservation Status

Original native forests have been reduced by 67 percent and moist forest and shrublands have been reduced by 61 percent (Noss and Peters 1995). Lowland mesic and dry forests, in particular, have been largely eliminated. Polynesians modified extensive areas of lowland and middle elevation habitat through clearing, fire, and agriculture, and their activities are hypothesized to have precipitated the extinction of over 40 species of birds (Cuddihy and Stone 1990; James and Olson 1991). Native habitats and species are currently under severe threat from habitat loss, fragmentation, and degradation due to development, agriculture, and fires, exploitation of native plants and wildlife, illicit marijuana cultivation, grazing by domestic livestock, and introduced plants (over 870 species, including aggressive grasses, guava, banana poka, blackberry, and *Clidemia hirta,* which is rapidly dominating rain forest understories) and animals such as feral pigs, mongoose, deer, sheep, ants, and mosquitoes hosting avian malaria (Gagné 1988; Cuddihy and Stone 1990; Noss and Peters 1995; Sohmer and Gon 1996). Over 30 percent of the flora is endangered or threatened and an estimated 10 percent is already extinct (Sohmer and Gustafson 1987). Nearly 50 percent of the flora is introduced, and the degree of habitat alteration, particularly in the lowlands, means that many visitors to the islands may never see a native plant species or community. Seven bird species have become extinct since 1963, and approximately 50 percent of the total avifauna has been eliminated (Chai, Cuddihy,

and Stone 1989). Fifty percent of native terrestrial landsnails have become extinct (Howarth, Nishida, and Asquith 1995). Hawaii has 25 percent of the United States' federally listed endangered species and 72 percent of the nation's recorded extinctions (Noss and Peters 1996). Unfortunately, the federal endangered species listing process is seriously backlogged and has not been able to keep pace with the rapid and pervasive threats to Hawaii's biodiversity.

Priority Activities to Enhance Biodiversity Conservation

The fragile nature of Hawaii's native species and communities requires strong protection of remaining habitats and aggressive programs to eliminate and control alien species, introduced epizootics, and fire. Federal, state, and private areas managed for the conservation of native habitats and species cover approximately 12 percent of the islands (Hawaii Natural Heritage Program 1987; Sohmer and Gon 1996). Four organizations are primarily involved in the acquisition and management of protected areas: the National Park Service, the U.S. Fish and Wildlife Service, the state Department of Land and Natural Resources, and the Hawaii Chapter of The Nature Conservancy (Gagné and Cuddihy 1990). Most coastal and lowland communities, dry forests, and mixed mesic forests are poorly represented in the existing protected-area system. Of 180 native terrestrial communities identified by Gagné (1988), 130 are considered to be globally imperiled, but only 28 enjoy some form of protection. Eighty-eight are predicted to be lost in the near future unless fully protected.

Relationship to Other Classification Schemes

The diverse array of Hawaiian habitats has been incorporated into three ecoregions for this study: tropical moist forests, tropical dry forests, and high shrublands. Küchler (1985) identifies seven different potential vegetation types for Hawaii, whose boundaries were used to delineate our ecoregions. The Hawaiian Moist Forest ecoregion [1] represents a lumping together of four of Küchler's vegetation types: Guava Mixed Forest (*Aleurites Hibiscus Mangifera Psidium Schinus,* relatively dry in comparison to other vegetation types), Ohia Lehua Forest (*Metrosideros cibotium*), Lama Manele Forest (*Diospyros sapindus*), and Koa Forest (*Acacia*). The tropical dry forests follow the boundaries of Küchler's (1985) Sclerophyllous Forest, Shrubland, and Grassland vegetation type. Some dry forest patches and elements likely occur in Küchler's Guava Mixed Forest and Lama Manele Forest vegetation classes, but these were generally classified as tropical moist broadleaf forests. The Hawaiian high shrublands ecoregion follows Küchler's vegetation type: Grassland, Microphyllous, Shrubland, and Barren. Bailey et al. (1994) have lumped all Hawaiian terrestrial habitats into the single province and section Hawaiian Islands (unit code: M423A). For more fine-grained patterns of biodiversity, the Hawaii Natural Heritage Program (1987, 1991) has classified around 100 native Hawaiian communities, including 48 forest types. Gagné (1988) identified 180 native terrestrial communities, while Gagné and Cuddihy (1990) recognized 106 plant communities, including 20 dominated by introduced species.

Key Number: **1**
Ecoregion Name: **Hawaiian Moist Forests**
Major Habitat Type: **1.1 Tropical Moist Broadleaf Forests**
Ecoregion Size: **6,944 km²**
Biological Distinctiveness: **Globally Outstanding**
Conservation Status: **Snapshot—Vulnerable Final—Endangered**

Introduction

Tropical moist forests of Hawaii are composed of mixed mesic forests (about 750–1,250 meters in elevation), rain forests (found above mixed mesic forests up to 1,700 m), wet shrublands, and bogs in swampy areas. Moist to wet forests are commonly found on the windward lowland and montane areas of the larger islands and on mountaintops of some of the smaller islands. Koa (*Acacia* spp.) and Ohia lehua (*Metrosideros* spp.) are common dominant canopy tree species.

Biological Distinctiveness

Mesic forests are the richest for many taxa and have the highest proportion of endemic tree species. Many of the honeycreepers, an endemic group of birds that displays many specialized

adaptations to different food and plant resources, were found in mesic and wet forests. Hawaiian moist forest is the main habitat for other forest birds, including the Hawaiian hawk, Hawaiian crow, Hawaiian honeyeaters (now extinct), and Hawaiian thrushes. This ecoregion was the center for adaptive radiation in honeycreepers, many plant species, Hawaiian *Drosophila,* and other invertebrates. These forests are also noted for the diverse assemblage of shrubs and trees that were found within the Koa and Ohia forests. Rain forests, which occur in montane areas with high rainfall, are largely dominated by the tree *Metrosideros polymorpha,* with other wet forest tree species commonly present (e.g., *Cheirodendron, Ilex, Antidesma, Melicope, Syzygium, Myrsine, Psychotria, Tetraplasandra*), and tree ferns (*Cibotium* spp.) and a variety of shrubs and epiphytic plants covering the forest floor and tree surfaces. *Clermontia, Cyanea, Gunnera, Labordia, Broussaisia, Vaccinium, Phyllostegia,* and *Peperomia* are some typical plant genera. Numerous ferns and mosses as well as Hawaii's three native orchids occur in the rain forest (Sohmer and Gustafson 1987). Bogs occur on montane plateaus and depressions and consist of a variety of sedges, grasses, ferns, mosses, small trees, and shrubs that form irregular hummocks.

Conservation Status

Habitat Loss

Lowland and foothill moist forests have been largely eliminated. Some relatively large blocks of montane forest still exist on the larger islands, but even here there is much degradation from feral ungulates, introduced weed species, development, and recreational activities.

Remaining Blocks of Intact Habitat

Several important areas of relatively intact tropical moist forests currently have incomplete protection, including:

- Waianae Mountains of Oahu
- East Molokai Mountains, West Maui Mountains
- Windward East Maui
- Lanaihale of Lanai
- Kohala Mountains
- Hamakua-Hilo subregion

- Kona subregion of Hawaii (Sohmer and Gon 1996)

Types and Severity of Threats

As stated above, recreation activities, including trampling by hikers, and the rooting of feral pigs seriously threaten remaining Hawaiian bogs and wet forests. Introduced weed species and other non-native species threaten native flora.

Suite of Priority Activities to Enhance Biodiversity Conservation

- Control of feral animals and alien plants at Kauai Summit.
- Establish protected areas and control alien species in the Koulau block.
- Control alien species and continue efforts of TNC preserves in the East Molokai block.
- Continue efforts of East Maui Watershed Partnership for control and monitoring of pigs and weeds in the East Maui watershed.
- Control alien species in the West Maui block.
- Continue efforts of natural area reserves, Hakalau NWR, and Havo National Park in windward Hawaii.
- Establish protected areas and engage in feral animal and weed control in the Kau, Hawaii, block.
- Enhance natural areas reserve system management and establish protected sites in central and north Kona and Hawaii blocks.

Conservation Partners

For contact information, please see appendix G.

- The Nature Conservancy
- The Nature Conservancy of Hawaii
- Hawaiian Natural Database

Relationship to Other Classification Schemes

The Hawaiian Moist Forest encompasses Küchler's (1985) units 2 (Guava Mixed Forest), 3 (Ohia Lehua Forest), and 4 (Lama Manele Forest). Omernik (1995b) did not classify Hawaii, and Bailey (1994) clumped all of Hawaii in one unit.

Prepared by S. Gon and D. Olson

Key Number:	**2**
Ecoregion Name:	**South Florida Rocklands**
Major Habitat Type:	**1.1 Tropical Moist Forests**
Ecoregion Size:	**2,157 km²**
Biological Distinctiveness:	**Regionally Outstanding**
Conservation Status:	**Snapshot—Critical**
	Final—Critical

Introduction

Extreme southern Florida is characterized by thin, droughty soils situated on outcroppings of limestone. Pinelands and tropical hardwood hammocks cover virtually all of these outcrops and are considered as rockland ecosystems. The best examples of the pine rocklands, an ecosystem that once covered about 1,500 km², are located at the southern tip of Florida above an extrusion of limestone known as the Miami Rock Ridge and along the lower Florida Keys. Pine rocklands are fire-maintained pine forests (dominated by slash pine, *Pinus elliottii*) with a mixture of tropical and temperate understory plants (Snyder, Herndon, and Robertson 1990). The associated hardwood hammocks contain a rare intrusion of tropical hardwoods more typical of the Bahamas and Greater Antilles than the adjacent southeastern conifer forests. We chose to put this ecoregion in the tropical moist forest MHT rather than in the subtropical and tropical coniferous forest MHT, in deference to the extraordinary richness of its tropical hardwood flora.

Biological Distinctiveness

The South Florida Rocklands support the only true tropical forest on the U.S. mainland. Tropical epiphytic orchids, bromeliads, and ferns festoon the trees of the hardwood hammocks and are restricted to southern Florida. At least 137 species of trees and shrubs occur here, 18 species of vine and scandent shrubs, and seven species of palms (Snyder, Herndon, and Robertson 1990). Some experts argue that while the tree flora is similar to a Bahamian forest flora, the South Florida hardwood hammock species have been isolated long enough to be considered distinct populations from the rest of the Caribbean forest species. Thirty-seven species of herbs are endemic to the pine rocklands, among the more than 250 species of herbaceous plants recorded. Unlike the flora, which maintains a strong Caribbean influence, the fauna is largely derived from southeastern temperate habitats.

Fire is essential for maintenance of rockland pine forest and determines the relative dominance of pine forest versus hammock, the latter being less likely to burn.

Conservation Status

Habitat Loss

The South Florida Rocklands were never widespread and are now virtually gone as a result of population growth and land clearing in the Miami area and the Keys. Only 2 percent of the original habitat is thought to remain, making the pine rocklands one of the most endangered of ecoregions (Noss and Peters 1995). Besides Long Pine Key in Everglades National Park, conversion of pine rocklands communities on the Miami Rock Ridge has left perhaps 2 percent intact within the other major block of this ecoregion. Hardwood hammocks are under much greater threat of development in the Keys, where they are mostly privately owned. Other reasons for habitat loss include conversion to agriculture through the use of rock plows to break up the limestone for planting, fire suppression, and introduction of exotic species such as Brazilian pepper (*Schinus*). The increase in water table from agricultural irrigation, beetle epizootics, and the effects of Hurricane Andrew damaged part of the remaining stands.

Remaining Blocks of Intact Habitat

The largest blocks of habitat occur in:

- Long Pine Key in Everglades National Park
- Big Pine Key in the Key Deer National Wildlife Refuge

Snyder, Herndon, and Robertson (1990) have reviewed the conservation status of the pine rocklands in detail. Only three remnant fragments are larger than fifty hectares. The remaining hammocks on the Rock Ridge are not in danger of being cleared, as they are owned by Dade County.

Degree of Fragmentation

Only small remnants are left, and the increased number of small fragments surrounded by other forms of land use, particularly urban areas, restricts fire management. Much of the remaining area is surrounded by water or cities, making restoration of corridors highly unlikely.

Degree of Protection

What little remains is represented by the existing reserves listed above. These habitats and the species they support are not secure because of fire suppression, edge effects, and high mortality of fauna from road networks and heavy vehicle traffic.

Types and Severity of Threats

Most of the privately owned lands containing rocklands have been converted, and little remains to be put under conservation management. Continued fire suppression and invasion of exotic species could drastically degrade these remaining blocks over the next two decades. Exotic plants pose a serious threat to the integrity of the rocklands.

Suite of Priority Activities to Enhance Biodiversity Conservation

A number of important areas on private land have been proposed for purchase under habitat conservation programs. Those efforts should be given top priority.

Snyder, Herndon, and Robertson (1990) concludes that while it is essential to put under protection the remaining sites of pine rocklands, just as important will be increased active management, particularly for controlled burning to maintain the integrity of these habitat patches and reduce the spread of invading species.

Conservation Partners

For contact information, please see appendix G.

- Conservation and Recreation Lands Program (CARL) of the state of Florida
- The Nature Conservancy
- The Nature Conservancy of Florida
- Florida Natural Areas Inventory

Relationship to Other Classification Schemes

This ecoregion is not delineated by Bailey or Omernik. Boundaries were taken from Küchler (1964).

Prepared by E. Dinerstein, A. Weakley, R. Noss, and K. Wolfe

Key Number:	**3**
Ecoregion Name:	**Puerto Rican Moist Forests**
Major Habitat Type:	**1.1 Tropical Moist Broadleaf Forests**
Ecoregion Size:	**7,544 km²**
Biological Distinctiveness:	**Not assessed**
Conservation Status:	**Snapshot—Endangered Final—Endangered**

Introduction

Lowland tropical moist forests historically occurred in the coastal flats all around Puerto Rico except in the dry southwest. The forests grew in the northern coastal plain up to 24 m (80 ft) in height but were lower in other regions. Several important tree species of the original 200 species include *Hymenaea courbaril, Acrocomia media, Nectandra coriacea,* and *Zanthoxylum martinicense.* Several species of the lowland flora were deciduous or semideciduous to adapt to periodic dry conditions. On drier limestone soils, forests were lower but supported a similar assemblage of plant species. Many land snails are known only from the lowland limestone areas.

Montane forests are restricted to the Luquillo Mountains and the higher peaks of the Cordillera Central. Middle elevation forests were spectacular, reaching 34 m (110 ft) with trees 2.5 m (8 ft) in diameter (Little and Wadsworth 1964). About 170 tree species were found in these habitats, the wettest in Puerto Rico. The Luquillo forests are isolated from those of the Central Cordillera, which has resulted in the presence of several species of invertebrates, plants, and frogs endemic to Luquillo Mountain. Common tree species in Luquillo include *Cyathea arborea, Euterpe globosa* (the Sierra palm), *Cecropia peltata,* and several *Ocotea* species. Cloud forests cloak the highest peaks and contain a number of interesting

species and endemics, including *Weinmannia pinnata, Brunellia comocladifolia,* and *Podocarpus coriaceus.* Puerto Rican forests, particularly the moist forests of Luquillo and the eastern half of the island, are strongly influenced by periodic hurricanes. Hurricane Hugo, an intense "50-year" storm, caused significant modification of natural and secondary forest habitats and affected populations of a wide range of native species (Walker 1991).

Conservation Partners

For contact information, please see appendix G.

- Conservation Trust of Puerto Rico

Relationship to Other Classification Schemes

Puerto Rican moist forest ecoregion boundaries were derived by combining the five moist, wet, and rain forest life zones of Ewell and Whitmore (1973). Bailey et al. (1994) consider all of Puerto Rico as a single province and section, M411A, Dry-Humid Mountains of Puerto Rico.

Prepared by D. Olson

Key Number:	**4**
Ecoregion Name:	**Hawaiian Dry Forests**
Major Habitat Type:	**1.2 Tropical Dry Broadleaf Forests**
Ecoregion Size:	**6,632 km²**
Biological Distinctiveness:	**Globally Outstanding**
Conservation Status:	**Snapshot—Critical Final—Critical**

Introduction

Tropical dry forests of Hawaii typically occurred on the leeward side of the main islands and once covered the summit regions of the smaller islands. Most native lowland forests of Hawaii are either seasonal or sclerophyllous to some degree (Sohmer and Gon 1996), and more mesic transition forests occur where conditions are favorable. These transition forests include mixed mesic forests that often contain patches and elements of dry forest communities.

Biological Distinctiveness

Dry forests vary from closed to open canopied forests, can exceed 20 meters in height in montane habitats, and are dominated by the tree genera *Acacia, Chamaesyce, Metrosideros, Sapindus, Sophora, Pritchardia, Pandanus, Diospyros, Nestegis, Erythrina,* and *Santalum.* (Sohmer and Gon 1996). Dry forests harbor a number of specialist species including native hibiscus trees of the genus *Hibiscadelphus, Kokia cookei, Caesalpinia kauaiense,* and *Santalum paniculatum* and several rare endemics such as *Gouania,* now represented by only a few individuals (Sohmer and Gustafson 1987; Cuddihy and Stone 1990). Around 22 percent of native Hawaiian plant species occur within this ecoregion, with lower habitat type endemism than tropical moist forests (Sohmer and Gustafson 1987). The palila (*Psittirostra bailleui*), an endangered finchlike bird, specializes on mamane trees that occur in dry forest habitats (Noss and Peters 1995). Several shrubland, grassland, and herbaceous formations occur within this ecoregion (Sohmer and Gon 1996). Lower Hawaiian dry forest was habitat for several forest birds, such as honeycreepers, fly catchers, flightless rails, other flightless birds (now extinct), and the Hawaiian owl (*Asio flammeus sandwicensis*).

Conservation Status

Habitat Loss

Tropical dry forests are globally threatened, and Hawaiian dry forests have been reduced by 90 percent (Noss and Peters 1995). Clearing and burning of lowland dry forests began with the arrival of Polynesians, and the last remnants are being destroyed today through development, expansion of agriculture and pasture, and burning. Most larger fragments of relatively intact dry forests are in montane areas.

Remaining Blocks of Habitat

A few relictual areas survive such as Puu Waawaa on Hawaii, Puu o Kali on Maui, Auwahi on Maui, Kanepuu on Lanai, and small stands (a few thousand square meters) in the Waianae mountains of Oahu that are currently surrounded by burned slopes or alien-dominated vegetation. Several other important dry forest conservation sites identified by Sohmer and Gon

(1996) include the Na Pali Coast of Kauai, East Molokai Mountains, West Maui Mountains, Leeward East Maui, Lanaihale-Kanepuu of Lanai, and the Kona subregion of Hawaii.

Degree of Fragmentation

What little habitat that remains is highly fragmented.

Degree of Protection

Remaining transition forests and dry forests are poorly represented in the existing protected-areas system. Strong protection and active management of the remaining remnants of Hawaiian dry forests are needed. Research on effective restoration methods is needed.

Types and Severity of Threats

Introduced plant species are widespread, and dense growth and competition for resources prevents the establishment of native plant seedlings. The African fountain grass (*Pennisetum setaceum*), the shrub *Lantana camara,* and molasses grass (*Melinis minutiflora*) are among the major problem species. Introduced rats, plants, and seed-boring insects, grazing by domestic livestock and introduced deer, goats, and pigs, as well as recurring fires inhibit almost any regeneration of native species in most altered habitats.

Conservation Partners

For contact information, please see appendix G.

- The Nature Conservancy
- The Nature Conservancy of Hawaii
- Hawaii Natural Heritage Program

Relationship to Other Classification Schemes

The Hawaiian dry forest corresponds to Küchler's (1985) units 1 (Sclerophyllous Forest, Shrubland, Grassland), 5 (Koa Forest), and 6 (Koa-Mamane Parkland). Omernik (1995b) did not classify Hawaii, and Bailey (1994) clumped all of Hawaii into one unit.

Prepared by S. Gon and D. Olson

Key Number: **5**

Ecoregion Name:	**Puerto Rican Dry Forests**
Major Habitat Type:	**1.2 Tropical Dry Broadleaf Forests**
Ecoregion Size:	**1,311 km²**
Biological Distinctiveness:	**Not assessed**
Conservation Status:	**Snapshot—Critical Final—Critical**

Introduction

Tropical dry forests are located along the south-central and southwestern coast of Puerto Rico and on adjacent islands such as Mona, Vieques, Culebra, Desecheo, Caja de Muertos, and portions of each of the larger U.S. Virgin Islands (see Little and Wadsworth 1964; Dansereau 1966; Ewell and Whitmore 1973). The vegetation displays a range of adaptations to the several-month dry season and low annual rainfall (600 mm to 1,000 mm on average), including deciduous leaves; waxy coatings on leaves, trunks, and branches; and water storage structures.

Biological Distinctiveness

Guaiacum officinale, Coccoloba venosa, Ceiba pentandra, and *Capparis cynophallophora* are common trees of the coastal dry forests, while on the dry limestone forests species such as *Pisonia albida, Guaiacum sanctum,* and *Plumeria alba* are characteristic. The Puerto Rican nightjar (*Caprimulgis noctitheris*) is known only from a few areas in the dry southwest portion of the island (Raffaele 1989b).

Conservation Status

The last large block of coastal dry forest and dry limestone forest (ca. 40 km²) occurs within the Guçnica Commonwealth Forest and Biosphere Reserve. Tropical dry forests are highly threatened globally, particularly within the Greater and Lesser Antilles, making this reserve a high conservation priority. Clearing for development, fires, and introduced species continue to threaten the remaining dry forests.

Conservation Partners

For contact information, please see appendix G.

• Conservation Trust of Puerto Rico

Relationship to Other Classification Schemes

Puerto Rican dry forest boundaries were derived from the subtropical dry forest life zone of Ewell and Whitmore (1973).

Prepared by D. Olson

Key Number: **6**

Ecoregion Name:	**Willamette Valley Forests**
Major Habitat Type:	**3.1 Temperate Grasslands/ Savanna/Shrub**
Ecoregion Size:	**14,850 km²**
Biological Distinctiveness:	**Bioregionally Outstanding**
Conservation Status :	**Snapshot—Critical Final—Critical**

Introduction

Cultivation and development have destroyed nearly all of the natural habitat in the Willamette Valley, once a prairie supporting oak stands and groves of Douglas fir and other trees. Just one-tenth of 1 percent of the valley's native grasslands and oak savannas remains (Noss and Peters 1995).

Fire shaped the Willamette Valley, as it did most of the northwest grassland and savanna communities. Possibly dating back to the Pleistocene era, periodic burning by Native Americans created ideal conditions for native perennial grasses. More recent fire suppression activities—with the concurrent spread of agriculture and development—have contributed enormously to the destruction of the natural habitats of the Willamette Valley. Without regular fires, forest is gradually replacing most of the savanna in the valley.

The Willamette Valley has nearly level to gently sloping floodplains bordered by dissected high terraces and hills. The climate is generally mild throughout the year, with moderate rainfall reaching its maximum in winter. Prior to cultivation, the valley had abundant swamp and bog communities in addition to the grasslands and oak savannas.

Biological Distinctiveness

The Willamette Valley provides the only habitat for Bradshaw's lomatium (*Lomatium bradshawii*), a yellow-flowered member of the parsley family (Noss and Peters 1995). The valley is also the sole wintering area of the dusky Canada goose (*Branta canadensis occidentalis*) (Bailey 1995).

Conservation Status

Habitat Loss

Less than 1 percent of the Willamette Valley remains as intact habitat due to conversion to agriculture, urbanization, and fire suppression. Practically no prairie remains, and the savanna is converting to forest. Most of the riparian areas have been lost, though some remain intact because their propensity for flooding makes them unsuitable for agriculture or development.

Remaining Blocks of Intact Habitat

The largest blocks of remaining habitat are no greater than 35 km², and management of these areas focuses on waterfowl. The existing private reserves are all smaller than 0.15 km².

Degree of Fragmentation

Remaining patches of habitat in the Willamette Valley are tiny, with effectively no connectivity in most areas and little core habitat due to edge effects. The individual fragments and clusters that remain are highly isolated, and the intervening suburban and agricultural landscape precludes dispersal for most taxa.

Degree of Protection

Management of the protected areas in the Willamette Valley seeks to provide waterfowl for hunters, not to maintain natural habitats or enhance natural values. Area managers do not allow natural disturbance events to proceed, nor do they seek to mimic those events through interventions, and much of the use of these areas may degrade the quality of remaining natural communities.

Types and Severity of Threats

So little natural habitat remains in the Willamette Valley that there are few conversion threats to the ecoregion. Degradation of the remaining fragments continues to be a problem, and there are still moderate levels of wildlife exploitation.

Suite of Priority Activities to Enhance Biodiversity Conservation

- Focus restoration activities on riparian areas to regenerate gallery forests and to connect corridors to the Cascade Range and the Coast Range foothills.
- Use conservation efforts to promote small prairie/savanna areas, but these will likely remain fragmented from the overall linked conservation strategy cited above.

Conservation Partners

For contact information, please see appendix G.

- Association of Forest Service Employees for Environmental Ethics
- Northwest Ecosystem Alliance
- National Wildlife Federation—Western Division
- Pacific Rivers Council
- Audubon Society

Relationship to Other Classification Schemes

Bailey (1994) combines the Willamette Valley with the Puget Sound Valley to form the Pacific Lowland Mixed Forest Province. We follow Omernik (1995b)in dividing this region into two separate regions and reclassifying the Willamette as a grassland/savanna. This more accurately reflects the communities originally present in the valley and creates smaller, more manageable ecoregions. Since conservation strategies will differ for the Willamette and Puget Sound Valleys, managing the two as a single large unit would be difficult.

Prepared by R. Noss, J. Strittholt, G. Orians, and J. Adams

7

Ecoregion Name:	**Western Great Lakes Forests**
Major Habitat Type:	**2.1 Temperate Broadleaf and Mixed Forests**
Ecoregion Size:	**273,623 km²**
Biological Distinctiveness:	**Nationally Important**
Conservation Status:	**Snapshot—Relatively Stable Final—Relatively Stable**

Introduction

The Western Great Lakes Forests stretch from northern Michigan and the Upper Peninsula through northern Wisconsin, much of northern Minnesota, and into the southern portion of northwestern Ontario and extreme southeastern Manitoba. Characteristic forests in this region are more closely identified with the warmer, more humid southeastern mixed forests than with the colder, drier boreal regions to the north.

This region is characterized by moist low-boreal and subhumid transitional low-boreal ecoclimates. The mean annual temperature ranges from 1°C to 2°C; the mean summer temperature ranges from 14°C to 15.5°C; and the mean winter temperature is –13°C. Mean annual precipitation ranges from 500 mm in the west to 700–800 mm in the east. Generally, this ecoregion experiences warm summers and cold winters.

The effects of the Great Lakes upon climate, commonly called the "lake effect," are of considerable importance to the biota. Lake effects occur along the shorelines of all the Great Lakes, increasing the length of the growing season and influencing average temperatures, extreme temperatures, and the amount and timing of precipitation. Lake effects play a lesser role in the western portions of this ecoregion than in northern Michigan because winds generally blow from the Great Plains to the southwest. As a result, the climate is considerably more "continental," with extreme minimum winter temperatures and short growing seasons. Because Lake Superior is the coldest of the Great Lakes, its ameliorating effect on temperature is less than that of the other Great Lakes.

Glaciers once covered this entire ecoregion, much of which is now mantled with thick deposits of glacial drift. As a result of the glaciation, nearly all of the soil has developed on glacial deposits and lake beds or on rock surface

scraped bare by the ice. Rock outcrops and glacial moraines provide a varied landscape, although the region is generally rolling, with little change in elevation. Paleozoic marine and nearshore deposits underlie much of the region. In the Upper Peninsula of Michigan, extensive exposures of basaltic bedrock occur on Isle Royale, the Keweenaw Peninsula, and the Porcupine Mountains. The flora of Isle Royale and the Keweenaw Peninsula is rich in disjunct species from the Pacific Northwest (Cordilleran region), presumably because of the chemical characteristics of the basalt, a bedrock type that is also abundant in the West. The resistant limestones and dolomites of the Niagara Escarpment form cliffs along Lake Michigan's northern shoreline and form the spine of the Bruce Peninsula on Georgian Bay's western shore.

Within the Canadian portion, this ecoregion takes in a portion of the Severn Upland and is underlain by massive, crystalline, acidic, Archean bedrock, forming hummocky, broadly sloping uplands and lowlands. Bedrock outcroppings are common throughout the region. Small to medium-sized lakes are numerous in the south-central area, and many are linked by bedrock-controlled networks of streams. Wetlands are also widespread and are characterized by bowl bogs that are treed and often surrounded by peat margin swamps (ESWG 1995).

Characteristic vegetation is a mixed forest that includes a succession from quaking aspen (*Populus tremuloides*), paper birch (*Betula papyrifera*), and jack pine (*Pinus banksiana*) to white spruce (*Picea glauca*), black spruce (*Picea mariana*), and balsam fir (*Abies balsamea*). Forest species assemblages are highly influenced by drainage characteristics and topography. In low, wet areas, black spruce is likely to be the dominant species. Northward into Canada, black spruce takes on a more dominant role and often grows in extensive stands covering thousands of acres on the uplands (Tester 1995).

Characteristic mixed forests are distinct from the predominantly deciduous boreal forests of the west and cooler boreal forests to the north. The region includes northern coniferous forest, northern hardwood forest, boreal hardwood-conifer forest, swamp forest, and peatland. Minnesota's peatlands are still relatively intact despite efforts in the early 1900s to ditch and drain large tracts in the north (Tester 1995).

Ecologists subdivide the pine forest of this ecoregion into Great Lakes pine forests of white pine and red pine (*Pinus resinosa*) with paper birch and aspen, and jack pine forests of jack pine, red pine, oak (*Quercus* spp.), and hazel (*Corylus cornuta*). Sandy, acid, nutrient-poor soils and high-frequency fires characterize all pine forest habitat in the northern Great Lakes. Jack pine historically occupied the most extreme, drought-prone sites, especially extensive out-wash plains. Northern pin oak (*Quercus ellipsoidalis*) was a common associate. Red pine occupied slightly more mesic sites, either on out-wash deposits, sandy end moraine, or sandy beach ridges and small transverse dunes. White pine was once common on sandy lake plains, occupying sites from poorly drained embayments to excessively drained sand dunes. Intensive logging of white pine and red pine between 1850 and 1900 and widespread fires resulted in the replacement of many pine forests by red maple and light-seeded, wind-dispersed aspens and paper birch (Albert 1994).

Wetlands are widespread in this ecoregion, with conifers dominating most of the wetlands, although hardwoods still grow along stream margins and well-aerated wetlands in stream headwater areas. The most extensive conifer swamps occur in the glacial-lake basins of eastern upper Michigan.

Common species of the northern hardwoods include sugar maple (*Acer saccharum*), red maple (*Acer rubrum*), American beech (*Fagus grandifolia*), hop hornbeam (*Ostrya virginiana*), basswood (*Tilia americana*), yellow birch (*Betula alleghaniensis*), and eastern hemlock (*Tsuga canadensis*). Some authorities separate this community into two types distinguished by the presence or absence of beech (*Fagus grandifolia*). The western half of Michigan's Upper Peninsula does not contain beech, probably because of extremely low winter temperatures. Fire, while rare, occurs often enough to maintain the presence of white pine and red oak. Fire is a more important factor in stands of spruce and balsam fir.

Biological Distinctiveness

Characteristic wildlife in the Western Great Lakes Forests include moose (*Alces alces*), black bear (*Ursus americanus*), wolf (*Canis lupus*), lynx (*Lynx canadensis*), snowshoe hare (*Lepus americanus*), white-tailed deer (*Odocoileus virginianus*),

and woodchuck (*Marmota monax*). Bird species include ruffed grouse (*Bonasa umbellus*), hooded merganser (*Lophodytes cucullatus*), pileated woodpecker (*Dryocopus pileatus*), bald eagle (*Haleaeetus leucocephalus*), turkey vulture (*Cathartes aura*), herring gull (*Larus argentatus*), and waterfowl. American black duck (*Anas rubripes*) and wood duck (*Aix sponsa*) occur in the eastern part of the ecoregion (ESWG 1995).

Isle Royale National Park supports an intact predator-prey system, including a breeding wolf population and large numbers of white-tailed deer, moose, snowshoe hare, and beaver (*Castor canadensis*). The park also contains perhaps as much as 350 km² (86,000 acres) of old-growth forest (Davis 1996). Wolves appear to be expanding their range to the southwest in Minnesota, and in 1994 the state population was estimated at two thousand (Tester 1995). This ecoregion contains several areas suitable for supporting wolf packs, and there is the potential for a rebound in the wolf population. Other endangered mammals include marten (*Martes americana*), cougar (*Felis concolor*), and lynx. Threatened and endangered birds include Kirtland's warbler (*Dendroica kitlandii*), merlin (*Falco columbarius*), and yellow rail (*Coturnicops noveboracensis*).

The southern shorelines of Lake Superior and Lake Huron are crucial for migrating birds, serving as migratory stopover points and major breeding areas. Terns and plovers nest on islands. The lakeshores are also important areas of aquatic insect diversity, while estuaries serve as important fish hatcheries for Great Lakes fisheries.

The acidic, Archean bedrock outcrops in many areas attract lightning strikes. As a result, fire is an important disturbance regime in the region, particularly on conifer-dominated dry sites.

Conservation Status

Habitat Loss

Only 20 percent of the western Great Lakes forests remains as intact habitat. Minnesota has some 2,630 km² (650,000 acres) of old-growth forest, more than any other state in the east (Davis 1996). Much of this habitat is concentrated in the Boundary Waters–Quetico area straddling the Minnesota-Ontario boundary. While in some areas, particularly in the northern reaches of the ecoregion, the coniferous forests still exist much as they did hundreds of years ago, much of the landscape is in transition from its presettlement status. Most of the original mature white and red pine forests have been logged and have been replaced by younger stands of birch and aspen, with only scattered pine. This has resulted in major forest conversion across much of the ecoregion, causing far-reaching changes to the biota found within the ecoregion.

In addition, extensive areas throughout the ecoregion have been converted to agricultural production or are increasingly being developed for new housing.

Remaining Blocks of Intact Habitat

Several relatively large blocks of more or less intact habitat and a number of smaller patches remain. Important blocks include:

- Boundary Waters Canoe Area Wilderness—northeastern Minnesota, 4,300 km² (American area contiguous to Quetico)
- Quetico Provincial Park—northwestern Ontario (Canadian area contiguous to Boundary Waters)
- Porcupine Mountains National Forest and Shequamagon National Forest—northern Wisconsin
- Nicollet National Forest—northeastern Wisconsin
- Superior National Forest—northeastern Minnesota
- Chippewa National Forest—northern Minnesota
- Ottowa National Forest—northwestern Michigan
- Hiawatha National Forest—northwestern Michigan
- Voyageurs National Park—northern Minnesota
- Isle Royale National Park—northern Wisconsin

Degree of Fragmentation

A forest matrix still characterizes this ecoregion, although parts of the ecoregion are fragmented by both public and logging roads. Within the United States, more than half of all fragments have some interaction with other intact habitat blocks, resulting in relatively low fragmentation pressure.

Degree of Protection

The Western Great Lakes forests include several large protected areas. The most important areas include:

- Boundary Waters Canoe Area Wilderness—northern Minnesota
- Quetico Provincial Park—northwestern Ontario, 4,758 km²
- Voyageurs National Park—northern Minnesota
- Isle Royale National Park—northern Michigan
- Apostle Islands National Park—northern Michigan
- Porcupine Mountains State Park—northern Michigan
- Turtle River Waterway Provincial Park—northwestern Ontario, 400 km²
- La Verendrye Provincial Waterway Park—northwestern Ontario, 183 km²
- Whiteshell Provincial Park—Manitoba (backcountry zones), 913 km²
- Nopiming Provincial Park—Manitoba (backcountry zones), 316 km²
- Lake of the Woods Provincial Park—northwestern Ontario, 129 km²
- Lola Lake Provincial Nature Reserve—northwestern Ontario, 65 km²
- Sandbar Lake Provincial Park—northwestern Ontario, 50 km²
- Winnange Lake Provincial Park—northwestern Ontario, 47 km²
- Portions of the national forests (Wilderness Areas, RNAs, etc.) also provide significant protection to the diversity of this ecoregion.

Types and Severity of Threats

The most significant conversion threat in this ecoregion is the conversion of pine to aspen forest. Logging is a significant cause of this conversion throughout the ecoregion. Paper mills and oriented strand board mills are now harvesting second-growth forests. Much of the forest outside core protected areas has been converted to young, successional stands of birch and aspen. Although aspen-dominated forests provide habitat for wildlife, they have crowded out the native white pine forests. Agriculture, scattered throughout the ecoregion, and development, especially for second homes along the lakeshores, also pose conversion threats.

Large deer populations threaten to degrade some southern areas of the ecoregion, as excessive browsing harms hemlock and white spruce. While changing water levels in the lakes once played an important ecological role, inlet and outlet controls have now stabilized the water levels, leading to significant changes in the lakeshore ecology. The cumulative effects of industrial toxins in the lakes and along the shorelines pose another threat to the health of the ecoregion. Overall, many populations of plants species are experiencing high mortality and low recruitment due to habitat degradation. Woodland caribou (*Rangifer caribou*) have essentially been extirpated from this ecoregion, with only infrequent sightings in the extreme northern areas bordering ecoregions 93 and 94.

Suite of Priority Activities to Enhance Biodiversity Conservation

- Maintain existing large, intact habitat blocks and corridors.
- End logging practices that result in forest conversion from pine-mixed hardwoods to aspen.
- Restore conifer forests where conversion has already occurred, including reestablishing the natural fire regime in the Boundary Waters–Quetico area.
- Protect the wetlands of Kakagen slews and the Door County Peninsula.
- Protect the undisturbed vegetation of the sea caves on Apostle Island.
- Conserve the rare species of the limestone grasslands (alvars) of northern Lake Huron.
- Protect the forests of Menomenee County, Wisconsin, which have never been cut and may be the most mature forests in the U.S. portion of the ecoregion.
- Upgrade the protection standards of provincial forests in Manitoba.
- Protect the wild and scenic Wolf River.
- Protect Aulneau Peninsula in Lake of the Woods, northwestern Ontario.

Conservation Partners

For contact information, please see appendix G.

- Endangered Spaces Campaign, Manitoba
- Environment North
- Federation of Ontario Naturalists
- Friends of the Boundary Waters Wilderness
- Manitoba Naturalists Society
- Michigan Environmental Council
- Michigan Natural Areas Council
- Michigan Nature Association
- Quetico Foundation
- Resource Conservation Manitoba
- The Nature Conservancy, Manitoba
- The Wildlands League
- World Wildlife Fund Canada

Relationship to Other Classification Schemes

The Western Great Lakes Forests include both hardwoods and conifers, unlike the deciduous forests to the south and the boreal forests to the north. This ecoregion largely corresponds to Omernik's (1995b) Northern Lakes and Forests ecoregion, though Omernik creates a separate ecoregion for the Northern Minnesota wetlands. The Western Great Lakes Forests of southeastern Manitoba and northwestern Ontario fall within the Boreal Shield Ecozone, incorporating the Lake of the Woods, Rainy River, and Thunder Bay–Quetico areas (TEC ecoregions 91–93) (ESWG 1995). In Rowe's (1972) classification, this ecoregion includes the Lower English River area in the boreal forest zone (B.14), and the Quetico and Rainy River areas within the Great Lakes–St. Lawrence forest region (GL.11, GL.12).

Prepared by S. Chaplin, A. Perera, S. Robinson, J. Adams, T. Gray, G. Whelan-Enns, K. Kavanagh, M. Sims, and G. Mann

Key Number:	**8**
Ecoregion Name:	**Eastern Forest/Boreal Transition**
Major Habitat Type:	**Temperate Broadleaf and Mixed Forests**
Ecoregion Size:	**346,724 km²**
Biological Distinctiveness:	**Nationally Important**
Conservation Status:	**Snapshot—Vulnerable Final—Vulnerable**

Introduction

This ecoregion includes most of the southern Canadian Shield in Ontario and Quebec. The shield, in fact, principally defines the southern boundaries of this ecoregion. It lies north and west of the St. Lawrence Lowlands except for a disjunct section comprised of the Adirondack Mountains in upper New York State.

The region has a humid mid-boreal ecoclimate in the northwest and a humid high cool temperate ecoclimate in the Algonquin area, and the mean annual temperature ranges from 1.5°C to 3.5°C, increasing toward the south. The mean summer temperature of this region is 15°C, and the mean winter temperature ranges from –8.5°C to –11°C. Temperatures are slightly cooler in the southern Laurentians. Mean annual precipitation ranges from 800 mm to 1,000 mm; however, along the shores of Lake Superior and Georgian Bay, and between Quebec City and the Saguenay River, annual precipitation is in excess of 1,000 mm. In general, this ecoregion experiences warm summers and cold, snowy winters (ESWG 1995).

The Ontario part of this ecoregion is underlain by massive, crystalline, acidic, Archean bedrock forming undulating, broadly sloping uplands and lowlands with outcroppings. The Cobalt Plain in the northeast section of the Ontario portion is composed of flat-lying clastic sediments with ridges and hills formed by gabbro sills or granitic rock inliers. The southern Laurentians in Quebec are composed mainly of Precambrian granites and gneisses and are incised by a number of southward-draining rivers through highlands (ESWG 1995).

The characteristic mixed forests of this ecoregion are distinct from the predominantly deciduous forests to the south and the cooler boreal forests to the north. In the northern reaches and in the Lake Temiskaming area, the forests transition into a more predominantly boreal forest characteristic of ecoregions to the north,

although on warmer, better-drained sites, deciduous species dominate.

Biological Distinctiveness

Mixed-wood forests characterize this region and include white spruce (*Picea glauca*), balsam fir (*Abies balsamea*), quaking aspen (*Populus tremuloides*), paper birch (*Betula papyrifera*), and yellow birch (*B. alleghaniensis*). Red (*Pinus resinosa*), white (*P. strobus*), and jack pine (*P. banksiana*) occur on drier sites in the northwest. To the south, in the Algonquin area, the mixedwood forest is characterized by stands of sugar maple (*Acer saccharum*), yellow birch, eastern hemlock (*Tsuga canadensis*), and eastern white pine, with beech (*Fagus grandifolia*) appearing on warmer sites. Poorly drained areas support tamarack (*Larix laricina*) and eastern white cedar (*Thuja occidentalis*), with black spruce (*Picea mariana*) in the north, and red maple (*Acer rubrum*) and black ash (*Fraxinus nigra*) in the Algonquin area (ESWG 1995). Wetlands occur throughout the ecoregion, usually in association with river systems and along parts of the Georgian Bay shoreline.

Moose (*Alces alces*), lynx (*Lynx canadensis*), black bear (*Ursus americanus*), snowshoe hare (*Lepus americanus*), wolf (*Canis lupus*), coyote (*Canis latrans*), white-tailed deer (*Odocoileus virginianus*), American black duck (*Anas rubripes*), wood duck (*Aix sponsa*), hooded merganser (*Lophodytes cucullatus*), and pileated woodpecker (*Dryocopus pileatus*) occur throughout, with chipmunk (*Tamias striatus*), mourning dove (*Zenaida macroura*), cardinal (*Cardinalis cardinalis*), and wood thrush (*Hylocichla mustelina*) in the Lake Nipissing–Algonquin area. The southern Laurentians provide habitat for fewer animal species (ESWG 1995).

Forest species assemblages are highly influenced by drainage characteristics and topography. Fire is an important disturbance regime in the ecoregion on spatial scales of up to 1,000 km², particularly in the northern parts of the ecoregion. Elsewhere, smaller fires are more common.

This is the southern limit of timber wolves in eastern North America, and there are emerging plans to reintroduce wolves into such areas as the Adirondacks. Recently, eastern cougar (*Felis concolor*) sightings have been increasing in this ecoregion. The most widespread old-growth red and white pine stands remaining in the world and one of the largest remaining areas of old-growth forest in the northeastern United States, Five Ponds Wilderness, is found here. A large percentage of the Great Lakes watershed headwaters remain as relatively intact (rare on a continental scale).

Conservation Status

Habitat Loss

It is estimated that only 10 percent of the ecoregion remains as intact habitat. Much of the area has been highly fragmented by forestry activities, settlements, summer homes and cottages, ski facilities, and agriculture.

Remaining Blocks of Intact Habitat

- Adirondack Park (although it is roaded and contains urban inclusions, such as Saranac Lake and Lake Placid)—northeastern New York
- Parc du Mont Tremblant—southwestern Quebec
- Parc Jacques Cartier—southern Quebec
- La Maurice National Park—southwestern Quebec
- Algoma Highlands
- Algonquin Provincial Park (highly roaded)—southern Ontario
- Bark Lake
- Mississagi Uplands
- Quirke-Whiskey Lakes
- South Ranger Lake

Degree of Fragmentation

The ecoregion is highly fragmented by public roads, logging roads, large-scale logging, and settlement patterns.

Degree of Protection

- Adirondack State Park
- Algonquin Provincial Park—protected portion (1,583 km^2). Logging permitted in an additional 6,069 km^2 of the park. Total park area: 7,652 km^2
- Lake Superior Provincial Park—southern Ontario, 15 km^2 (much of it previously logged)
- Parc du Mont Tremblant (provincial park)—1,490 km^2

- Lady Evelyn–Smoothwater Provincial Wilderness Park—central Ontario, 724 km^2
- Parc Jacques Cartier (provincial park)—670 km^2
- La Maurice National Park—543 km^2
- French River Waterway Provincial Park—central Ontario, 511 km^2
- Killarney Provincial Park—southern Ontario, 485 km^2
- Gatineau Park, National Capital Commission—southwestern Quebec, 344 km^2
- Parc des Grands Jardin (provincial park)—southern Quebec, 310 km^2
- Mississagi Provincial Waterway Park—central Ontario, 198 km^2
- Obabika River Provincial Waterway Park—central Ontario, 170 km^2
- Blackstone Harbour Provincial Natural Environment Park—southern Ontario, 119 km^2
- Wakami Lake Provincial Park—central Ontario, 88 km^2
- Chapleau-Nemegosenda Provincial Waterway Park—central Ontario, 81 km^2
- La Cloche Provincial Park—central Ontario, 74 km^2
- Bon Echo Provincial Park—southeastern Ontario, 66 km^2

Types and Severity of Threats

The timber industry continues to be very active in the ecoregion, particularly in the Canadian portion. There is increased mining potential throughout, and tourism is beginning to create significant impacts in parts of the ecoregion.

Suite of Priority Activities to Enhance Biodiversity Conservation

- Continue acquisition of private lands within Adirondack State Park.
- Protect Hautes Gorges in Quebec.
- Upgrade protection standards for La Verendrye, Rolland Germain, Papineau Labelle, and Laurentides Wildlife Reserves in Quebec and Algonquin Provincial Park in Ontario.
- Establish a major protected area in the Algoma Highlands of Ontario.
- Protect additional old-growth forest stands in the Temagami area in Ontario.

- Protect the Mississagi Uplands, Bark Lake, South Ranger Lake, and Rawhide Lake old-growth forest sites in Ontario.
- Upgrade protection standards for Chapleau, Nipissing, and Peterborough Crown Game Preserves in Ontario.
- Develop a protection plan for the Little Claybelt, shared by Ontario and Quebec.

Conservation Partners

For contact information, please see appendix G.

- Adirondack Council
- Ancient Forest Exploration and Research
- Association Touristique Régionale de Charlevoix (ATR)
- Canadian Parks and Wilderness Society, Quebec
- Earthroots
- Federation of Ontario Naturalists
- Muskoka Field Naturalists
- The Nature Conservancy, Eastern Regional Office
- The Nature Conservancy of Canada
- Nipissing Naturalists Club
- Northwatch
- Orillia Naturalists' Club
- Regroupement National des Conseils Régionaux de l'Environnement du Québec (RNCREQ)
- Residents Committee to Protect the Adirondacks
- Regroupement écologiste Val d'Or et environs (REVE)
- Union Québecoise pour la Conservation de la Nature (UQCN)
- Wild Earth, Vermont
- The Wildlands League (Canada)
- World Wildlife Fund Canada, Quebec Region

Relationship to Other Classification Schemes

This mixed-wood forest region is composed of the Lake Temiskaming Lowland, the Southern Laurentians, and the Algonquin–Lake Nipissing area (TEC ecoregions 97, 98, and 99) (ESWG 1995). Because this ecoregion is a transition zone, it is characterized by a variety of forest types from Rowe (1972), including the Laurentide-Onatchiway (B.1a), Chibougamau-Natashquan (B.1b), Gouin (B.3) and Missinaibi-Cabonga (B.7) within the boreal forest region. In the Great Lakes–St. Lawrence forest region, sections include the Laurentian, Algonquin-Pontiac, Middle Ottawa, Georgian Bay, Sudbury-North Bay, Saguenay, Haileybury Clay, Temagami, and Algoma (GL.4a, GL.4b, GL.4c, GL.4d, GL.4e, GL.7–GL.10).

Prepared by K. Kavanagh, L. Gratton, M. Davis, S. Buttrick, N. Zinger, T. Gray, M. Sims, and G. Mann

Key Number:	**9**
Ecoregion Name:	**Upper Midwest Forest/Savanna Transition Zone**
Major Habitat Type:	**2.1 Temperate Broadleaf and Mixed Forest MHT**
Ecoregion Size:	**166,122 km²**
Biological Distinctiveness:	**Nationally Important**
Conservation Status:	**Snapshot—Critical Final—Critical**

Introduction

One of the three ecotonal units separating the vast Great Plains grasslands from the forests of the eastern United States is the Upper Midwest Forest/Savanna Transition Zone (hereafter UMTZ). This is the smallest of the three transition units along the eastern and northern edge of the Great Plains. Much of the UMTZ is contained in central and southern Wisconsin and southwestern Minnesota. The UMTZ is distinguished from the adjacent Northern and Central Tall Grasslands ([59] and [60]) by the predominance of trees in a mosaic of forests, savannas, and woodlands, and from the Central Forest/Grassland Transition Zone to the south by differences in dominance of major tree species. The UMTZ is essentially an oak, maple, basswood woodland, forest, and savanna ecosystem (Küchler 1985). The boundaries of this ecoregion were heavily influenced by fire and drought. During periods of extensive burning this ecotone may have covered a much larger area to the east than currently estimated. Today, the oak savanna component of the UMTZ is one of the world's most endangered ecosystems. Much of the original ecoregion has been converted (see below).

Biological Distinctiveness

The UMTZ contains the last concentration of tall-grass prairie savanna (black soils) in the United States. It contains important sites used by geese during migrations, and it is home to populations of river otters extending out into the prairies. Riparian areas also support an interesting floodplain forest community.

Despite severe threats to its integrity, much of the fauna of the original savanna has persisted, largely because many species are generalists that can survive in altered habitats. The edges of forests often mimic savannas in terms of tree density, and vertebrates persist in these areas. The original oak savanna vegetation, an important community within this ecoregion, has not adapted as well. At least eleven species of herbaceous plants are now confined to much smaller areas and are considered to be threatened. Additional pressures may come from an overabundance of deer in this ecoregion. Several species of invertebrates are also considered threatened (Henderson and Epstein 1995).

Conservation Status

Habitat Loss

Today, less than 5 percent of the ecoregion is considered intact. The savannas found on the deep rich soils of the Midwest were completely fragmented and nearly entirely destroyed throughout the ecoregion by the early to mid nineteenth century (Henderson and Epstein 1995). Most of the ecoregion was affected by clearing, plowing, or overgrazing, and what remained in seminatural habitat suffered invasion by woody shrubs as a consequence of fire suppression. Recent estimates of remaining savanna fragments testify to the critical status of this ecoregion: only 2 km² of intact examples of oak savanna vegetation remain in Wisconsin, or less than 0.01 percent of the original 29,000 km².

Remaining Blocks of Intact Habitat

A 1985 survey of oak savanna across the ecoregion listed a mere 133 sites totaling 26 km², or 0.02 percent of the estimated presettlement

extent (Nuzzo 1986). The remaining examples are largely on either the wettest or driest sites. The intermediate sites on the most productive soils have been converted.

There are no national forests in this ecoregion. Remaining blocks of habitat include:

- Baraboo Hills, Devils State Park—southern Wisconsin
- Savanna River Depot (sandy grasslands and floodplain forests)—northwestern Illinois
- Upper Mississippi Wildlife Refuge (floodplain forests)—eastern Wisconsin, western Minnesota
- Mississippi Bluffs forests (relatively intact)—Minnesota
- Richard J. Dorer State Forest—Minnesota
- Whitewater Wildlife Management Area—Minnesota
- Neceda National Wildlife Refuge (protected and managed wetlands)—Wisconsin
- Horicon Marsh (stopover site for migratory geese)—southeastern Wisconsin
- Kettle Moraine State Forest—southeastern Wisconsin

Degree of Fragmentation

As stated above, remaining habitat is highly fragmented. Pioneering field research conducted in isolated forest remnants of this ecoregion contributed greatly to our understanding of the impact of habitat fragmentation on breeding success of songbirds. It was here that the severe impact of brown-headed cowbirds (*Molothrus ater*) on songbirds was well documented. The absence of large predators has given rise to high populations of smaller predators that prey on songbirds and their eggs and nestlings.

Degree of Protection

The state Departments of Natural Resources maintain several hundred very small protected sites, many less than 5 km². These pockets of vegetation are all isolated.

Types and Severity of Threats

Four main threats to the survival of this ecosystem have been identified: (1) loss of recovery opportunities as second-home and residential development spread into more "natural" areas, (2) lack of general awareness of the globally threatened status of oak-savanna vegetation,

(3) fire suppression and misunderstanding about the importance of burning in maintaining the integrity of the ecosystem, and (4) invasion by exotic plants such as honeysuckle (*Lonicera* sp.) and reed canary grass (*Phalaris arundinacea*) (Henderson and Epstein 1995). Grazing of wooded sites by cattle and deer continue to be a problem.

In some areas the forests containing maturing oaks and walnuts are big enough for a second cut. The need to protect these areas from logging is paramount.

If carefully managed, portions of the UMTZ have good potential for recovery. It is estimated that within a few decades thousands of hectares of overgrown oak savannas on public and private lands could be recovered. Restoration techniques involve thinning, removing brush, and burning.

Suite of Priority Activities to Enhance Biodiversity Conservation

- Identify and protect clusters of intact habitat (i.e., source pools) from further fragmentation.
- Implement savanna restoration, which has a high probability of success in many areas.
- Improve private timber management to prevent conversion and further degradation.

Conservation Partners

For contact information, please see appendix G.

- Illinois Department of Natural Resources
- Minnesota Department of Natural Resources
- The Nature Conservancy
- The Nature Conservancy, Iowa Field Office
- Wisconsin Department of Natural Resources

Relationship to Other Classification Schemes

The UMTZ is a combination of Omernik (1995b) ecoregions 51 (North Central Hardwood Forest), 52 (Driftless area), and 53 (Southeastern Wisconsin Till Plains). It corresponds closely to Küchler's (1985) vegetation type 72 (Oak Savanna), 90 (Maple-Basswood Forest, in part), and 97 (Northern Hardwoods, in part). It also corresponds generally with Bailey (1994), sections 222M in part (Minnesota and NE Iowa Morainal, Oak Savanna Section), 222L (North Central U.S. Driftless and Escarpment Section), and 222K (Southwestern Great Lakes Morainal Section).

Prepared by S. Chaplin, P. Sims, S. Robinson, and E. Dinerstein

Key Number:	**10**
Ecoregion Name:	**Southern Great Lakes Forests**
Major Habitat Type:	**2.1 Temperate Broadleaf and Mixed Forests**
Ecoregion Size:	**244,469 km²**
Biological Distinctiveness:	**Nationally Important**
Conservation Status:	**Snapshot—Critical Final—Critical**

Introduction

This ecoregion covers much of the industrial heartland of North America, including southern Michigan, much of Ohio and Indiana, extreme southwestern Ontario, including the lowlands of the south of Lake Ontario in Ontario and western New York State. The area is so heavily populated and developed that essentially no large blocks of natural habitat remain.

Most of this ecoregion is rolling, but some parts are nearly flat. The northern but not the southern parts were glaciated. The climate shows strong annual temperature cycles: summers are hot with frequent tornadoes, and winters are cold. In Ontario, the Lake Erie lowlands are marked by humid, warm to hot summers and mild, snowy winters. The mean annual temperature is 8°C; the mean summer temperature is 18°C, and in winter it is 2.5°C. Mean annual precipitation ranges from 750 to 900 mm and is distributed evenly throughout the year (ESWG 1995).

Much of the ecoregion is underlain by carbonate-rich, Paleozoic bedrock. Most of the region lies southwest of the Niagara escarpment, where the land surface slopes gradually toward the southwest through rolling topography. Bedrock outcrops are limited to only small areas (ESWG 1995).

The Southern Great Lakes Forests are dominated by deciduous forests that are distinct from the mixed forests in other parts of the St. Lawrence lowlands to the north and contain lower species diversity than ecoregions to the east and south. Historically, the ecoregion consisted of sugar maple (*Acer saccharum*) and beech (*Fagus grandifolia*), which together comprised 80 percent of the canopy in the region. Beech reaches its western limit in the transition zones between forest and grasslands and savannas (ecoregions [9] and [65]), and basswood (*Tilia americana*) becomes a more important species. Other forest types, dominated by oaks (*Quercus* spp.) and hickories (*Carya* spp.), occur on drier sites, whereas mixed swamp forest, especially elms (*Ulmus* spp.), ashes (*Fraxinus* spp.), and red maple (*Acer rubrum*), occupies the wettest soils. A tightly closed canopy and a thick layer of humus and leaf litter characterize these forests, encouraging the growth of spring perennial herbs and discouraging bryophytes (Greller 1988). Forest species assemblages are highly influenced by surficial materials (sands or clays). Fire and water table depth are important ecological processes in maintaining relict prairie grasslands and oak savannas. Forested areas on moist and wet sites are characteristically renewed largely through tree falls from wind and/or ice storms.

Biological Distinctiveness

Some characteristic mammals include white-tailed deer (*Odocoileus virginianus*), red fox (*Vulpes fulva*), eastern chipmunk (*Tamias striatus*), grey squirrel (*Sciurus carolinensis*), and red squirrel (*Tamiasciurus hudsonicus*). Breeding birds include northern cardinal (*Cardinalis cardinalis*), wood thrush (*Hylocichla mustelina*), screech owl (*Otus asio*), mourning dove (*Zenaida macroura*), green heron (*Butorides virescens*), pileated woodpecker (*Dryocopus pileatus*) and red-bellied woodpecker (*Melanerpes carolinus*), and wild turkey (*Meleagris gallopavo*) (reintroduced through parts of the ecoregion) (ESWG 1995).

Although little forest habitat remains, the Southern Great Lakes ecoregion contains some rare ecological phenomena. The Great Lakes include extensive interior wetlands and freshwater bodies with dune systems, such as Long Point, Presqu'ile, and Lake Erie's Rondeau and Point Pelee, which have large sand pits supporting unique plant communities. These are major staging areas for migrating birds. They are also of international significance as peninsular sand and dune complexes in a freshwater system. The sand dune complex on the shores of Lake Michigan is another important site. The southern extent of the Niagara escarpment has numerous ancient trees of varied species.

The ecoregion is the most northward distribution of many "Carolinian" species in North America. As well, it is the eastern extension of the Niagara escarpment and Niagara gorge. There is an archipelago of islands in western Lake Erie, including several endemic subspecies and varieties. This is the most eastern extension of midwestern prairies and savannas in North America.

Conservation Status

Habitat Loss

Agriculture and industrial and urban development are the predominant land uses in much of this ecoregion. Thus, the ecoregion is one of the most heavily impacted by human activities on the continent. Habitat loss is nearly complete in this ecoregion. Nearly 100 percent of the region was ranked as heavily altered. Wetland losses have been particularly severe; Ohio, for example, has lost 90 percent of its wetlands, and 80 percent of the southern tamarack swamp in Michigan has been destroyed (Noss and Peter 1995). Major urbans centers include: Toronto, Hamilton, Buffalo, Rochester, Syracuse, Detroit-Windsor, Erie, Cleveland, Cincinnati, Columbus, and Indianapolis.

Remaining Blocks of Intact Habitat

No habitat blocks of significant size remain, but the Long Point area of Ontario has approximately 40 km^2 of seminatural lands in various ownerships, much of it public.

Degree of Fragmentation

Remaining patches are tiny, with effectively no connectivity in most areas and little core habitat due to edge effects. The individual fragments and clusters that remain are highly isolated, and the intervening landscape precludes dispersal for most taxa.

Degree of Protection

This region has no protected areas larger than 500 km².

Types and Severity of Threats

The remaining tiny fragments of natural habitat in the Southern Great Lakes face intense conversion pressure from development and agricultural expansion. Agricultural conversion for corn, soybeans, tobacco, grains, canola, and tender fruit has occurred. Urban sprawl threatens this region. Agricultural land and woodlots are being severed to accommodate country homes. Habitat not being converted is being degraded by pollution and exotic species. Wildlife exploitation continues, and the elimination of most target species is imminent or complete.

Suite of Priority Activities to Enhance Biodiversity Conservation

Given the parlous state of natural habitats in this ecoregion, conservation requires:

• urgent protection of the remaining fragments
• planning for restoration of those degraded areas that have not been irreversibly altered

Conservation Partners

For contact information, please see appendix G.

• Carolinian Canada Program
• Essex County Field Naturalists
• Federation of Ontario Naturalists
• Friends of Point Pelee
• Hamilton Naturalists Club
• McIlwraith Field Naturalists of London
• Norfolk Field Naturalists
• Peninsula Field Naturalists
• Sydenham Field Naturalists
• The Wildlands League
• World Wildlife Fund Canada

Relationship to Other Classification Schemes

The Southern Great Lakes Forests ecoregion includes parts of Bailey's (1994) Laurentian Mixed Forest Province and Eastern Broadleaf Forest (Continental) Province. Omernik (1995b)

finds four distinct ecoregions here: the Eastern Corn Belt Plains, S. Michigan–N. Indiana Till Plains, Huron-Erie Plain, and Erie-Ontario Lake Plain. We believe that the forest types found in the Southern Great Lakes are sufficiently closely related to warrant a single ecoregion.

In Canada, the southern Great Lakes forests cover the Lake Erie lowland in southern Ontario and northern United States (ESWG 1995). In southern Ontario, this ecoregion overlies the Niagara deciduous forest region (Rowe 1972).

Prepared by K. Kavanagh, S. Buttrick, M. Davis, J. Adams, M. Sims, and G. Mann

Key Number:	**11**
Ecoregion Name:	**Eastern Great Lakes Lowland Forests**
Major Habitat Type:	**2.1 Temperate Broadleaf and Mixed Forests**
Ecoregion Size:	**115,538 km²**
Biological Distinctiveness:	**Nationally Important**
Conservation Status:	**Snapshot—Critical Final—Critical**

Introduction

This ecoregion includes lowland areas of New York and Vermont surrounding the Adirondacks, portions of Quebec along the St. Lawrence River, and much of southern Ontario between Lake Ontario and Lake Huron and Georgian Bay. Suburban development and pollution of the St. Lawrence have severely harmed natural areas here, and less than 5 percent of the ecoregion remains as intact habitat.

This ecoregion is characterized by warm summers and cold, snowy winters that are milder to the south. In the Canadian portion of the ecoregion, the climate ranges from humid, mid-cool temperate in the south to humid, high-cool temperate in the northeast. The mean annual temperature ranges from 4.5°C to 6°C, the mean summer temperature is approximately 16°C, and the mean winter temperature ranges from –4.5°C to –7°C. Mean annual precipitation ranges from 700 to 1,000 mm. Areas to the lee of the Great Lakes, both in the United States and

Canada, lie in major snowbelt areas (ESWG 1995).

The Eastern Great Lakes forests lie between the boreal forests and the broadleaf deciduous zones and are therefore transitional. Part of these forests consist of a few coniferous species—mainly eastern hemlock (*Tsuga canadensis*) and pine (*Pinus* spp.—and a few deciduous species—mainly yellow birch (*Betula alleghaniensis*), sugar maple (*Acer saccharum*), red maple (*Acer rubrum*), red oak (*Quercus rubra*), eastern hemlock, and American beech (*Fagus grandifolia*). The rest is a mosaic of pure deciduous stands in favorable habitats with good soils, and pure coniferous forests in less favorable habitats with poorer soils. In the northeast, beech is restricted to warmer sites. Drier sites contain red oak and red pine (*Pinus resinosa*) and white pine (*P. strobus*), as well as eastern white cedar (*Thuja occidentalis*). Moist sites are dominated by red maple, elms (*Ulmus* spp.), eastern cottonwood (*Populus deltoides*), and ashes (*Fraxinus* spp.). Eastern white cedar occurs in wet depressions and near streams. Early successional species include white pine, quaking aspen (*Populus tremuloides*), and paper birch (*B. papyrifera*). Pine trees are often the pioneer woody species that flourish in burned-over areas or on abandoned arable land. Fires started by lightning are common in this ecoregion, particularly where soils are sandy and there is a layer of dry litter in summer (Flader 1983).

The St. Lawrence lowlands are underlain by carbonate-rich Paleozoic bedrock. The landscape is a mix of bedrock outcrops and deeper marine and lacustrine clay deposits. The southern region is divided by the Niagara escarpment, which extends northeast to Manitoulin Island. The area to the west of the escarpment slopes to the southwest in rolling topography, and the area to the east of the escarpment rises from Lake Ontario north to Georgian Bay. Consequently, most of this ecoregion has low relief. Lakes, poorly drained depressions, morainic hills, drumlins, eskers, outwash plains, and other glacial features are typical of the area, which was entirely covered by glaciers during parts of the Pleistocene. The greatly varying soils include peat, muck, marl, clay, silt, sand, and gravel (ESWG 1995).

Biological Distinctiveness

This ecoregion contains numerous rare ecological or evolutionary phenomena. Pronounced is the mosaic of freshwater marshes and dunes, bogs and fens, hardwood and conifer swamps, as well as the rare and unique alvar communities (also called pavement barrens) restricted primarily to this ecoregion in North America. These alvar communities support a suite of prairie species that reach the eastern edge of their range here. These alvar communities are the most extensive in the world and represent a habitat type that is globally endangered. Elsewhere in the world, such communities are known to occur only on islands in the Baltic Sea of Sweden and in Estonia (Couchiching Conservancy 1996). Other rare phenomena include ancient eastern white cedar trees growing on the exposed limestone cliffs of the Niagara escarpment. Some trees have been aged at 700 to 800 years, making them among some of the oldest in eastern North America. The St. Lawrence lowlands of Quebec are the most northeastern distribution of many plant species, and the fresh tidal wetlands along the St. Lawrence are high in species endemism.

Many large mammals, such as black bear (*Ursus americanus*), moose (*Alces alces*), and wolf (*Canis lupus*) have been extirpated from much of this region. White-tailed deer (*Odocoileus virginianus*), coyote (*Canis latrans*), snowshoe hare (*Lepus americanus*), chipmunk (*Tamias striatus*), red squirrel (*Tamiasciurus hudsonicus*), and eastern gray squirrel (*Sciurus carolinensis*) represent a few of the more common species. Some characteristic breeding birds include northern cardinal (*Cardinalis cardinalis*), green-backed heron (*Butorides virescens*), mourning dove (*Zenaida macroura*), eastern screech owl (*Otus asio*), wood thrush (*Hylocichla mustelina*), pileated woodpecker (*Dryocopus pileatus*), wood duck (*Aix sponsa*), and American black duck (*Anas rubripes*) (ESWG 1995).

Conservation Status

Habitat Loss

Over 95 percent of the habitat in this ecoregion has been lost to suburban development and pollution of the St. Lawrence. Much of the remaining habitat consists of wetlands or abandoned farmlands undergoing reforestation. In some locations, recovery of abandoned agricultural land is beginning to occur, but these lands remain unprotected. Other areas continue to be converted to agriculture or are succumbing to widespread urban sprawl throughout this ecoregion.

Remaining Blocks of Intact Habitat

This ecoregion contains no blocks of intact habitat more than 250 km^2 in area. Important blocks include:

- Bald Mountain—Vermont, 4 km^2
- The Diameter, South Basin—New York
- Missisquoi National Wildlife Refuge—Vermont
- Split Rock—New York
- Chaumont Barrens—6.5 km^2 (1,600 acres) of alvar in northern New York
- Bruce Peninsula—south-central Ontario, between Georgian Bay and Lake Huron
- Alfred Bog—eastern Ontario
- Luther Marsh—south-central Ontario
- Ganaraska Forest—south-central Ontario
- Carden Plain—south-central Ontario, 200 km^2
- Mont St, Hilaire—southern Quebec, 11 km^2
- Lac St. François National Wildlife Area—southern Quebec, 13 km^2
- Cap Tourmente National Wildlife Area—eastern Quebec, 24 km^2

Degree of Fragmentation

The Eastern Great Lakes lowland forests are highly fragmented, with effectively no connectivity in most areas and little core habitat due to edge effects. The individual fragments and clusters that remain are highly isolated, and the intervening urban and suburban landscape precludes dispersal for most taxa.

Degree of Protection

None of the protected areas in this ecoregion exceed 250 km^2. Important areas include:

- Split Rock (New York State) and Coon Mountain (Adirondack Land Trust)
- Eastern Lake Ontario (A network of reserves here protects dunes, marshes, and fens.)
- Rome Sand Plains, 12 km^2 (3,000 acres) of upland and wetland pitch pine communities
- Albany Pine Barrens, a good example of pitch pine, scrub oak woodland containing occurrences of Karner Blue butterfly.

- Missisquoi National Wildlife Refuge—Vermont (This site protects ecologically important marshland, floodplain forests, and a large pitch pine bog—22 km^2.)
- Bald Mountain, Vermont, and South Bay, NY (The upland dry forests of this reserve are regionally important for their unusual plant communities.)
- Bruce Peninsula National Park—south-central Ontario, 266 km^2
- Cabot Head Provincial Nature Park—Ontario, 45 km^2
- Awenda Provincial Natural Environment Park—29 km^2
- Oka Provincial Park—southern Quebec, 23 km^2
- Rouge Valley Provincial Park Reserve—southern Ontario, 22 km^2
- Wasaga Beach Provincial Recreation Park—Ontario, 15 km^2
- Sandbanks Provincial Natural Environment Park—Ontario, 15 km^2
- Carillon Provincial Recreation Park—Ontario, 14 km^2
- Murphy's Point Provincial Natural Environment Park—Ontario, 12 km^2
- MacGregor Point Provincial Natural Environment Park—Ontario, 12 km^2
- Mont St. Hilaire Nature Reserve, McGill University—southern Quebec, 11 km^2
- Indian Point Provincial Natural Environment Park—Ontario, 9 km^2
- Presqu'ile Provincial Park—southeastern Ontario, 9 km^2
- Charleston Lake Provincial Natural Environment Park—Ontario, 9 km^2
- Iles de Boucherville Provincial Park—southern Quebec, 8 km^2
- St. Lawrence Islands National Park—eastern Ontario, 5 km^2
- Mont St. Bruno Provincial Park—Quebec, 5 km^2

Types and Severity of Threats

Development, particularly construction of summer homes and suburbanization, poses the greatest conversion threat to the Eastern Great Lakes lowland forests. Montreal (population greater than 2 million), Ottawa (population greater than 700,000), and Quebec City (population greater than 700,000) are some of the larger urban centers. Suburbs of other urban centers such as Toronto, Ontario, Syracuse, and Albany. spill out into this region as well, despite their city centers being in adjacent ecoregions. Widespread farming occurs on much of the rest of the landscape (along with smaller manufacturing centers); principal crops are corn, grains, soybeans, and apple.

Degradation due to pollution is another serious concern. The St. Lawrence is one of the most polluted waterways in North America, with high levels of mirex, PCBs, and DDT and its derivatives (Colborn et al. 1990). In spite of this, the area still supports a diversity of faunal populations, including breeding populations of common and black terns, caspian terns, and least and American bitterns. The shoreline of Lake Ontario and the St. Lawrence provides important migratory bird habitat for land birds, shorebirds, and waterfowl.

Suite of Priority Activities to Enhance Biodiversity Conservation

- Connect Split Rock in New York and Coon Mountain (Adirondack Land Trust), both of which were recently acquired.
- Protect South Bay area in New York.
- Acquire North and South Bouquet Mountains in Essex Country, New York, which contain valuable wildlife habitats, with wetlands, more than two hundred species of nesting birds, diverse plant communities, and black bear habitat on the west side of Lake Champlain in Adirondack State Park.
- Complete acquisition of Alfred Bog in Ontario to ensure protection of the area's moose population.
- Protect alvar sites in New York and Ontario.
- Protect the Albany Pine Barrens through land conservation and improved land management.
- Ensure representation of viable examples of all landscapes, natural community types, and native species in conservation areas throughout the ecoregion.
- Restore and conserve woodlots across the ecoregion.
- Protect Mont Rigaud in Quebec
- Protect Minising Swamp and Grenoch Swamp in Ontario
- Provide greater protection and enforcement of the entire Niagara escarpment in Ontario.

- Increase protection for the Oak Ridges Moraine in Ontario.
- Prevent further net loss of wetlands; several important sites exist between Drummondville and Quebec City.
- Protect the Carden Plain in Ontario.

Conservation Partners

For contact information, please see appendix G.

- Adirondack Council
- Association pour la Protection de l'Environnement de Rigaud (APER)
- Bereton Field Naturalists' Club
- Canadian Parks and Wilderness Society, Ottawa Valley Chapter
- Durham Region Field Naturalists
- Federation of Ontario Naturalists
- The Nature Conservancy, Great Lakes Program
- Le Centre de Données sur le Patrimoine
- McIlwraith Field Naturalists of London
- Ministère de l'Environnement
- Natural Heritage Information Centre
- Nature Action
- The Nature Conservancy of Canada
- The Nature Conservancy of New York, Central and Western New York Chapter
- The Nature Conservancy of New York, Eastern New York Chapter
- The Nature Conservancy of Vermont
- The Nature Conservancy, Adirondacks
- New York Natural Heritage Program
- Orillia Naturalists' Club
- Presqu'ile-Brighton Naturalists
- Quinte Field Naturalists
- Regroupement National des Conseils Régionaux de l'Environnement du Québec (RNCREQ)
- Residents Committee to Protect the Adirondacks
- Rideau Valley Field Naturalists
- Sierra Club, Northeast Regional Office
- Strategies Saint-Laurent
- Thousand Islands Land Trust
- Union Québecoise pour la Conservation de la Nature (UQCN)
- Vermont Nongame and Natural Heritage Program
- Wild Earth
- The Wildlands League

- World Wildlife Fund Canada, Quebec Region

Relationship to Other Classification Schemes

The Eastern Great Lakes Lowland Forests ecoregion represents a subdivision of Bailey's (1994) Laurentian Mixed Forest Province. This ecoregion extends Omernik's (1995b) Erie-Ontario Lake Plain to the north and west of Lake Ontario. The forests north of the lake are distinct from those along the lake's southern edge, so we chose to classify all forests south of Lake Ontario as part of a single ecoregion, the Southern Great Lakes Forests [10].

In Canada, the Eastern Great Lakes Lowland Forests are distributed through east-central Ontario and southern Quebec. This region encompasses the St. Lawrence Lowlands, the Frontenac Axis, and the Manitoulin–Lake Simcoe area (TEC ecoregions 132, 133, and 134) (ESWG 1995). This ecoregion includes a part of the Niagara section of the deciduous forest region N.1, and the Great Lakes–St. Lawrence regions GL.1–GL.3 and GL.4c: Huron-Ontario, Upper St. Lawrence, Middle St. Lawrence, and Middle Ottawa (Rowe 1972).

Prepared by K. Kavanagh, M. Sims, T. Gray, N. Zinger, and L. Gratton

Key Number:	**12**
Ecoregion Name:	**New England/Acadian Forests**
Major Habitat Type:	**2.1 Temperate Broadleaf and Mixed Forests**
Ecoregion Size:	**236,894 km²**
Biological Distinctiveness:	**Bioregionally Outstanding**
Conservation Status:	**Snapshot—Endangered Final—Endangered**

Introduction

Now increasingly forested, parts of the landscape in this ecoregion have changed dramatically over the past 350 years. Once covered by primeval forest, the land was cleared for agriculture at such a pace that by the middle of the

nineteenth century farm crops or pastures covered nearly three-quarters of the arable land in southern and central New England. One hundred years later, forests again blanketed 75 percent of New England, the result of an era of farm abandonment brought on by the opening of richer farmland to the west, the building of railroads, the Civil War, and even the California Gold Rush (Degraff 1991).

The New England/Acadian forests form a mosaic of forest types and nonforest habitats covering the Eastern Townships and Beauce regions of Quebec, approximately 50 percent of New Brunswick, most of Nova Scotia, northwestern Massachusetts and extreme northwestern Connecticut, all but the southwestern corner of Maine, the Champlain Valley of Vermont, and the coastal plain of New Hampshire. All of this area is hilly to mountainous, with the highest elevations occurring in the White Mountains of New Hampshire. The mountains of this region contain a number of forest types; northern hardwoods and spruce forests predominate and comprise roughly half of the forested landscape. Mature stands in many areas originated after extensive fires that were fueled by logging debris in the late nineteenth century. This led to fire protection policies and the decline of many fire-dependent ecosystems (Niering 1992).

Overall, this ecoregion can be described as a transition zone between the boreal spruce-fir forest to the north and the deciduous forest to the south, with the Atlantic Ocean strongly influencing vegetation dynamics of the ecoregion, especially in coastal areas. Along the Fundy Coast, high winds, cooler summers, and strongly broken topography with many areas of shallow soil result in a greater occurrence of conifer-dominated forests. On a few of the highest mountain peaks (which in New England were not separated out of this ecoregion as were the Cape Breton Highlands of Nova Scotia and the Christmas Mountains of New Brunswick in Canada), numerous arctic species occur as disjunct populations. This is true of the White Mountains in New Hampshire, and Mount Washington in particular, where a tundra-like alpine meadow occurs (Yahner 1995). Wide distribution of red spruce (*Picea rubens*) and red pine (*Pinus resinosa*) distinguish the ecoregion from the predominantly deciduous woodlands of the Great Lakes lowland forests and the mixed woods of the Eastern Forest/Boreal Transition area. Some combination of sugar maple (*Acer saccharum*), American beech (*Fagus grandifolia*), and yellow birch (*Betula alleghaniensis*) characterize most hardwood forests. The forests vary with elevation, with valleys containing hardwood forest with an admixture of eastern hemlock (*Tsuga canadensis*) and low mountain slopes supporting a mixed forest of red spruce, balsam fir (*Abies balsamea*), maple, beech, birch, white spruce (*P. glauca*), and red pine. Eastern hemlock and eastern white pine (*P. strobus*) are also present. Conifers also dominate low-elevation areas with shallow soils. The compensating effect of latitude is apparent in the altitudinal limits of zonation, which rise in elevation as one moves south.

Above the mixed-forest stands lie pure stands of balsam fir and red spruce, which devolve into krummholz at higher elevations. In Canada, this ecoregion encompasses part of the Appalachian Mountain complex. To the south of the St. Lawrence River, the Appalachian complex is dominated by folded Paleozoic sandstones and quartzites. The average elevation is 400 m asl, but peaks of 600 m asl are common. The Sutton Mountains in the south are a continuation of the Green Mountains of Vermont, and the Megantic Hills are a continuation of the White Mountains of New Hampshire. Toward the east, the uniform Chaleur Uplands and lower elevations of the New Brunswick Highlands range from 200 to 500 m asl. The uplands have developed on folded sedimentary and igneous Paleozoic strata and increase in elevation to the east, becoming more rugged and dissected. In the southern New Brunswick uplands, the terrain decreases in elevation and levels out to the west, where rolling and hummocky stony till plains predominate. The Fundy Coast bedrock is composed of Proterozoic, Paleozoic, and Mesozoic strata, which rises from sea level to 215 m asl. The Nova Scotia uplands consist of folded Paleozoic slates and quartzites that form broad, sloping plains. Toward central Nova Scotia, the uplands are elevated and underlain by granitic batholith. The Atlantic Coast is also underlain by Paleozoic metamorphics and granites. The ecoregion includes the Nova Scotia Highlands, which encompass the Cobequid Mountains to the west, Antigonish Highlands in the center, and in the northeast the dissected Cape Breton hills, which are remnant of a Cretaceous peneplain surface, composed of Paleozoic metamor-

phics and Proterozoic intrusives and volcanics (ESWG 1995).

Glaciers shaped the distinctive topography of mountains and plateaus characteristic of this ecoregion and also determined the mosaic of soil and forest types. The mountains and plateaus are underlain by granite and metamorphic rocks and are often thinly mantled by glacial till. Since soils did not develop in place, this ecoregion is not characterized by infertile uplands grading into fertile valleys. Often the best soils for forest development consist of till deposited on mid-slopes of hills and mountains (Degraff 1991). Many glacially broadened valleys have glacial outwash deposits with poor soils and contain numerous swamps and lakes.

The climate of this ecoregion is characterized by warm, moist summers and cold, snowy winters. Because maritime air masses have year-round access to the eastern seaboard, precipitation is evenly distributed throughout the year, unlike that of the Allegheny Highlands Forests [15] or the Eastern Great Lakes Lowland Forest [11]. Mean annual precipitation is relatively high, ranging from 1,000 to 1,600 mm, increasing toward the Atlantic Coast and at higher elevations. In the Canadian portion of the ecoregion, mean annual temperatures range from 3°C to 6.5°C, rising in the east, and mean summer temperature is 14.5°C. Mean winter temperature within this region ranges considerably, from −7.5°C in the northern New Brunswick Uplands to −1.5°C along the Atlantic Coast of Nova Scotia (ESWG 1995).

Fire plays a much less important role in the northern hardwood forests characteristic of this ecoregion, where spring and fall seasons are short, than in the oak-dominated forests of ecoregions farther to the south. Fire can be a crucial factor in areas where red spruce and balsam fir intermingle with the hardwoods, as in parts of northern New England and the Adirondacks, especially during dry periods (Niering 1992). Fire probably plays the most important role in the forest dynamics of the region. Fires tend to be on the order of 10 to 100 km^2 in New Brunswick, for example, although there has been active fire suppression for many decades. Periodic blowdowns, sometimes on very large scales, also play a role in forest dynamics. Where areas border the Atlantic Ocean, sea salt spray and wind strongly influence forest dynamics.

Biological Distinctiveness

The New England/Acadian forests are a moderately rich example of temperate broadleaf and mixed forests. The mosaic of forest types and habitats supports 222 bird species, making these forests the second-richest ecoregion within the temperate broadleaf and mixed forests MHT and among the twenty richest ecoregions in the continental United States and Canada. For example, mature northern hardwood stands in New England commonly contain softwoods—usually red spruce, eastern hemlock, or white pine—and as a result they also contain bird species associated with coniferous forests, such as red-breasted nuthatches (*Sitta canadensis*), golden-crowned kinglets (*Regulus satrapa*), and northern parula warblers (*Parula americana*) (Niering 1992). New England/Acadian forests contain fourteen species of conifers, more than any other ecoregion within this major habitat type, save for the Appalachian/Blue Ridge Forests [16] and the Southeastern Mixed Forests [22].

Characteristic mammals include moose (*Alces alces*), black bear (*Ursus americanus*), red fox (*Vulpes fulva*), snowshoe hare (*Lepus americanus*), porcupine (*Erithyzon dorsatum*), fisher (*Martes pennanti*), beaver (*Castor canadensis*), bobcat (*Lynx rufus*), marten (*Martes americana*), muskrat (*Ondatra zibethica*), and racoon (*Procyon lotor*), although some of these species are less common in the southern parts of the ecoregion. White-tailed deer (*Odocoileus virginianus*) have expanded northward in this ecoregion and displaced the woodland caribou (*Rangifer tarandus* ssp. *caribou*) from the northern parts of the ecoregion. Coyotes (*Canis latrans*) have recently replaced wolves, which were eradicated from this ecoregion in historical times. Numerous seabirds and migratory shorebirds inhabit the salt marshes and tidal flats along the coasts in the northern parts of the ecoregion.

This ecoregion contains several rare ecological or evolutionary phenomena including major areas of serpentine rocks and associated rare vegetation, raised peat bogs, ribbed fens, and coastal raised peatlands. Western Massachusetts and eastern New York have unusual fen ecosystems, which support populations of bog turtles (*Clemmys muhlenbergii*). The southernmost reticulated bog in eastern North America reportedly occurs in Frontenac Park, Quebec. Bald eagles (*Haliaeetus leucocephalus*) reach their highest breeding density in eastern North America (Nova Scotia). There are numerous Atlantic coastal plain plant species at their northern limits, and the northeastern limits of several deciduous tree species and forest communities with southern affinity can also be found within the ecoregion. Typical of the transitional nature of this ecoregion, the southernmost outliers of arctic vegetation in eastern North America also occur here. The ecoregion has many fast-flowing, cold rocky rivers with highly fluctuating water levels that give rise to interesting floral and faunal communities.

Conservation Status

Habitat Loss

Little intact habitat remains in this ecoregion, with only about 5 percent of the New England/Acadian forest in presettlement condition. Nearly all of the ecoregion shows some signs of human activity. In Canada, estimates were placed at less than 5 percent intact, with at least 50 percent of the ecoregion classed as heavily altered. Logging is the main cause of habitat loss. Many areas are now undergoing a third forest-cutting rotation. The most natural areas tend to be those that are difficult to access or occur at high elevations.

Agriculture is extensive in some jurisdictions such as western New Brunswick, Nova Scotia, and Vermont. Some farmland has been abandoned this century, and old-field succession is gradually returning these areas to forest cover. In areas of higher elevation in New England and Quebec, ski-hill development has had a severe impact on many mountains, while summer home development along with urban and suburban development (in the Halifax area) is increasing. These tend to impact the valley lands most significantly. Mining is a major land use in parts of the ecoregion in Quebec (of talc, marble, asbestos, granite), and interest remains high for the extensive serpentine areas of Quebec.

Remaining Blocks of Intact Habitat

Several relatively large blocks of more or less intact habitat remain, as do a number of smaller patches. Important blocks of old-growth forest in order of decreasing size include:

- Mahoosuc Mountains—Maine
- Tobeatic-Kejimikujik—Nova Scotia, more than 1,000 km²
- Baxter State Park—Maine, 97 km² (24,000 acres) of unlogged spruce and fir within a 810-km² (200,000-acre) park
- Big Reed Forest—Maine, 20 km² (5,000 acres) in Piscataquis County lowlands
- Nash Stream Forest—New Hampshire, 32 km² (8,000 acres)
- White Mountains—New Hampshire
- Green Mountains—Vermont
- Mont Orford—Quebec
- Frontenac Provincial Park—Quebec
- Mont Megantic—Quebec
- Bic Provincial Park—Quebec
- Fundy National Park—New Brunswick
- Cape Breton Highlands National Park—northern Nova Scotia

Degree of Fragmentation

Habitat fragmentation is relatively low in the New England/Acadian forests. More than half of all fragments are clustered to some degree and connected within a matrix of clearcut areas. The degree of fragmentation is probably higher in Nova Scotia and western New Brunswick and eastern Maine than in other areas of this ecoregion.

Degree of Protection

The most important protected areas in this ecoregion include:

- White Mountains National Forest—New Hampshire (Important areas within the White Mountains are the Great Gulf Wilderness, Dry River Wilderness, Crawford Notch, Sandwich Range Wilderness, and Nancy Brook RNA.)
- Franconia Notch State Park—New Hampshire
- Lafayette Brook Scenic Area—New Hampshire
- Baxter State Park—central Maine
- Big Reed Forest Reserve—Maine
- Shawangunk Mountains Dwarf Pine Plains—New York, mostly in the Mohonk Preserve and Minnewaska State Park
- Aiken, Lye Brook, and Bristol Cliffs Wilderness Areas in Green Mountain National Forest—Vermont

- Mt. Mansfield State Forest—Vermont
- Camels Hump State Forest—Vermont
- Putnam State Forest—Vermont
- Victory State Forest—Vermont
- Kejimkujik National Park and Tobeatic Provincial Protected Area—Nova Scotia, 1,371 km²
- Cape Breton Highlands National Park—Nova Scotia (75% of area is lowlands plus Polletts Cove–Aspy Forest and Margaree River), 1,050 km²
- Fundy National Park—New Brunswick, 205 km²
- Bonnet Lake Barrens and Canso Coastal Barrens—Nova Scotia, 193 km²
- Tidney River—Nova Scotia, 188 km²
- Tangier Grand Lake—Nova Scotia, 157 km²
- Frontenac Provincial Park—Quebec, 155 km²
- Clattenburgh Brook and Waverley–Salmon River Long Lake—Nova Scotia, 111 km²
- Cloud Lake—Nova Scotia, 108 km²
- Gabarus Provincial Protected Area and Louisbourg National Historic Park—Nova Scotia, 96 km²
- Economy River and Portapique River—Nova Scotia, 81 km²
- Boggy Lake, Alder Grounds and Big Bog—Nova Scotia, 72 km²
- French River—Nova Scotia, 72 km²
- Mont Orford Provincial Park—Quebec, 58 km²
- Middle River Framboise—Nova Scotia, 57 km²
- Ogden Round Lake—Nova Scotia, 57 km²
- Terence Bay—Nova Scotia, 56 km²
- Middle River—Nova Scotia, 54 km²
- Mont Megantic Provincial Park—Quebec, 54 km²
- Lake Rossignol—Nova Scotia, 52 km²
- White Lake—Nova Scotia, 45 km²
- North River—Nova Scotia, 43 km²
- Bowers Meadows—Nova Scotia, 43 km²
- Bic Provincial Park—Quebec, 33 km²
- Nash Stream Forest—New Hampshire, approx. 32 km²
- Liscomb River—Nova Scotia, 30 km²
- Cape Chignecto Provincial Park—Nova Scotia, 30 km²

Types and Severity of Threats

The major conversion and degradation threats to this ecoregion are development and logging. Development for second homes and ecotourism is a particular problem in Quebec and in the vicinity of other urban centers. Development and population growth are also a significant threat in northeastern Vermont. Logging remains an important industry in Maine and may alter large areas of habitat in that state as well as in the provinces of Quebec and New Brunswick. High-intensity recreational development (e.g., ski hills) and mining (especially in Quebec) combine to further reduce the remaining extent of natural habitat in this ecoregion.

Suite of Priority Activities Needed to Enhance Biodiversity Conservation

- Ensure that examples of all landscape types, all natural community types, and all native species are well represented, in all their natural variability, in conservation areas throughout the ecoregion.
- Expand North Woods preserve around Baxter State Park in north-central Maine through the establishment of the Maine Woods National Park. Together with Baxter State Park the area would create a park covering 3.2 million acres.
- Expand the preserve around the Cobscook Bay in Maine.
- Expand existing conservation activities in Maine's Mattagodus wetlands.
- Work to control housing development in Vermont's Northeast Kingdom.
- Work with private timber industries to promote conservation-minded activities.
- Encourage state and national governments to buy land that timber companies are willing to sell.
- Control exotics. During the last few decades certain introduced shrubs, vines, and trees, such as Japanese honeysuckle (*Lonicera japonica*), Oriental bittersweet (*Celastrus orbiculatus*), shrub honeysuckle (*Lonicera* spp.), multiflora rose (*Rosa multiflora*), autumn olive (*Elaeagnus umbellata*), and Norway maple (*Acer platanoides*), have become serious competitors in many post-agricultural and forest communities in New England. Some of these exotics may bring about the demise of native species and may also replace native plant cover. Exotics are a particularly severe threat in small, fragmented forests in urban and developed areas (Niering 1992).

- Pursue private stewardship for rich hardwood stands and white cedar wetlands along the Saint John River and its tributaries in New Brunswick.
- Increase conservation effort for riparian zones generally and the Bay of Fundy shoreline in New Brunswick particularly.
- Expand protection around Schenob Brook wetlands (in the Berkshires) in Massachusetts and Connecticut.
- Increase conservation efforts for the Mount Sutton mountains in Quebec.
- Protect Mount Gosford and Marble Hill (Twin Peaks on the Quebec-Maine border).
- Give serious conservation attention to the northern part of the Appalachian Mountains in Quebec.
- Pursue final designation of thirty-one new conservation sites recently announced in Nova Scotia.
- Create more coastal protected areas in Nova Scotia.
- Create better linkages (improved forestry practices) for lands between core protected areas in Nova Scotia.
- Increase the deciduous forest cover in Nova Scotia.
- Develop a protected-areas system plan that includes candidate sites in New Brunswick.
- Give more attention to conserving old-growth forests in New Brunswick.

Conservation Partners

For contact information, please see appendix G.

- Appalachian Mountain Club
- Canadian Parks and Wilderness Society, Nova Scotia Chapter
- Conservation Council of New Brunswick
- Ecology Action Centre
- Ford Alward Naturalist Association
- Friends of Nature Conservation Society
- Fundy Guild
- Le Centre de Données sur le Patrimoine
- Maine Natural Areas Program

- Margaree Environmental Protection Association
- Moncton Naturalists' Club
- The Nature Conservancy of Canada, Atlantic Canada
- The Nature Conservancy
- The Nature Conservancy, Eastern Regional Office
- The Nature Conservancy of Maine
- The Nature Conservancy of Massachusetts
- The Nature Conservancy of New Hampshire
- The Nature Conservancy of Vermont
- The Nature Conservancy, QuebecNature Trust of New Brunswick
- Naturel du Quebec
- New Brunswick Federation of Naturalists
- New Brunswick Protected Natural Areas Coalition
- New Hampshire Natural Heritage Inventory
- Northern Appalachian Restoration Project
- Nova Scotia Nature Trust
- Nova Scotia Wild Flora Society
- Parc d'environnement naturel de Sutton (PENS)
- Regroupement National des Conseils Régionaux de l'Environnement du Québec (RNCREQ)
- Restore the Northwoods
- Save Our Shores
- Society for the Protection of New Hampshire Forests
- Union Québecoise pour la Conservation de la Nature (UQCN)
- Vermont Nongame and Natural Heritage Program
- World Wildlife Fund Canada, Quebec Region

Relationship to Other Classification Schemes

The New England/Acadian forests are demarcated from the Northeastern Coastal Forests [14] to the south, and the Gulf of St. Lawrence and Eastern Canadian Forests [13] and [97] to the north by potential vegetation types (Küchler 1985) and elevation respectively. The Northeastern Coastal Forests ecoregion [14] is more dominated by oaks and occurs on the coastal plain. The New England/Acadian Forest ecoregion is largely similar to Omernik's (1995b) northeastern highlands, though it has been extended into similar areas of Canada.

Omernik includes the Adirondacks as a disjunct part of the northeastern highlands, while we believe them to be more similar to the Eastern Forests/Boreal Transition [8]. Bailey (1994) also extends this ecoregion farther west into the Adirondacks.

In Canada, the New England/Acadian Forest ecoregion extends through southern Quebec, New Brunswick, and Nova Scotia. This ecoregion incorporates a number of the Terrestrial Ecoregions of Canada: the southern part of the Appalachians, the Northern and Southern New Brunswick Uplands, the Saint John River Valley, the Southwest and South-Central Nova Scotia Uplands, the Nova Scotia Highlands, and the Atlantic and Fundy Coasts (TEC ecoregions 117, 118, 120, 121, 123–125, 127, and 128) (ESWG 1995). Forest cover here is primarily Acadian, including the Upper Miramichi-Tobique, Carleton, South Atlantic Shore, East Atlantic Shore, Cape Breton–Antigonish, Fundy Coast, Southern Uplands, Atlantic Uplands, and Cobequid sections [Rowe (1972) forest classes A.2, A.4, A.5a, A.5b, A.7, A.9, A.10, A.11, A.13].

Prepared by M. Davis, L. Gratton, J. Adams, J. Goltz, C. Stewart, S. Buttrick, N. Zinger, K. Kavanagh, M. Sims, and G. Mann

Key Number: **13**
Ecoregion Name: **Gulf of St. Lawrence Lowland Forests**
Major Habitat Type: **2.1 Temperate Broadleaf and Mixed Forests**
Ecoregion Size: **39,426 km²**
Biological Distinctiveness: **Nationally Important**
Conservation Status: **Snapshot—Critical Final—Critical**

Introduction

This ecoregion includes all of Prince Edward Island (PEI); Iles de la Madeleine, Quebec; most of east-central New Brunswick; and the Annapolis Valley and the Northumberland Strait coast of Nova Scotia.

The mean annual temperature ranges from 4.5°C to 6.5°C, and mean summer temperature ranges from 12°C to 15.5°C, with the lowest temperature in Iles de la Madeleine, Quebec. Mean winter temperature is colder in the Maritime lowlands than in the islands or the sheltered Annapolis-Minas lowlands. This ecoregion is marked by warm summers and mild, snowy winters and can be characterized primarily as an Atlantic high cool temperate eco-climate (ESWG 1995).

Prince Edward Island, Iles de la Madeleine, and the Maritime lowlands of New Brunswick and Nova Scotia are underlain by Carboniferous sandstones, shales, and conglomerates and rise inland from sea level to 200 m asl in elevation. Small bedrock outcrops stand as prominent hills. The eastern portion of this region is underlain by Mesozoic sandstone in the Annapolis Valley and Paleozoic shale, sandstone, gypsum, and limestone in the Minas lowlands (ESWG 1995).

The Gulf of St. Lawrence strongly influences climate and vegetation dynamics over much of the ecoregion. Lowland physiography and warmer summers allow for better growth of hardwoods than in much of the New England/Acadian Forests ecoregion.

Biological Distinctiveness

The closed mixed-wood forests of this ecoregion are strongly influenced by a maritime climate in which warm summers allow for good growth of hardwoods, which are often mixed with red spruce (*Picea rubens*) and balsam fir (*Abies balsamea*). Eastern hemlock (*Tsuga canadensis*) and eastern white pine (*Pinus strobus*) are found in the Maritime lowlands of New Brunswick and the Annapolis-Minas lowlands of Nova Scotia. Sugar maple (*Acer saccharum*), yellow birch (*Betula alleghaniensis*), and beech (*Fagus grandifolia*) are more common on well-drained, higher-elevation areas. Bogs and fens are significant in the Minas lowlands, while other wet sites support white elm (*Ulmus americana*), black ash (*Fraxinus nigra*), and red maple (*A. rubrum*). The original mixed-wood forest of PEI was characterized by red oak (*Quercus rubra*), sugar maple, yellow birch, and beech, while abandoned fields are returning to forests of white spruce (*P. glauca*). Although warm summers also characterize the climate of Iles de la Madeleine, poor drainage results in a conifer-dominated forest of black spruce (*P. mariana*) and balsam fir.

Characteristic mammals across the mainland portions of the ecoregion include black bear (*Ursus americanus*), moose (*Alces alces*), white-tailed deer (*Odocoileus virginianus*), red fox (*Vulpes fulva*), snowshoe hare (*Lepus americanus*), porcupine (*Erithyzon dorsatum*), fisher (*Martes pennanti*), beaver (*Castor canadensis*), bobcat (*Lynx rufus*), marten (*Martes americana*), racoon (*Procyon lotor*), and muskrat (*Ondatra zibethica*). On PEI, however, black bear, fisher, bobcat, and marten have been extirpated, and only red fox, snowshoe hare, beaver, and muskrat are native. Coyotes (*Canis latrans*) have now invaded most of the ecoregion, including PEI. A diversity of shorebirds and seabirds inhabits the salt marshes and tidal flats along the coast. This includes some of the highest breeding densities for the endangered piping plover (*Charadrius melodus*) in eastern North America.

Fires play a role in forest dynamics of the ecoregion, although probably on a relatively small scale (up to 10s of square kilometers). At present, they are actively suppressed. High winds, sea salt spray, and fog are additional strong influences on forest dynamics and vegetation communities, particularly in coastal areas.

Exhumed skeletal forests exist in sandy areas of Prince Edward Island. The dune system at Greenwich, Prince Edward Island, has a unique structure for northeastern North America. Unique forest communities composed of mixed hardwood and softwood species also occur. Several endemic species are found here, including the Laurentian aster (*Aster laurentiana*); maritime ringlet butterfly (*Coenonympha nepisiquit*), which is found in salt marshes; and the Bathurst salt marsh aster (*Aster subulatus* var. *obtusifolius*).

A continentally significant and large population of breeding great blue herons (*Ardea herodias*) is found on Prince Edward Island. In addition, perhaps the largest colony (twelve thousand pairs) of double-crested cormorants (*Phalacrocorax auritus*) in northeastern North America and a major breeding population of piping plover are found in this ecoregion.

Conservation Status

Habitat Loss

It is estimated that only about 3 percent of this ecoregion can be considered as intact habitat. More than 75 percent is considered to be heavily altered. Presently, the principal reasons for habitat loss are:

- conversion of land for agriculture (especially potatoes)
- forest harvesting and high-density road networks
- urban and rural housing
- recreational disturbance of dune vegetation in coastal areas
- extensive peat harvesting (especially in eastern New Brunswick)

Historically, agriculture and logging for shipbuilding destroyed much of the original forest cover of the ecoregion by the nineteenth century.

Remaining Blocks of Intact Habitat

The area encompassing Kouchibouguac National Park in New Brunswick constitutes the only substantial remaining block of intact habitat in this ecoregion.

Degree of Fragmentation

Habitat fragmentation is high in Prince Edward Island and Nova Scotia.

Degree of Protection

- Kouchibouguac National Park—eastern New Brunswick, 239 km^2
- Prince Edward Island National Park—northern Prince Edward Island, 25 km^2
- Dollar Lake Provincial Park—northern Nova Scotia, 12 km^2
- Ile Brion Ecological Reserve—eastern Quebec, 6 km^2
- Brudenell River Provincial Park—eastern Prince Edward Island, 6 km^2

Types and Severity of Threats

Threats to biodiversity protection in this ecoregion are very high. Continued logging (especially in New Brunswick , where the annual allowable cut is predicted to increase in the near future) and agriculture are the primary sources of habitat degradation and loss. Peat mining is a threat to bogs, and increased shoreline development threatens many coastal regions.

Suite of Priority Activities to Enhance Biodiversity Conservation

- Establish widespread planning for setting restoration priorities.
- Build and maintain incentives for private land protection.
- Increase protection for remaining woodlands.
- Increase protection for offshore islands.
- Encourage the government of New Brunswick to come forward with a protected-areas system plan that includes representative candidate sites.
- Give more attention to conserving old-growth forests and coastal habitat.

Conservation Partners

For contact information, please see appendix G.

- Annapolis Field Naturalists Society
- Blomidon Naturalists Society
- Canadian Parks and Wilderness Society, Nova Scotia Chapter
- Conservation Council of New Brunswick
- Environmental Coalition of Prince Edward Island
- Federation of Nova Scotia Naturalists
- Island Nature Trust
- Le Club de Naturalistes de la Peninsule Acadienne
- Moncton Naturalists' Club
- Natural History Society of Prince Edward Island
- Nature Conservancy of Canada, Atlantic Canada
- Nature Trust of New Brunswick
- New Brunswick Federation of Naturalists
- New Brunswick Protected Natural Areas Coalition
- Nova Scotia Nature Trust
- Parks and People PEI
- Prince Edward Island Salmon Association
- Prince Edward Island Wildlife Federation
- World Wildlife Fund Canada

Relationship to Other Classification Schemes

The Gulf of St. Lawrence Lowland Forests overlay the Maritime Lowlands, the Annapolis-Minas Lowlands, Prince Edward Island, and the Iles de la Madeleine (TEC ecoregions 122, 126,

130, and 131) (ESWG 1995). These ecoregions are all part of the Atlantic Maritime Ecozone, in the Acadian forest region. Forest sections include the Eastern Lowlands, Prince Edward Island, and the Central Lowlands (A.3, A.8, and A.12) Forest Regions (Rowe 1972).

Prepared by: B. Meades, C. Stewart, J. Goltz, K. MacQuarrie, K. Kavanagh, M. Sims, and G. Mann

Key Number:	**14**
Ecoregion Name:	**Northeastern Coastal Forests**
Major Habitat Type:	**2.1 Temperate Broadleaf and Mixed Forests**
Ecoregion Size:	**89,691 km²**
Biological Distinctiveness:	**Bioregionally Outstanding**
Conservation Status:	**Snapshot—Critical Final—Critical**

Introduction

Northeastern coastal forests of the Piedmont plateau and the coastal plain cover all or part of seven states, from northern Maryland to southern Maine. The ecoregion is dominated by Appalachian oak forests, characterized by white oak (*Quercus alba*) and northern red oak (*Q. rubra*). Until the early part of this century American chestnut (*Castanea dentata*) was so ecologically and commercially important a species in this ecoregion that these forests were known as oak-chestnut forests. A chestnut blight caused by *Endothia parasitica* and likely imported from Chinese chestnuts spread rapidly following its introduction in 1904, killing most chestnuts in New England in twenty years and at the southern end of the species' range by 1940. Although chestnuts still sprout profusely from root systems after more than half a century, few trees survive long enough to reach the main canopy of a mature forest stand (Barnes 1991).

Hurricanes and fires shaped the classic presettlement landscape of New England. Native Americans set most of the fires; they used fire to manage vegetation for some nine thousand years before the arrival of European settlers (Niering 1992). The fires in southern New England tended to create open, parklike forests that attracted wildlife, favored palatable ground cover plants, and facilitated movement and increased visibility for hunting (Cronon 1983). Severe storms, especially in the south, have also resulted in extensive blowdowns, which require a century or more to recover from. Ice and glaze storms are a significant but less obvious disturbance in these deciduous forests (Barnes 1991).

A transition occurs in this glaciated landscape from oak and oak-hickory communities to the south and east to hardwood communities to the north and west. Species of southern distribution, such as sweet gum (*Liquidambar styraciflua*), river birch (*Betula nigra*), Spanish oak (*Quercus falcata*), and red mulberry (*Moros rubra*), have reached or are reaching their northern limits. On disturbed sites a new set of species, including paper birch (*Betula papyrifera*), gray birch (*B. populifolia*), and quaking aspen (*Populus tremuloides*), becomes more important. The oak communities grade into mixed deciduous communities on the lower north slopes and ravines. The species composition varies widely, having either elements of the oak forest or those of the northern hardwood-conifer forest, depending on local climatic and soil conditions.

Biological Distinctiveness

The Northeastern coastal forests are relatively rich, with over 750 species and seven endemics. Of the eighteen ecoregions classified as temperate broadleaf and mixed forests, the northeastern coastal forests rank seventh in terms of richness and endemism and first in bird richness, with over 250 species. The region also supports numerous butterfly species. The northeastern coastal forests include part of Delaware Bay, a large-scale migratory corridor and feeding area for shorebirds and songbirds, where birds feed on horseshoe crab eggs. This ecoregion includes a significant portion of the range of the bog turtle (*Clemmys muhlenbergii*). The northern subspecies found in this area is a candidate for listing as a federally threatened species.

Conservation Status

Habitat Loss

Suburban sprawl has resulted in the loss of over 98 percent of the ecoregion's natural habitat. Remaining habitat is limited to fragments and degraded larger patches. The northeastern forests were the first on the continent to suffer from heavy logging pressure, and they may again come under the ax as loggers revisit the Northeast, for the fourth or fifth time, as western forests are depleted.

Remaining Blocks of Intact Habitat

Practically no old-growth forest remains in this ecoregion (Davis 1996), and there are no blocks of intact habitat more than 250 km² in area. Important blocks include:

- Devil's Den Nature Conservancy Preserve—Connecticut, 7 km² (1,700 acres)
- Cape May National Wildlife Refuge—southern New Jersey, 32 km² (7,956 acres)
- Great Swamp Management Area—Rhode Island, approx. 12 km² (3,000 acres)
- Crandall Swamp—Rhode Island, approx. 6 km² (1,500 acres)
- Great Bay Wildlife Refuge—southeastern New Hampshire
- Freetown State Forest and adjacent land—southeastern Massachusetts
- Shawangunk Mountains Dwarf Pine Plains—southern New York
- Quabbin Reservoir—Massachusetts
- Plum Island National Wildlife Refuge—northeastern Massachusetts
- Bare Mountain and Mt. Tom, Connecticut River Valley—west-central Massachusetts

Degree of Fragmentation

The northeastern coastal forests are highly fragmented, with effectively no connectivity in most areas and little core habitat due to edge effects. The individual fragments and clusters that remain are highly isolated, and the intervening urban and suburban landscape precludes dispersal for most taxa.

Degree of Protection

None of the protected areas in this ecoregion exceeds 250 km². The greatest challenges to conserving a fraction of this original ecosystem are to manage the public lands for multiple use and to marry those public lands to private reserves. Most of the sites suitable for biodiversity conservation are small and quite isolated. Within the ecoregion, the best opportunities for conservation in the short term occur in:

Delaware Bay Shores, Great Bay National Wildlife Refuge (which does not include the entire bay), Falls River State Park (where adjacent land is not protected), and New York's Mohonk Preserve and Minnewaska State Park, which protect the Shawangunk Mountains Dwarf Pine Plains.

Types and Severity of Threats

Development is the greatest threat and could significantly alter at least 25 percent of the remaining habitat within the next twenty years. Native plants are experiencing significant mortality due to shoreline erosion, the introduction of exotics, and overuse of natural resources. Collection of wild orchids and reptiles poses a threat to some species, and the recreational use of fragile shoreline constitutes a major threat to the wildlife of this ecoregion.

Suite of Priority Activities to Enhance Biodiversity Conservation

- Add to protection of the Delaware Bay. Land acquisition is ongoing, and a careful, well-managed approach to ecotourism is being developed. This includes linking existing preserves in the region through a system of site improvements and a unified interpretive plan.
- Protect Pennsylvania-Maryland serpentine barrens. This is the largest expanse of serpentine barrens in the eastern United States. The rare serpentine prairie, savanna, and woodland areas are sustained by a combination of fire and toxic minerals found in the serpentine rock. Recommendations include acquisition of remaining undeveloped land and improved management, particularly regarding the use of prescribed burns or other techniques for duff reduction.
- Protect bog turtle habitat. Significant threats to bog turtles include habitat destruction, metapopulation fragmentation, excessive groundwater withdrawal, and collection. Local habitats need to be reconnected and managed to slow succession to woodlands in order to restore metapopulation dynamics.
- Protect piping plover breeding areas along beaches in Connecticut, New York, Rhode Island, Massachusetts, Maine, and New Jersey. Protect natural processes of longshore drift, erosion, and overwash, which maintain coastal systems.

- Protect Connecticut River tidelands. The lower Connecticut River and its tributaries is one of the richest ecosystems in the Northeast, providing habitat for hundreds of species, seven of them globally rare or endangered, and containing an extraordinarily unsullied wetland complex. An ecosystem management approach is needed to conserve this nationally important natural resource.
- Improve management of Great Bay, acquire land nearby, and promote compatible development.

Conservation Partners

For contact information, please see appendix G.

- Association for Biodiversity Information
- Essex County Greenbelt Association
- Maine Coast Heritage Trust
- Massachusetts Audubon Society
- The Nature Conservancy
- The Nature Conservancy, Eastern Regional Office
- New Jersey Audubon Society
- New Jersey Conservation Foundation
- Society for the Protection of New Hampshire Forests
- Trustees of Reservations, Northeast Regional Office
- Trustees of Reservations, Southeast Regional Office

Relationship to Other Classification Schemes

The Northeastern Coastal Forests are demarcated from the New England/Acadian Forests [12] to the north by vegetation (Küchler 1975) and elevation. The latter ecoregion is dominated by hardwoods and occurs in the mountains rather than on the coastal plain. This ecoregion extends Omernik's (1995b) Northeastern Coastal Zone to the south and west, overlapping with parts of his Northern Piedmont and Middle Atlantic Coastal Plain ecoregions. We believe our classification is more in keeping with the ecological patterns of the region.

Prepared by M. Davis, W. Eichbaum, and J. Adams

Key Number:	**15**
Ecoregion Name:	**Allegheny Highlands Forests**
Major Habitat Type:	**2.1 Temperate Broadleaf and Mixed Forests**
Ecoregion Size:	**83,881 km²**
Biological Distinctiveness:	**Nationally Important**
Conservation Status:	**Snapshot—Endangered Final—Endangered**

Introduction

The presettlement forests of the Allegheny highlands consisted primarily of hemlock (*Tsuga canadensis*) and beech (*Fagus grandifolia*). Together the two species represent nearly 60 percent of all the trees observed in early land surveys of what is now Allegheny National Forest (Marquis 1975). Although hemlock and beech tolerate shade, the occurrence of both species can be traced to periodic catastrophes, particularly fire. Sugar maple (*Acer saccharum*), an associate of the beech-hemlock type, often replaced hemlock as the major component of presettlement forests. Beech-maple stands were more common on less moist sites, perhaps because fire had eliminated hemlock. Red maple (*Acer rubrum*), yellow and black birch (*Picea glauca* and *P. mariana*), white ash (*Fraxinus americana*), and black cherry (*Prunus serotina*) were common associates in both the hemlock-beech and the beech-maple forests. White pine (*Pinus strobus*) occurred in well-defined, rela- tively pure stands. These stands generally originated after fires or windthrow wiped out the preceding stands.

Between 1890 and 1920, loggers cleared most of the Allegheny plateau. Save for a few pockets of old growth, the current forests, which contain most of the presettlement species in different relative abundances and distribution, originated at that time. The heavy cutting favored hardwoods and created massive amounts of coniferous slash, providing ideal conditions for widespread and intense fires. The fires were a major factor in the virtual elimination of white pine and hemlock in the Allegheny forests. Repeated fires also tended to reduce the proportion of sugar maple, beech, and other typical hardwoods and increase such species as aspen, pin cherry, sedges, grasses, and honeysuckles.

These forests display large-scale patterning related to soil drainage, which segregates areas dominated by beech, hemlock, and white pine from areas dominated by hemlock and yellow birch. Smaller-scale patterning separates small areas of hemlock from yellow birch (Whitney 1990).

An expanding deer population plays an important role in these forests, particularly in old-growth areas. Heavy deer browsing since the 1930s has had a profound influence on the size-class distribution of stems. In one stand, for example, by 1978 deer had eliminated the smaller classes of once common trees other than beech (Whitney 1984).

Biological Distinctiveness

The Finger Lakes and Pocono Mountains are biologically important areas. The Finger Lakes contain a mixture of northern and southern species, and the deeply cut lakes reveal remarkable geological strata. The Pine Barrens in the Poconos are a unique formation.

Conservation Status

Habitat Loss

Less than 1 percent of this ecoregion remains intact, but once logged areas are now reforested. Agriculture, particularly in the western and central lowlands, is the leading cause of habitat loss, while recreation and development contribute to habitat loss in the northern parts of the ecoregion.

Remaining Blocks of Intact Habitat

Relatively few large habitat blocks remain. The most important blocks are:

- Pennsylvania State Forest—north-central Pennsylvania, Potter and Clinton counties, approx. 1,000 km²
- Allegheny National Forest—northwestern Pennsylvania, McKean and Warren counties, 16 km² (4,000 acres)
- Catskill State Park—central New York; 65 km² unlogged in one tract, 54,000–65,000 km² total unlogged is in thirty-eight tracts
- Allegany State Park—western New York

Degree of Fragmentation

The forests of the Allegheny highlands are moderately fragmented, with some connectivity, clusters of habitat fragments, and an intervening landscape that allows for dispersal of many taxa through some parts of the ecoregion.

Degree of Protection

This ecoregion contains no protected areas larger than 500 km². The most important protected areas are:

- Hammersley Fork Wilderness Area—north-central Pennsylvania, approx. 100 km²
- Cook State Forests—northwestern Pennsylvania, 6 unlogged km² (1,500 acres) in a 29-km² (7,200-acre) forest
- Catskills—New York, 219–263 km² (54,000–65,000 acres) of mostly state land
- Arbutus Peak Oak Barren Macrosite—northeastern Pennsylvania, Luzerne County 21 km² (5,313 acres) owned by the Pennsylvania Game Commission
- Lehigh Pond—northeastern Pennsylvania, Wayne County 15 km² (3,912 acres)
- Hemlock Lake and Canadise Lake—western New York, approx. 4 km² (1,000 acres)
- Bergen Swamp—New York, approx. 8 km² (2,000 acres)
- Allegheny National Forest—northwestern Pennsylvania
- Allegany State Park—western New York, includes 3 km² (700 acres) of old growth
- Woodbourne Forest—northwestern Pennsylvania, includes 2 km² (600 acres) of old-growth and second-growth forest

Types and Severity of Threats

Recreational and suburban development pose a significant threat to the forests of the Allegheny highlands, particularly in the Finger Lakes region and the Catskills. In the western portion of the ecoregion, a booming deer population is destroying herbaceous vegetation and preventing tree regeneration.

Suite of Priority Activities to Enhance Biodiversity Conservation

The decline in agriculture in much of this ecoregion presents an opportunity for conservation. A top priority should be either to acquire land taken out of production or to improve the management of that land. Key sites include the Finger Lakes, the Poconos, and French Creek, a pristine river valley in southwest New York and northwest Pennsylvania that harbors diverse flora and fauna. Deer must be controlled throughout the ecoregion and especially in Allegheny National Forest and the Catskills.

Conservation Partners

For contact information, please see appendix G.

- New York Natural Heritage Program
- Pennsylvania Natural Diversity Inventory, Central Bureau of Forestry
- Pennsylvania Natural Diversity Inventory, East
- Pennsylvania Natural Diversity Inventory, West
- The Nature Conservancy of Pennsylvania
- The Nature Conservancy, Central and Western New York Chapter
- The Nature Conservancy, Central/Western Office
- Western Pennsylvania Conservancy

Relationship to Other Classification Schemes

The Allegheny Highland Forests are demarcated from surrounding hardwood forests by elevation. Omernik (1995b) divides the region into four parts: the Northern Appalachian Plateau and Upland, the North Central Appalachians, the Western Allegheny Plateau, and the Erie/Ontario Lake Plain. We believe the communities here are similar enough to warrant classification into a single ecoregion that is still not so large as to be unmanageable.

Prepared by S. Buttrick

Key Number:	**16**
Ecoregion Name:	**Appalachian/Blue Ridge Forests**
Major Habitat Type:	**2.1 Temperate Broadleaf and Mixed Forests**
Ecoregion Size:	**159,266 km²**
Biological Distinctiveness:	**Globally Outstanding**
Conservation Status:	**Snapshot—Vulnerable Final—Vulnerable**

Introduction

The Appalachian/Blue Ridge Forests ecoregion encompasses major portions of Fenneman's (1938) Blue Ridge and Ridge and Valley physiographic provinces of the central and southern Appalachians. The region stretches north from northeastern Alabama and Georgia; through eastern Tennessee, western North Carolina, Virginia, and Maryland; and into central Pennsylvania.

The large variety of landforms, climate, soils, and geology, coupled with a long evolutionary history, has led to one of the most diverse assemblages of plants and animals found in the world's temperate deciduous forests (Stephenson, Ash, and Stauffer 1993). The Appalachians were once a large mountain range in the Paleozoic and have been relatively geologically stable since that period. Eroded ridges and valleys throughout the region run in a southwest-northeast direction. Climate differs across the ecoregion due to large variations in elevation, physiography, and latitude. Ancient limestones have eroded into extensive karst formations in some areas, creating a network of caves and unusual communities on limestone. During the Pleistocene glaciations, the Appalachians acted as a mesic and thermal refuge for a number of species and communities. In a similar manner, after the retreat of the glaciers, cold-adapted communities, such as cranberry bogs, remained in refugia in cooler portions of the Appalachians, well south of their usual range.

The Appalachian/Blue Ridge forests consist of two major community types, corresponding to elevational gradients. At lower elevations, between 250 and 1,350 m, mixed oak (*Quercus* spp.) forests dominate. Old-growth cove forests at mid elevations once supported massive tulip poplars (*Liriodendron tulipifera*), chestnuts (*Castanea dentata*), red spruce (*Picea rubens*), and oaks. Above 1,350 m, spruce-fir forests develop and dominate the landscape (Stephenson, Ash, and Stauffer 1993). Along high-elevation ridges, red spruce, the endemic Fraser fir (*Abies fraseri*), and balsam fir (*A. balsamea*) dominate.

Prior to 1890, the low-elevation dominant forest system of this region consisted of mixed oak and American chestnut communities. In the early 1900s the spread of the chestnut blight, caused by the fungus *Cryphonectria parasitica,* resulted in widespread loss of chestnut from the forest community. Chestnut trees once dominated much of the region's lower-elevation forest canopies. With their loss, red oak (*Quercus rubra*), hickory (*Carya* spp.), chestnut oak (*Q. prinus*), black oak (*Q. velutina*), locust (*Robinia pseudoacacia*), and birch (*Betula* spp.), as well as red maple (*Acer rubra*), pines (*Pinus* spp.), and additional hardwood species proliferated (Whitney 1994). The decline of chestnuts was likely associated with a major loss of mast for wildlife.

Biological Distinctiveness

This ecoregion, together with the adjacent Mixed Mesophytic Forests [17], represents one of the world's richest temperate broadleaf forests (only the temperate flora of central China is slightly richer). This ancient mesic forest type was formerly widespread in the Northern Hemisphere and is now represented by relictual ecosystems in eastern North America and eastern China. The related forests of the Appalachians and central and southwestern China share a large number of higher taxa and relict groups. Many genera and some species and families have disjunct distributions in these distant regions. Over fifty such genera of plants include magnolias, hickory, sassafras, ginseng, mayapple, skunk cabbage, several orchids, jack-in-the-pulpit, coffee-tree, stewartia, witch hazel, dogwoods, persimmons, hollies, sumacs, maples, and yellowwood. Several animal taxa also show unique affinities with East Asian relatives, including copperheads (*Agkistrodon* spp.), hellbender salamanders (Cryptobranchidae family), some land snails, and paddlefish (*Polyodon spathula*). The taxonomic similarities between these two regions are paralleled by ecological similarities. These include dominance by woody, broad-leaved, deciduous plants; nonwoody plants with underground storage structures and spring blooming (fourteen hundred species in the Great Smokies alone); early-leafing ephemeral herbs or shade-adapted perennials with buds that overwinter; and plant communities that vary predictably along environmental gradients (Constantz 1994).

The biodiversity of this ecoregion is also exceptional due to the broad range of microhabitats, the presence of numerous relict species and communities, and geologic stability over long periods of evolutionary history to allow for diversification within taxa. Indeed, a number of plants, invertebrates, salamanders, crayfish, freshwater mussels, and fish are restricted to single watersheds or peaks due to millions of years of isolation and favorable conditions. More than 158 tree species can be found within the region, ranking it among the highest ecoregions in North America for total floral diversity. In conjunction with the Appalachian Mixed Mesophytic Forests [17], it contains the highest total amount of endemic flora and fauna species in North America. The Appalachians are the world's center for plethodontid salamander diversity, harboring 34 species of these lungless salamanders. Geographic diversification is pronounced, with isolated and different species and populations on different peaks throughout the region. Several plethodontids exhibit a Mullerian mimicry complex in which several species have converged in appearance to warn predators of their general toxicity. Salamanders are often the most abundant vertebrate with the highest biomass in a given patch of forest, playing a significant role in ecological processes and food webs. Land snails and spiders are well represented and contain a number of relictual and ancient taxa, many with disjunct relatives in the forests of eastern Asia. More than 225 terrestrial vertebrates occupy this ecoregion. The breeding bird community is dominated by neotropical migrants, inhabiting all the successional stages of forest structure.

The prevalent limestone and karst formations in this ecoregion are associated with a cave fauna of salamanders, fish, and invertebrates. The diversity and distribution of these species are not well known, but they appear to rival cave faunas around the world in richness and endemism.

Conservation Status

Habitat Loss

Approximately 83 percent of the habitat in this ecoregion has been altered. Heaviest loss in habitat can be found in the ridge and valley provinces, particularly in limestone valleys that are most productive for agriculture. Habitat loss is greatest in low elevations and diminishes with increased elevation. Lower elevations have milder slopes and were preferentially selected for conversion to agriculture. In addition, suburban sprawl and urban development have occurred in the lower elevations. The vast majority of the region has been logged. Only a few blocks and patches of unlogged forest remain, with several larger blocks found in the Great Smokies region. Virtually all of Shenandoah National Park is regrowth, a situation repeated throughout the region where forests occur today. The extraordinary size of the trees and structural complexity of the ancient cove forests of the Great Smokies belie what has been lost throughout the region from centuries of deforestation and logging. The forests that have returned after logging have few, if any, large trees, no chestnuts, and low structural complexity, and they are uncharacteristically poor in wildflowers, land snails, salamanders, and other species normally abundant and diverse in undisturbed forests. Native wildflower communities have been severely reduced and altered throughout the region, with only a few areas that have a semblance of their former richness (J. Terborgh James B. Duke Professor of Environmental Science, Duke University, pers. comm., 1996).

The spruce-fir forests and portions of the mixed oak forest were subject to intensive logging in the early 1900s. In the wake of poor management, heath balds spread over many ridge tops, thwarting plant and tree regeneration (White et al. 1993). Forests were also cleared for agriculture and pastures. However, these clearings have slowly been abandoned and have subsequently begun to revert to forest communities (Stephenson, Ash, and Stauffer 1993). High-elevation forests consisting of red spruce, the endemic Fraser fir, and the more widespread balsam fir have been subject to environmental and anthropogenic stresses. These high-elevation communities are naturally subject to increased atmospheric moisture, cooler temperatures, and higher winds but also now suffer from the effects of acid rain deposition, which tend to be exacerbated in high-elevation communities, and from the depredations of an introduced homopteran insect, the wooly adelgid (*Adelges* spp.). In the early 1900s, short-sighted logging practices degraded these forests.

Remaining Blocks of Intact Habitat

Several blocks of more or less intact habitat remain as patches on the landscape. A large majority of them can be found within public lands. Larger and smaller patches of old-growth

forest have been inventoried by Davis (1993). Some of the larger blocks are:

- The Great Smoky Mountains National Park—western North Carolina and eastern Tennessee
- Pennsylvania mountain ridges—central Pennsylvania
- South Mountain—south central Pennsylvania
- Shenandoah Mountain and Allegheny Front Range—northern Virginia
- Massanutten Mountain—northern Virginia
- Shenandoah National Park—central Virginia
- Thomas Jefferson National Forests—western Virginia
- Seneca Rocks, Spruce Knob, Panther Knob—eastern West Virginia
- Allegheny Mountains—central Pennsylvania
- Mt. Rogers highlands—southwestern Virginia
- Amphibolite Mountains—northwestern North Carolina
- Blue Ridge escarpment—western North Carolina
- Brushy Mountains—northwestern North Carolina
- Roanoke Mountain highlands—western Virginia
- Nantahalah Mountains—western North Carolina
- Pisgah National Forest—western North Carolina
- Escarpment gorges—northwestern Georgia
- Ramsey's Draft Wilderness—western Virginia

Degree of Fragmentation

Fragmentation and isolation of remaining blocks of undisturbed habitat are severe. Secondary regrowth, however, has begun to offer opportunities for dispersal for a wide range of taxa, although significant movements of less vagile species such as salamanders, wildflowers, and land snails might require centuries. Regrowth areas are still highly fragmented from roads, power lines, and development. Opportunities for songbirds to nest in larger blocks of forests that are relatively undisturbed by cowbirds and mesopredators are few across the landscape.

Degree of Protection

Substantial acreage of forested land has been purchased by the federal and state governments for forests and parks during this century.

Terrestrial Ecoregions of North America

Government agencies, corporations, absentee owners, and 1 percent of the local population control 53 percent of the land (Stephenson, Ash, and Stauffer 1993). Management of these areas depends on both the owning party and the political climate of the day. For example, on lands federally or state controlled, management plans dictate multiple-use management (e.g., timber harvest, recreation, wildlife). The best opportunities for the protection of large portions of this ecoregion lie in national and state forests. These forests are currently being managed for timber extraction, but changes in public demand and Forest Service policy may eventually balance consumptive use (timber extraction) with an increase in nonconsumptive uses (viewsheds, hiking, camping, biodiversity protection) (Stephenson, Ash, and Stauffer 1993).

Types and Severity of Threats

The major types of conversion threats for the ecoregion are timber and mineral extraction, conversion to developed lands, fire suppression, air pollution, acid precipitation, high densities of deer, and the introduction of exotic pests and diseases.

Timber extraction is a serious threat to habitat protection. While the demand for forest products will increase in the future, the abandonment of farms, demand for recreation, and support of the general public for forested lands should offset the extraction of the hardwood forests (Stephenson, Ash, and Stauffer 1993). However, the forest will change, becoming a younger, disturbed forest, which will support different faunal species from those of mature hardwood stands. Trees in much of the regrowing forests are beginning to grow into age classes that attract logging interests. Conservation action must take place now before plans for logging preclude potential opportunities to conserve well-connected, larger blocks of forest across the landscape.

Mining of coal and minerals continues to be a signficant threat to large tracts of habitat through direct destruction and toxic runoff and deposition. Development also threatens the ecoregion. In the past two decades, natural lands have been increasingly developed for recreational resorts and second homes (Mardin and Schwartz 1981; Lovingood and Reiman 1985). This development has, for the most part, been unregulated at the government level in the past.

Most of this developed land was converted from abandoned agricultural fields.

A by-product of increased urban and suburban development, even in distant regions, is an increase in air pollution and acidic precipitation. The ecological effects associated with acid rain deposition include a degradation of fitness and growth of trees and shrubs, a loss of resilience to natural stresses, and direct mortality of sensitive animal species. In particular, the Appalachian high-elevation spruce-fir forest communities have experienced significant forest damage from air pollutants and are highly susceptible to future degradation (de Steiguer, Pye, and Love 1990).

The reduction and extirpation of large predators have caused rodents and deer to proliferate in abundance well above their estimated natural range. Browsing by deer and other abundant herbivores has been implicated in the extirpation of plant species and the alteration of communities throughout the region. Control programs must be intensified to reduce the impact of this pervasive threat. The reintroduction of large predators such as the red wolf and mountain lions could help reestablish an ecological balance to the ecoregion.

The introduction of exotic pests and diseases poses a serious threat to portions of the habitat. The introduction of the gypsy moth (*Lymantria dispar*), spruce budworm (*Choristoneura fumiferana*), hemlock wooly adelgid (*Adelges tsugae*), and balsam wooly adelgid (*A. piceae*), as well as dogwood anthracnost fungi is altering the forest composition and habitat composition.

A number of species in this ecoregion are classified as rare, endangered, or threatened. These include, but are not limited to, a number of land snails and salamanders, a variety of plant species including orchids and many herbaceous plants, the red-shouldered hawk (*Buteo lineatus*), the Virginia big-eared bat (*Plecotus townsendii virginianus*), the red wolf, (*Cnis niger*), the loggerhead shrike (*Lanius ludovicianus*), and perhaps the eastern cougar (*Felis concolor couguar*) (Stephenson, Ash, and Stauffer 1993).

Many cave fauna are threatened by altered hydrologic flows, toxicity and eutrophication from runoff, vandalization of caves and bat roosts, and alteration of cave openings, causing changes in environmental conditions.

Suite of Priority Activities to Enhance Biodiversity Conservation

- Identify and protect larger blocks (greater than 100 km²) of relatively unfragmented and undisturbed forests.
- Restore linkage zones of appropriate habitats between larger blocks.
- Free larger blocks of roads, powerline corridors, and other avenues for intrusions by cowbirds, racoons, and other predators of songbirds. Road closures and prohibition of road building are greatly needed, particularly on public lands.
- Identify and protect a regional system of large, undisturbed habitat blocks in order to maintain viable populations of migratory songbirds.
- Curtail poaching of black bears, freshwater mussels, ginseng, and other economically valuable native plants and animals for commercial purposes.
- Reduce deer populations everywhere to prevent the extirpation of populations of many plant species across their ranges.

Conservation Partners

For contact information, please see appendix G.

- Alabama Natural Heritage Program
- Georgia Natural Heritage Program
- Kentucky Natural Heritage Program
- National Park Service, Great Smoky Mountains National Park
- National Park Service, Shenandoah National Park
- The Nature Conservancy
- The Nature Conservancy, Southeast Regional Office
- The Nature Conservancy of Alabama
- The Nature Conservancy of Georgia
- The Nature Conservancy of Kentucky
- The Nature Conservancy of Maryland
- The Nature Conservancy of North Carolina
- The Nature Conservancy of Pennsylvania
- The Nature Conservancy of South Carolina
- The Nature Conservancy of Tennessee
- The Nature Conservancy of Virginia
- The Nature Conservancy of West Virginia
- Sierra Club
- U.S. Forest Service

- Virginia Division of Natural Heritage
- Western Pennsylvania Conservancy
- The Wildlands Project, SouthPAW

Relationship to Other Classification Schemes

The ecoregion was based on Omernik's (1995b) Central Appalachian Ridges and Valleys and Blue Ridge Mountains ecoregions. The ecoregion approximates Bailey's (1994) Northern Ridge and Valley section (M221A), the Blue Ridge Mountains section (M221D—although the northern sliver is left out), and the Central Ridge and Valley section (221J) and captures smaller portions of several other sections (231c, 231D, 221H, 221H, M221C, and M221B). The ecoregion covers a majority of Küchler's (1975) Appalachian Oak Forest as well as portions of the Oak-Hickory-Pine Forest found in West Virginia.

Prepared by C. Loucks, D. Olson, E. Dinerstein, A. Weakley, R. Noss, J. Stritholt, and K. Wolfe

Key Number:	**17**
Ecoregion Name:	**Appalachian Mixed Mesophytic Forests**
Major Habitat Type:	**2.1 Temperate Broadleaf and Mixed Forests**
Ecoregion Size:	**183,000 km²**
Biological Distinctiveness:	**Globally Outstanding**
Conservation Status:	**Snapshot—Endangered Final—Critical**

Introduction

The extraordinary forests of southeastern North America represent relicts of ancient mesic forests that once covered much of the temperate regions of the Northern Hemisphere. Today, examples of these forests can be found only in the southeast region of North America and in eastern and central China. The Appalachian Mixed Mesophytic Forests ecoregion encompasses the moist broadleaf forests that cover the plateaus and rolling hills west of the Appalachian Mountains. It extends southward into northwest Alabama and east central Tennessee. Moving north, the region includes eastern Kentucky, western North Carolina, most of West Virginia, southeastern Ohio, and southwestern Pennsylvania. Mixed mesophytic forests acted as a mesic refuge during drier glacial epochs for a wide range of taxa. The long evolutionary history of the region and wide range of topographic and edaphic conditions have contributed to the development of the rich biota and abundance of endemic species, particularly in freshwater communities.

Biological Distinctiveness

The Mixed Mesophytic Forest ecoregion represents one of the most biologically diverse temperate regions of the world. Forest communities often support more than thirty canopy tree species at a single site, as well as rich understories of ferns, fungi, perennial and annual herbaceous plants, shrubs, small trees, and diverse animal communities. Songbirds, salamanders, land snails, and beetles are examples of some particularly diverse taxa. Indeed, the ecoregion harbors some of the richest and most endemic land snail, amphibian, and herbaceous plant biotas in the United States and Canada. The ecoregion's freshwater communities are the richest temperate freshwater ecosystems in the world, with globally high richness and endemism in mussels, fish, crayfish, and other invertebrates.

The lower-elevation forests contain a variety of forest types, with magnolias (*Magnolia* spp.), oaks (*Quercus* spp.), hickories (*Carya* spp.), walnuts (*Juglans* spp.), elms (*Ulmus* spp.), birches (*Betula* spp.), ashes (*Fraxinus* spp.), basswoods (*Tilia* spp.), maples (*Acer* spp.), locusts (*Robinia* spp.), pines (*Pinus* spp.), the grand tulip poplar (*Liriodendron tulipifera*), black gum (*Nyssa sylvatica*), eastern hemlock (*Tsuga canadensis*), black cherry (*Prunus serotina*), sweet gum (*Liquidambar styraciflua*), American beech (*Fagus grandifolia*), and yellow buckeye (*Aesculus octandra*). The American chestnut (*Castanea dentata*) was a dominant canopy species but was extirpated at the turn of the century by the introduced chestnut blight fungus (*Cryphonectria parasitica*). Some endemic species include the Allegheny plum (*Prunus alleghaniensis*) and the Black Mountain salamander (*Desmognathus welteni*).

Higher-elevation forests toward the east have yellow birch, mountain maple, sugar maple, beech, and eastern hemlock, with extensive understories of mountain laurel (*Kalmia latifolia*) and rhododendron (*Rhododendron* spp.). A variety of restricted habitats occur within the forests including glades, heath barrens, shale barrens, and sphagnum bogs. Many of these communities support endemic plants and land snails. Cranberry bogs harbor a range of species that are normally associated with more northerly ecoregions such as cranberry (*Vaccinium* spp.), blueberry (*Vaccinium* spp.), bog rosemary (*Andromeda glaucophylla*), buckbean (*Menyanthes trifoliata*), northern goshawk (*Accipiter gentilis*), fisher (*Martes pennanti*), and black-billed magpie (*Pica pica*). Such bogs and glades are relicts that have survived with their disjunct populations of cool-adapted species since cooler glacial epochs. Surrounding high-elevation forests also support disjunct northern species such as the Canada yew (*Taxus canadensis*), eastern larch (*Larix laricina*), red pine (*Pinus resinosa*), and balsam fir (*Abies balsamea*).

Conservation Status

Habitat Loss

Over 95 percent of this habitat, perhaps more, has been converted or degraded at some point in the last two hundred years. Only a few very small and scattered fragments of undisturbed or old-growth forests remain, most less than a few hectares in size (Davis 1993). Forests were converted for agriculture, coal mining, logging for charcoal, dams, and road building. Most of the agricultural lands have subsequently failed and are being abandoned, with an increase in the growth of secondary, or pioneer, forests. These regrowing forests lack many of the features and much of the diversity of undisturbed, or old-growth forests, namely large trees; variable age classes of trees; structural complexity such as multiple canopy layers; and diverse and abundant wildflowers, salamanders, fungi, land snails, and other invertebrate taxa. Because of the intensity and broad extent of clearing of forests over the last two centuries, many forest-specialist species appear to have been extirpated over large portions of the landscape. If source populations in undisturbed forest fragments are not imbedded in or adjacent to regrowing tracts, large areas of secondary forests may remain depauperate into the future.

Secondary forests have the capacity to conserve a great deal of biodiversity and represent, in combination with the last fragments of undisturbed forest, the best opportunity to conserve the region's biodiversity over the long term. Larger, unroaded blocks of forest can also act as source pools for breeding migratory songbirds that are experiencing negative reproductive rates due to cowbird parasitism and nest predation by meso-predators in the mosaic of smaller forest fragments across the landscape. Trees within secondary forests are beginning to attain sizes that are attractive to logging interests. A landscape-scale conservation strategy for conserving large, interconnected blocks of mature forests urgently needs to be developed and implemented.

Remaining Blocks of Intact Habitat

Few remaining patches of undisturbed forest remain, although older pioneer forests (i.e., forests that have regrown from previously cleared land) can be relatively large. The larger habitat blocks that do exist are found primarily on public lands. Some of the larger extant blocks of relatively intact habitat can be found within the following areas:

- Daniel Boone National Forest—east-central and southeastern Kentucky
- Shawnee State Forest—southern Ohio
- Wayne National Forest—southern Ohio
- Big South Fork National Recreational Area—north-central Tennessee
- Savage Gulf State Natural Area—south-central Tennessee, Grundy County)
- Cranberry Wilderness—southeastern West Virginia
- Monongahela National Forest—eastern West Virginia
- Frozen Head State Natural Area—east-central Tennessee
- Cumberland Gap—southeastern Kentucky
- Pine Mountain—southeastern Kentucky, Letcher County
- Blanton Forest—southeastern Kentucky, Harlan County
- Sipsey Wilderness—north-central Alabama
- Talladega National Forest—east-central Alabama
- Scott State Forest—northeastern Tennessee

Degree of Fragmentation

Much of the existing forest, whether old-growth or regrowth forest, is still distributed in a highly fragmented mosaic throughout the region, broken by agriculture, roads, power lines, towns, and other forms of development. However, when one considers regrowth forests, the Appalachian Mixed Mesophytic Forests ecoregion has lower levels of fragmentation relative to other East Coast ecoregions. Fragmentation is highest in the northern part of the ecoregion, primarily in southwestern Pennsylvania and Ohio. The southern section of the ecoregion is comparatively less fragmented and has better potential for restoration into larger blocks within the context of a conservation strategy.

Degree of Protection

Most larger blocks of forest presently occur in federal and state forests, wilderness areas, and state natural areas. However, the management plans for federal forest lands do not strictly protect the forests, but reflect the multi-use management policy of the Forest Service. Present federal and state policies dictate intensive harvest of timber from national forests, usually accompanied by road building, fire suppression, thinning, application of herbicides and pesticides, and other ecologically damaging management practices. No effective formal process of identifying and protecting rare, distinctive, representative, or otherwise important communities, species, or ecosystems has been developed by federal or state agencies. Some small but highly distinctive or rare communities, such as bogs and glades, have been protected by private organizations such as The Nature Conservancy. Several landscape-level conservation systems have been proposed for this ecoregion and the adjacent Appalachian ecoregion, consisting of a network of core protected areas, corridors and linkage zones, and buffer zones (Mueller 1992; Leverett 1993).

Types and Severity of Threats

A primary threat is the increasing conversion and fragmentation of forests through logging and development. Hardwood forests are increasingly being exploited throughout the region as maturing forests become attractive to timber exploiters and production in West Coast forests declines. Multinational timber industries, as well as local chip mills in Kentucky and Tennessee, create demand for increased harvests on public and private lands.

Coal, copper, and ore mining in this ecoregion are a major cause of air and water pollution, causing widespread degradation and poisoning of ecosystems. The globally outstanding freshwater biodiversity of the ecoregion is highly imperiled from toxic pollution, acid runoff from mines, pesticides and herbicides, sedimentation, eutrophication from excess nutrient runoff, dams, dredging, channelization, and introduced species such as the zebra mussel. Acid rain deposition, from industrial and urban sources, continues to be a major problem in many sensitive ecosystems, particularly in higher-elevation forest communities.

Highways continue to cause high mortality in wildlife and are barriers to dispersal for many species. Numerous proposed highways, roads, and power lines cut across many of the larger blocks of forest in the ecoregion, particularly in the Monongahela National Forest (e.g., "Corridor H" transmission lines in the proposed Cherry River Wilderness). Road building into larger blocks of forests should be curtailed to reduce fragmentation and loss of source pool breeding sites for migratory songbirds. Off-road vehicle use and road building have severely degraded riparian communities and rare bogs and glades in many areas.

Abundant populations of deer, resulting from the eradication of large predators and poorly managed hunting programs, have been implicated in the extirpation and reduction of many understory plant species and the alteration of community structure (Alverson, Waller, and Solheim 1988). The nearly extirpated Canada yew (*Taxus canadensis*) of Monongahela National Forest is a classic example of this problem (Mueller 1992).

Many wild herbs and other plants are harvested for commercial purposes, and some, like wild ginseng, are threatened with extirpation over large areas of their range because of unregulated and illegal poaching. Large numbers of black bears are poached for their gall bladders for the Asian medicinal trade. Freshwater mussels are legally and illegally harvested for their shells to be used as nuclei for cultured pearls in Asia. A number of endangered species, including many plants and freshwater mussels and fish, occur within the ecoregion.

Suite of Priority Activities to Enhance Biodiversity Conservation

- Identify and protect large core areas of forest, linkage zones, and buffer zones, building upon existing protected sites. Examples include the proposed Cherry River Wilderness, Cranberry Wilderness expansion, Cheat Bridge corridor, Canaan Mountain Wilderness, Laurel Fork Wilderness, Kentucky River corridor, Cumberland River corridor, and the Gauley Mountain Wilderness (Mueller 1992; Leverett 1993).

- Identify, restore, and protect of large blocks of unfragmented forest habitat that can act as source pools for breeding migratory songbirds. Populations that breed in landscapes with only small, fragmented forest patches generally have negative reproductive rates because of cowbird parasitism and predation by racoons, crows, opossums, and other meso-predators. Recent studies suggest that the few remaining very large blocks of forest are maintaining populations of songbirds over vast regions.

- Protect and expand of existing large blocks and restoration of additional blocks distributed across the landscape. Conserving migratory songbirds will occur only if state and federal agencies can be persuaded to stop building roads and power line corridors and to begin to close existing roads to restore large contiguous blocks of forest. Plans to conserve larger blocks of forest for songbird conservation need to be implemented immediately before logging interests obtain concessions throughout the regions as regrowing forests becomes more lucrative.

- Implement plans to increase the connectivity of public and conserved private lands, particularly in Wayne State Forest and the Cumberland Plateau of Tennessee.

- Reduce and control acid precipitation, gypsy moths, wooly adelgids, and zebra mussels.

- Control poaching of black bears and other wildlife and commercially harvested herbs.

- Reevaluate fire suppression and management practices in light of maintaining native communities.

- Increase heritage inventories of the ecoregion to identify additional areas and species populations in need of protection and conservation action.
- Develop hunting management plans that would prevent overabundant deer populations from causing irreversible ecological damage. Reintroduce cougars and gray wolves and improve management of existing populations of black bear and mustellids to help reestablish ecological interactions that are sustainable and less damaging to the ecosystem than existing conditions.

Conservation Partners

For contact information, please see appendix G.

- Alabama Natural Heritage Program
- Cumberland Gap Friends of the Bankhead
- Georgia Natural Heritage Program
- Kentucky Heartwood
- Kentucky Natural Heritage Program
- The Nature Conservancy
- The Nature Conservancy, Southeast Regional Office
- The Nature Conservancy of Alabama
- The Nature Conservancy of Georgia
- The Nature Conservancy of Kentucky
- The Nature Conservancy of Ohio
- The Nature Conservancy of Pennsylvania
- The Nature Conservancy of Tennessee
- The Nature Conservancy of West Virginia
- Ohio Natural Heritage Data Base
- Tennessee Division of Natural Heritage
- U.S. Fish and Wildlife Service
- USDA Forest Service
- West Virginia Highlands Conservancy
- West Virginia Natural Heritage Program
- Western Pennsylvania Conservancy
- Wildlands Project

Relationship to Other Classification Schemes

The Mixed Mesophytic Forests ecoregion was based on an aggregation of several of Omernik's (1995b) level III ecoregions. Combined into our Mixed Mesophytic Forests ecoregion due to similarities in biodiversity characteristics and dynamics were Western Allegheny Plateau (70),

Central Appalachians (69), Southwestern Appalachians (68), and the extreme southwest portion of the Interior Plateau (71) that lies within Alabama. The Mixed Mesophytic Forests [17] covers a majority of Küchler's (1975) Mixed Mesophytic Forest as well as portions of the Oak-Hickory-Pine ecoregion in the south and the Appalachian Oak forest in southwestern Pennsylvania. This ecoregion roughly corresponds to Bailey's (1994) Southern Unglaciated Allegheny Plateau section (221E), Northern Cumberland Plateau section (221H), Southern Cumberland Plateau section (231C), and portions of the Southern Ridge and Valley section (231D).

Prepared by C. Loucks, D. Olson, E. Dinerstein, A. Weakley, R. Noss, J. Stritholt, and K. Wolfe

Key Number:	**18**
Ecoregion Name:	**Central U.S. Mixed Hardwood Forests**
Major Habitat Type:	**2.1 Temperate Broadleaf and Mixed Forests**
Ecoregion Size:	**296,019 km²**
Biological Distinctiveness:	**Regionally Outstanding**
Conservation Status:	**Snapshot—Endangered**
	Final—Endangered

Introduction

As in the ecoregions to the east, broadleaf deciduous trees dominate the Central U.S. Mixed Hardwood Forests. This region receives less precipitation than the more coastal areas, however, so drought-resistant oak-hickory forests predominate here. While other forests in the United States and Canada have both oak and hickory, this region was once the only one where both species occurred in abundance over a large area. Much of the natural habitat in this ecoregion has now been destroyed by development and agriculture. Most of the area covered by the Central U.S. Mixed Hardwood Forests is rolling, but some parts are nearly flat and the Ozark highlands reach approximately 1,000 m. Parts of Kentucky and Tennessee are marked by dissected plateaus and basins.

Biological Distinctiveness

The Central U.S. Mixed Hardwoods ecoregion is among the richest in North America for herbaceous plants and shrubs, with 2,527 species (Kartesz 1997). The tree flora is less diverse, dominated by only a few species. The oak-hickory forest becomes more savannalike in its northern reaches. In southern Illinois and Indiana, the forest forms a mosaic with prairie. Widespread dominants are white oak (*Quercus alba*), red oak (*Q. rubra*), black oak (*Q. velutina*), bitternut hickory (*Carya cordiformis*), and shagbark hickory (*C. ovata*). Flowering dogwood (*Cornus florida*) often occurs in the understory, along with sassafras (*Sassafras* spp.) and hop hornbeam (*Carpinus* spp.). The shrub layer is distinct, often with evergreens, and wildflowers are common. Intact wetter sites feature American elm (*Ulmus americana*), tulip tree (*Liriodendron tulipifera*), and sweet gum (*Liquidambar styraciflua*).

Conservation Status

Habitat Loss

Only about 1 percent of the central U.S. hardwoods remains as intact habitat. The majority of the ecoregion has been heavily altered by human activity, particularly conversion to agriculture, short-rotation silviculture, and pasture in some areas (e.g., bluegrass).

Remaining Blocks of Intact Habitat

All the remaining blocks of habitat in this ecoregion are smaller than 1,000 km². The most important are:

- Mammoth Cave—south-central Kentucky
- Land between the Lakes—southeastern Kentucky, northeastern Tennessee
- Hoosier National Forest, Yellowwood National Forest—southern Indiana
- Edge of Appalachia—central Tennessee
- Western Highland Rim—eastern Tennessee
- Lower Missouri Ozarks—southern Missouri
- Tennessee Cedar Glades—central Tennessee
- Missouri Ozarks and southern Mark Twain National Forest—southern Missouri, northern Arkansas
- Shawanee Hills—southern Illinois
- Big River Junction—eastern Illinois
- Wolf River—southeastern Tennessee
- Fort Campbell—northern Tennessee, southern Kentucky

Degree of Fragmentation

Most of this ecoregion has been highly fragmented, so large predators have disappeared. The remaining habitat blocks have some degree of connectivity, but few corridors exist between blocks. Degraded corridors within the blocks may be restorable, though the focus should be on improving the integrity of large blocks.

Degree of Protection

Mammoth Cave, Edge of Appalachia, and parts of the Missouri Ozarks and the Tennessee Cedar Glades are the only blocks of habitat in the ecoregion to enjoy substantial protection. These areas do not adequately protect the habitats of the central U.S. hardwoods.

Types and Severity of Threats

Urban sprawl and agriculture are the greatest conversion threats to the region. Invasion of exotic grasses, cave vandalism, overuse for recreation, fire suppression in fire-maintained systems, and loss of large ungulates (bison) are degrading the remaining natural habitats. Deer poaching continues to be a problem in Kentucky and Tennessee, and collection of wild herbs is ongoing across the region.

Suite of Priority Activities to Enhance Biodiversity Conservation

- Improve integrity of existing natural habitats
- Continue to identify existing biodiversity hot spots and protect them. These are important places that conserve the full range of communities and species within the ecoregion, some of which will be small, some large, some connectible, and some not.

Conservation Partners

For contact information, please see appendix G.

- Alabama Natural Heritage Program
- Arkansas Natural Heritage Commission
- Illinois Natural Heritage Division
- Indiana Natural Heritage Data Center

- Kentucky Heartwood
- Kentucky Natural Heritage Program
- Mississippi Natural Heritage Program
- Missouri Natural Heritage Database
- The Nature Conservancy
- The Nature Conservancy, Midwest Regional Office
- The Nature Conservancy, Southeast Regional Office
- Ohio Natural Heritage Data Base
- Oklahoma Natural Heritage Inventory
- Tennessee Division of Natural Heritage

Relationship to Other Classification Schemes

This ecoregion corresponds roughly to the southern half of Bailey's (1994) Eastern Broadlead Forest (Continental) Province. The northern portion of Bailey's province should be distinguished on the basis of climate, vegetation types, and ecosystem dynamics. Omernik (1995b) classifies the Ozark highlands as a separate ecoregion, while we believe these rightly belong with the central U.S. hardwoods.

Prepared by A. Weakley, R. Noss, E. Dinerstein, S. Robinson, and J. Adams

Key Number:	**19**
Ecoregion Name:	**Ozark Mountain Forests**
Major Habitat Type:	**2.1 Temperate Broadleaf and Mixed Forests**
Ecoregion Size:	**62,011 km²**
Biological Distinctiveness:	**Regionally Outstanding**
Conservation Status:	**Snapshot—Endangered**
	Final—Endangered

Introduction

The Ozark Mountain Forests are comprised chiefly of the Ouachita and Boston Mountains, whose forests are among the best developed oak-hickory forests in the United States. The primary species here are red oak (*Quercus rubra*), white oak (*Q. alba*), and hickory (*Carya* spp., especially *Carya texana*). Shortleaf pine (*Pinus echinata*) and eastern red cedar (*Juniperus virginiana*) are important on disturbed sites, shallow soils, and south- and west-facing slopes.

The Ouachita Mountains cover an area approximately 380 km east to west by 100 km north to south in western Arkansas and southeastern Oklahoma. The region has ridge and valley topography, and erosion has been the dominant geological force for the last 300 million years. Maximum average annual precipitation in the Ouachita Mountains is over 150 cm, and minimum average annual precipitation is less than 100 cm (Foti and Glenn 1991).

This ecoregion includes significant old-growth forest. Thousands of acres of stunted oaks on dry sites over 760 m (2,500 ft) in elevation in western Arkansas' Ouachita National Forest have never been cut. There is an abundance of acreage in cover types of pine-mixed, pine-hardwood, and hardwood seventy years of age and older, with relatively small amounts of forest over one hundred years old (Fryar 1991). In the Boston Mountains, scattered old growth may total 181 km² (Davis 1996).

Biological Distinctiveness

This ecoregion contains distinctive freshwater communities. The Ouachita and Boston Mountains also were minor Pleistocene refugia. The ecoregion is also home to an endemic earthworm, *Diplocardia meansi*. Restricted to the drier soils of Rich Mountain, this worm is the second-largest known earthworm in the United States and is bioluminescent. When tweaked or shocked, the worm secretes a fluid that glows in the dark (Robison and Allen 1995).

Conservation Status

Habitat Loss

Only about 3 percent of this ecoregion remains as intact habitat. While the upper-elevation forests are still in relatively good condition, riparian habitat has been severely degraded. The forests along the Arkansas River, for example, have been almost completely destroyed. Conversion to agriculture, logging, fire suppression, and grazing are the main causes of habitat loss.

Remaining Blocks of Intact Habitat

The Boston Mountains and Ouachita Mountains themselves are the only remaining large and intact habitat blocks in this ecoregion. There are no significant blocks of intact lowland habitats, and even the blocks on the Ouachita Mountains are broad and somewhat diffuse.

Degree of Fragmentation

Habitats in this ecoregion have been badly fragmented. There is some degree of connectivity within the habitat blocks, but little between the blocks. Degraded corridors within the two major habitat blocks (see above) may be restorable, but such is not the case elsewhere in the ecoregion. Similarly, ecosystem dynamics remain intact within the major habitat blocks but not outside them.

Degree of Protection

The Boston and Ouachita Mountains are the only adequately protected areas in this ecoregion.

Types and Severity of Threats

Conversion of hardwood to pine forests and construction of second homes and resorts pose the greatest conversion threats to this ecoregion. Fire suppression is a major cause of degradation. Wildlife exploitation in the form of bear poaching and herb extraction is also a significant concern.

Suite of Priority Activities to Enhance Biodiversity Conservation

- Improve biodiversity management of publicly owned areas.
- Identify and conserve important biodiversity areas.
- Continue major ecosystem restoration efforts in Ouachita National Forest.
- Complete a land exchange between Weyerhaeuser and the U.S. Forest Service.
- Conserve important old-growth sites (from Davis 1996) such as:
  - Blackfork Mountain
  - Clifty Canyons Special Interest Area (7 km^2)
  - Turkey Mountain and adjacent slopes (4 km^2 of apparently unlogged post oak and chinquapin oak savanna)
  - Cossatot River State National Park Natural Area (up to 8 km^2 of old-growth eastern red cedar glades and xeric pine or pine-hardwood forest on steep slopes)

- Cucumber Creek Watershed (72.85 km^2). The Oklahoma Nature Conservancy owns 6 km^2 of this watershed, which includes some large shortleaf pine on the steeper slopes that are 51 to 76 cm in diameter and black gum and hickory that may be up to 91 cm in diameter. Large sycamore trees occur along the stream. The upper part of the watershed has probably not been cut since the turn of the century.

Conservation Partners

For contact information, please see appendix G.

- Arkansas Natural Heritage Commission
- The Nature Conservancy
- The Nature Conservancy, Southeast Regional Office
- Oklahoma Natural Heritage Inventory
- The Ozark Society

Relationship to Other Classification Schemes

Omernik (1995b) creates separate ecoregions for the Boston and Ouachita Mountains, primarily on the basis of elevation. Bailey (1994) also has separate classifications for this region, placing the Boston Mountains in the Ozark Broadleaf Forest-Meadow Province and the rest of the region in the Eastern Broadleaf Forest (Continental) Province. We believe the similarities of ecological processes and communities justify placing the Boston and Ouachita Mountains in a single ecoregion. Those processes and communities also distinguish this region from the rest of the Ozarks and surrounding forest types.

Prepared by A. Weakley, E. Dinerstein, R. Noss, S. Robinson, J. Strittholt, and J. Adams

Key Number: **20**
Ecoregion Name: **Mississippi Lowland Forests**
Major Habitat Type: **2.1 Temperate Broadleaf and Mixed Forests**
Ecoregion Size: **112,284 km²**
Biological Distinctiveness: **Bioregionally Outstanding**
Conservation Status: **Snapshot—Critical Final—Critical**

Introduction

The Mississippi Lowland Forests ecoregion occurs in the floodplain of the Mississippi River and runs north from Louisiana to the southern tip of Illinois. It is separated from surrounding ecoregions by topography, floodplain dynamics, and vegetation. Surrounding ecoregions typically support more xeric conifer-dominated forests on upland landscapes. The Mississippi Lowland Forests were very similar in composition to the magnificent bottomland forests that occur along the rivers found in the Middle Atlantic Coastal Forests [50]. However, virtually all of the remaining riparian forests of this ecoregion are gone, making it one of the most heavily converted ecoregions in the United States.

Biological Distinctiveness

The Mississippi Lowland Forests serve as an important part of a major flyway route used by migratory birds. The rich bottomland forests once contained some of the most interesting hardwood communities in the United States, but these are virtually all cleared.

The natural vegetation is dominated by bottomland hardwood forests, primarily oak-hickory-pine forests (Küchler 1964). Küchler describes this region as "a dense medium tall to tall forest of broadleaf deciduous and evergreen trees and shrubs and needleleaf deciduous trees" that occurs along river and stream floodplains. During the last period of glaciation, it is likely that major river channels served as important pathways for migrating species. As a result, a mixing of upland and bottomland species occurred. Consequently, this ecoregion includes species from the mixed mesophytic forests, the Appalachian/Blue Ridge forests, and the central hardwoods (Sharitz and Mitsch 1993).

Bottomland hardwood forests represented the majority of the inland wetland acreage in the United States during European colonization.

Historical descriptions report that these areas grew cypress, gum, hickory, oak, and cedar as their major overstory components. Virgin stands of cypress were typically 400–600 years old at the time of European settlement. Over the last century most of these lands have been logged, and few individuals over 200 years old remain (Sharitz and Mitsch 1993).

The most important environmental condition in this ecoregion is the hydroperiod, which controls the amount of oxygen and moisture available to the forest communities. The floodplain forest communities can be segregated into different species assemblages, based on the hydroperiod. They include, in decreasing flood duration, river swamp forests, lower hardwood swamp forests, backwater and flats forests, and upland transitional forests. River swamp forests, which are adapted to continuous flooding, contain bald cypress (*Taxodium distichum*) and water tupelo (*Nyssa aquatica*), which often codominate the canopy. Associated with river swamp forests are button bush (*Cephalanthus occidentalis*), water ash (*Fraxinus caroliniana*), water elm (*Planera aquatica*), and black willow (*Salix nigra*). Lower hardwood swamp forests are similar to river swamp forests but have a more diverse woody community. Water hickory (*Carya aquatica*), red maple (*Acer rubrum*), green ash (*Fraxinus pennsylvanica*), and river birch (*Betula nigra*) increase in prevalence. Common herbs include butterweed (*Senecio glabellus*), jewelweed (*Impatiens capensis*), and royal fern (*Osmunda regalis*). Forests of backwaters or flats make up a third zone and are seasonally saturated. They support a greater richness of hardwood species. In addition to lower hardwood swamp species, sweet gum (*Liquidamber styraciflua*), sycamore (*Platanus occidentalis*), laurel oak (*Quercus laurifolia*), and willow oak (*Q. phellos*) are present. Woody vines increase in abundance, including poison ivy (*Toxicodendron radicans*), greenbriers (*Smilax* spp.), and trumpet-creeper (*Campsis radicans*). At the transition to upland forests are ridges and dunes formed during the Pleistocene, as well as natural levees. These areas are flooded for only brief periods of time. This area is similar to the oak-hickory-pine forests of the southeastern piedmont area (Sharitz and Mitsch 1993).

Conservation Status

Habitat Loss

About 91–95 percent of this habitat has now been converted to agriculture or other uses or is highly degraded. Soybean cultivation dominates land use. The habitats most affected are bottomland forests, which were cleared for agriculture or harvested for timber long ago.

Remaining Blocks of Intact Habitat

Remaining habitat is confined to the wettest sites, which are difficult to exploit economically or put under cultivation. The remaining blocks are not representative of the ecoregion's major habitats. Remaining fragments include:

- Atchafalaya area and surrounding lowlands (status and ownership uncertain)—southern Louisiana
- Crowley's Ridge (partly included in St. Francis National Forest)—northeastern Arkansas, southeastern Missouri
- Big Woods Conservation Area adjacent to USFWS refuge (TNC)—northern Louisiana
- Cache River Restoration Project—southern Illinois, southeastern Missouri, southwestern Kentucky
- Mingo NWR—southeastern Missouri
- several National Wildlife Refuges of uncertain biodiversity value

Degree of Fragmentation

There is no possibility at present of connecting the existing blocks mentioned above. This ecoregion has been greatly affected by fragmentation, levee construction, and the alteration of river flow. The long-term potential of corridor restoration is low.

Degree of Protection

This ecoregion contains minimal habitat protection. The current protected areas do not contain vegetation typical of the ecoregion.

Types and Severity of Threats

Because of the high degree of conversion, there is little left to conserve. Hydrologic alterations have the greatest impact. Logging remains a threat, as does continued exploitation of remaining forests. Pollutant effects in the lower section of the ecoregion are serious.

Suite of Priority Activities to Enhance Biodiversity Conservation

- Designate Atchafalaya area and surrounding lowlands and the Big Woods Conservation Area as protected areas and improve conservation management.
- Continue funding for interagency planning efforts to create corridors.
- Continue inventory and identification of important sites for biodiversity and protect remaining areas.

Conservation Partners

For contact information, please see appendix G.

- The Nature Conservancy
- The Nature Conservancy, Midwest Regional Office
- The Nature Conservancy, Southeast Regional Office
- Partners in Flight

Relationship to Other Classification Schemes

The Mississippi Lowland Forests corresponds to Küchler's (1975) unit 103 (Southern Floodplain Forest). This ecoregion is identical to Omernik's (1995b) unit 73 (Mississippi Alluvial Plain) with the exception of the coastal grasslands of Louisiana, which were separated and lumped with the Western Gulf Coastal Grasslands [68], based on Küchler (1975). Bailey's (1994) section 234A (Mississippi Alluvial Basin) approximates the boundaries of this ecoregion.

Prepared by A. Weakley, E. Dinerstein, R. Snodgrass, and K. Wolfe

Key Number:	**21**
Ecoregion Name:	**East Central Texas Forests**
Major Habitat Type:	**2.2 Temperate Broadleaf and Mixed Forests**
Ecoregion Size:	**52,637 km²**
Biological Distinctiveness:	**Bioregionally Outstanding**
Conservation Status:	**Snapshot—Critical**
	Final—Critical

Introduction

The East Central Texas Forests ecoregion is located entirely within the state of Texas and is one of the smallest ecoregions within the Temperate Broadleaf and Mixed Forests MHT. The potential natural vegetation of the ecoregion consists of oak-hickory forest, the "cross timbers" association of oaks and little bluestem, and juniper-oak savannas. Common hardwoods of the oak-hickory association include scarlet, post, and blackjack oaks (*Quercus coccinea, Q. stellata,* and *Q. marilandica,* respectively), and pignut and mockernut hickories (*Carya glabra* and *C. tomentosa*). Forests of elm (*Ulmus americana*), pecan (*Carya illinoensis*), and walnut (*Juglans regia*) occur along river courses. The ecoregion is distinguished from adjacent prairie units and the coastal plain grasslands by higher tree den- sity and from the Piney Woods forests by the more open nature of the habitat and the greater dominance of hardwoods. Fire and drought were historically principal disturbance factors.

Biological Distinctiveness

The East Central Texas Forests are character- ized by species associated with temperate, sub- humid forests. Jaguar (*Felis onca*) and bison (*Bison bison*) both occurred in the ecoregion pre- viously. The forests are notably rich in butterflies and reptiles.

Conservation Status

Habitat Loss

Like some of the surrounding habitat, this ecore- gion is being heavily altered through both ranch- ing and farming practices. Approximately 75 percent of the natural vegetation of this ecore- gion has been converted to agriculture.

Remaining Blocks of Intact Habitat

There are no large portions of intact habitat left in this ecoregion. A majority of the remaining habitat exists in small blocks of 100–200 acres, and the quality of habitat in these blocks is quite variable.

Degree of Fragmentation

The original habitat is highly fragmented, and while the original floral component of the ecore- gion still exists, the fragmented nature of the ecoregion has severely limited the flow of natural ecological processes.

Degree of Protection

There are no national forests located in this ecoregion, and similarly the level of protection is extremely minimal. The only partially protected piece of land, outside of state parks, is the Attwater Prairie Chicken National Wildlife Refuge.

Types and Severity of Threats

The major threats to the ecoregion are the con- version of forests to agriculture. In addition, the suppression of fire in the remaining blocks of habitat may threaten their biological integrity in the future.

Suite of Priority Activities to Enhance Biodiversity Conservation

- Work to protect the remaining blocks of intact habitat.
- Attempt to identify potential parcels of land for restoration.
- Work to promote greater awareness of living resources within the communities.

Conservation Partners

For contact information, please see appendix G.

- The Nature Conservancy of Texas

Relationship to Other Classification Schemes

The boundaries of the East Central Texas Forests are taken from Omernik (1995b) ecore- gion 33 (East Central Texas Plains). Both our classification and Omernik's unit are derived from Küchler (1975) unit 91 (Oak-Hickory forest). The unit is similar in location and shape to Bailey et al.'s (1994) subregion no. 255C (Oak Woods and Prairie section). Fire and drought were historically principal disturbance factors.

Prepared by C. Loucks

Key Number:	**22**
Ecoregion Name:	**Southeastern Mixed Forests**
Major Habitat Type:	**2.1 Temperate Broadleaf and Mixed Forests**
Ecoregion Size:	**347,803 km²**
Biological Distinctiveness:	**Globally Outstanding**
Conservation Status:	**Snapshot—Critical Final—Critical**

Introduction

The Southeastern Mixed Forests skirt the Appalachian/Blue Ridge Mountains, occupying the piedmont zone between upland forests and the Atlantic Coastal and Gulf Coastal plains. This ecoregion is by far the largest within the temperate broadleaf and mixed forests MHT, crossing nine states and running northeast to southwest from Maryland to Louisiana. The Southeastern Mixed Forests [51] are demarcated from the Southeastern Conifer Forests to the south by vegetation (Küchler 1975) and elevation (the fall line of the Atlantic piedmont). The latter ecoregion is more dominated by longleaf pine (*Pinus palustris*) and occurs on the coastal plain rather than on the piedmont. It is similarly separated from the Appalachian/Blue Ridge Forests [16] and Appalachian Mixed Mesophytic forests [17] by elevation and vegetation. The Southeastern Mixed Forests, lying between all of these species-rich ecoregions, is enriched by its proximity to these other units. However, this ecoregion is perhaps the most altered, having been heavily and repeatedly logged and now largely converted to agriculture.

Biological Distinctiveness

The Southeastern Mixed Forests are famous as the center of gastropod diversity for North America and perhaps the world, but many of the endemic taxa are extinct. The freshwater ecosystems found within this ecoregion are among the richest in the temperate latitudes. The Southeastern Mixed Forests rank among the top ten ecoregions in richness of amphibians, reptiles, and birds and among the top ten ecoregions in number of endemic reptiles, amphibians, butterflies, and mammals. Kartesz (1997) identifies 3,635 native herbaceous and shrub species, the highest in North America.

The natural vegetation is dominated by oak-hickory-pine forests (Küchler 1964). At the time of European settlement this ecoregion was dominated by stands of pure pines and stands of pure hardwoods, with mixtures of each between these extemes. Hardwoods were much more prevalent in distribution than they are today (Skeen, Doerr, and Van Lear 1993). The most dominant ecological force in shaping composition and structure of the Southeastern Mixed Forests, prior to European settlement, was fire. Fire disturbance provided good seed beds for pines and consequently maintained pine stands. Low-intensity frequent fires in hardwood stands favored oak regeneration over competing hardwoods.

With European settlement of the ecoregion, the natural vegetation pattern was significantly altered, as forests were converted to shifting agriculture. After World War II, many of the farms were abandoned. Pines outcompeted hardwoods for sun and nutrients and survived better in the extreme environmental conditions of the abandoned fields. Common pine species of this ecoregion include shortleaf pine (*Pinus echinata*), loblolly pine (*P. taeda*), and longleaf pine (*P. palustris*). Hardwood species grew in after the pines and established themselves prominently in the understory. Since the mid 1960s pine stands have been harvested, and hardwoods stands have taken their place, outcompeting juvenile pines for dominance in the overstory. In addition, the suppression of natural fire regimes has shifted vegetation to hardwood forests. This unnatural management practice has altered the plant communities across the ecoregion and threatens the long-term persistence of many fire-dependent species.

The understory and herbaceous layers of the ecoregion follow similar successional gradients. On open fields with high light, grasses and forbs dominate. As pines and hardwood tree species overtop them and the canopy begins to close, more shrubs and small trees fill in. They include species such as dogwood (*Cornus* spp.), red bud (*Cercis canandensis*), cedar (*Juniperus* spp.), and American holly (*Ilex opaca*). Common shrubs and herbaceous species include blackberry (*Rubus spp.*), sumac (*Rhus* spp.), honeysuckle (*Lonicera* spp.), and poison ivy (*Toxicodendron radicans*) (Skee, Doerr, and Van Lear 1993).

Conservation Status

Habitat Loss

About 99 percent of this habitat has now been converted to agriculture or other uses, or is highly degraded. Habitat loss is relatively uniform across the ecoregion. This is the most heavily settled ecoregion along the East Coast of the United States, and much of the land has been used for growing tobacco and peanuts. The once dense forests harvested long ago have never been allowed to regrow to a mature age. There are large amounts of tertiary forest that offers little biodiversity value. A remnant tallgrass prairie, the so-called Black Belt, has been completely converted. A few habitats are in relatively good condition, particularly on granite outcrops.

Remaining Blocks of Intact Habitat

Nine blocks of habitat have been identified by this analysis, but most are in relatively poor condition, fragmented, and poorly protected. These include:

- Sumter National Forest (in very poor shape)—western South Carolina
- Uwharrie National Forest—central North Carolina
- Bienville National Forest—east-central Mississippi
- Talladega National Forest (SW unit)—central Alabama
- Oconee National Forest and Piedmont National Wildlife Refuge—north-central Georgia
- Sauratown Mountains—north-central North Carolina
- Brushy Mountains—central North Carolina
- South Mountains—south-central North Carolina
- Tunica Hills—southwestern Mississippi, eastern Louisiana

Degree of Fragmentation

Fragmentation is very high, and creation of new corridors is unlikely except in riparian areas. The species most susceptible to fragmentation, such as black bears (*Ursus americanus*), have been largely extirpated.

Degree of Protection

Habitat protection is poor, and remaining habitat is in relatively poor condition.

Types and Severity of Threats

Because of the heavy rate of conversion, there is little left to conserve. Logging remains a threat, as does continued exploitation of remaining forests and conversion to pine plantations. The lack of fire management in remaining areas is viewed as a serious degradation threat.

Suite of Priority Activities to Enhance Biodiversity Conservation

- Promote panther reintroduction and recovery in the largest blocks of forest remaining in the piedmont areas.
- Enhance protection of Forest Service areas to increase biodiversity protection and move away from production in last remaining blocks.
- Improve fire management regimes.
- Inventory last remaining sites to identify biodiversity priorities that may have been overlooked.

Conservation Partners

For contact information, please see appendix G.

- Alabama Natural Heritage Program
- Georgia Natural Heritage Program
- Louisiana Natural Heritage Program
- Maryland Heritage and Biodiversity Conservation Programs
- Mississippi Natural Heritage Program
- The Nature Conservancy
- The Nature Conservancy, Southeast Regional Office
- The Nature Conservancy of Alabama
- The Nature Conservancy of Georgia
- The Nature Conservancy of Louisiana
- The Nature Conservancy of Maryland

- The Nature Conservancy of Mississippi
- The Nature Conservancy of North Carolina
- The Nature Conservancy of South Carolina
- The Nature Conservancy of Virginia
- North Carolina Heritage Program
- Pennsylvania Natural Diversity Inventory
- South Carolina Heritage Trust
- Tennessee Division of Natural Heritage
- Virginia Division of Natural Heritage
- West Virginia Natural Heritage Program
- Wilderness Society, Southeast Regional Office

Relationship to Other Classification Schemes

The Southeastern Mixed Forests ecoregion corresponds to the eastern piedmont and southwestern portions of Küchler's (1975) unit 101 (Oak-Hickory-Pine Forest) east of the Mississippi River. In addition, this ecoregion contains unit 80 (black belt). This ecoregion is in large part a combination of Omernik's (1995b) unit 65 (Southeastern Plains), portions of unit 74 (Mississippi Valley Loess Plains), and disjunct unit 68 (Southwestern Appalachians). Several portions of Bailey's (1994) sections 231A (Southern Appalachian Piedmont), 231B (Middle Coastal Plains), and the northern section of 232A (Middle Atlantic Coastal Plain) correspond to this ecoregion.

Prepared by A. Weakley, E. Dinerstein, R. Noss, and K. Wolfe

Key Number: **23**
Ecoregion Name: **Northern Pacific Coastal Forest**
Major Habitat Type: **2.2 Temperate Coniferous Forest**
Ecoregion Size: **60,894 km²**
Biological Distinctiveness: **Globally Outstanding**
Conservation Status: **Snapshot—Relatively Stable**
Final—Relatively Stable

Introduction

The Northern Pacific Coastal Forests occupy a narrow (about 160 km wide) coastal band extending from the southern portion of the Alexander Archipelago to the Prince William Sound region and eastern Kodiak Island. This ecoregion contains more than one-fourth of the world's coastal temperate rain forest and is one of the largest and most pristine temperate rain forest and shoreline ecosystems in the world (Alaback and Juday 1989; Ecotrust 1995). Most of the ecoregion lies within the Tongass National Forest (6.8 x 10⁴ km²), the Chugach National Forest (3.5 x 10⁴ km²), and Glacier Bay National Park (1.3 x 10⁴ km²) (Alaback and Juday 1989).

This ecoregion consists of thousands of small and several large mountainous islands (elevation to 1,500 m), long coastal valleys, and outburst flood fans (McNab and Bailey 1994). Some of the largest islands in North America are found here, including the eastern portion of Kodiak Island and Prince of Wales Island. Several long, narrow bays have been carved into mountainous terrain by glaciers, creating an extremely irregular coastline. The northern portion of the ecoregion consists of a mosaic of foothills, coastal lowlands of alluvial fans, uplifted estuaries, morainal deposits, dunes, river deltas, and terraces (Bailey 1995). Notably, limestone karst topography characterized by numerous sinkholes, caves, underground streams, and fractured bedrock is prominent in the southern portion of the region—Prince of Wales Island and Ketchikan area (DeMeo, Martin, and West 1993; USDA Forest Service 1994c). This karst landscape is a three-dimensional system that includes productive forests and peatlands on top of karst, the surface and subsurface interactions, and groundwaters originating from these systems (USDA Forest Service 1994c).

Glacial influences have shaped landform features throughout the region and are particularly evident from the numerous fjords in the southern portion of the region and the drumlin fields (small hills) of Prince of Wales Island (DeMeo, Martin, and West 1993). However, some areas have escaped glaciation and may provide refugia for remnants of ancient flora present before the period of last glacial advance (Heusser 1989; DeMeo, Martin, and West 1993). In addition, the presence of numerous islands of various sizes and distances from the mainland has influenced the distribution of local flora and fauna. For instance, brown bears (*Ursus arctos horribilis*) are present only on Admiralty, Baranof, and Chichagof Islands (the so-called ABC islands). In addition, the Alexander Archipelago (Samson et al. 1989) and Queen Charlotte Islands [25] to the south (Lindsey 1989; Kavanaugh 1989) contain endemic subspecies of invertebrates, birds, and mammals.

Climate in this region is generally characterized by a unique combination of moderate temperatures and high rainfall resulting from the warm Alaska Current. Annual precipitation averages 2,450 mm but is quite variable (range 762–5,588 mm) both within the region and on individual islands (Alaback 1988; DellaSala et al. 1994; Bailey 1995). These localized differences in climate are known to be important factors influencing site productivity and distribution of plant associations (DeMeo, Martin, and West 1993). High annual precipitation has resulted in the relative absence of fire throughout the region. Wind is the primary disturbance; however, landslides, avalanches, floods, and glacial disturbances also occur (Alaback and Juday 1989; Bailey 1994). The absence of fire in this region has resulted in few forests of intermediate ages—i.e., 50–150 years (DeMeo, Martin, and West 1993; DellaSala et al. 1994, 1996a).

Soils in the region are considered young (0–15,000 years) and vary from shallow and poorly developed to deeply weathered (DeMeo, Martin, and West 1993). Soil types consist of histosols and spodosols (McNab and Bailey 1994).

Plant species diversity varies extensively across the ecoregion, reflecting complex interactions of climate, geomorphology, history, and species interactions (Alaback 1993). The predominant forest type in the region is coastal Sitka spruce (*Picea sitchensis*)–hemlock (*Tsuga* spp.) (Bailey 1995). However, poorly drained sites contain muskegs, and low-lying areas along river channels contain alder (*Alnus* spp.), cottonwood (*Populus trichocarpa*), and Alaska paper birch (*Betula papyrifera*) (Bailey 1995). Twenty-one ecological provinces have been recognized in southeast Alaska (USDA Forest Service 1991) and seven vegetation series and forty-one associations have been identified in this area, including shore pine (*Pinus contorta*), mixed conifer, western hemlock (*T. heterophylla*)–western red cedar (*Thuja plicata*), western hemlock, western hemlock–yellow cedar (*Chamaecyparis nootatensis*), mountain hemlock (*T. mertensiana*), and Sitka spruce (DeMeo, Martin, and West 1993). In

general, species richness (conifers, plant associations, birds, mammals) declines with increasing latitude (DeMeo, Martin, and West 1993; DellaSala et al. 1996a). For instance, the southern portion of the region contains 60 percent of the species on only 20 percent of the land base (Alaback, in review).

Biological Distinctiveness

The extensive coastline juxtaposed with numerous mountainous islands, streams, estuaries, and dense forests makes this ecoregion one of the most productive (marine and terrestrial biomass) in North America. Because this ecoregion also contains one-fourth of the world's temperate rain forest, it was considered globally outstanding. Many of the few remaining intact (unlogged) watersheds in North America—primarily in the northern portion of the ecoregion (Ecotrust 1995) and relatively abundant old-growth forest further contribute to the global significance of this ecoregion. Also of global importance are the capacity of these forests to store carbon and the subsequent role they play in regional and global climates (Waring and Franklin 1979; Alaback 1991). Old-growth forests in particular are critically important fish and wildlife habitat and are characterized by unique structural attributes (e.g., multilayered canopies, diverse forb and shrub layers, coarsewoody debris, large-diameter trees). These attributes are usually present when a forest reaches 150 years (although this varies with plant association; see Capp et al. 1992). In addition, old-growth forests in this region are classified by the USDA Forest Service by timber volume classes ranging from low-volume muskeg (volume class 4) to high-volume and commercially productive forests (volume class 7) that are generally found along valley bottoms and low-elevation areas (USDA Forest Service 1991). High-volume old-growth forests are of greatest importance to wildlife because they contain relatively high levels of species richness, provide important winter refugia for birds and mammals, and support superabundant anadromous fish runs (Samson et al. 1989; Ecotrust 1995; DellaSala et al. 1994, 1996a).

Many species that are threatened in the lower forty-eight states are present in far greater numbers in this ecoregion. Some of the highest nesting concentrations of bald eagles (*Haliaeetus leucocephalus*) and marbled murrelets (*Brachyramphus marmoratus*) in North America occur in southeast Alaska, Prince William Sound, and the Kodiak Archipelago. Likewise, some of the highest concentrations of brown bear in North America occur in southeast Alaska and Kodiak Island. Similarly, the region is known for some of the most productive salmon (five species) runs in North America. However, when compared to other temperate coniferous forests within its MHT, this ecoregion contains relatively low species richness because its northern distribution lies outside the geographic range of most taxa. In particular, few herpetofauna (four amphibia and one reptile) and conifer species (eight) exist in these northern latitudes. Bird species represent the majority (59 percent) of taxa evaluated; however, many forest-dwelling families are poorly represented, particularly the woodpeckers and wood warblers. Although not included in this assessment, the region contains high levels of bryophytes and epiphytes; these taxa are the subject of ongoing investigation (Alaback and Juday 1989).

Conservation Status

Habitat Loss

On a coarse scale, old-growth losses (all volume classes combined) have not been as extensive as other temperate coniferous forests within this MHT. For instance, logging has eliminated 7 percent of productive old-growth since the early 1950s (USDA Forest Service 1991; DellaSala et al. 1994, 1996a). However, an additional 23 percent of old growth is scheduled for harvest over the next fifteen years (USDA Forest Service 1991). Such losses are taking place in some of the most productive (high-volume) forested ecosystems remaining in the ecoregion. For instance, roughly 10 percent of the high-volume old-growth remains on the Tongass National Forest and much of this is scheduled for harvesting (USDA Forest Service 1991; DellaSala et al. 1996a). In addition, logging and extensive road building in some ecological provinces (e.g., Prince of Wales Island northern and southern provinces) will eliminate up to 70 percent of the total old growth over the next 150 years (USDA Forest Service 1991; DellaSala et al. 1996a).

Projected logging levels in old-growth systems are expected to result in significant population declines in several species, including northern goshawk (*Accipiter gentilis laingi*), Alexander Archipelago wolf (*Canis lupus ligoni*), marten (*Martes americana*), northern flying squirrel (*Glaucomys sabrinus*), brown bear (Suring et al. 1993), and some neotropical and resident birds (DellaSala et al. 1996a). In addition, well-drained karst terrain has been particularly impacted by logging, as has species-rich estuarine and riparian fringes (protected on national forests by narrow—150–300-m—buffers). Village corporation lands (managed by Native Alaskan corporations) have been particularly impacted by extensive logging.

Remaining Blocks of Intact Habitat

Within the ecoregion, several blocks of relatively intact habitat remain. The Southeast Alaska Conservation Council (1993) has mapped seventeen intact but threatened forests in southeast Alaska, four areas in the Gulf of Alaska, nine areas in the Prince William Sound region, and five areas in the Kenai Peninsula and Kodiak Archipelago. Similarly, Ecotrust (1995) mapped several intact watersheds (greater than 1,000 km²) in the region, of which many are threatened by logging. Workshop participants during this assessment identified the following intact blocks:

- Admiralty Island—southeastern Alaska
- West side of Chichagof Island—southeastern Alaska
- South Baranof Island—southeastern Alaska
- South Kupreanof Island—southeastern Alaska
- Cleveland Peninsula (logging is occurring)—southeastern Alaska
- Misty Fjords National Monument—southeastern Alaska
- Southern half of Kuiu Island—southeastern Alaska
- Coastal areas of Glacier Bay National Park—southeastern Alaska
- Most of Prince William Sound (logging is occurring or planned on Native corporation lands)—southeastern Alaska
- Kachemak Bay State Park—southeastern Alaska
- Yakutat Forelands—southeastern Alaska

- Honker Divide and the Thorne River on Prince of Wales Island (logging is planned)—southeastern Alaska

Degree of Fragmentation

Because most of the ecoregion is distributed across several archipelagos, the system is naturally fragmented. In addition, drainage and soil development within the ecoregion are highly variable, resulting in a mosaic of various forest volume classes (muskeg to volume class 7), estuarine areas, beach fringe, and riparian areas. However, several large, relatively intact blocks of forest exist in many of the ecological provinces, providing good connections among related patch types.

Habitat fragmentation due to anthropogenic factors also varies across the region. Several ecological provinces have been substantially affected by logging, particularly those on Prince of Wales Island, where 70 percent of two ecological provinces are scheduled for logging over the next 150 years. Logging has produced the most extensive road system for all of the ecological provinces on the Tongass National Forest.

Degree of Protection

While landmark conservation acts were passed in the 1980s (e.g., Alaska National Interest Lands Conservation Act) protecting several important areas in this ecoregion, most of the biologically rich and productive lands remain vulnerable to logging, particularly high-volume old growth. The Tongass Land Management Plan (USDA Forest Service 1991) provides maps of these areas as well as of several large, roadless and wilderness complexes under land protection categories. Additional protected areas include: Glacier Bay National Park, Admiralty Island Wilderness, Kachemak Bay State Park, Misty Fjords Wilderness, west Chichagof Island, south Baranof Island, Kuiu Island, southern Prince of Wales Island, Southern Etolin, and the Stikine Delta. Additional areas may be protected in the near future through land acquisitions made possible by the Exxon Valdez oil spill settlement funds; many lands available for acquisition are threatened rain forest areas.

The greatest obstacle to conserving future lands in this ecoregion is political will. The Alaska delegation has blocked numerous attempts at protection of additional rain forest

lands on the Tongass and has proposed land exchanges that would remove or substantially weaken protection of rain forest areas. At a minimum, conservation in this region needs to focus on blocking such congressional measures while building support for additional protection of high-volume old growth.

Types and Severity of Threats

While many pristine areas remain in this ecoregion, old growth is on a trajectory similar to that of the Pacific Northwest (e.g., ecoregions [25], [32], [31], [38]), which has resulted in extensive degradation during the past fifty years. Population viability concerns already exist for several species associated with old growth, and trends in these species over the next century are alarming (several species have low persistence probabilities; see Suring et al. 1993). Other threats include expansion of roads and highways to accommodate the explosive growth in tourism, increases in cruise ships visiting popular tourism destinations such as Glacier Bay National Park, pollution from pulp mills and mining tailings, and potential long-term reductions in subsistence species (e.g., deer) due to logging in winter habitat areas.

Suite of Priority Activities to Enhance Biodiversity Conservation

The recommendations of the interagency viability population task force (Suring et al. 1993) are an excellent starting point for identifying lands in need of additional protection on the Tongass National Forest in order to achieve more stable population trajectories for old-growth associated species. Other conservation measures identified by the Southeast Alaska Conservation Council (1993) and Ecotrust (1995) should be adopted. In addition, conservation groups will need to balance protection of old-growth areas with ecologically appropriate logging (e.g., under the auspices of the Forest Stewardship Council principles and standards) while communities make the transition to more diversified economies. Recommendations for small-scale logging (outside high-volume old growth) are available from the Tongass Conservation Society, Alaback (in review), and DellaSala (1996a). The following suite of priority activities was identified by workshop participants:

- Protect all remaining high-volume old growth on Prince of Wales Island, Kuiu Island, southeast Chichagof Island, southeast Revillagigedo Island.
- Protect Wild and Scenic Rivers (e.g., Thorne River on Prince of Wales Island).
- Maintain connections across ecological provinces (e.g., south and north Prince of Wales Island, Stikine Delta, Copper River Delta).
- Institute habitat conservation areas, as proposed by Suring et al. (1993).
- Secure protection of roadless areas (Land Use Designation II).
- Continue to use Exxon Valdez oil spill settlement to purchase habitat in the oil spill region.
- Formally designate additional wilderness areas (e.g., several on the Chugach National Forest).

Conservation Partners

For contact information, please see appendix G.

- Alaska Rainforest Campaign
- Ecotrust
- The Nature Conservancy of Alaska
- Sierra Club, Alaska field office
- Southeast Alaska Conservation Council (SEACC)
- Tongass Conservation Society
- Wilderness Society, Alaska region

Relationship to Other Classification Schemes

The classification system used to delineate this ecoregion is similar to Omernik's (1995b) ecoregion 120 and Bailey's (1994) ecosections 245A–C. The ecoregion described here is also concordant with Ecotrust's classification for temperate rain forest in the subpolar and perhumid zones. We did not, however, include the seasonal and coastal redwood zones and instead split them into separate ecoregions (ecoregions [34] and [40]) out of concerns for species assemblages, climate, disturbance regimes, and physiography.

Prepared by D. DellaSala, L. Craighead, and R. Hagenstein

Key Number:	**24**
Ecoregion Name:	**Queen Charlotte Islands**
Major Habitat Type:	**2.2 Temperate Coniferous Forests**
Ecoregion Size:	**9,972 km²**
Biological Distinctiveness:	**Globally Outstanding**
Conservation Status:	**Snapshot—Vulnerable Final—Vulnerable**

Introduction

The Queen Charlotte Islands represent a major offshore archipelago of islands separated from the British Columbia and Alaskan mainland by Hectate Strait at a distance of approximately 75–100 km.

The climate in this ecoregion is considered oceanic and maritime South Pacific Cordilleran. The mean annual temperature is 7.5°C; mean summer temperature is 11.5°C, and mean winter temperature is 3.5°C. Annual precipitation in the islands is between 800 (in eastern areas) and 4,000 mm (on western slopes).

Physiographically, the Queen Charlotte Islands are characterized by irregular, steep slopes in the west and gently sloping lowlands in the east.

Biological Distinctiveness

This ecoregion has significant areas of old-growth, west coast rain forests. Watersheds on the islands are important for anadromous fish, while elevational gradients result in high terrestrial species richness and community variation.

Along the west coast of the islands, vegetation is composed of stunted, open-growing western red cedar (*Thuja plicata*), yellow cedar (*Chamaecyparis nootkatensis*), shore pine (*Pinus contorta* var. *contorta*), and western hemlock (*Tsuga heterophylla*). Better-drained sites also support Sitka spruce (*Picea sitchensis*). Wetlands are common in the islands and contain open western hemlock and shore pine.

Several species of introduced mammals are present on the island, including black-tailed deer (*Odocoileus hemionus*), elk (*Cervus canadensis*), racoon (*Procyon lotor*), rats (*Rattus* spp.), eastern gray squirrel (*Sciurus carolinensis*), and beaver (*Castor canadensis*). Native, common wildlife includes black bear (*Ursus americanus*), river otter (*Lutra canadensis*), seabirds, shorebirds, and marine mammals.

This ecoregion is a critical stopover for migratory waterbirds flying north to Alaska and south to Mexico. As one of the most isolated island archipelagos in western North America, the ecoregion harbors several endemic subspecies of plants, birds, and small mammals as well as an endemic subspecies of black bear. Critical nesting sites for colonial nesting birds and raptors are also found here.

Conservation Status

Habitat Loss

Approximately 50 percent of the habitat on the Queen Charlotte Islands has been altered, primarily as a result of clear-cut logging. The remaining habitat is relatively intact. High rainfall levels in this ecoregion have created serious erosion and landslide problems where logging has been extensive. The logging has provided habitat suitable for the introduced black-tailed deer, which in turn are causing other serious habitat impacts through selective overbrowsing of some conifer and forest understory species.

Remaining Blocks of Intact Habitat-

- Naikoon/Tlell Watershed—approx. 1,200 km²
- Gwaii Haanas National Park Reserve and Haida Heritage Site (1,450 km², of which 1,200 km² is not logged)
- Duu-Guusd—northwest Graham Island, 1,500 km²
- North-central Graham Island—approx. 800 km²
- Englefield Bay—approx. 700 km²

Degree of Fragmentation

Logging has been the principal land use responsible for the fragmentation of habitat. Since logging is directed principally at the valley bottom and lower slope forests, fragmentation occurs for upper-slope communities and impacts species movements.

Degree of Protection

- Gwaii Haanas National Park Reserve—1,470 km²
- Naikoon Provincial Park—726 km²

- Krajina Ecological Reserve—98 km²
- Drizzle Lake Ecological Reserve—8 km²
- Tow Hill Ecological Reserve—5 km²

Types and Severity of Threats

Logging and road building remain significant threats to mature forest habitat and to some species, such as marbled murrelets (*Brachyramphus marmoratus*), cavity nesters, and raptors. Introduced species to this island system are also a major threat to native biodiversity. Black-tailed deer are having a major impact on the regeneration of western red cedar, a dominant species in many of the islands' forested habitats. Damming by introduced beaver of small streams used by coho salmon (*Oncorhynchus kisutch*) for spawning threatens some stocks. The introduction of rats (*Rattus* spp.), squirrels (*Sciurus* spp.), and racoons has had a profound impact on seabird colonies.

Suite of Priority Activities to Enhance Biodiversity Conservation

- Protect Duu-Guusd—northwest Graham Island.
- Protect the Tlell River watershed.
- Protect Government Creek Watershed System.
- Reduce rate of forest harvest (especially clear-cutting of rainforest).
- Formulate plans to control (or remove) introduced species.

Conservation Partners

For contact information, please see appendix G.

- BC Wild
- Gowgaia Institute
- Laskeek Bay Conservation Society
- The Nature Conservancy, British Columbia
- Sierra Club of British Columbia
- Tlell Watershed Society
- World Wildlife Fund Canada

Relationship to Other Classification Schemes

The Queen Charlotte Islands are characterized by the Queen Charlotte Lowland and Ranges (TEC ecoregions 188 and 189). The ranges form the backbone of the islands, and the lowland in the north and east is primarily forested plain and wetlands. The islands fall within the Coast Forest Region, and have a rain forest type of vegetation.

Prepared by D. Demarchi, J. Broadhead, K. Kavanagh, M. Sims, and G. Mann

Key Number:	**25**
Ecoregion Name:	**Central British Columbia Mountain Forests**
Major Habitat Type:	**2.2 Temperate Coniferous Forests**
Ecoregion Size:	**1,747 km²**
Biological Distinctiveness:	**Nationally Important**
Conservation Status:	**Snapshot—Relatively Stable Final—Vulnerable**

Introduction

This ecoregion in north-central British Columbia occurs as a relatively narrow band oriented in a northwest-southeast direction. It encompasses part of the Rocky Mountain trench and most of the Hart ranges of the Rocky Mountains and the Omineca Mountains (ESWG 1995).

This ecoregion's climate is considered alpine and subalpine southern Cordilleran. Mean annual temperature is around 2°C; mean summer temperature is 12°C, and mean winter temperature is between −10°C and −7°C. Mean annual precipitation ranges from 500 to 700 mm, increasing from the northwest toward the southeast, and with elevation from east to west in the south. Climatic conditions in the valleys are characterized by warm, dry summers and mild, snowy winters. Subalpine summers are cool, wet, and prone to early frosts. Subalpine winters are cold and snowy (ESWG 1995).

The western section of this ecoregion encompasses the southern section of the Omineca Mountains, which form a bed of Paleozoic and Mesozoic sedimentary and massive crystalline rocks. The west also encompasses the eastern ranges of the Skeena Mountains, which are composed of Jurassic and Cretaceous sediments and volcanic strata. Both ranges have peaks around 2,400 m asl. The central section of the Rocky Mountains runs down the center of this region and is relatively subdued, yet rises above the prairie plains to the east. The Eastern Continental Ranges are linear, with great cliffs and thick sections of gray carbonate strata. Rock outcrops are found along most peaks and ridges (ESWG 1995).

Biological Distinctiveness

A vertically stratified complex of ecosystems characterizes this ecoregion: low-elevation forests of interior western red cedar (*Thuja plicata*) and western hemlock (*Tsuga heterophylla*) in the northwestern, Skeena Mountain, area; lodgepole pine (*Pinus contorta*), quaking aspen (*Populus tremuloides*), and white (*Picea glauca*) and black spruce (*P. mariana*) in the east; subalpine sections of Engelmann spruce (*Picea engelmannii*), alpine fir (*Abies lasiocarpa*), lodgepole pine, and white spruce; and alpine tundra consisting of heather (Ericaceae), heath (*Phyllodoce empetriformis*), sedge (*Carex* spp.), and mountain avens (*Dryas hookeriana*) at the highest elevations.

Wildlife species include woodland caribou (*Rangifer tarandus* ssp. *caribou*), elk (*Cervus canadensis*), moose (*Alces alces*), black-tailed deer (*Odocoileus hemionus*), black bear and grizzly bear (*Ursus americanus* and *U. arctos*) (very high populations), beaver (*Castor canadensis*), wolf (*Canis lupus*), red fox (*Vulpes fulva*), wolverine (*Gulo luscus*), marten (*Martes americana*), snowshoe hare (*Lepus americanus*), and ruffed grouse (*Bonasa umbellus*). Bighorn sheep (*Ovis canadensis*) and mountain goat (*Oreamnos americanus*) inhabit the rugged subalpine and alpine areas.

Among the rare ecological and evolutionary phenomena are large wetlands in valleys on the windward side of the Hart Ranges. The Rocky Mountain trench is a major north-south flyway for migratory birds. In addition, caribou (*Rangifer tarandus*) are found in southern parts of the ecoregion and Dall (*Ovis dalli*) and bighorn sheep are found within 150 kilometers of each other.

Conservation Status

Habitat Loss and Degradation

Approximately 75 percent of this ecoregion remains as intact habitat. Logging has been one of the principal activities resulting in habitat loss and has to date been focused primarily in the Rocky Mountain trench. It is now moving into all major valleys, and along with increased road access the rate of habitat loss is increasing significantly in some places. A major dammed reservoir (Williston Lake) now blocks east-west movement by wildlife across the Rocky Mountain trench and to some degree the north-south movement down the Rocky Mountains. A major hydroelectric transmission corridor transects Pine Pass. Mineral exploration and mines are also responsible for habitat loss and degradation.

Remaining Blocks of Intact Habitat

- Omineca-Bear—approx. 5,000 km² of alpine and subalpine
- Ospika—approx. 2,000 km² of alpine and subalpine
- Mount Selwyn—approx. 1,500 km² of alpine and subalpine
- Monkman Provincial Park—eastern British Columbia, 401 km² of alpine and subalpine

Note: No valley bottoms, sub-boreal spruce forests, large (natural) lakes or river habitats remain that are unroaded, unlogged, or without mineral exploration.

Degree of Fragmentation

Habitat fragmentation in this ecoregion has been caused primarily by road building and logging in valley bottom lands. Williston Lake (a reservoir) has created a major barrier to the movement of wildlife in the center of this ecoregion.

Degree of Protection

- Monkman Provincial Park—401 km2
- Gwillim Lake Provincial Park—91 km²
- Patsuk Creek Ecological Reserve—5 km²
- Sukunka Falls Provincial Park—3 km²
- Chunamon Creek Ecological Reserve—3 km²

Types and Severity of Threats

Aside from the significant impacts of the Williston Reservoir on large mammals, logging is the most serious threat. All commercially viable forests are slated to be logged within the next fifty years.

Suite of Priority Activities to Enhance Biodiversity Conservation

- Establish more protected areas representative of the region, especially mid and low elevations and valley bottoms.
- Ensure that critical wildlife movement corridors are maintained.

Conservation Partners

For contact information, please see appendix G.

- The Ecology Circle
- The Nature Conservancy, Alberta
- The Nature Conservancy, British Columbia
- Nechako Environmental Society
- Northwest Wildlife Preservation Society
- Prince George Naturalists
- World Wildlife Fund Canada

Relationship to Other Classification Schemes

Within the Montane Cordillera Ecozone are the Central British Columbia Mountain Forests, which are distributed throughout the Omineca and Rocky Mountains (TEC ecoregions 199 and 200) (ESWG 1995). As with other ecoregions in western Canada, the Northern BC forests are made up of different forest types. This ecoregion includes the Columbia forest region (CL.1), Montane Transition (M.4), Interior Subalpine (SA.2), and Forest Ecoregions (Rowe 1972).

Prepared by D. Demarchi, G. Smith, K. Kavanagh, M. Sims, and G. Mann

Key Number:	**26**
Ecoregion Name:	**Alberta Mountain Forests**
Major Habitat Type:	**2.2 Temperate Coniferous Forests MHT**
Ecoregion Size:	**9,769 km²**
Biological Distinctiveness:	**Nationally Important**
Conservation Status:	**Snapshot—Relatively Stable Final—Relatively Stable**

Introduction

This ecoregion lies almost wholly within Alberta but hugs the Alberta–British Columbia border from Banff northward to Jasper and Kakwa.

Mean annual temperature in the Eastern Continental Ranges is 2.5°C; mean summer temperature is 12°C, and mean winter temperature is –7.5°C. Precipitation increases from east to west with elevation, from 600 to 800 mm per year. Valley regions are marked by warm, dry summers and mild, snowy winters, and subalpine areas have cool, showery summers and cold, snowy winters (ESWG 1995).

This region covers the Rocky Mountains of Alberta, incorporating the eastern flanks of the Continental Ranges. The major peaks cluster around the Columbia Ice Field, the largest ice field in the Rocky Mountains. The ranges themselves are linear with great cliffs and precipitous faces of thick sections of gray carbonate strata, and peaked by rock outcrops (ESWG 1995).

Biological Distinctiveness

Vegetation in this ecoregion is composed of alpine and subalpine ecosystems characterized by mixed forests of lodgepole pine (*Pinus contorta*), Engelmann spruce (*Picea engelmannii*), and alpine fir (*Abies lasiocarpa*). Alpine fir is found at higher elevations. Alpine vegetation is also characterized by heather (Ericaceae) with sedges (*Carex* spp.) and mountain avens (*Dryas hookeriana*) on warmer sites (ESWG 1995).

Wildlife of this region includes bighorn sheep (*Ovis canadensis*), elk (*Cervus canadensis*), blacktailed deer (*Odocoileus hemionus*), wolf (*Canis lupus*), grizzly and black bear (*Ursus arctos* and *U. americanus*), caribou (*Rangifer tarandus*), and mountain goat (*Oreamnos americanus*). The ecoregion exhibits a high diversity of large mammals (ESWG 1995).

Conservation Status

Habitat Loss

Approximately 80 percent of this region was considered to remain as intact habitat. Major road corridors in valley lands with major outdoor recreation facilities and town sites are primarily responsible for loss of habitat.

Remaining Blocks of Intact Habitat

- Banff and Jasper National Parks—southwestern Alberta
- Willmore Provincial Wilderness Park and Kakwa Provincial Park—western Alberta and eastern British Columbia

Degree of Fragmentation

Road and travel corridors in the major valleys impede large carnivore and other wildlife movement.

Degree of Protection

- Jasper National Park—southwesthern Alberta, 10,878 km²
- Banff National Park—southwestern Alberta, 6,641 km²
- Willmore Provincial Wilderness Park—western Alberta, 4,596 km²
- Kakwa Provincial Park—southeasern British Columbia, 1,276 km²
- Whitegoat Provincial Wilderness Area—444 km²
- Bugaboo Alpine Provincial Park—southeastern British Columbia, 249 km²
- Ghost River Provincial Wilderness Area—southwestern Alberta, 153 km²
- Kootenay Plains Ecological Reserve—southwestern Alberta, 32 km²

Types and Severity of Threats

Expansion of road systems and recreational activities pose a major threat. A recent review (1996) of Banff National Park has recommended major changes to recreational and town site development in the park, which would redress some of the habitat loss in this ecoregion over the long term. These recommendations have generally been accepted by the federal government. Attempts are also underway to begin adding structural elements to the trans-Canada highway, which passes through Banff National Park, to lower wildlife-auto collisions, although the highway is undergoing a major expansion as well. New coal mining adjacent to Jasper National Park is a threat to the integrity of that protected area.

Suite of Priority Activities to Enhance Biodiversity Conservation

- Allow no net increase in town site size in the major protected areas.
- Increase control of recreation activities; limit access to trails and close off parts of the trail networks on a seasonal basis to provide less disturbance (and interaction) with wildlife such as grizzly bears.

Conservation Partners

For contact information, please see appendix G.

- Banff/Bow Valley Naturalists
- Canadian Parks and Wilderness Society, Calgary/Banff Chapter
- World Wildlife Fund Canada

Relationship to Other Classification Schemes

This ecoregion covers the Rocky Mountains of Alberta, incorporating the eastern portion of the Continental Ranges (TEC ecoregions 207) (ESWG 1995). The forests of this region cover the Subalpine East Slope Rockies (SA.1) and Montane Douglas-fir and Lodgepole Pine (M.5) (Rowe 1972).

Prepared by D. Demarchi, R. Usher, K. Kavanagh, M. Sims, and G. Mann

Key Number:	**27**
Ecoregion Name:	**Fraser Plateau and Basin Complex**
Major Habitat Type:	**2.2 Temperate Coniferous Forests**
Ecoregion Size:	**37,055 km²**
Biological Distinctiveness:	**Nationally Important**
Conservation Status:	**Snapshot—Vulnerable Final—Endangered**

Introduction

This ecoregion occupies much of the central interior of British Columbia. It is distinct from the drier "interior" to the south and southeast, the cooler interior forests to the north, and the coastal forests to the west.

Mean annual temperature for the region is 3°C; mean summer temperature is 12.5°C, and mean winter temperature is –7.5°C. Precipitation varies greatly; values of 250–300 mm per year occur in the area of the Chilcotin and Fraser River junction and rise to 600–800 mm per year at higher elevations in the west. In general, the climate in this ecoregion is boreal interior cordilleran (ESWG 1995).

This ecoregion covers the plateaus and plains of north-central British Columbia. The region is primarily underlain by flat-lying Tertiary and volcanic bedrock that is generally below 1,000 m asl in the northeast but rises to an elevation of 1,800 m asl in the west. It has a rolling surface covered by thick glacial drift, into which the Fraser River and its tributaries are incised (ESWG 1995).

Biological Distinctiveness

Vegetation of this ecoregion is transitional from the coastal forests to the west and the drier interior forests to the east. The subalpine zone is characterized by lodgepole pine (*Pinus contorta*), Engelmann spruce (*Picea engelmannii*), and alpine fir (*Abies lasiocarpa*). Montane forests include lodgepole pine, trembling aspen (*Populus tremuloides*), white spruce (*P. glauca*), and Douglas fir (*Pseudotsuga menziesii*). Bunchgrass-dominated grasslands occur at valley bottoms along the Fraser and Chilcotin Rivers. Alpine tundra vegetation is also present at the highest elevations. Fire is probably the most important disturbance regime (ESWG 1995).

Characteristic wildlife include caribou (*Rangifer tarandus*), coyote (*Canis latrans*), moose (*Alces alces*), bighorn sheep (*Ovis canadensis*), black-tailed deer (*Odocoileus hemionus*), wolf (*Canis lupus*), black bear (*Ursus americanus*), muskrat (*Ondatra zibethica*), lynx (*Lynx canadensis*), sandhill crane (*Grus canadensis*), blue grouse (*Dendragapus obscurus*), spruce grouse (*D. canadensis*), ruffed grouse (*Bonasa umbellus*), sharp-tailed grouse (*Tympanuchus phasianellus*), and migratory waterfowl (ESWG 1995).

The ecoregion contains western North America's southernmost stable (as of the present) herd of woodland caribou (*Rangifer tarandus* spp. *caribou*) (approximately 1,500 animals). There are also large concentrations of bighorn sheep, as well as the largest concentrations of breeding pairs of Barrow's goldeneye (*Bucephala islandica*) and bufflehead (*B. albeola*) in North America. One of the most northern extensions of dry grasslands in western North America, home to the Trirhabda beetle (*Trirhabda* spp.), occurs within the ecoregion.

Conservation Status

Habitat Loss

It is estimated that only about 25 percent of the ecoregion remains as natural, intact habitat. Much of the remaining habitat was described as "altered," although some parts of the ecoregion have been "heavily altered" due to agricultural activities (particularly along major valleys and southern uplands). Wetlands have also been seriously impacted by agricultural development in this ecoregion. Additional impacts have occurred from hydroelectric impoundments, pulpwood harvesting, and livestock grazing.

Remaining Blocks of Intact Habitat

- Tweedsmuir Park—west-central British Columbia—contains important alpine and subalpine habitat, as well as important habitat for caribou.
- Itcha Ilgachuz Provincial Park—central British Columbia—contains important alpine and subalpine habitat, as well as important habitat for caribou.
- Churn Creek Provincial Park—central British Columbia—contains habitat for bighorn sheep and black-tailed deer.

- Carp Lake Provincial Park—east-central British Columbia—contains wetland complexes
- Junction Sheep Range Provincial Park—central British Columbia—contains bighorn sheep range, northernmost bunchgrass ranges (including the northern distribution of the Trirhabda beetle).

Degree of Fragmentation

Habitat fragmentation is moderate but increasing rapidly in this ecoregion. Primary reasons for fragmentation include clear-cutting, wetland conversions, river impoundments, and highway and railway corridors in key lowland valleys. Low and mid-elevation habitats are far more fragmented than upper elevations.

Degree of Protection

- Tweedsmuir Provincial Park—9,609 km^2
- Itcha Ilgachuz Provincial Park—1,090 km^2
- Churn Creek Provincial Park—361 km^2
- Tweedsmuir Recreation Park—333 km^2
- Taseko Provincial Park—211 km^2
- Carp Lake Provincial Park—193 km^2
- Kluskoil Lake Provincial Park—156 km^2
- Nazco Lakes Provincial Park—80 km^2
- Junction Sheep Range Provincial Park—47 km^2
- Lang Lake School House Provincial Park—45 km^2

Types and Severity of Threats

Presently, rapidly expanding logging practices are the primary threat to biodiversity in the next twenty years. Between 15 and 20 percent of the total standing volume is scheduled for harvest in the short term. This will result in additional fragmentation to forest habitat and loss of critical winter habitat for woodland caribou. Access is increasing to much of the ecoregion as logging roads are built into most of the valley lands.

Suite of Priority Activities to Enhance Biodiversity Conservation

- Provide sufficient winter habitat protection to maintain stability for the woodland caribou herd in this ecoregion.
- Provide additional habitat protection for California bighorn sheep.
- Provide more protection for wetland habitats.

Conservation Partners

For contact information, please see appendix G.

- Canadian Parks and Wilderness Society, British Columbia Chapter
- The Nature Conservancy, British Columbia
- Nechako Environmental Society
- Northwest Wildlife Preservation Society
- Prince George Naturalists
- Williams Lake Environmental Society
- World Wildlife Fund Canada

Relationship to Other Classification Schemes

The Fraser Plateau and Basin Complex (TEC ecoregions 202 and 203) in central British Columbia occupies the Interior Plateau and foothills of the Coast Mountains, the Nechako Lowlands, the northern part of the Nechako Plateau, the southern portion of the Northern Rocky Mountain Trench, and the western flank of the MacGregor Plateau (ESWG 1995). This ecoregion incorporates a variety of Rowe's (1972) forest regions. Montane forest sections include Central Douglas-fir (M.2), Northern Aspen (M.3), and Montane Transition (M.4), and Interior Subalpine (SA.2).

Prepared by D. Demarchi, D. Neaves, J. Cooperman, K. Kavanagh, M. Sims, and G. Mann

Key Number: **28**

Ecoregion Name: **Northern Transitional Alpine Forests**

Major Habitat Type: **2.2 Temperate Coniferous Forests**

Ecoregion Size: **25,662 km²**

Biological Distinctiveness: **Nationally Important**

Conservation Status: **Snapshot—Relatively Stable Final—Vulnerable**

Introduction

This small ecoregion in west-central British Columbia lies just east of the coastal mountains in southernmost Alaska.

The transitional alpine forests' climate is characterized as moist montane southern cordilleran. The ecoregion is generally very wet, and precipitation ranges from 2,500 mm in the western coastal areas of the Nass River valley to 600 mm in the lower elevations of the interior Bulkley ranges. Temperature varies widely according to elevation but rests between 13°C in the summer and –9.5°C in the winter (ESWG 1995).

This ecoregion consists of the transitional physiography between the rugged and massive coastal ranges and the more subdued Omineca Mountains in the east. It is largely underlain by folded Jurassic and Cretaceous sediments, but there is extreme variation in elevation, ranging from the high, mountainous, and heavily glacierized areas of the Hazelton and Skeena Mountains (2,500–2,800 m asl), to the low Nass and Skeena River Valleys (less than 750 m asl). In the western portion of the ecoregion, these valleys are broad and gently rolling in the lower elevations, but they are increasingly deeply incised and steep as one moves east into the Skeena Mountains and away from the Nass Basin. The westernmost portion of the ecoregion contains several large, elongated lakes, which drain into the eastern interior of British Columbia. There are also isolated patches of permafrost in some of the higher elevations of the Skeena Range (ESWG 1995).

Biological Distinctiveness

Moist montane areas (low-elevation valley systems), particularly in the west, are dominated by western red cedar (*Thuja plicata*) and western hemlock (*Tsuga heterophylla*), and mountain hemlock (*T. mertensiana*) in the wetter lower subalpine. Lodgepole pine (*Pinus contorta*), Engelmann spruce (*Picea engelmannii*), and alpine fir (*Abies lasiocarpa*) dominate the subalpine throughout the ecoregion, although to a lesser extent as one nears the coast. The alpine vegetation is typically tundra and includes discontinuous patches of low-growing heather (Ericaceae), sedge (*Carex* spp.), and mountain avens (*Dryas hookeriana*) (ESWG 1995).

Grizzly bear (*Ursus arctos*), black bear (*U. americanus*), moose (*Alces alces*), red fox (*Vulpes fulva*), and wolf (*Canis lupus*) range throughout the ecoregion, and parts are populated by woodland caribou (*Rangifer tarandus* ssp. *caribou*), snowshoe hare (*Lepus americanus*), marten (*Martes americana*), beaver (*Castor canadensis*), grouse (*Dendragapus* spp.) and black-tailed deer (*Odocoileus hemionus*). Higher elevations support mountain goat (*Oreamnos americanus*), and wolverine (*Gulo luscus*) and ptarmigan (*Lagopus* spp.) can be found in the eastern ranges (ESWG 1995).

Among important ecological phenomena, the ecoregion supports a high density of grizzly bear and continues to have healthy salmon populations (*Oncorhynchus* spp.). Populations of mountain goat are found in the Bulkley Ranges. In addition, an unusual wet temperate rain forest, transitional between the coast and interior forests, exists in this ecoregion.

Conservation Status

Habitat Loss

It is estimated that approximately 75 percent of this ecoregion remains as intact natural habitat. Most of the remaining habitat has been altered by logging, particularly in the major valleys. Mineral exploration is occurring in the alpine and subalpine habitats, and a major transportation corridor separates the Bulkley Ranges from the Babine Ranges.

Remaining Blocks of Intact Habitat

- Babine Mountains—northwest-central British Columbia, alpine and subalpine habitat.
- Bulkley Ranges—northwest-central British Columbia, alpine and subalpine habitat used by mountain goats and caribou.

- Muckaboo Creek—northwest-central British Columbia, temperate rain forest with grizzly bears and salmon habitat.

Degree of Fragmentation

Most of this mountainous ecoregion remains intact. Some habitat is naturally fragmented due to the distribution of mountain blocks separated by fast-flowing rivers. Additional habitat fragmentation has occurred as the result of clear-cut logging in some of the valleys and a major transportation corridor in the Bulkley Valley.

Degree of Protection

- Babine Mountains Provincial Park—northwest-central British Columbia, 324 km²
- Ningunsaw River Ecological Reserve—northwest-central British Columbia, 20 km²

Types and Severity of Threats

Threats are high for this ecoregion. All remaining forests (including temperate old growth) are slated to be cut in the next sixty years. As low-elevation areas are encroached upon for logging, access is becoming a concern as logging roads open up remote areas where ungulates such as caribou are present. Loss of low-elevation forests creates significant habitat loss for both permanent low-elevation species and others that use these areas as critical winter habitat.

Suite of Priority Activities to Enhance Biodiversity Conservation

- Establish additional protected areas in alpine and subalpine habitats to protect mountain goat and caribou habitat. Also establish protected-area candidate in the Muckaboo Creek area to conserve temperate rain forest with prime grizzly bear and salmon habitat.
- Provide better protection for the Babine River corridor.

Conservation Partners

For contact information, please see appendix G.

- Canadian Parks and Wilderness Society, British Columbia Chapter
- The Nature Conservancy, British Columbia
- Nechako Environmental Society

- Northwest Wildlife Preservation Society
- Prince George Naturalists
- Williams Lake Environmental Society
- World Wildlife Fund Canada

Relationship to Other Classification Schemes

This ecoregion covers the western portion of the Skeena Mountains and the Bulkley Ranges in the far south (TEC ecoregions 198 and 201). The Bulkley Ranges have an elevation of up to 2,500–2,800 m asl (ESWG 1995). Montane Transition (M.4), Interior Subalpine (SA.2), and tundra forest regions are found in this ecoregion (Rowe 1972).

Prepared by D. Demarchi, K. Kavanagh, M. Sims, and G. Mann

Key Number: **29**
Ecoregion Name: **Alberta/British Columbia Foothills Forests**
Major Habitat Type: **2.2 Temperate Coniferous Forests**
Ecoregion Size: **120,460 km²**
Biological Distinctiveness: **Nationally Important**
Conservation Status: **Snapshot—Endangered Final—Critical**

Introduction

This ecoregion occupies the Rocky Mountain foothills in western Alberta and a small section of east-central British Columbia. Two disjunct geographic areas comprise this ecoregion.

Mean annual temperature ranges from −0.5°C in the north to 2°C in the southern region. Mean summer temperature ranges from 13°C in the north to 15°C in the south, and mean winter temperature ranges significantly, from −17.5°C in the north to −10°C in the south. Annual precipitation is even throughout the entire ecoregion (400–600 mm). This climate is classified as boreal southern cordilleran (ESWG 1995).

The foothills of the Western Alberta Upland rise above the plains and are mainly linear ridges, rolling plateau remnants, and broad valleys.

Higher elevations range from 700 to 1500 m asl. The Clear Hills Upland in the north has elevations in the range of 550–1,050 m asl and has steep slopes, some rolling plateau remnants, and broad, gently undulating valleys (ESWG 1995).

Biological Distinctiveness

Vegetation is characterized by mixed forests of lodgepole pine (*Pinus contorta*), quaking aspen (*Populus tremuloides*), and white spruce (*Picea glauca*), with balsam poplar (*P. balsamifera*), paper birch (*Betula papyrifera*), and balsam fir (*Abies balsamifera*). Aspen and open stands of lodgepole pine occur on drier sites; black spruce (*P. mariana*) and tamarack (*Larix laricina*) are associated with wet sites. The Alberta/British Columbia Foothills Forests are transitional, marking mixed vegetation between boreal and Cordilleran vegetation. Fire is probably the most important disturbance regime, although seasonal and year-to-year changes in precipitation will also alter species assemblages (ESWG 1995).

Beaver (*Castor canadensis*), black bear (*Ursus americanus*), moose (*Alces alces*), muskrat (*Ondatra zibethica*), wolf (*Canis lupus*), snowshoe hare (*Lepus americanus*), sandhill crane (*Grus canadensis*), ruffed grouse (*Bonasa umbellus*), spruce grouse (*Dendragapus canadensis*), and waterfowl are characteristic of this ecoregion (ESWG 1995).

Among rare ecological and evoutionary phenomena are some of the highest population densities of moose in North America, and a variety of eastern warbler species (Parulinae) that reach their northwestern limits in North America here. In addition, Halfway Valley is a major stopover for migrating sandhill cranes. The northern portion of this ecoregion contains an area known as the Chinchauga Hills, which contains two subspecies of caribou (*Rangifer tarandus*), one a mountain subspecies and the other a woodland subspecies.

Conservation Status

Habitat Loss

There is virtually no completely undisturbed habitat in this ecoregion. Most of the ecoregion was ranked as "altered" with a smaller portion ranked as "heavily altered."

A variety of extensive human use is prevalent in this ecoregion, including logging, agricultural expansion from the south that is converting aspen forests to cereal and hay crops, livestock grazing in community pastures, intensive seismic exploration, and pipelines.

Remaining Blocks of Intact Habitat

There are no completely unaltered large habitat blocks. The largest site with the least disturbance appears to be the Chinchauga Hills.

Degree of Fragmentation

Agriculture, roads, and logging have fragmented the ecoregion. To a lesser extent, an extensive network of seismic lines creates divisions, but their narrow character is less disruptive to some species than roads. They do, however, create opportunities for increased access to otherwise more remote areas.

Degree of Protection

- Goose Mountain Ecological Reserve—western Alberta, 12 km²
- Whitecourt Mountain Natural Area—western Alberta, 5 km²

Types and Severity of Threats

Much of the territory in this ecoregion has been allocated to various kinds of industrial resource uses several times over. Forestry, oil and gas, and agriculture combine to place a high human demand on the landscape. Predator control of "nuisance" animals, under the guise of wildlife management, is a serious threat to both grizzly bear (*Ursus arctos*) and wolf (*Canis lupus*) populations in this ecoregion. The caribou populations in the Chinchauga Hills are severely threatened due to habitat loss and degradation.

Suite of Priority Activities to Enhance Biodiversity Conservation

- Protect the Chinchauga Hills
- Protect riparian habitats throughout the ecoregion

Conservation Partners

For contact information, please see appendix G.

- Alberta Wilderness Association
- Canadian Parks and Wilderness Society, Calgary and Banff Chapter
- Canadian Parks and Wilderness Society, Edmonton Chapter
- The Nature Conservancy, Alberta
- World Wildlife Fund Canada

Relationship to Other Classification Schemes

The Western Alberta Upland (TEC ecoregions 145 and 146) occurs as two separate areas in west-central Alberta. The small southern region extends into the foothills, and the larger area crosses into British Columbia. The Clear Hills Upland, which comprises the northern section of the Foothills Forests (TEC ecoregion 137), spans the British Columbia–Alberta boundary north of the Peace River district (ESWG 1995). This ecoregion falls within the Lower and Upper Foothills sections of the Boreal forest region (B.19a and B.19c) (Rowe 1972).

Prepared by D. Demarchi, R. Usher, K. Kavanagh, M. Sims, and G. Mann

Key Number:	**30**
Ecoregion Name:	**North Central Rockies Forests**
Major Habitat Type:	**2.2 Temperate Coniferous Forests**
Ecoregion Size:	**245,430 km²**
Biological Distinctiveness:	**Bioregionally Oustanding**
Conservation Status:	**Snapshot—Relatively Stable**
	Final—Vulnerable

Introduction

The North Central Rockies Forest is an extensive montane ecoregion stretching roughly six hundred miles from north to south. It is virtually contiguous from southeastern British Columbia and the extreme southwestern corner of Alberta to west-central Montana and south-central Idaho. The ecoregion contains some of North America's best-known wildlands: Alberta and British Columbia's Rocky Mountain parks (Mt. Revelstoke-Glacier, Yoho, and Waterton Lakes),

Montana's Glacier Park and Bob Marshall Wilderness complex, and Idaho's Selway-Bitterroot Wilderness. This ecoregion is distinct from the drier "interior" to the west, the cooler boreal forests to the north, and the drier foothills to the east.

Climate varies extensively from west to east in the ecoregion, with the western edge experiencing the moderating effects of maritime influence and the eastern edge experiencing a harsher, more continental regime. Climate likewise varies from north to south, with local topographic change. The mean annual temperature in the Canadian portion of this ecoregion ranges from 3.5°C in the east to 5.5°C in the west. Mean summer temperature ranges from 12.5°C to 14.5°C, and mean winter temperature ranges from −3.5°C to −6.5°C, following the pattern of warmer temperatures in the west and cooler temperatures in the east. Mean annual precipitation ranges from 500 to 800 mm in the valleys, to over 1,000 mm at higher elevations. In the southeast region, frequent chinooks moderate winter temperatures. Permafrost occurs in isolated patches in some alpine areas. Valleys are characterized by warm, showery summers and mild, snowy winters, and subalpine summers are cool, showery, and prone to frosts, with moderately cold and snowy winters. Generally, the climate in the North Central Rockies Forests is a combination of alpine, subalpine, and montane southern cordilleran (ESWG 1995).

The Columbia Mountains, which cover most of this ecoregion, contain many peaks higher than 3,000 m asl and have rugged rock and glacier outcrops. The highest mountain in the Canadian Rocky Mountains is Mount Robson, at just over 3,900 m asl. This region is composed of folded sedimentary and volcanic strata and massive metamorphic rocks of Paleozoic and Mesozoic age. Glaciation has shaped great valleys and filled them with glaciofluvial and morainal sediments (ESWG 1995).

Biological Distinctiveness

The dominant vegetation type in the ecoregion is coniferous forest. Species composition and associations reflect the influence of maritime weather systems that penetrate from the Pacific. Thus, tree species found in the Cascades and Pacific coastal ranges are also strongly represented here. Hemlocks (*Tsuga* spp.), Pacific yew (*Taxus brevifolia*), and larch (*Larix* spp.) are found here, yet are absent from other Rocky Mountain forests (Peet 1988). Montane forests include western hemlock (*T. heterophylla*) and western red cedar (*Thuja plicata*), with white spruce (*Picea glauca*) and alpine fir (*Abies lasiocarpa*) forests more prevalent to the south. Montane areas also include stands of lodgepole pine (*Pinus contorta*) and Douglas fir (*Pseudotsuga menziesii*), with some western white pine (*P. monticola*) and western larch (*L. occidentalis*). The subalpine forests are characterized by Engelmann spruce (*P. engelmannii*) and alpine fir, and stands of lodgepole pine that develop after fire. Fire is probably the most important disturbance regime, although the rugged terrain results in high precipitation and flash floods and landslides in some areas.

In addition to expansive conifer forests, the ecoregion contains several other vegetation communities. Mountain meadows, foothill grasslands, riparian woodlands, and upper treeline/alpine communities exist throughout the ecoregion. The ecoregion is characterized by dramatic vertical zonation of vegetation and associated fauna. This zonation is a consequence of abrupt elevational gradients between flatlands and mountains. Secondary climatic effects of topographic relief (e.g., rain-shadow effects, exposure to or shelter from prevailing winds, and thermal inversions) likewise influence zonation (Peet 1988).

The area has noteworthy populations of large carnivores, including wolves (*Canis lupus*), grizzly bears (*Ursus arctos*), and wolverines (*Gulo luscus*). There are also populations of rare woodland caribou (*Rangifer tarandus* ssp. *caribou*), the only caribou to live in areas of deep snow. Other wildlife include: black bear (*Ursus americanus*), mountain goat (*Oreamnos americanus*), grouse (*Dendragapus* spp.), waterfowl, black and white-tailed deer (*Odocoileus hemionus* and *O. virginianus*), and moose (*Alces alces*). In the southeast, marten (*Martes americana*) and bobcat (*Lynx rufus*) occur, while in the northern part of the southern Rocky Mountain trench, coyote (*Canis latrans*) and cougar (*Felis concolor*) are found.

Conservation Status

Habitat Loss

Logging, hard-rock mining, oil and gas development, and recreational-residential construction are all major anthropogenic threats to the ecoregion. Domestic livestock grazing and introduction of exotic species are altering species compositions. Burgeoning recreational use of remote areas is also affecting the ecoregion.

Remaining Blocks of Intact Habitat

- Bob Marshall Wilderness Complex—northwestern Montana
- Glacier National Park—northern Montana
- Selway-Bitterroot Wilderness—northeastern Idaho
- Cabinet Mountains Wilderness—northwestern Montana
- Wells-Gray Provincial Park—British Columbia, 5,157 km^2
- Mount Robson Provincial Park—British Columbia, 2,195 km^2
- Kootenay National Park—British Columbia, 1,406 km^2
- Glacier National Park—British Columbia, 1,349 km^2
- Purcell Mountain Provincial Wilderness Park—British Columbia, 1,315 km^2
- Yoho National Park—British Columbia, 1,313 km^2
- Bowron Lake Provincial Park—British Columbia, 1,231 km^2
- Mitchell Lake/Niagara Provincial Park—British Columbia, 1,105 km^2
- Upper Elbow Sheep Provincial Park—Alberta, 906 km^2
- Goat Range Provincial Park—Alberta, 795 km^2
- Waterton Lakes National Park—southwestern Alberta and southeastern British Columbia, 505 km^2

Degree of Fragmentation

Transportation corridors such as Canada Highway 3, U.S. Highway 2, and I-90 are major fragmentation areas that may reduce the long-term viability of the ecoregion's carnivore populations. Major mining sites, large clear-cuts, and other high-impact resource extraction activities

have also reduced connectivity within the ecoregion and among neighboring ecoregions.

Degree of Protection

The ecoregion has a substantial amount of protected areas (see "Remaining Blocks of Intact Habitat" above). However, these are not optimally configured for long-term viability. Thus, major changes and disturbances in relatively small areas could have the disproportionately large effect of isolating these protected areas from one another.

Types and Severity of Threats

The major threats are loss of connectivity among habitat blocks due to resource extraction and development, and increased human activity within habitat blocks as more people occupy the region.

Suite of Priority Activities to Enhance Biodiversity Conservation

The highest-priority activity should be to implement a comprehensive large carnivore conservation strategy for the ecoregion. Key sites for action include:

- In the North Fork of the Flathead River, reduce road access to prevent population declines for grizzlies and other large mammals.
- In the Cabinet-Yaak ecosystem, limit intensive logging and associated road building as they are greatly reducing habitat effectiveness.
- In the Crowsnest Pass area, where Canada Highway 3 crosses the Continental Divide, take measures to preserve important linkage habitat, so as to conserve the international populations of carnivores in the area.

Conservation Partners

For contact information, please see appendix G.

- Alliance for the Wild Rockies
- American Wildlands
- Craighead Wildlife-Wildlands Institute
- Canadian Parks and Wilderness Society, British Columbia Chapter
- Canadian Spirit of Ecotourism Society
- Corporation for the Northern Rockies

- Craighead Environmental Research Institute
- Defenders of Wildlife
- East Kootenay Environmental Society, Cranbrook-Kimberley Branch
- East Kootenay Environmental Society, Creston Valley Branch
- East Kootenay Environmental Society, Elkford Branch
- East Kootenay Environmental Society, Golden Branch
- East Kootenay Environmental Society, Invermere Branch
- East Kootenay Environmental Society, Sparwood Branch
- Ecology Center
- Friends of Yoho
- Great Bear Foundation
- Montana Wilderness Association
- The Nature Conservancy, British Columbia
- Northern Rockies Conservation Coop
- Northwest Wildlife Preservation Society
- Pro Terra-Kootenay Nature Allies
- Rocky Mountain Naturalists
- Shuswap Environmental Action Society
- Shuswap Naturalists
- Waterton Natural History Association
- The Wilderness Society
- Williams Lake Environmental Society
- World Wildlife Fund Canada
- Yellowstone Institute
- Yellowstone Ecosystem Studies

Relationship to Other Classification Schemes

The ecoregion boundary follows the northern part of Omernik's (1995b) Northern Rockies. This ecoregion, however, reaches its southern limit at the Clearwater River in Central Idaho, rather than stretching to the Snake River shrub steppe as in Omernik. There is no closely corresponding Bailey ecoregion. Küchler (1975) classifies the same area as 10, 11, 12, 13, and 45.

The North Central Rockies Forests run along the southeastern boundary of British Columbia and into the United States. The Columbia Mountains and Highlands terrestrial ecoregion (TEC ecoregion 205) forms most of this region, along with the Western Continental Ranges to the east (TEC ecoregion 206). The Selkirk-Bitterroot Foothills to the southwest, the northern half of the Southern Rocky Mountain Trench, and the Northern Continental Divide

(TEC ecoregions 212–214) are also included in this ecoregion (ESWG 1995). This ecoregion includes various forest types: Northern and Southern Columbia (CL.1, CL.2), Montane Douglas-fir and Lodgepole Pine (M.5), Interior Subalpine (SA.2), East Slope Rockies Subalpine (SA.1), and Tundra (Rowe 1972).

Prepared by S. Primm, D. Demarchi, K. Kavanagh, and M. Sims

Key Number:	**31**
Ecoregion Name:	**Okanagan Dry Forests**
Major Habitat Type:	**2.2 Temperate Coniferous Forests**
Ecoregion Size:	**53,210 km²**
Biological Distinctiveness:	**Nationally Important**
Conservation Status:	**Snapshot—Endangered Final—Critical**

Introduction

This ecoregion occupies the southern interior of British Columbia and adjacent Washington state between the Rocky Mountains to the east and the coastal ranges to the west.

Mean annual temperature is around 7°C; mean summer temperature ranges from 15°C to 16.5°C, and mean winter temperature is around –2.5°C. A strong elevational gradient in precipitation occurs, ranging from 250 mm to over 1,000 mm per year. On average, precipitation is in the range of 400 mm on the plateaus. This ecoregion is characterized by very warm to hot, dry summers and moderately cool winters with little snowfall (ESWG 1995).

This ecoregion is composed of flat-lying Tertiary sediments and volcanic rocks 1,220–1,525 m asl in elevation. Its gently rolling surface is incised by many rivers below the general surface (ESWG 1995).

Biological Distinctiveness

Vegetative cover is quite diverse in this region. It encompasses alpine, forests, and grasslands. Forest cover ranges from lodgepole pine (*Pinus contorta*) with quaking aspen (*Populus tremuloides*), white spruce (*Picea glauca*), and Douglas fir (*Pseudotsuga menziesii*) on the plateau to Douglas fir and pine grass (*Calamogrosis rubescens*) at moderate mid-slope elevations. In subalpine areas, Engelmann spruce (*P. engelmannii*), subalpine fir (*Abies lasiocarpa*), and lodgepole pine grow. Valley bottoms contain ponderosa pine (*P. ponderosa*) with bluebunch wheatgrass (*Agropyron* spp.), blue grass (*Poa predensis*), June grass (*Koelaria* spp.), and sagebrush (*Artemisia tridentata*) (ESWF 1995).

Characteristic wildlife of this ecoregion includes black bear (*Ursus americanus*), elk (*Cervus canadensis*), northern river otter (*Lutra canadensis*), bighorn sheep (*Ovis canadensis*), black and white-tailed deer (*Odocoileus hemionus* and *O. virginianus*), coyote (*Canis latrans*), bobcat (*Lynx rufus*), cougar (*Felis concolor*), American badger (*Taxidea taxus*), California quail (*Callipepla californica*), waterfowl, blue grouse (*Dendragapus obscurus*), and long-billed curlew (*Numenius americanus*) (ESWG 1995).

This ecoregion contains the Northern Continental Range extensions of many species of reptiles, amphibians, insects, and plants.

Conservation Status

Habitat Loss

It is estimated that only about 20 percent of this ecoregion remains as intact habitat. Some parts of the ecoregion have been heavily altered due to growing urban expansion and conversion of land into agricultural production. This is particularly true of the valleys and basins. Upper elevations have been impacted by livestock grazing, logging, open-pit mines, agriculture, and transmission and pipeline corridors. Some grasslands have been seriously overgrazed by livestock.

Remaining Blocks of Intact Habitat

Few unprotected blocks of habitat remain in this ecoregion.

Degree of Fragmentation

Connectivity of grassland habitats for reptiles and amphibians is highly impaired in the major valleys. The connectivity of forest species is intermediate in most forest habitats, except for the major travel corridors across the Thompson Plateau, where there are median dividers and 2–3-meter-high fences for great lengths.

Degree of Protection

- Dunn Peak Provincial Park—southern British Columbia, 198 km²
- Lac Du Bois Provincial Park—southern British Columbia, 145 km²
- Bonaparte Provincial Park—southern British Columbia, 118 km²
- Okanagan Mountain Provincial Park—southern British Columbia, 106 km²
- Arrowstone Provincial Park—southern British Columbia, 104 km²
- Silver Star Provincial Park—southern British Columbia, 87 km²
- Guichon Provincial Park—southern British Columbia, 51 km²
- Toweel Provincial Park—southern British Columbia, 46 km²
- Roche Lake Provincial Park—southern British Columbia, 21 km²
- Porcupine Provincial Park—southern British Columbia, 19 km²

Types and Severity of Threats

Many of the valley bottoms, including bunchgrass and bunchgrass-sagebrush, are under intensive pressure from agriculture and urban development. Most of the remaining patches are slated for conversion. Upland forest habitats have either been logged in the past or are slated for logging in the future. Many of these areas lie on public lands and will remain under forest cover, although forest communities may be significantly altered or degraded.

Suite of Priority Activities to Enhance Biodiversity Conservation

- Restore and increase long-term protection for the grassland habitats.
- Promote private stewardship and nature trust activities to conserve the biodiversity in this ecoregion.

Conservation Partners

For contact information, please see appendix G.

- Canadian Parks and Wilderness Society, British Columbia Chapter
- Friends of Okanagan Mountain Park

- Kamloops Naturalists
- The Nature Conservancy, British Columbia
- Nature Trust of BC
- North Okanagan Naturalists Club
- Northwest Wildlife Preservation Society
- Okanagan Silkameen Parks Society
- World Wildlife Fund Canada

Relationship to Other Classification Schemes

The Thompson-Okanagan Plateau (TEC ecoregion 209) comprises most of the area of the Okanagan Dry Forests, and the Okanagan Highland (TEC ecoregion 211) in the south is also included in the Canadian portion of the ecoregion (ESWG 1995). In terms of forest regions, this ecoregion incorporates the Strait of Georgia Coast forest (C.1), Grassland, Interior Subalpine (SA.2) and Montane forests of Ponderosa Pine and Douglas-fir (M.1) and Central Douglas-fir (M.2) (Rowe 1972).

Prepared by D. Demarchi, K. Kavanagh, M. Sims, and G. Mann

Key Number:	**32**
Ecoregion Name:	**Cascade Mountains Leeward Forests**
Major Habitat Type:	**2.2 Temperate Coniferous Forests**
Ecoregion Size:	**46,626 km²**
Biological Distinctiveness:	**Nationally Important**
Conservation Status:	**Snapshot—Relatively Stable Final—Relatively Stable**

Introduction

This ecoregion stretches as a narrow band along a north-south axis on the leeward side of the Cascade Mountains that straddle the Canada-U.S. border in the province of British Columbia and the state of Washington.

In the southern half of the ecoregion the mean annual temperature is 6°C; mean summer temperature is 15°C, and mean winter temperature is −3.5°C. In the northern half of the ecoregion, temperatures are approximately 2.5°C cooler. A strong climatic gradient exists from the moist coastal climate to the semiarid continental climate of the southern interior. Precipitation ranges from 300 mm in the south to 500 mm in the north, and from 600 mm in the east to 1,200 mm in the west (ESWG 1995).

The southern half of the ecoregion is made up of the Cascade Ranges, a mountainous upland within the southern Pacific Ranges. The Chilcotin Ranges make up the northern half of the ecoregion and reach elevations of up to 2,700 m asl. The Okanagan Range in the south runs along the Canada–United States border (ESWG 1995).

Biological Distinctiveness

Alpine tundra communities consist of Engelmann spruce (*Picea engelmanii*), subalpine fir (*Abies lasciocarpa*), and lodgepole pine (*Pinus contorta*). Montane forests are made up of lodgepole pine, quaking aspen (*Populus tremuloides*), white spruce (*P. glauca*), and Douglas fir (*Pseudotsuga menziesii*) with a pine grass understory. At the lowest elevations in the eastern region there is a parkland of scattered ponderosa pine (*P. ponderosa*) in a matrix of bluebunch wheatgrass (*Agropyron* spp.) and sagebrush (*Artemisia tridentata*) grasslands (ESWG 1995).

This ecoregion is home to bighorn sheep (*Ovis canadensis*), mountain goat (*Oreamnos americanus*), grizzly and black bear (*Ursus arctos* and *U. americanus*), black-tailed deer (*Odocoileus hemionus*), coyote (*Canis latrans*), blue grouse (*Dendragapus obscurus*), cougar (*Felis concolor*), and various raptors in the southern part of the region (ESWG 1995). Large numbers of salmon (*Oncorhynchus* spp.) use the Fraser River as a major migration corridor. The ecoregion contains habitat for the endangered spotted owl (*Strix occidentalis*).

Conservation Status

Habitat Loss

Nearly 70 percent of the ecoregion is considered to remain as intact habitat. A variety of human activities occur in the ecoregion, including logging, base-metal mining, cattle grazing in lower-elevation areas, and some sheep grazing in alpine areas and some significant transportation corridors. There are small areas of agriculture in the lower elevations.

Remaining Blocks of Intact Habitat

Few large blocks of habitat remain in this ecoregion. Some remaining blocks include:

- South Chilcotin-Spruce Lake
- Stein Valley
- Manning Park
- Snowy Mountain
- TS42-OS
- Cathedral

Degree of Fragmentation

The Fraser River is a natural movement barrier to the north-south movement of small mammals, ungulates, and carnivores. However, the construction of several major east-west highways and railways, complete with high fencing in some cases, has further disrupted the north-south movement of species such as grizzly and black bears, wolverine, and fisher.

Degree of Protection

- Ts'yl-os Provincial Park—British Columbia, 2,332 km²
- North Cascades National Park—north-central Washington
- Big Creek-South Chilcotin Provincial Park—British Columbia, 868 km²
- Manning Provincial Park—British Columbia, 658 km²
- Cathedral Provincial Park—British Columbia, 332 km²
- Lower Stein Wilderness Area—British Columbia, 330 km²
- Skagit Valley Recreation Park—British Columbia, 325 km²
- Marble Range Provincial Park—British Columbia, 180 km²
- Cascade Recreation Park—British Columbia, 166 km²
- Edge Hills Provincial Park—British Columbia, 130 km²
- Upper Stein Wilderness Area—British Columbia, 100 km²

Types and Severity of Threats

Outside of protected areas, there are significant threats to forests on both the U.S. and Canadian sides of the border. There are additional threats from mineral exploration, and rural and recre- ational developments adjacent to some of the lakes in the ecoregion (e.g., Carpenter Lake).

Suite of Priority Activities to Enhance Biodiversity Conservation

- Establish new protected areas at Spruce Lake and Snowy Mountain in British Columbia.
- Restore suitable corridors to enable dispersal of mammalian species in a north-south direction.

Conservation Partners

For contact information, please see appendix G.

- Canadian Parks and Wilderness Society, British Columbia Chapter
- The Nature Conservancy, British Columbia
- Nature Trust of BC
- Northwest Wildlife Preservation Society
- Sierra Club, Western Canada
- World Wildlife Fund Canada

Relationship to Other Classification Schemes

The Chilcotin Ranges (TEC ecoregion 204) make up the northern part of the Cascade Mountains Leeward Forests; the Interior Transition Ranges (TEC ecoregion 208) lie in the southern part; and the small Okanagan Range area (TEC ecoregion 210) in the far south is included as well (ESWG 1995). Ponderosa Pine and Douglas-fir Montane forest (M.2), Interior and Coastal Subalpine forest (SA.2, SA.3), Grassland, and Tundra all characterize this south-central area of British Columbia (Rowe 1972).

Prepared by K. Kavanagh and M. Sims

Key Number:	**33**
Ecoregion Name:	**British Columbia Mainland Coastal Forests**
Major Habitat Type:	**2.2 Temperate Coniferous Forests**
Ecoregion Size:	**142,153 km²**
Biological Distinctiveness:	**Globally Outstanding**
Conservation Status:	**Snapshot—Vulnerable Final—Endangered**

Introduction

This ecoregion occupies all of the British Columbia mainland coast and a small part of Washington state. The ecoregion extends inland about 150 km to the crest of the Coast Mountains.

A climatic gradient exists across the Coast Mountains to slightly drier and cooler conditions toward the interior. In the valleys of the southern half of this ecoregion, the mean annual temperature is 6.5°C, but it is higher in the far south. The summer mean temperature is approximately 14°C, and the winter mean is around –1°C, but also is higher in the far south. Precipitation varies greatly with elevation, from 1,500 mm in the lower elevations to 3,400 mm at higher elevations. Mean annual temperature in the northern half of the ecoregion is around 5°C; mean summer temperature is 13°C, and mean winter temperature ranges from –4.5°C to –0.5°C. Temperatures are significantly lower in the Nass Basin in the far north of the region. Precipitation ranges anywhere from 1,200 to 4,500 mm, depending on elevation (ESWG 1995).

This ecoregion incorporates the northern part of the Cascade Mountains, the Pacific Ranges and the Kitimat Ranges of the Coast Mountains, the Nass Ranges, and the Nass Basin. The Pacific Ranges in the south vary from sea level to 4,000 m asl, and the highest peaks are surrounded by expansive ice fields. Many large, steep-sided, transverse valleys, inlets, or fjords dissect this mountainous coastal region. The Kitimat Ranges in the northern part of the ecoregion form a saddle between the Pacific Ranges and the higher Boundary Ranges to the north (ESWG 1995).

Biological Distinctiveness

The Pacific Ranges encompass three different vegetation zones. Low-elevation coastal forests include productive stands of western hemlock (*Tsuga heterophylla*), western red cedar (*Thuja plicata*), and amabilis fir (*Abies amabilis*). The subalpine zone is dominated by forests of mountain hemlock (*T. mertensiana*) and amabilis fir, with some yellow cedar (*Chamaecyparis nootkatensis*). Alpine tundra also includes sedge (*Carex* spp.)-dominated meadows and lichen-colonized rock fields.

Characteristic wildlife of this ecoregion includes black-tailed deer (*Odocoileus hemionus*), black and grizzly bear (*Ursus americanus* and *U. arctos*), mountain goat (*Oreamnos americanus*), wolf (*Canis lupus*), mink (*Mustela vison*), northern river otter (*Lutra canadensis*), blue grouse (*Dendragapus obscurus*), and waterfowl, and moose (*Alces alces*), woodland caribou (*Rangifer tarandus* ssp. *caribou*), beaver (*Castor canadensis*), red fox (*Vulpes fulva*), marten (*Martes americana*), and snowshoe hare (*Lepus americanus*) in the far north.

This ecoregion contains one of the most extensive mountain-fjord complexes in the world. A large number of species reach their northern or southern limit along this extensive coastal ecoregion in western North America, and an unusual white form of the black bear (Kermode or Spirit bear) occurs on Princess Royal Island.

Conservation Status

Habitat Loss

It is estimated that approximately 40 percent of this ecoregion remains as fully intact habitat. Most of the disturbed habitat is in the low to mid elevation forests, particularly in the southern and northern ends of the ecoregion where logging has accessed many of the valleys. Transportation corridors in the valleylands are present, particularly south of Squamish, British Columbia.

Remaining Blocks of Intact Habitat

In addition to existing protected areas, there remain blocks of habitat in the central portion of the ecoregion. In the southern part of the ecoregion, the Skagit Valley on the British

Columbia–Washington border is a somewhat smaller block of remaining habitat.

Degree of Fragmentation

The numerous coastal inlets and islands, combined with the mountainous terrain, create a certain level of natural fragmentation of habitat in this ecoregion. Valley bottoms and old-growth coastal rain forest have been significantly fragmented in more than 50 percent of this region by logging activities and logging roads. Such fragmentation and loss of habitat are of concern for species such as the spotted owl and grizzly bear, which require large tracts of undisturbed habitat.

Degree of Protection

- Kitlope Provincial Park—west-central British Columbia, 3,172 km²
- Garibaldi Provincial Park—southwestern British Columbia, 1,946 km²
- Hakai Recreation Park—western British Columbia, 1,229 km²
- Fiordland Recreation Park—western British Columbia, 910 km²
- Gitnadoix River Recreation Park—western British Columbia, 580 km²
- Golden Ears Provincial Park—southwestern British Columbia, 555 km²
- Khutzeymateen Provincial Park—west-central British Columbia, 449 km²
- Pinecove Lake–Burke Mountain Provincial Park—southwestern British Columbia, 368 km²
- Homathko River–Tatlayoko Provincial Park—southwestern British Columbia, 340 km²
- Swan Lake Wilderness Area—west-central British Columbia, 190 km²

Types and Severity of Threats

The greatest threat to habitat in this ecoregion is logging, which is occurring or soon scheduled to occur throughout most of the valleys. South of 50 degrees latitude, urbanization and recreational development, especially in valleys, are locally significant threats.

Suite of Priority Activities to Enhance Biodiversity Conservation

- Create more protected areas that represent the middle- and lower-elevation coastal rain forest communities.
- Properly enforce the B.C. Forest Practices Code. Avoid further damage to remaining salmon habitat and restore other riparian habitats.

Conservation Partners

For contact information, please see appendix G.

- Canadian Parks and Wilderness Society, British Columbia Chapter
- The Nature Conservancy, British Columbia
- Nature Trust of BC
- Northwest Wildlife Preservation Society
- Sierra Club, Western Canada
- World Wildlife Fund Canada

Relationship to Other Classification Schemes

The British Columbia Mainland Coastal Forests encompass, from north to south, the Nass Basin, the Nass Ranges, the Coastal Gap, the Pacific Ranges, and the Cascade Ranges (TEC ecoregions 187, 190–192, and 197) (ESWG 1995). This ecoregion includes Southern and Northern Pacific Coast forest (CL.2, CL.3), Coastal Subalpine (CL.3), Montane Transition (M.4), Northern Aspen (M.3), and Tundra forests (Rowe 1972).

Prepared by D. Demarchi, J. Nelson, K. Kavanagh, M. Sims, and G. Mann

34

Ecoregion Name:	**Central Pacific Coastal Forests**
Major Habitat Type:	**2.2 Temperate Coniferous Forests**
Ecoregion Size:	**73,696 km²**
Biological Distinctiveness:	**Globally Outstanding**
Conservation Status:	**Snapshot—Endangered Final—Endangered**

Introduction

The Central Pacific Coastal Forests stretch from southern Oregon to the northern tip of Vancouver Island. Major habitats of this diverse region include sea stacks, sandy beaches, rocky coastal cliffs, coastal headlands, tide pools, mud flats, salt marshes and estuaries, streams and rivers of various sizes, grass balds, and many forest types (Noss 1993).

Influenced by cool moist air from the ocean, the Central Pacific Coastal Forests experience frequent clouds and fog, with most precipitation occurring in the winter. Generally, precipitation is greater in the western half of the ecoregion. Annual rainfall ranges from 2,000 to 4,080 mm, with higher coastal mountain areas receiving the bulk of this precipitation. Mean annual temperature is around 13°C. Vancouver Island experiences a mean annual temperature of 8.5°C, averaging 13.5°C in the summer and 3.5°C in the winter. The climate on the island is marked by warm summers and mild winters; it is one of the mildest areas in Canada. Rainfall varies between 1,500 and 3,500 mm per year. Precipitation on the eastern side of the island ranges from 800 to 2,500 mm per year. Climate in this ecoregion is a combination of alpine, sub-alpine, and maritime South Pacific cordilleran (ESWG 1995).

The forests of the Central Pacific Coast are among the most productive in the world, characterized by large trees, substantial woody debris, luxuriant growths of mosses and lichens on trees, and abundant ferns and herbs on the forest floor. The major forest complex consists of Douglas fir (*Pseudotsuga menziesii*) and western hemlock (*Tsuga heterophylla*), encompassing seral forests dominated by Douglas fir and massive old-growth forests of fir, hemlock, western red cedar (*Thuja plicata*), and other species. These forests occur from sea level up to elevations of 700–1,000 m in the Coast Range and Olympic

Mountains. This forest type occupies a wide range of environments with variable composition and structure and includes such other species as grand fir (*Abies grandis*), Sitka spruce (*Picea sitchensis*), and western white pine (*Pinus monticola*) (Franklin 1988).

On Vancouver Island, low-elevation coastal forest cover includes stands of western hemlock, Douglas fir, and amabilis fir (*A. amabilis*). Western hemlock, Douglas fir, and grand fir characterize most forests of eastern Vancouver Island. Drier sites support stands of western hemlock and western red cedar. The driest areas in eastern Vancouver Island are comprised of mixed stands of Douglas fir and western hemlock, with occasional Garry oak (*Quercus garryana*), Pacific dogwood (*Cornus nuttallii*), and arbutus (*Arbutus menziesii*). The subalpine forests are composed of mountain hemlock (*T. mertensiana*) and amabilis fir, with some yellow cedar (*Chamaecyparis nootkatensis*) and western hemlock in high-elevation areas on eastern Vancouver Island. Coastal areas of British Columbia are best characterized as hydroriparian forest ecosystems, such that the hydrologic cycle is the most important ecological process influencing forest dynamics. High winds, sea salt spray, and fog strongly influence forest dynamics. Fires are not as important as in other areas (ESWG 1995).

Although Douglas fir is the most abundant species at lower elevations in the region, western hemlock is the major climax species. Douglas fir typically dominates young forests because of its relatively large and hardy seedlings and rapid growth rate; however, in mature forests Douglas fir cannot regenerate because western hemlock and several other species of fir are more tolerant of shade.

While hemlock and fir dominate much of the ecoregion, the cool, wet conditions along the coast create a narrow band of forests distinguished by Sitka spruce. With its high tolerance of salt spray, in areas near the ocean Sitka spruce may form nearly pure forests or codominate with lodgepole pine (*Pinus contorta*). The Sitka spruce zone—in which hemlocks also occur in large numbers—may be only a few kilometers in width and generally occurs below 150 m. Where mountains abut the coast, however, Sitka spruce forests may extend up to 600 m (Noss 1993). The alluvial rain forests of the western Olympic

Peninsula are outstanding examples of the spruce-hemlock forest.

Riparian forests of this ecoregion are quite distinct from the Douglas fir–hemlock forests. Broadleaf species such as black cottonwood (*Populus trichocarpa*) and red alder (*Alnus rubra*) replace the otherwise ubiquitous conifers along the many rivers and streams of the Pacific Northwest. Occasional grasslands, sand dune and strand communities, rush meadows and marshes, and western red cedar and alder swamps, these last often formed by beaver activity, break up the conifer forests.

Biological Distinctiveness

The Central Pacific Coastal Forests are among the richest temperate coniferous forests in North America for amphibians and birds. Gray's Harbor in Washington, for example, is a critical migratory stopover site for shorebirds. Rare and endangered species found in this ecoregion include California condor (*Gymnogyps californianus*), hoary elfin butterfly (*Incisalia polios obscurus*), North Pacific plantain (*Plantago macrocarpa*), and possibly the Pacific fisher (*Martes pennanti*) and California wolverine (*Gulo luscus luteus*) (Noss 1993). Other characteristic wildlife include elk (*Cervus canadensis*), black-tailed deer (*Odocoileus hemionus*), wolf (*Canis lupus*), black bear (*Ursus americanus*), mink (*Mustela vison*), racoon (*Procyon lotor*), grouse (*Dendragapus* spp.), seabirds, shorebirds, and waterfowl.

Conservation Status

Habitat Loss

Human activity, particularly clear-cut logging, plantation forestry, road building, agriculture, and development have heavily altered the Central Pacific Coastal Forests. Only about 4 percent of the region remains as intact habitat. Some ecosystem types, such as the coastal temperate rain forests in Oregon, have been virtually destroyed (Kellogg 1992).

Remaining Blocks of Intact Habitat

Several relatively large blocks of more or less intact habitat remain, as do a number of smaller patches. Important blocks include:

- Strathcona Provincial Park—southwestern British Columbia, 2,193 km²
- Olympic National Park—northwestern Washington, 1,200 km² (assuming that roughly half the park is forested) N.B.: Forests at the upper elevations of the park are Cascadian in nature and fall within the North Cascades Forest Ecoregion [33].
- Brooks Peninsula Provincial Park—southwestern British Columbia, 380 km²
- Tahsish watershed—southwestern British Columbia

Degree of Fragmentation

Although heavily altered, most of the ecoregion remains forested. More than half of the remaining fragments have some degree of interaction with other intact habitat blocks.

Degree of Protection

The only protected habitats in this ecoregion are in national and provincial parks, listed below. Several of those protected areas, however, are relatively large and contain undisturbed forests. Human activities have significantly altered nearly all habitats outside the parks.

- Olympic National Park—northwestern Washington
- Strathcona Provincial Park—southwestern British Columbia, 2,193 km²
- Pacific Rim National Park—southwestern British Columbia, 499 km²
- Checleset Bay Ecological Reserve—southwestern British Columbia, 346 km²
- Brooks Peninsula Provincial Recreation Park—southwestern British Columbia, 287 km²
- Pacific Rim-West Coast Trail National Park—southwestern British Columbia, 263 km²
- Megin Watershed Provincial Park—southwestern British Columbia, 213 km²
- Brooks-Nasparti Provincial Park—southwestern British Columbia, 212 km²
- Cape Scott Provincial Park—southwestern British Columbia, 150 km²
- Pacific Rim–Broken Islands Group National Park—southwestern British Columbia, 106 km²
- Tahsish-Kwois Provincial Park—southwestern British Columbia, 106 km²

Types and Severity of Threats

So little intact habitat remains outside national and provincial parks that conversion of the Central Pacific Coastal Forests has ceased to be an important issue for conservationists. Degradation from pollution, grazing, burning, introduced species, road building, and excessive recreational impacts is causing significant mortality in some native plant communities.

Suite of Priority Activities to Enhance Biodiversity Conservation

- Establish protected areas where ecological gaps remain in the protected-areas system, for example, in Clayoquot Sound.
- Extend Olympic National Park to connect coastal and upland sections.
- Create more protected areas that represent the middle- and lower-elevation coastal rain forest communities.
- Properly enforce the B.C. Forest Practices Code. Avoid further damage to remaining salmon (*Oncorhynchus* spp.) habitat and restore other riparian habitats.

The following areas in the Oregon Coast Range are high priorities for protection and restoration (Noss 1993). All require consolidation, road closures, and revegetation. These areas contain substantial old-growth and other late-successional forests, are of high value to fisheries and aquatic biodiversity, support numerous spotted owl (*Strix occidentalis*) nesting sites, and form important natural linkages in a network of significant sites:

- Cummins Creek/Rock Creek—427 km^2
- Drift Creek—229 km^2
- Mt. Hebo/Nestucca River—506 km^2
- Mary's Peak/Grass Mountain—141 km^2
- Elliott State Forest—364 km^2

These five areas are among thirty-one proposed high-priority reserves that together cover 5,065 km^2, or over 23 percent of the Oregon Coast Range Bioregion. The high-priority reserves by themselves are too small to conserve viable populations of wide-ranging species or to maintain natural disturbance regimes. They must be linked and insulated by reserves under slightly less strict protection and by multiple-use buffer zones. Together, these three classes of reserves would form a continuous network through the Coast Range.

Conservation Partners

For contact information, please see appendix G.

- Canadian Parks and Wilderness Society, British Columbia Chapter
- Clayoquot Biosphere Project
- Friends of Ecological Reserves
- Friends of Strathcona Park
- Friends of the Stikine
- Islands Trust
- The Nature Conservancy, British Columbia
- Nature Trust of BC
- Northwest Wildlife Preservation Society
- Sierra Club, Western Canada
- Victoria Natural History Society
- Western Canada Wilderness Committee
- World Wildlife Fund Canada

Relationship to Other Classification Schemes

Bailey (1994) combines the Central Pacific Coastal Forests and the Cascades into a single province. We follow Omernik (1995b) by distinguishing the regions on the basis of climate, elevation, and dominant communities. We have further divided the area by delineating two ecoregions within the Cascades: the British Columbia Mainland Coastal Forests [33], which run from central Washington to northern British Columbia; and the Central and Southern Cascades Forests [36], which run from Washington almost to the Oregon-California border. Although the mountains of the Olympic Peninsula fall entirely within the coastal ecoregion, like Omernik we include the higher-elevation forests with the North Cascades ecoregion. Conservation strategies for the coastal and montane forests would be unwieldy if managed as a single large unit stretching nearly from California to the Yukon Territory.

The Canadian portion of the Central Pacific Coastal Forests rests on Vancouver Island (TEC ecoregions 193 and 194) (ESWG 1995). In addition to Southern and Northern Pacific Coast vegetation (CL.2, CL.3), Vancouver

Island has Coastal Subalpine forests (CL.3) (Rowe 1972).

Prepared by R. Noss, J. Strittholt, G. Orians, J. Adams, K. Kavanagh, M. Sims, and G. Mann

Key Number: **35**
Ecoregion Name: **Puget Lowland Forests**
Major Habitat Type: **2.2 Temperate Coniferous Forests**
Ecoregion Size: **22,488 km²**
Biological Distinctiveness: **Nationally Important**
Conservation Status: **Snapshot—Critical Final—Critical**

Introduction

The Puget Lowland Forests occupy a north-south depression between the Olympic Peninsula and the western slopes of the Cascade Mountains, extending from across the Canadian border to the lower Columbia River along the Oregon border. In British Columbia, this ecoregion includes the Fraser Valley lowlands, the coastal lowlands locally known as the Sunshine Coast, and several of the Gulf Islands. It is distinct from the mountainous hydroriparian systems to the west and the drier areas to the north and east. The Puget Sound Valley is a depressed glaciated area consisting of moderately dissected tableland covered by glacial till, glacial outwash, and lacustrine deposits (Franklin and Dyrness 1973, 16; Bailey 1995, 38). Relief in the valley is moderate with elevation ranging from sea level to 460 m and seldom exceeding 160 m (Franklin and Dyrness 1973, 16). North of the U.S.-Canada border, bedrock outcrops of Mesozoic and Paleozoic origin form rolling hills on the Lower Mainland that reach an elevation of 310 m asl. The Fraser River dominates this ecoregion. The majority of soils in the valley are formed in glacial materials under the influence of coniferous forest (ESWG 1995). Haplorthods (brown podzolic soils) are most common and contain moderately thick forest floor layers with well-developed humus (Franklin and Dyrness 1973, 17).

This ecoregion has a Mediterranean-like climate, with warm, dry summers and mild, wet winters. The mean annual temperature is 9°C; the mean summer temperature is 15°C, and the mean winter temperature is 3.5°C. Annual precipitation averages 800 to 900 mm but may be as high as 1,530 mm (Franklin and Dyrness 1973, 17; Bailey 1995, 38). Only a small percentage of this precipitation falls as snow. However, rainfall amounts on the San Juan Islands can be as low as 460 mm because of rain-shadow effects caused by the Olympic Mountains (Franklin and Dyrness 1973, 88). This local rain-shadow effect results in some of the driest sites encountered in the region. Varied topography on these hilly islands results in a diverse assemblage of plant communities arranged along topographically determined moisture gradients (Franklin and Dyrness 1973, 313). Open grasslands with widely scattered trees dominate the exposed southern aspects of the islands, while moister dense forests occur on northern sheltered slopes characterized by western red cedar (*Thuja plicata*), grand fir (*Abies grandis*), and sword fern (*Polystichum munitum*) communities.

Before cultivation and European settlement, the Puget Sound Lowland Forests were dominated by dense coniferous forests most commonly made up of western red cedar, western hemlock (*Tsuga heterophylla*), and Douglas fir (*Pseudotsuga menziesii*) (Bailey 1995, 38). Mixed stands of Douglas fir with some Garry oak (*Quercus garrayana*), Pacific dogwood (*Cornus nuttallii*), and arbutus (*Arbutus menziesii*) are common on drier sites. Moist sites support stands of western hemlock and western red cedar. Periodic flooding and infrequent fires were once the predominant disturbance regimes in the region (McNab and Bailey 1994). Long intervals (centuries) between large-scale fire events were more typical of moister forest types (Agee 1993, 58), with drier forests (*Quercus* spp., *Pinus ponderosa*) and prairies experiencing frequent fires.

Biological Distinctiveness

Characteristic wildlife includes racoon (*Procyon lotor*), sea otter (*Enhydra lutris*), mink (*Mustela vison*), coyote (*Canis latrans*), black-tailed deer (*Odocoileus hemionus*), and harbour seal (*Phoca vitulina*). There is a rich diversity of birds, which includes turkey vulture (*Cathartes aura*), bald eagle (*Haliaeetus leucocephalus*), blue grouse

(*Dendragapus obscurus*), seabirds, numerous shorebirds, and waterfowl (ESWG 1995).

In general, this ecoregion is intermediate in species richness (429 species) and conifer richness (11 species) compared to the thirty other ecoregions within its MHT. Birds (200 species) make up the majority (47 percent) of taxonomic groups represented. Notably, 8 of the total species are endemics, and 6 of these are snails. Although plant communities within the region are similar to others in the *T. heterophylla* zone of western Washington, large areas once contained prairie, oak woodland, and pine (*Pinus* spp.) forest types (Franklin and Dyrness 1973, 88). Other notable features that are uncommon elsewhere in western Washington include the following: (1) *Pinus contorta, P. monticola,* and *P. ponderosa* as major constituents, along with *P. menziesii* and *Glautheria shallon;* (2) *Q. garryana* groves (relict examples on the Ft. Lewis Air Force Base); (3) extensive prairies often invaded by *P. menziesii* and associated with groves of *Quercus* (Ft. Lewis); (4) abundant and poorly drained sites with swamp or bog communities (relict examples in the Seattle area); (5) occurrence of species rarely or never found elsewhere in western Washington, such as *Juniperus scopulorum, Populus tremuloides, P. ponderosa,* and *Betula papyrifera.* Prairies were significant features south of Puget Sound and once included the Tacoma Prairies near Tacoma, Washington, and Wier Prairie near Olympia, Washington (Franklin and Dyrness 1973, 89). Since settlement the extent of these prairies has been substantially reduced by urbanization, cultivation, and invasion by Douglas fir and Garry oak communities caused by fire suppression and livestock grazing.

Remaining riparian forests in the region provide important spawning areas for salmonids (*Oncorhynchus* spp.), habitat for amphibians and snails, roost sites for bats, perching and nesting sites for bald eagles, and travel corridors for wildlife (e.g., black-tailed deer, neotropical migratory birds).

Conservation Status

Habitat Loss

This ecoregion lies within the most densely populated area of Washington and British Columbia, encompassing the cities of Vancouver, Victoria, Bellingham, Seattle, Tacoma, and Olympia. Consequently, only 5 percent of the original habitat within the region remains, and most remaining areas have been heavily (90–100 percent) altered. Small, isolated "islands" of original habitat (e.g., old-growth forest, bogs, prairie-oak woodlands) are surrounded by urbanization and agriculture.

Among the plant communities in this region, the prairie–*Q. garryana* woodlands, riparian forests and wetlands, and old-growth lowland forests (almost completely gone) are most threatened.

Remaining Blocks of Intact Habitat

No sizable blocks of intact habitat remain in the region. A few relict examples of prairie-oak communities occur on Ft. Lewis and are managed by the base to maintain characteristic plant composition. The remaining forests have been largely converted to tree farms or exist as small city or state parks. Remaining areas should be used to restore degraded habitats through the use of prescribed fire (prairies) and long-rotation timber harvest (plantations). Burn's Bog wetland complex, the southernmost domed peatbog in western North America, is found adjacent to Vancouver.

Degree of Fragmentation

Fragmentation has been extensive throughout the region; almost no native habitat remains.

Degree of Protection

Very little of the natural communities in this region has been protected. Opportunity exists to use local conservation easements to protect remaining riparian and wetland areas and to restore a portion of the old-growth forests. However, this potential is limited by urbanization that is projected to increase substantially over the next decade; the region is one of the fastest-growing areas in the United States. A case can be made to better manage riparian areas and nearby watersheds, to maintain city watersheds and open spaces, and to provide connections to adjacent ecoregions.

Types and Severity of Threats

The ecoregion was given a critical ranking because of threats to remaining native habitats from urbanization, agriculture, fire suppression, invasive species, flood control and hydroelectric dams, and logging. A serious threat to native plant communities, especially prairie-oak woodlands, is invasion by scotchbroom (*Cytisus scoguarius*) and encroachment by trees as a result of fire suppression in these prairies.

Suite of Priority Activities to Enhance Biodiversity Conservation

- Establish one or more forest connections between the Cascade and Coast Range forests south of Olympia (if restored, the Skookimchuck River valley would be a prime candidate for acting as a connection between populations of northern spotted owls (*Strix occidentalis caurina*) on the Olympic Peninsula and in the Cascade Mountains).
- Establish several riparian habitat corridors along streams draining into Puget Sound from the Cascades (e.g., Nisqually, Skykomish, Nooksack Rivers).
- Protect Burn's Bog, British Columbia.
- Maintain the remaining prairie-oak woodlands on the Ft. Lewis base through the use of prescribed fire management.

Conservation Partners

For contact information, please see appendix G.

The region is a hot-bed for local conservation, with several regional and national groups having offices in the Seattle area. Some examples include:

- Burn's Bog Conservation Society
- Canadian Nature Federation
- Friends of Caren
- Galiano Conservancy Association
- National Audubon Society
- The Nature Conservancy, British Columbia
- Nature Trust of BC
- Northwest Ecosystem Alliance
- Pender Harbour and District Wildlife Society
- Sierra Club
- Sierra Club of British Coumbia
- Tetrahedron Alliance
- White Rock and Surrey Naturalists
- The Wilderness Society
- World Wildlife Fund Canada

Relationship to Other Classification Schemes

In general, there is a high concordance with Omernik's (1995b) classification of this ecoregion (Puget lowlands ecoregion 2). Bailey (1994), however, delineated this ecoregion farther south of the lower Columbia River to the Willamette Valley of central Oregon, and thus the boundary we used in this analysis corresponds more closely to Bailey's northern portion of the Pacific Lowland Region (i.e., above the lower Columbia River). Justification for splitting the ecoregion is provided by Franklin and Dyrness (1973, 89), who indicate that the Puget lowlands may be recognized as a separate vegetative zone similar to the coastal Douglas fir zone in British Columbia because both the Puget Sound and British Columbia areas were glaciated and are influenced by large oceanic bodies. Such climatic influences are much less dramatic in the Willamette Valley. Therefore, our delineation more closely approximates these suggested changes.

Only a small portion of this ecoregion lies in Canada. The Lower Mainland of British Columbia (TEC ecoregion 196) extends westward from the foothills of the Cascade Range at Chilliwack to the Fraser River delta at Richmond and northward to include the narrow Georgia Lowland along the Sunshine Coast. The Georgia-Puget Basin (TEC ecoregion 195) incorporates the numerous Gulf Islands of the Strait of Georgia off the coast of British Columbia (ESWG 1995); hence, its forest region names: Coastal Strait of Georgia and Southern Pacific Coast (C.1, C.2) (Rowe 1972).

Prepared by D. DellaSala, G. Orians, K. Kavanagh, and M. Sims

Key Number:	**36**
Ecoregion Name:	**Central and Southern Cascades Forests**
Major Habitat Type:	**2.2 Temperate Coniferous Forests**
Ecoregion Size:	**44,848 km²**
Biological Distinctiveness:	**Bioregionally Outstanding**
Conservation Status:	**Snapshot—Vulnerable Final—Vulnerable**

Introduction

The Central and Southern Cascades span several physiographic provinces in Washington and Oregon, including the southern Cascades, the western Cascades, and the High Cascades (Franklin and Dyrness 1973, 21–26). This ecoregion extends from Snoqualmie Pass in Washington to just north of the California border. The region is characterized by accordant ridge crests separated by steep, deeply dissected valleys strongly influenced by historic and recent volcanic events (Mt. Saint Helens). Ridge elevations in the northern section are as high as 2,000 m, with three dormant volcanoes ranging from 2,550 m (Mt. Saint Helens) to 4,392 m (Mt. Rainier) (Franklin and Dyrness 1973, 21). The stratigraphy dates back to Precambrian-Cenozoic epochs (McNab and Bailey 1994). Pleistocene glacial activity has been widespread, creating numerous lakes and mountain valleys. However, most glaciers were restricted to small alpine areas (Franklin and Dyrness 1973, 22).

The region is characterized by numerous perennial streams and lakes maintained by abundant annual precipitation (1,270–3,048 mm; McNab and Bailey 1994). Soils are generally andisols and spodsols; however, soil series vary considerably across the region in association with local climatic differences and edaphic processes (Franklin and Dyrness 1973, 21–26; McNab and Bailey 1994). Potential vegetation includes the western hemlock (*Tsuga heterophylla*), Pacific silver fir (*Abies amabilis*), and western red cedar (*Thuja plicata*) series. Predominant natural disturbances include fire, wind, floods, and volcanoes (McNab and Bailey 1994). Extensive logging and fire suppression have substantially altered natural disturbance regimes, shifting regional landscapes from those with a full range of seral stages to highly fragmented landscapes where late-seral stages are rapidly being replaced by monocultural tree

plantations (see DellaSala et al. 1995 for basin-wide declines, and USDA Forest Service and USDI Bureau of Land Management 1996 for ecoregional declines).

Biological Distinctiveness

When compared to other ecoregions within the Temperate Coniferous Forest major habitat type, this ecoregion contains intermediate levels of biodiversity (e.g., total species richness = 493). Birds represent the majority (41 percent) of taxa evaluated, followed by butterflies (26 percent), and mammals (13 percent). This ecoregion contains one of the highest levels of endemic amphibians (five of eleven endemics are amphibians) of any ecoregion within its major habitat type. Several taxa, including salamanders (e.g., Pacific giant salamander [*Dicamptodon ensatus*] and *Ensatina* spp.), frogs (e.g., tailed frog [*Ascaphus truei*]), fishes (e.g., chinook salmon [*Oncorhynchus tshawytscha*], and bull trout [*Salvelinus confluentus*], and birds (e.g., northern spotted owl [*Strix occidentalis caurina*] and northern goshawk [*Accipiter gentilis*]), have been the focus of conservation attention in this region because of their close association with declining habitat types such as aquatic areas, seeps, talus slopes, old growth, and riparian forests (Corn and Bury 1990, USDA Forest Service 1994b). The threatened northern spotted owl has been used as an indicator species in environmental impact assessments because its range overlaps with 39 listed or proposed species (10 of which are late-seral associates) and 1,116 total species associated with late-seral forests (USDA Forest Service 1994b, S-15). Late-seral forests in general are of national and global importance because they provide some of the last refugia for dependent species and perform vital ecological services, including sequestration of carbon, cleansing of atmospheric pollutants, and maintenance of hydrological regimes.

At finer mapping scales, the USDA Forest Service and USDI Bureau of Land Management (1996, 98) have identified two hot biodiversity spots within the southern Cascades and Upper Klamath Ecological Reporting Units (overlaps with this ecoregion). These areas contain unusually high levels of species richness for diverse taxa (e.g., invertebrates, bryophytes, lichens, fungi, vascular plants, and vertebrates) relative to other ecological reporting areas within the

Columbia River Basin. One of the two hot spots also received a relatively high composite ecological integrity ranking, meaning that the area still consists of a mosaic of plant and animal communities maintained by well-connected, high-quality habitats (USDA Forest Service and USDI Bureau of Land Management 1996, 115).

Conservation Status

Habitat Loss and Degradation

Habitat loss within late-seral forests and aquatic areas has been extensive throughout this ecoregion (see USDA Forest Service and USDI Bureau of Land Management 1996 for portions of this ecoregion that lie within the Columbia River Basin analysis area). In particular, late-seral multi- and single-story forests have declined basin wide to as much as 40 percent of their original extent (see USDA Forest Service and USDI Bureau of Land Management 1996, 81). Late-seral forests in general have experienced sharp declines in many other ecoregions within the temperate coniferous forest major habitat type and are therefore a national as well as an ecoregional priority (e.g., see ecoregions [23, 24, 29–35, 37–39, 42, 44]). In addition, mid-seral subalpine forests have experienced a 35 percent decrease while late-seral montane multistory forests have increased by 35 percent (USDA Forest Service and USDI Bureau of Land Management 1997, 2–88). Moreover, many terrestrial and aquatic areas within this ecoregion have relatively low composite ecological integrity rankings due primarily to cumulative impacts of extensive logging, road building, and hydroelectric development. Declines in salmonid populations have been severe throughout the ecoregion and basin wide; however, a few aquatic strongholds and areas of very low road densities persist within the ecoregion (USDA Forest Service and USDI Bureau of Land Management 1996, 109, 110, 118, 121, 124, 127).

In addition to the above declines, ecological processes in this region have been altered by a century of fire suppression activities that have increased the severity and extent of fires and altered fire-dependent plant communities (see DellaSala, Olson, and Crane 1995a; Dellasala et al. 1996 for general discussions of basin-wide fire suppression effects, and USDA Forest Service and USDI Bureau of Land Management

1996, 90, 91, 127, for regional discussions). The absence of periodic fires also has resulted in declines in rangeland integrity and increases in exotic species invasions (USDA Forest Service and USDI Bureau of Land Management 1996, 88).

Remaining Blocks of Intact Habitat

Both the Oregon and the Washington GAP projects provide more detailed information on remaining intact blocks of habitat and areas of high biological importance outside protected areas. In addition, the USDA Forest Service and USDI Bureau of Land Management (1996) identified other key areas of importance to fish and terrestrial biodiversity, particularly the aquatic strongholds and biodiversity hot spots discussed above. The workshop participants provided the following local knowledge of relatively intact habitats that should be added to those sources:

- Mt. Rainier National Park—central Washington, 947 km^2
- Three Sisters Wilderness Area—central Oregon, 1,107 km^2
- Crater Lake National Park—south-central Oregon, 614 km^2
- Sky Lakes Wilderness Area—south-central Oregon, 453 km^2
- Mt. Jefferson Wilderness Areas—central Oregon, 427 km^2
- Goat Rocks Wilderness Area—south-central Washington, 355 km^2
- Aquatic diversity areas and late-seral owl reserves in each national forest in the ecoregion (see Henjum et al. 1994, USDA Forest Service 1994b)

Degree of Fragmentation

Fragmentation has been extensive throughout this region, as indicated by the few remaining intact late-seral forests, high road densities, and low ecological integrity rankings discussed above.

Degree of Protection

Protected sites in this ecoregion primarily include the remaining intact habitat blocks identified above. In addition, several late-seral forest reserves have been administratively protected under President Clinton's Northwest Forest Plan (USDA Forest Service 1994b) for forests within the range of the northern spotted owl.

However, many of these administratively protected areas, as well as the aquatic diversity areas, remain vulnerable to congressionally mandated salvage logging under the timber salvage rider.

Types and Severity of Threats

Logging is the primary threat to biodiversity in this region. Fire suppression, exotic species invasions, and road building have caused extensive damage to key watersheds, late-seral forests, and rangelands in this ecoregion.

Suite of Priority Activities to Enhance Biodiversity Conservation

- Restore forest and aquatic integrity to the region by protecting all remaining late-seral forests, native grassland/shrub communities, aquatic strongholds, roadless areas, and biodiversity hot spots. These areas provide building blocks and source pools for colonization into degraded landscapes. The preferred alternative recently released for the Columbia River Basin EIS (USDA Forest Service and USDI Bureau of Land Management 1997) fails to provide adequate protection for these areas and instead proposes "aggressive" restoration for most of the basin, despite the recommendations of conservationists for a comprehensive restoration plan that includes a network of core reserves, buffers, landscape connectivity, and restoration areas (DellaSala et al. 1996b, DellaSala and Olson 1996).
- Reintroduce fire through prescribed fire management in fire-dependent rangelands and forests.
- Integrate biodiversity conservation objectives with more sustainable and diversified regional economies. This ecoregion as well as many other places in the Columbia River Basin is experiencing a shift in human demographics and jobs from single-resource dependency to more diversified economies (see USDA Forest Service and USDI Bureau of Land Management 1996, 130). Basinwide lumber and wood products production, for instance, represent only 2.5 percent of the basin employment, while recreation represents 15 percent (ibid., 42). In addition, timber harvest in the basin currently accounts for 10 percent of the U.S. harvest, down from

17 percent since 1986 and expected to decline to 5 percent by the new century. In particular, this ecoregion was given a relatively high socioeconomic resiliency ranking, meaning (in part) that it is not as dependent on any one resource or job sector as other places within the basin (ibid., 130). Thus, the opportunities have probably never been better in this ecoregion for integrating conservation with sustainable economies.

Conservation Partners

For contact information, please see appendix G.

- Association of Forest Service Employees for Environmental Ethics
- Audubon Society
- Inland Empire Public Lands Council
- Klamath Forest Alliance
- National Wildlife Federation, Western Division
- Northwest Ecosystem Alliance
- Pacific Rivers Council
- Selkirk-Priest Basin Association

Relationship to Other Classification Schemes

This ecoregion corresponds to Omernik's (1995b) Level III ecoregion 4 (Cascades). There is high concordance in the centroids of this ecoregion when compared to McNab and Bailey's (1994) classification (ecoregion 242B). However, their delineation of this ecoregion continues farther north, encompassing the entire Cascade Mountain range within the United States (i.e., includes the southern portion of ecoregion [33]). In addition, this ecoregion overlaps with three physiographic provinces identified by Franklin and Dyrness (1973), including the southern, western, and high Cascades.

Prepared by D. DellaSala, J. Strittholt, R. Noss, and G. Orians

Key Number:	**37**
Ecoregion Name:	**Eastern Cascades Forests**
Major Habitat Type:	**2.2 Temperate Coniferous Forest**
Ecoregion Size:	**55,200 km²**
Biological Distinctiveness:	**Bioregionally Outstanding**
Conservation Status:	**Snapshot—Endangered Final—Endangered**

Introduction

The Eastern Cascades Forests span the eastern slopes of the Cascade Mountains in Oregon and Washington, from the southern reaches of the Cascade Mountains Leeward Forests [32] to Northern California. Vegetation is highly variable throughout this ecoregion and is influenced primarily by edaphic processes and disturbance regimes (Franklin and Dyrness 1973, 160). Several ecotones exist, particularly along the Cascade crest, where western Cascade forest types overlap with eastern Cascade forests (e.g., the Wenatchee National Forest in Washington has conifer species present on both sides of the Cascades), and along the lower timberline, where forest species mix with shrub and shrub-steppe communities (Franklin and Dyrness 1973, 160).

The geomorphology of the region is characterized by a series of steep, rugged mountains with the interior Cascade range rising to 2,700 m and volcanic peaks extending to 4,300 m (e.g., Mt. Rainier in Washington). Soil types are primarily andisols underlain by volcanic ash and other dry soils (Bailey 1995). Serpentine soils also occur in some areas (e.g., Wenatchee National Forest) and may support rare community types. Climate is generally mild, with precipitation averaging less than 511 mm annually for the region.

The natural vegetation of the region is a complex mosaic of shrublands, grasslands, and coniferous forests (Küchler 1964; Franklin and Dyrness 1973, 160; Bailey 1995). The dominant forest type along the eastern slopes of the Cascades is ponderosa pine (*Pinus ponderosa*) (Franklin and Dyrness 1973, 160). Within forested landscapes, species composition (forest type) varies along environmental gradients defined by physical factors such as temperature and moisture (DellaSala et al. 1996b). Topographic-moisture gradients (e.g., from sheltered valleys to exposed ridges) and soil conditions further determine the distribution of vege-

tation types. Fire resistance among different communities varies considerably (Habeck and Mutch 1973). Seven forest zones and numerous plant associations have been recognized, including *Juniperus occidentalis* (driest type, receiving 200–250 mm precipitation), *P. ponderosa* (dry, warm areas 600–2,000 m in elevation), *Pseudotsuga menziesii* (mesic areas), *Abies grandis* (predominates mid slopes), *P. contorta* (wide ecological amplitude, receiving 1,200–1,525 mm precipitation), *Tsuga heterophylla* (eastern extension of west Cascade forest), and *Abies lasiocarpa* (subalpine zone, coolest and wettest forest type) (Franklin and Dyrness 1973, 161). In addition, *A. concolor* and *A. magnifica shastensis* associations, which are widespread in California and southwestern Oregon, occur along the eastern slopes of the southern Oregon Cascades, reflecting another ecotonal zone in this region (Franklin and Dyrness 1973, 160).

Prior to European settlement (pre-1850), a wide variety of disturbances characterized the region, ranging from frequent small-scale and localized events such as treefall gaps to rare, large-scale events such as stand-replacing fires and epizootic outbreaks (DellaSala, Olson, and Crane 1995; DellaSala et al. 1996b). Such disturbances resulted in a dynamic equilibrium between patch creation and loss (Everett et al. 1994). This active disturbance regime has resulted in a larger proportion of younger seral stages than in areas west of the Cascade Mountains (Hejl 1992). However, the low-elevation (900–1,500 m) forests, which experienced frequent low-intensity fires, were predominantly (up to 90 percent) old-growth ponderosa pine (Henjum et al. 1994). In general, forest ecosystems in this region are adapted to more frequent fire disturbances than mesic westside forests. Fire cycles range from periodic (5–15 years) surface fires in dry and warm ponderosa pine and Douglas fir types, to infrequent (more than 100 years and up to 900+ years) stand-replacement crown fires (greater than 1,000 km²) in mesic and cool western red cedar (*Thuja plicata*), western hemlock, and cedar-spruce forest types (Agee 1993, 13). Such disturbances played a crucial role in maintaining inland forest structure, species composition, and ecosystem processes (e.g., species interactions, epizootics, plant species adaptations to fire, nutrient cycling, succession) (DellaSala, Olson, and Crane 1995; DellaSala et al. 1996b). A new anthropogenic regional landscape mosaic has now replaced this dynamic equilibrium that was once maintained by natural forces. Logging and fire suppression have shifted disturbance regimes and landscape dynamics to less frequent and more intense fires, and frequent and large-scale anthropogenic disturbances have disrupted natural processes and led to declines in various ecosystem types and species (Henjum et al. 1994; DellaSala et al. 1996b).

Biological Distinctiveness

This ecoregion contains important habitat for up to 268 taxa that have federal listing status, including 45 native fish (recognized by federal and state agencies as sensitive or of special concern) and 10 listed or candidates for listing under the Endangered Species Act (Franklin and DellaSala 1996). The total number of species in this region, however, is intermediate when compared to other coniferous forests within its MHT. Birds make up the majority (42 percent) of the taxa evaluated, followed by butterflies (28 percent), and mammals (14 percent). Conifer richness is intermediate compared to other coniferous forests in this MHT; however, beta-diversity is locally high in ecotones along the crest of the Cascades (e.g., Wenatchee National Forest). Within this ecoregion, elements from many adjacent regions, such as the Klamath-Siskiyou, Great Basin, and Sierra Nevadas, intermingle in a complex mosaic of communities.

Conservation Status

Habitat Loss

The forests of eastern Oregon and Washington have experienced dramatic changes in the past fifty years (Henjum et al. 1994; DellaSala, Olson, and Crane 1995; DellaSala et al. 1996b). Of particular concern is the loss of old-growth forest types such as low-elevation ponderosa pine, western larch (*Larix occidentalis*), and Douglas fir (Henjum et al. 1994). Only about one fourth of the remaining late-seral/old growth (LS/OG) has been protected administratively or by statute (in some areas less than 3 percent of LS/OG remains), and 75–90 percent of remaining patches in the region are too small (less than 0.4 km²) to conserve LS/OG dependent species

or processes (Henjum et al. 1994). Continued logging in these areas could further reduce LS/OG types to 7–13 percent of their original extent. Other losses identified in this ecoregion include degradation of shrub-steppe caused by extensive livestock grazing and invasion of exotic species, degradation of riparian, wetland, and aquatic ecosystems, and reductions in habitat quality caused by invasive species (DellaSala, Olson, and Crane 1995; DellaSala et al. 1996b). While habitat loss has been extensive within the ecoregion, the ecoregion was given a vulnerable rating compared to other coniferous forests within its MHT that had higher losses (e.g., [32] and [38]).

Remaining Blocks of Intact Habitat

Maps of remaining LS/OG, wilderness, roadless areas, aquatic diversity areas, and designated old growth (*Note*: Not all areas designated as old growth by the USDA Forest Service meet old-growth criteria) are available in Henjum et al. (1994). These maps, together with state GAP analyses of vegetation coverage, provide regional information needed in conservation reserve planning (see DellaSala et al. 1996b for models). In addition, the workshop participants identified the following intact blocks:

- Gearhart Mt. Wilderness and surrounding intact forest areas—160 km² wilderness and intact forest
- Deschutes National Forest—west of Bend, Oregon, approx. 800 km²
- Bear Valley National Wildlife Refuge—south-central Oregon, 43 km²
- Tule Lake National Wildlife Refuge—northern California, 170 km² (primarily wetlands)
- Clear Lake National Wildlife Refuge—northeastern California, 176 km² (primarily wetlands)
- Oak Creek Washington State Recreation Area—central Washington, 137 km²
- South Warner Wilderness Area—northeastern California, 217 km² (the area has been grazed but not logged)
- Lake Murry Washington State Recreation Area—central Washington, 128 km²
- Indian Heaven Wilderness Area, Gifford Pinchot National Forest—south-central Washington
- Mt. Adams Wilderness Area, Gifford Pinchot National Forest—south-central Washington

- William O. Douglas Wilderness Area, Wenatchee National Forest—central Washington
- Norse Peak Wilderness Area, Wenatchee National Forest—central Washington

Degree of Fragmentation

Fragmentation has been extensive in the region, particularly in the southern portion because of agriculture and clear-cut logging (Henjum et al. 1994). Few blocks greater than 20 km^2 remain.

Degree of Protection

The greatest threat in the region is continued logging, particularly salvage logging. Several areas of ecological importance were proposed for salvage logging, including old-growth forests with live trees greater than 2 m diameter breast height; late-seral reserves and key watersheds previously protected under the President Clinton's Northwest Forest Plan; habitat previously protected because of concerns over listed or candidate species; areas abutting wilderness; sacred Native American sites; Wild and Scenic Rivers; roadless areas; and steep slopes previously removed under forest plans (DellaSala, Olson, and Crane 1995; DellaSala et al. 1996b; DellaSala and Olson 1996). The Western Ancient Forest Campaign (Washington, D.C.), local Audubon Society, and Sierra Club have more detailed information on the location of salvage logging operations in the region.

Types and Severity of Threats

The major types of threat in the ecoregion include the following: (1) logging, (2) livestock grazing in riparian areas and native shrub-steppe, (3) fire suppression, (4) spread of noxious weeds exacerbated by road building and fragmentation, and (5) hydroelectric dams and flood control (Henjum et al. 1994; DellaSala, Olson, and Crane 1995; DellaSala et al. 1996b). Congressionally imposed salvage logging levels remain the greatest threat to protecting biodiversity in the region.

Suite of Priority Activities to Enhance Biodiversity Conservation

Most of the forests in this region are publicly owned (USDA Forest Service lands). Thus, the key to conservation of biological diversity lies in

how the national forests are managed. The Forest Service and Bureau of Land Management are currently preparing an environmental impact statement (EIS) that will determine the fate of many remaining roadless areas and LS/OG in the region. This EIS should remain a priority for conservation organizations, particularly those concerned about the need for a conservation reserve network and restoration of degraded ecosystems (see DellaSala, Olson, and Crane 1995; DellaSala et al. 1996b). The following conservation activities were recommended by DellaSala et al. (1996) and generally apply to this ecoregion:

- Determine where the most natural conditions persist (e.g., existing parks, wilderness, and roadless areas).
- Assemble or develop the necessary geo-referenced data and information to determine the status of ecosystem representation (e.g., GAP analysis).
- Identify hot spots of species richness and endemism for the region (e.g., Wenatchee National Forest, crest of the Cascades).
- Determine areas most threatened by human impact.
- Describe existing spatial patterns of regional biodiversity and attempt to reconstruct historic patterns where and when possible.
- Examine historic records of vegetation at appropriate ecological scales (e.g., across regional landscapes) to determine which ecosystems have lost critical components or have been most degraded.
- Determine necessary size, level of redundancy, and distribution of core areas to meet well-defined conservation goals.
- Conduct population viability studies for forest carnivores (e.g., wolverine [*Gulo luscus*]; marten [*Martes americana*], fisher [*Martes penanti*], lynx [*Lynx lynx*], flammulated owl [*Otus flammeolus*], boreal owl [*Aeoius funereus*], and great gray owl [*Strix nebulosa*]) (Ruggiero et al. 1994; Hayward and Verner 1994).
- Examine landscape pattern and linkages, paying particular attention to the interfaces between patches.
- Conduct watershed assessments so as to complement the greater regional planning effort.

- Consider both aquatic and terrestrial conservation requirements and their interaction.
- Integrate public lands management with conservation of important biodiversity features on private lands through the use of economic incentives and cooperative agreements.
- Restore fire-suppressed ecosystems through ecologically appropriate silviculture (e.g., thinning from below the canopy) and prescribed fire (DellaSala, Olson, and Crane 1995; DellaSala et al. 1996b).

Conservation Partners

For contact information, please see appendix G.

- Inland Empire Public Lands Council
- Kettle Range Conservation Group
- National Wildlife Federation, Western Division
- Northwest Ecosystem Alliance
- Oregon Natural Resources Council
- Selkirk-Priest Basin Association

Relationship to Other Classification Schemes

The Eastern Cascades Forest ecoregion delineated in this assessment is concordant with Omernik's (1995b) ecoregion 9. However, we split this ecoregion from the northern portion of Bailey's (1994) ecoregion M242 and put this portion into the Cascade Maountains Leeward Forests [32]. In addition, Bailey's ecoregion includes both slopes of the Cascades, while we felt the plant series and climate were different enough to warrant splitting along the Cascade crest.

Prepared by D. DellaSalla

Key Number: **38**
Ecoregion Name: **Blue Mountains Forests**
Major Habitat Type: **2.2 Temperate Coniferous Forests**
Ecoregion Size: **44,848 km²**
Biological Distinctiveness: **Bioregionally Outstanding**
Conservation Status: **Snapshot—Vulnerable Final—Endangered**

Introduction

The Blue Mountains Forests ecoregion lies within northeastern Oregon and extreme southeastern Washington. This ecoregion consists of several basin and range areas, alluvial fans at the base of mountain ranges, and floodplains along streams draining valleys (Bailey 1995, 77). Mountainous areas include the Strawberry, Greenhorn, Elkhorn, Aldrich, and Maury Ranges, and the Ochoco, Blue, and Wallowa Mountains (Franklin and Dyrness 1973, 27). Relief is highly variable, ranging from moderate slopes in the Blue and Ochoco Mountains to the deeply dissected and glaciated Wallowa Mountains. Elevations in mountainous areas range from 2,100 to 2,900 m, with basin areas ranging from 750 to 900 m (Franklin and Dyrness 1973, 28). Hells Canyon is an exceptionally deep (1,660 m, deeper than the Grand Canyon) and wide (24 km) canyon along the eastern boundary of the ecoregion where the Snake River cuts across it (Bailey 1995, 77). The western Blue Mountains contain some of the oldest rocks in Oregon, dated to the Paleozoic formation (Franklin and Dyrness 1973, 28). Much of the area in the central and northern portions of the ecoregion has experienced volcanic activity at some time. Soil types include regosols, fragiorthods, argixerolls, haploxerolls, and palexerolls (Franklin and Dyrness 1973, 28). Major hydrologic regimes primarily influencing basin areas include the John Day, Grande Ronde, Powder, and Malheur Rivers; however, low annual precipitation (305–1,270 mm) has limited formation of perennial streams (McNab and Bailey 1994). The ecoregion contains 11 of 116 forest and range ecosystems, including sagebrush (*Artemisia tridentata*), Pinyon-juniper (*Juniperus occidentalis*), ponderosa pine (*Pinus ponderosa*), Douglas fir (*Pseudotsuga menziesii*), western larch (*Larix occidentalis*), spruce-fir (*Picea* spp. and *Abies* spp.), lodgepole pine (*P. contorta*), chaparral-mountain shrub, mountain meadows, mountain grasslands, and alpine (Garrison et al. 1977).

Primary disturbance events include fire (various intensities and frequencies; see USDA Forest Service 1994a; Agee 1994; DellaSala et al. 1995) and epizootics (USDA Forest Service 1994a; DellaSala et al. 1995). A century of intensive management has created an anthropogenic landscape that bears little resemblance to the natural ecology or disturbance patterns that once shaped the region's ecosystems (USDA Forest Service 1994a; DellaSala et al. 1995).

Biological Distinctiveness

In general, this ecoregion contains intermediate levels of species richness (466 species) when compared to other ecoregions within the Temperate Coniferous Forest major habitat type. Bird species make up the majority (42 percent) of taxa evaluated, followed by butterflies (25 percent), and mammals (14 percent). Total endemism (9 species) was relatively low compared to other ecoregions within this major habitat type.

Riparian and old-growth forests represent important habitats in this region that receive disproportionate use by many fish and wildlife species. For instance, 285 of 378 (75 percent) terrestrial vertebrate species known to occur in the Blue Mountains either directly depend on riparian areas or use them more than other habitat types (Thomas, Maser, and Rodiek 1979, 41). Wildlife use riparian areas as travel corridors (e.g., for seasonal movements of ungulates between summer and winter ranges), as areas with favorable microclimates (e.g., shading and cooling properties benefit herpetofauna and fish), for perch and nest sites (e.g., bald eagles [*Haliaeetus leucocephalus*]), and as feeding and roosting areas (e.g., bats, carnivores, neotropical migrants). Similarly, 174 species reproduce and 192 species feed in old-growth mixed-conifer forests in the region (Thomas, Maser, and Rodiek 1979, 33). While the total number of species in old growth is lower than in some other seral stages (e.g., grass-forb), these forests provide important refugia for 19 vertebrate species, including bald eagle, northern spotted owl (*Strix occidentalis caurina*), flammulated owl (*Otus flammeolus*), boreal owl (*Aegolius funereus*), Vaux's swift (*Chaetura vauxi*), northern goshawk (*Accipiter gentilis*), black-backed woodpecker (*Picoides arcticus*), pileated woodpecker (*Dryocopus pileatus*), white-headed woodpecker (*P. albolarvatus*), three-toed woodpecker (*P. tridactylus*), red-breasted nuthatch (*Sitta canadensis*), pygmy nuthatch (*S. pygmae*), brown creeper (*Certhia americana*), golden-crowned kinglet (*Regulus satrapa*), Swainson's thrush (*Catharus ustulatus*), hermit thrush (*C. guttatus*),

Townsend's warbler (*Dendroica townsendi*), pine marten (*Martes americana*), and fisher (*M. pennanti*) (Henjum et al. 1994, 184). Many of these species feed on insects and depend on large snags found primarily in older forests (DellaSala et al. 1995; DellaSala, Olson, and Crane 1995). Old growth forests also provide important source pools for species recolonization of disturbed areas and are more likely to contain intact ecological processes (e.g., nutrient cycling, hydrologic and fire regimes) and complex structural attributes (e.g., snags, down logs, diverse tree-size classes; see Henjum et al. 1994) than intensively managed forests (DellaSala et al. 1995; DellaSala, Olson, and Crane 1995). Both riparian and old-growth forest types are being rapidly depleted throughout the region because of extensive logging, livestock grazing, flood abatement, hydroelectric dams, and fire suppression (Henjum et al. 1994).

Conservation Status

Habitat Loss

Habitat loss has been extensive, with approximately 10 percent of the region still intact (based on workshop participants' evaluation of threats). Losses have been greatest in old-growth forest types, fire-dependent forests (e.g., parklike ponderosa pine forests), riparian zones, and stream areas (Henjum et al. 1994). Based on comparisons between current and 1936 timber inventories in the Blue Mountains, Henjum et al. document nearly a 90 percent decline in old-growth ponderosa pine. Originally, virgin forests of this type extended the length of Oregon along the east slopes of the Cascade Range, from within a few kilometers of the summit to the desert's edge, interrupted only by small openings of nonforest land (Henjum et al. 1994). Fire suppression and livestock grazing have resulted in shifts in species composition from parklike (open, grassy understories) stands of large ponderosa pine to dense fir understories with accumulations of fuel loadings and more frequent conflagrations (USDA Forest Service 1994b; Agee 1994; DellaSala et al. 1995, 1996). Additional losses in low-elevation old-growth Douglas fir and mixed-conifer forest types have been nearly as dramatic, with most (75–90 percent) of remaining old growth restricted to national forests averaging less than 0.4 km^2

(Henjum et al. 1994). Notably, less than 8 percent of 2,888 km^2 of forested roadless areas have been administratively protected. Because of their relative size (roadless areas defined by the Rare II roadless area inventory are greater than 20 km^2) and composition, roadless areas represent one of the last remaining opportunities to conserve large-scale ecological processes, intact hydrologic regimes, and species that are intolerant of road-related impacts (e.g., large carnivores). Roadless areas are also important benchmarks for gauging the effectiveness of restoration and ecosystem management activities (Henjum et al. 1994; DellaSala et al. 1995, 1996b).

In addition to losses in the above forested types, aquatic areas have been highly degraded throughout the region. Forty-three taxa of resident fish have been identified as at risk of extinction within the region due to extensive habitat degradation (Henjum et al. 1994, 115). For instance, since 1941, 60 percent of the pool habitat for fish on the Grande Ronde River in Oregon has been degraded, and more than 70 percent of the streams in the Wallowa-Whitman National Forest on the Upper Grande Ronde fail to meet forest plan standards (Henjum et al. 1994). On the John Day River in Oregon, fall chinook salmon (*Oncorhynchus tshawytscha*) and coho salmon (*O. kisutch*) appear to be extinct, and spring chinook salmon and steelhead (*O. mykiss*) numbers have dropped precipitously (Henjum et al. 1994; Pacific Rivers Council 1995). The bull trout (*Salvelinus confluentus*) has experienced extensive declines in spawning habitat and is a "species of concern" and a candidate for listing under the Endangered Species Act. Additional threats in this region include exotic species invasions, shifts in plant community composition related to fire suppression and livestock grazing, and changes in hydrologic and stream temperature regimes that have disrupted food webs and ecological processes within riparian zones, aquatic areas, and adjacent uplands (Henjum et al. 1994).

Remaining Blocks of Intact Habitat

As discussed, few old-growth areas in this region are large enough to sustain dependent species or ecological processes; however, roadless areas represent the best opportunities for maintaining examples of relatively intact ecosystems. Workshop participants listed the following priority areas for conserving representative blocks of intact habitat:

- Eagle Gap Wilderness, Wallowa-Whitman National Forest—northeastern Oregon, 1,280 km^2
- Hells Canyon National Recreation Area—northeastern Oregon, 1,370 km^2
- North Fork of the John Day River—north-central Oregon, 454 km^2
- Wenaha-Tucannon area, Umatilla National Forest—northeastern Oregon and southeastern Washington, 714 km^2
- Strawberry Mountain, Malheur National Forest—eastern Oregon, 280 km^2

Additional roadless areas, wilderness areas, and old-growth forests requiring protection are provided in Henjum et al. (1994). Henjum et al. provide the foundation for restoration and conservation planning activities within the region (also see DellaSala et al. 1996 for a conservation approach applicable to this region that is based on Henjum et al. 1994 and Noss and Cooperrider 1994).

Degree of Fragmentation

Habitat fragmentation has been extensive in this ecoregion; however, the southern portion is more heavily fragmented and altered than the northern portion (based on workshop participants' evaluation of threats). Most forests in this ecoregion could be restored if logging were curtailed (especially in old growth and other declining types), livestock grazing were greatly reduced (especially in riparian areas and parklike ponderosa pine forests), some fires were allowed to burn (e.g., in wilderness areas and forests that have not had fire regimes significantly altered by fire suppression; DellaSala et al. 1995), and prescribed fires were used to return fire cycles to historic conditions (see Agee 1994; DellaSala et al. 1995, 1996b). Restoration activities should use remaining native habitats (e.g., roadless areas, old growth) as benchmarks in adaptive management approaches combined with prescribed fire and limited thinning in fire-suppressed forest types (e.g., thinning from below the dominant canopy; DellaSala et al. 1995, 1996b). Such activities would be useful in directing landscape trajectories to recreate patch and disturbance patterns more characteristic of pre-European settlement conditions.

Degree of Protection

According to Henjum et al. (1994), unless logging ceases in old-growth ponderosa pine, this forest type will decline below 10 percent of its historic extent within the foreseeable future. Although President Clinton commissioned the Forest Service and the Bureau of Land Management to conduct a basin-wide, scientific assessment and environmental impact statement (Columbia River basin planning includes the Blue Mountains), salvage logging and fire suppression activities continue to threaten remaining old growth and roadless areas, further jeopardizing opportunities for restoration and conservation. While some remaining roadless areas and old growth have ostensibly been evaluated for their contribution in maintaining biodiversity (A. Brunelle, Interior Columbia Basin Ecosystem Management Team Leader, pers. comm., 1996), it is unclear whether these areas will be administratively protected or will be subject to congressional attempts to overturn forest planning (e.g., forest health legislation proposed by Senator Craig continues to threaten roadless and old-growth areas).

Type and Severity of Threats

Workshop participants identified the rapid rate of logging in remaining old growth and roadless areas as the greatest threat to biodiversity in the ecoregion, followed by overgrazing. Forest health measures legislated by Congress (e.g., salvage logging law, PL 104-19) and conducted by federal agencies as part of ecosystem management approaches continue to focus narrowly on tree mortalities at the expense of ecosystem functions, processes, and integrity (DellaSala et al. 1995). Large-scale salvage logging as a treatment for perceived forest health emergencies continues to pose the greatest threat to ecosystem integrity in the region. The Blue Mountains have received the most attention regarding forest health because of recent increases in conflagrations and epizootics. However, many of the forest health problems in this region have been caused by a century of intensive management that has reduced ecosystem resiliency, elimi- nated complex structural features important to insectivorous species (e.g., snags, large trees) and fire-resistant trees, and substantially altered fire and hydrologic regimes (DellaSala et al. 1995). Other threats to biodiversity in the region include exotic species invasions and potential regional shifts in climatic patterns related to predicted global climate change. These threats are similar to threats identified in surrounding ecoregions within this major habitat type (e.g., [29], [34], [35]).

Suite of Priority Activities to Enhance Biodiversity Conservation

Workshop participants identified the following priority activities that along with recommendations of Henjum et al. (1994) and DellaSala et al. (1996b) provide a basis for restoration and conservation planning in this ecoregion:

- Protect all remaining large, relatively intact blocks, including Hells Canyon and Eagle Cap; John Day wilderness and surrounding areas; Wenaha-Tucannon wilderness; Aldrich Mountain (northwest corner of Malheur National Forest has old growth but most is unprotected); and Ochoco National Forest (northern block has unprotected old growth).
- Protect all old growth, roadless areas, and aquatic diversity areas identified by Henjum et al. (1994) and do not construct roads in any roadless areas, particularly those of more than 2.5 km^2.
- Cut no trees of any species older than 150 years or with a diameter-at-breast height *greater than* 50 cm, and no ponderosa pine dominant or codominants from any forest (Henjum et al. 1994).
- Conduct restoration activities, including thinning from below and prescribed fire management in areas where fire suppression has caused excessive accumulation of fuels and dense fir understories (e.g., fire-suppressed ponderosa pine forests).
- Reduce disturbance in riparian areas, including road closures, road obliteration, and reductions in livestock grazing pressure (see Henjum et al. 1994; DellaSala et al. 1995; DellaSala, Olson, and Crane 1995).
- Establish protected corridors along streams, rivers, lakes, and wetlands and restrict logging and grazing activities within these corridors (Henjum et al. 1994).
- Halt salvage logging activities in ecologically sensitive areas, including areas recovering from intense fire, areas with unstable slopes,

areas with fragile soils, riparian areas, and old-growth areas (Beschta et al. 1995).

- Base forest health assessments on the health and integrity of entire ecosystems and not only tree mortalities (DellaSala et al. 1995).
- Establish a forest health monitoring program that is based on appropriate indices of ecological integrity (e.g., degree of fragmentation, presence of intact terrestrial and hydrological processes; see Angermeier and Karr 1994; DellaSala and Olson 1996) and biodiversity conservation (Henjum et al. 1994; DellaSala et al. 1995, 1996b).

The above recommendations also apply to surrounding ecoregions within this major habitat type (e.g., [34], [35], [29]).

Conservation Partners

For contact information, please see appendix G.

- Association of Forest Service Employees for Environmental Ethics
- Clearwater Forest Watch
- Inland Empire Public Lands Council
- Kettle Range Conservation Group
- Klamath Forest Alliance
- National Wildlife Federation
- Northwest Ecosystem Alliance
- Pacific Rivers Council, Oregon
- Selkirk-Priest Basin Association

Relationship to Other Classification Schemes

The Blue Mountain ecoregion mapped in this assessment was based on Omernik's (1995b) ecoregion 11. This ecoregion is roughly concordant with Bailey's classification (ecoregion 332G), with the exception of the southeast and northwest corners, which were split into different ecoregions by Bailey (1994). The ecoregion generally corresponds to Franklin and Dyrness's (1973, 27) description of the Blue Mountains Province and Thomas et al's. (1979) description of wildlife and plant communities in the Blue Mountains.

Prepared by D. DellaSala, J. Strittholt, and G. Orians

Key Number:	**39**
Ecoregion Name:	**Klamath-Siskiyou Forests**
Major Habitat Type:	**2.2 Temperate Coniferous Forests**
Ecoregion Size:	**50,299 km²**
Biological Distinctiveness:	**Globally Outstanding**
Conservation Status:	**Snapshot—Vulnerable Final—Endangered**

Introduction

In northwestern California and southwestern Oregon, complex terrain, geology, climate, and biogeographic history have created one of the earth's most extraordinary expressions of temperate biodiversity in the Klamath and Siskiyou Mountains. Although well known among biologists, few North Americans realize the uniqueness and importance of the species and communities in this ecoregion. Indeed, logging, mining, road building, and grazing continue to be intensive and pervasive threats to this area. To date no strict protected areas have been established in this rich region.

Biological Distinctiveness

The Klamath-Siskiyou ecoregion is considered a global center of biodiversity (Wallace 1983), has been named an IUCN Area of Global Botanical Significance (one of seven in North America), and is proposed as a World Heritage Site and UNESCO Biosphere Reserve (Vance-Borland et al. 1995). The biodiversity of these rugged coastal mountains of northwestern California and southwestern Oregon has garnered this acclaim because the region harbors one of the four richest temperate coniferous forests in the world (along with the southeastern conifer forests of North America; the forests of Sichuan, China; and the forests of the Primorye region of the Russian Far East), with complex biogeographic patterns, high endemism, and unusual community assemblages. A variety of factors contribute to the region's extraordinary living wealth. The region escaped extensive glaciation during recent ice ages, providing both a refuge for numerous taxa and long periods of relatively favorable conditions for species to adapt to specialized conditions. Shifts in climate over time have helped make this ecoregion a junction and transition zone for several major biotas, namely those of the Great Basin, the Oregon Coast

Range, the Cascades Range, the Sierra Nevada, the California Central Valley, and the Coastal Province of Northern California. Elements from all of these zones are currently present in the ecoregion's communities. Temperate conifer tree species richness reaches a global maximum in the Klamath-Siskiyous with thirty species, including seven endemics, and alpha diversity (single-site) measured at seventeen species within a single square mile (2.59 km^2) at one locality (Vance-Borland et al. 1995). Overall, around thirty-five hundred plant species are known from the region, with many habitat specialists (including ninety serpentine specialists) and local endemics.

The great heterogeneity of the region's biodiversity is due to the area's rugged terrain, very complex geology and soils (giving the region the name "the Klamath Knot"), and strong gradients in moisture decreasing away from the coast (e.g., from more than 300 cm [120 in.] per annum to less than 50 cm [20 in.] per annum). Habitats are varied and range from wet coastal temperate rain forests to moist inland forests dominated by Douglas fir (*Pseudotsuga menziesii*), *Pinus ponderosa*, and *P. lambertiana* mixed with a variety of other conifers and hardwoods (e.g., *Chamaecyparis lawsoniana, Lithocarpus densiflora, Taxus brevifolia*, and *Quercus chrysolepis*); drier oak forests and savannas, with *Quercus garryana* and *Q. kelloggii;* serpentine formations with well-developed sclerophyllous shrubs; higher-elevation forests with Douglas fir, *Tsuga mertensiana, Abies concolor*, and *A. magnifica;* alpine grasslands on the higher peaks; and cranberry and pitcher plant bogs. Many species and communities have adapted to very narrow bands of environmental conditions or to very specific soils such as serpentine outcrops. Local endemism is quite pronounced, with numerous species restricted to single mountains, watersheds, or even single habitat patches, tributary streambanks, or springs (e.g., herbaceous plants, salamanders, carabid beetles, land snails; see Olson 1992). Such fine-grained and complex distribution patterns mean that any losses of native forests or habitats in this ecoregion can significantly contribute to species extinction. Several of the only known localities for endemic harvestmen, spiders, land snails, and other invertebrates have been heavily altered or lost through logging within the last decade, and the current status of these species is unknown

(Olson 1992). Unfortunately, many invertebrate species with distribution patterns and habitat preferences that make them prone to extinction, such as old-growth specialist species, are rarely recognized or listed as federal endangered species. Indeed, eighty-three species of Pacific Northwest freshwater mussels and land snails with extensive documentation of their endangerment were denied federal listing by the U.S. Fish and Wildlife Service in 1994 (J. Belsky, Oregon Natural Resources Council, pers. comm. 1994).

Rivers and streams of the Klamath-Siskiyou region support a distinctive fish fauna, including nine species of native salmonids (salmon and trout) and several endemic or near endemic species such as the tui chub (*Gila bicolor*), the Klamath small-scale sucker (*Catostomus rimiculus*), and the coastrange sculpin (*Cottus aleuticus*). Many unusual aquatic invertebrates are also occur in the region.

Conservation Status

Habitat Loss

Only 25 percent of the ecoregion is considered to be relatively intact.

Remaining Blocks of Intact Habitat

Larger blocks of relatively intact habitat include the Trinity Alps (1,910 km^2), Marble Mountain (769 km^2), Whiskeytown-Shasta (657 km^2), Kalmiopsis (649 km^2), Eel River (586 km^2), Siskiyou (569 km^2), Yolla Bolly–Middle Eel River National Forest Wilderness areas (555 km^2), and the Klamath Wild and Scenic River (669 km^2). Most of these larger blocks encompass higher-elevation habitats. Few large blocks of lower-elevation undisturbed habitat remain. The Dillon Creek watershed represents one of the last remaining lowland forest blocks but is still unprotected.

Degree of Fragmentation

The biological communities of this ecoregion are naturally heterogeneous, but logging has greatly increased fragmentation. Species that are found only within narrow altitudinal belts are particularly sensitive to the isolation, fragmentation, and reduction of their habitat.

Degree of Protection

Higher-elevation habitats are well represented within a system of protected areas (e.g., wilderness areas), but middle- and lower-elevation sites are poorly protected. The ecological integrity of habitats within Wild and Scenic River areas are often compromised by land use in surrounding areas and linear habitat configurations that do not favor viable ecological and population dynamics.

Types and Severity of Threats

The Klamath-Siskiyou's native communities have been heavily impacted by commercial logging, fire suppression, mineral extraction, road building, and grazing of sheep and cattle. Many of the larger rivers and tributaries have been severely degraded by dredging for gold, toxic runoff from mining operations, and sedimentation and associated changes from excessive and poorly managed logging. Introduced species, particularly in aquatic habitats, have altered many natural communities.

Suite of Priority Activities to Enhance Biodiversity Conservation

- Strictly protect remaining roadless areas, lowland forest habitats, the last remaining intact watersheds, and riparian habitats immediately.
- Because of the globally outstanding biodiversity of this region, halt all logging of undisturbed native forests and initiate restoration efforts to regain ecological viability for remaining native habitats.
- Implement controlled burns to reverse problems associated with decades of fire suppression.
- Immediately protect and restore lowland habitats that act as linkages between higher-elevation protected areas.
- Eliminate cattle and sheep grazing from alpine and high-elevation habitats, as this contributes to significant damage of *Darlingtonia* bogs and other fragile habitats.
- Complete conservation analyses for the region to ensure adequate representation and landscape-level integrity in proposed conservation strategies. Some specific activities include:

- Preserve and protect existing roadless areas, particularly those providing linkages between existing protected areas such as the Kalmiopsis, Siskiyou, Marble Mountain, and Trinity Alps wilderness areas. About 60 percent of them have been designated late-successional reserves under the Clinton forest plan but need legislative action for permanent protection.
- Protect all remaining intact blocks of low-elevation forests, an increasingly rare habitat type due to excessive logging. The Dillon Creek watershed is one of the last larger blocks of lowland forest and also maintains one of the last intact altitudinal gradients in the region.
- Establish a biodiversity reserve system centered on Klamath National Forest, or a system of parks and/or reserve areas, including a Siskiyou National Park in southern Oregon. Reserve system should conform to watersheds critical to salmon. Spawning streams are being degraded by sediment deposition and exposure to solar radiation caused by logging. Add key watersheds, such as the Illinois and North Fork Smith Rivers, South Fork Chetco, and Eagle Creek to the reserve system.
- Take steps to protect downstream habitat for salmon and other species on public and private lands. End overfishing of salmon stocks, particularly wild coho and south coastal Oregon chinook. Place adequate fish screens on irrigation diversions. Prevent chronic dewatering of streams in the Illinois Valley and elsewhere that kill thousands of juvenile salmon. Remove or modify earthen dams that block access to headwater streams. Correct hatchery practices damaging to wild populations.
- Preserve the Port Orford cedar (*Chamaecyparis lawsoniana*), an ecoregional endemic conifer, by stopping the spread of an exotic waterborne root fungus that is threatening this species with extinction. The fungus is spread by vehicles traveling on roads and by logging activity.
- Reduce accumulated fuel loads (built up by decades of fire suppression) in the Siskiyou and Klamath Mountains to reduce the chance of catastrophic wildfire. Salvage logging is not an ecologically sound prescription to achieve this goal.
- Bring BLM forested land under the jurisdiction of the Forest Service. Focus the Forest Service's ecosystem management policy by

establishing a panel of independent scientists to develop standards for harvesting timber that are in accord with conservation of the biological diversity of the system.

- Encourage the federal government to purchase private lands that harbor or are needed to sustain endangered and threatened species.
- Encourage the federal government to promote improved forest practices on private land as a complement to ecologically sustainable management of federal forests.
- Undertake recovery of declining species such as fisher (*Martes pennanti*) and wolverine (*Gulo luscus*), and reintroduce extirpated species such as wolf (*Canis lupus*) and grizzly (*Ursus horribilis*). The relatively intact condition of the Klamath-Siskiyou ecoregion provides a rare opportunity for recovery of large carnivores in the West.
- Declare a moratorium on logging and road building in the Blue, Somes, Red Cap, Perch, Sucker, Monte, and Biose Creek watersheds. These areas contain key salmonid habitat and unprotected low-elevation forests and roadless areas. The forest supervisor should be directed to enforce the roadless area directive by restoring the roadless area boundaries originally identified under the RARE II process in these areas.
- Prohibit road reconstruction activities in the Dillon Creek watersheds where the New Year's Day storm of 1997 caused a number of significant mudslides in watershed already damaged by years of logging and road building. Roads in this area should be decommissioned (not repaired) to assist in recovery of threatened coho and to allow watersheds to recover from cumulative impacts. The administration should use federal highway funding available to these forests as a result of storm disaster relief to decommission roads in watersheds at high risk of subsequent flooding and mudslides.
- Withdraw all timber sales in roadless areas, late-successional reserves, and key salmonid watersheds in the Siskiyou National Forest, including the Mineral Fork, Quosatana Roadless Area, and Smith River National Recreation Area. The Forest Service has offered "reserve" timber in these areas, which would greatly diminish the ecological integrity of the reserves and undermine the credibility of President Clinton's Northwest Forest Plan.

Conservation Partners

For contact information, please see appendix G.

- California Native Plant Society
- Headwaters
- Humboldt State University
- Klamath Forest Alliance
- The Nature Conservancy
- The Nature Conservancy of California
- Northcoast Environmental Center
- Siskiyou Regional Education Project
- The Wildlands Project

Relationship to Other Classification Schemes

The Klamath-Siskiyou ecoregion corresponds to Omernik's (1995b) Klamath Mountains ecoregion and roughly overlaps with Bailey's (1994) Klamath Mountains and Southern Cascades sections (M261A, M261D).

Prepared by D. Olson, R. Noss, G. Orians, J. Strittholt, C. Williams, and J. Sawyer

Key Number:	**40**
Ecoregion Name:	**Northern California Coastal Forests**
Major Habitat Type:	**2.2 Temperate Coniferous Forests**
Ecoregion Size:	**13,274 km²**
Biological Distinctiveness:	**Globally Outstanding**
Conservation Status:	**Snapshot—Critical Final—Critical**

Introduction

The Northern California Coastal Forests are largely defined by two features, the largely persistent moist environments provided by Pacific storms in the winter and coastal fogs in the summer, and the distribution of the redwood (*Sequoia sempervirens*). Redwoods range from central California to the Oregon border and are typically found within 65 km (40 miles) of the coast. Redwood groves are patchily distributed among a variety of natural communities found within this coastal belt, including Douglas fir–tan oak forests, oak woodlands, closed-cone pine forests, bogs, and coastal grasslands (Sawyer 1996)

Biological Distinctiveness

The coastal forests of Northern California are in many ways an extension of the temperate rain forests that hug the coasts in Washington and Oregon, except that, in California, redwoods and Douglas fir–tan oak forests dominate many lowland areas. These ancient and spectacular conifers are among the biggest, tallest, and oldest trees in the world, often exceeding 200 feet in height (some individuals are more than 369 feet), 15 feet in diameter, and 2,200 years of age. Redwood groves have the greatest biomass accumulation known for any terrestrial ecosystem. They are globally unique forests, and only a few other forests in the world have a similar assemblage and structure of ancient, giant conifers (e.g., the giant sequoia groves of the Sierra Nevada, the Sitka spruce and Douglas fir forests of the Pacific Northwest, and the Alerce forests of southern Chile). Distribution is patchy, but redwoods generally occur in the fog belt ranging from five to thirty-five miles wide along the coast and from 100 to 2,000 feet in elevation (Barbour et al. 1993). Redwood-dominated forests tend to occur in valley bottoms, where there is abundant fog drip, alluvial soils, and floods about every thirty to sixty years. On the uplands, where fire was a recurring disturbance, a more diverse assemblage of trees, includes Douglas fir, grand fir, western hemlock, Sitka spruce, western red cedar, tan oak, bigleaf maple, California bay, and Port Orford cedar, as well as redwoods. Without periodic disturbances, some ecologists suspect that redwood groves may eventually be replaced by western hemlock (Zinke 1995; but see Viers 1982).

Drier slopes within this ecoregion are dominated by Douglas fir and tan oak. They are co-dominates of a "Douglas fir–tan oak forest" madrone and contain Garry oak, black oak, interior live oak, and coast live oak. Eight conifer species are endemic to the ecoregion. A rich understory of herbs, shrubs, treelets, ferns, and fungi is found under the towering redwood and other conifers.

Redwood forests harbored a diversity of animal life including bears, fishers, pine martens, numerous warblers, and the endangered mar-

bled murrelet (*Brachyramphus marmoratus*), which nests in mature forest canopies. A number of amphibians live here including the Pacific giant salamanders (*Dicamptodon ensatus*), red-bellied newts (*Taricha rivularis*), and tailed frogs (*Ascaphus truei*). Silver salmon and steelhead trout breed in coastal rivers and streams. Also found in this ecoregion is the extraordinary bright yellow-orange banana slug (*Ariolimax columbianus*), a mature forest specialist and a candidate for California's state invertebrate. A number of other invertebrate species, including beetles, harvestmen, spiders, millipedes, and freshwater mussels, are specialists on habitats modified by old redwood and other conifer forests and maintain very local distributions (Frest and Johannes 1991; Olson 1992). Given the propensity of species in these invertebrate groups for very restricted ranges and the virtual elimination of mature forests in this ecoregion, the probability is high that numerous species extinctions have already occurred (Olson 1991).

North Coast grasslands, often called bald hills and coastal scrubs, can be found close to the sea in some areas, particularly on coastal terraces below redwoods. Grasses, spring wildflowers, and shrubs dominate these habitats, which have now been largely converted to farmland and pasture. Unusual closed-cone pine forests, sphagnum bogs, and pygmy forests also occur on coastal terraces, often above the redwood belt in some areas.

Conservation Status

"Redwoods rank among the most resilient trees on Earth. Adapted to naturally occurring floods and fires and able to resprout, they can turn a burned or cut over hillside into a green forest within a century. But during that century, the plants and animals dependent on the shade, soil moisture, shelter, and interrelated life of the old-growth ecosystem have died and vanished. And the second-growth forest replacement cannot begin to compare with the primeval beauty, the ambience, and the biodiversity of the 1,000+-year-old forest gem long gone" (Johnston 1994, 27).

Habitat Loss

Less than 4 percent of the original extent of virgin redwood forests remains, and only 2.5 percent of this is protected. Unfortunately, protected lower-elevation groves can be threatened by severe flooding and sedimentation caused by logging in surrounding watersheds. Several large groves of old-growth redwoods on the Eel, Klamath, and Van Duzen Rivers were lost due to severe floods in the 1950s exacerbated by extensive and ecologically damaging logging in surrounding watersheds. Most of the coastal grasslands have been converted to agriculture or rangelands. All vegetation types have been reduced due to urbanization.

Remaining Blocks of Intact Habitat

One large remaining block of redwood forest occurs at Redwood Creek within the national park. Redwood National Park is largely surrounded by plantations, and its core habitat area is thus influenced by pronounced edge effects.

It is possible that all remaining blocks of redwoods are too small to maintain viable redwood ecosystems into the future. However, for a low-elevation forest type, redwood forest has more acreage than any other coastal forest type in California, Oregon, or Washington. The larger patches of coastal grassland occur in:

- Mendocino County
- Marin County
- Point Reyes National Seashore
- Bodega Bay area

In addition, there are large areas of old-growth redwood forest in:

- Jedediah Smith State Park
- Prairie Creek Redwoods State Park
- Humboldt State Park

A far more critical kind of forest for the ecoregion, which was once quite extensive, is the Douglas fir–tan oak forest (often called mixed evergreen forest). All but a few acres exist in this ecoregion as old growth. Most was cut in the 1950s and 60s before ecologists even recognized this specific type of forest (Sawyer 1996).

Degree of Fragmentation

Redwood forests were naturally patchy to some degree, but wholesale logging has greatly isolated remaining patches and caused much fragmentation.

Degree of Protection

Redwood National Park is the only hope for survival of functioning redwood ecosystems, yet even this is questionable given their size and surrounding land use. Jedediah Smith and Del Norte Redwood State Parks are two smaller reserves. Muir Woods and Big Basin toward the south are too small for realistic prospects of long-term conservation of this unique community.

King's Range Conservation Area, considered at one point for national park status, lacks the better-known redwood but does contain remnants of the Douglas fir–tan oak forest, although it has been mostly cut.

Types and Severity of Threats

The last redwood groves on private land, mostly in the Headwater Forest area near the Van Duzan River, are under imminent threat of cutting by Pacific Lumber. Compromising agreements between state and federal agencies and this company leave in doubt the survival of these last remnants. It is unfathomable with the knowledge and resources we have today that there would be any question of total protection of the last remaining groves of these globally unique ecosystems, and unconscionable that the government and citizens of this country have let the destruction continue to this point.

Many remaining groves, both protected and unprotected, are threatened by significant alteration of surrounding watersheds from development and logging, which can increase the frequency and severity of floods, fires, and sedimentation. Fire suppression can lead to hot fires that kill many individual trees and other species or allow other tree species to eventually out-compete young redwoods. Local conditions of rainfall and understory moisture can be changed due to landscape-level deforestation. Selective logging in redwood forests has significant impacts on native invertebrate biodiversity, even after fifteen years, and may be responsible for whole extinctions of some taxa (Hoekstra et al. 1995). Marbled murrelets and other old-growth specialist species continue to be threatened through deforestation and subsequent ecological changes in remaining habitats.

Ecological change from other human activities has also contributed to the loss of species. Strobeen's parnassian butterfly , a redwood subspecies of *Parnassius clodius,* disappeared thirty-three years ago from former redwood habitats of the Santa Cruz Mountains due to the loss of its host plant (*Dicentra* spp.) from fire and timber management practices (Murphy 1993). The lotis blue butterfly, a local endemic of a coastal sphagnum bog in Mendocino, was lost in 1983 as forest cover replaced the wetland, perhaps due to changes in disturbance regimes or the loss of alternate habitats. Salmon migrations have been severely compromised through stream destruction from logging and mining as well as overfishing. The native flora competes with the highest percentage (34 percent) of introduced plant species for any ecoregion in the continental United States and Canada.

Urbanization is also a major threat to this ecoregion, due to the spread of urban areas between Monterey and San Francisco and the area north of San Francisco.

Suite of Priority Activities to Enhance Biodiversity Conservation

- Provide strict and unequivocal protection for the groves and surrounding watersheds of all remaining patches of redwood forest.
- Implement long-term management plans to ensure that appropriate disturbances occur to facilitate long-term regeneration and ecological viability.
- Restore degraded redwood habitat, particularly in areas adjacent to undisturbed patches.
- Carry out surveys and conservation analyses of terrestrial and aquatic invertebrates. Conserve remaining coastal prairies, scrubs, and pygmy forests on coastal terraces.

Conservation Partners

For contact information, please see appendix G.

- California Native Plant Society
- Department of Interior, National Park Service
- Humboldt State University
- Mattole River Coalition
- Sierra Club
- U.S. Army
- The Wildlands Project

Relationship to Other Classification Schemes

This ecoregion is based on the southern section of Omernik's (1995b) Coast Range ecoregion (1). The northern boundary was determined by the northernmost extent of redwoods. There is approximate congruence with Bailey's (1994) Northern California Coast Section (263A), which is broader toward the north and does not include the redwood areas south of San Francisco Bay. The ecoregion generally corresponds to Küchler's (1975) vegetation classes, Redwood Forest, Mixed Evergreen Forest with Rhododendron, and Coastal Prairie and Coastal Scrub.

Prepared by D. Olson and J. Sawyer

Key Number:	**41**
Ecoregion Name:	**Sierra Nevada Forests**
Major Habitat Type:	**2.2 Temperate Coniferous Forests**
Ecoregion Size:	**52,832 km²**
Biological Distinctiveness:	**Globally Outstanding**
Conservation Status:	**Snapshot—Endangered Final—Endangered**

Introduction

The Sierra Nevadas run northwest to southwest and are approximately four hundred miles long and fifty miles wide. The range is highest toward the south, with several peaks over fourteen thousand feet. Several large river valleys dissect the western slope with steep canyons. The eastern escarpment is much steeper than the western slope, in general. The range supports a diverse set of natural communities with many endemic species and extraordinary habitats.

Biological Distinctiveness

The Sierra Nevada ecoregion harbors one of the most diverse temperate conifer forests on earth, displaying an extraordinary range of habitat types and supporting many unusual species. Fifty percent of California's estimated seven thousand species of vascular plants occur in the Sierra Nevada, with four hundred Sierra endemics and two hundred rare species (CWWR 1996). The southern region has the highest concentration of species and rare and endemic species, but pockets of rare plants occur throughout the range. The eastern slope west of the Owens River valley is also noted for many unusual plant species. Geographic patterns of plant diversity are complex, with the species compositions of communities changing dramatically with altitude and between watersheds along a north-south gradient (i.e., high beta-diversity).

Different communities are distributed in elevational belts on both sides of the range, with counterpart communities occurring at higher elevation on the eastern slope. Above the chaparral and foothill woodlands on the west slope of the Sierra Nevada, part of the Interior chaparral and woodland ecoregion, occurs a series of forests at montane and subalpine elevations. The montane elevation is itself sorted into several extensive forest types. First is the ponderosa pine forest, which replaces chaparral and woodland, then the mixed-conifer forest dominates. Here five conifers, ponderosa pine, sugar pine, giant sequoia (*Sequoiadendron giganteum*), Douglas fir, and white fir, mix on the most productive soils in the Sierra. Sugar pine is the world's tallest and largest pine species of this forest type. Some seventy-five groves of giant sequoia exist within the mixed-conifer forest. Giant sequoias are the most massive trees on earth; some trees are 273 feet tall, over 36 feet in diameter, and over 3,200 years old. Sequoias and many other tree, shrub, and herbaceous plant species tolerate frequent fires, and some require fire for their regeneration. Sequoias benefit from fire's reduction of fungal pathogens and carpenter ants, which weaken structural integrity in larger trees; clearing of understory trees and brush that compete with seedlings; and opening of cones through heat (Schoenherr 1992). Fires in the past were often frequent and of low intensity because fuel loads were generally low. A diverse assemblage of herbaceous plants and shrubs is also found in this zone, including numerous local endemics and species specialized on particular soil types such as serpentine. Above the mixed-conifer forest grows simple forests dominated first by white fir and then by red fir. They occur in the zone of deepest snow.

From 7,000 to 9,000 feet, in the subalpine zone, lodgepole pine forms extensive forests mixed with meadows and montane chaparral.

Jeffrey pine, western white pine, mountain juniper, pine, and aspen are locally common. Between 9,000 and 10,000 feet mountain hemlock, whitebark pine, foxtail pine, and limber pine are characteristic. Above the timberline, alpine meadows, talus slopes, and rocky outcrops cover the land. On the eastern slope, the alpine zone extends down to around 11,800 feet, lodgepole pine– or red fir–dominated forests occur down to 8,000 feet, and Jeffrey pine forests to 6,000 feet. Below this elevation the pinyon-juniper woodlands grade into sagebrush scrubs of the Great Basin. Extensive talus slopes, meadows, montane chaparral, lakes, and rock outcrops occur throughout the Sierra.

Approximately four hundred terrestrial vertebrate species occur in the Sierra Nevada, although around one hundred of these are largely distributed elsewhere. Around 60 percent of California's vertebrate species are found here. Thirteen vertebrate species (e.g., salamanders, frogs, rodents, birds) are endemic to the range and include some of the highest levels of mammal endemism in the United States and Canada. Some endemics include the Yosemite toad, Mount Lyell salamander, limestone salamander, Kern salamander, long-eared chipmunk, alpine chipmunk, heather vole, Walker Pass pocket mouse, yellow-eared pocket mouse, and golden trout. A diverse vertebrate predator assemblage once occurred in the ecoregion, including grizzly bear, black bear, coyote, cougar, ringtail, bobcat, fisher, pine marten, wolverine, and several large owls, hawks, and eagles.

The Sierras support a diverse invertebrate fauna with a number of endemics, including Behr's colias butterfly (*Colias behrii*), restricted to a small area around Tioga Pass. Many other invertebrate species have very local distributions. With further investigations in the region, new plant and invertebrate species are being added to lists. Between 1968 and 1986 sixty-five new plant species were described for the Sierras (CWWR 1996).

Conservation Status

Habitat Loss

A century of intensive logging, mining, railroad building, development, fire suppression, and grazing by sheep and cattle have left only around 25 percent "intact" natural habitat in the Sierra

Nevada. Much of this intact habitat occurs at higher elevations, often in nonforested alpine or less productive forests and woodlands. More than 60 percent of the ponderosa pine or mixed-conifer forests have been altered, and many remaining forests have been degraded through logging and fire suppression. Overall, around 19 percent of the entire mid-elevation conifer belt is still relatively intact (CWWR 1996). Between 7 percent and 30 percent of late-successional old-growth forests of middle elevations remain, the percentage depending on the forest type. One third of the original extent of giant sequoia groves have been harvested, with the U.S. Forest Service allowing these rare and unique trees to be cut in some of the northern groves as recently as the 1980s. The greatest percentage loss of habitat has occurred in late-successional forests, foothill woodlands, and riparian habitats.

Remaining Blocks of Intact Habitat

Four national parks, Lassen, Yosemite, Sequoia, and Kings Canyon, and some national forest wilderness areas, harbor the largest remaining blocks of relatively intact montane mixed-conifer forests. Most of the rest of this zone has been cut over at least once. There are extensive high montane and subalpine forests and meadows, and alpine habitats in national park and Forest Service wilderness areas. Much of the higher elevations were once heavily grazed, and Forest Service wilderness areas are still subjected to grazing. Most remaining habitats, including late-successional forests, suffer from significant alterations of historic fire regimes and consequent changes to processes (e.g., dense growth of shade-tolerant species). A variety of changes resulting from intensive exploitation and current management of remaining forests contribute to the lower resiliency of forests to fire and epizootic disturbances. The Sierra Nevada Ecosystem Project (SNEP) analysis came to the conclusion that historic logging, creating simple forests, was more important in creating the fire problems in the Sierra Nevada than fire suppression itself. These problems include forest simplification; removal of older trees, with their structural and genetic resistance; loss of natural firebreaks such as old-growth patches with sparse understories and moist riparian vegetation; replanting schemes using genetically similar seedlings of a single tree species; and intensive application of

pesticides, which also destroys natural predators on epizootics.

Degree of Fragmentation

Fragmentation varies inversely with altitude. The last remaining less disturbed habitats at lower elevation are severely fragmented.

Degree of Protection

The national parks share a high degree of protection, including the few remnant lower-elevation habitats they encompass. Recent changes in fire management policies within national parks are helping to restore natural disturbance regimes and successional processes. Wilderness areas, which usually are located at higher elevations, are protected from commercial logging, but they are still intensively grazed by domestic livestock, causing significant damage to riparian habitats and other vegetation types.

Types and Severity of Threats

The vast majority of native forests have already been largely converted to tree plantations. Intensive forestry practices have simplified forest structure and composition in most remaining forests, causing reduced ecologic resilience, genetic variability, and impaired function. Forest simplification and fire suppression together contribute to greatly increased probabilities of catastrophic fires and increased frequencies and severity of widespread mortality from epizootics such as bark beetles and fungal pathogens (CWWR 1996; DellaSala et al. 1996b). Most remaining fragments of unlogged forests outside national parks are threatened by logging, fire suppression, and grazing. Meadows throughout the range outside the national parks continue to be significantly damaged by sheep and cattle grazing and pack animals. Soil compaction from logging and development activities is altering natural succession patterns.

Introduced pathogens from livestock harm native species. The entire herd of sixty-five bighorn sheep in the Warner Range, a small range northwest of the Sierra Nevada, was extirpated by an introduced virus from a single sheep in 1988 (Jensen, Torn, and Harte 1993). Sugar pines in many areas are being killed by the introduced white pine blister rust (*Cronartium ribicola*), which is rapidly spreading. Continued cutting of larger and potentially resistant sugar pine trees may preclude any effective conservation programs. The creation of clear-cuts, a commonly used timber extraction practice, promotes the growth of gooseberry and currants, vectors of the rust (Johnston 1994). Tree mortality from Anosus root rot fungus (*Heterobasidion annosum*), a native pathogen, continues to increase from logging, which creates stumps that are a foci for the fungus, and from dense stands of young trees resulting from fire suppression (Johnston 1994).

In the middle-elevation conifer belt, oxidant-induced air pollution damage caused by ozone and other chemicals from air pollution from coastal and valley cities, is damaging many tree species, including *Abies concolor, Pinus ponderosa,* and *P. jeffreyi*. Lichens have also suffered significant damage and reduction due to air pollution.

Across the range, 218 endemic plant species are considered rare or threatened, and 3 plant species (*Monardella leucocephala, Mimulus whipplei, Erigeron mariposanus*) are believed to be extinct. Sixty-nine terrestrial vertebrate species (17 percent of the fauna) are considered at risk by government agencies. Many amphibian species at all elevations have severely declined and are disappearing in many areas throughout the range (Drost and Fellers 1996). Introduced fish appear to be a major cause of amphibian decline at higher elevations, although increased UV radiation, viruses, loss of habitat, and acid rain have been suggested as additional causes of this dramatic decline. Dams and impoundments, as well as grazing and logging in surrounding catchments, have degraded most aquatic habitats. Only 10 percent of the original habitat of anadromous fish in the Sierra are reachable for spawning.

Suite of Priority Activities to Enhance Biodiversity Conservation

*   Stop all logging of all remaining late-successional forest blocks.
*   Establish a series of core reserves, buffer zones with limited-use guidelines, and linkage habitats to enhance landscape-level ecological integrity. The Sierra Biodiversity Institute and the Center for Water and Wildland Resources (CWWR 1996) have conducted several analyses to identify priority areas for conservation (e.g., ecologically significant areas and SNEP—significant areas inventory) in the Sierra Nevada. The SNEP

analysis estimates that approximately 500,000 acres of federal and state lands require protection to conserve a minimum of 10 percent of remaining habitat of each Sierran plant community. Several candidate biodiversity management areas identified in the CWRR (1996) analysis include the lower elevations of Calaveras County and portions of the Consumnes River basin, the middle elevations of Sierra County north of Highway 49, parts of Plumas County east of Highway 89 and south of Highway 70, portions of Mariposa County along Highway 49, the South Fork of the Kern River to Walker Pass, and along the Greenhorn Mountains. High rates of beta-diversity (i.e., changes in species along gradients and over distance) in the Sierran biota require a system of protected areas well distributed over the landscape in order to represent the full range of biodiversity. Several challenges for effective representation include (1) the need to protect sufficiently large landscapes to promote long-term persistence of natural communities and biodiversity, (2) the distribution of many distinctive areas on privately owned lands, and (3) the reluctance of federal agencies, such as the Forest Service, to protect natural communities in areas designated for resource exploitation.

*   Identify and restore linkage zones to maintain and restore altitudinal corridors for seasonal migration of wildlife.
*   Abolish clear-cuts as a logging practice in order to improve forest health and reduce the spread of white pine blister rust.
*   Remove domestic livestock from all meadows, riparian areas, wilderness areas, and national forests.
*   Expand programs to restore ecologically sound fire regimes through revised fire suppression policies and prescribed burns and reduction of high fuel loads, in some cases instead of salvage logging.
*   Stop destruction of riparian habitats for development, logging, and grazing.
*   Establish strong conservation measures in areas recognized for concentrations of endemic and rare plants and invertebrates (see CWWR 1996).

Conservation Partners

For contact information, please see appendix G.

- Bureau of Land Management
- California Native Plant Society
- Center for Water and Wildland Resources
- Friends of the Earth
- National Park Service
- Sierra Biodiversity Institute
- Sierra Club
- U.S. Fish and Wildlife Service

Relationship to Other Classification Schemes

The Sierra Nevada ecoregion used in this analysis is based on Omernik's (1995b) Sierra Nevada ecoregion. This ecoregion generally covers the same habitats as Bailey's (1994) Sierra Nevada Section (M261E), except that Bailey identifies the northern fifth of the range as the Southern Cascades and has an extension of the Sierras running southwest toward the Tehachapi Mountains. This ecoregion covers several of Küchler's (1977) vegetation classes, including Big Trees, North Jeffrey Pine, Sierran Montane, Alpine, Upper Montane Subalpine Forests, Sierra Yellow Pine, Juniper-Pinyon Woodland, and Yellow Pine-Shrub Forests.

Prepared by D. Olson and J. Sawyer

Key Number:	**42**
Ecoregion Name:	**Great Basin Montane Forests**
Major Habitat Type:	**2.2 Temperate Coniferous Forests**
Ecoregion Size:	**5,786 km²**
Biological Distinctiveness:	**Bioregionally Outstanding**
Conservation Status:	**Snapshot—Vulnerable Final—Vulnerable**

Introduction

The Great Basin Montane Forests rise steeply out of the semiarid sage-covered plains of the Great Basin Shrub Steppe [76] in Nevada and southeastern California. This ecoregion consists of montane vegetation differentiated into distinct life zones that are roughly organized along elevational gradients (Hall 1946, 34). Within the basin, there is a distinct east-west pattern of declining species richness that correlates with distance from the western Rockies (Hamrick, Schnabel, and Wells 1994, 148). Geographic isolation on mountain tops has resulted in a high degree of genetic variation among populations of conifers and other taxa. These isolated mountain tops represent Holocene refugia and thus are believed to have a high degree of genetic diversity relative to "mainland" areas. Evidence of long-term shifts in species distributions is present in the fossil record and has been related to past climatic changes (Hamrick, Schnabel, and Wells 1994,149). The stratigraphy is Precambrian and Cenozoic with complex volcanics (McNab and Bailey 1994).

There are few perennial streams in low-lying areas in this dry ecoregion. Annual precipitation in the basin ranges from 130 to 490 mm, with most precipitation falling in the mountains as snow (Bailey 1995). At higher elevations, forest soils are present and include entisols (mesic-cryic) and aridisols (xeric) (McNab and Bailey 1994). Vegetation from valley bottoms to mountain tops includes shrub-steppe, woodlands, pinyon pine (*Pinus* spp.), juniper (*Juniperus* spp.), Douglas fir (*Pseudotsuga menziesii*), and subalpine communities (Bailey 1995). Woodlands represent a transition between moister coniferous forests of higher elevations and drier grasslands and deserts of the basin. They tend to be more open than forests and contain smaller trees. The dominant woodland trees are drought-tolerant pines and junipers derived from Mexican sources, while conifers are mainly cold tolerants derived from boreal regions (Whitney 1985). Four major woodland and forest types have been identified in this region: montane white fir (*Abies concolor*) forest, subalpine woodland, limber pine (*P. flexilis*) woodland, and Great Basin bristlecone pine (*P. longaeva*) woodland (Vasek and Thorne 1988, 822). Great Basin bristlecone is a true timberline species of highest (up to 3,540 m) desert elevations. The species typically forms pure open stands; otherwise, it mixes with limber pine and is found on the poorest and driest soils and grows in some of the harshest climates within the region (e.g., mountain tops). Bristlecones are among the oldest living organisms on earth, with one individual dated at more than 4,600 years. Ironically, this ancient individual was cut down in the dating process. Subalpine woodlands are among the highest-elevation woodlands in the region and consist of white pines (*P. flexilis* and *P. longaeva*). These trees are usually too short and widely spaced to be considered "forests" (Vasek and Thorne 1988, 822).

Montane white fir forests are widely scattered open groves that often mix with pinyon and other trees in steep mesic, north-facing ravines and slopes below the crests of mountains. Limber pine woodlands are desert-oriented white pines that rarely form forests and are usually mixed with bristlecone. Limber pine and white fir mix in groves at the heads of east-facing canyons where white bark pine (*P. albicaulis*) and mountain hemlock (*Tsuga mertensiana*) occur along west-facing canyons. In the White Mountains, limber pine is most abundant on moist, granitic soils at 2,900–3,355 m, and at lower elevations this species forms pure stands (Vasek and Thorne 1988, 825). Other less prominent woodland types include lodgepole pine (*P. contorta*) and aspen (*Populus tremuloides*). Lodgepole pine is distributed sporadically within some north-facing canyons, and aspen groves occur on moist slopes and meadows on east slopes. Fire is the chief disturbance throughout the region (Whitney 1985, 119).

Biological Distinctiveness

The biogeography of the forested mountaintops of the Great Basin has often been compared to that of an archipelago of oceanic islands. This "continental" island system has allowed biologists to explore many interesting ecological and evolutionary questions, such as the rate of differentiation among isolated taxa. The cool coniferous forests occur on widely separated mountaintops in a sea of Great Basin shrub and desert habitats, resulting in pronounced isolation for many forest species on individual ranges or peaks (Harper et al. 1994). Thus, several species and subspecies of plants, invertebrates, and some vertebrates occur only on one or a few peaks. The mosaic of forest and woodland types occurring along elevation gradients in this region results in high local beta-diversity. However, when compared with other temperate coniferous forests within its major habitat type, the region contains intermediate levels of biodiversity (total

richness = 395, total endemism = 9). Birds constitute the greatest number of species (40 percent) of the various taxa evaluated, followed by butterflies (31 percent), and mammals (14 percent). Although not evaluated in this assessment, the region is estimated to contain up to 26,000 species of insects (Nelson 1994). The numerous mountain ranges and riparian areas in the basin act as important refugia and corridors for geographic expansion of species currently found more commonly in more northern regions (Nelson 1994).

Conservation Status

Habitat Loss

Habitat loss in this region has been moderate, with 25–50 percent of the remaining area still intact (based on workshop participants' conservation status assessment). However, fire suppression and livestock grazing have resulted in shifts in species composition, particularly in lowland areas, where invasive species such as Russian thistle (*Salsola australis*), cheatgrass (*Bromus tectorum*), and exotic wheatgrass (*Agropyron* spp.) have prospered (Whitney 1985, 119).

Remaining Blocks of Intact Habitat

Few large (greater than 250 km^2) areas are present in this ecoregion. This is, in part, an artifact of the ecoregion delineation; most of the ecoregion is defined by relatively small ecological units.

Habitat Fragmentation

Workshop participants did not recognize habitat fragmentation as a significant factor affecting biodiversity in this region. Again, this was an artifact of the delineation; most of the area is naturally isolated because of its confinement to isolated mountain ranges.

Degree of Protection

Most protected lands and wilderness areas are managed by the USDA Forest Service and Great Basin National Park. Many of the larger ranges are part of National Forest Service lands, especially the Humboldt and Toiyabe National Forests.

Types and Severity of Threats

The major types of conversion threats in this region are livestock grazing, gold mining, and microwave communication sites.

Suite of Priority Activities to Enhance Biodiversity Conservation

The highest-priority activities in this ecoregion include restoring native communities impacted by livestock grazing and invasive species.

Conservation Partners

For contact information, please see appendix G.

- Nevada Association of Conservation Districts
- Nevada Wildlife Federation
- Wildlife Society, Nevada Chapter

Relationship to Other Classification Schemes

This ecoregion corresponds to the montane regions of Omernik's (1995b) ecoregion 13 and Bailey's (1994) ecoregion 341A. Because beta-diversity created drastically different communities between montane forests and basin types, we split the montane portion of the Great Basin into its own ecoregion and treated the basin separately [76].

Prepared by D. DellaSala

Key Number: **43**
Ecoregion Name: **South Central Rockies Forests**
Major Habitat Type: **2.2 Temperate Coniferous Forests**
Ecoregion Size: **159,348 km²**
Biological Distinctiveness: **Bioregionally Outstanding**
Conservation Status: **Snapshot—Relatively Stable Final—Vulnerable**

Introduction

The South Central Rockies Forest is centered primarily on the Yellowstone Plateau and the mountain ranges radiating outward from the plateau. This unit lies mainly in western Wyoming, extending into eastern Idaho and central Montana. A second large unit includes the mountains of central and eastern Idaho south of the Clearwater River. The ecoregion also exists in two additional isolated units: the Bighorn Mountains of north-central Wyoming and south-central Montana, and the Black Hills of western South Dakota and northeastern Wyoming.

Biological Distinctiveness

The ecoregion is characterized by dramatic vertical zonation of vegetation and asssociated fauna. This zonation is a consequence of abrupt elevational gradients between flatlands and mountains. Topographic relief is quite dramatic; for example, the Bighorn Mountains rise 2,794 m over the surrounding lowlands (Knight 1994). The range of biotic zones is greater in the higher mountains (e.g., Wind River and Teton Ranges in Wyoming; Madison Range in Montana). Secondary climatic effects of topographic relief (e.g., rain-shadow effects, exposure to or shelter from prevailing winds, and thermal inversions) likewise influence zonation (Peet 1988).

The Black Hills unit of the ecoregion, being the lowest in elevation and having relatively gentle topographic relief, exhibits the least amount of zonal variation. This unit has distinctive floristic diversity, however, containing flora representative of the Great Basin, Eastern Deciduous, Boreal, Rocky Mountain, and Southern Great Plains (Knight 1994).

The dominant vegetation type in the ecoregion is coniferous forest. Küchler (1985) classifies the potential vegetation type as Douglas fir–spruce-fir forest, which would be dominated by Englemann spruce (*Picea englemannii*), subalpine fir (*Abies lasiocarpa*), and Douglas fir (*Psuedotsuga menziesii*). As Peet (1988) points out, however, most forests in the Rocky Mountains "are in some stage of recovery from prior disturbance . . . climax stands [are] less common than seral communities." Thus, instead of one or all of the expected fir species, large areas of the ecoregion are dominated by lodgepole pine. To some extent, the preponderance of lodgepole pine reflects the greatly altered (via fire suppression) disturbance regime in the ecoregion (Knight 1994). Whitebark pine (*Pinus albicaulis*) is an important species at the upper treeline–krummholz zone.

In addition to expansive conifer forests, the ecoregion contains several other vegetation communities. Mountain meadows, foothill grasslands, riparian woodlands, and upper treeline–alpine communities exist throughout the ecoregion. In the Yellowstone unit, unique biotic communities occur in association with geothermal features (e.g., geysers, hot springs), due to the micro-environments created by varying chemical compositions and relatively warm temperatures (Knight 1994).

Relative to other Rocky Mountain ecoregions, the South Central Rockies are dry, experiencing a predominantly continental climate. Summers are brief and winters long and cold. Significant precipitation occurs in the higher elevations, typically as snow.

Fire, snow avalanches, major seismic disturbances, and wind are major disturbance patterns in this ecoregion. Prevailing winds alter distribution and morphology of tree species at higher elevations, while periodic "blowdown" events can topple several hundred hectares of mature forest at a time. These blown-down areas in turn can fuel stand-replacing fires during dry seasons. Herbivory is also a significant influence, particularly on aspen and riparian willow communities (Knight 1994).

Conservation Status

Habitat Loss and Degradation

Logging, hard-rock mining, oil and gas development, and recreational-residential construction are all major anthropogenic threats to the ecoregion. Domestic livestock grazing and spread of exotic species are altering species compositions. Burgeoning recreational use of remote areas is also affecting the ecoregion.

Remaining Blocks of Intact Habitat

Many of the ecoregion's mountain ranges are still relatively intact, though most have been altered somewhat by historic mining, logging, grazing, and fire suppression. The mountains in and immediately adjacent to Yellowstone National Park have seen only minor human influence in the last century.

The following are largely intact:

- Frank Church Wilderness—central Idaho
- Lemhi and Lost River Ranges—eastern Idaho
- Beaverhead National Forest—southwestern Montana
- Anaconda-Pintler Wilderness—southwestern Montana
- Pioneer Range—southwestern Montana
- Tobacco Root Mountains—southwestern Montana
- Snowcrest Mountains—southwestern Montana
- Centennial Mountains—southwestern Montana
- Madison Ranges—southwestern Montana
- The Bridger Mountains—south-central Montana
- Big Belt Mountains—south-central Montana
- Little Belt Mountains—south-central Montana
- Crazy Mountains—south-central Montana

But these areas show the effects of being isolated ranges near large population centers. Mountains to the south of Yellowstone National Park, including the Wind River, Wyoming, and Salt River Mountain Ranges, are rugged, remote, and relatively intact.

Degree of Fragmentation

Intensive development in valley bottoms, combined with existing transportation corridors, is beginning to disrupt connectivity within the ecoregion. Low-elevation development (often in ecoregions [57], [75], and [77]) has eliminated winter range or blocked off migratory routes for ungulates. Massive clear-cuts, particularly on the Targhee National Forest, have similarly caused fragmentation. Development of the high alpine environment has been limited to date, with only four major downhill ski resorts throughout the entire area.

Degree of Protection

The degree of protection in the ecoregion is fairly high, relative to some other Rocky Mountain ecoregions. Parks like Yellowstone and Grand Teton embrace a broad elevational and climatic gradient. Combined with adjacent wilderness areas, they form large blocks of protected habitat for many species. More work, however, still needs to be done to meet the needs of wider-ranging species like grizzly bears.

Types and Severity of Threats

Indiscriminate logging, and especially associated road building, are major problems. Existing road networks in nonwilderness areas are quite dense and contribute to an overall loss of habitat security. Rapid development of low-elevation areas is another threat, although concentrated mainly in other ecoregions. Mortality to grizzly bears and possibly to wolves through ungulate hunters in the fall is unacceptably high and could well be making the difference between a growing and a declining grizzly population in Greater Yellowstone.

Suite of Priority Activities to Enhance Biodiversity Conservation

- Identify and maintain critical linkage habitats within the ecoregion and among other ecoregions.
- Prevent building of any more forest roads and implement seasonal closures and obliteration projects to reduce motorized access.
- Increase management of visitor and recreationist behavior in parks and wildernesses. Yellowstone should consider abandoning and obliterating part of its extensive paved road network.

Conservation Partners

For contact information, please see appendix G.

- Alliance for the Wild Rockies
- American Wildlands
- Craighead Environmental Research Institute
- Craighead Wildlife-Wildlands Institute
- Defenders of Wildlife
- Ecology Center
- Great Bear Foundation, Greater Yellowstone Coalition
- Montana Wilderness Association
- Northern Rockies Conservation Co-op
- Predator Project
- Wild Forever
- The Wilderness Society, Northern Rockies Regional Office
- The Wildlands Project
- Yellowstone Institute
- Wyoming Outdoor Council
- Yellowstone Ecosystem Studies

Relationship to Other Classification Schemes

This ecoregion closely matches Omernik's (1995b) Middle Rockies, with the addition of the Frank Church and mountain ranges to the south and east of that area. It corresponds roughly to Bailey's (1994) M331A, D, E but does not include the Black Hills as part of this ecoregion. Küchler (1975) classifies the area as 11, 14, 16, 45, and 49.

Prepared by S. Primm

Key Number:	**44**
Ecoregion Name:	**Wasatch and Uinta Montane Forests**
Major Habitat Type:	**2.2 Temperate Coniferous Forests MHT**
Ecoregion Size:	**41,465 km²**
Biological Distinctiveness:	**Bioregionally Outstanding**
Conservation Status:	**Snapshot—Vulnerable Final—Endangered**

Introduction

The Wasatch and Uinta Montane Forests ecoregion is a distinct block of high montane habitat stretching from southeastern Idaho and extreme southwestern Wyoming to the isolated ranges of the Colorado Plateau in southern Utah. The ecoregion includes the Wasatch Range, a major north-south range; and the Uintas, one of a very few major east-west ranges.

Biological Distinctiveness

The dominant vegetation of the ecoregion is coniferous forests of varying composition. Ponderosa pine (*Pinus ponderosa*), Douglas fir (*Pseudotsuga menziesii*), subalpine fir (*Abies lasiocarpa*), and Englemann spruce (*Picea engelmanni*) communities all exist in these mountains in various associations. Limber pine (*Pinus flexilis*) also occurs but is relatively limited (Mauk and Henderson 1984). A major distinguishing feature of this ecoregion is its large areas of gambel oak (*Quercus gambelii*).

The Wasatch and Uinta Rockies differ climatically from other Rocky Mountain ecoregions in their relative aridity, a function of the extensive rain shadow cast by the Sierra Nevada five hundred miles to the west. Moist air from the southwest or southeast does not penetrate this far. The higher peaks nevertheless receive a good deal of snow, which is notably consistently dry. This uniformly dry snowpack accounts for Utah's lack of snow avalanches in the mountains. Disturbances consist mainly of fire.

Conservation Status

Habitat Loss and Degradation

Most of the ecoregion has been impacted by grazing, logging, mining, and recreational use. Large predators are fully extirpated, and ungulates like bighorn sheep (*Ovis canadensis c.*) are apparently in decline.

Remaining Blocks of Intact Habitat

The High Uinta Primitive Area represents a fairly intact block of high-elevation habitat. The northern extension of the Wasatch supports large numbers of mule deer and is relatively intact. The Aquarius Plateau in the southeast portion of the ecoregion is an important remnant block of aspen, ponderosa pine, and spruce-fir forests at higher elevation.

Degree of Fragmentation

Motorized recreation and widespread livestock grazing have had major fragmentation effects on the ecoregion. Elk and deer appear to be little affected by these processes, but most of the native vegetation is fragmented by converted and degraded areas from intensive use.

Degree of Protection

Protection of the ecoregion overall is very poor. The High Uintas Primitive Area protects mainly high-alpine habitats rather than a broad elevational gradient. Very little of this montane system is protected at all.

Types and Severity of Threats

Although some popular native fauna have survived and even thrived in the ecoregion, the outlook for its long-term viability is not optimistic. Increased motorized recreation in the mountains may compromise ungulate habitat security beyond a critical threshold if allowed to expand. Domestic livestock grazing continues unabated.

The downhill ski industry poses some of the same threats here as it does in Colorado.

Suite of Priority Activities to Enhance Biodiversity Conservation

- Expand the protected area in the High Uintas to encompass a broader elevational gradient, as this remote mountain range still has high conservation potential.
- Maintain representation of native vegetation types, particularly gambel oak—an important consideration in designing a reserve network. Ultimately, this ecoregion could functionally connect with Greater Yellowstone, facilitating movement of carnivores south into the southern deserts. For this to occur, restoration of habitat quality and security would need to begin.

Conservation Partners

For contact information, please see appendix G.

- Southern Utah Wilderness Alliance

Relationship to Other Classification Schemes

The ecoregion boundary is taken from Omernik (1995b). It approximates the boundaries of Bailey's (1994) M331E and M341C. Küchler (1975) classifies the area as 11, 14, 19, 21, and 31.

Prepared by S. Primm

Key Number:	**45**
Ecoregion Name:	**Colorado Rockies Forests**
Major Habitat Type:	**2.2 Temperate Coniferous Forests**
Ecoregion Size:	**132,740 km²**
Biological Distinctiveness:	**Bioregionally Outstanding**
Conservation Status:	**Snapshot—Relatively Stable**
	Final—Relatively Stable

Introduction

The Colorado Rockies Forest is a massive ecoregion dominated by the highest mountains of the Rockies. It extends from Casper, Wyoming, to Santa Fe, New Mexico, arrayed as both linear ranges (e.g., Laramie Mountains and Sangre de Cristo) and compex spatial expanses of peaks and massifs (e.g., San Juan Mountains).

Biological Distinctiveness

The ecoregion is characterized by dramatic vertical zonation of vegetation and asssociated fauna. This zonation is a consequence of abrupt elevational gradients between flatlands and mountains. Secondary climatic effects of topographic relief (e.g., rain-shadow effects, exposure to or shelter from prevailing winds, and thermal inversions) likewise influence zonation (Peet 1988). The overall elevation of the ecoregion is very high and reaches greater altitudes than any other part of the Rockies. It therefore exhibits the full range of life zones seen in other ecoregions farther north, while having much greater spatial extent of certain zones than other ecoregions. In addition to this distinctive physiography, the ecoregion's latitude and contiguity distinguish it as a clear ecological unit.

The dominant vegetation type in the ecoregion is coniferous forest. Species composition and associations are similar to those of the south-central Rockies, with a few exceptions. Bristlecone pine (*Pinus aristata*) replaces whitebark pine (*Pinus albicaulis*) as the predominant upper treeline–krummholz species (Peet 1988). Lodgepole pine (*Pinus contorta*) is relatively rare, while there is a good deal more ponderosa pine (*Pinus ponderosa*) woodland in the Colorado Rockies than in more northerly sites in the cordillera (Peet 1988). Extensive stands of aspen (*Populus tremuloides*) are prominent in the Colorado Rockies, apparently out-competing lodgepole pine as a major post-fire seral species (Peet 1988). In addition to expansive conifer forests, the ecoregion contains several other vegetation communities. Mountain meadows, foothill grasslands, riparian woodlands, and upper treeline–alpine tundra communities exist throughout the ecoregion.

Large, important herds of elk (*Cervus elaphus*) and mule deer (*Odocoileus hemionus*) inhabit the ecoregion. Mountain lion (*Puma concolor*) and black bear (*Ursus americanus*) are also abundant. It is likely that a remnant population of grizzly bears (*Ursus arctos*) still survives in the San Juan Mountains in southwest Colorado and northwest New Mexico. The Colorado Rockies may in fact have nearly all the species that were present prior to European settlement. Wolverine (*Gulo gulo*), lynx (*Lynx canadensis*), and American marten (*Martes americana*) all occur in the ecoregion, reaching perhaps the southern limit of their peninsular distribution here. As such, the ecoregion may hold long-term evolutionary potential for these species.

Fire, snow avalanches, and wind are major disturbance patterns in this ecoregion. Prevailing winds alter distribution and morphology of tree species at higher elevations. Alpine tundra systems are some of the most extensive in the Western Hemisphere.

Conservation Status

Habitat Loss and Degradation

Logging, hard-rock mining, oil and gas development, and recreational-residential construction are all major anthropogenic threats to the ecoregion. Domestic livestock grazing and introduction of exotic species are altering species compositions. The Medicine Bow–Snowy Range mountains in southern Wyoming have been subject to particularly intense commercial logging and sheep grazing. Low-elevation areas in the San Juans have been logged extensively, including late-seral and old-growth stands of ponderosa pine.

Burgeoning recreational use of remote areas is also affecting the ecoregion. Expansion of existing downhill ski resorts, and development of new resorts, is a pressing threat to the alpine and subalpine environments. Additionally, human settlement and transportation impacts associated with ski developments at Vail and Aspen are rapidly consuming low-elevation habitat and dramatically increasing vehicle traffic. Environmental pollution from the Denver metropolitan area and urbanizing areas in the mountains is also a concern. Densely concentrated recreational use near the Front Range metroplex threatens fragile alpine environments; unregulated off-trail travel in easily accessible tundra can cause damage that will remain for decades.

Remaining Blocks of Intact Habitat

Rocky Mountain National Park, Indian Peaks Wilderness, Comanche Peak Wilderness, and Rawah Wilderness form one large, relatively contiguous block of wildlands in the northern part of the ecoregion. The San Juan wilderness complex in the southwest is a massive conglomerate of high, wild mountains. Large wilderness areas around Aspen and Vail, centered around the Maroon Bells and Collegiate Peaks, and the Flat Tops area north of I-70 protect extensive habitat areas in the central part of the ecoregion. The Sangre de Cristo Range, a long, narrow range extending into New Mexico, occupies the southeastern part of the ecoregion.

Degree of Fragmentation

The ecoregion is fragmented by several major roads. Along these roads, development tends to creep outward in an ever-expanding zone. The Gunnison River corridor and I-70 are particularly problematic. Continued development is disrupting elk and bighorn sheep movements; the bighorns are particularly susceptible to stress. Ski area development and expansion represent another source of fragmentation. Disrupting everything from treeline to valley bottom in many locales, downhill skiing is a particularly damaging form of recreation—perhaps worse than off-road recreational vehicle (ORV) use.

Degree of Protection

There are several major protected areas, corresponding to many of the intact habitat blocks listed above. These include:

- Weminuche Wilderness
- The wild core of the southwestern part of the ecoregion—southern Colorado (protected by south San Juan and associated roadless areas)
- The Rocky Mountain National Park and adjacent wilderness areas in the northeast,

which protect a wide elevational range. However, river valleys and foothills are greatly under-represented—north-central Colorado.

The same could be said of many other wilderness areas in the ecoregion, which do an adequate job of protecting alpine and subalpine zones but do not embrace the full complement of habitat types. Lower-elevation old growth, particularly in the San Juans, is poorly represented in the protected-area system.

Types and Severity of Threats

Loss of valley bottoms (and increasingly, hillsides) to home building has major impacts on ungulate movements in many parts of the ecoregion. Downhill ski resort expansion is another source of habitat destruction. Concentrated, heavy recreational use of mountain areas near resort towns and the urban Front Range is another threat. Logging of low-elevation forests remains a threat as well.

Suite of Priority Activities to Enhance Biodiversity Conservation

- Halt any further old-growth logging in the ecoregion.
- Maintain lynx, wolverine, and marten by conserving adequate areas of the forest habitats they require, which includes old growth to some extent. Efforts to restore natural predator-prey relations, as well as natural fire processes, are both possible and desirable in an area of such expansive wildlands.
- Manage recreational use to avoid long-term degradation of sensitive alpine communities. Ending public subsidies for ski resort expansion would likely bring an end to further downhill ski development, as demand has remained flat over the past decade or more.

Conservation Partners

For contact information, please see appendix G.

- Colorado Environmental Coalition
- Southern Rockies Ecosystem Project
- Sierra Club, Southwest Regional Office
- The Wilderness Society

Relationship to Other Classification Schemes

The ecoregion boundary is taken from Omernik (1995b). It corresponds closely to Bailey's (1994) M331F, G, H, and I, and Küchler's (1975) 14, 17, 20, and 45 for the same geographic area.

Prepared by S. Primm

Key Number: **46**
Ecoregion Name: **Arizona Mountain Forest**
Major Habitat Type: **2.2 Temperate Coniferous Forests**
Ecoregion Size: **109,080 km²**
Biological Distinctiveness: **Regionally Outstanding**
Conservation Status: **Snapshot—Relatively Stable Final—Relatively Stable**

Introduction

The Arizona Mountain Forest extends from the Kaibab Plateau in northern Arizona to south of the Mogollon Plateau into portions of southwestern Mexico and eastern Arizona. This area consists mainly of steep foothills and mountains but includes some deeply dissected high plateaus (Bailey 1995, 64). Elevations range from 1,370 to 3,000 m with some peaks to 3,840 m. Soils types have not been well defined; however, most soils are entisols with alfisols and inceptisols in upland areas (Bailey 1995, 65). Stony land and rock outcrops occupy large areas on the mountains and foothills.

Vegetation zones in this ecoregion resemble the Rocky Mountain Life Zones but at higher elevations (Bailey 1995, 64). Although forests in this ecoregion are too far south to support distinct alpine communities, they do have a well-defined Transition Zone at 1,980–2,440 m where a cool, moist climate supports pine forests above the drier pinyon-juniper-oak woodlands of lower elevations (Kricher and Morrison 1993, 252). These forests are both wet and cold, averaging 635 mm to 1,000 mm, with annual precipitation increasing in the upper elevation Canadian Zone (Lowe and Brown 1994, 10). The growing season is typically shorter than 75 days, with occasional nighttime frosts.

The Transition Zone in this region comprises a strong Mexican fasciation, including Chihuahua pine (*Pinus leiophylla*) and Apache pine (*P. engelmannii*) and unique varieties of ponderosa pine (*P. ponderosa* var. *arizonica*) (Kricher and Morrison 1993, 253). Such forests are open and parklike and contain many bird species from Mexico seldom seen in the United States. The Canadian Zone (above 2,000 m) includes mostly Rocky Mountain species of mixed-conifer communities such as Douglas fir (*Pseudotsuga menzeisii*), Engelmann spruce (*Picea engelmannii*), subalpine fir (*Abies lasiocarpa*), and white fir (corkbark variety, *A. lasiocarpa* var. *arizonica*). Dwarf juniper (*Juniperus communis*) is an understory shrubby closely associated with spruce-fir forests. Exposed sites include southwestern white pine (*P. strobiformis*, a variety of limber pine), while disturbed north-facing sites consist primarily of lodgepole pine (*P. contorta*) or quaking aspen (*Populus tremuloides*).

Virgin forests in this region often exceed 25 m in height and are commonly layered in two or more age classes. Below 2,900 m one or more of the age classes may be composed solely of quaking aspen, an important wildlife habitat component and pioneer species following fire (Pase and Brown 1994, 37). Wetter sites contain Rocky Mountain maple (*Acer glabrum*), Bebb willow (*Salix bebbiana*), scouler willow (*S. scouleriana*), blueberry elder (*Sambucus glauca*), thin-leafed alder (*Alnus tenuifolis*), and bitter cherry (*Prunus emarginata*). Dry windy sites may be occupied by limber pine (*P. flexis*) and bristlecone pine (*P. aristata*). At lower elevations (less than 2,600 m) Douglas fir intermingles with ponderosa pine and white fir (*A. concolor*).

Biological Distinctiveness

In general, this ecoregion was considered regionally outstanding because of its relatively high levels of species richness (2,817 species) and endemism (132 species). Plants were the richest taxa (78 percent of the total species), followed by birds (7 percent) and butterflies (7 percent), snails (3 percent) and mammals (3 percent) and other taxa. Most endemics (26 percent) were also plants.

This ecoregion was also the southern extent of spruce-fir forests and the northern extent of many Mexican wildlife species, including tropical birds and reptiles. The Gila Wilderness in southwestern New Mexico contains perhaps the largest and "healthiest" ponderosa pine forest in the world. The region also has outstanding subterranean biodiversity with an extensive cave fauna in Guadalupe. In addition, there is great potential for restoring Mexican wolf (*Canis lupus*) and grizzly bear (*Ursus arctos horribilis*) populations in the area because of its remoteness and juxtaposition to other ecoregions where these species were formerly prevalent.

Of local conservation importance is the status of riparian areas. Riparian areas, in general, represent less than 1 percent of southwestern landscapes, yet they are critically important to wildlife, water quality, and fish habitat (Ohmart 1995).

Conservation Status

Habitat Loss and Degradation

In general, the ecoregion was considered relatively stable, with approximately 25 percent of it still intact. However, several threats to the ecoregion were identified by workshop participants and the published literature, including the following:

- logging and related fragmentation of old-growth and roadless areas
- severe overgrazing in wilderness areas
- heavily degraded stream channels and loss of habitat for the endangered Gila trout (*Oncorhynchus gilae*) and southwestern willow flycatcher (*Empidonax traillii extimus*) (Ohmart 1995)
- global endangerment of Freemont cottonwood (*P. fremontii*) and Goodding willow (*S. gooddingii*). These trees grow primarily in wet soils along streams.
- timber harvest in mature and old-growth forests preferred by the Mexican spotted owl (*Strix occidentalis lucida*) (Grubb, Ganey, and Masek 1997) and northern goshawk (*Accipiter gentalis*), particularly on the Kaibab Plateau (Crocker-Bedford 1990)

Remaining Blocks of Intact Habitat

Several sizable blocks of intact habitat remain in this ecoregion, including the following areas:

- Aldo Leopold–Gila Wilderness—southwestern New Mexico
- Blue Range Primitive Area—eastern Arizona
- Guadalupe-Carlsbad area—southeastern New Mexico and western Texas
- Kaibab Plateau and National Forest—north-central Arizona
- Grand Canyon National Park—northwestern Arizona
- Chuska Mountains on Navaho lands—northeastern Arizona and northwestern New Mexico
- Mazatzal Complex—central Arizona
- Superstition Mountains—central Arizona
- El Malpais National Monument and Conservation Area—western New Mexico

In addition, Foreman and Wolke (1992, 298) identified several large (40,000 ha) roadless areas, including:

- Sycamore Canyon–Secret Mountains—north-central Arizona
- Hellsgate—central Arizona
- Four Peaks—central Arizona
- Salt River, Baldy Bill—central Arizona
- Eagle Creek, Gila Mountains—eastern Arizona
- Galiuro Mountains—eastern Arizona

Degree of Fragmentation

Habitat fragmentation in this region has been primarily related to timber harvest in roadless areas and old-age forest classes (McClellan 1992; Pase and Brown 1994, 45). Several areas were recommended by workshop participants as potential corridors for minimizing fragmentation and insularization effects, including connecting the Gila complex with the Sky Islands to the south for future wolf movements; connecting the Gila complex with the Mataxal complexes at Showlow, New Mexico; and connecting riverine habitat through stream buffers designed to restore degraded fish populations.

Degree of Protection

The large wilderness areas identified above are largely protected from logging; however, grazing continues to be a concern in some wilderness areas. In general, about 9 percent of the ecoregion is in areas permanently protected from industrial logging and mining (DellaSala et al. 1997).

Types and Severity of Threats

There are several threats to the ecoregion, including conversion (road building, timber harvest); degradation (fire suppression, mining, ORV use, logging, fuel gathering); and potential wildlife losses (high threat to future Mexican wolf reintroduction from poaching) (McClellan 1992). Additional threats to riparian areas (Ohmart 1995; Randall 1995) and old-growth forests and roadless areas (McClellan 1992; Pase and Brown 1994; Grubb, Ganey, and Masek 1997) were identified in the literature.

Suite of Priority Activities to Enhance Biodiversity Conservation

- Control overgrazing on the Diamond Bar allotment in the Gila Wilderness.
- Protect and restore degraded native fish populations, particularly endemic trout, through habitat restoration in degraded riparian areas.
- Protect remaining old-growth and roadless areas.
- Promote reintroduction of the Mexican wolf and maintain habitat connections across ecoregions of suitable occupation.
- Designate Blue Range Wilderness area.
- Restore fire to fire-suppressed forest types.

Conservation Partners

For contact information, please see appendix G.

- Sierra Club
- Sky Island Alliance
- Southern Rockies Ecosystem Project
- Southwest Center for Biological Diversity

Relationship to Other Classification Schemes

This ecoregion corresponds to Omernik's (1995b) ecoregion 23 (Arizona/New Mexico Mountains), and there is a fair degree of overlap with Bailey's (1995, 64) M313 (Arizona–New Mexico Mountains Semi-Desert-Open Woodland-Coniferous Forest-Alpine Meadow Province).

Prepared by D. DellaSala

Key Number: **47**
Ecoregion Name: **Madrean Sky Islands**
Major Habitat Type: **2.2 Temperate Coniferous Forests**
Ecoregion Size: **11,420 km²**
Biological Distinctiveness: **Globally Outstanding**
Conservation Status: **Snapshot—Relatively Stable Final—Relatively Stable**

Introduction

The Madrean Sky Islands ecoregion comprises roughly twenty-seven small mountain ranges in southern Arizona, southwestern New Mexico, and northwestern Mexico. These ranges form a transition between the southern end of the Rocky Mountain cordillera and the northern end of Mexico's Sierra Madre Occidental. Although we treat them as a separate ecoregion in this analysis for ecoregions of the United States and Canada, we consider them the northern extension of the Sierra Madre Occidental and are following this interpretation in the ongoing analysis of ecoregions of Mexico.

Biological Distinctiveness

The biodiversity of the ecoregion is diverse and complex since it harbors both subtropical and temperate flora and fauna and is comprised of an archipelago of mountaintops surrounded by xeric Chihuahuan grasslands and scrub. Significant past changes in climate from moist to more xeric conditions have caused many communities to shift upward in elevation, creating a complex mosaic of communities and unusual distribution patterns. The dynamic history of the region and the isolation of forested peaks in a desert ecosystem make this ecoregion rich in unusual ecological and evolutionary phenomena. The island-like nature of the forest and woodland communities has promoted much differentiation in various taxa among different peaks and ranges. The mixing of subtropical and temperate plants and animals also creates unusual ecological interactions and assemblages. In general, the lower elevations of the Sky Islands include many subtropical species at their northernmost limit, while higher elevations support many montane species at their southern limit (McLaughlin 1995). Associations, community structure, and diversity vary from range to range, according to location, elevational range, and physiographic structure of each range (Warshall 1995).

The dominant vegetation community in the Sky Islands is pine-oak woodland, consisting of *Pinus leiophylla, P. cembroides, P. ponderosa arizonica, Quercus hypodleucodes, Q. arizonica, Q. emoryi, Q. rugosa, Juniperus deppeana,* and *Arctostaphylos pungens* (Peet 1988). Some ranges also have a zone of spruce-fir forest, containing *Picea englemannii, Abies lasiocarpa, A. concolor,* and *Pseudotsuga menziesii.* The lower slopes may consist of chaparral, grasslands, or oak woodlands (Warshall 1995).

The Sky Islands are clearly distinct from the surrounding desert and semidesert grassland lowlands. Some biogeographers also consider them distinct from the nearby major mountain systems (i.e., Sierra Madre Occidental, Arizona Mountains, and Colorado Plateau), as they combine elements from both major systems, and refer to the biogeographic region as Apachean. However, at a continental scale, we interpret the Sky Islands as primarily Madrean in character and refer to them as a northern extension of the Sierra Madre Occidental.

Fire and climatic variation are the major natural disturbances in the Sky Islands (Warshall 1995).

Conservation Status

Habitat Loss and Degradation

Overgrazing, fuelwood gathering, recreational pressure, military use, and development are serious anthropogenic threats to the ecoregion (Warshall 1995). Kit Peak, Mount Graham, and the Catalina Mountains have experienced major high-elevation development. Riparian areas and lower-elevation oak woodlands are specific areas that have seen heavy impacts from grazing and development. Relatively intact, lower-elevation riparian woodland is now extremely rare throughout the region. Much of the Sierra Madre Occidental forests in Mexico have been cleared through logging and agricultural expansion. A recent WWF survey of forests in the Sierra Madre Occidental identified only two relatively large, intact blocks of forest remaining throughout the whole ecoregion.

Remaining Blocks of Intact Habitat

Although logging has had an impact on many larger blocks, some remain because remoteness, steep slopes, and aridity have limited incursions into many areas. Some larger blocks of relatively intact habitat can be found in the following areas:

- Chiricahuas—southeastern Arizona
- Rincon Mountains—southern Arizona
- Catalina Mountains—southern Arizona
- Huachuca and Canelo Hills—southern Arizona
- Pajaritos-Atascosta Mountains—southern Arizona
- Animas Range—southwestern New Mexico
- Peloncillo Mountains—southeastern Arizona and southwestern New Mexico
- San Aritas—southern Arizona
- Whetstone Mountains—southern Arizona

Degree of Fragmentation

The mountain ranges are naturally isolated habitats to some extent. Valley bottoms, lower slopes, and riparian zones at lower elevations have been relatively easy to access and develop, thus impeding movements of fauna among the Sky Islands.

Degree of Protection

Important protected areas in the Sky Islands are:

- Chiricahua National Monument and Wilderness Area
- Galiuso Wilderness Area
- Saguaro National Monument East
- Rincon Wilderness Area
- Huachuca Mountains Wilderness Area
- Pusch Ridge Wilderness Area
- Dos Cabezas Wilderness Area
- Santa Teresa Wilderness Area
- Pajarito Wilderness Area
- Gray Ranch

Types and Severity of Threats

Intensive and inappropriate recreational use is a potentially severe threat as the Tucson-Phoenix metropolitan area grows in population. Fuelwood harvest is a continuing source of habitat degradation. Logging pressure can rapidly reduce and alter natural communities. Intensive collection of snakes, lizards, and birds for the pet trade may be causing local extinctions. The fragmented nature of the ecoregion makes populations of such species highly sensitive to extirpation from exploitation. Poaching pressure on the few jaguars (*Felis onca*) in the region could suppress recolonization from Mexico. Black bears (*Ursus americanus*) may also be targeted by poachers for the international trade in bear parts.

Suite of Priority Activities to Enhance Biodiversity Conservation

- Designate more of the Sky Islands as wilderness and identify or restore functional linkage habitat among the various ranges.
- Protect groundwater and flowing water.
- Protect and restore native large carnivores and protect other species from collecting and exploitation to ensure full faunal assemblages.

Conservation Partners

For contact information, please see appendix G.

- Defenders of Wildlife
- Sky Island Alliance
- Southwest Center for Biological Diversity
- Wildlands Project

Relationship to Other Classification Schemes

Neither Omernik (1995b) nor Bailey (1994) specifically delineates the Sky Islands in his classification. Küchler (1975) identifies the ecoregion variously as Arizona pine forest and oak-juniper woodland.

Prepared by S. Primm

Key Number:	**48**
Ecoregion Name:	**Piney Woods Forests**
Major Habitat Type:	**2.2 Temperate Coniferous Forests**
Ecoregion Size:	**140,892 km²**
Biological Distinctiveness:	**Regionally Outstanding**
Conservation Status:	**Snapshot—Endangered Final—Critical**

Introduction

The Piney Woods Forests stretch across eastern Texas, northwestern Louisiana, and southwestern Arkansas. This ecoregion includes parts of what is commonly known as the Big Thicket region of east Texas. The Piney Woods occupies the western extent of the southeastern coastal plain, and its vegetation reflects similarities with the communities found within the Southeastern Mixed Forests [22] and the Southeastern Conifer Forests [51]. Despite its name, Küchler (1985) classified this ecoregion as oak-hickory-pine forest. Little of the longleaf pine forests that once dominated this ecoregion remain. Pine plantations are widespread, and the effects of fire suppression have caused considerable ecological damage to this ecoregion.

Biological Distinctiveness

Sandhill pine forests are one of the communities characteristic of the Piney Woods. Longleaf pine (*Pinus palustris*) shares dominance with shortleaf pine (*P. echinata*) and loblolly pine (*P. taeda*). In this flatwoodlike habitat, pines dominate the overstory with a well-developed woody understory (Christensen 1988). Pine density is low, the herb layer is sparse, and exposed sandy tracts are common. Common associated trees are bluejack oak (*Quercus incana*) and post oak (*Q. stellata*), with a characteristic understory of yaupon (*Ilex vomitoria*) and flowering dogwood (*Cornus florida*). Savannalike areas occur on poorly drained soils and contain scattered individuals of longleaf and loblolly pine along with tupelo (*Nyssa sylvatica*), sweet gum (*Liquidambar styraciflua*), and magnolia (*Magnolia virginiana*). The interaction of moisture and fire frequency determines vegetation structure and composition (Ware, Frost, and Doerr 1993). In other sections oaks and hickories are mixed in with pines.

Conservation Status

Habitat Loss and Degradation

About 3 percent of the remaining habitat is considered intact. Bottomland forests around the Red River have been completely converted. Longleaf pine areas have been converted to loblolly or slash pine plantations or severely fire suppressed. Urban development was a major cause of habitat loss in the early part of this century, as was logging. Today, fire suppression is a major cause of habitat loss for fire-dependent species, as is conversion to pine plantation.

Remaining Blocks of Intact Habitat

Remaining small fragments include:

- Fort Polk and Vernon District of the Kisatchie National Forest—central Louisiana
- Remainder (other districts) of Kisatchie National Forest—central and northern Louisiana
- Big Thicket National Reserve (very linear-shaped habitat block offering little core area)—eastern Texas
- Texas National Forests (Sabine, Angela, Davy Crockett, Sam Houston; all with a lower level of protection and the last two heavily used for timber production)
- Sandylands (TNC reserve of 2,500 hectares)—East Texas
- Smaller fragments in adjacent areas of Arkansas and Louisiana

Degree of Fragmentation

Fire dynamics in this ecoregion are very much affected by fragmentation, as are large carnivores (e.g., black bears). Expert assessment is that it will be difficult to restore corridors in this ecoregion, and perhaps greater attention should be given to expanding existing areas.

Degree of Protection

Poor and protected areas do not adequately represent typical vegetation. One of the major conservation initiatives in this ecoregion is the Piney Woods Conservation Initiative that seeks to maintain nearly 3,200 hectares of longleaf pine forest.

Types and Severity of Threats

Conversion threats are continued conversion to pine plantations and, to a lesser extent, urban areas. The main degradation threat is considered to be fire suppression.

Suite of Priority Activities to Enhance Biodiversity Conservation

- Ensure better representation of major community types in protected areas (representation is poor at present).
- Promote fire management and promulgate right-to-burn laws.
- Work to conserve wilderness areas by removing fire-suppression policies.
- Continue inventory and identification of important sites for biodiversity and protect remaining areas.

Conservation Partners

For contact information, please see appendix G.

- Arkansas Natural Heritage Commission
- Louisiana Natural Heritage Program
- The Nature Conservancy of Arkansas
- The Nature Conservancy
- The Nature Conservancy of Louisiana
- The Nature Conservancy, Piney Woods
- The Nature Conservancy of Texas
- The Nature Conservancy, Southeast Regional Office
- Oklahoma Natural Heritage Inventory
- Texas Biological and Conservation Data System

Relationship to Other Classification Schemes

The Piney Woods Forests include parts of Omernik's (1995b) ecoregion 35 (South Central Plains), Bailey's (1994) subregions 231E and 232F (Mid Coastal Plains and Coastal Plains and Flatwoods, Western Gulf Section), and Küchler's (1975) unit 101 (Oak-Hickory-Pine Forest).

Prepared by A. Weakley, T. Cook, E. Dinerstein, and K. Wolfe

Key Number:	**49**
Ecoregion Name:	**Atlantic Coastal Pine Barrens**
Major Habitat Type:	**2.2 Temperate Coniferous Forests**
Ecoregion Size:	**8,975 km²**
Biological Distinctiveness:	**Nationally Important**
Conservation Status:	**Snapshot—Relatively Stable Final—Relatively Stable**

Introduction

The Atlantic Coastal Pine Barrens cover the coastal plain of New Jersey, much of the southern half of Long Island, and Cape Cod, Massachusetts, an areal extent of nearly 9,000 km². The largest block of this disjunct ecoregion occurs in the coastal plain of New Jersey. A smaller, inland example of a Pine Barrens ecosystem can be found near Albany, New York. The Pine Barrens are underlain by sandy, nutrient-poor soils that typically support a stunted forest of pitch pine (*Pinus rigida*) and blackjack oak (*Quercus marilandica*) maintained by frequent fires. Hydrology, soils, disturbance regimes, and vegetation combine to separate this ecoregion from surrounding units. In particular, fire regimes and sandy, droughty soils distinguish this unit from surrounding ecoregions.

Biological Distinctiveness

Three major types of pine-dominated habitats are widely recognized, each distinguished by canopy height and composition (Christensen 1988). Pine-oak forest is the tallest type, with a well-developed tree layer composed of post oak (*Quercus stellata*) and blackjack oak. The dwarf pine plains represent the other extreme, dominated by a scrub version of pitch pine and blackjack oak that often is less than 3 m tall. Shrubby oaks such as *Q. ilicifolia* are also common. The pine–shrub oak forest grades between the two other types with pitch pines commonly reaching a height of more than 10 m. Given the dry, sandy soils of the ecoregion, pine barrens do not support particularly rich floras. About eight hundred species and varieties of plants have been recorded for the New Jersey Pine Barrens (Collins and Anderson 1994).

Fire history is the major disturbance factor that influences stand height and composition (Good, Good, and Andresen 1979). Evidence

suggests that severe, frequent fires have selected for a distinct genotype of pitch pine, with reduced apical dominance of trees and highly serotinous cones. The herb layer of the pine barrens includes forty to fifty species found among the three habitat types.

A number of rare endemic plants are common in parts of the New Jersey Pine Barrens, which is also home to the endemic pine barrens frog. Perhaps most outstanding is the number of relatively rare community types contained within such a small ecoregion. They include: the pitch pine–scrub oak barrens; the coastal plain ponds, an important habitat for the American burying beetle; and maritime grasslands native to Martha's Vineyard and the eastern tip of Long Island, the only sites in the eastern United States with this type of grasslands. The beaches adjacent to this ecoregion are critical breeding habitat for piping plover (Cape Cod, Martha's Vineyard, and Long Island) and for roseate terns (approximately one-third of the U.S. population breeds on Bird Island).

Conservation Status

Habitat Loss and Degradation

Intact habitat is limited to about 10 percent of the original areal extent of the pine barrens. Habitat loss in the three disjunct units is a result of urbanization (New Jersey) and suburban sprawl. The rapid expansion of housing developments, retirement communities, and vacation homes has affected all three units. Extensive development in some areas has led to longer intervals for habitats to return from fire episodes, tilting succession to greater dominance of oaks and reduction of dwarf pine areas.

Remaining Blocks of Intact Habitat

Intact habitat is distributed among the following blocks (presented in decreasing order of size):

- New Jersey Pine Barrens—New Jersey
- Long Island Pine Barrens (only a small part remains)—New York
- Miles Standish State Park—Massachusetts
- Cape Cod National Seashore—Massachusetts
- State Forest, Martha's Vineyard—Massachusetts

Degree of Fragmentation

Pine barrens are naturally fragmented and are historically dependent on the substrate. However, road networks have further fragmented parts of the pine barrens and caused mortality among vertebrates.

Degree of Protection

For its size, this ecoregion represents one of the best conserved habitats in the eastern United States, largely because its poor soils have precluded intensive agriculture. Farsighted planning, leading to the establishment of the Cape Cod National Seashore, has preserved an important stretch of this habitat in Massachusetts.

The five remaining blocks of habitat listed above also constitute the most important protected areas in this ecoregion.

Types and Severity of Threats

The major conversion and degradation threats are development and fire suppression. These are considered to be moderate over the next decade. Exploitation of wildlife is not viewed as a major threat to biodiversity conservation, unlike in other ecoregions in the eastern United States.

Suite of Priority Activities to Enhance Biodiversity Conservation

None of the sites listed above is completely protected in terms of land area or management for biodiversity conservation. Increased conservation effort is needed in:

- Peconic Area on Long Island
- Cape Cod National Seashore (largely management issues)
- Most of Nantucket
- Southern edge of Martha's Vineyard

Several conservation plans for the New Jersey Pine Barrens are reviewed by Berger and Stinton (1985).

Like many ecoregions within the Temperate Coniferous Forest MHT, the Pine Barrens are seriously threatened by fire suppression. Improved zoning and planning must be implemented to reduce development in order to allow areas to burn in controlled fashion.

Conservation Partners

For contact information, please see appendix G.

- Environmental Defense Fund (Long Island)
- Long Island Chapter
- Massachusetts Audubon Society
- National Park Service
- The Nature Conservancy
- The Nature Conservancy of New Jersey
- New Jersey Pinelands Commission
- Sierra Club, Atlantic Chapter
- Sierra Club, New Jersey Chapter
- The Pinelands Commission

Relationship to Other Classification Schemes

Boundaries for the Atlantic Coastal Pine Barrens are taken from Küchler (1975) (unit 100). This ecoregion is subsumed under the Upper Atlantic and Lower New England Coastal Plain (sections 221C and 221A, respectively) in Bailey (1994) and the Middle Atlantic Coastal Plain and Northern Atlantic Coastal Plain (ecoregions 63 and 59) in Omernik (1995b).

Prepared by E. Dinerstein, S. Buttrick, M. Davis, and B. Eichbaum

Key Number:	**50**
Ecoregion Name:	**Middle Atlantic Coastal Forests**
Major Habitat Type:	**2.2 Temperate Coniferous Forests**
Ecoregion Size:	**133,855 km²**
Biological Distinctiveness:	**Regionally Outstanding**
Conservation Status:	**Snapshot—Endangered Final—Endangered**

Introduction

The Middle Atlantic Coastal Forests stretch from the eastern shore of Maryland and Delaware to just south of the Georgia–South Carolina border. The western border of this ecoregion is the juncture of the Atlantic coastal plain and the edge of the Piedmont, whereupon it gives way to the Southeastern Mixed Forests

[22]. Although much of the ecoregion shares the same topography, the flat Atlantic coastal plain, it is one of the most heterogeneous units from a vegetation perspective. Some of the most majestic plant communities of the United States occur along rivers of this ecoregion. The Atlantic white cedar (*Chamaecyparis thyoides*) swamps once occurred along the entire eastern seaboard and were common in this ecoregion. Perhaps most famous are the river swamp or bottomland forests, admired for their towering bald cypress and gum trees. The Middle Atlantic Coastal Forests contain the most diverse assemblage of freshwater wetland communities in North America and perhaps of all temperate forest ecoregions. Nonalluvial wetlands, including freshwater marshes, shrub bogs, white cedar swamps, bayheads, and wet hammocks are particularly prominent. The coastline is bordered by barrier or sea islands that protect extensive estuaries, lagoons, and sounds. Many of these wetlands have already been converted, and others are highly threatened.

Biological Distinctiveness

The mosaic of plant communities observed is a result of dramatic gradients in soil structure and chemistry and hydrology (Christensen 1988). Habitats are dynamic, with fire being a major source of disturbance in drier areas, and hurricanes and floods in the bottomlands, coastal plains, and maritime habitats. The interaction of moisture and fire frequency influences species richness in the herbaceous layer with frequent (1–3 yr) summer fires favoring a herbaceous-dominated savanna and less frequent (5–10 yr) fires leading to a dense shrubby understory.

River swamp forests, or bottomland forests, were once prominent in this ecoregion and are one of the most visually appealing habitats in North America. This forest type is dominated by bald cypress (*Taxodium distichum*) and swamp tupelo (*Nyssa sylvatica* var. *biflora*). Eastern, or Atlantic, white cedar (*Chamaecyparis thyoides*) occurs along blackwater rivers, most commonly on organic substrates underlain by sand (Wharton et al. 1982). Christensen (1988) describes in detail the variety of river forest and nonalluvial wetland communities found throughout the ecoregion. Some of the more interesting types include the pocosins and Carolina bays. Pocosins, from an ancient

Algonquin term for "swamp on a hill," are characterized as extensive, flat, damp, sandy or peaty areas far from streams, with a scattered growth of pond pine (*Pinus serotina*) and a dense growth of mostly evergreen shrubs (often gallberry, *Ilex glabra*), that taken together, resemble a heath scrub community. Pond pine is especially prevalent in coastal North Carolina on poorly drained organic soils and where wildfire is common. Carolina bays are ovate-shaped, shallow depressions and occur abundantly across a broad band of the coastal plain from southern North Carolina to the South Carolina–Georgia border (Richardson and Gibbons 1993). They represent a type of bog or bog-lake complex unique to the southeastern coastal plain and are thought to have been formed by a meteor or comet impact. Maritime communities are also an important feature of this ecoregion (Christensen 1988).

The ecoregion as a whole ranks among the top ten ecoregions of the United States and Canada in reptiles, birds, and tree species. Within this ecoregion, the bottomland or floodplain forests are some of the most biologically important habitats in North America. They occur on rich alluvial soils, maintain moderate climates and microhabitats that provide seasonal refugia for many species, maintain an abundance of arthropods and mast from canopy species to sustain overwintering migratory birds, and offer abundant rotting logs, bole, and branch cavities for detritivores and hole-nesting species, respectively (Harris 1989, in Sharitz and Mitsch 1993). Many of the remaining pockets of forested habitat are crucial as seasonal habitat for songbirds and migratory waterfowl. Bottomland forests also provide a distinct contrast to the pine-dominated uplands, serve as a resource sink for upland aquatic communities, support aquatic food webs when flooded and terrestrial food chains during the dry season. Finally, the linear distribution of floodplain forests facilitates local and regional movements of species, and river flow maintains effective water dispersal of seeds and larvae (Harris 1989, in Sharitz and Mitsch 1993). Other habitats, such as bogs, provide habitat for rich assemblages of herbaceous species, including many endemics. Carnivorous plants, such as Venus fly-traps (*Dionea* spp.), pitcher plants (*Saracena* spp.), and sundews (*Drosera* spp.), are restricted to very small areas. The ecoregion was a Pleistocene refugia, although this section is out on the continental shelf.

Conservation Status

Habitat Loss and Degradation

Approximately 12 percent of the ecoregion contains habitat that meets the definition of intact used in this assessment. The highest levels of conversion are in the western part of this ecoregion, the upper coastal plain, where upland vegetation on loamy soils has been nearly completely converted. Longleaf pine communities have largely disappeared and are now absent in Virginia. Much of the cypress forests of the Middle Atlantic Coastal Forests has been lost to logging (Christensen 1988). Stands where cypress has been high-graded often revert to bay forests. Where logging and fire have occurred, cypress is extremely slow to recover. The Great Dismal Swamp in Virginia was one of the strongholds of Atlantic white cedar swamps and is now virtually gone (Noss and Peters 1995).

One of the greatest threats is to the diverse wetlands communities and in particular bottomland forests. These were once extensive: in the mid-1970s, 47 percent (188,000 km^2) of remaining wetlands in the lower forty-eight states of the United States were in the Southeast. Sixty-five percent of the pallustrine forested wetlands (pocosins, swamps, bottomland hardwoods, and bogs) in the United States occur in the Southeast (Hefner and Brown 1984). Pocosins originally covered 9,080 km^2 of the forty-one coastal plain counties of North Carolina. By 1979, some 6,080 km^2 of natural or slightly altered pocosins remained. Of this amount only 2,810 km^2 were still considered in a natural state as of 1980 (Richardson and Gibbons 1993). The number of Carolina bays is uncertain, although at least six thousand once occurred in North and South Carolina. Few natural bays remain, the majority having already been modified by agricultural or urban development. The South Carolina Trust Program provides details on the numbers and proportions of altered and natural bays (Bennett and Nelson 1989). The least affected communities in this ecoregion are the coastal marshes and deep peatlands.

The main reasons behind conversion are agriculture, fire suppression, urbanization, coastal development (including resorts), ditching and draining of wetlands, and damming of rivers, which affects hydrology.

Remaining Blocks of Intact Habitat

Numerous blocks of habitat are scattered about the ecoregion, but all are relatively small in size. Those that include at least a fraction of intact habitat are:

- Savannah River bottomlands—southern South Carolina–Georgia border
- C.E. basin—southern South Carolina
- Francis Marion National Forest—eastern South Carolina
- Winyah Bay—eastern South Carolina
- Lake Waccamaw and River—southeastern North Carolina, northeastern South Carolina
- Brunswick County Pinelands—southeastern North Carolina
- Bladen Lakes—southern North Carolina
- Holly Shelter Gamelands—southeastern North Carolina
- Camp Lejeune—southeastern North Carolina
- Croatan National Forest—eastern North Carolina
- Outer Banks—coastal North Carolina
- Pamlimarle Peninsula—eastern North Carolina
- Roanoke River—eastern North Carolina
- North and Northwest River—northeastern Carolina
- Great Dismal Swamp—Virginia–North Carolina border
- Assateague and Chincoteague—Maryland–Virginia Atlantic Coast
- Virginia coast reserve—Virginia
- Cape Romain—South Carolina coast
- Fort Bragg—southeastern North Carolina
- Sandhills Gamelands—southeastern North Carolina
- Sandhills NWR—northern South Carolina
- Fort Jackson—central South Carolina
- Fort Stewart—eastern Georgia

Degree of Fragmentation

Fragmentation is an important threat in this ecoregion because it exacerbates the main problem of fire suppression. An area where much work remains to be done is in planning and creating corridors along the coast. In the uplands, a corridor between Fort Bragg and the Sandhills Gamelands represents the only possibility at present for linking upland areas.

Other opportunities for establishing corridors include links among the Croatan National Forest, Camp Lejeune, and the Holly Shelter Gamelands. Another possibility is among the Brunswick County Pinelands, and Lake Waccamaw.

Degree of Protection

This ecoregion contains the longest undammed river sections in the country. However, there is very poor protection of blackwater bottomlands. Forested wetlands, such as bottomland forests, are undergoing rapid reduction in area and alteration of composition. Many are being converted to farmland, used for industrial parks, or modified by urban and suburban expansion. Other forested wetlands are being managed for timber production, which typically reduces their value as wildlife habitat for sensitive species (Sharitz and Mitsch 1993). North Carolina and South Carolina were estimated at having 12,950 km^2 and 12,790 km^2, respectively, of bottomland hardwood forest in 1952. Projections for the year 2000 show losses of 20 percent in North Carolina and 28 percent in South Carolina bottomland hardwood forests.

Other communities that are not represented in the protected areas of this ecoregion are areas underlain by loamy soils and upland blocks of habitat.

The most important protected areas for biodiversity conservation are:

- Francis Marion National Forest—southeastern South Carolina
- Brunswick County Pinelands—North Carolina
- Holly Shelter Gamelands—North Carolina
- Croatan National Forest—eastern North Carolina
- Outer Banks—eastern North Carolina
- Pamlimarle Peninsula—eastern North Carolina
- Roanoke River—southwestern Virginia
- Sandhills Gamelands—northeastern South Carolina
- Sandhills NWR—northeastern South Carolina
- Fort Stewart—eastern Georgia

The North Carolina Heritage Program, in conjunction with the state, has purchased more than a dozen bays totaling less than 20 km^2. The South Carolina Trust Program has acquired or is negotiating the purchase of more than thirty Carolina bays (Richardson and Gibbons 1993).

Thirty-seven species of southeastern songbirds are known to require extensive forest areas for maintaining viable populations. The continued fragmentation of these forests has put greater stress on these species.

The wetlands of the Middle Atlantic Coastal Forests are keystone habitats for a variety of reptile, amphibian, and shrub species that require the moist conditions available to complete stages of their life cycle (herps) or maintain viable populations (shrubs). The loss of a significant portion of wetlands in this ecoregion will have a dramatic impact on many species that are either native or reach high densities within this ecoregion. Bays and pocosins are also important reservoirs of carbon, and their continued conversion will compound the rise in atmospheric CO_2 (Richardson 1983).

Types and Severity of Threats

Conversion threats include development along the coast and pine plantations on the outer coastal plain (but not the coast itself). The major degradation threats are fire suppression, dams, and ditching. There is no legal fire management regime in this ecoregion. Poaching of carnivorous plants and black bear is judged as moderate.

Suite of Priority Activities to Enhance Biodiversity Conservation

- Increase protection in Winyah Bay, Lake Waceamah, and Waceamah River; Brunswick County Pinelands; and Bladen Lakes.
- Improve management for biodiversity conservation, particularly fire management, in Francis Marion National Forest, Great Dismal Swamp, Fort Bragg, Sandhills Gamelands, Sandhills National Wildlife Refuge, and Fort Stewart.
- Establish corridors among units listed above.
- Protect blackwater river systems in northeast Cape Fear area, northeast Black River, and the Little Perdu.
- Strengthen right-to-burn laws.
- Identify and inventory sites with loamy soils and last bits of longleaf pine for inclusion in protected areas.
- Develop strong state land-use management systems that include the protection of biodi-

versity as a fundamental goal. One example of such a program is the Chesapeake Bay Critical Area Program of the State of Maryland.

- Enact and implement effective forest management programs at the state level. These programs should include minimum regulatory standards as well as a mix of economic incentives for good forest practices. This is particularly important because a high proportion of forest resources in this region is located on private lands.
- Improve implementation of state programs for the protection of both tidal and nontidal wetlands.
- Undertake significant restoration efforts in selected areas. For example, substantial work needs to be undertaken to remove physical barriers obstructing movement of aquatic species from the many rivers flowing into the coastal waters.
- Pay special attention to retaining the natural characteristics of the barrier island systems that remain in the region. Acquisition programs such as those carried out by The Nature Conservancy in the Virginia Barrier Islands are vital but should be supplemented by development that is more sensitive where it is allowed.

A revision to the policy issues is also needed. This would include:

- The effects of transportation corridors have been devastating in certain areas and threaten much of the region. High priority should be placed on development of a long-range strategy for increased reliance on mass transit alternatives to the automobile.
- Throughout many of the rural parts of the ecoregion there is a growing animal livestock industry, such as those of chickens in the Del-Mar-Va Peninsula and hogs in North Carolina. Conversion of open space to these uses and the associated potential water pollution impacts are serious and need to be controlled more effectively. Of special concern is the very high density of animals in some areas.
- Widespread use of drainage systems to allow for agricultural activities has substantially altered much of the low-lying habitat of the region. Action should be undertaken to remove

the government subsidies for these systems, and where high-value habitat is at stake, restoration actions should be undertaken.

Conservation Partners

For contact information, please see appendix G.

- Adkins Arboretum
- Chesapeake Bay Foundation
- EPA Chesapeake Bay Liaison Office
- Maryland Ornithological Society
- The Nature Conservancy
- The Nature Conservancy of Maryland
- The Nature Conservancy of Virginia
- The Nature Conservancy, Southeast Regional Office
- North Carolina Coastal Federation

Relationship to Other Classification Schemes

The Middle Atlantic Coastal Forests cover the same area as Küchler (1985) unit 101 (Oak-Hickory Pine Forest) until the ecoregion abuts the Piedmont (fall line). Omernik (1995b) unit 63 (Middle Atlantic Coastal plain) is virtually the same except that we chose to separate southern New Jersey into the Atlantic Coastal Pine Barrens. Bailey (1994) delineates this unit as part of sections 232A and 232C (Middle Atlantic Coastal Plains and Atlantic Coastal Flatlands, respectively).

Prepared by A. Weakley, E. Dinerstein, R. Noss, B. Eichbaum, and K. Wolfe

Key Number:	**51**
Ecoregion Name:	**Southeastern Conifer Forests**
Major Habitat Type:	**2.2 Temperate Coniferous Forests**
Ecoregion Size:	**236,759 km²**
Biological Distinctiveness:	**Globally Outstanding**
Conservation Status:	**Snapshot—Endangered Final—Critical**

Introduction

The Southeastern Conifer Forests span the coastal plain of the southeastern United States, stretching across southeastern Louisiana, southern Mississippi, southern Alabama, central and southern Georgia, and the Florida panhandle and upper peninsula. This ecoregion is the largest conifer forest ecoregion east of the Mississippi and the second-largest coniferous ecoregion in the continental United States (second to ecoregion [30], North Central Rockies Forests). Most of the area of this ecoregion falls within the state of Florida.

The Southeastern Conifer Forests were dominated by relatively open tall stands of longleaf pine (*Pinus palustris*) with an understory of wiregrass (*Aristida stricta*). The open nature of the mature longleaf pine stands and the frequency of understory fires helped maintain perhaps the richest temperate herbaceous flora on earth. Unfortunately, virtually all of the mature longleaf pine forests are gone, being restricted to a few isolated sites (Ware, Frost, and Doerr 1993).

Because of the human-induced disappearance of the longleaf pine forests and their subsequent replacement by mixed-hardwood forest, some controversy surrounds the classification of this ecoregion. We follow Ware, Frost, and Doerr (1993) and many others in viewing this ecoregion as a conifer forest maintained by fire. In fact, this ecoregion is part of a mosaic of vegetation types that grade from longleaf pine forests (or sandhill communities) to pine savannas, flatwood habitats (pine forests with woody understories), and xeric hardwood communities (Christensen 1988; Myers 1985). Fire regimes largely determine the areal extent of these communities; high-frequency, low-intensity fire regimes favor the maintenance of the once-dominant longleaf pine and wiregrass communities.

Biological Distinctiveness

The biological diversity of this ecoregion is virtually unparalleled in North America. The Southeastern Conifer Forests were a refugium during the Pleistocene and a major area of adaptive radiation of amphibians, reptiles, and vascular plants. Tree diversity and endemism is highest in this ecoregion, totaling 190 species with 27 endemics. The Southeastern Conifer Forests also rank among the top ten ecoregions in richness of amphibians, reptiles, and birds and among the top ten ecoregions in number of endemic reptiles, amphibians, butterflies, and mammals. The wiregrass understory contains some of the richest herbaceous floras in the world, with a single stand containing as many as 200 species (Noss and Peters 1995). Kartesz lists 3,417 species of native herbaceous and shrub species, among the highest levels of endemism found in North America.

Mature longleaf pine forests are of particular importance to endangered species such as the red-cockaded woodpecker (*Picoides borealis*) and the declining gopher tortoise (*Gopherus polyphemus*). The gopher tortoise is an important keystone species of this ecoregion—nearly 400 species of animals use the burrows of tortoises during various stages of their life cycle (Cox et al. 1994). Not surprisingly, many federally listed species are native to longleaf pine forest communities (Noss and Peters 1995).

Suppression of natural fire regimes has shifted vegetation to hardwood forests. This unnatural management practice has altered plant communities across the ecoregion and threatens the long-term persistence of many fire-dependent species.

Conservation Status

Habitat Loss and Degradation

Over 98 percent of this habitat is now gone in the southeastern section of this ecoregion, much having been converted to agriculture or tree farms. Remaining habitat is limited to fragments and degraded larger patches. Since the middle of this century, longleaf pine forests have been cut at a rate of 525 km²/yr (130,000 acres/year) and replaced with monoculture plantations of slash pine (*Pinus elliottii*) (Noss and Peters 1995). Slash pine plantations support less diverse species assemblages than the original longleaf pine habitats.

Among the plant communities found within this ecoregion, the longleaf pine forests are among the most heavily converted and threatened (Myers and Ewel 1990a). Close behind are the mixed conifer and hardwood forests and all upland communities. The least affected habitats are tidal marshes; most other wetland habitats have been moderately affected by conversion. Besides agriculture and pine plantations, fire suppression and suburban sprawl have also resulted in habitat conversion.

Remaining Blocks of Intact Habitat

Several relatively large blocks of intact habitat remain, as well as a number of smaller patches. Some of the best conserved are on military bases. Important blocks in order of decreasing size include:

- Eglin U.S. Air Force Base—western Florida panhandle
- Apalachicola National Forest—northern Florida
- Osceola National Forest—northern Florida
- Ordway Preserve (University of Florida)—northern Florida
- Desoto National Forest—southern Mississippi
- Okefenokee National Wildlife Reserve—southern Georgia
- St. Marks National Wildlife Refuge—northern Gulf Coast of Florida
- Lower Suwannee National Wildlife Refuge—northern Gulf Coast of Florida
- Ocala National Forest (only a small part is natural)—north-central Florida
- Conecuh National Forest—southern Alabama
- Blackwater State Forest—northwestern Florida panhandle
- Withlacoochee State Forest—central Gulf Coast of Florida
- Grand Bay Savannah—southwestern Alabama Gulf Coast.

Restoration of habitat in these units would boost the total area of intact habitat to about 12 percent of the ecoregion.

Degree of Fragmentation

Although much of the ecoregion has been converted and exists as isolated fragments, there are a number of undammed rivers with undamaged riparian corridors because flood regimes have prevented the development of riparian areas. The entire coastline extending into the adjacent Western Gulf Coastal Grasslands [68] is critical for migratory birds.

Degree of Protection

According to Ware, Frost, and Doerr (1993), "unless concerted action is taken, the end is in sight for this old growth community that has evolved over the millennia and dominated the southeastern Coastal Plain for at least 5,000 years. Longleaf pine, wiregrass, the associated oaks, and herbaceous species, and the animals uniquely adapted to taking advantage of the fire-adapted vegetation . . . may not be with us into the next century. Red-cockaded woodpeckers are considered by some to be the 'flagship' animal species of this community, as longleaf pine and wiregrass are the flagship plants of the woody and herbaceous communities." The authors point out that only 25 percent of the small amount of intact longleaf pine forests is on public lands under some form of protection.

The greatest challenge to conserving a fraction of this original ecosystem is to manage the 2 to 3 percent of the landscape for maintenance of 100 percent of the original diversity of species and habitats once contained within longleaf pine forests. Ware et al. (1993) illustrate the locations of thirty-seven sites containing remnants of the longleaf pine communities and mature southern mixed-hardwood forests. Most of these sites are small and quite isolated. Within the ecoregion, the best opportunities for conservation in the short term occur in: Blackwater River State Forest; Eglin Air Force Base; several Florida state parks (especially Wekiwa Springs, Torreya, Gold Head Branch, and San Felasco Hammock); the University of Florida's Ordway Preserve; and the Nature Conservancy's Janet Butterfield Brooks Preserve. Cox et al. (1994) propose a series of strategic habitat conservation areas, covering approximately 20,000 km², that contain critical habitats and species assemblages that currently lack formal protection.

Types and Severity of Threats

The major types of conversion threats are fire suppression due to concerns about air quality, highway development, and urban sprawl and suburban development. Florida remains one of the fastest-growing states in the United States. The major degradation threat is from introduced species. Perhaps the most serious problem is the spread of kogon grass (*Imperata cylindrica*), an Asian native that is replacing the native herbaceous understory and is virtually impossible to eradicate. The major threats in the form of wildlife exploitation are illegal hunting of fox squirrels (*Sciurus niger*) and black bears (*Ursus americanus*), collection of herpetofauna, and destruction of poisonous snakes. In sum, this ecoregion is considered to be under high threat. We chose not to elevate the conservation status of this ecoregion from endangered to critical because its score for snapshot conservation status was on the edge of endangered/vulnerable. There are still large areas that, while altered, could become part of an effective conservation program if properly managed for biodiversity.

Suite of Priority Activities to Enhance Biodiversity Conservation

- Prevent remaining sites from being logged or converted to pine plantations, and work toward ensuring that areas designated as strategic habitat conservation areas be granted formal protection.
- Give greater emphasis in an ecoregion conservation plan to create at least one large longleaf pine forest reserve of at least 5,000 km^2 in which this ecosystem can be properly restored and maintained through ecologically sound burning practices.
- Manage remaining sites properly by using controlled burns to maintain fire-dependent communities. Federal agency land managers, especially U.S. Forest Service, Department of Defense, and U.S. Fish and Wildlife Service, will be required to design and implement burning practices that may vary from site to site in order to support the full spectrum of species that require different burning regimes. The right to burn natural vegetation should be assured for land managers and owners to encourage more natural fire management techniques and procedures. Related

recommendations are: phase out human-induced winter burns and replace them with growing-season burning, minimize fire lane building, restore fire lanes after fire has occurred, and relocate or reduce the soil-destructive aspects of fire lane building.
- Identify core areas and corridors. An important acquisition is the Pinhook Swamp (60,000–70,000 acres), which would be purchased to connect the Osceola National Forest (Florida) and the Okefenokee National Wildlife Refuge (Georgia).
- Track recovery of the red-cockaded woodpecker in the region. Survey forests in which the woodpecker is found and promote appropriate burns in these areas to improve woodpecker habitat. Monitor the emergency ban that was placed on clear-cutting within 1.2 km (0.75 miles) of active and inactive woodpecker colonies.
- Identify additional biodiversity hot spots and develop and implement plans to protect them.
- Track and influence the management of public lands in the region via public land trustees, who work with state resource agencies to make management decisions for state lands and revisions of national forest plans. National forest plans govern the management of federal lands and are revised every decade. Groups that review and comment upon these plans should promote such issues as endangered species recovery, biodiversity maintenance, ground cover maintenance and management, restoration and maintenance of native species, and fire management.
- Develop a strategy to address and combat chip milling in the southeast forests. This process is used to make paper and involves the clear-cutting of large tracts of forest, which are processed in newly constructed chip-processing factories. Until recently, this process was a problem only in the western United States.
- Identify the type and level of recreation that is compatible with the continued survival and health of wildlife.
- Map the location of roads, obliterate unused forest roads, and prevent unnecessary road building.

Conservation Partners

For contact information, please see appendix G.

- Alabama Natural Heritage Program
- Conservation and Recreation Lands (CARL) Acquisition Program (in Florida)
- Coastal Plains Institute and Land Conservancy
- Florida Natural Areas Inventory
- Georgia Department of Natural Resources
- Georgia Natural Heritage Program
- Louisiana Natural Heritage Program
- Mississippi Natural Heritage Program
- National Audubon Society, Southeast Regional Office
- The Nature Conservancy
- The Nature Conservancy, Southeast Regional Office
- The Nature Conservancy of Alabama
- The Nature Conservancy of Florida
- The Nature Conservancy of Georgia
- The Nature Conservancy of Louisiana
- The Nature Conservancy of Mississippi
- Sierra Club, North Florida Representative
- Sierra Club, Southeast Regional Office
- The Wilderness Society, Southeast Regional Office
- Wildlife Resources Division

Relationship to Other Classification Schemes

The Southeastern Conifer Forests are demarcated from the Southeastern Mixed Forests [22] to the north by vegetation (Küchler 1975) and elevation (the fall line of the Atlantic piedmont). The latter ecoregion is more dominated by oaks and occurs on the piedmont rather than the coastal plain. Omernik (1995b) and Bailey (1994) both extend this ecoregion farther west along the Gulf Coastal Plain and north along the Atlantic Coastal Plain, respectively. We divided these coastal plain units into three because: (1) the Western Gulf Coastal Grasslands [68] and the Middle Atlantic Coastal Forests [50] contain a mosaic of habitat types that distinguish them from the longleaf pine–dominated forests; (2) the biologically distinct wire grass communities disappear in the coastal areas west of the boundary we have delineated, replaced by big bluestem grass understories more typical of the central

grasslands ecoregions; and (3) conservation strategies for these three units would be unwieldy if managed as a single large unit that would essentially span the entire coastline from eastern Texas to Virginia.

Prepared by E. Dinerstein, A. Weakley, R. Noss, R. Snodgrass, and K. Wolfe

Key Number:	**52**
Ecoregion Name:	**Florida Sand Pine Scrub**
Major Habitat Type:	**2.2 Temperate Coniferous Forests**
Ecoregion Size:	**3,389 km²**
Biological Distinctiveness:	**Globally Outstanding**
Conservation Status:	**Snapshot—Critical Final—Critical**

Introduction

Most ecoregions described in this conservation assessment are ecosystems of regional extent covering at least 10,000 km². An important exception, and one of the smallest ecoregions in the continental United States, is the Florida Sand Pine Scrub (hereafter scrub), which occurs as a small archipelago of sandy ridges and limestone areas restricted to central and southern Florida and a narrow stretch along the Gulf Coast of Florida. The extraordinary biodiversity found within these units, a significant portion of which is restricted to these habitats, warrants their elevation to ecoregion status. The biodiversity of these islands is greatly threatened by development, but the state of Florida has made major efforts to purchase remaining scrub habitats.

Biological Distinctiveness

Scrub is frequently cited as Florida's most distinct ecosystem; physiognomy and composition are quite distinct from surrounding habitats, and 40–60 percent of scrub species are considered to be endemic. Scrub contains a biological treasure house of plants and animals adapted to life on scattered ridges of sand; the ancient origins of these sand dune communities date back to the Pliocene savannas and provide a relic example of an extremely old and formerly extensive ecosystem (Deyrup and Eisner 1993).

Scrub vegetation represents an exceedingly xeric plant community in an otherwise lush, subtropical belt. Scrub is often defined botanically as a xeromorphic shrub community dominated by a layer of evergreen or nearly evergreen oaks (*Quercus geminata, Q. myrtifolia, Q. inopina, Q. chapmanii*) or Florida rosemary (*Ceratiola ericoides*) or both, in the presence or absence of an overstory of sand pine (*Pinus clausa*) and occupying well-drained, infertile, sandy soils. Three units of scrub can be defined: inland peninsula, coastal peninsula, and coastal panhandle (Myers and Ewel 1990b). The scrub is home to about one hundred plant species, about one-third endemic to this ecoregion (Fergus 1993). An even higher number of invertebrates (forty-five plus) are endemic to the harsh desertlike conditions. Perhaps the flagship animal species of this part of the ecoregion is the scrub jay (*Aphelocoma coerulescens coerulescens*), a distinct race recently elevated to species rank, which has evolved an interesting social system in the scrub islands. High genetic variability among scrub species is likely, as many are poor dispersers.

The largest blocks of scrubs are inland scrubs, and the best studied is the Lake Wales Ridge, home to the Archbold Biological Station. Many of the vertebrate and invertebrate species endemic to the scrub have been studied here. Other islands of scrub are less well studied and under great threat from development. Scrub is dependent upon fire to maintain its unique assemblage of species. Left alone, scrub patches will burn at five- to forty-year intervals. Scrub fires are mediated by the surrounding landscape, which can either be more fire prone than scrub (e.g., high pine) or nonflammable (e.g., swamps). Fires can be spectacular; the 1935 fire in the Ocala National Forests burned 140 km² in only four hours. Scrub seems to be remarkably resilient and persistent in the face of these natural disturbances. The absence of the driving disturbance event, fire, may ultimately spell the greatest threat to the persistence of scrub.

Conservation Status

Habitat Loss and Degradation

Today only about 10–15 percent of the scrub habitat remains, the rest replaced by citrus groves and housing developments. There has been a relatively uniform loss of scrub habitat across the region, but the most severe loss is in the south (Lake Wales Ridge), which also has the highest biodiversity value. Along with increased urbanization and the planting of citrus groves, fire suppression is an important factor that has led to the conversion of scrub. Proximity of housing developments and other facilities near scrub preclude allowing fires to burn naturally or at all. Specifically, conversion has most affected sand pine scrub, sand pine islands, and scrubby flatwoods. A review of the current conservation status of remaining scrub can be found in Myers and Ewel (1990b).

Remaining Blocks of Intact Habitat

The largest block of scrub habitat is the Ocala National Forest and the Archbold Biological Station, both part of the Central Ridge section of this ecoregion. In decreasing order of size, other blocks include (all blocks are found in southern Florida):

- Jonathan Dickinson State Park (coastal scrub section)
- Avon Park Air Force Base (Central Ridge section)
- Arbuckle State Forest (Central Ridge section)
- Cedar Key Scrub State Preserve (coastal scrub section)
- Merritt Island/Cape Canaveral (coastal scrub section)

A more complete list of remaining blocks is listed in Myers and Ewel (1990b, 142–143).

Degree of Fragmentation

The Lake Wales Ridge section is not naturally highly fragmented, but residual sections have become fragmented by citrus and residential development. Corridors are not as relevant for scrub as for other ecoregions because scrub occurs in naturally disjunct blocks of habitat, although some species require relatively undisturbed intervening habitat for their effective dispersal. Fragmentation and urbanization reduce the ability to maintain fire susceptibility. Existing corridors and linkage zones would be restorable if intense burning is permitted.

Degree of Protection

The most promising development is that relatively large sums of money have been made available for conservation of scrub by the state of

Florida through the Conservation and Recreation Lands (CARL) Acquisition program and other avenues. Although scrub occurs in national forests, these areas are not being managed in a manner that promotes maintaining scrub over the long term. A number of important areas on private land have been proposed for purchase under habitat conservation programs. Overall, the variation in scrub habitat is adequately represented by many small reserves.

Types and Severity of Threats

The major conversion threats in this ecoregion are citrus production and residential development. Fire suppression remains the most pervasive degradation threat. Unlike other Florida ecoregions, there are no serious exotic species problems.

Suite of Priority Activities to Enhance Biodiversity Conservation

- Take action to properly conserve the important remnants of scrub habitat that have been identified, especially on the Lake Wales ridge.
- Thoroughly inventory remaining scrub habitats.
- Promote greater national awareness of the biological value of scrub.

Conservation Partners

For contact information, please see appendix G.

- Archbold Biological Station
- Florida Natural Areas Inventory
- The Nature Conservancy
- The Nature Conservancy of Florida
- U.S. Fish and Wildlife Service (has planned sand pine refuge)

Relationship to Other Classification Schemes

This ecoregion is not delineated by Bailey or Omernik. Boundaries were taken from Küchler (1964) and Myers and Ewel (1990b).

Prepared by E. Dinerstein, A. Weakley, R. Noss, R. Snodgrass, and K. Wolfe

Focus on the Great Plains Grasslands

North American visitors to the great savannas and grasslands of East Africa return home filled with awe after witnessing the spectacular assemblages of large grazing mammals and the abundant large predators. The magnificent drama of large mammal migrations across unbroken wild vistas is an increasingly rare phenomenon worldwide. Less than two hundred years ago, huge populations of grazing ungulates offered a similar spectacle over the vast extent of the Great Plains of North America. The presettlement population of American bison or buffalo (*Bison bison*) was estimated to be between 30 and 70 million. Bison were the prey base for North America's natural predators—wolves and grizzly bears—and were also at the center of Native American tribal cultures.

As a backdrop to the large mammal migrations was a rich temperate grassland fauna and flora. The grasses that flourished in these habitats were clearly adapted to large mammalian grazers and the ecological effects of trampling and fire. Rather than relying solely on sexual reproduction and formation of seeds, perennial grasses also spread via underground rhizomes, plant tissues unaffected by thundering hooves. In turn, buffalo and other wild grazers and predators were able to prosper in the prairie ecosystem despite its wide temperature swings and frequent droughts, blizzards, wildfires, and tornados. Bison, through their grazing behavior, wallowing, and migrations, were one of the keystone species of this ecosystem. Others were black-tailed prairie dogs (*Cynomys ludovicianus*) in the western portions of the Great Plains, which formed extensive underground colonies and whose feeding behavior and diggings maintained diverse assemblages of prairie plants, and pocket gophers (*Geomys bursarius*), whose burrowing behavior turned over soil in large areas of the Great Plains.

Today, only minute fragments of this glorious ecosystem remain. The large herds of buffalo are gone, as are the wolves and grizzly bears and other large ungulates such as elk. Virtually all of the tallgrass prairies, relatively moist grasslands on deep organic soils, have been converted to farmland. Most of the drier, mixed-grass prairies have also been converted to farmland or planted for hay. Much of the still drier short-grass prairies are now grazed intensively by domestic livestock. Woody plants have invaded much of the grasslands as a result of fire suppression, water management schemes, and the planting of farm and ranch shelterbelts. Thus, in nearly all of the Great Plains, woody plants are now far more abundant than before the advent of farming. It is estimated that non-native species now account for 30–60 percent of all grassland species. This estimate includes agricultural crops and many aggressive invading weeds from Eurasia. Overall, some experts estimate that less than 1 percent of the original grasslands remains unaffected by human activities (Klopatek et al. 1979). Taken as a whole, the complex of Great Plains ecoregions, particularly the more mesic eastern units, represents some of the most altered ecoregions on earth.

The conversion and alteration of this habitat type are staggering when one considers the original extent of the Great Plains in North America. Grasslands are the largest of the vegetation formations found in North America, most occurring in the Great Plains. Grasslands once ranged over almost 3 million km² of the continental United States (39 percent) (Küchler 1964) and a much smaller area of Canada and Mexico (about 500,000 and 200,000 km², respectively) (Rowe 1972; Rzedowski 1978).

For the purpose of this analysis, the Great Plains grasslands can be divided into seven distinct grassland ecoregions bordered by three grassland-savanna transition zones. Each is described below. Strong longitudinal gradients in precipitation, a result of the rain-shadow effect of the Rocky Mountains, persist across the Great Plains. The western edge receives about half the rainfall as the eastern and southeastern edge, which in turn grade into oak-hickory savanna. A strong latitudinal gradient also prevails with more severe winters and shorter growing seasons in the northern ecoregions than in the southern units. The dominance of grass species derived from the cool temperate zone vs. warm temperate areas influences composition and further delineates ecoregions. Finally, the composition of grasslands also affects structure. The tallest grasslands occur on the more mesic sites and the shorter grasslands on the drier areas.

Plant species richness in Great Plains grasslands is correlated with increased rainfall during the growing season, increased topographic relief, and low human impact. In general, the relatively recent history of North American grasslands restricts the level of plant endemism to quite low levels in comparison with other habitat types. Endemism in animal taxa is also relatively low. Ten bird species are endemic to the complex of ecoregions that constitutes the Great Plains, and seven of these have shown major declines over the past quarter century (Knopf 1995). Grassland nesting birds in general have shown a steeper decline on a continental scale than even the more heralded neotropical migrants of the forested ecosystem (see essay 11). The Prairie Potholes region, an amalgam of three different Great Plains ecoregions, once supported vast populations of breeding waterfowl. Populations for five of these species are at their lowest levels since the early 1970s (Shaffer and Newton 1995). Native prairie fishes have also suffered declines in recent years (Willson 1995).

Despite being the most converted of all the North American MHTs, the restoration of the Great Plains grasslands and its distinct fauna has become a topic of great popular and research interest. Across the Great Plains, experiments are underway to replace domestic cattle operations with buffalo ranching. Other experiments use controlled burning, among other techniques, to restore grassland floras. Without question, the single greatest threat to most native grasslands in the western Great Plains has been unsustainable levels of cattle grazing. While there is some controversy surrounding the impact of stocking rates and the idea that some species perform better in response to grazing, it seems clear that some grassland species are tolerant of grazers at appropriate densities, while other species do poorly under any grazing regime (see Bock and Bock 1995). In the eastern Great Plains, habitat destruction has been the major threat historically, but now threats arise from a variety of factors (see individual ecoregions for a more detailed discussion of this topic).

The challenge for the future will be to devise grazing regimes appropriate to maintaining a mosaic of habitats that allow both grazing-tolerant species and grazing-intolerant species to survive in suitably sized landscapes under restoration management. Bock and Bock (1995) offer three important suggestions for restoration that apply to all of the Great Plains ecoregions described below.

First, continue a modified Conservation Reserve Program that reduces the risk that grasslands currently being kept out of cultivation will be subjected to the plow. Furthermore, stop the planting of non-native species in fields under the CRP program. This practice does nothing to restore the Great Plains flora.

Second, Bock and Bock (1995) recommend dedicating a portion of each federal grazing lease as a livestock exclosure. At least 860,000 km² of public federal land is being grazed by domestic livestock in seventeen western states (Sabadell 1982). They prescribe creation of a system of federal livestock exclosures whereby 20 percent of each parcel leased to a livestock grazer would be set aside as permanently ungrazed reserve. This practice would provide habitat over more than 150,000 km² of previously unavailable habitat for plants and animals intolerant of large ungulate grazing. Bock and Bock describe a series of tradeoffs to maintain higher grazing levels on appropriate leased lands as compensation. Much restoration would have to occur within the exclosures.

The third proposal is to remove all livestock from U.S. national grasslands. These areas are divided among nineteen administrative units, seventeen of which occur in the western Great Plains from Montana to Texas. Together, they account for more than 15,000 km², and some parcels are large enough to maintain populations of bison and pronghorn antelope. The remainder are mixed into a matrix of private landholdings.

We add two more proposals: (1) Create a certification program for beef producers that set aside portions of land as grazing exclosures. This program could be similar to others that focus on marketing of sustainably harvested tropical hardwoods. (2) Create large-scale bison-ranching cooperatives where ranch owners would share in profits proportionally to the amount of land they contribute to the cooperative.

Restoration experiments and the aesthetic desire to reclaim some fraction of the Great Plains come at a time when human demographics and economic and social conditions are shifting the debate from an emotional longing to a pragmatic argument. For example, the Great Plains include about 20 percent of the land mass of the continental United States, but only 2 percent of the population. This population is declining as a result of an emigration shift from farms and ranches to urban centers. The decline in social, educational, and health services in sparsely populated areas will likely drive this shift even more. In some areas of the Great Plains where restoration efforts are underway, nature tourism and game ranching have the potential to become the major income earners in otherwise economically depressed areas. Perhaps the era of restoration of North America's ecoregions will eventually include significant parts of the Great Plains.

Key Number:	**53**
Ecoregion Name:	**Palouse Grasslands**
Major Habitat Type:	**3.1 Temperate Grasslands/ Savanna/ Scrub**
Ecoregion Size:	**46,826 km²**
Biological Distinctiveness:	**Regionally Outstanding**
Conservation Status:	**Snapshot—Endangered Final—Endangered**

Introduction

Grasslands and savannas once covered large areas of the intermountain West, from southwest Canada into western Montana (Sims 1988). Today, areas like the great Palouse prairie of eastern Washington, northeastern Oregon, and northwestern Idaho are virtually gone. The Palouse, formerly a vast expanse of native blue-bunch wheatgrass (*Agropyron spicatum*), Idaho fescue (*Festuca idahoensis*), and other grasses, has been plowed and converted to wheat fields or is covered by cheatgrass (*Bromus tecturum*) and other exotic plant species (Noss and Peters 1995).

The Palouse lies in the rain shadow of the Cascades and has a generally semiarid climate. This climate is similar to that of the annual grasslands of California, yet the Palouse historically resembled the mixed-grass vegetation of the central grasslands, except for the absence of short grasses. Such species as *Agropyron spicatum*, *Festuca idahoensis*, and *Elymus condensatus*, and the associated species *Poa scabrella*, *Koeleria cristata*, *Elymus sitanion*, *Stipa comata*, and *Agropyron smithii* originally dominated the Palouse prairie grassland (Sims 1988).

Fire was a major force in shaping the Palouse. For thousands of years, Native Americans set periodic, cool-burning fires that did not damage perennial grasses. Without frequent burning to reduce fuel levels, conditions were ideal for rare but intense fires that destroyed the native perennial species and allowed exotic grasses and annual forbs to invade. Excessive grazing also resulted in the demise of many of the perennial grasses (Sims 1988).

The Palouse prairie is now intensive agricultural land with patches of shrub-steppe grassland. Once a rich ecosystem year-round, the species that invaded the Palouse following the increase in grazing and the change in fire regime transformed these grasslands to seasonal rangelands.

Biological Distinctiveness

Except for the presence of shrubs, the Palouse grassland resembles the Great Plains shortgrass prairie. The once dominant species, however— bluebunch wheatgrass, Idaho fescue, and bluegrass—are distinctive.

Conservation Status

Habitat Loss and Degradation
Conversion to agriculture has destroyed more than 99 percent of the Palouse grasslands.

Remaining Blocks of Intact Habitat
Only two relatively large blocks of more or less intact habitat remain:

- Hell's Canyon National Recreation Area— eastern Washington, 795 km^2
- Coulee Dam National Recreation Area— eastern Washington, 279 km^2

Degree of Fragmentation
The Palouse Grasslands are highly fragmented. Remaining patches are tiny, with effectively no connectivity in most areas and little core habitat due to edge effects. The individual fragments and clusters that remain are highly isolated, and the intervening landscape precludes dispersal for most taxa.

Degree of Protection
Two protected areas in this ecoregion exceed 250 km^2 (see above).

Types and Severity of Threats

The Palouse does not face serious threats from agriculture because nearly all of the habitat has already been converted. Degradation of the remaining fragments continues to be a problem, however, and there are still moderate levels of wildlife exploitation.

Suite of Priority Activities to Enhance Biodiversity Conservation

Given that nearly all of this ecoregion has already been destroyed, conservation activities are largely futile. Acquisition and preservation of the remaining fragments should nevertheless be a priority, as should restoration of any abandoned agricultural land.

Conservation Partners

For contact information, please see appendix G.

- Idaho Conservation Data Center
- The Nature Conservancy of Washington
- The Nature Conservancy, Western Region
- The Nature Conservancy of Idaho
- The Nature Conservancy of Oregon
- Oregon Natural Heritage Program
- Washington Natural Heritage Program

Relationship to Other Classification Schemes

Bailey (1994) includes the Palouse in the Great Plains–Palouse Dry Steppe Province. Omernik's (1995b) Columbia Basin ecoregion extends somewhat farther to the west, whereas we classify the area west of the Palouse as part of the Snake/Columbia Shrub Steppe [75].

Prepared by R. Noss, J. Strittholt, G. Orians, and J. Adams

Key Number:	**54**
Ecoregion Name:	**California Central Valley Grassland**
Major Habitat Type:	**3.1 Temperate Grasslands/ Savanna/Shrub**
Ecoregion Size:	**48,965 km²**
Biological Distinctiveness:	**Regionally Outstanding**
Conservation Status:	**Snapshot—Critical**
	Final—Critical

Introduction

The valley runs northwest to southeast for 430 miles in central California, paralleling the Sierra Nevada Range to the east and the coastal ranges to the west (averaging seventy-five miles apart) and stopping abruptly at the Tehachapi Range in the south. Two rivers flow from opposite ends and join around the middle of the valley to form the extensive Sacramento–San Joaquin Delta that flows into San Francisco Bay. Desert grasslands occur only in the southern end of the valley because of increasing aridity. The valley is ringed by oak woodlands and chaparral of the California Interior Chaparral and Woodland ecoregion [70].

Biological Distinctiveness

California's Great Central Valley once supported a diverse array of perennial bunchgrass ecosystems including prairies, oak-grass savannas, and desert grasslands, as well as a mosaic of riparian woodlands, freshwater marshes, and vernal pools. In its original state, it comprised one of the most diverse, productive, and distinctive grasslands in temperate North America.

Perennial grasses that were adapted to cool-season growth dominated the habitats. *Stipa pulchra* was particularly important, although *Stipa cerua, Elymus* spp., *Poa scabrella, Aristida* spp., *Koeleria cristata, Muhlenbergia rigins,* and *Melica imperfecta* occurred in varying proportions (Sims 1988). Most growth occurred in the late spring after winter rains and the onset of warmer and sunnier days. Interspersed among the bunchgrasses were a rich array of annual and perennial grasses and forbs, the latter creating extraordinary flowering displays during certain years. Some extensive mass flowerings of the California poppy (*Eschscholtzia californica*), lupines (*Lupinus* spp.), and purple owl clover (*Orthocarpus purpurascens*) still occasionally

occur in several areas and are now best known from Antelope Valley in the Tehachapi foothills (Schoenherr 1992).

Native grasslands supported several herbivores including pronghorn antelope (*Antilocarpa americana*), elk (including a valley subspecies, the Tule Elk [*Cervus elaphus nannodes*]), mule deer (*Odocoileus hemionus*), California ground squirrels, gophers, mice, hare, rabbits, and kangaroo rats. Several rodents are endemics or near-endemics to southern valley habitats including the Fresno kangaroo rat (*Dipodomys nitratoides exilis*), Tipton kangaroo rat (*Dipodomys nitratoides nitratoides*), San Joaquin antelope squirrel (*Ammospermophilus harrisi*), San Joaquin pocket mouse (*Perognathus inornatus*), and giant kangaroo rat (*Dipodomys ingens*). Predators once included grizzly bear, gray wolf, coyote, mountain lion, ringtail, bobcat, and the San Joaquin Valley kit fox (*Vulpes macrotis mutica*), another southern valley–foothill endemic.

The valley and deltas once supported enormous populations of wintering waterfowl in extensive freshwater marshes. Riparian woodlands acted as important migratory pathways and breeding areas for many neotropical migratory birds. Three species of bird are largely endemic to the valley, the surrounding foothills, and portions of the southern coast ranges, namely, the yellow-billed magpie (*Pica nuttalli*), the tri-colored blackbird (*Agelaius tricolor*), and Nuttall's woodpecker (*Picoides nuttalii*).

The valley contains a number of reptile and amphibian species, including several endemic or near-endemic species or subspecies such as the San Joaquin whipsnake (*Masticophis flagellum ruddocki*), the blunt-nosed leopard lizard (*Gambelia sila*), Gilbert's skink (*Eumeces gilberti*), and the giant aquatic garter snake (*Thamnophis couchii*).

Various invertebrates are known to be restricted to valley habitats. These include the delta green ground beetle (*Elaphrus viridis*), known only from a single vernal pool site, and the valley elderberry longhorn beetle (*Desmocerus californicus dimorphus*), found only in riparian woodlands of three counties.

Vernal pool communities occur throughout the valley in seasonally flooded depressions and contain many endemic plant species (e.g., solano grass [*Tuctoria mucronata*]). Several types are recognized, including valley pools in basin areas that are typically alkaline or saline, terrace pools on ancient flood terraces of higher ground, and pools on volcanic soils. Vernal pool vegetation is ancient and unique, with many habitat and local endemic species (Holland and Jain 1995). During wet springs, the rims of the pools are encircled by flowers that change in composition as the water recedes. Several aquatic invertebrates are restricted to these unique habitats, including a species of fairy shrimp and tadpole shrimp.

Riparian forests once bordered many of the valley's major rivers and their tributaries. Willows, western sycamore, box elder, Fremont cottonwood, and the valley oak (*Quercus lobata*) were dominant tree species. Some riparian forests and associated woodlands were up to 30 km wide along the lower reaches of the San Joaquin and Sacramento Rivers. These are unusual forests for California because the trees are deciduous in winter, perhaps because of the cool, foggy conditions in many parts of the valley (Barbour et al. 1993). The California hibiscus (*Hibiscus californicus*) and the valley elderberry longhorn beetle are restricted to limited areas of remaining woodlands. Many neotropical migratory birds use these forests for dispersal pathways or breeding habitat. The western yellow-billed cuckoo (*Coccyzus americanus*) and the yellow-breasted chat (*Icteria virens*), as well as several other riparian specialists, have all declined significantly in California due to loss of riparian forests.

The extensive rivers and lakes of the Central Valley supported vast freshwater marshlands dominated by rushes, bulrushes (tules), sedges, cattails, willows, and floating plants. The delta and several large lakes such as Tulare, Buena Vista, and Kern were bordered by vast marshes, which, in turn, supported huge populations of fish, waterfowl, shorebirds, and blackbirds. Tulare Lake was the largest lake west of the Mississippi River (Schoenherr 1992). The native fish fauna was continentally distinctive with four separate Chinook salmon runs and a number of endemic species (and higher taxa) such as the Tule perch, delta smelt, Sacramento blackfish, Sacramento splittail, Sacramento perch, and Sacramento sucker (P. Moyle, Professor of Wildlife Fisheries Biology, UC–Davis, pers. comm., 1996).

Conservation Status

Habitat Loss and Degradation

Virtually all Central Valley habitats have been altered. Introduced annual grasses now dominate grassland habitats. In most habitats native taxa make up less than 1 percent of the standing grassland crop. Agricultural development, urban expansion, alteration of hydrologic regimes and channelization, grazing by domestic livestock, fires, and introduced plants and animals have all contributed to the pervasive destruction of native habitats.

An estimated 11,310 km^2 (2.8 million acres), over 66 percent, of vernal pools have been destroyed with the most intact pools left on the higher terraces. Agriculture, conversion to pastureland, water diversion and channelization, and draining have all taken their toll on these unique habitats. The USFWS has just begun an ecosystem recovery plan for over forty species dependent upon vernal pools (R. F. Holland, pers. comm., 1996).

Out of 416 km^2 of remaining riparian woodlands, only about 40 km^2 or 1 percent of original riparian woodlands can be considered intact, down from an estimated 4,000 km^2 (Schoenherr 1992). Channelization, dams, clearing for pasture, flood control, alien plants, overgrazing by domestic livestock, fires, and logging have all taken their toll.

Intensive agricultural development has left few freshwater marshlands (less than 6 percent of their original extent), and those that are left are generally degraded and heavily managed for duck production, water impoundments, or runoff and effluent storage. Dams, channelization of rivers, and pollution continue to threaten the unusual and productive freshwater biodiversity of the region.

Remaining Blocks of Intact Habitat

Scattered large vernal pool sites are functionally intact, including vine plains and Jepson Prairie. The largest blocks of desert grassland and scrub are owned by oil companies in the southern portion of the valley. Small patches of riparian forests, woodlands, and marshlands are scattered throughout the valley, especially along the Sacramento River and some of its tributaries. The Kaweah Oaks Preserve near Visalia and the Creighton Ranch near Corcoran conserve remnant oak savanna (Johnston 1994). The Nature

Conservancy's Cosumnes River Preserve between Stockton and Sacramento conserves one of the best remaining examples of valley oak riparian woodlands. A few notable remnant freshwater marsh sites include the Creighton Ranch Reserve, a relict of Tulare Lake, Gray Lodge, and Butte Sink in the northern valley, portions of the delta, and various managed National Wildlife Refuges scattered throughout the valley. The huge Carrizo Plain natural area (3,180,000 acres), located just west of the valley proper but representative of the San Joaquin Valley, embraces extensive stands of saltbush scrub, desert grassland, alkali scrub, and wetlands.

Degree of Fragmentation

Remaining patches of relatively undisturbed native habitats are all severely fragmented and isolated. Loss of habitat linkages are likely to be most significant for species that rely on contiguous riparian woodlands for dispersal corridors, such as migratory warblers, cuckoos, and reptiles and amphibians. Populations of threatened mammals and reptiles in the southern desert grasslands are also effected by fragmentation. The foothill areas are more intact than the valley floor, but fragmentation has altered them as well.

Degree of Protection

A number of important vernal pool sites have been protected within small reserves, such as The Nature Conservancy's Pixley Reserve in Tulare County, Jepson Prairie Reserve near the Delta, and Vine Plains Preserve near Chico. Several of the larger blocks of desert grasslands in the southern portion of the valley are privately owned by oil companies that work with federal and state agencies and conservation organizations to manage the areas for biodiversity conservation. Consumnes River Preserve protects some of the last blocks of riparian woodland, marshland, and vernal pools in the valley. A system of managed wetlands owned by private landholders and state and federal agencies protects large populations of wintering waterfowl of the Pacific Flyway, as well as associated marshland species such as the giant garter snake.

Types and Severity of Threats

Remaining native habitats are threatened by continuing habitat clearance for agriculture, alteration of hydrologic regimes, dams, channelization, fires, grazing by domestic livestock, and alien species. Over 526 introduced plant species exist in California, and many of these occur in the annual grasslands of the Central Valley (Heady 1995). Several federally endangered species occur in the valley, including the San Joaquin kit fox, the blunt-nosed lizard, the Delta green ground beetle (*Elaphrus viridis*), the valley elderberry longhorn beetle, the Delta smelt, and a number of vernal pool plant species. Populations of the tri-colored blackbird are also declining precipitously in recent years due to loss of breeding habitat and changes in agricultural practices. Many amphibian species show dramatic declines associated with habitat loss and introduced predators (Fisher and Shaffer 1996). Extensive riparian forests and woodlands have been destroyed by decades of tree cutting, channelization, and flood-control projects. Salinization, toxic runoff, and erosion from ecologically unsound agricultural practices increasingly degrade habitats.

Suite of Priority Activities to Enhance Biodiversity Conservation

Strict protection and restoration of the last remnants of native communities are needed for all valley habitats.

- Extensive restoration of marshlands and riparian woodlands. Introduced annual grasses dominate grassland and savanna communities; intensive management is required for possible restoration of the original bunchgrass prairies.
- Protect and restore linkages between lowland valley habitats and foothill habitats. The Nature Conservancy has identified 385 priority sites in the northern half of the valley.

Conservation Partners

For contact information, please see appendix G.

- Bureau of Land Management
- California Department of Fish and Game

- California Native Grass Association
- California Native Plant Society
- California Oak Foundation
- The Nature Conservancy
- The Nature Conservancy, California Region
- Sacramento River Preservation Trust
- Sierra Club

Relationship to Other Classification Schemes

The Central Valley ecoregion was originally based on Omernik's (1995b) Central California Valley ecoregion (7), although boundaries in some areas have been revised by Bob Holland, an expert on vernal pools and valley vegetation. This ecoregion generally corresponds to Bailey's (1994) Great Valley Section (262A) and Küchler's (1975) California Prairie, Blue Oak–Digger Pine, and riparian forest vegetation classes in the region.

Prepared by D. Olson and R. Cox

Key Number:	**55**
Ecoregion Name:	**Canadian Aspen Forest and Parklands**
Major Habitat Type:	**3.1 Temperate Grasslands/ Savanna/Shrub**
Ecoregion Size:	**394,376 km²**
Biological Distinctiveness:	**Nationally Important**
Conservation Status:	**Snapshot—Endangered Final—Critical**

Introduction

This ecoregion stretches in an arc from the Manitoba–North Dakota border in the east to central Alberta in the west with a disjunct occurrence in northwestern Alberta crossing the British Columbia border in the Peace River area.

This ecoregion is classified primarily as having a subhumid low-boreal ecoclimate, which distinguishes it from the warmer, drier areas to the south and the cooler boreal forests to the north. It also has a transitional grassland ecoclimate. Summers are short and warm, and winters are cold and long. The mean annual temperature ranges from 0.5°C to 2.5°C; the mean summer temperature ranges from 13°C to 16°C, and the mean winter temperature ranges from –14.5°C to –12.5°C. The Peace River Lowland area of the region generally represents the coolest temperatures for each range, while the Southwest Manitoba Uplands region represents the warmest temperatures. Mean annual precipitation ranges from 375 mm to just under 700 mm, with the driest area being in the northwestern section of the Mid-Boreal Lowlands (ESWG 1995). Fire is probably the most important natural disturbance regime.

Much of the region is underlain by Cretaceous shale and covered by undulating to kettled, calcareous, glacial till with significant areas of level lacustrine and hummocky to ridged fluvioglacial deposits. Associated with the rougher hummocky glacial till are a large number of small lakes, ponds, and sloughs occupying shallow depressions. The gently undulating or sloping lands associated with the Peace River are underlain by Tertiary sandstone and shale strata, covered mainly by imperfectly drained clayey lacustrine sediments with some fine-textured tills and sandy fluvioglacial deltas associated with the major river systems. Finally, the Interlake Plain area is underlain by flat-lying Paleozoic limestone and covered by broadly ridged, extremely calcareous glacial till and by shallow, level, lacustrine sands, silts, and clays (ESWG 1995).

Biological Distinctiveness

Vegetation in this ecoregion is characterized by a cover of quaking aspen (*Populus tremuloides*), with secondary quantities of balsam poplar (*P. balsamifera*), together with an understory of mixed herbs and tall shrubs. White spruce (*Picea glauca*) and balsam fir (*Abies balsamea*) are the climax species but are not well represented because of fires. Jack pine (*Pinus banksiana*) stands may be present on drier, sandy sites. Poorly drained sites are usually covered with sedges (*Carex* spp.), willow (*Salix* spp.), some black spruce (*P. mariana*), and tamarack (*Larix laricina*). In the Turtle Mountain and Spruce Woods areas (TEC 163, 164), quaking aspen dominates with secondary quantities of balsam poplar, although white spruce and balsam fir are the climax species if fires do not occur frequently (ESWG 1995).

Characteristic wildlife include moose (*Alces alces*), white-tailed deer (*Odocoileus virginianus*), black bear (*Ursus americanus*), wolf (*Canis lupus*), beaver (*Castor canadensis*), coyote (*Canis latrans*), marten (*Martes americana*), mink (*Mustela vison*), red fox (*Vulpes fulva*), snowshoe hare (*Lepus americanus*), northern pocket gopher (*Thomomys talpoides*), Franklin's ground squirrel (*Citellus franklinii*), sharp-tailed grouse (*Tympahuchus phasianellus*), ruffed grouse (*Bonasa umbellus*), black-billed magpie (*Pica pica*), cormorant (*Phalacrocorax* spp.), gull (*Larus* spp.), tern (*Sterna* spp.), American white pelican (*Pelecanus erythrorhynchos*), and many neotropical migrant bird species (ESWG 1995).

Of ecological significance, the Aspen Parkland and Forests ecoregion represents the most extensive boreal-grassland transition in the world. This ecoregion contains the northern-most breeding distribution for many warbler species (Parulinae) and has some of the most productive and extensive waterfowl breeding habitat on the continent. White-tailed and black-tailed deer (*Odocoileus virginianus* and *O. hemionus*) reach their northern continental limit here.

Conservation Status

Habitat Loss and Degradation

It is estimated that less than 10 percent of the natural habitat in this region remains intact. Of the 90 percent disturbed, most has been converted to agricultural cropland, including canola (*Brassica napsus*), alfalfa (*Medicago sativa*), and wheat (*Triticum aestivum*). Cultivation of land for grazing purposes is also widespread. In those parts of the ecoregion where forest cover was historically more widespread, forest harvesting continues in the remaining farm woodlots.

Remaining Blocks of Intact Habitat

Few significant blocks of habitat remain. These include:

- Moose Mountain Provincial Park—south-eastern Saskatchewan, approx. 266 km^2
- Elk Island National Park—central Alberta, 194 km^2
- Bronson Forest—Saskatchewan
- Wainwright Military Reserve—Alberta

Degree of Fragmentation

There are very high levels of habitat fragmentation due to agriculture. This is particularly true for most forest species.

Degree of Protection

- Moose Mountain Provincial Park—Saskatchewan, 265 km^2
- Spruce Woods Provincial Park (backcountry zones)—southern Manitoba, 201 km^2
- Elk Island National Park—Alberta, 194 km^2
- Turtle Mountain Provincial Park (backcountry zones)—southwestern Manitoba, 118 km^2
- Rumsey Ecological Reserve—south-central Alberta, 34 km^2
- Sand Lakes Natural Area—central Alberta, 28 km^2
- Wainwright Dunes Ecological Reserve—eastern Alberta, 28 km^2
- Jack Pines Natural Area—central Alberta, 18 km^2
- Silver Valley Ecological Reserve—western Alberta, 18 km^2

Types and Severity of Threats

Ongoing agricultural conversion of remnant patches of natural habitat, often for grazing and haying, is an important threat, as is logging for aspen in many of the remaining forested areas. In fact, aspen pulpwood harvests are expanding in the ecoregion. Extensive use of agricultural pesticides is a major concern for wildlife populations. Predator control is still occurring in some areas and is becoming more problematic with the start-up of game farms. Seismic oil and gas exploration is widespread in many areas.

Suite of Priority Activities to Enhance Biodiversity Conservation

- Protect and restore habitat throughout the ecoregion. Some specific sites include:
  - Porcupine Forest, Bronson Forest, and Nisbet Forest in Saskatchewan
  - Major Sand Hills throughout the ecoregion
  - Shilo Defense Base in Manitoba
  - Rumsey Block of Alberta
  - Beaverhill Lake in Alberta

- Upgrade protection standards for select wildlife management areas in Manitoba.
- Establish Manitoba Lowlands National Park.

Conservation Partners

For contact information, please see appendix G.

- Alberta Wilderness Association
- Brandon Naturalists' Society
- Canadian Parks and Wilderness Society, Calgary-Banff Chapter
- Canadian Parks and Wilderness Society, Edmonton Chapter
- Critical Wildlife Habitat Program
- Ducks Unlimited Canada
- Endangered Spaces Campaign, Manitoba
- Endangered Spaces Campaign, Saskatchewan
- Federation of Alberta Naturalists
- Friends of Elk Island Society
- Friends of Prince Albert National Park
- Lower Fort Garry Volunteers
- Manitoba Heritage Habitat Corporation
- Manitoba Naturalists Society
- Meewasin Valley Authority
- Nature Saskatchewan
- The Nature Conservancy, Alberta
- The Nature Conservancy, British Columbia
- Red Deer River Naturalists
- Resource Conservation Manitoba
- Saskatchewan Forest Conservation Network
- Sierra Club
- TREE (Time to Respect the Earth's Ecosystems)
- Watchdogs for Wildlife
- The Wildlife Society
- World Wildlife Fund Canada

Relationship to Other Classification Schemes

The Canadian Aspen Forest and Parklands weave across four Canadian provinces and encompass eight terrestrial ecoregions: the Peace Lowland, Western Boreal, Boreal Transition, Interlake Plain, Aspen Parkland, and Southwest Manitoba Uplands (TEC ecoregions 138, 143, 149, 155, 156, 161, 163, and 164). These ecoregions lie in both the Boreal Plains Ecozone and the Prairies Ecozone (ESWG 1995). The Boreal

sections are Manitoba Lowlands, Aspen-Oak, Aspen Grove, Mixedwood, and Lower Foothills (B.15, B.16, B.17, B.18a, and B.19a). This ecoregion also overlaps some of the grasslands adjacent to the south of this ecoregion (Rowe 1972).

Prepared by S. Primm, J. Shay, S. Chaplin, K. Carney, E. Dinerstein, P. Sims, A. G. Appleby, R. Usher, K. Kavanagh, M. Sims, and G. Mann

Key Number:	**56**
Ecoregion Name:	**Northern Mixed Grasslands**
Major Habitat Type:	**3.1 Grasslands, Savannas, and Shrublands**
Ecoregion Size:	**218,897 km²**
Biological Distinctiveness:	**Nationally Important**
Conservation Status:	**Snapshot—Critical Final—Critical**

Introduction

The Northern Mixed Grasslands cross Alberta, Saskatchewan, and southern Manitoba in Canada and run from North Dakota to northern Nebraska in the United States, covering about 270,000 km². Essentially an ecotone, this transitional belt separates the three tallgrass prairie ecoregions to the east (Northern, Central, and Flint Hills Tall Grasslands) from the Northwestern Mixed Grasslands. In addition, the ecoregion separates the shortgrass prairie to the south from the cooler boreal forests to the north. This ecoregion is intermediate in growing-season length, vegetation structure, and rainfall from the drier units to the west and the more mesic tallgrass prairies to the east. It is separated from the Central and Southern Mixed Grasslands by climatic factors, the southern unit having a much warmer climate and a longer growing season. In the Canadian portion of this ecoregion, the mean annual temperature is approximately 3°C; mean summer temperature is 15.5°C, and mean winter temperature is −10°C. This ecoregion is considered to be a transitional grassland ecoclimate with semiarid moisture conditions. Mean annual precipitation is 325–450 mm (ESWG 1995).

This ecoregion is arguably the most disturbed among all grassland ecoregions, with only a few remnant patches remaining, none of which is considered to be intact. Conversion to agriculture has completely changed the landscape in an ecoregion that once supported large populations of bison (*Bison bison*). Restoration of representative examples of this ecoregion will be a major challenge.

Biological Distinctiveness

The mixed-grass prairie was recognized by Clements (1920) as a mixture of the tallgrass and shortgrass prairies. The dominant grasses found here include grama (*Bouteloua gracilis*), little bluestem (*Schizachrium scoparium*), needle-and-thread grass (*Stipa comata*), wheatgrass (*Agropyron smithii*), *Carex filifolia*, junegrass (*Koeleria cristata*), and *Poa secunda*. Küchler (1964) classified the dominant vegetation as wheatgrass-bluestem-needlegrass. Essentially, the floristic composition of this unit was determined largely by drought and grazing pressure, with fire having a less prominent role than in the tallgrass prairie ecoregions.

Farther north in the ecoregion, the transitional grassland ecoclimate supports a vegetation of quaking aspen (*Populus tremuloides*), bur oak (*Quercus macrocarpa*) groves, mixed tall shrubs, and intermittent fescue grasslands. Generally, quaking aspen and shrubs occur on moist sites, while bur oak and grass species occur on increasingly drier sites. The southeast portion of the ecoregion tends to be warmer and drier. Remaining native vegetation in this portion of the ecoregion is dominated by spear grass and wheatgrass. In addition, local saline deposits support alkali grass, wild barley, red samphire, and sea blite. In the separated Cypress Upland, fescue and wheatgrass grow below 1,000 m, and mixed montane forests of lodgepole pine, deciduous trees, and shrubs grow at higher elevations. Larkspur (*Delphinium* spp.), death camas (*Zigadenus elegans*), and wild lupine (*Lupinus* spp.) are found here and nowhere else in the prairies.

The topography is broken by many glacial pothole lakes, making this ecoregion the most productive breeding area for waterfowl in the United States. Other wildlife characteristic of the moist mixed grassland are black-tailed and white-tailed deer (*Odocoileus hemionus* and *O. virginianus*), pronghorn antelope (*Antilocapra americana*), coyote (*Canis latrans*), short-horned lizard (*Phrynosoma douglassi*), western rattlesnake (*Crotalus viridis*), rabbit (*Sylvilagus* spp.), ground squirrel (*Spermophilus saturatus*), and sage grouse (*Centrocercus urophasianus*) (ESWG 1995). Yellow-rumped warbler (*Dendroica coronata*) is also found only in this part of the prairies. Bison were once a common feature of this ecoregion. Grazing by ungulates, fire, and drought were major sources of disturbance.

The ecoregion provides continentally significant waterfowl production and is a major staging area. It is estimated that up to 80 percent of the wetlands, however, have been lost or degraded. The Cypress Uplands (Alberta-Saskatchewan), which are believed to have escaped the last glaciation, are located in this ecoregion. A large number of disjunct species populations of both flora and fauna are found here.

Conservation Status

Habitat Loss

Virtually no major areas of intact habitat remain. It is estimated that more than 75 percent of the ecoregion has been heavily altered. Agricultural conversion of natural prairie is the prime reason for the high rate of habitat loss. Most of the less altered patches of habitat, however, contain representatives of the native flora.

Remaining Blocks of Intact Habitat

No completely intact blocks of habitat remain. Restoration of some areas is just beginning. The following sites may be characterized more as large fragments or clusters of fragmented habitat.

- Turtle Mountain—southern Manitoba, on the border of the ecoregion
- Prairie Coteau—northeastern South Dakota; only part, also included in the central tallgrasslands
- Last Mountain Lake—southern Saskatchewan, an important prairie restoration site and a huge migratory bird stopover area
- Missouri WSR—324 km² of habitat (unprotected)
- J. Clark Salyer National Wildlife Refuge—north-central North Dakota, 269 km²
- Pembina Gorge—northeastern North Dakota
- Prairie Coteau
- Sheyenne Delta—North Dakota

- Cypress Hills/Uplands—Alberta and Saskatchewan
- Dundurn Military Reserve—Saskatchewan

Degree of Fragmentation

As stated above, the small patches of altered habitat worthy of conservation consideration are essentially isolated and highly fragmented.

Degree of Protection

Few sites are protected. These include:

- Rumsey Ecological Refuge—Alberta
- Cypress Hills Provincial Park—Saskatchewan, 183 km^2
- Rumsey South—Alberta, 125 km^2
- Douglas Provincial Park—Saskatchewan, 61 km^2
- Hand Hills Ecological Reserve—Alberta, 22 km^2
- Danielson Provincial Park—Saskatchewan, 21 km^2
- Blackstrap Provincial Park—Saskatchewan, 5 km^2
- Pike Lake Provincial Park—Saskatchewan, 5 km^2

Types and Severity of Threats

Virtually all of the ecoregion has already been converted or heavily altered. Introduction of prairie exotics and intensification of habitat conversion (including altered habitat to more heavily altered habitat) for agricultural purposes are major threats. Disturbance to major water-fowl populations is a concern.

Suite of Priority Activities to Enhance Biodiversity Conservation

There is a need to identify as many remnant habitat patches as possible and to review their status to determine priority areas for restoration. Opportunities include the Prairie Coteau area, further work on Last Mountain Lake (Saskatchewan), a review of Wildlife Habitat Protection lands, and lands held by Ducks Unlimited.

The following steps are recommended to restore representative examples of this ecoregion:

- Systematically identify remnants.

- Protect and restore these areas to prevent critical loss of this ecosystem.
- Design an ecoregion strategy based on bringing back small pieces of this unit.

Conservation Partners

For contact information, please see appendix G.

- Alberta Wilderness Association
- Canadian Parks and Wilderness Society, Calgary-Banff Chapter
- Critical Wildlife Habitat Program
- Ducks Unlimited Canada
- Endangered Spaces Campaign, Manitoba
- Endangered Spaces Campaign, Saskatchewan
- Federation of Alberta Naturalists
- Manitoba Heritage Habitat Corporation
- Manitoba Naturalists Society
- Meewasin Valley Authority
- The Nature Conservancy of Nebraska
- The Nature Conservancy, Alberta
- The Nature Conservancy, British Columbia
- The Nature Conservancy, Tallgrass Prairie Office
- Nature Saskatchewan
- Prairie Conservation Forum
- Resource Conservation Manitoba
- Society of Grassland Naturalists
- U.S. Fish and Wildlife Service
- World Wildlife Fund Canada

Relationship to Other Classification Schemes

The boundary for the Northern Mixed Grasslands is derived from Sims (1988). It corresponds to Omernik's (1995b) ecoregion 46 (Northern Glaciated Plains) in the United States and Küchler's (1975) unit no. 60 (wheatgrass, bluestem, needlegrass). The comparable Bailey (1994) sections south of the Canada-U.S. border are: 332A (Northeast Glaciated Plains) 332B (Western Glaciated Plains), 251B (North Central Glaciated Plains—extreme western part).

In Canada, most of this ecoregion is Moist Mixed Grassland (TEC ecoregion 157), surrounding Fescue Grassland. Separated from the larger portion of the ecoregion is the Cypress

Upland (TEC ecoregion 160), which is an outlier of the montane vegetative zone that occurs on the lower slopes of the Rocky Mountains (ESWG 1995). This ecoregion includes both grassland and boreal forest regions, specifically Aspen-Oak, Aspen Grove, and Northern Foothills Boreal forest sections (B.16, B.17, and B.19a) (Rowe 1972).

Prepared by J. Shay, G. Whelan-Enns, A. Appleby, R. Usher, K. Kavanagh, M. Sims, S. Chaplin, P. Sims, K. Carney, and E. Dinerstein

Key Number:	**57**
Ecoregion Name:	**Montana Valley and Foothill Grasslands**
Major Habitat Type:	**3.1 Temperate Grasslands/ Savanna/Shrub**
Ecoregion Size:	**81,666 km²**
Biological Distinctiveness:	**Nationally Important**
Conservation Status:	**Snapshot—Vulnerable Final—Endangered**

Introduction

The Montana Valley and Foothill Grasslands ecoregion occupies high valleys and foothill regions in the central Rocky Mountains of Montana and Alberta. The ecoregion occupies the Rocky Mountain Front, the uppermost flatland reaches of the Missouri River drainage, and extends into the Clark Fork–Bitterroot drainage of the Columbia River system.

The Canadian component of this ecoregion is characterized by undulating to rolling topography, and surface deposits are composed of loamy glacial till and clayey lacustrine deposits (ESWG 1995). It should be noted that there are smaller outliers of this region nested within some of the major river valleys (e.g., Bow River valley) west of the ecoregion that are not shown on the coarse-scaled North American map.

Located in the Chinook belt, this ecoregion is characterized by dry, warm summers and mild winters. Mean annual temperature is 3.5°C; mean summer temperature is 14°C, and mean winter temperature is –8°C. Annual precipitation is approximately 425 mm (ESWG 1995).

Biological Distinctiveness

The dominant vegetation type of this ecoregion varies somewhat but consists mainly of wheatgrass (*Agropyron* spp.) and fescue (*Festuca* spp.). Certain valleys, notably the upper Madison, Ruby, and Red Rock drainages of southwestern Montana, are distinguished by extensive sagebrush (*Artemisia* spp.) communities as well. This is a reflection of semiarid conditions caused by pronounced rain-shadow effects and high elevation. Thus, near the Continental Divide in southwestern Montana, the ecoregion closely resembles the nearby Snake-Columbia shrub steppe.

While other sites may experience similar rain-shadow effects, lower elevation, greater overall precipitation, and edaphic variations impede sagebrush growth. Along the Rocky Mountain Front, glaciated potholes in the foothills prairie create extensive wetlands and much more mesic grassland conditions than in the high valleys. The western segments of the ecoregion are within the range of moist airmasses from the Pacific.

In Canada, the grassland community is dominated by rough fescue, with lesser quantities of Parry oat grass, junegrass (*Koelaria* spp.), and wheatgrass (*Agropyron* spp.). Forbs are also abundant and include sticky geranium (*Geranium viscosissimum*), bedstraw (*Galium* spp.), yellow bean (*Thermopsis* spp.), and wheatgrass. Moist sites support shrub communities. Drier sites have an increased amount of needle-and-thread grass (*Stipa comata*). Trees are found only in very sheltered locations along waterways (ESWG 1995).

Historically, heavy grazing by native herbivores—mainly bison (*Bison bison*)—was a major influence on most of this ecoregion. Periodic fires were also part of the disturbance regime. Some characteristic wildlife species of the foothill grasslands include white-tailed deer (*Odocoileus virginianus*), pronghorn antelope (*Antilocapra americana*), coyote (*Canis latrans*), rabbit (*Sylvilagus* spp.), grouse (*Dendragapus* spp.), and ground squirrel (*Spermophilus saturatus*). The ecoregion supports endemic and relict fisheries: Westslope cutthroat trout (*Oncorhynchus lewisi clarki*), Yellowstone cutthroat (*O. clarki*), and fluvial arctic grayling (*Thymallus arcticus*), a relict species from past glaciation.

The moderating effects of strong chinook winds result in a high diversity of vegetation communities in close proximity to one another. Certain sites contain relatively high levels of species richness. The Centennial Valley of southwestern Montana supports 487 vascular plant taxa (Mahr 1996). This valley supports large breeding populations of trumpeter swans (*Cygnus bucinator*) and sandhill cranes (*Grus canadensis*). Owing to its highly productive wetlands and juxtaposition to nearby montane habitats, the Centennial Valley may also be a candidate site for grizzly bear (*Ursus arctos*) recolonization of grassland habitat. This valley and other undeveloped valleys provide critical linkage habitat for grizzlies and other species moving between separate mountain ranges. Such areas also provide critical seasonal range for ungulates like elk (*Cervus canadensis*) and bighorn sheep (*Ovis canadensis*).

Conservation Status

Habitat Loss and Degradation

Approximately 25 percent of the ecoregion remains as intact habitat. Less than 10 percent of the Canadian portion of this ecoregion is estimated to remain as intact habitat. Most of the ecoregion has been heavily altered. Domestic stock grazing, draining of wetlands, and conversion to row crops have been major anthropogenic changes to the ecoregion. Replacement of native species with exotic grasses and noxious weed invasions are serious problems.

More recently, development for residential homes has been a major threat to the ecoregion. Most of the tremendous population growth in the Rocky Mountains naturally ends up in this ecoregion, because it is largely privately owned and lends itself to home building.

Long-term environmental pollution from hard-rock mining is a major concern. The ecoregion contains the nation's largest Superfund toxic waste site, stretching along the Clark Fork and Blackfoot Rivers from the Continental Divide west to Missoula, Montana.

Remaining Blocks of Intact Habitat

There are still major tracts of relatively intact habitat in the ecoregion. Along the East Front of the Rockies, from Great Falls to near Calgary, it is possible to find large tracts of grassland that have experienced only grazing. Key areas include:

- Pine Butte Swamp Reserve, a TNC site near Choteau—southwestern Montana
- extensive wetland areas on the Blackfeet Indian Homeland
- Whaleback—Alberta

In the southwestern corner of Montana, there are vast expanses of undeveloped foothills and valley bottoms. In these remote areas, home building is not an immediate large-scale threat, although some key riparian areas are being subdivided and developed.

The Centennial Valley, the Big Hole Valley, and undeveloped parts of the Madison Valley are relatively pristine sites that are tremendously important to the health of adjacent montane systems.

Degree of Fragmentation

Rapid subdivision in some locales is contributing to wholesale fragmentation of habitat. With subdivision and home building come new beachheads for exotic plant invasions, whether intentional or not. Home building also leads to loss of travel routes and winter range for ungulates and other fauna. This process is most serious in the Paradise Valley, the Bitterroot Valley, and the Gallatin Valley around Bozeman. Similar trends may take hold in the Madison Valley in the near future.

Possible oil and gas development on the Rocky Mountain Front in Montana and Alberta could have direct habitat impacts as well as impair connectivity between mountains and grasslands for ungulates, large predators, and other species.

Degree of Protection

The following areas are protected:

- Red Rock Lakes National Wildlife Refuge and Wilderness—southwestern Montana
- Pine Butte Swamp Reserve—southwestern Montana
- National Bison Range—western Montana
- various BLM Wilderness Study Areas in southwestern Montana
- Ross Lake Natural Area—Alberta, 48 km^2

Types and Severity of Threats

Conversion of native habitat to home sites is a serious localized threat, although the boom appears to be leveling off. The loss of undeveloped riparian forest linkages across valley bottoms could have serious impacts on the long-term viability of sensitive species like the wolverine (*Gulo gulo*) and grizzly. Spread of exotic and noxious plant species could be a major long-term threat. Conversion of forest to grazing lands has a major impact on forest species.

Suite of Priority Activities to Enhance Biodiversity Conservation

• Identify priority sites. Three essential components should be protected: (1) rare plant communities and associations (dune communities, fen bogs, wetlands); (2) linkage habitat for large carnivores and other species; and (3) seasonal range for ungulates.
• Encourage and preserve the phenomenon of grizzlies using grassland or wetland habitat. These opportunities are present on the Rocky Mountain Front and in the Centennial Valley, in particular.
• Protect the Whaleback in Alberta.

Conservation Partners

For contact information, please see appendix G.

• Alberta Wilderness Association
• Canadian Parks and Wilderness Society, Calgary-Banff Chapter
• Castle Crown Wilderness Coalition
• Corporation for the Northern Rockies
• Federation of Alberta Naturalists
• Greater Yellowstone Coalition
• The Nature Conservancy of Montana
• The Nature Conservancy, Alberta
• Waterton Natural History Association
• The Wilderness Society
• World Wildlife Fund Canada

Relationship to Other Classification Schemes

The ecoregion boundary closely follows Omernik (1995b). Bailey (1994) lacks a distinctive counterpart to this ecoregion, since his classification of the same geographic area amalgamates the entire elevational gradient, with the exception of M332C, corresponding to the Rocky Mountain Front portion of this ecoregion. This ecoregion follows Küchler's (1975) 56 and 57; he classifies the Beaverhead, Red Rock, and Big Hole drainages as 49, distinguishing it markedly from the other Montana valleys.

The Canadian portion of the Montana Valley and Foothill Grasslands is made up of Fescue Grassland (TEC ecoregion 158) along the face of the Rocky Mountain foothills (ESWG 1995). This narrow region is characterized by Grassland, Aspen Grove Boreal Forest (B.17), and Douglas-fir and Lodgepole Pine Montane forests (M.5) (Rowe 1972).

Prepared by S. Primm, R. Usher, K. Kavanagh, M. Sims, and G. Mann

Key Number:	**58**
Ecoregion Name:	**Northwestern Mixed Grasslands**
Major Habitat Type:	**3.1 Grasslands, Savannas, and Shrublands**
Ecoregion Size:	**638,300 km²**
Biological Distinctiveness:	**Nationally Important**
Conservation Status:	**Snapshot—Endangered Final—Endangered**

Introduction

The Northwestern Mixed Grasslands is the largest grassland ecoregion in North America, covering almost 640,000 km². This ecoregion covers parts of southeastern Alberta and southwestern Saskatchewan, much of the area east of the Rocky Mountains, central and eastern Montana, western North and South Dakota, and northeastern Wyoming. Four major features distinguish this unit from other grasslands: the harsh winter climate, with much of the precipitation falling as snow; short growing season; periodic, severe droughts; and vegetation. Two environmental gradients determine species composition in mixed and shortgrass prairies: increasing temperatures from north to south and increasing rainfall from west to east. With increasing latitude, the shortgrass prairies take on an aspect more similar to mixed grass, such as in this ecoregion, where many cool-season species predominate (Sims 1988). Mean annual temperature in this part of the ecoregion ranges from 3.5°C but rises as high as 5°C in the west. Mean summer temperature is 16°C, and mean winter temperature is −10°C. In late summer, moisture deficits occur, due to low precipitation and high evapotranspiration. In general, this ecoregion has an arid grassland ecoclimate.

In pre-settlement times, drought, fire, and grazing were probably the major disturbance factors, with fire playing less of a role than in other grassland ecoregions. Today, virtually all of this ecoregion is converted to wheat farms or rangelands. However, the potential for large-scale restoration is perhaps greater in this ecoregion than in almost any other in North America. Attempts to reestablish populations of native black-footed ferrets (*Mustela nigripes*) and bison (*Bison bison*) are well underway.

Biological Distinctiveness

The dominant grass communities include grama-needlegrass (*Bouteloua* spp.–*Stipa* spp.) and wheatgrass (*Agropyron* spp.), and wheatgrass-needlegrass (Küchler 1964). Farther north into Canada the natural vegetation of this area is characterized by spear grass (*Poa annua*), blue grama grass, wheatgrass, and, to a lesser extent, junegrass (*Koelaria* spp.) and dryland sedge (*Carex* spp.). A variety of shrubs and herbs also occurs, but sagebrush (*Artemesia tridentata*) is most abundant, and on drier sites yellow cactus and prickly pear (*Opuntia* spp.) can be found. On shaded slopes of valleys and river terraces, scrubby aspen (*Populus* spp.), willow (*Salix* spp.), cottonwood (*Populus* spp.), and box-elder (*Acer negundo*) occur. Saline areas support alkali grass (*Puccinellia* spp.), wild barley (*Hordeum* spp.), greasewood (*Sarcobatus vermiculatus*), red samphire (*Salicornia rubra*), and sea blite (*Suaeda depressa*).

Prior to 1850, the Northwestern Mixed Grasslands contained some of the last extensive habitat for bison in the United States and Canada. Bison populations likely had a major impact on the structure and composition of this ecoregion. Their numbers are increasing again, as herds are growing on Native American lands

and private ranches. Black-footed ferrets were also once common here, and reintroductions should eventually capitalize on abundant prairie dog (*Cynomys* spp.) populations. Recovery efforts are underway for swift fox (*Vulpes velox*) in the northern part of the ecoregion.

The Northwestern Mixed Grasslands are surprisingly rich in mammals for an ecoregion so far north. Much of the bird fauna is composed of species typically associated with the prairie potholes: ferruginous hawk (*Buteo regalis*) and Swainson's hawk (*Buteo swainsoni*), golden eagle (*Aquila chrysaetos*), sharp-tailed grouse (*Tympahuchus phasianellus*) and sage grouse (*Centrocercus urophasianus*), mountain plover (*Charadrius montanus*), and clay-colored sparrow (*Spizella pallida*). Black-tailed and white-tailed deer (*Odocoileus hemionus* and *O. virginianus*), bobcat (*Lynx rufus*), and cougar (*Felis concolor*) are typical large mammals. Short-horned lizard (*Phrynosoma douglassi*) and western rattlesnake (*Crotalus viridis*) occur here as well.

The Northwestern Mixed Grasslands contain the largest breeding population of endangered piping plovers (*Charadrius melodus*) around alkaline lakes in all of North America, and important breeding populations for rails (*Rallus* spp.) and more threatened species of sparrows. Conservation efforts are increasing around threatened birds such as the burrowing owl (*Athene cunicularia*) and ferruginous hawk.

Conservation Status

Habitat Loss and Degradation

More than 85 percent of the ecoregion is now grazed by livestock or converted to dryland farming. In the Canadian portion, it is estimated that only about 2 percent of the ecoregion remains as natural, intact habitat. Considerable potential exists for habitat recovery in some areas by only partially grazing the lands. However, oil and gas development and the creation of road networks are very significant factors on the Canadian side of the border, and tame grazing and hay crops are increasingly replacing more native grasslands.

Essentially no unaltered habitat remains in this ecoregion. However, there is extraordinary potential for rapid recovery, as much of it is degraded rather than converted. A few exotic species have invaded, but most of the dominant plant species persist on rangelands. Plant species characteristic of the vegetation of this ecoregion evolved to withstand intense grazing by bison. Thus, it is not surprising to see that many previously dominant plants persist and are likely to become reestablished where restoration efforts are carefully managed.

Remaining Blocks of Intact Habitat

None of the following blocks is considered intact, but many are only partly altered and quite large. These include:

*   Western North Dakota
*   Roosevelt National Park, within the Little Missouri National Grassland—western North Dakota
*   Missouri Coteau—south-central North Dakota
*   Little Missouri National Grassland
*   Badlands National Park–Buffalo Gap National Grassland—southwestern South Dakota
*   Thunder Basin National Grasslands—eastern Wyoming
*   Lower Yellowstone River—eastern Montana (the largest section of intact Missouri River, undammed and with a population of endangered paddlefish)
*   Charles M. Russell National Wildlife Refuge—northwestern Montana
*   Suffield National Wildlife Area—southeastern Alberta, 420 km²
*   Butala Ranch—southwestern Saskatchewan, more than 100 km²
*   Matador Grasslands—southwestern Saskatchewan
*   Grasslands National Park and surrounding area—southern Saskatchewan
*   Great Sand Hills—southwestern Saskatchewan
*   South Saskatchewan River riparian zones—southwestern Saskatchewan
*   Sage Creek area—southeastern Alberta

Degree of Fragmentation

Among the rangelands that are "relatively intact," fragmentation is low because they occur in close proximity. Habitat fragmentation is relatively higher in some areas on the Canadian side of the border. A combination of oil and gas pipelines and road network densities contributes to the greater dissection of the landscape.

Degree of Protection

What little remains of fully intact habitat is under protection in the following:

*   Charles M. Russell National Wildlife Refuge—east-central Montana
*   Badlands National Park—western South Dakota
*   Grasslands National Park—southern Saskatchewan

And in smaller patches in:

*   Bittercreek Mountain Wilderness Study Area
*   Sage Creek (Badlands)—western South Dakota
*   Saskatchewan Landing Provincial Park—southwestern Saskatchewan, 58 km²
*   Prairie Coulees South, Ecological Reserve—southeastern Alberta, 14 km²
*   Kennedy Coulee Ecological Reserve—southeastern Alberta, 11 km²
*   Matador Grasslands Provincial Park—southwestern Saskatchewan, 8 km²

Types and Severity of Threats

The major threat is conversion of altered habitat (rangeland) to wheat production. Major degradation threats are exotic species such as leafy spurge (*Euphorbia* spp.) and yellow sweet clover. There is increased industrial activity (particularly oil and gas), road expansion (with associated access issues), and widespread application of pesticide and herbicide in agricultural production.

Suite of Priority Activities to Enhance Biodiversity Conservation

*   Restore the following important sites for the Northwestern Mixed Grasslands:
    *   Matador Grassland—home to a former International Biome Program research site
    *   Sydney, Montana
    *   Bozeman, Montana, Rangelands Insect Research Site
    *   Old Wives Lake—important for migratory birds

- community pastures in Saskatchewan and Manitoba
- Little Missouri Badlands
- Lower Yellowstone River
- Charles M. Russell NWR—possibility of adding 640 km² to area
- Improve contact with ranchers to encourage model grazing programs.
- Increase efforts to restore the ecological integrity of this ecoregion through carefully managed programs, including bison and black-footed ferret reintroductions and prairie dog recovery.
- Encourage the development of ecotourism as an alternative to ranching in ecologically valuable areas and in areas that have been depopulated over the past few decades.
- Increase protection standards for Suffield National Wildlife Area, Alberta.
- Give stronger conservation attention to Prairie Farm Rehabilitation Administration (PFRA) pasture lands in Saskatchewan. Increase protection standards for the most significant PFRA pasture lands and Wildlife Habitat Protection Act lands.
- Complete Grasslands National Park in Saskatchewan.
- Encourage the province of Saskatchewan to lend greater support to private initiatives to protect the Great Sand Hills.

Conservation Partners

For contact information, please see appendix G.

- Alberta Wilderness Association
- Canadian Nature Federation
- Canadian Parks and Wilderness Society
- Ducks Unlimited Canada
- Endangered Spaces Campaign, Saskatchewan
- Federation of Alberta Naturalists
- Great Sand Hills Planning District Commission
- Natural Resources Conservation Service
- The Nature Conservancy, Alberta
- Nature Saskatchewan
- Prairie Conservation Forum, Calgary-Banff Chapter
- Prairie Farm Rehabilitation Administration
- Saskatchewan Wildlife Federation

- Society of Grassland Naturalists
- World Wildlife Fund Canada
- several Native American tribes active in restoration of bison range

Relationship to Other Classification Schemes

The boundaries of the Northern Short Grasslands are taken from an amalgamation of Omernik's (1995b) ecoregions 41 (Northern Montana Glaciated Plains), 42 (Northwestern Glaciated Plains), 43 (Northwestern Great Plains), and 45 (Northeastern Great Plains) in the United States. The ecoregion corresponds to Küchler (1975) units 57 (Grama-needlegrass-wheatgrass) and 59 (Wheatgrass-needlegrass). The boundary overlaps with Bailey (1994) sections 331D (Northwestern Glaciated Plains), 331E (Northern Glaciated Plains section), and 331F (Northwestern Great Plains section).

The Northern Short Grasslands ecoregion corresponds to the Mixed Grassland terrestrial ecoregion (TEC ecoregion 159) (ESWG 1995). Aspen Grove Boreal forest (B.17) and Grassland characterize this region of south Alberta and Saskatchewan (Rowe 1972).

Prepared by S. Primm, J. Shay, S. Chaplin, K. Carney, E. Dinerstein, P. Sims, A. G. Appleby, R. Usher, K. Kavanagh, M. Sims, and G. Mann

Key Number:	**59**
Ecoregion Name:	**Northern Tall Grasslands**
Major Habitat Type:	**3.1 Grasslands, Savanna, and Shrublands**
Ecoregion Size:	**73,148 km²**
Biological Distinctiveness:	**Nationally Important**
Conservation Status:	**Snapshot—Endangered Final—Endangered**

Introduction

The Northern Tall Grasslands is the northernmost extension of the true tallgrass prairie in North America. It follows the Red River valley from Lake Manitoba in Manitoba south into eastern North Dakota and western Minnesota. The Northern Tall Grasslands can be distinguished from the mixed grasslands to the west by the dominance of tallgrass species—a feature once relatively uniform across their range—and by the highest levels of rainfall (100 cm/yr). This ecoregion is one of the warmest and most humid regions in the Canadian prairies, with an annual temperature of around 2.5°C. Mean summer temperature is 16°C, and mean winter temperature is –12.5°C. Mean annual precipitation ranges from 450 to 700 mm.

The Northern Tall Grasslands ecoregion is separated from the Central Tall Grasslands [60] to the south by the more depauperate flora and fauna of this northernmost section and the dominance of species more characteristic of the boreal regions to the north. In contrast, the species assemblages characteristic of the Central Tall Grasslands are derived from the eastern deciduous forest/transition to the east. Historically, fire and drought and grazing by bison and other ungulates have been the principal sources of habitat disturbance in this ecoregion. Today, very little of the natural vegetation remains in this unit. It is one of the most highly converted of all ecoregions and a major area of agricultural activity.

Much of the ecoregion was significantly influenced by glacial Lake Agassiz, and the present landforms represent former beach ridges, dunes, and ancient lake bottoms. On the Canadian side of the border, the major physiographic feature is classified as the Lake Manitoba Plain, an extensive area underlain by limestone bedrock. This is a low-relief region, with elevations ranging from 218 to 410 m asl. Thick deposits of clays, sands, and gravels cover the rock surface in many areas. The southern two thirds of this ecoregion encompass much of the Red River valley.

Biological Distinctiveness

As in the other tallgrass ecoregions, the dominant grass species in this ecoregion are big bluestem (*Andropogon gerardii*), switchgrass (*Panicum virgatum*), and Indian grass (*Sorghastrum nutans*) (Küchler 1975). The richness of the herbaceous cover is less than in the other two tall grasslands (Kartesz, pers. comm.). The ecoregion is transitional between the aspen (*Populus* spp.) parkland to the north and northwest, mixed-grass prairie to the west, and prairie-forest transitional ecoregions to the east. The tallgrass prairie is most dominant on well-drained, drier sites and is often mixed with quaking aspen (*Populus tremuloides*), oak (*Quercus macrocarpa*) groves, and rough fescue grasslands.

As in other ecoregions of this section of North America, bison (*Bison bison*) and elk (*Cervus canadensis*) once roamed these tallgrass prairies, where they were hunted by the coyote (*Canis latrans*). These species are now gone, although bison are slowly being reintroduced to the area, and wolves (*Canis lupus*) occasionally enter the ecoregion from the east. Common wildlife species in this ecoregion include white-tailed deer (*Odocoileus virginianus*), rabbit (*Sylvilagus* spp.), ground squirrel (*Spermophilus saturatus*), and significant waterfowl populations.

This is the northernmost grassland-savannah complex in eastern North America.

Conservation Status

Habitat Loss and Degradation

Approximately 5 percent of the ecoregion remains as intact habitat, mostly on former beach ridge sites that include coarse sands, are droughty and less valuable for crop production, and are found along riparian corridors. Habitat loss is extensive in this ecoregion given the high rate of conversion to agriculture. More than 75 percent of the ecoregion is considered to be heavily altered.

Remaining Blocks of Intact Habitat

The following blocks of habitat remain in the Northern Tall Grasslands:

- Manitoba Tall Grass Prairie Preserve—southern Manitoba, an area 20 km² and highly fragmented
- Agassiz Beach Ridges—northwestern Minnesota, highly fragmented glacial lake ridges in an area of 159 km²
- Sheyenne Delta—southeastern North Dakota, stabilized sand dunes in an area about 465 km²
- Delta Marsh—southern Manitoba
- Shilo Military Reservation—southern Manitoba
- Caribou-Beaches Parklands—northern Minnesota, 356 km², mostly contiguous area immediately sourth of the Tolstoi-Gardenton Prairie

Degree of Fragmentation

Remaining blocks of habitat are highly fragmented. Most of the remaining habitat blocks are separated by agricultural lands and a relatively dense network of roads.

Degree of Protection

Only parts of the remaining fragments are under some form of conservation management, as follows:

- The Tolstoi Prairies are about 50 percent protected.
- The Agazzi Ridge is about 30 percent protected.
- The Sheyenne Delta is about 70 percent public owned, but heavily grazed.
- Many smaller TNC properties in Minnesota and North Dakota are under some form of conservation management.

Types and Severity of Threats

The major threat for the ecoregion is inappropriate management of lands with biodiversity value. Agricultural conversion of native habitat and development pressures continue. Noteworthy threats include potato farming in the Sheyenne Delta, mining in the Lake Agassiz beach ridges and dunes, and drainage of moist prairie wetlands throughout.

Suite of Priority Activities to Enhance Biodiversity Conservation

- The Shilo Military Reservation contains 48 km^2 of mixed-grass prairie and supports over two hundred plant species. It is a former International Biological Program site. Ensure future protection of this important site.
- The Delta Marsh is an important waterfowl stopover area and the site of intensive research. Piping plovers nest on the beach. Upgrade the protection status of this important site.
- Reduce habitat fragmentation in areas where habitat blocks remain.
- Restore fire cycles in prairie management.
- Secure more land as protected areas, including reviews of existing conservation and wildlife management areas to determine where protection standards need to be increased to prohibit inappropriate activities and further habitat conversion.
- Establish community pastures and public grazing lands where natural vegetation is maintained.

Conservation Partners

For contact information, please see appendix G.

- Brandon Naturalists' Society
- Critical Wildlife Habitat Program
- Ducks Unlimited Canada
- Endangered Spaces Campaign, Manitoba
- Manitoba Heritage Habitat Corporation
- Manitoba Naturalists' Society
- Minnesota Department of Natural Resources
- The Nature Conservancy
- The Nature Conservancy, Midwest Regional Office
- The Nature Conservancy of Canada
- The Nature Conservancy, Manitoba
- Red River Basin Water Management Consortium
- Resource Conservation Manitoba
- Sierra Club
- TREE (Time to Respect the Earth's Ecosystems)
- U.S. Forest Service, Region 3
- U.S. Forest Service, Region 6
- The Wildlife Society
- World Wildlife Fund Canada

Relationship to Other Classification Schemes

Tallgrass prairie is derived from Sims (1988). It corresponds to Küchler's (1975) unit no. 66 (Bluestem Prairie) and Omernik's (1995b) ecoregion 48 (Red River Valley). The tallgrass prairie defined above corresponds to Bailey (1994) as section 251A (Red River Valley section).

This ecoregion corresponds to the Lake Manitoba Plain terrestrial ecoregion (TEC ecoregion 162) (ESWG 1995). Aspen-Oak Boreal forest (B.16) and Grassland identify this region (Rowe 1972).

Prepared by S. Chaplin, J. Shay, J. Moore, P. Sims, G. Whelan Enns, K. Kavanagh, M. Sims, and G. Mann

Key Number:	**60**
Ecoregion Name:	**Central Tall Grasslands**
Major Habitat Type:	**3.1 Grasslands, Savanna, and Shrublands**
Ecoregion Size:	**248,435 km²**
Biological Distinctiveness:	**Globally Outstanding**
Conservation Status:	**Snapshot—Critical**
	Final—Critical

Introduction

The tallgrass prairie of the United States and Canada is divided into three ecoregions: the Central, Northern, and Flint Hills tall grasslands. The Central Tall grasslands cover southern Minnesota, most of Iowa, a small section of eastern South Dakota, and extend as a narrow finger through eastern Nebraska and northeastern Kansas. The Central Tall Grasslands are the most mesic of the grasslands of the Central Plains (Risser et al. 1981). They can be distinguished from other grassland associations by the dominance of tallgrass species—a feature once relatively uniform across their range—and by the highest levels of rainfall (100 cm/yr). The Central Tall Grasslands also have the longest growing season of the ecoregions of the Great Plains. How abundant bison (*Bison bison*) were is uncertain, but elk (*Cervus canadensis*) were probably very important to this ecoregion.

The Central Tall Grasslands are separated from the Northern Tall Grasslands by a much higher diversity of species and by the presence of more northerly species in the Northern Tall Grasslands. The Central Tall Grasslands are demarcated from the Flint Hills Tall Grasslands to the south by soil type. The rocky, limestone soils of the Flint Hills made that ecoregion unsuitable for intensive agriculture. The soils of the Central Tall Grasslands presented no such problems to intensive farming and are now virtually converted to corn and soybean production. Historically, fire and drought and grazing by bison and other ungulates were the principal sources of habitat disturbance in this ecoregion.

Biological Distinctiveness

The Central Tall Grasslands must have been one of the most visually appealing ecoregions of North America in their original state. Before being settled and converted, this ecoregion was the largest tallgrass prairie on earth. The large

number of brightly flowering herbaceous plants added greatly to the plant diversity as well as to its physical beauty. The dominant grass species in this ecoregion are big bluestem (*Andropogon gerardii*), switchgrass (*Panicum virgatum*), and Indian grass (*Sorghastrum nutans*) (Küchler 1975). The richness of the herbaceous cover is obvious across this ecoregion: about 265 species constitute the bulk of the tallgrass prairie in Iowa; 237 species were recorded in a square mile near Lincoln, Nebraska; and 225 species were recorded from the Missouri Valley (Weaver 1954). Many of the plant species found here originated from several different regions; having been exposed to a wide range of climates over the long term, they exhibit relatively wide ecological ranges and thus are widespread throughout the Great Plains.

Because the Great Plains grasslands attained their current extent only in the post-glacial period, these associations are characterized by very low endemism in plants and animals. Low endemism is also characteristic of adjacent grassland ecoregions. As in other ecoregions of this section of North America, bison and elk once roamed these tallgrass prairies, where they were hunted by the prairie wolf (*Canis lupus*). These species are now gone.

Conservation Status

Habitat Loss and Degradation

The Central Tall Grasslands ecoregion is now the corn belt of the United States. Nearly all of this ecoregion has been converted to tilled cropland, and the rest is used for haying and pasture.

Remaining Blocks of Intact Habitat

Essentially no sizable blocks of intact habitat exist in this ecoregion. Remnants of the Central Tall Grasslands in southern Iowa and adjacent Missouri are restricted to twenty patches, all less than 0.08 km² (20 acres) in size (USDA Forest Service 1994a). The Loess Hills in western Iowa (16 km²) and the Prairie Coteau in eastern South Dakota contain important remnants, although the former is rather linear in shape and grazed by livestock.

Degree of Fragmentation

Fragmentation is high among the few, widely scattered parcels of tallgrass prairie.

Degree of Protection

None of the remaining fragments has any formal protection, although restoration is underway at Walnut Creek National Wildlife Refuge in central Iowa.

Types and Severity of Threats

Because this ecoregion is virtually converted, future threats to the remaining tiny fragments are low.

Suite of Priority Activities to Enhance Biodiversity Conservation

Working with appropriate agencies, make efforts to protect and restore the remaining fragments. This will be an uphill effort in most areas and perhaps futile in others because of the high value of the land for agricultural purposes. An example of an appropriate target for this approach would be the Prairie Coteau, which could be restored to an area as large as 162 km².

Conservation Partners

For contact information, please see appendix G.

- Dr. Clinton Owensby, world authority on the tallgrass prairie
- The Nature Conservancy, Iowa Field Office
- U.S. Fish and Wildlife Service

Relationship to Other Classification Schemes

Tallgrass prairie is derived from Sims (1988). It corresponds to Küchler's (1975) unit no. 66 (Bluestem Prairie) and Omernik's (1995b) ecoregion 47 (Western Corn Belt plains). The tallgrass prairie defined above corresponds to Bailey (1994) as parts of sections 251B (North Central Glaciated Plains section), 251C (Central Dissected Till Plains section), and 222M (Minnesota and NE Iowa Morainal, Oak Savanna section).

Prepared by P. Sims, S. Chaplin, T. Cook, J. Shay, S. Smith, E. Dinerstein, and K. Carney

Key Number:	**61**
Ecoregion Name:	**Flint Hills Tall Grasslands**
Major Habitat Type:	**3.1 Grasslands, Savanna, and Shrublands**
Ecoregion Size:	**29,612 km²**
Biological Distinctiveness:	**Globally Outstanding**
Conservation Status:	**Snapshot—Vulnerable**
	Final—Vulnerable

Introduction

The Flint Hills Tall Grasslands cover the Flint Hills of Kansas and the Osage Plains of northeastern Oklahoma. The Flint Hills Tall Grasslands constitute the smallest grassland ecoregion in North America. It can be distinguished from other grassland associations by the dominance of tallgrass species and from the Central Tall Grasslands to the north by its more depauperate biota and a thin soil layer spread over distinct beds of limestone. These flinty beds of limestone, from which the name of this ecoregion is derived, rendered large areas unsuitable for corn or wheat farming. Today, the Flint Hills Tall Grasslands ecoregion is an anomaly—an essentially unplowed (although heavily grazed) remnant of the tallgrass prairie. Historically, fire, drought, and grazing by bison (*Bison bison*) and other ungulates were the principal sources of habitat disturbance in this ecoregion. A new tallgrass prairie national park has been established covering about 44 km². This ecoregion offers the best opportunity for restoration of tallgrass prairie in the United States (Madson 1993).

Biological Distinctiveness

The Flint Hills and adjacent Osage Hills contain the last large pieces of tallgrass prairie in the world. The Flint Hills are less rich in species than the Central Tall Grasslands. The dominant grass species in this ecoregion are big bluestem (*Andropogon gerardii*), switchgrass (*Panicum virgatum*), and Indian grass (*Sorghastrum nutans*) (Küchler 1975). As in other ecoregions of this section of North America, bison and elk (*Cervus canadensis*) once roamed these tallgrass prairies, where they were hunted by the prairie wolf (*Canis lupus*). These species are now gone, although bison are being reestablished in this ecoregion. Greater prairie chickens (*Tympanuchus cupido*) are common on the Tallgrass Prairie National Reserve.

Conservation Status

Habitat Loss and Degradation

Much of the Flint Hills and the Osage Hills is in relatively good condition. Although grazed by livestock, native common plant species still occur.

Remaining Blocks of Intact Habitat

This unit contains the largest blocks of "relatively intact" tallgrass prairie among the three tallgrass units and some of the largest blocks in all of the Great Plains ecoregions. Several important blocks of habitat in this ecoregion include:

- Barnard Ranch (TNC-owned ranch in Osage Hills)—northeastern Oklahoma
- Tallgrass Prairie National Preserve, Flint Hills region—central Kansas, approximately 44 km^2
- Konza Prairie—northeastern Kansas (important research site for tallgrass prairie ecosystem), 35 km^2

Degree of Fragmentation

Habitat fragmentation is relatively low, considering existing land use and level of grazing.

Degree of Protection

The creation of the Tallgrass Prairie National Park in 1996 adds to the amount of habitat protected in this unit. The Nature Conservancy has also established a large conservation unit in the Osage Hills (Barnard Ranch).

Types and Severity of Threats

Because this ecoregion is too difficult to farm, future threats of conversion to agriculture remain low relative to other grassland ecoregions. Grazing could become a more severe threat if not properly managed. Some rangeland has been converted to non-native cool-season grass. There is increasing fragmentation due to smaller homestead units being incorporated into large ranches.

Suite of Priority Activities to Enhance Biodiversity Conservation

- Working with appropriate agencies, make efforts to increase protection of the Flint and Osage Hills.
- Reconnect and restore the remaining blocks of habitat. This will be an uphill effort in most areas and perhaps futile in others. The extent of habitat conversion of this ecoregion is discouraging, and support for conservation is weak in parts of this unit.
- Alter range practices to manage for biodiversity value rather than maximum beef production.

Conservation Partners

For contact information, please see appendix G.

- Dr. Clinton Owensby, world authority on the tallgrass prairie
- Kansas Natural Heritage Inventory
- The Nature Conservancy, Tallgrass Prairie Office
- U.S. Fish and Wildlife Service

Relationship to Other Classification Schemes

Tallgrass prairie is derived from Sims (1988). It corresponds to Omernik's (1995b) ecoregion 28 (Flint Hills) and Bailey's (1994) section 251F (Flint Hills).

Prepared by S. Chaplin, P. Sims, T. Cook, and E. Dinerstein

Key Number:	**62**
Ecoregion Name:	**Nebraska Sand Hills Mixed Grasslands**
Major Habitat Type:	**3.1 Grasslands, Savanna, and Shrublands**
Ecoregion Size:	**61,117 km²**
Biological Distinctiveness:	**Nationally Important**
Conservation Status:	**Snapshot—Relatively Stable Final—Relatively Stable**

Introduction

The Nebraska Sand Hills (hereafter, Sand Hills), located almost entirely within the state of Nebraska, is considered by Küchler (1964) to be one of three major grassland associations that he grouped as the tallgrass prairie. One of the smallest of the Great Plains ecoregions, the Sand Hills cover an area of about 61,100 km². The Sand Hills can be distinguished from other tallgrass and mixed-grass ecoregions by a combination of physiography, soils, elevation, surface water characteristics, and natural vegetation. The Sand Hills consist of various dune types and interdunal valleys, dominance of Cenozoic sands, and a distinct soil type (entisols). The irregular dunes and sandy soils of the Sand Hills are so distinct within the Great Plains as to warrant elevation of the Nebraska Sand Hills to its own ecoregion. The western and portions of the northern Sand Hills are dotted with small lakes and wetlands. Precipitation and temperature are average among the Great Plains ecoregions. Historically, fire and grazing by migratory herds of bison played a major role in shaping the landscape. Today, the major disturbance factor is cattle grazing on the large ranches that cover much of the ecoregion.

Biological Distinctiveness

The Sand Hills contain the most intact natural habitat of the Great Plains ecoregions. The Sand Hills contain a distinct grassland association dominated by sand bluestem (*Andropogon hallii*), *Calamovilfa longifolia,* and needle-and-thread (*Stipa comata*).

Conservation Status

Habitat Loss and Degradation

It is estimated that as much as 85 percent of the Sand Hills is still intact. Toward the eastern section of this ecoregion, habitat loss and degradation are higher than in the rest of the Sand Hills due to center-pivot irrigation and development. Degradation of the tallgrass prairie is less in the Sand Hills than in other grassland ecoregions; fragility of soils has dissuaded excessive overgrazing and cropping (Sims 1988). In overgrazed areas, decreases in little bluestem (*Schizachyrium scoparium*), needle-and-thread, and *Andropogon hallii* and increases in hairy grama (*Bouteloua hirsuta*), *Calamovilfa longifolia,* and dropseed (*Sporobolus cryptandrus*) are apparent.

Remaining Blocks of Intact Habitat

Very little of the Sand Hills has been plowed; remaining blocks cover large parts of the ecoregion.

Degree of Fragmentation

The absence of farming in the region and other developments contribute to an extraordinarily low level of fragmentation for a Great Plains ecoregion.

Degree of Protection

The most important protected areas in the Sand Hills are:

- Valentine National Wildlife Refuge—north-central Nebraska, 285 km²
- Crescent Lake—west-central Nebraska, 175 km²
- Niobrara Valley Preserve, with Wild and Scenic designation of the Niobrara River—north-central Nebraska, 202 km²
- Fort Niobrara National Wildlife Refuge—central-northern Nebraska

Types and Severity of Threats

Because of the sandy soils and topography, farming is less conspicuous than in other units. Center-pivot irrigation was implemented in the past but has largely proven unfeasible. If it were attempted again, conversion to agriculture might become a more serious threat. The pumping of the vast reservoirs of groundwater is also a potential threat. Drainage and haying of wet prairies and meadows between the dunes potentially could have a major influence. Whether wetland species are being impacted is unknown.

Suite of Priority Activities to Enhance Biodiversity Conservation

Priority activities should focus on creating dialogues with landowners. Key conservation messages to stress are the need to limit stocking rates to promote the health of grazing lands, and where and where not to graze.

The Sand Hills contain only a few protected areas. Creation of new areas is not considered to be an urgent need because of the relative degree of intactness of the landscape and a potentially high level of awareness of ecosystem sensitivities among landowners.

Conservation Partners

For contact information, please see appendix G.

- Institute of Agriculture and Natural Resources
- West Central Research and Extension Station
- The Nature Conservancy of Nebraska
- Natural Resources Conservation Service
- Nebraska Natural Heritage Program
- University of Nebraska
- U.S. Fish and Wildlife Service

Relationship to Other Classification Schemes

The boundary of the Sand Hills is taken from Omernik (1995b) and is very similar to the boundary described in Bailey's (1994) unit 332C, Nebraska Sand Hills, and Küchler's (1975) unit 67, Sand Hills Prairie.

Prepared by S. Chaplin, P. Sims, E. Dinerstein, K. Carney, R. Schneider, and T. Cook

Key Number:	**63**
Ecoregion Name:	**Western Short Grasslands**
Major Habitat Type:	**3.1 Grasslands, Savannas, and Shrublands**
Ecoregion Size:	**435,154 km²**
Biological Distinctiveness:	**Regionally Outstanding**
Conservation Status:	**Snapshot—Endangered**
	Final—Endangered

Introduction

The Western Short Grasslands is the second-largest grassland ecoregion of North America, covering slightly more than 435,000 km². This unit ranges over portions of western Nebraska and southeastern Wyoming, across much of eastern Colorado, southwestern Kansas, West Texas, the Oklahoma panhandle, and into eastern New Mexico. This grassland ecoregion is distinguished from other grassland units by low rainfall, relatively long growing seasons, and warm temperatures. From a structural standpoint, the short stature of the dominant sod-forming grasses, grama (*Bouteloua gracilis*) and buffalo grass (*Buchloe dactyloides*), separate the Western Short Grasslands from other units.

Küchler (1964) classified this ecoregion as grama–buffalo grass prairie, bluestem-grama prairie, sandsage-bluestem prairie, and wheatgrass-bluestem-needlegrass prairie. Under more natural conditions, major sources of ecosystem disturbance were drought and grazing by wildlife, rather than fire as in some of the other Great Plains grassland ecoregions to the east. Unlike its value in the more mesic grassland ecoregions, fire is thought to be detrimental to shortgrass prairie plants (Wright and Bailey 1980, 1982). Today, livestock grazing is the major form of disturbance, but much of this ecoregion is in better condition as far as grazing impact than other ecoregions. Part of the explanation is that many of the sod-forming perennial grassland plants, the dominant species in this ecoregion, evolved under intense grazing and trampling by migratory herds of bison (*Bison bison*).

Biological Distinctiveness

The Western Short Grasslands are among the richest ecoregions in the United States and Canada for species of butterflies, birds, and mammals (see appendix C). Part of this pattern can be explained by the closer proximity of this ecoregion to the subtropics. This ecoregion once supported one of the most impressive migrations of a large ungulate species anywhere in the world—the American bison migration. Today, bison no longer migrate, but bison ranching is becoming increasingly popular. The Western Short Grasslands also contain the fastest declining bird populations on the continent—those of the endemic birds of the short grasslands of the Great Plains. These species are declining faster than many neotropical migratory birds, whose plight receives much more attention.

Conservation Status

Habitat Loss and Degradation

Much of this area was severely affected by largely unsuccessful efforts to develop dryland cultivation. In western Kansas and eastern Colorado this damaging activity continues. The dustbowl of the 1930s was centered in this ecoregion and stands as proof of the unsuitability of this area for farming, unless heavily irrigated. Areas in the southern part of this ecoregion, in Texas, have been invaded by mesquite (*Prosopis glandulosa*) and thorny shrubs (e.g., *Opuntia* spp., *Zizyphus obtusifolia, Berberis trifoliata*), forming a savanna or shrubland with a shortgrass prairie understory (Sims 1988).

Nearly all of this ecoregion is in farms and ranches. Cropland cover varies between 30 and 60 percent across the ecoregion (even higher in western Kansas and western Nebraska), with grazing lands occupying the remainder. The amount of irrigated farmland varies across the ecoregion. In the northern section in Colorado, almost all of the land is in farms or ranches with a buildup of urban areas along the eastern edge of the Rockies. About 68 percent is used for grazing domestic livestock, with about 15 percent of the area planted in dry crops.

In the extreme southern section of the ecoregion, grazing covers more than 75 percent of the area. Overgrazing has allowed the spread of woody shrubs and trees and the near permanent conversion of plains grasslands to desert scrublands (Dick-Peddie 1993). Approximately 40 percent of the remaining habitat in this ecoregion is considered to be intact, one of the highest percentages among North American grasslands, and the highest among grassland

ecoregions greater than 70,000 km². In the western section of this ecoregion, rangelands are moderate to heavily grazed.

Remaining Blocks of Intact Habitat

Six of the national grasslands in this ecoregion contain intact blocks of habitat:

- Black Kettle—western Oklahoma
- Comanche—southeastern Colorado
- Pawnee—northeastern Colorado
- Cimarron—southeastern Kansas (one of the strongholds of the lesser prairie chicken)
- Rita Blanca—along the borders of New Mexico, Texas, and Oklahoma
- Kiowa—panhandle of Texas

Other significant areas of intact habitat include:

- Arkansas River Sandsage—southwestern Kansas
- Tule and Palo Duro Canyon areas—northwestern Texas

Degree of Fragmentation

The species that occur in the Western Short Grasslands are relatively widespread and good dispersers. Aside from areas that have been plowed, the native fauna is relatively unaffected by the levels of fragmentation found in this ecoregion.

Degree of Protection

Although there is considerable rangeland and grassland worthy of conservation, few sites are formally protected. National grasslands have considerable potential for biodiversity conservation, but grazing on these units by domestic livestock must be modified.

Types and Severity of Threats

The main threat to this ecoregion is conversion to agriculture. New technologies, including four-wheel-drive tractors, "precision farming," herbicides, and irrigation, make farming more productive in areas that were previously difficult to cultivate. The Conservation Reserve Program (CRP) in the eastern and southern part of this unit in the Oklahoma and Texas panhandles has kept some areas under good conservation management. If the CRP program were legislatively modified or discontinued, some of these areas would be under great threat of conversion. Overgrazing is less of a threat in most parts of the ecoregion because of the abundance of grazing-tolerant plant species. The invasion of tree species is a problem in some areas.

Suite of Priority Activities to Enhance Biodiversity Conservation

- Make national grassland management more sensitive to biodiversity.
- Work with conservation associations to maintain rangelands.
- Increase the use of rotational grazing to mimic natural grazing patterns.
- Improve relationships with private landowners.
- Improve management of irrigation systems.
- Restoring bison populations.

Conservation Partners

For contact information, please see appendix G.

- Kansas Natural Heritage Inventory
- National Cattlemen's Association
- Natural Resources Conservation Service
- Nebraska Natural Heritage Program
- Society for Range Management
- U.S. Forest Service

Relationship to Other Classification Schemes

The Western Short Grasslands ecoregion is derived from Omernik's (1995b) ecoregions 25 (Western High Plains) and 26 (Southwestern Tablelands). It corresponds to Küchler's (1975) unit no. 58 (Grama-Buffalo grass). Bailey (1994) classifies this as parts of six different sections (331H, 331C, 331I, 331B, 315A, and 315B).

Prepared by T. Cook, E. Dinerstein, P. Sims, S. Chaplin, S. Smith, and K. Carney

Key Number:	**64**
Ecoregion Name:	**Central and Southern Mixed Grasslands**
Major Habitat Type:	**3.1 Grasslands, Savannas, and Shrublands**
Ecoregion Size:	**282,139 km²**
Biological Distinctiveness:	**Bioregionally Outstanding**
Conservation Status:	**Snapshot—Critical**
	Final—Critical

Introduction

The Central and Southern Mixed Grasslands ecoregion has a north-south orientation, spanning central Nebraska, central Kansas, western Oklahoma, and north-central Texas. It separates the Central Tall Grasslands [60] and the Central Forest/Grassland Transition Zone [65] from the Western Short Grasslands [63]. Essentially, this region is a broad ecotone that covers slightly more than 282,000 km². It is distinguished from the Northern Mixed Grasslands by warmer temperatures and a much longer growing season, and from the adjacent tallgrass and short grasslands by the intermediate stature of the grassland layer. It is distinguished from the Central Forests/Grassland Transition Zone to the east by the relative scarcity of trees and shrubs. The major disturbance regimes are drought, the degree and frequency of grazing by domestic animals and wild ungulates, and fire.

Biological Distinctiveness

The mixed-grass prairie contains the floristic elements of the tall- and shortgrass prairies and, combined with a rich forb flora, contains the highest floral complexity of any North American grassland ecoregion (Barbour, Burk, and Pitts 1980). Typical grasses include little bluestem (*Schyzachrium scoparium*), western wheatgrass (*Agropyron smithii*), and grama grasses (*Bouteloua cartipendala*). These species mix with taller grasses in the wetter areas and give way to shorter grasses in the drier areas (e.g., *Bouteloua, Buchloe, Muhlenbergia,* and *Aristida*) (Sims 1988). The effects of drought cycles and grazing intensity shift floristic composition to favor drought-tolerant species during dry periods and more shallow-rooted, mesic-loving plants during wetter periods.

The Central and Southern Mixed Grasslands ecoregion is among the top ten ecoregions in the number of reptile species and is an important breeding area for endemic Great Plains bird species. It also contains very important stopover sites for migratory birds, particularly on wetland sites scattered throughout this ecoregion. The Platte River Valley in Nebraska is a prominent area for sandhill cranes, and the Cheyenne Bottoms in Kansas for populations of shorebirds during spring migration.

Conservation Status

Habitat Loss and Degradation

Overall, only about 5 percent of the remaining habitat is considered to be intact. During the dustbowl of the 1930s, basal cover of grasses on even moderately grazed and heavily grazed grasslands declined from 80 percent or more to less than 10 percent in a period of three to five years (Sims 1988), but the region has since mostly regained its cover. Natural vegetation has been converted to cropland or pasture on about 90 percent of this ecoregion in Oklahoma and Texas. In Kansas and Nebraska, about 60 percent is in cropland and about 35 percent is grazed (USDA Forest Service 1994a).

Remaining Blocks of Intact Habitat

Most of the remaining blocks of intact habitat are quite small. Some of the most prominent include:

- Wichita Mountains Wildlife Refuge, a 90-km^2 site for reestablishment of bison and an important area for conservation of black-capped vireos—southeastern Oklahoma
- Platte River Valley, a 45-km^2 important restoration site, a major stopover for migratory sandhill cranes—southern Nebraska
- Rainwater Basins, a 45-km^2 area managed in part by USFWS, which includes a series of clay bottom wetlands that are highly fragmented—southern Nebraska
- Central Kansas wetlands, including Cheyenne Bottoms and Quivira National Wildlife Refuge—central Kansas
- Great Salt Plains—north-central Oklahoma
- Red Hills—Oklahoma and Kansas
- Smokey Hills River Breaks—west-central Kansas, 45 km^2

Degree of Fragmentation

Much of the ecoregion that occurs in Kansas and Oklahoma still contains fragments of intact grasslands associated with farms. The remaining habitat in the rest of the ecoregion is fragmented.

Degree of Protection

With the exception of a high level of protection in the Wichita Mountains and Salt Plains, remaining habitat in the ecoregion is essentially unprotected.

Types and Severity of Threats

The major threat is conversion to agriculture. High wheat prices in the mid 1990s encouraged land conversion in western portions of the ecoregion. Center-pivot irrigation has also caused conversion. Water flow into streams due to diversions is another problem. Fire apparently increases forage production in the eastern portion of this ecoregion, makes grasses more palatable, eliminates undesirable annuals, and suppresses the invasion of mesquite, juniper, and cacti (Wright and Bailey 1980). Thus, fire suppression constitutes another threat. Overgrazing by domestic stock, particularly in riparian areas, is a localized serious threat.

Suite of Priority Activities to Enhance Biodiversity Conservation

- Manage and preserve wetlands, primarily by seeking to maintain water flows to wetland areas.
- Improve grazing management to make it more compatible with biodiversity conservation, such as by encouraging rotation grazing.
- Restore and enlarge the best representative areas for biodiversity.

Conservation Partners

For contact information, please see appendix G.

- Kansas Natural Heritage Inventory
- Natural Resources Conservation Service
- The Nature Conservancy of Nebraska
- U.S. Fish and Wildlife Service—wildlife refuges
- U.S. Forest Service—national grasslands border this ecoregion

Relationship to Other Classification Schemes

The Central and Southern Mixed Grasslands correspond to Omernik's (1995b) ecoregion 27 (Central Plains Grasslands) and most of Küchler's (1985) units 62 (Bluestem-grama prairie) and 76 (Mesquite-buffalograss). The ecoregion also corresponds to Bailey's (1994) sections 332E (South Central Great Plains), 311A (Redbed Plains), and 315C (Rolling Plains).

Prepared by S. Chaplin, T. Cook, E. Dinerstein, P. Sims, and K. Carney

Key Number:	**65**
Ecoregion Name:	**Central Forest/Grassland Transition Zone**
Major Habitat Type:	**3.1 Grasslands, Savannas and Shrublands**
Ecoregion Size:	**406,989 km²**
Biological Distinctiveness:	**Regionally Outstanding**
Conservation Status:	**Snapshot—Critical Final—Critical**

Introduction

The Central Forest/Grassland Transition Zone (hereafter, the CTZ) extends from northern Illinois, across much of Missouri, and into eastern Kansas, Oklahoma, and Texas. The CTZ is one of the larger "savanna-type" ecoregions, covering more than 380,000 km², and along with the Upper Midwest Forest/Savanna Transition Zone [9] separates the eastern deciduous forests from the tallgrass and mixed-grass prairies. The CTZ can be distinguished from the Central U.S. Hardwood Forests [18] and other forested ecoregions to the east by its mixture of savanna, prairie, and woodlands. It can be delineated from the tallgrass prairie and Central and Southern Mixed Grasslands [64] to the west by the higher tree and shrub densities. Annual precipitation ranges from 600 to 1,040 mm, with wetter areas supporting a more closed tree canopy. The uniform soil type (mollisol) unites this wide-ranging ecoregion. The major disturbance regimes were clearly fire and drought. The intensity, frequency, and areal extent of

fires, combined with drying periods, probably kept the boundaries of this ecoregion in a state of flux. Unfortunately, virtually no intact habitat remains in the CTZ because this ecoregion is one of the most converted of U.S. ecoregions. Almost all of this unit is intensively farmed for corn and soybeans.

Biological Distinctiveness

The CTZ is one of the richer ecoregions in North America due to its size and location as the ecotone between the Great Plains and the eastern deciduous forest. It is also much richer than the other two transition ecoregions, the Upper Midwest Forest/Savannah Transition Zone, and the Canadian Aspen Forest and Parklands [55]. The CTZ ranks among the top ten ecoregions for reptiles, birds, butterflies, and tree species. It shares a strong affinity with the adjacent grassland ecoregions in that many of the tallgrass prairie species can be found in the understory layer. The CTZ also shares much of the fauna of the adjacent grassland ecoregions; these species persist in the ecotones and openings within the ecoregion. Oaks and hickories are the dominant tree species throughout the unit but often occur at low to moderate densities. Typical oaks are blackjack oak (*Quercus marilandica*) and post oak (*Q. stellata*) in the southern part of this ecoregion. Bison (*Bison bison*) were abundant in this ecoregion in the presettlement period.

Conservation Status

Habitat Loss and Degradation

Less than 1 percent of the remaining habitat is considered to be intact. Conversion to intensive production of corn and soybeans is almost complete.

Remaining Blocks of Intact Habitat

The extremely high level of habitat loss translates into few remaining blocks of intact habitat. All of the remaining units are small. The most important example is the Emiquon floodplain forests in western Illinois, an important wetland and migratory stopover.

Other sites with high potential include:

- Goose Lake Prairies and the Midewin National Grassland—northeastern Illinois

- Palos Savanna, 52.5 km² of moderately fragmented savanna—northeastern Illinois
- Kankakee Sands, a savanna-wet prairie that is the site of a TNC restoration program, a 32 km² area with "macro-site" expansion to 80 km²—on the Illinois-Indiana border
- Osage Plains prairie fragments, an important site for the prairie chicken (*Tympanuchus cupido*), the richest and largest fragments of tallgrass prairie in this ecoregio—Missouri
- Cross Timbers area, an oak savanna with tallgrass prairie understory that is highly fragmented—Oklahoma and Kansas
- Arbuckle Uplift native grassland—southeastern Oklahoma
- Indiana Dunes Lake Shore grassland savanna—northern Indiana
- Emiquon floodplain forest—western Illinois

Degree of Fragmentation

Degree of fragmentation is extremely high in this ecoregion among remaining pieces of natural vegetation.

Degree of Protection

There are many small protected areas and larger ones along the Mississippi and Illinois Rivers. Prairie State Park is protected, as are some scattered remnants of the Osage Plains. Willow Slough and Goose Lake Prairie receive partial protection.

Types and Severity of Threats

Because almost all of this ecoregion has already been converted to corn and soybean production, there is little chance for further extensive conversion to occur. The Osage Plains faces some threats of degradation.

Suite of Priority Activities to Enhance Biodiversity Conservation

Restoration and immediate protection of remaining fragments must be the key activities in an ecoregion like the CTZ that has been so dramatically altered from its original state. Important activities include:

- Taking advantage of high potential for acquiring floodplain forests as a result of recent floods

- Identifying clusters of land to acquire to connect and restore aggregates of large substantial areas

Conservation Partners

For contact information, please see appendix G.

- Illinois Department of Natural Resources
- Indiana Department of Natural Resources
- Midwest Science Center
- Missouri Department of Conservation
- Missouri Resource Assessment Project (MoRAP)
- The Nature Conservancy
- The Nature Conservancy, Midwest Regional Office
- The U.S. Fish and Wildlife Service
- The U.S. Forest Service, Midewin Tallgrass Prairie

Relationship to Other Classification Schemes

The CTZ combines Omernik's (1995b) ecoregions 54 (Central corn belt plains), 40 (Central irregular plains), and 29 (Central Oklahoma/Texas plains). It corresponds to Küchler's (1975) unit 73, a mosaic of two other units (66, bluestem prairie, and 91, oak hickory forest). The CTZ corresponds roughly to an area that spans about eight sections in Bailey (1994).

Prepared by S. Robinson, T. Cook, S. Chaplin, and E. Dinerstein

Key Number:	**66**
Ecoregion Name:	**Edwards Plateau Savannas**
Major Habitat Type:	**3.1 Grasslands, Savannas, and Shrublands**
Ecoregion Size:	**61,800 km²**
Biological Distinctiveness:	**Regionally Outstanding**
Conservation Status:	**Snapshot—Critical**
	Final—Critical

Introduction

The Edwards Plateau Savannas form an important part of the Texas hill country, a moderately sized ecoregion separated from adjacent units by a distinct soil type (mollisol) and a vegetation type distinguished by juniper-oak savanna and mesquite-*Acacia* savanna underlain by mid to short grasslands (Küchler 1964). In other aspects, the Edwards Plateau Savannas ecoregion is intermediate among the dry grassland and savanna ecoregions in terms of rainfall, temperature, and length of growing season. It is estimated that up to 90 percent of this ecoregion has been converted to pasture, urban areas, and agricultural crops. Previously, major disturbance regimes were dominated by fire, drought, and perhaps grazing by bison (*Bison bison*).

Biological Distinctiveness

The limestone bedrock of the Edwards Plateau helps to contribute to the distinctiveness of the biota. An array of species are specialists on limestone habitats, including caves. Some of the largest assemblages of cave-dwelling bats anywhere in the world, indeed the largest aggregations of mammals anywhere in the world, roost in several of the large caves of this ecoregion. Millions of Mexican free-tailed bats (*Tadarida brasiliensis*) use these caves as maternity roosts, creating a globally outstanding phenomenon. The high number of endemic invertebrate cave species also qualifies this ecoregion as a global hot spot for cave-dwelling species. The aquatic vertebrates endemic to the plateau include the widemouth blindcat (*Satan eurystomus*), San Marcos salamander (*Eurycea nana*), Comal blind salamander (*Eurycea tridentifera*), Texas blind salamander (*Typhlomolge rathbuni*), Blanco blind salamander (*Typhlomolge robusta*), and the Texas salamander (*Eurycea neotenes*).

This ecoregion also ranks among the top ten ecoregions for reptiles and birds. The Edwards Plateau Savannas contain most of the breeding habitat for an endemic migratory warbler, the golden-cheeked warbler (*Dendroica chrysoparia*), which nests only in mature oak-juniper savannas, or cedar brakes. Some important breeding habitat for the black-capped vireo (*Vireo atricapillus*), a species endangered by habitat loss, occurs in this ecoregion. A disjunct population of maples (*Acer* spp.) occurs in mesic pockets, an unusual example of a relict population in a presently dry region.

Conservation Status

Habitat Loss and Degradation

Only about 2 percent of the remaining habitat in this important ecoregion is considered intact. Important ecological processes such as fire have been suppressed, leading to changes in vegetation structure and composition. Extensive soil loss has contributed to changes in the disturbance regimes of riparian communities. Overgrazing has fragmented and eliminated native grasslands and contributed to the expansion of woody species.

The southern and western portions of the Edwards Plateau are heavily altered by major encroachment of shrubs as a result of fire suppression. The eastern section of the plateau is in better condition and contains much second growth.

Remaining Blocks of Intact Habitat

The Edwards Plateau contains only a few relatively small areas of intact habitat. These occur around Austin, where about 30 percent of a 182 km² tract has been acquired and put under conservation management. About 50 percent of a 122 km² tract in the Balcones Canyonlands has been acquired. Other intact blocks remain in Real County along the Frio River and in small patches outside San Antonio in northwest Bexar County.

Degree of Fragmentation

All remaining blocks of habitat are highly fragmented.

Degree of Protection

The Balcones Canyonland National Wildlife Refuge near Austin and the Balcones Canyonlands have improved protection of biodiversity. However, many of the area's threatened species and habitats face considerable risk.

Types and Severity of Threats

Urban and suburban sprawl around Austin and San Antonio pose a high threat to remaining habitats. Invasion of exotic species and overgrazing will also increase habitat degradation unless mitigated by conservation measures. "Cedar" choppers in the hill country have removed much of the old-growth juniper-oak woodland. Change in riparian disturbance regimes affects the dynamics of this system and threatens species dependent on this habitat.

Suite of Priority Activities to Enhance Biodiversity Conservation

Priority sites for conservation include:

- Areas along the Balcones Escarpment
- Areas in western Travis County and northwest Bexar County that preserve the best remaining contiguous habitat of the golden-cheeked warbler
- Karst areas that contain the highest concentration of Mexican free-tailed bat maternity colonies in the world
- Fort Hood, which contains the largest population of golden-cheeked warblers and black-capped vireos under a single management unit, as well as significant karst features
- Areas of high salamander diversity and endemism (Some individual springs, streams, and artesian wells are the only known localities for some species.)
- Karst areas rich in endemic invertebrates

Conservation Partners

For contact information, please see appendix G.

- Bat Conservation International
- Bexar Land Trust
- Department of Texas Parks and Wildlife
- Land Trust Alliance
- Audubon Society, national, state, and local chapters
- Native Plant Society of Texas
- The Nature Conservancy
- The Nature Conservancy, Southeast Regional Office
- The Nature Conservancy of Texas
- Save Our Springs Association
- Sierra Club, Lone Star Chapter

Relationship to Other Classification Schemes

The boundary of the Edwards Plateau Savannas is taken from Omernik (1995b). It generally corresponds to Küchler's (1975) juniper-oak savanna (87) and Bailey's (1994) section 315D (Edwards Plateau).

Prepared by T. Cook and D. Olson

Key Number: **67**
Ecoregion Name: **Texas Blackland Prairies**
Major Habitat Type: **3.1 Grasslands, Savannas, and Shrublands**
Ecoregion Size: **779,092 km²**
Biological Distinctiveness: **Regionally Outstanding**
Conservation Status: **Snapshot—Endangered Final—Critical**

Introduction

The Texas Blackland Prairies ecoregion spans approximately 6.1 million hectares from the Red River on the north to near San Antonio in South Texas. It is part of a tallgrass prairie continuum that stretches from Manitoba to the Texas coast. The region consists of the main belt, covering 4.3 million hectares, and two smaller islands of prairie, southeast of the main belt. These are the Fayette Prairie, 1.7 million hectares, and the San Antonio Prairie, 0.7 million hectares. A significant part of the main belt is bounded by oak woodland and savanna on the north, east, and northwest. The San Antonio and Fayette Prairies are imbedded in and separated from the main belt by the oak woodlands of the East Central Texas Forests [21].

The main belt of the Blackland Prairies is divided into four narrow, geomorphic areas aligned in a north-south direction. These include—from west to east—the Eagle Ford Prairie, the White Rock Cuesta, the Taylor Black Prairie, and the Eastern Marginal Prairie (Montgomery 1993). The soils of the Eagle Ford and Taylor Black Prairies are primarily clays of the order vertisol, while the soils of the White Rock Cuesta are mollisols and the Eastern Marginal prairie of the order alfisol. Alfisols are the important soil order in the San

Antonio Prairie, while both alfisols and vertisols are important in the Fayette Prairie. Microtopography such as gilgai on vertisols and mima mounds on alfisols are important microhabitats. Gilgai are shallow microdepressions 1 to several meters across formed by pedoturbation of montmorillonitic clays. Mima mounds are small circular hills that are variable in size but may be more than a meter high and 1 to 14 meters across. The origins of mima mounds are not clear and are probably variable (Diamond and Smeins 1993). The climate is warm temperate to subtropical and humid. Precipitation ranges from 762 mm on the western edge to 1,016 mm on the east.

The natural vegetation of the region was dominated by tallgrass prairie on uplands. Deciduous bottomland woodland and forest were common along rivers and creeks (Diamond and Smeins 1993). The Blackland Prairie is characterized by a high degree of plant community diversity. This diversity, which is in part represented by four major prairie community types, is attributable to the ecoregion's variety of soil orders and their variation in texture and soil pH (Diamond, Riskind, and Orzell 1987; Diamond and Smeins 1985). Little bluestem (*Schizachyrium scoparium*) and Indian grass (*Sorghastrum nutans*) are frequently dominants on Blackland Prairie alfisols and vertisols. Big bluestem (*Andropogon gerardii*) is of variable importance on vertisols and is frequently a dominant on Blackland Prairie mollisols. Gamma grass–switchgrass (*Tripsacum dactyloides–Panicum virgatum*) prairies are associated with bottomland sites throughout the region and are also found on upland sites of the northern main belt vertisols, where they are especially associated with gilgai microtopography. Silveanus dropseed–mead's sedge (*Sporobolus silveanus–Carex meadii*) prairies are found over low pH soils of the northern main belt. Little bluestem–brownseed paspalum (*S. scoparium–Paspalum plicatulum*) prairie is associated with Fayette Prairie alfisols. Each community differs further in secondary florae. For example, eastern forb species such as *Liatris pycnostachya* and *Coreopsis grandiflora* are largely limited to the alfisols of the Eastern Marginal Prairies, while grasses such as *Bouteloua hirsuta* and *Muhlenbergia reverchonii,* as well as a diversity of species in the genus Dalea are generally found on the mollisols of the White Rock Cuesta.

The Blackland Prairie was a disturbance-maintained system. Prior to European settlement (pre-1825 for the southern and pre-1845 for the northern half), important natural landscape-scale disturbances included fire and periodic grazing by large herbivores, primarily bison and to a lesser extent pronghorn antelope. Fire and infrequent but intense, short-duration grazing suppressed woody and invigorated herbaceous prairie species. The latter were adapted to fire and grazing by virtue of maintaining perenniating tissues below ground. It has been suggested that second only to climate, fire has been the most important determinant of the spread and maintenance of grasslands (Anderson 1990). Fire frequency in the pre-settlement Blackland Prairie is unclear, but fires may have occurred at intervals of five to ten years (Wright and Bailey 1982). Both natural (e.g., lightning strike) and anthropogenic ignition sources are recognized. Bison herds, though reported for the Blackland Prairie, were far smaller than those found farther west in the mixed- and shortgrass prairies (Strickland and Fox 1993). Their impact was probably local, with long intervals between grazing episodes. Bison were probably extirpated from the region by the 1850s.

Biological Distinctiveness

Compared to its northern tallgrass counterpart, the original community diversity of the Blackland Prairie was greater, by virtue of its greater representation of soil orders. Among the four plant community series, six major plant associations have been identified. On vertisols and alfisols, microhabitat processes are important. Mima mounds, circular mounds several meters across and tens of centimeters to more than a meter tall, occur on all undisturbed alfisols of the region (Diamond and Smeins 1993). Three distinct vegetation zones and as many as four different habitats can occur within a single mound (Collins 1975). Gilgai microtopography is associated with the dramatic shrink-swell capacity characteristic of vertisols. On level ground, this pedoturbation creates microdepressions often more than 40 centimeters deep. On sloping topography trough-like depressions of up to 20 centimeters in depth run perpendicular to the contour of the slope (Miller and Smeins

1988). Xeric, mesic, and hydric moisture regimes correspond to microtopographic position. The Blackland Prairie is habitat for more than 500 native faunal taxa, including 327 species of bird (Schmidly, Scarbrough, and Horner 1993). Seven reptile and 15 bird species are considered imperiled (ranging from state "watch listed" to federally endangered). Most officially listed mammals have been extirpated from the Blacklands.

Conservation Status

Habitat Loss and Degradation

By the second half of the nineteenth century, row-crop agriculture was well established in the Blackland Prairie. By the middle 1920s more than 80 percent of the original vegetation had been lost to cultivation (Bland and Jones 1993). In the second half of the century urbanization continued to reduce the remaining prairie. Today less than 1 percent of the original vegetation of the Blackland Prairie remains, in scattered parcels across the region (Smeins and Diamond 1983). Most of the remaining Blackland Prairie survived by virtue of the value of the cattle forage that it produced. Over the past century, annual hay crops were taken from a majority of the prairie remnants. This practice continues today. Annual mowing, with and without the addition of fire, provided valuable disturbance. However, it is possible that long-term mowing at the same time of year, while not shifting overall composition, has changed the numeric relationships of the species. The greatest portion of remnant prairie is found in the northern part of the main belt, where hay production was an important agricultural business.

Remaining Blocks of Intact Habitat

The following sites are protected or under a volunteer registry program. Voluntarily protected lands are identified only by area and county to protect the privacy of the owners.

Protected Lands (land under conservation easement, public land, or ownership by conservation organization; all found within eastern Texas) include:

- Clymer Meadow, Hunt County—106 ha
- Tridens Prairie, Lamar County—41 ha
- Leonhardt Prairie, Falls County—16 ha
- County Line Prairie, Hunt County—8 ha
- Mathews Prairie, Hunt County—41 ha
- Parkhill Prairie, Collin County—21 ha
- Kachina Prairie, Ellis County—16 ha
- Rosehill Prairie, Dallas County—30 ha
- Cedar Hill State Park, Dallas County—8 ha

Voluntarily Protected Lands (listed by county) include:

- Hunt—five tracts totaling 159 ha
- Lamar—two tracts totaling 970 ha
- Franklin—one tract totaling 41 ha
- Van Zandt—two tracts totaling 117 ha
- Kaufman—one tract totaling 81 ha
- Rockwall—one tract totaling 16 ha
- Denton—one tract totaling 4 ha

Degree of Fragmentation

Fragmentation is extreme due to agricultural development and urbanization. Few patches greater than 16 ha remain.

Degree of Protection

Almost all of the remaining Blackland Prairie is under private ownership. Approximately 12 percent of the remaining Blackland Prairie is currently under protection by The Nature Conservancy of Texas, the Texas Chapter of The Nature Conservancy, a private, nonprofit conservation organization. Almost 60 perent is voluntarily protected under a private land registry program administered by that organization.

Types and Severity of Threats

Primary threats to prairie remnants are urbanization, row-crop agriculture, invasion by exotic plant species, fragmentation, and loss of landscape-scale processes, especially fire and grazing by large native herbivores. The most threatened community type is the *Schizachyrium-Andropogon-Sorghastrum* association typical of the mollisol prairie of the White Rock Cuesta. In this case, the single greatest threat is urbanization, as this narrow, elongated strip follows a line of urban development stretching between Dallas, Austin, and San Antonio.

Suite of Priority Activities to Enhance Biodiversity Conservation

The Blackland Prairie remains today as small islands of native diversity in a matrix of primarily oldfield, tame pasture, and urban development. Over the past forty years much of the former cropland has reverted to marginally productive oldfield. Land use in these rural portions of the region has begun to shift to small-scale cow-calf operations and low-density residential development. Customarily, these marginally productive oldfields are lightly stocked or are "improved" with exotic forages requiring fertilization to realize maximum productivity. Concurrently, one of the greatest limitations in conservation of the Blackland Prairie is lack of adequate remnant size to reestablish landscape-scale processes such as fire and grazing of bison. One site conservation strategy under development by The Nature Conservancy of Texas and other partners, particularly Texas A&M University, is the expansion of existing native grasslands and reconnection of prairie fragments by the introduction of a planted native forage system based on genetic materials obtained from remnant prairies. Forage quality produced by such systems may be comparable to that produced by exotic systems but without chemical management and associated environmental impacts. In order to implement this strategy, the following will be undertaken:

- Complete and update inventory of remaining prairie.
- Identify and prioritize sites by threats, ecological condition, and potential for long-term viability.
- Fully protect critical tracts.
- Evaluate opportunities for landscape-scale restoration by identification of linkages between remnant patches.
- Develop sustainable planted native forage systems using locally obtained native genetic material.
- Identify profitability relative to exotic systems and develop establishment technologies.
- Identify and develop markets for prairie products (e.g., seed and forage).
- Introduce native forage grazing strategies to private-partner grazing lands within critical ecological sites.

- Provide native genetic material to CRP and other revegetation programs with critical ecological sites.
- Identify appropriate management strategies for planted native forage systems.

Conservation Partners

For contact information, please see appendix G.

- Collin County Open Space Program
- The Nature Conservancy of Texas
- Natural Areas Preservation Association
- Native Prairies Association of Texas
- Texas A&M University
- Texas Parks and Wildlife Department

Relationship to Other Classification Schemes

The boundary of the Texas Blackland Prairies is taken from Omernik (1995b). It generally corresponds with K(chler's (1975) bluestem prairie (68).

Prepared by J. Eidson and F. E. Smeins

Key Number:	**68**
Ecoregion Name:	**Western Gulf Coastal Grasslands**
Major Habitat Type:	**3.1 Grasslands, Savannas, and Shrublands**
Ecoregion Size:	**77,425 km²**
Biological Distinctiveness:	**Regionally Outstanding**
Conservation Status:	**Snapshot—Critical Final—Critical**

Introduction

Tallgrass coastal prairie was the primary plant community from southeastern Louisiana to the mouth of the Rio Grande River and even into the northern portion of Tamaulipas. Presently, less than 1 percent of this community type remains in pristine condition (Smeins, Diamond, and Hanselka 1991). Only 607 km² remain in Texas. In Louisiana, tallgrass coastal prairie has been essentially extirpated.

Conversion to row-crop and rice production, overgrazing, introduction of tame pasture grasses, urbanization, exotic plant establishment and expansion, and a lack of fire management have all contributed to the loss and degradation of this ecoregion. Fragmentation of remaining tallgrass prairie sites limits their function and quality for grassland bird species and other vertebrate as well as invertebrate endemics.

Diamond (1993) described the little bluestem–brownseed paspalum (*Schizachyrium scoparium–Paspalum plicatulum*) series, the most representative classification on upland sites, as globally imperiled with only six to twenty occurrences recorded. Climax grasses include tall bunch grasses such as seacoast bluestem (*Andropogon scoparuium* var. *littoralis*), eastern gamma grass (*Tripsacum dactyloides*), gulf muhly (*Muhlenbergia capiallris* var. *filipes*), and several species of panicum. As you move toward the Gulf Coast, the topography shifts to lower elevations and more saline soils. Concomitantly, the prairie becomes more intermixed with gulf cordgrass (*Spartina spartinae*), sedges (*Carex* spp., *Cyperus* spp.), rush (*Junicus* spp.), bulrush (*Scirpus* spp.), and salt grass (*Distichlis spicata*). Occasional shrublands consisting of mesquite (*Prosposis glandulosa*), huisache (*Acacia farnesiana*), lime prickly ash (*Zanthoxylum fagara*), and Texas persimmon (*Diospyros texana*) may be found in the lower one third of the Texas coast and into Tamaulipas (Gould 1962; Johnston 1963).

The tallgrass coastal prairie region of Texas is generally thought of as a continuum of the north-south range of tallgrass communities in Texas (Diamond and Smeins 1984). In contrast, the coastal sand plain of Texas is distinct enough that it cannot be considered an extension of the true prairie continuum (Diamond and Fullbright 1990). In general, grasslands eventually meld into freshwater and intertidal marsh habitat at the interface with Gulf bays and estuaries.

During predevelopment time, fire, grazing, and occasional tropical storms were the primary natural forces acting upon this ecoregion. While those forces still act upon remaining representative examples of this ecosystem, in general, overgrazing, conversion to tame grasses, fragmentation, and woody encroachment are the most critical disturbances at this time.

Biological Distinctiveness

The Western Gulf Coastal Grasslands (WGCG), while linked in some ways to the Texas Blackland Prairies, is a distinct ecosystem due to its more temperate climate, proximity to the Gulf of Mexico and associated natural processes (e.g., tropical storms), and geological origin with subsequent succession since the Pleistocene era inundation. This region is critical to many species of vertebrates, especially grassland birds, both resident and migrant, that are declining on a continental scale (Knopf 1994). Several endemic species of the WGCG have become extinct in the wild (e.g., red wolf) or are critically imperiled (e.g., Attwater's prairie chicken, whooping crane). Intertidal, estuarine marsh habitats are less threatened. However, emergent palustrine wetlands within the WGCG are the most threatened type along the Gulf Coast.

Conservation Status

Habitat Loss and Degradation

As already mentioned, less than 1 percent of the WGCG remains in near pristine condition (Smeins, Diamond, and Hanselka 1991). Conversion to agricultural production has caused the greatest loss. Fragmentation of remaining habitat via subdividing large tracts into more marketable "ranchettes" leads to other degrading factors such as overgrazing, exotic plant expansion, lack of fire as a natural or prescribed process, and modification of local hydrological features by means of land leveling. Urbanization around larger metropolitan areas (e.g., Houston) has been another direct cause of habitat loss. This will proceed with the continued proliferation of suburban development.

Coastal wetlands are less suited, in most cases, for high-density development and agricultural conversion. However, channelization projects, with all their associated damage to overland sheet flow and hydrological function, continue to impact this portion of the WGCG. Subsidence, erosion, and loss of emergent wetlands are serious problems that will continue.

Remaining Blocks of Habitat

The National Wildlife Refuge System possesses several important sites within the WGCG, although the main emphasis of their acquisition was development of refuges specifically for waterfowl. Tallgrass prairie existing on these refuges was generally not considered key to acquisition priorities with few exceptions (e.g. Attwater's Prairie Chicken National Wildlife Refuge). In the list below, stars indicate the presence of significant grassland areas. The refuges include:

- Delta National Wildlife Refuge
- Lacassine National Wildlife Refuge
- Cameron Prairie National Wildlife Refuge
- Rockefeller National Wildlife Refuge
- Sabine National Wildlife Refuge
- Texas Point National Wildlife Refuge
- McFaddin National Wildlife Refuge
- Anahuac National Wildlife Refuge
- Moody National Wildlife Refuge
- Attwater's Prairie Chicken National Wildlife Refuge★★
- Brazoria National Wildlife Refuge
- San Bernard National Wildlife Refuge
- Big Boggy National Wildlife Refuge
- Whitmire Division of Aransas National Wildlife Refuge
- Aransas National Wildlife Refuge★★
- Matagorda Island National Wildlife Refuge★★
- Laguna Atascosa National Wildlife Refuge
- Santa Ana National Wildlife Refuge
- Padre Island National Seashore (National Park Service)★★

Other significant areas owned or managed by state or NGOs include:

- Paul J. Rainey Sanctuary (Audubon)
- Russell Sage Refuge (State of Louisiana)
- Little Pecan Island Preserve (TNC)
- J.D. Murphree Wildlife Management Area (WMA) (State of Texas)
- High Island Sanctuary (Audubon)
- Candy Abshier WMA
- Armand Bayou Preserve (private)★★
- Coastal Center (University of Houston)★★
- Galveston Bay Prairie Preserve (TNC)★★
- Pierce Marsh Preserve (TNC)
- Peach Point WMA (State of Texas)★★
- Mad Island WMA (State of Texas)
- Clive Runnells Family Mad Island Marsh Preserve (TNC)★★
- Mustang Island State Park (State of Texas)★★

In addition, there are approximately ten known sites on private land that contain excellent representative examples of the WGCG.

Degree of Fragmentation

Native, nonmigratory endemics are very much affected by fragmentation of the WGCG. Fragmentation has led to population isolation and presumed negative impacts to genetic heterozygosity for Attwater's prairie chickens. Less is known at this time regarding fragmentation effects upon small mammals and invertebrates.

Degree of Protection

A great deal of protection work has been done in this region. However, most protected sites are clustered along the immediate Gulf Coast wetland portion of the region. Priorities for protection were related to waterfowl protection and conservation, not tallgrass coastal prairie preservation. The best coastal prairie in existence occurs on private land.

Types and Severity of Threat

Agricultural conversion has basically consumed all land suitable for agriculture. Fragmentation of remaining sites, urbanization, lack of fire, exotic plant expansion, and overstocking of cattle are the primary present threats. The difficulty of private landowners to keep family lands intact because of the existing tax structure (especially as it relates to inheritance taxes) poses a prime problem for conserving tallgrass coastal prairie on private lands. Restoration potential is high for many sites; however, lack of concern by the general public (and therefore, minimal financial and legislative support) for tallgrass prairie conservation is a significant problem.

Suite of Priority Activities to Enhance Biodiversity Conservation

- Develop baseline inventory of all remaining significant tallgrass coastal prairie tracts.
- Devise triage approach to conserving remaining sites to identify highest-priority tracts for conservation action.
- Work with state and federal agencies, NGOs, and private landowners that presently possess significant habitat to determine if stewardship practices in place are maintaining and enhancing tallgrass prairie habitat.

- Establish relationships with private landowners and discuss significance of their grasslands when possible.
- Work on developing private lands initiative in cooperation with the USFWS and state agencies.
- Develop grazing and fire management guidelines that will assist small landowners (>80 ha) in assessing how they can improve their grasslands.
- Develop and implement conservation education projects specifically related to West Gulf Coastal Grasslands, their ecology, significance, and importance.
- Establish a network of native seed sources that could be used for prairie restoration efforts along the Gulf Coast.

Conservation Partners

For contact information, please see appendix G.

- Louisiana Natural Heritage Association
- National Biological Service, South Science Center
- Native Prairies Association of Texas
- The Nature Conservancy of Louisiana
- The Nature Conservancy of Texas
- Society of Ecological Restoration, Texas Chapter
- Southwestern Cattle Raisers Association
- Texas Farm Bureau
- Texas Parks and Wildlife Department
- U.S. Fish and Wildlife Service

Relationship to Other Classification Schemes

The Western Gulf Coastal Grasslands basically overlays with the Gulf Coast Prairies and Marshes region as delineated for Texas by Gould et al. (1960), Bailey (1994), and The Nature Conservancy (1997). On the lower Texas coast, Diamond and Fullbright (1990) separate the coastal sand plain from Gulf Coast Prairies and Marshes.

Prepared by J. Bergan

Key Number:	**69**
Ecoregion Name:	**Everglades**
Major Habitat Type:	**3.2 Flooded Grasslands**
Ecoregion Size:	**20,117 km²**
Biological Distinctiveness:	**Globally Outstanding**
Conservation Status:	**Snapshot—Vulnerable**
	Final—Vulnerable

Introduction

The Everglades, located at the southern tip of peninsular Florida, is the most famous wetland in the United States and one of the most distinct in the world. The Everglades is unique among the world's large wetlands because it derives its water from rainfall. Other large and famous wetlands, such as the Pantanal of South America, the Okavango of Botswana, and the Llanos in Venezuela and Colombia, derive most of their water and nutrient inputs from river flooding. The unique "sheet flow," the slow flow of water over shallow, broad tracts of marsh, inspired Douglas (1947) to name the Everglades "River of Grass." As important as sheet flow is, the groundwater connections of the Everglades to Lake Okeechobee, the second-largest freshwater lake entirely within the United States, are also essential for the maintenance of the wetland. The linkages between the Everglades, Lake Okeechobee, and the Kissimmee River, which provides 80 percent of the surface flow into Lake Okeechobee, illustrate the importance of connectivity among ecoregions to maintain integrity.

The boundaries of this ecoregion extend to include the Big Cypress Swamp to the northwest, the southern edge of Lake Okeechobee to the north, and the Atlantic coastal ridge to the east (i.e., the edge of ecoregion [52]). The Everglades climate has been classified as subtropical, featuring hot humid summers when 80 percent of rainfall occurs and mild winters. Rainfall varies spatially across southern Florida so that the inland marshes and Lake Okeechobee receive only about 60 percent of the rainfall levels recorded in the coastal areas (Gunderson and Loftus 1993). The most important climatic feature is also the most important natural disturbance factor: the recurrent hurricanes that strike most frequently from August through October. Extensive habitat destruction can occur from high winds, storm surge, and rainfall. Frosts also limit the northern distribution of many tropical species to this ecoregion and help to further define its boundaries.

Many observers have identified the Everglades as one of the most endangered of North American ecoregions as a result of clearing for agriculture, diversion of water flow, and other developments. Recovery efforts are now underway, supported by a broad association of environmentalists active in the region.

Biological Distinctiveness

The extraordinary biological richness of the Everglades has been well documented, particularly the spectacular wading birds, alligators, crocodiles, snail kites, and mangrove species. The habitats that support this rich assemblage of species include: ponds, sloughs, graminoid (grasslike wetlands), and forested wetlands (Gunderson and Loftus 1993). The forested uplands of the Everglades also harbor distinct assemblages of species, dominated by many trees with tropical Caribbean affinities. Ponds occur throughout the Everglades on the lowest elevational sites and are important areas for alligators. Typical aquatic vegetation may include water lilies (*Nymphaea* spp.) and spatter dock (*Nuphar advena*). Sloughs are found on the wettest sites, which experience nearly constant inundation and are underlaid by peat soils. Floating aquatics such as white water lily (*Nymphaea odorata*) are common and a submerged complex of plants is dominated by bladderwort (*Utricularia* spp.). Perhaps the most famous habitat of the Everglades is the sawgrass (*Cladium jamaicense*, in reality a member of the sedge family rather than a true grass), which dominates marshes in terms of abundance and biomass. Sawgrass is well adapted to fire and fluctuating water levels but is adversely affected by prolonged high water (Davis 1990). Wet prairies are dominated by rushes.

Also of great biological interest is the diversity of forest types also called tree islands due to being surrounded by a "sea" of sawgrass. They include swamp forest dominated by red bay (*Persea borbonia*), pond apple (*Annona glabra*) forests, cypress (*Taxodium ascendans*) forests, and hardwood hammocks with a variety of tropical species, including palms, that occur in the United States only in southern Florida. The tropical hammocks contain a diverse assemblage

of local endemic tree snails of the genus *Liguus*, and tropical butterfly species are also common.

Fire is an important feature in the Everglades that along with water flow helps to maintain early successional habitats. Increases in soil surface elevation lead to greater dominance of hardwoods and scrub.

As with other peninsulas, animal species richness tends to decline in most taxa from the mainland connection to the tip; Southern Florida hosts reduced numbers of vertebrates and invertebrates from north to south. About seventy breeding bird species occur, seventeen species of mammals, thirty species of reptiles, and fourteen species of amphibians. Ponds and creeks are important as dry-season refugia for many of these species. Sufficient water levels are crucial for a number of taxa to carry out breeding.

Conservation Status

Habitat Loss and Degradation

After floods in the middle of this century, a vast area of 2,830 km² was converted to agriculture just south of Lake Okeechobee. Today, sugar cane is the most commonly cultivated crop along with truck crops. The effects of fertilizer and pesticide application have reverberated throughout the ecoregion, causing widespread degradation. The limestone-based (marl) prairies of the southern Everglades have also been largely converted to agriculture. Eutrophication from agricultural runoff and urbanization has caused the invasion of cattails (*Typha* spp.).

Most insidious has been disruption of natural water flow through the construction of canals and water management practices. In order to mitigate droughts, water levels were kept abnormally high, leading to a decline in a number of plant communities including sawgrass. Another aspect of water management is excessive drainage of wetlands, which affects foraging areas for wading birds and has led to increased levels of fires, which shifts plant succession (Gunderson and Loftus 1993). The southern prairies have been irreversibly altered as a result of plowing of the substrate and farming. When abandoned, rather than returning to native assemblages, these areas are invaded by exotics such as Brazilian pepper (*Schinus terebinthifolius*). Between 1940 and 1980, the populations of Dade, Broward, and West Palm Beach Counties increased by 830 percent, to 3.2 million people (U.S. Department of Commerce 1980). This rapid population growth has led to the conversion of the eastern 12 percent of the Everglades.

Habitat alteration and outright loss have endangered a number of plant and animal species. There has been an approximate 90 percent decline in wading bird abundance over the last century. The invasion by the Australian tree *Melaleuca* has been a minor ecological disaster. Many species of introduced fish and lizards have displaced native species. Fourteen plants and nine vertebrates are federally listed.

In sum, it is estimated that no more than 2 percent of the original Everglades ecosystem is truly intact. However, about 30 percent of this unit remains in an altered state that could be restored with proper management as described below.

Remaining Blocks of Habitat

The most important blocks of habitat (all in southern Florida) are:

- Everglades National Park
- Big Cypress National Preserve
- Arthur Marshall Loxahatchee National Wildlife Refuge
- Florida Panther Wildlife Reserves
- Fakahatchee Strand
- Corkscrew Swamp
- Water Conservation Areas
- Biscayne National Park

Degree of Fragmentation

Highways and roads pose a barrier to some species (mammals and herpetofauna), but underpasses constructed on I-75 have relieved barrier effects for panthers and others. Critical corridors on a microscale are well documented. Large blocks of moderately altered habitat (Big Cypress, Everglades) are contiguous or nearly so.

Degree of Protection

Within the historic Everglades, there are three conservation areas:

- Everglades National Park—2,140 km^2
- Big Cypress National Preserve—222 km^2
- Arthur Marshall Loxahatchee National Wildlife Refuge—572 km^2

Other water management areas are managed in part for biodiversity conservation. Despite protection of a significant portion of the historic Everglades and recognition as an international biosphere reserve, the ecosystem faces severe threats from the impact of surrounding urban sprawl, ecologically unsound water management, agricultural development, invasion of exotic species, and fire.

Types and Severity of Threats

Assuming that restoration efforts will succeed, degradation threats will decline. Otherwise, degradation threats will intensify and irretrievably affect this ecosystem. Poaching of deer, alligators, and panthers continues to be a threat.

Suite of Priority Activities to Enhance Biodiversity Conservation

The restoration of the South Florida/Everglades ecosystem is dependent on a number of strategies and priorities that need to be addressed as the federal and state governments develop a comprehensive plan for restoration of the ecosystem. These include:

- major changes in water flow, delivery, and management in South Florida intended to benefit Everglades National Park, Florida Bay, the Everglades Water Conservation Areas, and other key areas of the Everglades system
- improvements in water quality throughout the system, and especially in meeting standards for phosphorus reduction from sugar cane lands in the Everglades Agricultural Area
- substantial additional land acquisition by both the federal and state governments in order to complete the boundaries of existing conservation lands, or to provide the land base necessary to restore more natural water flows through the system
- a very significant investment in state and federal funds for restoration of at least $500 million/year for the next five or six years for research, monitoring, land acquisition, and construction of water delivery and water-quality improvement projects.
- cost-sharing for restoration by the federal government, the state of Florida, and private interests that have contributed to the decline

of the Everglades system, including sugar cane growers and other agricultural interests.

Conservation Partners

For contact information, please see appendix G.

- Everglades Coalition
- Florida Audubon
- Florida Natural Areas Inventory
- National Audubon Society
- National Audubon's Everglades Ecosystem Restoration
- The Nature Conservancy of Florida

Relationship to Other Classification Schemes

Ecoregion boundaries are derived from Omernik's (1995b) unit no. 76 (Southern Florida Coastal Plain) and combines Küchler's (1975) units 82 (Cypress savannas), 83 (Everglades), and 96 (Mangroves). Bailey (1994) identifies a similar ecoregion (section 411A), which differs from our unit in that we separate the Pine Rocklands as part of a separate ecoregion (Florida Sand Pine Scrub [52]).

Prepared by E. Dinerstein, A. Weakley, R. Noss, and K. Wolfe

Focus on the Mediterranean Habitat Type

Mediterranean climates are characterized by cool, wet winters and hot, dry summers. Only five regions in the world experience Mediterranean conditions: the Mediterranean region of Europe, the kwonga of southwestern Australia, the fynbos of southern Africa, the Chilean matorral, and the Mediterranean ecoregions of California. Relative to other major habitat types, Mediterranean ecosystems are quite rare, are widely scattered around the earth, and cover only a small area of the earth's surface. Yet the biodiversity of these ecosystems is extraordinary. They harbor approximately 20 percent of the world's plant species, very high levels of local and regional endemism, high rates of beta-diversity (species turnover across distance or gradients), and numerous species and communities

with unusual adaptations, ecologies, and evolutionary histories. Habitats within Mediterranean zones are varied and include shrublands, chaparral, grasslands, desert grasslands, savannas, woodlands, wetlands, and coastal and montane forests. Regular inundation by fog during the long, dry summers creates fog-dependent scrub or forest communities narrowly distributed along coasts in most Mediterranean regions. Coastal communities are often highly distinctive from those found farther inland in terms of their species assemblages, structure, and disturbance regimes, and harbor many endemic species.

Plants and animals of Mediterranean areas must adapt to the stressful conditions of long, hot summers with little rain. Plants have developed adaptations to reduce moisture loss during the dry season, such as sclerophyllous leaves, twigs, and trunks; pronounced pilosity; deciduous leaves; and underground structures for storing water. Animals can aestivate, retreat to permanent water sources, migrate to other habitats, be nocturnal, or have the ability to subsist solely on water from dew, seeds, or plant parts. Mediterranean communities typically experience recurring fires that shape successional processes and select for fire-adapted biotas.

California has four ecoregions that experience Mediterranean conditions: California Central Valley Grasslands [54], California Interior Chaparral and Woodlands [70], California Montane Chaparral and Woodlands [71], and California Coastal Sage and Chaparral [72]. Almost the full range of habitat types that can be found within Mediterranean zones occurs within these ecoregions. Topography, proximity to coastal fog, soil type, slope, aspect, latitude, rainfall, and length of the dry season all influence the particular communities that develop in different areas. Mediterranean habitats are found from the north end of the Central Valley southward, in a band extending from the foothills of the Sierra Nevada to the coast, to the coastal area of northern Baja California. The coastal ranges and Mojave and Sonoran deserts narrow the width of the Mediterranean zone in southern California. Fog belts extend along the coast from the southern redwood forests in northern California to northern Baja California. The foothills of the Sierra Nevada occasionally experience summer rains from mountain thunderstorms. Summer rains from the Southwest

Monsoon may occur irregularly in southern and central areas.

California Mediterranean ecoregions are quite threatened, as are all of the Mediterranean regions around the world. The favorable climate attracts human settlement and agriculture. Many Mediterranean communities have low resiliency to human disturbance, are readily invaded by alien species, and can be significantly altered through suppression or increase of fire. A much greater conservation effort is needed in these highly threatened and biologically outstanding ecosystems.

Key Number:	**70**
Ecoregion Name:	**California Interior Chaparral and Woodland**
Major Habitat Type:	**4.1 Mediterranean Scrub and Savanna**
Ecoregion Size:	**71,658 km²**
Biological Distinctiveness:	**Globally Outstanding**
Conservation Status:	**Snapshot—Vulnerable Final—Vulnerable**

Introduction

This ecoregion forms a nearly continuous ellipse of oak woodland and chaparral around the California Central Valley, ranging from 300 to 3,000 ft in elevation (Barbour et al. 1993). The ecoregion continues across the coast ranges to the Pacific from Point Reyes to Santa Barbara, with breaks around the redwood belt south of San Francisco Bay and the montane communities of the Santa Lucia Range that parallel the coast south of Monterey Bay.

Biological Distinctiveness

Within the California Interior Chaparral and Woodland ecoregion, one finds a mosaic of grasslands, chaparral shrublands, open oak savannas, oak woodlands, serpentine communities, closed-cone pine forests, pockets of montane conifer forests, wetlands, salt marshes, and riparian forests. Oak savannas and chaparral are the most widespread and characteristic communities. Valley communities at lower elevation are characterized by foothill pine (*Pinus sabiniana*) and blue oak (*Quercus douglasii*), as well as a

range of woodland and chaparral plants such as California buckeye (*Aesculus californica*), manzanita (*Arctostaphylos* spp.), redbud (*Cercis occidentalis*), chamise (*Adenostoma fasciculatum*), and scrub oak (*Quercus dumosa*). Coast live oak, canyon live oak, golden-cup oak, valley oak, interior live oak, and maul oak are also part of the diverse oak flora of this ecoregion. A number of endemic and relict pines and cypresses occur within the ecoregion, often restricted to single ranges or soil types such as serpentine (e.g., Sargent cypress, *Cupressus sargentii*, and McNab cypress, *Cupressus macnabiana*). Many of the closed-cone pines are dependent upon periodic fires to open their cones and prepare the understory for seedlings.

A diverse assemblage of shrubby and herbaceous plants occurs within the savannas and chaparral, with many local and habitat endemics. Kartesz (pers. comm., 1996) estimates that the ecoregion harbors 2,036 species of herbaceous plants, vines, and shrubs. Maritime chaparral around Monterey Bay is noted for a number of endemics, including Hooker manzanita (*Arctostaphylos hookeri*), Monterey manzanita (*A. monteryensis*), pajaro manzanita (*A. pajorensis*), sandmat manzanita (*A. pumila*), Monterey ceanothus (*Ceanothus rigidis*), and Monterey goldenbush (*Ericameria fasciculata*) (Schoenherr 1992). Vegetative communities on serpentine are particularly rich in endemic plants and invertebrates, including leather oak (*Quercus durata*), interior silktassel (*Garrya condonii*), milkwort streptanthus (*Streptanthus polygaloides*), and Muir's hairstreak (*Mitoura nelsoni muiri*). Sandstone-derived soils near Coalinga support a number of disjunct populations of desert species, including Mojave sand verbena (*Abronia pogonatha*) and narrowleaf goldenbush (*Ericameria linearifolius*) (Schoenherr 1992).

Sixty species of mammals occur in these habitats, with six endemic and near-endemic species—the giant kangaroo rat (*Dipodomys ingens*), Heermann kangaroo rat (*D. heermani*), Santa Cruz kangaroo rat (*D. venustus*), Sonoma chipmunk (*Eutamias sonomae*), Suisun shrew (*Sorex sinuosus*), and salt marsh harvest mouse (*Reithrodontomys raviventris*)—the highest number of endemic mammal species for U.S. and Canadian ecoregions. Plethodontid salamanders are diverse, with five endemic species. Among the 100 species of birds that occur in this ecoregion, scrub jays (*Aphelocoma coerulescens*), acorn woodpeckers (*Melanerpes formicivorus*), and wrentits (*Chamaea fasciata*) are a few of the most characteristic species. The fresh and salt marshes of San Francisco Bay and other large estuaries along the coast once supported enormous populations of ducks, geese, shorebirds, and wading birds, particularly during the winter and migration seasons. Army ants (*Neivamyrmex* spp.) and primitive bristletails and land snails are among the ecoregion's large number of relict and unusual invertebrate species.

Many species and communities within the ecoregion are adapted to periodic fires; indeed, many species depend on fires for regeneration. An entire guild of annual herbaceous plants that occur in chamise chaparral has seeds that lie dormant for long periods until fires trigger their germination approximately every twenty to twenty-five years. Closed-cone pine communities historically burned about once every twenty-five to fifty years, and some species, such as the knobcone pine (*Pinus attenuata*), require fire to open their cones.

In summary, this ecoregion harbors a number of unique communities, with many species whose distributions clearly illustrate the ecological islands, specialization, relict nature, unique geologic history, and endemism of Californian biodiversity.

Conservation Status

Habitat Loss and Degradation

Approximately 30 percent of this ecoregion can be described as intact, concentrated in steeper foothill and montane areas. Virtually all native bunchgrass communities have been replaced by annual grassland understories in woodland areas. Some woodland areas above 915 m (3,000 ft) in the Sierra Nevada belt still have larger blocks of relatively intact habitat. Valley oak savannas and woodlands, particularly those on valley bottoms and gentler slopes, have been largely eliminated. Valley oak savannas are presently much rarer than blue oak savannas. Habitat for a number of species has been lost through centuries of grazing, alteration of fire regimes, harvesting of oaks, and introduction of aggressive alien species. Only 10 percent of all wetlands in the state of California remain, and much of what is left is highly threatened.

Remaining Blocks of Intact Habitat

The largest blocks of chaparral and sclerophyll-oak woodlands occur in the inner north and south coast ranges. Larger blocks of freshwater and salt marshes occur around Suisun and San Pablo Bays. Military bases such as Vandenberg and Hunter Liggett encompass extensive high-quality habitat.

Degree of Fragmentation

Habitat fragmentation is high, given the extent of range modification, road building, developments, and natural watershed topography. Many communities are naturally disjunct and fragmented, and human fragmentation exacerbates their isolation.

Degree of Protection

Few good examples of this habitat type are adequately protected. The Nature Conservancy, BLM, California Department of Fish and Game, and USFWS are working cooperatively to restore the vast Carrizo Plains. Additional reserves are Cold Canyon Nature Reserve managed by the University of California, and Pinnacles National Monument near Monterey. Unfortunately, many of these reserves are too small for long-term viability, and most suffer from decades of fire suppression, invasion of exotic species, and overgrazing by deer and rodent populations elevated through predator control. Much of the remaining salt marsh habitats in San Francisco Bay are protected to some degree, although external changes to hydrologic conditions and pollution may ultimately degrade these habitats.

Types and Severity of Threats

Rural residential development is a major threat to remaining low-elevation oak woodlands. Fire suppression represents a significant and pervasive threat in many chaparral and woodland habitats. Accumulation of combustible material, a natural process in these fire-dominated ecosystems, will inevitably lead to hot burns that will kill seedbanks, seedlings, and normally resistant larger trees. Plant species that specialize on regeneration after fires can also be extirpated by prolonged periods of dormancy between fire events. In the highly altered landscape, many species may not be able to find refuge in natural habitats during the inevitable fires, precluding their function as source populations.

Overgrazing by domestic livestock, including sheep and cattle, and wild deer and rodents is a serious problem. Blue oaks are not regenerating throughout their range because of high seedling mortality from grazing, seed predation, and, in some part, competition from introduced grasses. Cutting down trees for firewood and pasture is a persistent threat. The revival of wood-burning stoves has increased demand for oak wood in recent years. Clearing open woodlands for pasture also reduces forage quality (Barbour et al. 1993). Riparian areas and water sources, formerly critical sites for maintaining wildlife and rare plant species during the dry summers, have been extensively degraded or destroyed by domestic animals and water diversion. Expansion of agriculture, such as vineyards, and development projects, such as housing and golf courses, is swiftly converting many habitats in the Sierra foothills and areas near larger cities on the coast.

The continuing practice of type conversion of chaparral into grasslands through repeated burning and planting of annual grasses reduces habitat for species of herbaceous plants, invertebrates, birds, mammals, and reptiles that specialize on chaparral habitats. The loss of chaparral over the landscape is also thought to increase erosion and reduce water storage capacity of the habitat (Barbour et al. 1993). Introduced species are a serious and pervasive problem for native species. This ecoregion has 2,105 species of introduced plants, constituting 30 percent of the flora, the highest for any ecoregion in the United States and Canada.

The Bureau of Land Management has been involved in a number of large conservation projects in the ecoregion, including the Carrizo Plains, Cosumnes River, Sacramento River, and Panoche Hills. However, the bureau's authorization of off-road vehicle use on the serpentine communities of the Clear Creek Recreation Area of the inner Coast Range threatens a globally unique habitat (Schoenherr 1992).

Wetlands throughout the region continue to be threatened by development, draining, diversions, alteration of hydrologic conditions, and pollution. Introduced rats, cats, opossums, foxes, and other predators also cause mortality in wetland specialists such as salt marsh harvest mice and clapper rails.

Suite of Priority Activities to Enhance Biodiversity Conservation

- Take action to reduce the loss of oak woodlands. Management practices to promote regeneration of valley and blue oaks need to be widely implemented, including fencing of areas, planting of seedlings, and control of overabundant herbivores and alien plant species. Ranching techniques compatible with biodiversity conservation need to be devised and implemented.
- Restore fire cycles within their natural range of variation through controlled burns, fuel reduction, and control of alien plants. Areas identified as supporting relatively intact natural communities would be immediate targets for intensive fire management.
- Curtail or carefully manage grazing of cattle and other livestock around riparian areas and in sensitive habitats (e.g., some edaphic communities and areas where locally endemic species persist) to mimic natural processes. Cessation of off-road vehicle activity is also needed for conservation in many areas.
- Protect and restore wetlands, wherever they occur. Restoration of hydrologic regimes is particularly important.

Conservation Partners

For contact information, please see appendix G.

- Audubon Society
- Bureau of Land Management
- California Department of Fish and Game
- California Native Grass Association
- California Native Plant Society
- California Oak Foundation
- Department of Defense
- The Nature Conservancy
- The Nature Conservancy of California
- Sierra Club
- U.S. Fish and Wildlife Service
- state parks, county parks, and local land trusts

Relationship to Other Classification Schemes

The ecoregion boundaries used in this analysis closely match that of Omernik's (1995b) northern section of the Southern and Central California Plains and Hills (6). The interface between the Central Valley Grasslands and this ecoregion has been slightly modified by Robert Holland from Omernik's original boundaries. When taken together, Bailey's (1994) Northern California Interior Coast Ranges Section, Sierra Nevada Section, and Central California Coast Ranges Section roughly approximate this ecoregion. However, none of these sections continues to the coast, and the Santa Lucia Range and parts of the southern transverse ranges are incorporated into the latter section. Moreover, there is a break in the ring of sections at the northernmost area of the Central Valley. Küchler (1975) notes a variety of vegetation types within this ecoregion, including California oakwoods, California steppe and prairie, yellow pine forest, and chaparral.

Prepared by D. Olson, reviewed by R. Cox

Key Number:	**71**
Ecoregion Name:	**California Montane Chaparral and Woodlands**
Major Habitat Type:	**4.1 Mediterranean Scrub and Savanna**
Ecoregion Size:	**20,408 km²**
Biological Distinctiveness:	**Globally Outstanding**
Conservation Status:	**Snapshot—Vulnerable**
	Final—Vulnerable

Introduction

The montane habitats of Southern California share many species with the Sierras to the north and the lower-elevation Mediterranean woodlands and chaparral. Their communities, however, are distinctive in structure and composition, in addition to supporting a number of endemic and relict species. The ecoregion encompasses most of the Transverse Range that includes the San Bernardino Mountains; San Gabriel Mountains; portions of the Santa Ynez and San Rafael Mountains; Topatopa Mountains;

San Jacinto Mountains; the Tehachapi, Greenhorn, Piute, and Kiavah Mountains that extend roughly northeast and southwest from the southern Sierra Nevada; and the Santa Lucia Range (part of the Coast Range) that parallels the coast southward from Monterey Bay to Morro Bay. Several of the mountain ranges in this ecoregion are complex and high, with peaks ranging up to 3,500 m (11,485 ft) elevation in the Transverse Range. Such topography creates conditions for a wide range of natural communities, ranging from chaparral to mixed-conifer forests and alpine habitats.

Biological Distinctiveness

The California Montane Chaparral and Woodland ecoregion consists of a complex mosaic of coastal sage scrub, lower chaparral dominated by chamise, upper chaparral dominated by manzanita, desert chaparral, pinyon-juniper woodland, oak woodlands, closed-cone pine forests, yellow pine forests, sugar pine–white fir forests, lodgepole pine forests, and alpine habitats. The prevalence of drought-adapted scrub species in the flora of this ecoregion helps distinguish it from similar communities in the Sierras and other portions of northern California. Many of the shared Sierra Nevadan species typically are adapted to drier habitats in that ecoregion, Jeffrey pine (*Pinus jeffreyi*) being a good example.

Some coastal sage scrub occurs on the southern slopes of the Transverse Range, although chamise (*Adenostoma fasciculatum*) chaparral and scrub oak chaparral cover most lower habitats. Higher up, cold chaparral dominated by manzanitas are interspersed with closed-cone pine forests, Coulter pine (*Pinus coulteri*) woodlands, and endemic bigcone Douglas fir (*Pseudotsuga macrocarpa*) communities. Oak species are an important component of many chaparral and forest communities throughout the ecoregion. Canyon live oak, interior live oak, tan oak, Engelmann oak, golden-cup oak, and scrub oak are some examples. Mixed-conifer forests are found between 1,371 and 2,896 m (4,500 to 9,500 ft) elevation with various combinations and dominance of incense cedar, sugar pine and white fir, Jeffrey pine, ponderosa pine, and mountain juniper. Subalpine forests consist of groves of limber pine (*Pinus flexilis*), lodgepole pine, and Jeffrey pine. Very old individual trees are commonly observed in these relict subalpine forests. Within this zone are subalpine wet meadows, talus slope herbaceous communities, krummholz woodlands, and a few small aspen groves. Herbaceous and shrubby species are very diverse and share affinities with the Sierras, Mojave Desert, and coastal and interior chaparral and woodlands. Numerous endemic plant species occur in many different communities.

In addition to these general vegetation patterns, this ecoregion is noted for a variety of ecologic islands, communities with specialized conditions that are widely scattered and isolated and typically harbor endemic and relict species. Examples include two localities of knobcone pine (*Pinus attenuata*) on serpentine soils, scattered vernal pools with a number of endemic and relict species, and isolated populations of one of North America's most diverse cypress floras, including the rare Gowen cypress (*Cupressus goveniana goveniana*), restricted to two sites on acidic soils in the northern Santa Lucia Range; Monterey cypress (*C. macrocarpa*), found only at two coastal localities near Monterey Bay; and Sargent cypress (*C. sargentii*), restricted to serpentine outcrops. Monterey pine (*Pinus radiata*) is also restricted to three coastal sites near Monterey Bay. The ecoregion, together with the Northern California Coastal Forests ecoregion [40], supports eight endemic conifer species, the highest number for any ecoregion in the United States and Canada.

The Santa Lucia Range supports scattered populations of redwoods limited to fog-inundated coastal valleys. Coast live oaks and madrone form coastal evergreen communities intermixed with coastal sage and chamise chaparral. Higher up, one finds tan oak and canyon live oak woodlands eventually grading into forests of ponderosa pine, sugar pine, Jeffrey pine, Coulter pine, and serpentine-associated knobcone pine. The range also harbors the unusual and endemic Santa Lucia or bristlecone fir (*Abies bracteata*) between 610 and 1,525 m (2,000 and 5,000 ft) elevation.

The ecoregion is also home to a few endemic or near-endemic vertebrates, such as the white-eared pocket mouse (*Perognathus alticolis*) and five endemic and near-endemic amphibians, largely Plethodontid salamanders. California condors once inhabited much of the ecoregion, with the western Transverse Range acting as a refuge for the last wild population. Winter aggre-gations of monarch butterflies occur at several localities near Monterey Bay and southward along the coast. Some larger vertebrate predators still occur in the ecoregion, including puma, bobcat, coyote, and ringtails.

Conservation Status

Habitat Loss and Degradation

Approximately 30 percent of the ecoregion is relatively intact habitat, with the caveat that virtually all bunchgrass elements have been replaced by introduced annual grasses, and fire suppression, grazing, and loss of riparian and aquatic habitats are a major problem everywhere.

Remaining Blocks of Intact Habitat

The Ventana wilderness area in the Santa Lucia Range and the Ventura region of the Transverse Range have some of the larger intact habitat blocks. The northern extension of the ecoregions toward the Sierra Nevada has some large blocks, although much of this region is significantly impacted by grazing, cement mining, cotton, and windmill farms. Some larger, more intact blocks occur on Forest Service lands and the San Emigidio Ranch, recently purchased by the Wildlands Conservancy. The few blocks of conifer forests on the mountain peaks of the Transverse Range are all disturbed by development, grazing, logging, and fire suppression. Most of the designated wilderness areas are small and heavily used. Some of the best examples of native blue and valley oak woodlands occur in the inland valleys of the northern Santa Lucia Range, near Hunter Liggett and the Ventana Wilderness.

Degree of Fragmentation

The entire region is heavily roaded, and valley bottoms are largely developed. Fragmentation and isolation of intact habitat blocks are relatively high.

Degree of Protection

Much of the ecoregion falls within the Los Padres National Forest. The forests and chaparral of this national forest suffer from intensive logging of low-productivity ecosystems, overgrazing, air pollution, loss of aquatic habitats, heavy recreational use, and decades of fire suppression.

Types and Severity of Threats

Fire suppression is a severe problem throughout the ecoregion, allowing fuel loads to build up and increase the probability of ecologically devastating hot fires. Few large predators remain due to centuries of hunting and predator control. The high densities of deer, rodents, and other herbivores that have resulted from predator extirpation contribute to intensive grazing and seed predation. This effect results in dramatic changes to plant and animal communities throughout the ecoregion. Many springs, streams, rivers, and other aquatic habitats have been highly disturbed through land development, overgrazing, sedimentation, introduced species (18 percent of the flora), and water diversions. Extensive development around lakes and streams for resorts and vacation homes has altered many montane aquatic systems. The wide range of terrestrial species that depend on critical water resources to survive, such as amphibians, has been severely impacted by the loss and alteration of aquatic habitats. High-impact recreational activities, such as driving off-road vehicles and hunting, cause significant damage of plant communities and mortality and disturbance of wildlife.

Mixed conifer and closed-cone pine forests are heavily impacted by air pollution from urban centers. Ozone from smog causes ponderosa pine and other species of conifer, shrubs, and lichen to weaken and die.

Suite of Priority Activities to Enhance Biodiversity Conservation

- Restore fire events to frequencies and intensities within their natural range of variation through management prescriptions such as controlled burns and fuel reduction (not salvage logging).
- Strictly protect ecologic islands and populations of rare species such as Gowen cypress and knobcone pine forests, vernal pools, and bigcone Douglas fir and Santa Lucia fir groves. For example, strong protection needs to be given to Huckleberry Hill and its environs near Monterey to conserve several rare tree species.

- Protect the last blocks of foothill oak woodlands, a habitat type severely threatened throughout its range.
- Prohibit off-road vehicle use and grazing in fragile and rare serpentine plant communities.
- Prohibit further cutting of the last remaining groves and trees used by overwintering monarchs. These sites may experience rare environmental conditions that are necessary for the butterflies to survive. Some city governments near Monterey Bay recently voted to allow cutting of some of the butterfly trees for development, showing a disregard for the global rarity of this phenomenon and the fragile nature of the scattered populations. Multiple butterfly sites may be necessary to allow long-term persistence of the butterflies in the face of natural weather events.
- Encourage the U.S. Forest Service to reduce timber harvest in these habitats, which are characterized by low productivity and experience significant periods of drought. Grazing and continued destruction of riparian and aquatic habitats must be curtailed on both federal and private lands.

Air pollution is also a significant problem, yet reduction in smog emissions from Los Angeles and environs is a very challenging hurdle.

Conservation Partners

For contact information, please see appendix G.

- California Native Plant Society
- The Nature Conservancy
- The Nature Conservancy of California
- Sierra Club
- U.S. Fish and Wildlife Service
- U.S. Forest Service

Relationship to Other Classification Schemes

This ecoregion matches Omernik's (1995b) Southern California Mountain ecoregion (8). Bailey (1994) does not distinguish the montane areas of Southern California and incorporates the Transverse Range into four ecoregions that share boundaries in the region: Southern California Mountains and Valley section (M262B), Central California Coast Ranges section (M262A), Southern California Coast section (261B), and Sierra Nevadas Foothill section (261F, largely encompassing the Tehachapis and associated ranges). Bailey splits the Santa Lucia Range longitudinally into the Central California Coast Ranges section and the Central California Coast section. Küchler (1975) maps a variety of vegetation types within the boundaries of the ecoregion, including chaparral, southern oak woodland, southern subalpine forests, Jeffrey pine forests, mixed-hardwood forests, oak savannas, and mixed-hardwood and redwood forests of the Santa Lucia Range.

Prepared by D. Olson, reviewed by R. Cox

Key Number:	**72**
Ecoregion Name:	**California Coastal Sage and Chaparral**
Major Habitat Type:	**4.1 Mediterranean Scrub and Savanna**
Ecoregion Size:	**23,403 km²**
Biological Distinctiveness:	**Globally Outstanding**
Conservation Status:	**Snapshot—Critical Final—Critical**

Introduction

The California Coastal Sage and Chaparral ecoregion encompasses the coastal terraces, plains, and ranges south of the Transverse Range and is bounded to the east by the Colorado Desert and the Sierra de Juarez and Sierra San Pedro Martir of northern Baja California. The Santa Rosa Mountains of the Peninsular Range are included, although the San Jacinto Mountains just to the northwest are considered under the adjacent ecoregion. The coastal extent of the ecoregion ranges from Santa Barbara southward to Punta Baja in Baja California, Mexico. The eight Channel Islands are also part of this ecoregion.

Biological Distinctiveness

The California Coastal Sage and Chaparral supports a diversity of habitats including montane conifer forests, Torrey pine woodlands, cypress woodlands, southern walnut woodlands, oak woodlands, riparian woodlands, chamise chap-

arral, inland and coastal sage scrub, grasslands, vernal pools, and freshwater and salt marshes. Coastal sage scrub, chamise chaparral, and oak woodlands predominate over much of the landscape.

Coastal sage scrub, also known as soft chaparral, is characterized by low, aromatic, and drought-deciduous shrublands of black sage (*Salvia mellifera*), white sage (*S. apiana*), California buckwheat (*Eriogonum fasciculatum*), California sage (*Artemisia californica*), golden yarrow (*Eriophyllum confertiflorum*), toyon (*Heteromeles arbutifolia*), lemonade-berry (*Rhus integrifolia*), and a diverse assemblage of other shrubs, herbaceous plants, cacti, and succulents. *Opuntia*, *Yucca*, and *Dudleya* are some of the most common succulent genera, the latter represented by several locally endemic species. This diverse and globally rare habitat type occurs in coastal terraces and foothills below around 1,000 m, interspersed with chamise chaparral, oak woodland, grasslands, and salt marsh. Five major coastal sage shrub associations are found within this ecoregion. The relatively xeric Vizcainan and Martirian associations are restricted to Baja California, each characterized by abundant cacti and succulent elements; the Diegan covers coastal habitats from Ensenada to San Pedro Bay; the Riversidian encompasses inland habitats from the U.S.-Mexico border to the northern Los Angeles basin; and the Venturan ranges from San Pedro Bay to Point Arguello, bounded by the Transverse Range (Westman 1983). The Diablan, a sixth association outside of this ecoregion, continues north toward San Pablo Bay of the San Francisco Bay area.

Like other chaparrals, coastal sage scrub is a fire-adapted community with many species that resprout quickly from root crowns or rapidly germinate after burns. A number of plants lie dormant in the seed bank for decades, only germinating and flowering after periodic fires. Fire frequencies in sage and chamise chaparral habitats are estimated to have ranged between twenty and forty years. Unlike chamise chaparral, coastal sage is primarily active during the cool, wet winters and largely sheds leaves during the dry summers. Chamise chaparral generally occurs on the lower slopes of the ecoregion's ranges but can also be found near the coast in areas with deeper soils and increased moisture. A large number of specialist and endemic species of plants and animals are found in coastal sage and chamise chaparrals, including the Quino checkerspot butterfly (*Euphydryas editha quino*), Hermes copper butterfly (*Lycaena hermes*), San Diego thorn mint (*Acanthomintha ilicifolia*), San Diego ambrosia (*Ambrosia pumila*), San Diego barrel cactus (*Ferocactus viridescens*), San Diego pocket mouse (*Perognathus fallax*), San Diego horned lizard (*Phrynosoma coronatum blainvillei*), Stephens kangaroo rat (*Dipodomys stephensi*), red-diamond rattlesnake (*Crotalus ruber*), California gnatcatcher (*Polioptila californica*), San Diego banded gecko (*Coleonyx variegatus abbotti*), Merriam kangaroo rat (*Dipodomys merriami*), and the coastal populations of the cactus wren (*Campylorhyncus brunneicapillus*).

Southern oak woodlands once covered much of the foothills and plains of the ecoregion. The Los Angeles basin and San Fernando Valley were noted for their extensive savannas of coast live oak (*Quercus agrifolia*) and valley oak (*Q. lobata*). Canyon live oak (*Q. chrysolepis*) is more common at higher elevations. California walnut (*Juglans californica*) woodlands once occurred in foothills around inland valleys in the northern portion of the ecoregion. A few vernal pools are scattered among the oak savannas and grasslands.

Riparian woodlands once lined streams and supported several species of willow, cottonwoods, sycamore, coast live oak, ash, white alder, and a diverse flora of herbaceous plants, shrubs, and vines. Many species of wildlife depend on riparian habitats for resources and habitat.

Some of the inland higher ranges and peaks support many of the same conifer communities found in the Transverse Range, with bigcone Douglas fir, sugar pine (*Pinus lambertiana*), white fir (*Abies concolor*), ponderosa pine (*Pinus ponderosa*), Jeffrey pine (*Pinus jeffreyi*), Coulter pine (*Pinus coulteri*), and incense cedar (*Calocedrus decurrens*). Black (*Quercus velutina*) and canyon live oaks and manzanitas (*Arctostaphylos* spp.) are common associates in these montane forests. Cuyamaca cypress (*Cupressus arizonica arizonica*) and Tecate cypress (*C. forbesii*) occur on only a few isolated peaks.

Near San Diego, the unique Torrey pine (*Pinus torreyana*) woodland occurs in Torrey Pine State Reserve. This tree occurs only here in a small population and on Santa Rosa Island.

The ecoregion supports between 150 and 200 species of butterflies, has the highest species richness of native bees in the United States, and has a number of relict species such as the Riverside fairy shrimp (*Streptocephalus woottoni*) of vernal pools. The rosy boa (*Lichanura trivirgata*), California legless lizard (*Aniella pulchra*), and several relict salamanders are some examples of the unusual and distinctive herpetofauna.

Much of the Channel Islands is covered in coastal sage and chamise chaparral, with some oak woodlands on the larger islands. The islands are noted for endemic and relict plant and animal species and subspecies, many being restricted to single islands. Buckwheats, locoweeds, oaks, and the succulent *Dudleya*'s are noted for several island endemics. The Catalina ironwood (*Lyonothamnus floribundis*) is a good example of a relict species once found throughout the mainland. The island night lizard (*Xantusia riversiana*), island fox (*Urocyon littoralis*), and Santa Catalina shrew (*Sorex willetti*) are other endemic species.

Conservation Status

Habitat Loss and Degradation

More than 85 percent of coastal sage scrub has been lost to urban and agricultural development. Some scrub communities, such as riversidian alluvial fan scrub, are now confined to remnant patches along unaltered streams and washes. The vast majority of oak savannas and walnut woodlands have been destroyed. Most coastal habitats have been highly altered throughout the ecoregion. The Channel Islands have experienced widespread loss of original habitats and degradation from grazing and introduced species.

Remaining Blocks of Intact Habitat

Isolated blocks of coastal sage scrub occur in Camp Pendleton Marine Corps Base, Santa Monica Mountains parklands, San Joaquin Hills near Laguna Beach, and Irvine Ranch in Orange County. Patches elsewhere are quite small and highly fragmented. Chamise chaparral still occurs in relatively large blocks on some inland foothills.

California walnut woodlands were formerly most abundant in the Puente Hills, but now the last remaining patches occur in the San Jose

Hills south and east of Covina. Tonner Canyon and Soquel Canyon once had well-developed walnut forests, but these are being rapidly destroyed. The few remaining remnant groves represent one of the most endangered forest communities in the United States.

Some of the best opportunities to conserve large blocks of coastal sage exist in northern Baja California, Mexico. An increase in home building along the Baja coast is threatening many of these larger blocks. Conservationists and government agencies must act quickly to conserve habitat in this region.

Degree of Fragmentation

Coastal sage habitats are extremely fragmented and isolated in areas of intensive development, precluding effective dispersal of most species. Field studies suggest that isolated fragments of less than 1 km² (10–100 ha) will lose their native vertebrate species within a few decades (Fleishman and Murphy 1993). Isolated canyons in southern California lose at least half of their bird species within twenty to forty years after isolation, although many have been observed in narrow corridors.

Degree of Protection

Torrey Pine State Reserve protects one of only two known populations of this rare conifer. The fragile coastal sage understory is often damaged by visitors. Native grasslands, vernal pools, and oak savannas with Engelmann oak (*Quercus engelmanii*) are conserved in the Santa Rosa Plateau Reserve near Elsinore, which is managed by Riverside County California Department of Fish and Game (CDFG), and The Nature Conservancy. Existing laws offer only minimal protection; these include the Natural Communities Conservation Planning Program (NCCP) of 1991 that restricts destruction of some coastal sage scrub, and the Endangered Species Act listing of the California gnatcatcher, which created restrictions on destruction of habitat for extant birds. Native habitats on Santa Cruz Island are currently being restored by The Nature Conservancy, which manages the island as a reserve.

Types and Severity of Threats

Native habitats continue to be cleared for housing, golf courses, orchards, and other forms of development. Much of the remaining habitat,

particularly near the coast, occurs only in very small patches and is highly isolated, fragmented, and surrounded by development, which is generally hostile environment for most native species. Small habitat blocks face numerous threats, including invasion of alien species, predation by introduced animals and people, frequent fires from human activity, dumping, and pollution. Type conversion of chaparral and shrub communities to grassland is carried out through burning, grazing, and herbicide eradication of shrub species. Consequently, grasslands dominated by introduced species have greatly increased in area. Decades of fire suppression in chaparral have allowed the accumulation of senescent vegetation, creating conditions for catastrophic hot fires that kill dormant seed beds and individual plants that are usually resistant to low-intensity fires.

Grazing by domestic livestock, particularly sheep, can have a serious impact on coastal sage scrub communities. Santa Cruz Island has been heavily damaged by 130 years of feral sheep grazing, with only 6 percent of the island still covered by degraded coastal sage scrub. Cattle, rabbits, deer, and pigs also contribute to similar situations throughout the Channel Islands. Plowing for agriculture or development destroys root crowns, thereby reducing any opportunities for regeneration. After fires and intense grazing, several invasive species can dominate, including foxtail fescue, red brome, and wild oats. Planting of ryegrass after fires as a short-term measure to prevent erosion can impede regeneration of native species. Controlled burns and brush removal are not necessary, because coastal sage scrub does not accumulate high fuel loads. Oxidants from smog have been implicated in damage and reduction in growth of coastal sage scrub. Numerous species of alien plants and animals compete for resources and kill native species. Introduced plant species make up 23 percent of the flora. Riparian woodlands are impacted by clearing, grazing, and the reduction of flow in streams, which alters successional patterns.

Seventy-seven species in Southern California are currently listed under the Endangered Species Act, and another 378 are under consideration. Some of the federally recognized threatened and endangered species include the California gnatcatcher, San Diego banded gecko, cactus wren, Merriam kangaroo rat, flannel-mouthed sucker (*Catostomus latipinnis*), western patch-nosed snake (*Salvadora hexalepis*), and

cheese-weed moth lacewing (*Chrysoperla* spp.).
Thirteen plant species of the coastal sage scrub
are recognized as threatened or endangered.
Unfortunately, many other species are threatened with extinction and extirpation, but the
conservative listing practices of federal and state
agencies preclude their formal recognition by the
government.

Suite of Priority Activities to Enhance Biodiversity Conservation

- Curtail further destruction of coastal sage,
  oak woodlands, walnut woodlands, and riparian woodlands.
- Protect all remaining patches of threatened
  native habitats to conserve more intact examples and to provide eventual source pools for
  landscape-level restoration efforts.
- Restore larger blocks of remaining habitat
  and linkage zones among habitat patches.
- Restore fire regimes to their natural range of
  variation within native habitats through controlled burning and fuel reduction prescriptions.
- Eliminate type conversion and sowing of
  alien plant species for erosion control.

Several factors that impede effective conservation of coastal sage have been identified by
Reid and Murphy (1995): Coastal sage is difficult to define for regulatory purposes; remaining
patches of sage are highly fragmented and surrounded by hostile environments; private
landownership predominates; multiple jurisdictions are involved; and existing laws offer little
protection. The Natural Community
Conservation Planning Act of 1991 was applied
to the coastal sage and chaparral habitats of this
ecoregion in 1992. Nonregulatory agreements
between the USFWS, California Resources
Agency, environmental organizations, developers, and private landowners are designed to promote both conservation and development in the
region. The resulting habitat conservation plans
are supposed to incorporate conservation biology
guidelines and address the conservation needs of
endangered species. Accurate mapping efforts of
remaining habitats in several counties have been
conducted and provide an invaluable data layer
for conservation planning and analyses.
Representation analyses are being conducted to
assess how well different habitats, communities,

associations, and species are conserved within
protected areas.

Habitat conservation plans have the potential
to develop into effective conservation strategies
for many of the ecoregion's critical habitats.
However, political and economic pressures have
led to reserve and development plans that significantly compromise the long-term effectiveness
of proposed conservation objectives. Many of
the reserves that have been designated to date
are based on maintaining viable populations of
target species and certain structural characteristics. However, they appear to be too few, small,
isolated, and surrounded by hostile environments for many native species, and the habitat
plans allow for further conversion of the last
remnants of this globally unique ecosystem for
housing, golf courses, roads, and shopping
malls. One major problem is the eventuality of
fire within each of the proposed reserves. For
native biodiversity to persist over the long term
within reserves, they need to be of sufficient size
to absorb inevitable fires; that is, they need to be
large enough such that the entire patch is not
burned during a single fire event. Unburned
patches need to be present to provide refuge for
mobile species and source pools for reinvasion of
burned areas after the fire. Small reserves will
blink out over the next several decades as fires
occur, eventually eliminating species from all but
the largest blocks of habitat. Small blocks of
habitat are commonly proposed as reserves by
developers and local governments. In addition to
the serious fire problem, the biodiversity in small
reserves faces extreme threats from small population sizes; introduced species; dumping; off-
road traffic; toxins from urban runoff; and high
mortality by feral cats, rats, racoons, small children, and other predators. Few patches of habitat close to the coast, habitat that is often opti-
mal for many coastal sage species due to
abundant fog, are proposed as reserves because
of their real estate potential. Biologists and
resource agencies must be firm in ensuring that
reserve designs will be effective over appropriate
spatial and temporal scales and fully recognize
the full range of threats present. Conservation
strategies that realistically address the impact of
fire and the minimum requirements of sensitive
species and ecological processes must be immediately developed and implemented.

Conservation Partners

For contact information, please see appendix G.

- Bureau of Land Management
- California Resources Agency
- County governments; Metropolitan Water District
- Department of Defense
- Endangered Habitats League
- The Nature Conservancy
- The Nature Conservancy of California
- Rancho Santa Ana Botanic Gardens, Claremont
- Southern California Botanists
- Southern California Natural Resources Defense Council
- U.S. Fish and Wildlife Service
- U.S. Forest Service (Cleveland National Forest)

Relationship to Other Classification Schemes

This ecoregion matches Omernik's (1995b) Southern and Central California Plains and Hills (6). Bailey (1994) maps the coastal portion of this ecoregion as the Southern California Coast section (261B) and the interior portion as the Southern California Mountains and Valley section (M262B). The latter section extends farther inland than Omernik's boundaries, encompassing more montane and desert habitats. Küchler (1975) identifies several vegetation types within the ecoregion, including valley oak savanna, southern oak woodland, coastal sagebrush, chaparral, California prairie, southern Jeffrey pine forest, and coastal salt marsh.

Prepared by D. Olson, reviewed by R. Cox

Key Number: **73**
Ecoregion Name: **Hawaiian High Shrublands**
Major Habitat Type: **4.2 Xeric Shrublands/Deserts**
Ecoregion Size: **1,853 km²**
Biological Distinctiveness: **Bioregionally Outstanding**
Conservation Status: **Snapshot—Vulnerable**
Final—Vulnerable

Biological Distinctiveness

Hawaiian high shrublands range from open shrublands to alpine grasslands and deserts. The upper slopes of the high volcanoes, Mauna kea, Mauna loa, Hualalai, and Haleakala, support shrubland habitats with species such as *Chenopodium oahuense, Vaccinium reticulatum, Dubautia menziesii,* and *Santalum haleakalae.* Subalpine grasslands, patchily distributed within and adjacent to the shrub zone, are dominated by tussock-forming species such as *Deschampsia nubigena, Eragrostis atropioides, Panicum tenuifolium,* and *Trisetum glomeratum* (HINHP 1996; Sohmer and Gustafson 1987). On the highest peaks, cold and dry conditions create alpine deserts inhabited by silversword (*Agryroxiphium sandwicense*), *Dubautia* spp., and other alpine-adapted plants, as well as alpine-adapted invertebrate species. The Hawaiian nene goose (*Branta sandvicensis*) lives in high shrubland areas, and endangered Hawaiian dark-rumped petrels nest in burrows in subalpine and alpine cinderlands.

Conservation Status

Several important areas for conserving high shrublands that have no or incomplete protection have been identified by Sohmer and Gon (1996): Leeward East Maui, Haleakala Summit of Maui (protected), Alpine summits (Hawaii), and the Pohakuloa-Saddle area of Hawaii.

Types and Severity of Threats

Although some large blocks of relatively intact high shrublands and alpine deserts still exist, overgrazing by domestic and feral livestock, wildfires, trampling from recreational activities, competition from introduced plants, removal of plants such as silverswords, and introduced ants that kill native invertebrate pollinators all pose significant threats to native species and communities (Gagne and Christensen 1985; Sohmer and Gustafson 1987). Alpine grasslands have been reduced in range (Sohmer and Gustafson 1987), but still occupy more than 50 percent of their presumed original range (Jacobi 1985; HINHP 1996).

Conservation Partners

For contact information, please see appendix G.

- Hawaii Natural Heritage Program
- The Nature Conservancy
- The Nature Conservancy of Hawaii

Relationship to Other Classification Schemes

The Hawaiian High Shrublands corresponds to Küchler's (1985) unit 7 (Grassland, microphyllus shrubland, and barren). Omernik (1995b) did not classify Hawaii, and Bailey (1994) clumped all of Hawaii into one unit.

Prepared by S. Gon

Key Number:	**74**
Ecoregion Name:	**Hawaiian Lowland Shrublands**
Major Habitat Type:	**4.2 Xeric Shrublands/ Deserts**
Ecoregion Size:	**1,522 km²**
Biological Distinctiveness:	**Bioregionally Outstanding**
Conservation Status:	**Snapshot—Critical Final—Critical**

Biological Distinctiveness

Coastal and lowland dry shrublands occur on the lowest leeward slopes of the higher Hawaiian Islands and on all but the summit regions of the islands of Lanai, Kahoolawe, and Niihau. This ecoregion also includes the terrestrial portions of all of the northwestern Hawaiian Islands, mostly composed of atolls and small basalt remnants. Vegetation includes grasslands of *Eragrostis, Fimbristylis, Sporobolus,* and *Lepturus* and mixed shrublands dominated by one or more of *Sida, Dodonaea, Scaevola, Heliotropium, Gossypium, Chamaesyce, Chenopodium, Myoporum, Vitex, Anthium,* and *Styphelia.* Non-tree plant diversity of this ecoregion is high (more than two hundred species) and highly endemic (more than 90 percent endemic). Tree diversity is relatively low.

Conservation Status

Over 90 percent of the Hawaiian Low Shrublands have been lost to development or displacement by alien vegetation. Small, degraded examples of the natural communities of the ecoregion remain. Kahoolawe Island is a natural/cultural reserve, and Moomomi Preserve on Molokai is managed by The Nature Conservancy of Hawaii. The northwestern Hawaiian Islands constitute a USFWS refuge.

Types and Severity of Threats

Fire, weed invasions, feral animals (especially goats and deer), and continued development threaten this ecoregion.

Conservation Partners

For contact information, please see appendix G.

- Hawaii Natural Heritage Program
- The Nature Conservancy
- The Nature Conservancy of Hawaii

Relationship to Other Classification Schemes

The Hawaiian Low Shrublands encompasses a portion of Küchler's (1985) unit 1 (Schlerophyllous forest, shrubland, and grassland). Omernik (1995b) did not classify Hawaii, and Bailey (1994) clumped all of Hawaii into one unit.

Prepared by S. Gon

Key Number:	**75**
Ecoregion Name:	**Snake-Columbia Shrub Steppe**
Major Habitat Type:	**4.2 Xeric Shrublands/ Deserts**
Ecoregion Size:	**218,111 km²**
Biological Distinctiveness:	**Bioregionally Outstanding**
Conservation Status:	**Snapshot—Vulnerable Final—Endangered**

Introduction

The Snake-Columbia Shrub Steppe is a vast, mostly arid ecoregion. Its easternmost limit is the Continental Divide in eastern Idaho. From there, the ecoregion follows the arc of the Snake River plain as far as Hell's Canyon. The ecoregion then spreads throughout southeastern Oregon, extending along the Deschutes River basin to the Columbia River. It also includes, following hydrographic lines, parts of northern Nevada and the extreme northeast of California. To the north, the ecoregion dominates the western portion of the Columbia Basin in Washington.

The ecoregion is largely in the rain shadow of the Cascade Mountains and thus receives little precipitation. Latitude and physiography are influential factors in distinguishing this ecoregion from other similar ecoregions, such as the Wyoming Basin and Great Basin Shrub Steppes. The Snake-Columbia Shrub Steppe is lower in elevation than the Wyoming Basin, and the two are separated by mountainous areas.

Biological Distinctiveness

The dominant vegetation in the ecoregion is sagebrush (*Artemisia* spp.), typically associated with various wheatgrasses (*Agropyron* spp.), Idaho fescue (*Festuca idahoensis*), or other perennial bunchgrasses (Franklin and Dyrness 1988). The ecoregion contains a number of isolated mountain ranges in the southern parts of Idaho and Oregon, and here the sagebrush shrublands grade into bunchgrasses and juniper woodlands. Some parts of these ranges contain areas of Douglas fir (*Pseudotsuga menziesii*), subalpine fir (*Abies lasiocarpa*), and aspen (*Populus tremuloides*) (Wuerthner 1986). Riparian zones in the ecoregion typically contain cottonwoods (*Populus* spp.) and willows (*Salix* spp.) (West 1988). The ecoregion exhibits more desertlike vegetation in southeastern Oregon, where elevation is considerably higher and precipitation lower (Franklin and Dyrness 1988). The same area, however, contains extensive wetlands, which provide vital waterfowl habitat in the Pacific Flyway.

The Great Basin is hotter and drier than the Snake-Columbia, lacks distinct major watersheds, and exhibits vegetation associations indicative of its proximity to true desert ecoregions like the Mojave. The Snake-Columbia ecoregion lacks the floristic diversity found in the Great Basin ecoregion. Situated as it is in a distinct major river system with consistent climatic and physiographic features, the Snake-Columbia ecoregion forms a logical single unit.

Fire, grazing, and variations in precipitation and temperature are the major disturbances in the ecoregion. Burning may encourage grass growth and impede sagebrush. Sagebrush, on the other hand, adapts to drought conditions that kill grasses (West 1988).

The Owyhee drainage (southwest Idaho and southeast Oregon) once supported salmon runs, making it one of the few high desert anadromous spawning areas.

Conservation Status

Habitat Loss and Degradation

Overgrazing by domestic livestock, fire suppression and resultant high-intensity blazes, and spread of exotic grasses are the major anthropogenic changes to the ecoregion. Loss of perennial grasses is a major problem in this ecoregion. Irrigation in the Washington and Idaho portions of the ecoregion have also caused major changes and have encouraged the spread of exotic species. Wuerthner (1988) notes the loss of bluebunch wheatgrass (*Agropyron spicatum*) communities in the Owyhee Uplands portion of the ecoregion due to overgrazing in arid conditions.

Remaining Blocks of Intact Habitat

Large intact areas, though degraded by overgrazing and exotic grass invasion, remain in southwestern Idaho, southeastern Oregon, and northwestern Nevada. Defense installations, such as the Yakima Firing Range and the Hanford Nuclear Reservation, also protect extensive habitat areas in Washington.

Degree of Fragmentation

Conversion of native habitat to agriculture, particularly in eastern Idaho and in Washington, are major sources of fragmentation. Exotic grass and noxious weed invasions are likewise becoming serious enough to cause fragmentation.

Degree of Protection

Defense installations, national wildlife refuges, national monuments, and wilderness study areas provide a fair degree of protection. However, without control of exotic grass and noxious weed invasions, protected areas may have a limited benefit in the long run. Important protected areas include:

- Yakima Firing Range—southern central Washington
- Hanford Reservation—southern central Washington
- Hart Mountain National Antelope Refuge—southern Oregon
- Malheur National Wildlife Refuge—southeastern central Oregon
- Craters of the Moon National Monument—southern Idaho
- Birch Creek Fen Preserve (TNC)—eastern Idaho
- Sheldon National Wildlife Refuge—northwestern Nevada

Types and Severity of Threats

Livestock grazing, invasion of exotic plants, irrigated agriculture, and recreation, especially through use of off-road vehicles, are the major threats to the ecological integrity of this ecoregion. As there are no highly effective methods for checking or reversing the tide of exotic and noxious plant invasions, the spread of these species—and consequent elimination of native plant communities—is a serious threat. Combined with overgrazing and continued conversion to row crops, it may lead to serious alteration and degradation of the ecoregion in the near future.

Suite of Priority Activities to Enhance Biodiversity Conservation

- Designate wilderness study areas as actual wilderness areas.
- As defense installations are phased out, maintain these sites as protected areas.
- Target specific sites and vegetation communities for protection from severe overgrazing.
- Develop effective techniques for combating invasive species.
- Restore salmon fisheries. This is an interesting prospect that could have major implications for certain systems within the ecoregion.
- Make major reductions in livestock grazing on public lands.

Conservation Partners

For contact information, please see appendix G.

- Idaho Conservation League
- Oregon Wildlife Federation
- Oregon Natural Desert Association
- Oregon Natural Resources Council

Relationship to Other Classification Schemes

The ecoregion boundary follows Omernik (1995b). It encompasses Bailey's (1994) units 342 B, C, D, H, and I. It corresponds to Küchler's (1975) 49, 44, 45, 39, and some of 34 for the same geographic area.

Prepared by S. Primm

Key Number:	**76**
Ecoregion Name:	**Great Basin Shrub Steppe**
Major Habitat Type:	**4.2 Xeric Shrublands/**
	Deserts
Ecoregion Size:	**335,868 km²**
Biological Distinctiveness:	**Bioregionally Outstanding**
Conservation Status:	**Snapshot—Relatively Stable**
	Final—Relatively Stable

Introduction

The Great Basin is the most northerly of the four American deserts. Unlike the other three, which have almost exclusive ties to warm-temperate and tropical-subtropical vegetation types, the Great Basin has affinities with cold-temperate vegetation. The Chihuahuan and Sonoran Deserts may in fact be more closely related to the Argentine "Monte" than they are to the Great Basin (Turner 1994a).

The Great Basin is composed of a series of uplifted mountain ranges and their associated intervening valleys. There are approximately one hundred internally drained basins within this ecoregion. The basin is bounded to the east by the Colorado Plateau and central Rocky Mountains, to the north by the Columbia Plateau, and to the west by the Cascade-Sierra Range. The southern boundary is generally placed at the confluence of the Colorado River drainage and the Mojave Desert of southern California and southernmost Nevada (Morris and Stubben 1994).

Dominant species in the region include such distinctly cold-temperate species as sagebrushes (*Artemisia*), saltbrushes (*Atriplex*), and winterfat (*Ceratoides lanata*). These scrub species are much-branched, nonsprouting, aromatic semi-shrubs with soft wood and evergreen leaves. The Great Basin also contains species with evolutionary ties to warmer climates, such as rabbitbrush (*Chrysothamnus*), blackbrush (*Coleogyne*), hopsage (*Grayia*) and horsebursh (*Tetradymia*). The region contains few cacti, however, in numbers of either individuals or species, and also lacks characteristic desert plants in minor waterways (Turner 1994a).

Sagebrush communities are often considered steppe or shrub steppe because of the role of grasses, and in parts of the Great Basin grasses are important understory elements in distinctly shrub-steppe communities. Toward the south of the ecoregion, however, sagebrush may grow to the virtual exclusion of grasses even in the absence of grazing. The near absence of grasses may be due to the fact that the peak precipitation occurs in winter (Turner 1994a).

Grazing and fire are the two most important forces in the sagebrush communities of the Great Basin. Neither domestic nor native ruminants eat sagebrush, resulting in reduction of more palatable grasses and forbs. Native annuals suffered under heavy grazing, and since approximately 1990 introduced annuals have become increasingly conspicuous throughout the region. These annuals, such as cheatgrass brome (*Bromus tectorum*) and Russian thistle (*Salsola paulensii*), have become important in arresting succession in many areas. The highly flammable cheatgrass brome also greatly increases the incidence of fire. After a fire, sprouting species of such genera as *Chrysothamnus, Tetradymia*, and *Gutierrezia* may take on the role of the non-sprouting sagebrush.

The other dominant vegetation communities in the Great Basin are shadscale (*Atriplex confertifolia*) and blackbrush (*Coleogyne ramosissima*). Shadscale is a wide-ranging species, but it is found as a dominant only within and adjacent to the Great Basin. Blackbrush occurs primarily in southern Nevada, southeastern California, north-central Arizona, and southeastern Utah.

The Great Basin lies mostly to the north of the thirty-sixth parallel. Part of the ecoregion occurs to the south of that line along the Little Colorado River drainage in Arizona and New Mexico, where in some places it is referred to as the Painted Desert. The region generally receives less than 250 mm of precipitation per year. Mean monthly precipitation shows a strong winter-dominated pattern in the west, with a gradual shift eastward toward a stronger summer influence with wet and dry seasons less distinct than in the other deserts (Turner 1994a).

Biological Distinctiveness

The Great Basin is the largest arid area in the United States. It is a true basin and range, with completely self-contained drainage. The region supports numerous threatened and endangered species—Nevada is third in the nation in listed species—as well as an endemic species of greasewood (*Larrea* spp.) and an endemic kangaroo mouse (*Microdipodops pallidus*).

This ecoregion also contains significant evolutionary and ecological phenomena. The size and orientation of Goshute Mountain, for example, concentrate hawk migrations. The small lakes in the basin contain endemic shrimp and other species.

Conservation Status

Habitat Loss and Degradation

Virtually the entire basin has been grazed and browsed, and less than 10 percent remains as intact habitat. Exotic species have become established across the ecoregion. Irrigation for alfalfa has increased salinization in the region, while mining has led to pollution by heavy metals. Urban areas in the region, particularly Las Vegas, Reno, and Salt Lake City, are growing rapidly.

Remaining Blocks of Intact Habitat

The Great Basin contains several large intact habitat blocks. The most important are:

- Black Rock Desert—northwestern Nevada
- Sheep Range—southern Nevada
- Desert National Wildlife Range—southern Nevada
- Nevada test site—southern Nevada
- Great Basin National Park—eastern Nevada

Degree of Fragmentation

High-elevation, lush habitats in the Great Basin are badly fragmented, but low-elevation habitat is relatively intact. Much of the fragmented area can be restored.

Degree of Protection

The most important protected areas in the Great Basin are:

- Desert National Wildlife Reserve—southern Nevada
- Dugway Proving Grounds—northwestern Utah
- National Electronic Warfare Center—Utah
- Hill Air Force Base, Wendover Range—northwestern Utah
- Arc Dome Wilderness Area—Nevada
- Still Water National Wildlife Refuge—Nevada

- Important lakes of conservation significance: Pyramid, Walker, Mono, Topaz

Types and Severity of Threats

Grazing poses the greatest conversion threat to the Great Basin. While levels of grazing are unlikely to increase, grazing creates ongoing opportunities for invasion by non-native species. The expansion of cities like Salt Lake City and Provo also may lead to increased development and further conversion of habitat.

Invasive species and fires are degrading the habitats of the Great Basin, while hunting threatens populations of fur-bearing mammals.

Suite of Priority Activities to Enhance Biodiversity Conservation

- Protect the desert experimental range near Milford, Utah. This range has the most detailed historical grazing records in the United States. These provide the opportunity for research to determine the impact of grazing, the extent of the return of native species, and so on.
- Encourage BLM to make wilderness designations in Utah and Nevada.
- Encourage the Department of Defense to increase stewardship activities in Dugway Proving Grounds and Hill Air Force Range, Utah.
- Restore riparian areas.
- Find and protect representative pinyon-juniper sites.

Conservation Partners

For contact information, please see appendix G.

- Sierra Club
- Southern Utah Wilderness Alliance

Relationship to Other Classification Schemes

The Great Basin corresponds to Omernik's (1995b) Northern Basin and Range ecoregion. The Great Basin falls within Bailey's (1994) Intermountain Semi-Desert and Desert Province and covers parts of several Bailey sections, some of which extend as far east as central Colorado. We believe that vegetation and climatic patterns distinguish the Great Basin from the Colorado Plateau and the intervening Wasatch and Uinta montane forests.

Prepared by B. Holland, G. Orians, and J. Adams

Key Number:	**77**
Ecoregion Name:	**Wyoming Basin Shrub Steppe**
Major Habitat Type:	**4.2 Xeric Shrublands/ Deserts**
Ecoregion Size:	**132,361 km²**
Biological Distinctiveness:	**Nationally Important**
Conservation Status:	**Snapshot—Vulnerable Final—Vulnerable**

Introduction

The Wyoming Basin Shrub Steppe is an expansive ecoregion of high, open, arid country. The ecoregion is nearly surrounded by mountain ecoregions. It is drained by three major river systems: the Green River to the south, the Wind-Bighorn River to the north, and the North Platte to the east. A fourth "watershed," the Great Divide Basin, has no surficial connection to any river system; it is also the driest area of the ecoregion.

The ecoregion is generally arid or semiarid. It is in the rain shadow of the Rocky Mountains and thus receives little precipitation. Latitude and physiography are influential factors in distinguishing this ecoregion from other similar ecoregions, such as the Snake-Columbia and Great Basin Shrub Steppes. The Snake-Columbia Shrub Steppe is lower in elevation than the Wyoming Basin, and the two are segregated by mountainous areas. Unlike the other two, the Wyoming Basin receives at its eastern limit some summer moisture from the south. This precipitation does not penetrate into either the Bighorn Basin or the Great Divide Basin, the most desertlike portions of the ecoregion (Knight 1994).

Biological Distinctiveness

The dominant vegetation in the ecoregion is sagebrush (*Artemisia* spp.), often associated with various wheatgrasses (*Agropyron* spp.) or fescue

(*Festuca* spp.). Knight (1994) describes the great variability of the sagebrush-steppe mosaic in the ecoregion. Elevation, aridity, snow accumulation, prevailing winds, and other factors affect the species composition, morphology, and density of sagebrush communities in the ecoregion. Ecotones between sagebrush steppe and adjacent mountain ecoregions may appear at elevations as high as 3,000 m. The sagebrush steppe is also interspersed with desert shrublands, dunes, and barren areas in more arid regions (e.g., Red Desert) and with mixed-grass prairie at the eastern limit of the ecoregion (Knight 1994).

Fire, wind, grazing, and variations in precipitation and temperature are the major disturbances in the ecoregion. Burning may encourage grass growth and impede sagebrush. Sagebrush, on the other hand, adapts to drought conditions that kill grasses (West 1988). Excessive moisture along riparian areas or in places where snow accumulates can also encourage grass growth, which displaces many sagebrush species (Knight 1994). Seasonal browsing and grazing by native herbivores, including elk (*Cervus elaphus*), mule deer (*Odocoileus hemionus*), and pronghorn (*Antilocapra americana a.*), constitute another influential disturbance. Native herbivory is a somewhat distinctive factor in this shrub steppe ecoregion, since most of the ecoregion is in close proximity to extensive mountain ecoregions supporting large herds of migratory ungulates.

A particularly interesting aspect of the ecoregion is the presence of remnant prairie dog "ecosystems," developed in this case by white-tailed prairie dogs (*Cynomys leucurus*). Knight (1994, 87–88) states that prairie dogs "maintain biological diversity by creating habitat for other organisms through their small-scale disturbances. . . . [Their influence] far exceeds the biomass they constitute in the ecosystem." Raptors, mesopredators (e.g., coyote [*Canis latrans*] and swift fox [*Vulpes velox*]), and ungulates disproportionately use prairie dog towns when available. Presumably, ungulates like pronghorn and bison (*Bison bison*) key on these sites because forage quality is higher, due to greater concentrations of nutrients from animal waste. Prairie dog towns (near Meteetse) in this ecoregion maintained the last known black-footed ferrets (*Mustela negripes*) when they were thought to have gone totally extinct, and have served as the first reintroduction site (near

Medicine Bow) for captive-reared ferrets. The persistence of these ecosystems may be due to the harsh climate and remote location of many areas of the ecoregion, impeding agricultural conversion and persecution.

Conservation Status

Habitat Loss and Degradation

Habitat loss, degradation, and outright conversion are severe in some locales. Conversion of sagebrush habitat types to grasslands for domestic grazing has taken place in areas where the climate will support grass production. Large-scale energy and mineral developments have had major long-term impacts in certain areas.

Native ungulates that use the shrub steppe either year-round or as seasonal range are often intolerant of industrial activities and are thus displaced from otherwise productive habitat. Displacement tends to concentrate ungulates, leading to overutilization of winter range and rapid transmission of pathogens. Additionally, the road networks associated with resource extraction have allowed hunters greater access to ungulates, resulting in dramatically reduced herd sizes in some areas (Debevoise and Rawlins 1996).

Land reclamation has had a mixed success rate at restoring native vegetation (Knight 1994; Debevoise and Rawlins 1996). Reclamation in some cases has relied on non-native grass species, in addition to unintentional introduction of noxious weeds. Sagebrush and other important ungulate forage plants have, on the whole, been difficult to restore on reclamation sites (Debevoise and Rawlins 1996).

The combined effects of heavy livestock grazing and fire suppression have also altered the structure and composition of some areas of the ecoregion. Heavy grazing removes potential grass fuels, thus minimizing the likelihood of periodic fires. Woody plant species like sagebrush are quite robust under such conditions and can attain heights of up to 1.5 m. At ecotones within the surrounding montane ecoregions, conifer invasions occur over time, especially along relatively mesic riparian zones.

Remaining Blocks of Intact Habitat

Many areas of the ecoregion are remote from human population centers and have remained relatively intact. Scattered resource extraction sites and dispersed livestock grazing have occurred over most areas but have left little long-term impact on many habitat blocks (Knight 1994; Debevoise and Rawlins 1996).

Degree of Fragmentation

Outside of major resource extraction sites, fragmentation is relatively limited. Since most of the ecoregion is far too arid for cultivation, there has been little effort at irrigation and row cropping. The exception to this is the Bighorn Basin, which contains several extensive cultivated areas. Fragmentation through localized conversion or degradation may affect specific vegetation types and smaller fauna.

With respect to megafauna, fragmentation is locally severe for species like elk and mule deer, where topography, climate, and development-related displacement severely restrict access to crucial winter range. Large predators like grizzly bears (*Ursus arctos*), which existed at low densities and probably used the habitat only seasonally, have been eliminated from the ecoregion. This extirpation probably has more to do with the loss of bears in more productive source habitats in the surrounding mountains than with any particular change in the shrub steppe itself.

Eliminating individual bears that frequented the basins was accomplished easily, since the open country afforded little visual cover, and since poisons were freely available for predator control. Recolonization or sporadic use of the shrub steppe by grizzly bears may occur in the near future as the Greater Yellowstone population shifts and expands its range; this phenomenon could be suppressed by widespread use of M-44 cyanide traps set for coyote control, and by uncontrolled human access.

Degree of Protection

Overall, protected areas in the shrub steppe are lacking. The largest designated area is the Flaming Gorge National Recreation Area, which consists of a dammed reservoir on the Green River. Arguably, the shoreline areas and surrounding terrain are protected, but a large impoundment clearly amounts to wholesale habitat conversion that may even alter noninundated habitats nearby in southwestern Wyoming

There are many wilderness study areas (WSA) that may soon receive protection. A lower level of protection is afforded at some sites, administratively designated by the Bureau of Land Management as Areas of Critical Environmental Concern (ACEC). These include:

- Steamboat Mountain ACEC—south-central Wyoming
- the Greater Sand Dunes ACEC, which, along with the Sand Dunes WSA, could potentially constitute the core of a large reserve network in the Great Divide Basin—south-central Wyoming

Types and Severity of Threats

Overgrazing by domestic livestock, fire suppression, wholesale replacement of sagebrush with grasses, and spread of exotic grasses are the major anthropogenic threats to the ecoregion. Oil and gas exploration, along with mining (strip mining for coal and hard-rock mining for trona and other industrial minerals), has the potential to cause considerable disturbance, wildlife displacement, and environmental pollution. Debevoise and Rawlins (1996) report that 6,000 to 11,000 new oil and gas wells could be sunk in southwestern Wyoming in the next twenty years. Unbridled industrial expansion could have severe long-term impacts on the ecoregion.

Suite of Priority Activities to Enhance Biodiversity Conservation

- Carefully review the oil and gas leasing process, undertaking new exploration only after thorough assessment of possible impacts and development of effective mitigation measures.
- Make a concerted effort to develop and enact a wilderness bill that would provide full protection to existing WSAs and ACECs.

To help focus both activities, conservationists should undertake the following:

- Conduct a rapid assessment of the ecoregion.
- Identify and document rare plant communities and assemblages. Catalog and prioritize dune communities and other geologic-vegetative associations.
- Carefully evaluate prairie dog ecosystems, ungulate habitat, and potential large-predator recolonization areas. Much of this information is already available for southwestern Wyoming (Debevoise and Rawlins 1996).

This information could be used to identify areas in which it may be acceptable to concentrate impacts, as well as areas that are at the threshold of systemic collapse (such as the elimination of elk herds or plant communities). Large undisturbed areas should receive high priority for protection.

Conservation Partners

For contact information, please see appendix G.

- Greater Yellowstone Coalition
- Sierra Club, Northern Plains Field Office
- Wyoming Outdoor Council
- Wyoming Wildlife Federation

Relationship to Other Classification Schemes

The ecoregion boundary is taken from Omernik (1995b) and closely matches Bailey (1994) units 342 A, E, F, and G; it amalgamates Küchler (1975) units 50, 34, 31, 57, and part of 49.

Prepared by S. Primm

Key Number:	**78**
Ecoregion Name:	**Colorado Plateau Shrublands**
Major Habitat Type:	**4.2 Xeric Shrublands/ Deserts**
Ecoregion Size:	**326,390 km²**
Biological Distinctiveness:	**Regionally Outstanding**
Conservation Status:	**Snapshot—Relatively Stable Final—Relatively Stable**

Introduction

The Grand Canyon epitomizes the Colorado Plateau, an area that has been called the "land of color and canyons." The plateau can be thought of as an elevated, northward-tilted saucer. It is characterized by its high elevation and arid to semiarid climate. The Colorado Plateau has developed great relief through the erosive action of high-gradient, swift-flowing rivers that have downcut and incised the plateau. Approximately

90 percent of the plateau is drained by the Colorado River and its tributaries.

The region has conspicuous but irregular vegetation zones. The woodland zone is the most extensive, dominated by what is often called a pygmy forest of pinyon pine (*Pinus edulis*) and several species of juniper (*Juniperus* spp.). Between the trees the ground is sparsely covered by grama, other grasses, herbs, and various shrubs, such as big sagebrush (*Artemisia tridentata*) and alderleaf cercocarpus (*Cercocarpus montanus*).

The mountain zone extends over considerable areas on the high plateaus and mountains but is actually much smaller than the pinyon-juniper zone. The vegetation varies considerably, from ponderosa pine in the south to lodgepole pine and aspen farther north. Northern Arizona contains four distinct Douglas fir habitat types (Alexander et al. 1984). The lowest zone has arid grasslands but with many bare areas, as well as xeric shrubs and sagebrush. Several kinds of cacti and yucca are common at low elevations in the south.

Monoclines—local steepening in otherwise uniform, gently dipping strata—are the region's single most distinctive structural feature. The plateau also has igneous laccoliths, flat-bottomed igneous intrusive bodies that dome up over the sedimentary rocks, such as the Henry, La Sal, Navajo, Abajo, Ute, and Carrizo Mountains of southeastern Utah and northern Arizona (Morris and Stubben 1994). The plateau is bounded on the east by the southern Rocky Mountains, on the north by the central Rocky Mountains, and on the south and west by the Basin and Range Province.

Elevations in the plateau are generally over 1,525 m (5,000 ft) and in some areas are as high as 3,960 m (13,000 ft). The climate is thus characterized by cold winters and summers with hot days and cool nights. Average annual precipitation is about 510 mm, but some parts of the region receive less than 260 mm.

Biological Distinctiveness

The Colorado Plateau is the only area in the United States and Canada where large mountain rivers run through exposed sandstone. That unique juxtaposition created the Grand Canyon, an internationally important ecotourism site, and other spectacular canyons in the region.

The Colorado River fish fauna display distinctive adaptive radiations. The humpback chub (*Gila cypha*), for example, is a highly specialized minnow that lives in the upper Colorado. It adapted to the water's fast current and its extremes of temperature and volume. Dams and water diversion, however, have created a series of placid, still-water lakes and side streams, and the humpback chub may not be able to adapt to these altered conditions. The species, along with other native Colorado River fishes, including the bonytail (*Gila elegans*), squawfish (*Ptychocheilus lucius*), and perhaps the flannelmouth sucker (*Catostomus latipinnis*), may not survive to the end of this century (Sigler and Sigler 1994).

This ecoregion is also rich in certain species of insects. The portion of the Colorado Plateau in the state of Colorado harbors 61 of the 131 species of grasshopper found in the state. The plateau is also rich in ant species and supports several endemic leafhopper species (Nelson 1994).

Conservation Status

Habitat Loss and Degradation

Approximately 15 percent of the Colorado Plateau remains as intact habitat. While little of the remaining 85 percent has been heavily altered by human activity, all shows some signs of stress. Riparian areas and areas with mineral resources have been the hardest hit. The main reason for habitat loss in the region is grazing, and there is widespread grazing damage in the ecoregion. Other important causes of habitat loss include mining for coal and uranium, agriculture, invasion of exotics following heavy grazing, oil and gas exploration, dams, and urbanization, particularly around Albuquerque and Santa Fe, New Mexico.

Habitat loss is concentrated along Interstate 40 in the Four Corners region, along the Colorado River, and in the coal mining region in the northwest corner of the ecoregion.

Remaining Blocks of Intact Habitat

Several relatively large blocks of more or less intact habitat remain. Important blocks (mostly concentrated along the Colorado River) include:

- Grand Canyon National Park—northwestern Arizona
- Canyonlands National Park—eastern Utah
- Escalante–Capitol Reef–Kaiparowits Plateau—south-central Utah
- Desolation Canyon—Utah
- San Raphael Swell—Arizona
- Navajo Mountain—southern Colorado
- Arches National Park—eastern Utah
- Cebolleta Mesa complex—New Mexico
- Sevilleta National Wildlife Refuge—New Mexico

Degree of Fragmentation

The southern and eastern parts of the Colorado Plateau have been fragmented by urbanization, mining, and agriculture. The Colorado River and Upper Rio Grande riparian corridors are also highly fragmented.

Degree of Protection

This ecoregion has numerous large protected areas and is representative of most habitat types. Xeric shrubland is not well protected, but this habitat appears to be doing well outside of the reserves. The most important protected areas in this ecoregion include:

- Grand Canyon National Park—Arizona
- Canyonlands National Park—Utah
- Arches National Park—Utah
- Dinosaur National Monument—western Colorado and eastern Utah
- Mesa Verde National Park—southwestern Colorado
- Sevilleta National Wildlife Refuge—New Mexico
- Capital Reef National Park—Utah
- Rio Grande National Wild and Scenic River
- Petrified Forest National Park—eastern Arizona

Types and Severity of Threats

While conversion of natural habitat to other uses poses a problem for only a small portion of the Colorado Plateau, urban and suburban development, strip mining, and other activities threaten some of the most sensitive habitats in the region. Development is a particular problem near Albuquerque, Santa Fe, and Farmington, New Mexico. Strip mining and the construction of a

power plant threaten the Kaiparowits Plateau. Air pollution from uranium and coal mining poses the greatest degradation threat to the ecoregion, along with off-road vehicles, over-grazing, and excessive impacts of recreation around Moab. The greatest threat to wildlife on the Colorado Plateau is the destruction of native fish by dam building and other forms of development.

Suite of Priority Activities to Enhance Biodiversity Conservation

- Repeal RS 2477. This federal law allow counties to establish rights-of-way across old cattle trails and horse trails on federal lands. Counties have made thousands of claims to rights-of-way within the Colorado Plateau. By exercising these rights, counties fragment public lands and thus disqualify them for wilderness designation, often leading to development.
- Control the impact of livestock grazing. Livestock grazing has serious impacts on riparian areas, which are the arteries of the plateau.
- Control exotics. Tamarisk, cheat grass, and other exotics threaten native plants and animals. This will require management plus research into the most cost-effective control methods.
- Protect Colorado River endangered fishes: Change reservoir operations to benefit fishes. For example, manage flows so as to mimic the natural hydrograph (floods) and release water from the top rather than the cold bottom of Glen Canyon reservoir (certain fishes need this warm water).
- Inventory and monitor biodiversity. Little is known about biodiversity in this region. In particular, there is a need to determine if and why amphibians are declining
- Protect neotropical migratory birds. There is a need to protect their nesting habitat in this region and determine the cause of their population decline. Is the problem on their wintering grounds or their migration routes?
- Protect threatened and endangered species. The Colorado Plateau has some localized threatened and endangered species. The Mexican spotted owl (*Strix occidentalis lucida*) tends to occur in areas where people like to recreate, leading to conflicts.

- Repeal the salvage logging rider. A number of southern Utah forests have been targeted for logging under the provisions of the salvage rider.

Conservation Partners

For contact information, please see appendix G.

- Grand Canyon Trust
- The Nature Conservancy of Arizona
- The Nature Conservancy of Colorado
- The Nature Conservancy of New Mexico
- The Nature Conservancy
- The Nature Conservancy of Utah
- The Nature Conservancy, Western Region
- Navaho Nation Natural Heritage Program
- Southern Utah Wilderness Alliance
- Utah Wilderness Coalition

Relationship to Other Classification Schemes

Bailey's (1994) Colorado Plateau Semi-Desert Province begins farther to the south than this ecoregion. Much of what we classify as Colorado Plateau shrublands Bailey includes in the Intermountain Semi-Desert and Desert Province. That province, however, extends far to the west and covers most of the Great Basin. Including northeastern Utah and northwestern Colorado but not western Utah in the Colorado Shrub Steppe provides a more consistent and manageable ecoregion. This ecoregion corresponds more closely to two of Omernik's (1995b) ecoregions, the Colorado Plateaus and the Arizona–New Mexico Plateaus.

Prepared by S. Primm

Key Number:	**79**
Ecoregion Name:	**Mojave Desert**
Major Habitat Type:	**4.2 Xeric Shrublands/ Deserts**
Ecoregion Size:	**130,634 km²**
Biological Distinctiveness:	**Regionally Outstanding**
Conservation Status:	**Snapshot—Relatively Stable Final—Relatively Stable**

Introduction

The Mojave Desert is the smallest of the four American deserts. While the Mojave lies between the Great Basin Shrub Steppe and the Sonoran Desert, its fauna is more closely allied with the lower Colorado division of the Sonoran Desert. Dominant plants of the Mojave include creosote bush (*Larrea tridentata*), all-scale (*Atriplex polycarpa*), brittlebush (*Encelia farinosa*), desert holly (*Atriplex hymenelytra*), white burrobush (*Hymenoclea salsola*), and Joshua tree (*Yucca brevifolia*), the most prominent endemic species in the region (Turner 1994b).

The Mojave is bounded to the north by the Great Basin Shrub Steppe, to the west by the Sierra Nevada and California montane scrub, and to the east by the Colorado Plateau. To the south, the Mojave blends into the Sonoran Desert, with a combination of species typical of the two deserts plus species most often found in the scrublands or conifer woodland of the Great Basin. Mojave species generally favor the colder plains, while Sonoran species are found on hillsides.

The Mojave's warm temperate climate defines it as a distinct ecoregion. Species that serve to separate the Mojave from the Sonoran Desert include such widespread Sonoran species as ironwood (*Olneya tesota*), blue palo verde (*Cercidium florium*), and chuparosa (*Justica californica*). Mojave indicator species include spiny mendora (*Mendora spinescens*), desert senna (*Cassia armata*), Mojave dalea (*Psorothamnus arborescens*), and goldenhead (*Acamptopappus shockleyi*) (Turner 1994b).

The Mojave supports numerous species of cacti, including several endemics, such as silver cholla (*Opuntia echinocarpa*), Mojave prickly pear (*O. erinacea*), beavertail cactus (*O. basilaris*), and many-headed barrel cactus (*Echinocactus polycephalus*). The region is also rich in ephemeral plants, with perhaps eighty to ninety endemics (Turner 1994b).

The Mojave Desert contains a range of elevations not found in other North American deserts. Elevation ranges from below sea level in Death Valley (−146 m) to over 1,600 m on some mountains. Most of the region lies between 610 and 1,220 m, giving rise to the term "high desert." The Mojave receives little precipitation (between 65 and 190 mm annually), and dry lakes are a common feature of the landscape.

Biological Distinctiveness

While the Mojave Desert is not as biologically distinct as the other desert ecoregions, distinctive endemic communities occur throughout it. For example, the Kelso Dunes in the Mojave National Preserve harbor seven species of endemic insects, including the Kelso Dunes jerusalem cricket (*Ammopelmatus kelsoensis*) and the Kelso Dunes shieldback katydid (*Eremopedes kelsoensis*). The Mojave fringe-toed lizard (*Uma scoparia*), while not endemic to the dunes, is rare elsewhere (Schoenherr 1992).

The native range of California's threatened desert tortoise (*Gopherus agassizii*) includes the Mojave and Colorado Deserts. The desert tortoise has adapted for desert existence by storing up to a liter of water in its urinary bladder. The tortoise feeds on ephemeral plants in the spring and accumulates enough reserves of water to carry it through the remainder of the year (Schoenherr 1992).

Other endemic fauna include the Mojave ground squirrel (*Spermophilus mojavensis*) and Amargosa vole (*Microtus californicus scirpensis*),

The Mojave Desert is rich in ephemeral plants, most of which are endemic. Of the approximately 250 Mojave taxa with this life form, perhaps 80–90 are endemic (Shreve and Wiggins 1964). During favorable years, the region supports more endemic plants per square meter than any location in the United States. Most of these species are winter annuals. Flowering plants also attract butterflies such as the Mojave sooty-wing (*Pholisora libya*) and the widely distributed painted lady (*Vanessa cardui*) (Schoenherr 1992).

A majority of the fauna found in the Mojave Desert also extends into the Sonoran or Great Basin Desert as well. However, the following avifauna and herpetofauna are characteristic of the Mojave region in particular: LeConte's thrasher (*Toxostoma lecontei*), banded gecko (*Coleonyx variegatus*), desert iguana (*Dipsosaurus dorsalis*), chuckwalla (*Sauromalus obesus*), and regal horned lizard (*Phrynosoma solare*). Snake species include the desert rosy boa (*Lichanura trivirigata gracia*), Mojave patchnose snake (*Salvadora hexalepis mojavensis*), and Mojave rattlesnake (*Crotalus scutulatus*) (Brown 1994).

Conservation Status

Habitat Loss and Degradation

Roughly half of the Mojave Desert remains as intact habitat, and the remaining half has not been heavily altered by human activity. The main reasons for habitat loss in the region include urbanization and suburbanization from Los Angeles and Las Vegas, the increasing demand for landfill space (Los Angeles and San Diego are proposing a large landfill in the region), agricultural development along the Colorado River, grazing, off-road vehicles, and military activities. Areas under particular pressure include Ward Valley (near Mojave National Park) and Riverside County, west of Joshua Tree National Monument. A falling water table also threatens Death Valley National Park.

Remaining Blocks of Intact Habitat

The most important remaining habitat blocks include:

*   Death Valley National Monument—eastern California
*   Desert National Wildlife Range—southern Nevada
*   Joshua Tree National Monument—southeastern California
*   Lake Mead—southeastern Nevada and northwestern Arizona
*   Nevada test site—southern Nevada

Degree of Fragmentation

Habitats in the Mojave Desert are generally contiguous, with a high degree of connectivity. Roads, however, have fragmented habitat for certain species, such as desert tortoise and some species of snakes. Bighorn sheep migration routes also are not adequately protected between reserves.

Degree of Protection

By historical accident and the California Desert Protection Act, the Mojave Desert is the one of the best-protected ecoregions in the United States. The full range of habitats is included in reserves, although riparian areas need more protection. In addition to the habitat blocks listed above, important protected areas include:

*   Kingston Range—southeastern California
*   Mojave National Park—southeastern California

- Sheephole Wildlife Area—southeastern California
- Greenwater Range—southeastern California
- lands covered under the California Desert Protection Act

Types and Severity of Threats

Conversion threats to the Mojave are concentrated in the southwest and east-central portions of the ecoregion. Lower-elevation valleys are largely in private hands and lack protection. Off-road vehicles and development threaten these valleys, and development also is harming creosote bush areas. Tamarisk is invading springs, and grazing is damaging mid-elevation pastures. Wildlife trade threatens chuckwallas, Gila monsters, and desert tortoises.

Suite of Priority Activities to Enhance Biodiversity Conservation

The most important conservation activity in the Mojave Desert is to protect riparian areas and low-elevation valleys.

Conservation Partners

For contact information, please see appendix G.

- California Desert Protection League
- California Native Plant Society
- Sierra Club
- The Wilderness Society

Relationship to Other Classification Schemes

This ecoregion roughly corresponds to Bailey's (1994) Mojave Desert section, which falls within the American Semi-Desert and Desert Province, though the Bailey section extends somewhat farther to the east, into the Sonoran Desert and Colorado Plateau, and to the west into the Southern California montane scrub. Omernik (1995b) places the Mojave in the Southern Basin and range, a region that also includes much of the area we classify as the Sonoran Desert. We feel the communities within these desert ecosystems are distinct enough to warrant separate classifications and conservation planning.

Prepared by B. Holland, G. Orians, and J. Adams

Key Number:	**80**
Ecoregion Name:	**Sonoran Desert (U.S.)**
Major Habitat Type:	**4.2 Xeric Shrublands/ Deserts**
Ecoregion Size:	**116,770 km²**
Biological Distinctiveness:	**Globally Outstanding**
Conservation Status:	**Snapshot—Relatively Stable Final—Relatively Stable**

Introduction

The Sonoran Desert in the United States reaches from extreme southeastern California across the western two thirds of southern Arizona. In Mexico, the desert stretches south to encompass much of the state of Sonora, as well as the eastern shore of Baja California to the town of Loreto. Unless otherwise noted, this ecoregion description will be confined to the portion of the Sonoran Desert within the United States. It is one of three warm desert ecoregions in the United States and, taken together with the Mojave to the west and the Chihuahuan to the east, forms a desert region that stretches from the Sierra Nevadas in California to the Edwards Plateau in Texas. The bulk of the American Sonoran is in the state of Arizona.

The American Sonoran Desert can be divided into two sections, the distinctiveness of each section dictated by the availability of water. The lower Colorado River valley section to the west is characterized by creosote (*Larrea divaricata*) and white bursage (*Ambrosia dermosa*) (McNab and Avers 1994). Temperatures are relatively high in winter and summer. Rainfall is infrequent, and its pattern is highly irregular. The Arizona upland section to the north and east is more mesic, resulting in greater species diversity and richness. Rainfall totals between 100 and 300 mm, and it is distributed biseasonally in roughly equal amounts. Lower-elevation areas are dominated by dense communities of creosote and white bursage, but on slopes and higher portions of bajadas, subtrees such as paloverde (*Cercidium floridum, C. microphyllum*) and ironwood (*Olneya tesota*), saguaros (*Carnegiea gigantia*), and other tall cacti are abundant (Turner, Bowers, and Burgess 1995).

Biological Distinctiveness

The Sonoran Desert supports a wide variety of plants and animals. Despite their relative abundance within the desert, cacti primarily occur in the Nearctic. By far the most recognizable of Sonoran species is the saguaro cactus. This largest of all cacti is truly an amazing plant, but it is only one of numerous cactus species found in this rich desert. Other plant species include the cholla cactus (*Opuntia fulgida*), organ pipe cactus (*Lemaireocereus thurberi*), the silver-dollar cactus (*Opuntia chlorotica*), and the jojoba (*Simmondsia chinensis*). Another remarkable plant found in the Sonoran Desert is the ocotillo (*Fouquieria splendens*), which may remain leafless during the coldest months of winter but experience five or six leafy periods throughout the year. The brilliant red conical flowers are often triggered by the first cool-season rain in the spring, flowering within as few as forty-eight hours and attracting hummingbirds and other nectar feeders.

These plants form the foundation of habitats for numerous animal species. The Sonoran Desert is recognized as an exceptional birding area within the United States. Forty-one percent (261 of 622) of all terrestrial bird species found in the United States can be seen here during some part of the year. The Sonoran Desert, together with its eastern neighbor the Chihuahuan Desert, is the richest area in the United States for birds, particularly hummingbirds. Among the bird species found here are the saguaro-inhabiting cactus wren (*Campylorhynchu brunneicapillus*), black-tailed gnatcatcher (*Polioptila melanura*), phainopepla (*Phainopepla nitens*), Gila woodpecker (*Melanerpes uropygualis*), and Costa's hummingbird (*Calypte costae*). Perhaps the most well-known Sonoran bird is the roadrunner (*Geococcyx californianus*), distinguished by its preference for running over flying as it hunts scorpions, tarantulas, rattlesnakes, lizards, and other small animals.

The Sonoran Desert harbors two endemic bird species, the highest level of bird endemism in all of the United States. The rufous-winged sparrow (*Aimophila carpalis*) is fairly common along the central portion of the Arizona-Mexico border, seen in desert grass mixed with brush. Rare in extreme southern Arizona along the Mexican border, the five-striped sparrow (*Amphispiza quinquestriata*) lives predominantly in canyons on hillsides and slopes among tall, dense shrubs.

Fifty-eight species of reptiles, including six species of rattlesnake, are found in the U.S. por-tion of this rich desert. The threatened desert tortoise (*Gopherus agassizii*) makes its home in burrows and frequents desert oases, washes, and riverbanks throughout the Sonoran. Perfectly camouflaged for life in the Sonoran, the giant Gila monster (*Heloderma suspectum*) is one of several lizards inhabiting this region. The widespread tiger salamander (*Ambystoma tigrinum*), the world's largest land-dwelling sala-mander, is able to survive in the Sonoran by burrowing underground during the dry season and emerging during wet times. The tiger sala-mander is one of only a dozen amphibians able to live in this arid ecoregion.

Mammals that are able to withstand the extreme climate of the Sonoran include the California leaf-nosed bat (*Macrotus californicus*), ring-tailed cat (*Bassariscus astutus*), black-tailed jack-rabbit (*Lepus californicus*), desert kangaroo rat (*Dipodomys deserti*), and the endemic Bailey's pocket mouse (*Perognathus baileyi*), round-tailed ground squirrel (*Spermophilus tereticaudus*). Other wider-ranging large mammals also found here are the mountain lion (*Felis concolor*), bighorn sheep (*Ovis canadensis*), coyote (*Canis latrans*), and the endangered pronghorn (*Antilocapra americana sonoriensis*).

Conservation Status

Habitat Loss and Degradation

About 60 percent of the habitat in the Sonoran Desert ecosystem has been altered by agriculture, grazing, excessive groundwater pumping, and urbanization. Riparian habitats throughout the ecoregion are severely degraded, particularly along the Salt, Verde, and Gila Rivers. Residential development on bajadas is eliminating the habitat of bajadas-dependent species such as cholla cacti (*Opuntia*) and columnar cacti such as saguaro (MacMahon 1988). Rocky habitat areas preferred by Gila monsters and bighorn sheep are prime real estate for development. The last population of Sonoran pronghorn is confined to the Cabeza area in southern Arizona, isolated by lack of connectivity with other habitat areas. Riparian woodlands in the region are now one of the rarest habitat types in North America because of widespread destruction. Threats to riparian areas are numerous, including trampling, grazing, and fouling by domestic livestock; water diversion and dam building; conversion for

agricultural or urban development; and introduction of species such as the tamarisk tree
(*Tamarix chinensis*) (Noss and Peters 1995).

Remaining Blocks of Intact Habitat

There are several large blocks of relatively intact
habitat in the ecoregion, much of which includes
wildlife refuges, national monuments, and military installations. Important blocks include:

- Organ Pipe Cactus National Monument,
  Cabeza Prieta National Wildlife Refuge, and
  Barry Goldwater Bombing Range complex,
  which if coupled with Pinacate National Park
  and El Gran Desierto area in Mexico, consti-
  tutes a virtually roadless intact area of over
  25,000 km² (6 million acres).
- Kofa Mountains
- Lower Colorado River (includes Imperial
  National Wildlife Refuge, Trigo Mountain
  Wilderness, several BLM wilderness study
  areas in southwest Arizona and southeast
  California, and the Yuma Proving Grounds
  Marine Base)
- Arrastra Mountains
- Chuckwilla Mountains
- Eagletail Mountains
- Turtle Mountain
- Hassayampa Riparian Reserve (TNC), which
  conserves some of the last riparian woodland

In addition, there are large areas of intact
habitat within the Tohono O'odham Indian
Reservation.

Degree of Fragmentation

There are several large areas of relatively intact
habitat remaining in the American Sonora, but
they are separated by areas of intense agriculture
and urbanization. Desert tortoises and other
herpetofauna are adversely impacted by habitat
fragmentation due to roads. Lack of connectivity
between Cabeza Prieta and other habitat areas
has effectively isolated the Sonoran pronghorn
population. Habitat fragmentation due to roads
adversely impacts the desert tortoise (*Gopherus
agassizii*).

Riparian and freshwater habitats are extremely fragmented. Dams on the Salt, Gila, and
Verde Rivers have heavily altered the habitat in
and around those streams. Excessive groundwater pumping and agricultural runoff in the
Imperial Valley are contributing to the problem.

Degree of Protection

Approximately 17 percent of natural habitat of
the American Sonora enjoys some form of protection, as wilderness, wildlife refuge, national
monument, or other administrative unit.
Though this represents a large and valuable network of core habitat areas, lack of connectivity
between these areas remains a serious problem,
particularly for large, wide-ranging species such
as Sonoran pronghorn, bighorn sheep, Mexican
wolves (assuming reintroduction), and jaguars.
Perhaps the greatest opportunity exists in the
western Sonora. There is currently good connectivity between the Cabeza Prieta–Organ
Pipe–Goldwater complex and the Pinacate–
El Gran Desierto in Mexico. The integrity of
this area should be maintained and intact habitat
expanded by reestablishing linkage zones
between Cabeza Prieta and the Kofa Mountains.
Riverine systems are virtually unprotected.

Types and Severity of Threats

The major conversion threats are urbanization,
suburban sprawl, agricultural expansion, and
resource extraction. The urban and suburban
areas of Phoenix and Tucson continue to expand
rapidly. This in turn is pushing agricultural
operations farther into the desert and along
riparian areas such as the Gila River, with
tremendous impacts on wildlife habitat.

Desert habitats have not evolved to withstand
heavy grazing pressure. Thus, grazing by domestic livestock continues to have severe impacts on
natural communities, especially riparian habitats. A major degradation threat comes from
introduced species such as buffelgrass
(*Pennisetum ciliare*), originally planted as cattle
forage but rapidly spreading and forcing out
native plants. Wild burros, protected by federal
law and supported by a vocal constituency, are
extremely destructive to desert plant communities and compete directly with wildlife. Off-road
vehicle activity is increasing outside protected
areas, such as the military lands north of Cabeza
Prieta.

Collecting of herpetofauna, cacti, and ironwood is increasing to a level that could threaten
native wildlife and plant populations.
Coordinated efforts to enforce and regulate
wildlife laws, and protect populations of sensitive
species, particularly cacti, are greatly needed.

Suite of Priority Activities to Enhance Biodiversity Conservation

- Maintain connectivity between the Pinacate Desert and El Gran Desierto in Mexico and the Cabeza Prieta–Organ Pipe complex. Prevent further roading within and between these areas. Explore the option of combining these areas in a 2.5-million-hectare (6-million-acre) international wilderness protected area.

- Restore connectivity between Cabeza Prieta and the Kofa Mountains.

- Take advantage of the conservation opportunities presented by the vast Department of Defense holdings in the ecoregion. The Barry Goldwater Bombing Range is a 1-million-hectare (2.6-million-acre) area in the heart of the Sonoran Desert. Military activities have kept ranching, agriculture, and extractive industries out of the area, and much of the range is virtually intact desert habitat. The Yuma Proving Grounds Marine Base in southwest Arizona contains almost a quarter million acres of roadless desert landscape. Ensure that the DOD manages these areas to conserve native plants and wildlife and that land stewardship and research are included in their operating budgets.

- Protect and restore riparian habitat. Riparian habitat is greatly degraded throughout the ecoregion. Establish protection for habitat along the lower Gila, lower San Pedro, Santa Cruz, and Verde Rivers. Research promising areas for riparian restoration and the establishment of riparian corridors on, for example, the Bill Williams River and the Hassayampa River.

- Reestablish natural flow regimes in the ecoregion's rivers by controlling overpumping of river and groundwater, leaving enough water in the rivers to keep aquatic habitats intact and restore habitats that have been degraded.

- Coordinate existing scientific data and future research on both sides of the international border. Areas for coordination of future research include: the pervasive problem of non-native vegetation, including Sahara mustard (family: Cruciferae) and buffelgrass (*Pennisetum ciliare*), species that are causing large-scale ecological change in the ecoregion; trans-border wildlife species such as the jaguar, jaguarundi, ocelot, and Sonoran pronghorn; and likely areas for riparian restoration.

- Build the capacity of the Tohono O'Odham Nation to manage its extensive natural resources. The Tohono O'Odham reservation covers almost 800,000 hectares (2 million acres) of largely intact habitat adjacent to the Cabeza Prieta–Organ Pipe complex.

- Control collecting of native plants and herpetofauna with greater enforcement of existing laws and regulations and promulgation of new ones as necessary.

- Develop a comprehensive plan for conservation of the Lower Colorado River area, currently a jumble of jurisdictions and management regimes. This plan should include additional wilderness designations as appropriate; Wild and Scenic River status for the Colorado River below Picacho; steps to conserve the many endangered and threatened fish, wildlife, and plant species that occur in the area; and steps to curtail the increasing agricultural pollution of the river with salt and heavy metals.

Conservation Partners

For contact information, please see appendix G.

- Arizona Sonora Desert Museum
- The Nature Conservancy of Arizona
- The Nature Conservancy of California
- The Sonoran Institute
- Tucson Audubon Society
- The Wildlands Project

Relationship to Other Classification Schemes

The Sonoran Desert ecoregion corresponds to Küchler's (1975) units 36 (Creosote bush-bur sage) and 37 (Palo verde–cactus shrub). Omernik's (1995b) unit 14 (Southern Basin and Range) was split into the Mojave Desert [79] and Sonoran Desert [80] ecoregions, based on differences in community structure. The boundary location was based on Bailey (1994) and MacMahon (1988). This ecoregion incorporates Bailey's sections 322B (Sonoran Desert) and 322C (Colorado Desert).

Prepared by C. Williams, D. Olson, and P. Hurley

Key Number:	**81**
Ecoregion Name:	**Chihuahuan Desert (U.S.)**
Major Habitat Type:	**4.2 Xeric Shrublands/**
	Deserts
Ecoregion Size:	**205,439 km²**
Biological Distinctiveness:	**Globally Outstanding**
Conservation Status:	**Snapshot—Vulnerable**
	Final—Vulnerable

Introduction

The Chihuahuan Desert in the United States stretches from the extreme southeastern corner of Arizona (where it mingles with the Madrean Sky Islands and the eastern edge of the Sonoran Desert) across southern New Mexico and west Texas to the Edwards Plateau. The desert reaches deep into central Mexico as far south as San Luis Potosi; it is bounded by the Sierra Madre Occidental to the west and the Sierra Madre Oriental to the east. Unless otherwise noted, this ecoregion description will be confined to the portion of the Chihuahuan Desert within the United States. It is one of three warm desert ecoregions in the United States and, together with the Sonoran and the Mojave to the west, forms a vast desert region that stretches from the Sierra Nevada in California to central Texas.

Due to its generally higher elevation, the Chihuahuan Desert is cooler and has more rainfall than our other warm desert ecoregions, averaging 235 mm annually. Common throughout the Chihuahuan Desert, characterizing it in much the same way that large cacti characterize the Sonoran, are shrubs: creosote (*Larrea divarecata*), tarbush (*Florensia cernua*), mesquite (*Prosopis articulata*), and acacia (*Acacia* spp.). As one moves north from central Mexico, the desert grades from a landscape of multi-stemmed cacti, yucca, and shrubs to the dry grassland ecosystem that dominates much of the Chihuahuan Desert within the United States (MacMahon 1988). It is the cooler, more mesic climate that probably accounts for the extensive grass component of the American Chihuahuan ecoregion. Here the desert floor is often covered with grasses such as bush muhly (*Muhlenbergia porteri*), blue grama (*Bouteloua gracilis*), and, in the bottomlands, big sacaton (*Sporobolus wrightii*) (McClaran 1995). It was probably the latter that the early Spanish explorers encountered when they excitedly reported grasses that were "belly high to a horse" (Tweit 1995).

Biological Distinctiveness

The Chihuahuan Desert may be the most biologically diverse desert ecoregion in the world, whether measured by species richness or measured by endemism. It features 250 species of butterflies, over 100 mammals, over 250 bird species, 100 reptile species, 20–25 amphibian species, and over 100 species of cacti. The American Chihuahuan is home to 5 endemic reptile species, over 25 endemic snails, and 5 endemic butterflies.

Because of its recent origin, few warm-blooded vertebrates are restricted to the Chihuahuan Desert scrub. However, the Chihuahuan Desert supports a large number of wide-ranging mammals, such as the pronghorn antelope (*Antilocapra americana*), jaguar (*Felis onca*), collared peccary or javelina (*Dicotyles tajacu*), desert cottontail (*Sylvilagus auduboni*), and kangaroo rat (*Dipodomys* spp.). Common bird species include the greater roadrunner (*Geococcyx californianus*), curve-billed thrasher (*Toxostoma curvirostra*), scaled quail (*Callipepla squamata*), and Scott's oriole (*Icterus parisorum*). In addition, numerous raptors inhabit the desert and include the red-tailed hawk (*Buteo jamaicensis*), the great horned owl (*Bubo virginianus*), and the rare zone-tailed hawk (*Buteo albonotatus*).

The Chihuahuan Desert herpetofauna is more strongly associated with the region. Several lizards are centered in the Chihuahuan Desert and include the Texas banded gecko (*Coleonyx brevis*), reticulated gecko (*C. reticulatus*), greater earless lizard (*Cophosaurus texanus*), several species of spiny lizards (*Scelpoprus* spp.), and marbled whiptails (*Cnemidophorus tigris marmoratus*). Two other whiptails (*C. neomexicanus* and *C. tesselatus*) occur as all-female parthenogenic clones in select disturbed habitats (Wright and Vitt 1993). Representative snakes include the trans-Pecos rat snake (*Elaphe subocularis*), Texas blackheaded snake (*Tantilla atriceps*), and whipsnakes (*Masticophis taeniatus* and *M. flagellum lineatus*) (Brown 1994).

Conservation Status

Habitat Loss and Degradation

The Chihuahuan ecoregion is heavily degraded. Historical accounts report that in the mid-1800s the native grasses were lush and the landscape was relatively free of shrubs. Riparian areas were lined with galeria forests, and unchanelled streams often spread out to form wetland systems (cienegas). Antelope (*Antilocapra americana*), prairie dog (*Cynomys ludovicianus*), and Mexican wolf (*Canis lupus baileyi*) were abundant. Today, brush and shrubby trees dominate the landscape. Native grasses, overgrazed, under siege from aggressive non-native species, and deprived of their natural, fire-based disturbance regime, are struggling. In southern New Mexico around Las Cruces, for example, the desert floor is covered with creosote and little else. Most of the riparian forests and cienegas have disappeared, victims of overgrazing, heavy erosion, and excessive water diversion. The few free-flowing streams that remain are in deeply cut (often human-built) channels. Antelope and prairie dogs are scarce, and the Mexican wolf has been extirpated (Bahre 1995).

Remaining Blocks of Intact Habitat

There are several large blocks of relatively intact habitat in federal, state, and private ownership:

- Big Bend National Park, Big Bend Ranch State Park, Rio Grande Wild and Scenic River complex, which, coupled with adjacent state, private, and Mexican roadless areas, constitutes a relatively intact habitat block of over 1 million hectares (2.5 million acres)—southwestern Texas
- Guadalupe Escarpment—southwestern New Mexico and western Texas
- White Sands National Monument and Missile Range—southern New Mexico
- West Potrillo Mountains—southern New Mexico
- Gray Ranch—southwest New Mexico
- Ladder Ranch—southwest New Mexico
- Pedro Arminderos Ranch—southwest New Mexico
- Davis Mountains—Texas
- Big Hatchet Alamo-Hueco Mountains—west Texas and southern New Mexico
- Hueco Tanks Mountains—Texas

Degree of Fragmentation

Virtually the entire ecoregion has been heavily grazed. Even some of the relatively intact habitat areas above have in the past been significantly altered by grazing. Heavily grazed areas are characterized by increasing dominance of shrubs such as creosote bush, mesquite, tarbush, and acacia and drastic alteration in the species make-up of the grasslands, with native grasses often replaced by non-native species (Brown 1995). In some areas, little grass remains.

The grassland ecosystem of the Chihuahua evolved under conditions of frequent fire. Overgrazing and active fire suppression have drastically changed the frequency and, consequently, the ecological function of fire in the system. As the grasslands have degraded, the fuel load has become inadequate to sustain the type of wildfire necessary to clear invading brush. Non-native species such as Lehmann lovegrass (*Eragrostis lehmannia*) and buffelgrass (*Pennisetum ciliare*), planted to control erosion or provide forage for livestock, often enjoy competitive advantages over native species that result in the latter's decline. Studies have shown that areas of mixed native grasses support greater biodiversity than monocultures of lovegrass (Bahre 1995).

Fields cleared for irrigated agriculture have destroyed thousands of hectares of native grassland. Growing urban areas and overpumping of groundwater for agriculture are severely affecting flows of Chihuahuan rivers, including the San Pedro, Pecos, and Rio Grande. The dewatering of rivers and streams, coupled with damage done by grazing, has severely degraded much of the freshwater and riparian habitat in the ecoregion.

In general, connectivity is poor, with intact habitat areas separated from one another by long distances.

Degree of Protection

Though there are some large areas that enjoy official or de facto protection, such as the Big Bend wilderness areas and the White Sands complex, the Chihuahuan ecoregion as a whole suffers from a lack of protection. For example, only a small fraction of the publicly owned habitat blocks described above are protected as wilderness, and others are vulnerable to road building, resource extraction, off-road vehicle use, and other threats. Much of the habitat described above, particularly in Texas, is in private ownership. Though some is well managed and in good ecological condition, some remains under threat.

In the New Mexico section of the Chihuahuan, a large amount of publicly owned land is administered by federal agencies, but only a small fraction has been set aside as wilderness or park land. In the Chihuahuan region of Texas, there is virtually no federal land outside the Big Bend and Guadalupe National Parks.

Types and Severity of Threats

The major conversion threats are urbanization, agricultural expansion, and resource extraction. Residential development is increasing, and urban and suburban expansion around Las Cruces, New Mexico; El Paso, Texas; and other cities is threatening surrounding areas. Urban and agricultural expansion is putting ever greater strain on groundwater supplies, threatening the flows (and thus the riverine and riparian habitats) of the Pecos, San Pedro, Rio Grande, and other rivers and streams.

Degradation threats include increasing off-road vehicle use in some areas, invasions of non-native species, increasing dominance of native shrub species in areas historically characterized by open grasslands, water pollution in the Rio Grande from the growing El Paso–Ciudad Juarez metropolitan area, and air pollution from coal-fired power plants just over the border in Mexico. Although overgrazing remains a concern in some areas, the worst of the damage that can be inflicted by grazing has already taken place. Riparian areas throughout the ecoregion are threatened with dewatering and overuse by livestock.

Collecting of Chihuahuan flora and fauna is rising rapidly. Among the favorites of collectors (and thus the most threatened) are cacti, peyote, and certain herpetofauna such as the gray-banded king snake (*Lampropeltis mexicana alterna*) and the trans-Pecos rat snake (*Elaphe subocularis*).

Suite of Priority Activities to Enhance Biodiversity Conservation

- Expand the network of protected areas in New Mexico and, particularly, in Texas through wilderness designations, land acquisition, and agreements with private landowners. This includes shoring up protection in existing habitat blocks and identifying and protecting other areas as habitat blocks or linkage zones.

- Maintain connectivity between the Big Bend National Park, the Big Bend Ranch State Park, the Rio Grande WSR complex, and adjacent state, private, and Mexican roadless areas. Extend protection to heretofore unprotected areas through wilderness designation, land acquisition, and agreements with private landowners.

- Take advantage of the conservation opportunities presented by the White Sands Missile Range, a de facto protected area rich in native biodiversity. Ensure that the Department of Defense manages the area to conserve native wildlife and plants, and that land stewardship and research are included in the Range's management budget.

- Restore the proper flow and prevent further pollution of the Rio Grande. Restore the riverine and riparian habitats of the Rio Grande WSR.

- Protect key habitat areas on private land, through agreements with landowners or, when necessary and appropriate, acquisition. Examples include the Hueco Tanks Mountains near El Paso and the Davis Mountains north of Big Bend.

- Coordinate existing scientific data and future scientific research on both sides of the border. Areas for coordination and future research include: trans-border wildlife species such as jaguar, jaguarundi, and ocelot; the pervasive problem of non-native species; and promising areas for riparian restoration.

- Improve the management of the ecoregion's federal and state parks, including the Big Bend complex, Guadalupe National Park, and White Sands National Monument, putting greater emphasis on the conservation of native biodiversity.

- Increase the amount and quality of water in the Rio Grande drainage by convincing U.S. and Mexican water management agencies to release more water at regular intervals for instream flows.

- Improve air quality by working with Mexican power companies to reduce emissions of sulfur dioxide from coal-fired plants along the border.

- Improve federal grazing policy to prevent damaging grazing practices in Chihuahuan desert grasslands and riparian areas.

- Control collecting of native plants and herpetofauna with education and greater enforcement of existing laws and regulations or promulgation of new ones as necessary.
- Foster collaboration with Mexican federal and state agencies on binational conservation issues such as the Rio Grande watershed and trans-border wildlife species. Provide scientific and technical assistance to Mexican authorities as they devise and implement management plans for protected areas in the Mexican Chihuahua.

Conservation Partners

For contact information, please see appendix G.

- Chihuahuan Desert Research Institute
- The Conservation Fund, Texas Chapter
- National Parks and Conservation Association, Southwest Regional Office
- The Nature Conservancy
- The Nature Conservancy of New Mexico
- The Nature Conservancy of Texas
- The Nature Conservancy, Western Region
- Sierra Club, Lone Star Chapter
- Texas Center for Policy Studies

Relationship to Other Classification Schemes

The Chihuahuan Desert ecoregion incorporates Küchler's (1975) unit 27 (Oak-juniper woodland), 48 (Grama-tobosa prairie), 52 (Grama-tobosa shrubsteppe), 53 (trans-Pecos shrub savanna), and 64 (Shinnery). The ecoregion boundary is the same as Omernik's (1995b) unit 24 (Southern Deserts), with the exception of the Madrean Sky Islands [47], which were separated from the surrounding ecoregion. This ecoregion corresponds to Bailey's (1994) sections 321A (Basin and Range) and 321B (Stockton Plateau).

Prepared by C. Williams and D. Olson

Key Number: **82**
Ecoregion Name: **Tamaulipan Mezquital**
Major Habitat Type: **4.2 Xeric Shrublands/ Deserts**

Ecoregion Size: **57,776 km²**
Biological Distinctiveness: **Regionally Outstanding**
Conservation Status: **Snapshot—Critical Final—Critical**

Introduction

The Tamaulipan Mezquital region of southern Texas and northeastern Mexico has unique plant and animal communities containing tree- and brush-covered dunes, wind tidal flats, and dense native brushland. These communities contain such species as ocelot (*Felis pardalis*), jaguarundi (*Felis yagouaroundi*), reddish egret (*Egretta rufescens*), Texas indigo snake (*Drymarchon corais erebennus*), and over four hundred species of birds (USFWS 1993).

The Rio Grande flows through this ecoregion and once formed a broad and meandering waterway that produced numerous resacas or oxbows within its floodplain and an extensive marshy environment at its mouth. Some of the earliest Spanish explorers called the river Rio de las Palmas, because of the abundant palm trees that grew along its banks. These Texas palms once extended upriver as much as eighty miles. Today, the valley's only remaining native palm grove is the 74-hectare (172-acre) Sabal Palm Grove Sanctuary. Most of the Rio Grande floodplain has been cleared for agriculture, grazing, and development.

Both banks of the Rio Grande are now crowded with homes, businesses, and farms. The only remaining natural areas south of the river are the salt marshes and mud flats east of the city of Matamoros. The Rio Corona floodplain was relatively intact until recently, but since the clearing of some lands for agriculture, there has been increased erosion, pollution from agrochemicals, invasion of exotics, and general loss of native wildlife.

The native vegetation type covering much of northeastern Mexico and parts of southern Texas is mesquite-grassland, an important element of the region that plant ecologists classify as the Tamaulipan biotic province. The Tamaulipan province extends south of the border for almost two hundred miles between the coast and the deciduous woodlands on the slopes of the Sierra Madre Oriental. The Tamaulipan thornscrub, a subtropical, semiarid vegetation type, occurs on either side of the Rio Grande. The slightly higher, drier, and rockier sites originally had vegetation of chaparral and cacti, whereas the flat, deep soils supported mesquite as well as taller brush and a few drought-resistant trees, often rather openly spaced and savannalike in a grassland matrix.

Spiny shrubs and trees dominate Tamaulipan thornscrub, but grasses, forbs, and succulents are also prominent (Crosswhite 1980). This region also includes elements of pastizal, a combination of grassland, savanna, and páramo-like communities (Rzedowski 1994). Leguminous shrubs and trees constitute one third of the diverse woody flora, which the rural population uses for extensive grazing of livestock, fuelwood, and timber for fencing and construction (Reid, Marroquin, and Beyer-Münzel 1990).

The two species that characterized the Mezquital communities before habitat alteration are mesquite itself (*Prosopis glandulosa*) and the curly mesquite grass (*Hilaria belangeri*) that grows under it. The most common shrubs are probably chaparro (*Zizyphus obtusifolia*) and jazmincillo (*Aloysia gratissima*). Parts of this region consist of open woods of mesquite with a pronounced understory of grasses that often contain a layer of taller species such as hooded finger grass (*Chloris cucullata*) and a layer of shorter grasses such as grama (*Bouteloua* spp.) In some places dense stands of prickly pear or nopal (*Opuntia lindheimeri*) take the place of many of the shrubs and grasses (Crosswhite 1980).

Montezuma bald cypress (*Taxodium mucronatum*) was once common along the Rio Grande for 160 km (100 miles) from the Gulf of Mexico. The tree is now rare along the Rio Grande, though it is still the dominant tree along the Rio Corona to the south. Perhaps the most striking endemic along the lower Rio Grande is the palma de Micharos (*Sabal texana*), which was once common along resacas in the floodplain and held a dominant position over mesquite under certain conditions.

A few species of plants account for the bulk of the brush vegetation of the Tamaulipan Mezquital and give it a characteristic appearance. The most important of these plants include: mesquite, various species of acacia including *Acacia smallii* and *A. tortuosa*, desert hackberry, javelina bush, cenizo (sometimes called purple sage), common bee-brush or white brush (*Aloysia wrightii*), Texas prickly pear, and tasajillo or desert Christmas cactus. The only exceptions to the rather arid, shrub-covered landscapes are the lines of riparian vegetation within the few river valleys.

Parts of this region support grasslands not unlike those of the Great Plains, though the grasslands here are not as uniform due to the highly variable soil and moisture conditions. The flora of the Tamaulipan Mezquital differs dramatically from the nearest desert, the Chihuahua, largely because of the Chihuahua's higher elevations and generally colder winters. The number of species in the Tamaulipan Mezquital that are frankly desert plants, many showing affinities with the Sonoran Desert, is startling (Crosswhite 1980).

Elevation increases northwesterly from sea level near the Gulf Coast to a base of about 300 m (1,000 ft) near the northern boundary of the ecoregion, from which a few hills and mountains protrude. The Anacacho Mountains and Turkey Mountain (550 m, or 1,805 ft) in Kinney County, Texas, as well as Blue Mountain (389 m, or 1,277 ft) in Uvalde County, occur within this region.

Rainfall tends to increase from west to east, with desert species such as creosote bush (*Larrea tridentata*) tending to grow in the western part, and nondesert species such as American elm (*Ulmus americana*) and black hickory (*Carya taxana*) native to the eastern part.

Biological Distinctiveness

The diversity of the Tamaulipan Mezquital ranks it among the four richest examples of this major habitat type in the United States and Canada. While not quite as diverse as the Chihuahuan Desert to the northwest, the Tamaulipan Mezquital supports over six hundred species of plants and animals. The region is particularly rich in tree species (including two endemics) and birds. The region has a high diversity and density of the elements tracked by the Texas Heritage Program.

Conservation Status

Habitat Loss and Degradation

Clearing and conversion of shrubland for agriculture have had the greatest impact on altering the patterns and processes of the landscape of South Texas and northeastern Mexico. Only about 2 percent of this ecoregion remains as intact habitat, and most of the remaining area has been heavily altered by human activity.

Remaining Blocks of Intact Habitat

This ecoregion contains no habitat blocks larger than 250 km^2.

Degree of Fragmentation

Only small patches of the original landscape remain. These remnants are largely isolated and provide little opportunity for species dispersal. The surrounding matrix, however, is composed of elements of the original landscape, and some movement is possible.

Degree of Protection

The Tamaulipan Mezquital contains no blocks of protected habitat greater than 250 km^2. However, new mapping activities targeting a 100-km buffer zone along the U.S.-Mexico border have begun.

Types and Severity of Threats

Conversion to agriculture poses the greatest threat to this ecoregion. Agricultural expansion and clearing for development may significantly alter more than a quarter of the remaining habitat over the next two decades. Degradation of habitat due to pollution and exploitation of wildlife are significant but less pressing threats.

Suite of Priority Activities to Enhance Biodiversity Conservation

- Survey and inventory the South Texas landscape.
- Continue mapping the buffer zone along the U.S.-Mexico border.
- Restore private lands, such as TNC's shrub restoration project.
- Restore to the Rio Grande and tributaries flow regimes that enhance biodiversity conservation (i.e., natural regeneration of native forests, conservation of native and endemic fish and amphibians).

- Representative communities needing protection:
  - Big Sacaton
  - Texas Palmetto
  - Plateau Live Oak
  - Texas Ebony
  - Ceniza
  - Blockbrush
  - Guajilo
- Protect and conserve the following areas in the Mexican portion of the Tamaulipan Mezquital:
  - Tamaulipan Shrubland Frontier Zone (Coahuila, Nuevo León, Tamaulipas). This region of Tamaulipan dry shrubland is far less degraded than that in Texas; roughly a third of the land remains intact, compared to less than 10 percent in Texas. The area supports about sixty species of threatened or endangered plants.
  - Sierra Pichachos (Nuevo León). This sierra is an "island" in northeastern Mexico that provides a stopover point for migratory fauna. The sierra contains both puma and black bear. Most of the land here is in private hands, and conservation agencies have had little involvement, although the Fish and Wildlife Service has shown a particular interest in studying the region.
  - Rio Bravo Delta (Tamaulipas). The low-lying, saline grasslands of the Rio Bravo Delta include large systems of dunes. The region supports various endemic plants, including an endemic genus. The delta also contains several species with interesting disjunct distributions, such as the eastern American mole (*Scalopus aquaticus*), a kangaroo rat (*Dipodomys campactus*), and a pocket gopher (*Geomys personatus*).
  - North Laguna Madre (Tamaulipas). This region is an important breeding area for numerous aquatic birds, supports a productive shrimp fishery and oyster beds, and includes a barrier island with dunes.
  - Sierra de San Carlos (Tamaulipas). This isolated sierra on the coastal plateau supports high shrublands and intact pine and pine-oak forests.
  - Rancho Nuevo (Tamaulipas). Rancho Nuevo is a nesting areas for sea turtles and shares floristic and physionomic characteristics with the barrier islands off the Texas coast. The estuary of the Rio Barberena includes man-

groves, and the region as a whole includes several endangered species.

- Tamaulipan Tropical Oak Forest (Tamaulipas). A mixture of tropical forest, oak forest, and grassland.

Conservation Partners

For contact information, please see appendix G.

- local Audubon chapters
- local land trust
- The Nature Conservancy
- The Nature Conservancy, Southeast Regional Office
- The Nature Conservancy of Texas
- Texas Center for Policy Studies

Relationship to Other Classification Schemes

The Tamaulipan Mezquital ecoregion corresponds to Küchler's (1975) units 54 (Mesquite-acacia savanna) and 55 (Mesquite–live oak savanna). The boundary is the same as Omernik's (1995b) unit 31 (Southern Texas Plains) and is similar to Bailey's (1994) section 315E (Rio Grand Plain), part of the Southwest Plateau and Plains Dry Steppe and Shrub Province.

Prepared by T. Cook and J. Adams

Key Number:	**83**
Ecoregion Name:	**Interior Alaska/Yukon Lowland Taiga**
Major Habitat Type:	**6.1 Boreal Forest/Taiga**
Ecoregion Size:	**443,319 km²**
Biological Distinctiveness:	**Bioregionally Outstanding**
Conservation Status:	**Snapshot—Relatively Intact Final—Relatively Intact**

Introduction

This large ecoregion covers much of the interior Alaskan and Yukon forested area, extending from the Bering Sea on the west to the Richardson Mountains in the Yukon Territory on the east and bounded by the Brooks Range

on the north and the Alaska Range on the south. The ecoregion includes an extensive patchwork of ecological characteristics due to local differences in topography, micro-climate, and drainage, but this finer mosaic of habitats is united into a single ecoregion by a predominance of spruce and hardwood forests, continental climate, and lack of Pleistocene glaciation. The terrain consists of rolling hills and lowlands dissected by nearly flat bottomlands along major rivers. Elevations range from sea level to approximately 600 m, and slope gradients are usually less than five degrees. The unglaciated landscape of the Canadian portion of this ecoregion is generally flat to gently rolling, with low relief (300–600 m asl). The highest peak is 925 m asl and occurs in the Eagle Plains in the south (ESWG 1995).

Spruce-dominated coniferous forests cover the majority of the ecoregion and occupy a variety of site conditions. White spruce (*Picea glauca*) forests occur on warmer, drier sites on hillsides, in timberline areas, and along rivers. Black spruce (*Picea mariana*) is found in similar areas but has higher tolerance for poorly drained soils and extends into bottomlands and other wet areas. River meanders support a continuous succession of colonizing willow (*Salix* spp.) and alder (*Alnus* spp.), followed by balsam poplar (*Populus balsamifera*) and quaking aspen (*Populus tremuloides*), which are replaced by spruce. Recently disturbed sites, areas near timberline, north-facing slopes, and wetter areas support scrub communites dominated by willow, alder, and dwarf birch (*Betula* spp.) Bottomland bogs and other extremely wet areas are occupied by scrub-graminoid communities, including willow, dwarf birch, Labrador-tea (*Ledum decumbens*), bush cinquefoil (*Potentilla fruticosa*), and sedges (*Eriophoum vaginatum* and *Carex* spp.). Wildfire is very common throughout the ecoregion and keeps a continuous mosaic of successional communities present, including herbaceous communities, scrub communities, and broadleaf, coniferous, and mixed forests (Bailey et al. 1994). Wetlands dominate the Old Crow Flats portion of the bioregion, composed of polygonal peat plateau bogs with basin fens and locally occurring shore fens. Permafrost is continuous, with medium to high ice content in the form of ice wedges and massive ice bodies.

The Interior Alaska/Yukon Taiga experiences a continental climate, with short, warm summers and long, cold winters. In general, annual temperature variance and precipitation increase from west to east. Average annual precipitation ranges from 250 mm to 550 mm, falling to 170 mm in the upper Yukon flats. Average daily winter minimum temperatures range from −35°C to −18°C, and average daily summer maximum temperatures range from 17°C to 22°C (Gallant et al. 1995). The climate of the Canadian portion of this ecoregion is strongly continental, considering its proximity to the Beaufort Sea. Mean annual temperature ranges from −10°C to −6.5° increasing toward the south. Mean winter temperature ranges from −27°C to −23.5°C, and summer temperature ranges from 7.5°C to 10°C. Mean annual precipitation is around 250 mm in the Old Crow areas but increases up to 600 mm in the south. In general, this ecoregion has a high subarctic ecoclimate (ESWG 1995).

A wide range of soils occupy the many different ecological environments (for details, see Gallant et al. 1995). All are shallow above continuous to discontinuous permafrost, except along rivers. Parts of this ecoregion were left unglaciated in the Pleistocene and formed part of the extensive Bering Sea Pleistocene refugium (Pielou 1994; Gallant et al. 1995).

Biological Distinctiveness

The Interior Alaska/Yukon Lowland Taiga ecoregion has retained intact ecosystems, with healthy populations of all natural top predators. This is becoming increasingly rare in North America. The Porcupine, Central Arctic, and Western Arctic caribou herds migrate across, and winter in, this ecoregion. The rivers and wetlands of this ecoregion support breeding populations of many birds, including grebes, loons, and goldeneyes. Grouse (*Dendragapus* spp.) and flycatchers (*Empidonax* spp. and *Contopus* spp.) also breed in the river valley forests. Beaver (*Castor canadensis*), moose (*Alces alces*), caribou (*Rangifer tarandus*), and snowshoe hare (*Lepus americanus*) are common, as are mink (*Mustela vison*), river otter (*Lutra canadensis*), marten (*Martes americana*), and muskrat (*Ondatra zibethica*) along the major rivers. Other wildlife in this ecoregion includes grizzly bear (*Ursus arctos*), black bear (*Ursus americanus*), wolf (*Canis lupus*), red fox (*Vulpes fulva*), raven (*Corvus corax*), osprey (*Pandion haliaetus*), bald

eagle (*Haliaeetus leucocephalus*), golden eagle (*Aquila chrysaetos*), rock ptarmigan (*Lagopus mutus*), willow ptarmigan (*L. lagopus*), spruce grouse (*Dendragapus canadensis*), waterfowl, and chinook salmon (*Oncorhynchus tshawytscha*), which spawn in the Porcupine River and its tributaries.

The Yukon Flats area in particular, in the northeastern corner of the ecoregion, is thought by McNab and Avers (1994) to be the most productive Arctic wildlife habitat in North America. In addition to supporting most of the above wildlife in dense numbers, it is home to lesser scaups (*Aythya affinis*), pintails (*Anas acuta*), scoters (*Melanitta* spp.), and widgeons (*Anas* spp.), as well as 15–20 percent of the canvasback (*Aythya valisineria*) population. Sandhill cranes (*Grus canadensis*) are common, and lynx (*Lynx canadensis*) are abundant. Minto Flats contains similar features and qualities to those of Yukon Flats but on a smaller scale.

A number of rare, narrowly endemic, or disjunct plant species also occur throughout this ecoregion, often associated with south-facing river bluffs. These species, including *Cryptantha shakletteana, Erysimum asperum,* and *Eriogonum flavum* are thought to be relicts from more widespread late-Pleistocene tundra-steppe communities.

Conservation Status

Habitat Loss and Degradation

This ecoregion is almost entirely intact, with little habitat loss or fragmentation. Habitat loss has occurred mainly surrounding human communities, especially Fairbanks, and in the Tanama Valley State Forest, which has experienced some clear-cutting. Other forms of human disturbance include subsistence and recreational hunting (birds, terrestrial mammals) and fishing. Metallic element ore and sand and gravel deposits have been mined, and there has been limited agricultural use along major rivers. There has been little historic fragmentation, but this ecoregion, like most others in the Arctic, experiences enormous natural disturbances from fire, so blocks of intact habitat need to be large.

Occurrence of lightning-ignited wildfire is common throughout the ecoregion, and individual burns average about 20 km² in area (6.85 km² in the upper Yukon Flats). Soils in this ecoregion are very susceptible to wildfire alteration, due to the relatively warm (–1.5°C) and shallow permafrost. Organic mat disturbance from wildfire can warm soils, significantly lower permafrost tables, alter soil properties and hydrology, and change vegetation composition (Gallant et al. 1995).

Degree of Protection

The protected areas in Alaska were all designated by the Alaska National Interest Lands Conservation Act of 1980 and include large, intact areas of important lowland habitat. These areas generally contain significant inholdings owned by Alaska Native corporations. Important protected areas include the following:

- Yukon Flats National Wildlife Refuge—east-central Alaska
- Kanuti National Wildlife Refuge—central Alaska
- Koyukuk National Wildlife Refuge—west-central Alaska
- Innoko National Wildlife Refuge—western Alaska
- Nowitna National Wildlife Refuge—central Alaska
- Arctic National Wildlife Refuge—northeastern Alaska
- Vuntut National Park—northwestern Yukon Territory

Types and Severity of Threats

Major threats include:

- high likelihood of timber harvest in Tanana Valley State Forest and on native corporation lands (These harvests will not cover a significant portion of the ecoregion but will impact the highest-volume stands disproportionately.)
- opening of Yukon Flats to timber harvests and oil and gas development in the future
- oil and gas development in Eagle Plains, Yukon Territory
- high threat of exploitation of porcupine caribou herd from the Dempster Highway
- potential overharvest or overemphasis of management on game/commercial wildlife species outside of natural range of variation.

Suite of Priority Activities to Enhance Biodiversity Conservation

- Encourage sustainable harvest at lower levels than currently planned in Tanama Valley State Forest.
- Acquire timber rights to critical areas or encourage sustainable practices in Yukon Flats.
- Conserve the Sucke, Bonnett-Plume, and Wind Rivers, which flow north into Peale River, then MacKenzie River.

Conservation Partners

For contact information, please see appendix G.

- Alaska Boreal Forest Council
- Canadian Parks and Wilderness Society, Yukon Chapter
- Friends of Yukon Rivers
- Northern Alaska Environmental Center
- World Wildlife Fund Canada
- Yukon Conservation Society

Relationship to Other Classification Schemes

This ecoregion combines Gallant et al.'s (1995) ecoregions 104, 106, and 107 with the Ecological Stratification Working Group's (ESWG 1995) ecoregions 166, 167, and 169. It was determined that these three regions contained similar habitat mosaic elements, only in different proportions, so the aggregate consists of similar habitat linked by dominant commonalities. The same reasoning was used for combining the three Gallant regions. In addition, much of the wildlife moves freely through and among the three component regions. Major delineation features are the dominance of forest cover, lowland to hilly terrain, and lack of Pleistocene glaciation. Combination decisions are based on Gallant et al. (1995), McNab and Avers (1994), Ecoregions Working Group (1989), Viereck et al. (1992), and Küchler (1985).

Although most of this ecoregion lies in the state of Alaska, it encompasses the Old Crow Basin, the Old Crow Flats, and the Eagle Plains (TEC ecoregions 166–167 and 169) in the Yukon Territory on the Canadian side of the border (ESWG 1995). Vegetation is considered Boreal Alpine Forest–Tundra (B.33) (Rowe 1972).

Prepared by R. Hagenstein, T. Ricketts, J. Peepre, M. Sims, K. Kavanagh, and G. Mann

Key Number:	**84**
Ecoregion Name:	**Alaska Peninsula Montane Taiga**
Major Habitat Type:	**6.1 Boreal Forest/ Taiga**
Ecoregion Size:	**47,885 km²**
Biological Distinctiveness:	**Nationally Important**
Conservation Status:	**Snapshot—Relatively Intact**
	Final—Relatively Intact

Introduction

The Alaska Peninsula Montane Taiga ecoregion stretches along the southern, Pacific Ocean side of the Alaska Peninsula, from the mouth of Cook Inlet westward to (and including) Unimak Island. It also includes the majority of the Kodiak Island archipelago. Most of the ecoregion consists of rounded ridges ranging between sea level and 1,200 m in elevation, but several steep, rugged volcanic peaks rise to 1,400–2,600 m. The slopes are covered by dwarf scrub communities on upper slopes and in exposed areas and low scrub in lower, more protected sites. The dwarf scrub communities are dominated by crowberry (*Empetrum nigrum*) and include other ericads (*Vaccinium* spp.), arctic willow (*Salix arctica*), and white mountain avens (*Dryas octopetala*). Low scrubs are dominated by willows (*Salix* spp.), along with dwarf shrub species and some forbs. Tall scrub communities of alder (*Alnus sinuata*) and willow also occur in the lower elevations, and some floodplains and south-facing slopes support stands of balsam poplar (*Populus balsamifera*) (Gallant et al. 1995).

The climate in the Alaska Peninsula Montane Taiga ecoregion is dominated by maritime influences, with high precipitation and moderate temperature ranges. Precipitation varies greatly, ranging from 600 mm to 3,300 mm along the coasts, and reaching more than 4,000 mm in high elevations. Winter temperatures average between −11°C and 1°C, and summer tempera-

tures average between 6°C and 15°C. The maritime climate has kept the region generally free from permafrost, but it was heavily glaciated in the Pleistocene and retains glaciers on the higher peaks. Soils are mostly Typic Haplocryands and Typic Vitricryands. They have formed from volcanic ash and cinders and erode easily in the heavy rains. This can hinder vegetation development (Gallant et al. 1995).

Main activities are commercial fishing and processing, mining, and subsistence fishing and hunting. Indigenous communities rely mostly on fishing and the hunting of marine mammals. Gold, silver, lead, and copper have been mined on a small scale, and some coal and petroleum extraction occurs (Bailey et al. 1994).

No information is available on fire occurrence in this ecoregion. Perhaps the most regular disturbance factors are the frequent and enormously violent winter storms that buffet the region and the more infrequent volcanic eruptions and resulting ash falls. The combination of ash slopes and heavy storms creates a very readily eroded landscape, which continually disturbs the growth of vegetation.

Biological Distinctiveness

This ecoregion provides important seasonal staging and migration habitat for many waterfowl and supports populations of caribou, moose, ground squirrel, and hare (Bailey et al. 1994). Large numbers of brown bears inhabit the ecoregion, including the largest brown bears on earth—the Kodiak brown bears on Kodiak Island. The bears congregate for the large salmon runs that occur in summer and fall at the short, steep rivers, especially the McNeil River and those in Katmai National Park. The bears and other top-level predators exist in population numbers within their natural range of variation, with predator-prey relationships intact.

Exceptionally large sea bird colonies exist along the coastlines as well. For example, Unimak Island supports over 500,000 tufted puffins. Colonies in Stepovak Bay support 200,000 murres and 300,000 puffins. The Semidi Islands have 500,000 fulmars and 650,000 murres (Sowls, Hatch, and Lensink 1978).

Conservation Status

Habitat Loss and Degradation

The Alaska Peninsula Montane Taiga is almost entirely intact, with habitat loss almost exclusively restricted to the localized effects of development surrounding the small communities and settlements along the coastline.

Remaining Blocks of Habitat

As there is localized and minimal habitat loss in this ecoregion, the entire ecoregion remains essentially intact.

Degree of Fragmentation

The long, thin shape of this ecoregion and its many islands creates a naturally fragmented landscape, but anthropogenic fragmentation is very slight.

Degree of Protection

Important protected areas include:

- Alaska Peninsula National Wildlife Refuge—southwestern Alaska
- Izembek National Wildlife Refuge—southwestern Alaska
- Kodiak National Wildlife Refuge—southwestern Alaska
- Aniakchak National Monument and Preserve—southwestern Alaska
- Becharof National Wildlife Refuge—southwestern Alaska
- Katmai National Park and Preserve—southern Alaska
- McNeil River State Game Sanctuary—southwestern Alaska

Types and Severity of Threats

Major threats include:

- continued habitat damage from existing ranching, feral cattle on some islands, and predation effects on some islands from feral foxes
- some potential for overharvest of game species, especially for brown bear populations near McNeil River

Suite of Priority Activities to Enhance Biodiversity Conservation

- Protect brown bear populations around McNeil River from unsustainable harvest just outside the state wildlife refuge.
- Continue to consolidate Kodiak National Wildlife Refuge holdings by purchasing properties with Exxon Valdez settlement funds.

Conservation Partners

For contact information, please see appendix G.

- Conservation Fund
- Friends of McNeil River

Relationship to Other Classification Schemes

This ecoregion is identical to Gallant et al.'s (1995) ecoregion 113. Major delineation features are the high relief and scrub vegetation communities. The Bering Sea side of the peninsula (Beringia Lowland Tundra ecoregion) is lower and flatter and supports wet and moist tundra communities instead.

Prepared by R. Hagenstein and T. Ricketts

Key Number: **85**
Ecoregion Name: **Cook Inlet Taiga**
Major Habitat Type: **6.1 Boreal Forest/ Taiga**
Ecoregion Size: **27,963 km²**
Biological Distinctiveness: **Nationally Important**
Conservation Status: **Snapshot—Relatively Intact**
Final—Relatively Intact

Introduction

This ecoregion surrounds the upper reaches of Cook Inlet in south-central Alaska and is in turn surrounded by the mountains of ecoregions [103] and [104]. Its relatively mild climate, level to rolling topography, and coastal position have made it the focus of most of the human activity in Alaska. These factors have also contributed to the wide variety of vegetation communities found in the ecoregion. The most widespread are coniferous, broadleaf, and mixed forests, dominated in differing combinations by black spruce (*Picea mariana*), white spruce (*P. glauca*), Sitka spruce (*P. sitchensis*), quaking aspen (*Populus tremuloides*), balsam poplar (*P. balsamifera*), black cottonwood (*P. trichocarpa*), and paper birch (*Betula papyrifera*) (Gallant et al. 1995). Other important communities include low scrub, tall scrub, low scrub bog, mesic graminoid, graminoid herbaceous, and wet forb herbaceous. For more complete descriptions of these communities, see Gallant et al. (1995).

The Cook Inlet Taiga enjoys a generally mild climate in comparison with interior and arctic Alaska. Average annual precipitation ranges from 380 mm to 680 mm across the region. Average daily minimum temperature in winter is −15°C, while average daily maximum temperature in summer is 18°C. Soils are formed with windblown loess from the glacial floodplains and with volcanic ash from mountains to the west. The soils lie on top of glacial deposits. Unlike the majority of the other relatively low elevation ecoregions in Alaska, the Cook Inlet Taiga was extensively glaciated during the Pleistocene.

Biological Distinctiveness

This ecoregion probably has experienced the most extensive human disturbance and alteration in Alaska. Nevertheless, it remains approximately 90 percent intact and still supports all of the top-level terrestrial predators within, or close to, their natural ranges of variation. (These predators include brown bear, wolf, wolverine, and coyote.) The ecoregion also produces all five species of Pacific salmon, which support a wide range of terrestrial species as well as large commercial, sport, and subsistence fisheries.

The Kenai River watershed is worthy of special note for its biological values. It supports all five species of Pacific salmon, including a unique stock of the world's largest king salmon. The Kenai River also supports the second-highest concentration of overwintering American bald eagles in Alaska. Virtually the entire population of Wrangell Island snow geese uses the mouth of the Kenai River and Trading Bay (on the west side of Cook Inlet) as a migratory staging area each spring. Finally, populations of wolf, bear, lynx, and other animals on the Kenai Peninsula are separated from the rest of Alaska by water,

glaciers, and development and are subject to local extirpation as a result of development, exploitation, and habitat changes.

Wildfire occurrence is moderate to high (especially in dry years), and fires range in area from 1 ha to 22.7 km^2, averaging 1.6 km^2 (Gallant et al. 1995). Spruce bark beetle is also a common disturbance in the forests of this ecoregion. A current infestation has reached all parts of the ecoregion; up to 80 percent of the mature spruce in many stands have been killed. The spruce bark beetle is naturally occurring and may be the most important cause of stand renewal in the ecoregion.

Conservation Status

Habitat Loss and Degradation

As stated above, although this ecoregion is the most impacted by human activity of any in Alaska, only approximately 10 percent of its area is altered or heavily altered. Most human disturbance is concentrated in the urban and residential development of the lower Kenai River, Anchorage Basin, and Palmer-Wasilla area. Some agriculture occurs in Palmer and Point McKenzie, across Knik Arm from Anchorage. Other forms of human land use include timber harvest and oil and gas exploitation on the Kenai Peninsula and across Cook Inlet from Anchorage.

Both areas have high potential for resource exploitation, however, with timber harvest occurring throughout most parts of the ecoregion (especially in response to an ongoing spruce bark beetle epidemic), oil and gas development occurring in large parts of the Kenai Peninsula and on the west side of Cook Inlet, and potential coal mining on the west side of the inlet in the future. Recreational and subsistence hunting and fishing are generally managed well, although potential for unsustainable hunting exists.

Remaining Blocks of Habitat

The ecoregion is naturally divided into two large blocks of relatively intact habitat by Cook Inlet. One block, covering the Susitna Valley and the west side of Cook Inlet, is larger than 10,000 km^2. The second block, on the Kenai Peninsula, is subdivided into two smaller blocks of about

4,000 km² each by the Kenai River and the human development along it.

Degree of Fragmentation

The ecoregion is fragmented into three major blocks, but this is due to major water bodies.

Degree of Protection

Important protected areas include:

- Kenai National Wildlife Refuge—southern Alaska
- Chugach State Park—southern Alaska
- Nancy Lake State Recreation Area—southern Alaska
- Redoubt Bay Critical Habitat Area—southern Alaska
- Susitna Flats, Trading Bay, and Palmer Flats state game refuges—southern Alaska

Types and Severity of Threats

Major threats include:

- continued development along the Kenai River and overuse of the Kenai River, restricting wildlife movement, removing habitat, and disturbing hydrologic functioning
- unsustainable timber harvest in the extensive forests throughout the ecoregion
- expansion of oil, gas, and coal development on the Kenai Peninsula and the west side of Cook Inlet, with potential for localized but severe impacts on surrounding habitat
- overexploitation of fish and game stocks
- killing of bears that come into conflict with growing human residential areas, for defense of life and property

Suite of Priority Activities to Enhance Biodiversity Conservation

- Balance community needs and wildlife requirements in the Kenai River watershed.
- Implement planning to prevent unsustainable timber harvest from occurring in the Susitna Valley forests.
- Monitor coal and oil and gas development activities.

Conservation Partners

For contact information, please see appendix G.

- Alaska Center for the Environment
- The Great Land Trust
- Kachemak Heritage Land Trust
- The Nature Conservancy, Alaska

Relationship to Other Classification Schemes

This ecoregion is identical to Gallant et al.'s (1995) ecoregion 115. Major delineation features are the low (less than 600 m), relatively gentle terrain, forest-dominated vegetation, and mild climate.

Prepared by R. Hagenstein and T. Ricketts

Key Number:	**86**
Ecoregion Name:	**Copper Plateau Taiga**
Major Habitat Type:	**6.1 Boreal Forest/Taiga**
Ecoregion Size:	**17,170 km²**
Biological Distinctiveness:	**Nationally Important**
Conservation Status:	**Snapshot—Relatively Intact**
	Final—Relatively Intact

Introduction

The Copper Plateau Taiga ecoregion occupies the site of a large Pleistocene lake. The low (420 m to 900 m), flat or gently rolling plain is completely surrounded by the high mountains of ecoregions [103] and [104]. A shallow permafrost table and poor soil drainage have resulted in a landscape dotted with many lakes and wetlands. The vegetation reflects these wet conditions. Coniferous forests and woodlands dominated by black spruce (*Picea mariana*) are the most common communities. The many wetland areas support low scrub bog communities dominated by birch (*Betula glandulosa* and *B. nana*) and ericaceous shrubs. Also found in wetland areas are wet graminoid herbaceous communities, dominated by sedges (e.g., *Eriophorum anjustifolium* and *Carex* spp.) or codominated by sedges and herbs (e.g., *Menyanthes trifoliata*, *Petasites frigidus*, and *Potentilla palustris*).

Exceptionally well-drained sites support coniferous forests dominated by white spruce (*Picea glauca*) or broadleaf forests dominated by black cottonwood (*P. trichocarpa*), and/or quaking aspen (*Populus tremuloides*) (Gallant et al. 1995).

This ecoregion experiences a continental climate. Annual precipitation ranges from 250 mm to 460 mm and increases from south to north. Winter daily minimum temperatures average –27°C, while summer daily maximum temperatures average 21°C. Soils, generally poorly drained and shallow to permafrost, include histic pergelic cryaquepts, aquic cryochrepts, typic cryochrepts, pergelic cryaquolls, and typic cryoborolls. This region was extensively glaciated during the Pleistocene, in addition to containing the lake.

Wildfire occurrence in this ecoregion is low, and fires range in area from less than 1 ha to approximately 40 ha, averaging 5 ha. Because the permafrost is discontinuous and shallow, wildfire disturbance to the organic mat can significantly raise soil temperature, increase permafrost depth, and result in changes in soil hydrology and structure.

Biological Distinctiveness

The many thaw lakes and wetlands of this ecoregion provide excellent nesting habitat for a variety of migratory bird species. The north-central portion in particular supports very high numbers of breeding trumpeter swans. For the most part, top-level predators are still present in numbers close to their natural variations. The western part of the ecoregion is used by the Nelchina caribou herd as part of its annual migration route.

The Copper River supports strong runs of salmon, particularly king and sockeye salmon. The large lakes in the central part of the ecoregion, including Lake Louise, Tyone Lake, and Susitna Lake, support nesting waterfowl and fish and serve as the headwaters for the Susitna River.

Conservation Status

Habitat Loss and Degradation

Total habitat loss is estimated to be 10 percent, due mostly to development in the Glenallen area and to timber harvesting in the Copper River valley and on Native corporation lands near Chitina.

Remaining Blocks of Habitat

The Copper Plateau Taiga ecoregion essentially consists of one large block of contiguous and generally intact habitat.

Degree of Fragmentation

The roads connecting the areas of development, the major rivers, and the areas of timber harvest are responsible for the majority of habitat fragmentation in the ecoregion. Probably the most important of the fragmenting features is the timber harvesting near Chitina, where the ecoregion is narrowed naturally.

Degree of Protection

Important protected areas include:

- Wrangell–St. Elias National Park and Preserve, the largest block of protected habitat
- The state-managed Nelchina Caribou Special Management Area, which affords protection for caribou and generally for maintaining intact habitat but may not sufficiently protect top-level predators, given a management emphasis throughout the region on meat-producing species

One gap in the protected-area system for this ecoregion is the lake country in the north-central portion of the ecoregion.

Types and Severity of Threats

This is a likely timber harvest area, and there is potential for the use of unsustainable methods on the bottomlands of the Copper River, especially those owned by Native corporations. Recreation, tourism, and recreational hunting are increasing as Park Service policies deflect tourism use to this area. Sport and subsistence hunting is generally well managed, but policies are generally designed to increase meat species (caribou, moose) at the expense of predators. This is especially problematic for the brown bear populations along the Denali Highway in the northwest part of the ecoregion.

Suite of Priority Activities to Enhance Biodiversity Conservation

- Work with Native corporations to limit timber harvest areas and to change logging practices to sustainable levels and methods.
- Monitor and manage impacts of increased tourism.
- Create protected areas for trumpeter swan nesting habitats near Paxson Lake.

Conservation Partners

For contact information, please see appendix G.

- Copper Country Alliance
- Copper River Watershed Forum
- National Parks and Conservation Association

Relationship to Other Classification Schemes

This ecoregion is identical to Gallant et al.'s (1995) ecoregion 117. Delineation features include low elevation (less than 900 m), relatively level or gentle terrain, and major vegetation type.

Prepared by R. Hagenstein and T. Ricketts

Key Number:	**87**
Ecoregion Name:	**Northwest Territories Taiga**
Major Habitat Type:	**6.1 Boreal Forest/Taiga**
Ecoregion Size:	**345,833 km²**
Biological Distinctiveness:	**Globally Outstanding**
Conservation Status:	**Snapshot—Relatively Stable**
	Final—Relatively Stable

Introduction

This ecoregion comprises the northern Mackenzie River valley and areas to both the west and east of the lower Mackenzie Valley in the Northwest Territories. Much of the area to the south and west of Great Bear Lake is also included.

The climate in this ecoregion is marked by short, cool summers and long, cold winters. The mean annual temperature ranges between −10°C in the Mackenzie Delta region and −1°C in southern regions. Mean winter temperature ranges considerably from −26.5°C in the north to −1°C in the south, and mean summer temperature ranges from 6.5°C to 14°C. Mean annual precipitation is low, 200–400 mm but reaching 500 mm in the southwest. This ecoregion is characterized by both low subarctic and high subarctic ecoclimates (ESWG 1995).

This ecoregion is the northern extension of the flat Interior Plains, which dominate the Prairie and Boreal Plains ecoregions to the south. The broad lowlands and plateaus are incised by major rivers. The area is underlain by horizontal sedimentary rock and is nearly level to gently rolling in topography (ESWG 1995).

Biological Distinctiveness

Vegetation consists of open, very stunted stands of black spruce (*Picea mariana*). The shrub component includes dwarf birch (*Betula* spp.), Labrador tea (*Ledum decumbens*), and willow (*Salix* spp.). Understory species include bearberry, mosses, and sedges. Upland and foothills areas, and southern regions, are often better drained, are warmer, and support mixed-wood forests characterized by white and black spruce (*Picea glauca*), lodgepole pine (*Pinus contorta*), tamarack (*Larix laricina*), white birch (*Betula* spp.), quaking aspen (*Populus tremuloides*), and balsam poplar (*P. balsamifera*). White spruce and balsam poplar grow to greater heights along large rivers. Low-lying wetlands cover 25–50 percent of the ecoregion (ESWG 1995).

Characteristic wildlife of the Taiga ecoregion are woodland caribou (*Rangifer tarandus* ssp. *caribou*), moose (*Alces alces*), bison (*Bison bison*), wolf (*Canis lupus*), black bear (*Ursus americanus*), marten (*Martex americana*), lynx (*Lynx canadensis*), and Arctic ground squirrel (*Citellus parryi*). Caribou overwinter in the northwest corner of the region. Bird species include the common redpoll (*Carduelis flammea*), gray jay (*Perisoreus canadensis*), common raven (*Corvus corax*), red-throated loon (*Gavia stellata*), northern shrike (*Lanius excubitor*), sharp-tailed grouse (*Tympanuchus phasianellus*), fox sparrow (*Passerella iliaca*), bald eagle (*Haliaeetus leucocephalus*), peregrine falcon (*Falco peregrinus*), osprey (*Pandion haliaetus*), and migratory waterfowl (ESWG 1995).

Seasonal flooding and forest fire are important natural disturbance features in this ecoregion.

This ecoregion was a possible Pleistocene refugia for some plant species. Intact vertebrate population densities in this ecoregion are in the natural range of variation.

Conservation Status

Habitat Loss and Degradation

Approximately 90 percent of the habitat remains intact. Most habitat loss is the result of disturbance around small communities, seismic lines throughout the region, and oil and gas development in the Norman Wells area. Small-scale logging is occurring in the south end of the ecoregion.

Remaining Blocks of Intact Habitat

Most of the ecoregion is intact.

Degree of Fragmentation

Fragmentation of habitat resulting from human disturbance is low. A year-round travel route planned for the MacKenzie Valley could cause disruption to some seasonal wildlife movement. Wildfire, a major natural disturbance in this ecoregion, creates a mosaic of vegetation communities across the landscape.

Degree of Protection

- Nahanni National Park—western Northwest Territories, 4,765 km^2

Types and Severity of Threats

Mining, oil and gas development, and the associated exploration phases of these industries are serious threats, in part due to the increased level of access created by the road networks and seismic lines. The proposed northward extension of the Mackenzie Highway may create new pressures for wildlife populations in the northern portion of this ecoregion. Like many other northern ecoregions, a recent series of abnormal winters may mean that climate change is already creating significant seasonal changes to weather patterns.

Suite of Priority Activities to Enhance Biodiversity Conservation

- Create additional protected areas that represent the ecological variation characteristic of this ecoregion.
- Designate candidate protected areas in the Peel River Plateau in the Yukon portion of this ecoregion.

Conservation Partners

For contact information, please see appendix G.

- Canadian Arctic Resources Committee
- Canadian Nature Federation
- Canadian Parks and Wilderness Society, Yukon Chapter
- Ecology North
- World Wildlife Fund Canada
- Yukon Conservation Society

Relationship to Other Classification Schemes

The Northwest Territories Taiga lies within the Taiga Plains Ecozone and includes the following terrestrial ecoregions: Mackenzie Delta, Peel River Plateau, Great Bear Lake Plain, Fort MacPherson Plain, Colville Hills, Norman Range, Grandin Plains, Franklin Mountains, Keller Lake Plain, Great Slave Lake Plain, Nahanni Plateau, and Sibbeston Lake Plain (TEC ecoregions 50–55 and 57–62) (ESWG 1995). Forest region types here are Boreal Upper Mackenzie, Northwest Transition, and Alpine Forest-Tundra (B.23a, B.27, and B.33) and Tundra (Rowe 1972).

Prepared by A. Gunn, J. Shay, S. Smith, R. Hagenstein, J. Peepre, C. O'Brien, K. Kavanagh, M. Sims, and G. Mann

Key Number:	**88**
Ecoregion Name:	**Yukon Interior Dry Forests**
Major Habitat Type:	**6.1 Boreal Forest/Taiga**
Ecoregion Size:	**62,370 km²**
Biological Distinctiveness:	**Nationally Important**
Conservation Status:	**Snapshot—Vulnerable**
	Final—Vulnerable

Introduction

This ecoregion lies predominantly within the Yukon Territory. A small portion dips into extreme northwestern British Columbia. This ecoregion contains much of the human population in the Yukon, including parts of most major highways in the territory.

The climate in this ecoregion in the Yukon interior can be characterized as cold and semi-arid. Mean annual temperature is around –3°C; mean summer temperature is 11°C, and mean winter temperature ranges between –16.5°C and –19°C. The southern portion of the ecoregion lies within the rain shadow of the St. Elias Mountains. Mean annual precipitation is in the range of 225–400 mm, increasing with elevation and in the northeast (ESWG 1995).

The Yukon Plateau is the dominant physiographic feature. It is composed of groups of rolling hills and plateaus separated by deeply cut, broad valleys. Elevations in this ecoregion are generally above 1,000 m asl, with some smaller peaks in the south. Low-ice-content permafrost occurs in a sporadic discontinuous pattern here (ESWG 1995).

Biological Distinctiveness

White and black spruce (*Picea glauca* and *P. mariana*) form the most common forest types. Lodgepole pine (*P. contorta*) often invades very dry sites and burnt areas. South-facing slopes at low elevations are often characterized by grassland communities. Scrub birch (*Betula* spp.) and willow (*Salix* spp.) occur up to the tree line, which is usually defined by the presence of alpine fir (*Abies lasocarpa*). In the colder alpine regions, mountain avens (*Dryas hookeriana*), dwarf shrubs, forbs, grasses, and lichens constitute the main vegetative cover. Due to the relatively dry interior climate, forests are frequently renewed from recurring natural fires such that young successional communities are most common (ESWG 1995).

Characteristic wildlife species include caribou (*Rangifer tarandus*), moose (*Alces alces*), mountain goat (*Oreamnos americanus*), Stone's and Dall sheep (*Ovis dalli* spp.), grizzly and black bear (*Ursus arctos* and *U. americanus*), wolf (*Canis lupus*), coyote (*Canis latrans*), beaver (*Castor canadensis*), ground squirrel (*Spermophilus saturatus*), hare (*Lepus* spp.), raven (*Corvus corax*), ptarmigan (*Lagopus* spp.), and golden eagle (*Aquila chrysaetos*) (ESWG 1995).

This ecoregion contains one of the most northern areas on the continent with grassland communities. The juxtaposition of these grasslands with northern boreal forest creates an unusual association of plant communities.

Conservation Status

Habitat Loss and Degradation

It is estimated that approximately 75 percent of the ecoregion remains intact. Most of the disturbed habitat is considered to be altered, with only a small amount of area heavily altered. Factors leading to habitat loss include forestry and mining activities, urban growth around Whitehorse, and major transportation corridors.

Remaining Blocks of Intact Habitat

Most of the ecoregion, particularly the uplands, can be considered as intact habitat. Valley bottoms have been most widely developed.

Degree of Fragmentation

Habitat fragmentation in this ecoregion has principally resulted from the construction of major transportation corridors through the lower elevations. Susceptibility is high for some species with seasonal migrations or movements between habitat types, especially large carnivores, woodland caribou, and Dall sheep.

Degree of Protection

- Charlie Cole Creek Ecological Reserve— southern Yukon Territory, 1.6 km²
- Nisutlin River Delta National Wildlife Area (not fully protected)—southern Yukon Territory, 5.3 km²

There are no large protected areas yet established in this ecoregion.

Types and Severity of Threats

Threats to biodiversity are increasing rapidly in this ecoregion. Timber harvesting has begun and is yet to be well regulated or planned in the Yukon. Mining activity is significant in some parts of the ecoregion, and a number of wildlife species are in serious decline. A government-sanctioned wolf kill has been occurring in this area for the past few years.

Suite of Priority Activities to Enhance Biodiversity Conservation

- Most important, establish a new protected area following the Kusawa Lake territorial park proposal.
- Permanently end control programs for carnivores.

Conservation Partners

For contact information, please see appendix G.

- Canadian Nature Federation
- Canadian Parks and Wilderness Society, Yukon Chapter
- Friends of Yukon Rivers
- World Wildlife Fund Canada
- Yukon Conservation Society

Relationship to Other Classification Schemes

The Yukon Plateau–Central and the Yukon Southern Lakes (TEC ecoregions 175 and 177) in the Boreal Cordillera ecozone correspond to the Yukon Interior Dry Forests (ESWG 1995), which include Rowe's (1972) Dawson and Central Yukon Boreal forest regions (B.26b and B.26c) and tundra vegetation.

Prepared by S. Smith, J. Peepre, K. Kavanagh, M. Sims, and G. Mann

<table>
<tr><td>Key Number:</td><td>**89**</td></tr>
<tr><td>Ecoregion Name:</td><td>**Northern Cordillera Forests**</td></tr>
<tr><td>Major Habitat Type:</td><td>**6.1 Boreal Forest/Taiga**</td></tr>
<tr><td>Ecoregion Size:</td><td>**262,780 km²**</td></tr>
<tr><td>Biological Distinctiveness:</td><td>**Globally Outstanding**</td></tr>
<tr><td>Conservation Status:</td><td>**Snapshot—Relatively Stable**</td></tr>
<tr><td></td><td>**Final—Vulnerable**</td></tr>
</table>

Introduction

This ecoregion represents a combination of alpine, subalpine, and boreal mid-Cordilleran habitats across much of northern British Columbia and southeastern Yukon. The northern Cordillera forests extend across northern British Columbia and southern Yukon Territory and cover a minute area in the Northwest Territories.

The mean annual temperature for this ecoregion is generally –2°C; mean summer temperature is 10°C, and mean winter temperature ranges from –13°C to –18.5°C. Mean annual precipitation is approximately 350–600 mm but increases up to 1,000 mm at higher elevations (ESWG 1995).

This ecoregion includes a number of different physiographic features: the northern Rocky Mountains in northern British Columbia; the Hyland Highland in southeastern Yukon north of the Liard River; the Liard Basin, a broad, rolling, low-lying area; the complex, rugged Boreal Mountains and Plateaus; the Yukon-Stikine Highlands in the rain shadow of the Coast Mountains; and the Pelly and northern Cassiar Mountains. Discontinuous permafrost with low ice content occurs throughout the ecoregion, usually confined to lower, north-facing slopes (ESWG 1995).

Biological Distinctiveness

Vegetation associations in this ecoregion follow elevational gradients. Alpine communities include dwarf ericaceous shrubs (Ericaceae), dwarf birch (*Betula* spp.), willow (*Salix* spp.), grass (Gramineae), lichen, and bare bedrock at elevations above the tree line. Subalpine forests are characterized by alpine fir (*Abies lasiocarpa*), black spruce (*Picea mariana*), and white spruce (*P. glauca*) together with deciduous shrubs, and occasional Engelmann spruce (*P. englemannii*). Closed boreal forests at lower, warmer elevations include white and black spruce, lodgepole pine

(*Pinus contorta*), and some paper birch (*Betula papyrifera*) and aspen (*Populus tremuloides*). Lodgepole pine and aspen regenerate following fires, which is the principal form of renewal for forests in this ecoregion (ESWG 1995).

Characteristic wildlife include moose (*Alces alces*), wolverine (*Gulo luscus*), snowshoe hare (*Lepus americanus*), black bear (*Ursus americanus*), grizzly bear (*U. arctos*), mountain goat (*Oreamnos americanus*), pika (*Ochotona* spp.), bison (*Bison bison*), Stone's sheep (*Ovis dalli* spp.), Dall sheep (*Ovis dalli* spp.), weasel (*Mustela* spp.), red fox (*Vulpes fulva*), beaver (*Castor canadensis*), muskrat (*Ondatra zibethica*), Arctic ground squirrel (*Citellus parryi*), spruce grouse (*Dendragapus canadensis*), ptarmigan (*Lagopus* spp.), snowy owl (*Nyctea scandiaca*), raptors, waterfowl, crane (*Grus canadensis*), and ruffed grouse (*Bonasa umbellus*). Many species of wildlife reach their continental southern or northern range limits in this ecoregion (ESWG 1995). A large and intact predator-prey system includes wolves (*Canis lupus*), grizzly bears, caribou (*Rangifer tarandus*), and moose (*Alces alces*). There are especially high concentrations of grizzly bears in some of the valley lands.

Conservation Status

Habitat Loss and Degradation

Much of the region (85 percent) remains as intact habitat. New and increasing natural resource developments are increasing human pressure on this ecoregion. This includes major mining (open-pit) sites, hydroelectric impoundments, logging, and transportation corridors.

Remaining Blocks of Intact Habitat
Most of this ecoregion remains intact.

Degree of Fragmentation
Road networks, including logging and mineral-oil exploration roads, are the principal causes of habitat fragmentation. These are of increasing concern throughout the ecoregion, as they create access to new areas and disrupt the movement patterns of large carnivores and ungulates.

Degree of Protection

- Spatsizi Plateau Wilderness Provincial Park—northern British Columbia, 6,082 km²

- Atlin Provincial Park—northern British Columbia, 2,326 km²
- Mount Edziza Provincial Park—northern British Columbia, 2,287 km²
- Stikine River Recreation Park—northern British Columbia, 2,170 km²
- Kwadacha Wilderness Provincial Park—northern British Columbia, 1,144 km²
- Tatlatui Provincial Park—northern British Columbia, 1,058 km²
- Muncho Lake Provincial Park—northern British Columbia, 884 km²
- Gladys Lake Ecological Reserve—northern British Columbia, 485 km²
- Kwadacha Recreation Park—northern British Columbia, 440 km²
- Atlin Recreation Park—northwestern British Columbia, 384 km²

Types and Severity of Threats

The types and severity of threats are both increasing. Timber harvesting is heaviest in riparian spruce and poplar areas and upland lodgepole pine areas. Wildlife exploitation is now considered high in southeastern Yukon and moderate in the British Columbia portion of the ecoregion.

Suite of Priority Activities to Enhance Biodiversity Conservation

- Make the following protected areas in northern and northwestern British Columbia:
  - Muskwa-Kechika Wildlife Area
  - Jennings Plateau
  - Lower Stikine River Corridor
  - Taku
  - Tutshi
  - Kawdy Plateau/Lord Mountain
  - Liard Eskers
- Make the following protected areas in Yukon:
  - Coal River Watershed
  - examples of LaBiche and Beaver River watersheds
  - Wolf Lake–Meister River headwaters

Conservation Partners

For contact information, please see appendix G.

- Canadian Nature Federation
- Canadian Parks and Wilderness Society, Yukon Chapter
- Canadian Parks and Wilderness Society, British Columbia Chapter
- Friends of Yukon Rivers
- Northwest Wildlife Preservation Society
- The Nature Conservancy, British Columbia
- World Wildlife Fund Canada
- Yukon Conservation Society

Relationship to Other Classification Schemes

In the north of this ecoregion are the Pelly Mountains (TEC ecoregion 178), and along the British Columbia–Yukon border are the Liard Basin (TEC ecoregion 181) and Hyland Highland (TEC ecoregion 182). The Yukon-Stikine Highlands (TEC ecoregion 179) make up the western extension of the ecoregion, and the Boreal Mountains and Plateaus and Northern Canadian Rocky Mountains extend across northern British Columbia (TEC ecoregions 180 and 183) (ESWG 1995). This ecoregion primarily relates to the Tundra and Boreal forest regions of the Upper Liard and the Stikine Plateau (B.24 and B.25). Interior Subalpine (SA.2) forest is also present here (Rowe 1972).

Prepared by D. Demarchi, J. Peepre, K. Kavanagh, M. Sims, and G. Mann

Key Number: **90**
Ecoregion Name: **Muskwa/Slave Lake Forests**
Major Habitat Type: **6.1 Boreal Forest/Taiga**
Ecoregion Size: **262,295 km²**
Biological Distinctiveness: **Globally Outstanding**
Conservation Status: **Snapshot—Relatively Stable**
Final—Relatively Stable

Introduction

This ecoregion occurs in the northeastern corner of British Columbia, the northwestern cor-

ner of Alberta, and much of the Mackenzie River valley in the southwestern Northwest Territories.

This ecoregion is classified as having a sub-humid mid to high boreal ecoclimate with cool summers and very cold winters. Mean annual temperature ranges from –2.0°C to –6.5°C, mean winter temperature ranges from –18°C to –24.5°C, and mean summer temperature is around 12.5°C. Annual precipitation is between 250 mm and 500 mm on average (ESWG 1995).

The Mackenzie River Plain is a narrow northern extension of the boreal forest along the east side of the Mackenzie River. The plain is a broad, rolling area lying between Mackenzie and Franklin Mountains. The Horn Plateau extends from the Horn River west along the Willowlake River to the Mackenzie River. To the northeast and south, the plateau rises above the Great Slave Lake Plain and the Hay River Lowland regions. The Northern Alberta Uplands include the Caribou Mountains in northern Alberta and the Cameron Hills uplands that span the border with British Columbia and Northwest Territories (ESWG 1995).

Biological Distinctiveness

Vegetation consists of medium to tall, closed mixed stands of quaking aspen (*Populus tremuloides*), white spruce (*Picea glauca*), and balsam fir (*Abies balsamea*) with lesser amounts of balsam poplar (*Populus balsamifera*) and black spruce (*Picea mariana*). Organic blankets support open stands of stunted black spruce and some birch and shrubs. Sporadic, discontinuous permafrost with low to medium ice content is common throughout and associated with organic deposits. Wetlands and bogs cover 25–50 percent of the land. Fire is widespread among the forest and shrubland communities in this ecoregion (ESWG 1995).

One of the continent's most diverse and intact large mammal systems occurs in this ecoregion. Woodland caribou (*Rangifer tarandus* ssp. *caribou*), moose (*Alces alces*), lynx (*Lynx canadensis*), grizzly bear (*Ursus arctos*), black bear (*U. americanus*), wolf (*Canis lupus*), snowshoe hare (*Lepus americanus*), grouse (*Dendragapus* spp.), waterfowl, deer (*Odocoileus* spp.), and elk (*Cervus canadensis*) are all found in the Muskwa Plateau in the southeast part of this ecoregion.

Large-scale mammalian migrations occur within their natural range of variation. There are relatively undisturbed predator-prey systems, although human impacts are beginning to affect these systems. Bison (*Bison bison*) herds are found in the Athabasca Basin–Wood Buffalo National Park area.

Conservation Status

Habitat Loss and Degradation

Approximately 75 percent of the habitat is considered to remain intact. Seismic lines dissect the landscape every 10 km in northern British Columbia, parts of Yukon Territory, and Alberta. Logging of riparian habitats has been extensive in some areas and is increasing.

Remaining Blocks of Intact Habitat

Most of the ecoregion is intact, but there are local watersheds that have been very heavily impacted by logging.

Degree of Fragmentation

Habitat fragmentation is increasing due to the cumulative impact of seismic lines, logging, and the MacKenzie Valley highway and pipeline transportation corridors.

Degree of Protection

- Wood Buffalo National Park (a part)—southern Northwest Territories and northeastern Alberta, 31,364 km^2
- Maxhamish Lake Provincial Park—northeastern British Columbia, 5.2 km^2

Types and Severity of Threats

There is high potential for intensive logging in the Alberta portion of the ecoregion. Logging allocations are also being considered throughout most of the forested sections of this ecoregion. Upgrades to the transportation corridor following the Mackenzie Valley may increase access to wildlife in adjacent areas.

Suite of Priority Activities to Enhance Biodiversity Conservation

Establish the following as protected areas:

- Grayling River—British Columbia
- Hay-Zama wetland complex—Alberta
- Mills Lake—Northwest Territories
- Liard River area—Northwest Territories

Conservation Partners

For contact information, please see appendix G.

- Alberta Wilderness Association
- Canadian Nature Federation
- Canadian Parks and Wilderness Society, British Columbia Chapter
- Canadian Parks and Wilderness Society, Edmonton Chapter
- Ecology North
- Federation of Alberta Naturalists
- The Nature Conservancy, Alberta
- The Nature Conservancy, British Columbia
- Northwest Wildlife Preservation Society
- World Wildlife Fund Canada

Relationship to Other Classification Schemes

Within the Taiga Plains ecozone, the Muskwa/Slave Lake Forests extend across: MacKenzie River Plain, Horn Plateau, Hay River Lowland, Muskwa Plateau, and Northern Alberta Uplands (TEC ecoregions 56 and 63–67). The MacKenzie River Plain (TEC ecoregion 56) winds far into the northern Northwest Territories parallel to its border with Yukon Territory (ESWG 1995). The forests in this region are boreal, with the following sections: Hay River, Lower Foothills, Northern Foothills, Upper Mackenzie, and Upper Liard (B.18b, B.19a, B.19b, B.23a, and B.24) (Rowe 1972).

Prepared by A. Gunn, J. Shay, S. Smith, J. Peepre, R. Usher, G. Smith, K. Kavanagh, M. Sims, and G. Mann

91

Ecoregion Name:
Major Habitat Type:
Ecoregion Size:
Biological Distinctiveness:
Conservation Status:

**Northern Canadian Shield
Taiga**
6.1 Boreal Forest/Taiga
609,238 km²
Regionally Outstanding
Snapshot—Relatively Intact
Final—Relatively Intact

Introduction

This ecoregion covers a large area of the Northwest Territories, extreme northeastern Alberta, northern Saskatchewan, and northwestern Manitoba.

Short, cool summers and very cold winters typify this ecoregion. In fact, Yellowknife, on the north shore of Great Slave Lake, has the lowest mean annual temperature of any major city in all of Canada (–5°C). Generally, however, mean annual temperature ranges from –8°C to –5°C; mean summer temperature ranges from 8°C to 11°C, and mean winter temperature is between –24.5°C and –21.5°C. Mean annual precipitation is in the range of 200–400 mm. This ecoregion is classified as having a low to high subarctic ecoclimate (ESWG 1995).

Massive, crystalline Archean rocks form broad, sloping uplands and lowlands, with numerous small lakes and eskers that drain into Great Slave Lake. Bedrock outcrops are common, and maximum elevation reaches about 490 m asl. Permafrost is discontinuous to continuous, with low to medium ice content with sparse ice wedges throughout (ESWG 1995).

Biological Distinctiveness

The limit of tree growth is reached along the northern boundaries of this ecoregion. The ecoregion as a whole constitutes the tundra and boreal transition zone. Vegetation consists of open, very stunted stands of black spruce (*Picea mariana*) and tamarack (*Larix laricina*), with secondary quantities of white spruce (*Picea glauca*) and ground cover of dwarf birch (*Betula* spp.), ericaceous shrubs (Ericaceae), cottongrass (*Eriophorum* spp.), lichen, and moss. Drier sites can support open stands of white spruce, ericaceous shrubs, and a ground cover of mosses and lichens. Poorly drained sites support tussock vegetation of sedge (*Carex* spp.), cottongrass, and sphagnum moss (*Sphagnum* spp.).

Vegetation in the Tazin Lake upland in the southwestern region is characterized by medium to tall closed stands of quaking aspen (*Populus tremuloides*) and balsam poplar (*P. balsamifera*), with white spruce, balsam fir (*Abies balsamea*), and black spruce occurring in late successional stages. Fire is an important renewal agent for upland communities in this ecoregion (ESWG 1995).

Species inhabiting this ecoregion include moose (*Alces alces*), grizzly bear (*Ursus arctos*), black bear (*U. americanus*), wolf (*Canis lupus*), barren-ground caribou (*Rangifer tarandus* ssp. *arcticus*), wolverine (*Gulo luscus*), weasel (*Mustela* spp.), mink (*M. vison*), muskrat (*Ondatra zibethica*), otter (*Lutra canadensis*), beaver (*Castor canadensis*), snowshoe hare (*Lepus americanus*), brown lemming (*Lemmus trimucronatus*), red-backed vole (*Clethrioomys gapperi*), willow ptarmigan (*Lagopus lagopus*), rock ptarmigan (*L. mutus*), sandhill crane (*Grus canadensis*), osprey (*Pandion haliaetus*), raven (*Corvus corax*), spruce grouse (*Dendragapus canadensis*), and waterfowl (ESWG 1995).

This ecoregion represents nearly one quarter of the linear extent of the tree line in North America. Wolf denning is common along the treeline. The combination of forests and tundra in this ecoregion results in significant overlap of woodland and barren-gound caribou (*Rangifer tarandus* ssp. *arcticus*) herds. Major caribou herds include the Bathurst herd and the Beverley-Quaminirauq herds. This is perhaps the most frequented area in northwestern Canada for winter range of caribou.

Conservation Status

Habitat Loss and Degradation

It is estimated that 90–95 percent of the natural habitat in this ecoregion remains intact. The most widespread form of habitat disturbance relates to mining and mineral exploration. This activity is growing rapidly in the western half of this ecoregion. Uranium, diamonds, nickel, and copper are some of the important minerals that constitute the mining industry in this area. Hydroelectric development is of concern in northern Manitoba and parts of the Northwest Territories. Urban development along the north shore of Great Slave Lake is having local impacts on natural habitats.

Remaining Blocks of Intact Habitat
Most of the ecoregion remains intact.

Degree of Fragmentation
Greatest fragmentation is occurring in the western portion of the ecoregion where permanent and winter roads for mining and mineral exploration are increasing land-based access to the area. Some additional habitat fragmentation is possible with respect to flooding associated with extensive, future hydroelectric projects, particularly in eastern portions of the ecoregion.

Degree of Protection
Protected areas are primarily clustered in the southeastern portion of the ecoregion. Part of Thelon National Wildlife Sanctuary occurs in the north-central portion of this ecoregion in the Northwest Territories (shared with ecoregion [110]).

- Sand Lakes Provincial Park—northwestern Manitoba, 8,475 km²
- Caribou River Provincial Park—northwestern Manitoba, 7,515 km²
- Numaykoos Lake Provincial Park—northwestern Manitoba, 3,695 km²
- Baralzon Lake Ecological Reserve—northwestern Manitoba, 390 km²

Types and Severity of Threats
Fly-in hunt camps need to be monitored to ensure that there are not unsustainable levels of caribou harvest occurring. The grizzly bear population in this ecoregion is at relatively low densities due to the carrying capacity of the landscape. Hence, even low numbers of animals being destroyed as a result of becoming "problem bears" near mining camps is of concern. There may be loss of terrestrial habitat resulting from flooding associated with proposed hydroelectric projects.

Suite of Priority Activities to Enhance Biodiversity Conservation
- Establish more protected areas. Some candidates that have been identified include:
  - East Arm of Great Slave Lake, Northwest Territories, as a national park
  - Hasbala Lake, Saskatchewan
  - Arctic Butte, Saskatchewan
  - Grease River, Saskatchewan
  - Tazin Lake, Saskatchewan
  - Cracking Stone Peninsula, Saskatchewan
  - Tribal Park proposals put forward by First Nations in the Northwest Territories
- Improve protection of critical habitat and management for the Beverley-Quaminirauq caribou herd.
- Implement recommended actions in Manitoba with respect to this ecoregion according to the schedule established in the 1996–98 Action Plan.
- Identify and designate protected-area candidates under the Protected Areas Strategy for the Northwest Territories.

Conservation Partners
For contact information, please see appendix G.

- Canadian Nature Federation
- Ecology North
- Endangered Spaces Campaign, Manitoba
- Endangered Spaces Campaign, Saskatchewan
- Manitoba Future Forest Alliance
- Manitoba Naturalists Society
- The Nature Conservancy, Manitoba
- Nature Saskatchewan
- Resource Conservation Manitoba
- World Wildlife Fund Canada

Relationship to Other Classification Schemes
The Northern Canadian Shield Taiga includes the Coppermine River Upland, the Tazin Lake Upland, the Kazan River Upland, and the Selwyn Lake Upland (TEC ecoregions 68–71) (ESWG 1995). This ecoregion is characterized primarily as boreal forest but is transitional between boreal and tundra vegetation. Forest sections include Northwestern Transition and Forest-Tundra (B.27 and B.32) (Rowe 1972).

Prepared by A. Gunn, A. G. Appleby, G. Whelan Enns, C. O'Brien, K. Kavanagh, M. Sims, and G. Mann

Key Number:
Ecoregion Name:

Major Habitat Type:
Ecoregion Size:
Biological Distinctiveness:
Conservation Status:

92
**Mid-Continental Canadian
Forests**
6.1 Boreal Forest/Taiga
361,759 km²
Bioregionally Outstanding
Snapshot—Relatively Stable
Final—Vulnerable

Introduction

The mid-continental Canadian forest ecoregion extends from southern Great Slave Lake in the Northwest Territories to encompass most of northeastern Alberta, central Saskatchewan, and parts of west-central Manitoba.

This ecoregion is classified as having a sub-humid mid-boreal ecoclimate. It is marked by short, cool to warm summers and long, cold winters. The mean annual temperature ranges from −2°C to 1°C; the mean summer temperature ranges from 13°C to 15.5°C; and the mean winter temperature ranges from −17.5°C to −13.5°C. Mean annual precipitation ranges from 300 mm to 625 mm, the wettest areas being in the southeastern portions of the Mid-Boreal Lowland (ESWG 1995).

This ecoregion consists of both lowland and upland areas, which may be grouped into three regions: the Slave River Lowland in northeastern Alberta; the Mid-Boreal Lowland, which includes the northern section of the Manitoba Plain; and the Mid-Boreal Uplands, which occur as a group of upland areas south of the Canadian Shield stretching from north-central Alberta to southwestern Manitoba. The lowland areas are underlain by relatively flat, low-relief Paleozoic carbonates forming undulating sandy plains, or flat-lying Paleozoic limestone bedrock covered by level to ridged glacial till, lacustrine silts and clays, and extensive peat deposits. The upland areas, for the most part, consist of Cretaceous shales and are covered entirely by kettled to dissected, deep, loamy to clayey-textured glacial till, lacustrine deposits, and inclusions of coarse, fluvoglacial deposits (ESWG 1995).

Elevations in the uplands range from about 400 to over 800 m asl. Both upland and lowland areas have sporadic, discontinuous permafrost with low ice content, but the occurrence of permafrost is much rarer in the upland regions and is found only in peatlands. Wetlands and peat-

lands are prevalent in this ecoregion, especially in the lowland areas, while small lakes, ponds, and sloughs fill numerous shallow depressions associated with rougher morainal deposits in the upland regions (ESWG 1995).

Biological Distinctiveness

The ecoregion forms part of the continuous mid-boreal mixed coniferous and deciduous forest extending from northwestern Ontario to the foothills of the Rocky Mountains. The mixed coniferous and deciduous forest is characterized by medium to tall closed stands of quaking aspen (*Populus tremuloides*) and balsam poplar (*P. balsamifera*) with white and black spruce (*Picea glauca* and *P. mariana*), and balsam fir (*Abies balsamea*) occurring in late successional stages. Cold and poorly drained fens and bogs are covered with tamarack (*Larix laricina*) and black spruce and may also include ericaceous shrubs (Ericaceae) and mosses. Deciduous stands in the uplands have a diverse understory of shrubs and herbs, while coniferous stands tend to promote feathermoss (ESWG 1995). Fire, drainage, and topography are probably the most significant factors influencing species assemblages.

Characteristic wildlife species include moose (*Alces alces*), black bear (*Ursus americanus*), wolf (*Canis lupus*), lynx (*Lynx canadensis*), white-tailed deer (*Odocoileus virginianus*), elk (*Cervus canadensis*), beaver (*Castor canadensis*), muskrat (*Ondatra zibethica*), snowshoe hare (*Lepus americanus*), ducks, geese, American pelican (*Pelecanus erythrorhynchos*), sandhill crane (*Grus canadensis*), ruffed grouse (*Bonasa umbellus*), common loon (*Gavia immer*), and many other bird species. The Interlake Plain to the south is home to moose, coyote (*Canis latrans*), and eastern cottontail (*Sylvilagus floridanus*) as well (ESWG 1995). Wood Buffalo National Park—within the Slave River Lowland—is populated by the world's largest bison (*Bison bison*) herd.

The principal breeding range for whooping cranes (*Grus americana*) in North America is found within this ecoregion. The wetlands found here are of hemispheric significance to waterfowl migration, including many staging sites. The Cumberland delta—a very large wetland complex—is one such example. This "inland" delta was formed in post-glacial times on a lake that no longer exists. This ecoregion also has a rich diversity of large ungulate species: bison, elk, woodland caribou (*Rangifer tarandua* ssp. *caribou*), moose, and deer. In addition, the northernmost bat hibernacula in North America are located within this ecoregion (Manitoba).

Conservation Status

Habitat Loss and Degradation

It is estimated that 50 percent of the ecoregion remains as intact habitat. Most of the remaining habitat is considered as altered, with only a very small percentage defined as heavily altered. Most habitat disturbance has resulted from large-scale forestry operations, oil and gas development in western Saskatchewan, and localized areas of mining activity.

Remaining Blocks of Intact Habitat

- Wood Buffalo National Park—northern Alberta and southern Northwest Territories
- Cold Lake–Primrose Lake Air Weapons Range—eastern Alberta and western Saskatchewan
- Prince Albert National Park—central Saskatchewan
- Dore-Smoothstone Lakes—central Saskatchewan
- northern half of Cumberland Delta—eastern Saskatchewan
- Riding Mountain National Park—southwestern Manitoba
- Porcupine Hills—western Manitoba
- Duck Mountain—western Manitoba

Degree of Fragmentation

Habitat fragmentation is occurring primarily due to extensive forestry activity across the landscape. Large areas of boreal forest are under forest licenses. Logging roads and clear-cuts are significant in parts of the ecoregion.

Degree of Protection

- Wood Buffalo National Park—northern Alberta and southern Northwest Territories, 31,364 km²
- Prince Albert National Park—central Saskatchewan, 3,874 km²

- Riding Mountain National Park—southwestern Manitoba, 2,950 km²
- Clearwater River Provincial Park—Saskatchewan, 2,240 km²
- Meadow Lake Provincial Park—western Saskatchewan, 1,653 km²
- Narrow Hills Provincial Park—central Saskatchewan, 536 km²
- Duck Mountain Provincial Park (backcountry zones)—western Manitoba, 480 km²
- Duck Mountain Provincial Park—eastern Saskatchewan, 261 km²
- Wildcat Hill Provincial Park—eastern Saskatchewan, 217 km²
- Greenwater Lake Provincial Park—eastern Saskatchewan, 207 km²

Types and Severity of Threats

Existing and proposed logging throughout the ecoregion is a serious threat. In some areas, the potential for increased oil and gas development is also a concern.

Suite of Priority Activities to Enhance Biodiversity Conservation

- Establish protected areas in the following locations:
  - Cold Lake–Primrose Lake Air Weapons Range—Alberta and Saskatchewan
  - Cumberland Delta—Saskatchewan
  - Dore-Smoothstone Lakes Wilderness Area—Saskatchewan
  - Manitoba Lowlands National Park—Manitoba
- Make protection-standard upgrades to Saskeram and Tom Lamb Wildlife Management Areas in Manitoba.
- Restore the flood regime to the Peace and Athabasca Rivers in Alberta, which are currently being regulated by the Bennet Dam in British Columbia. There are floodplain communities of international significance along these rivers that are becoming degraded as a result of flood control.
- Protect the area where wood bison are being reintroduced in the Chitek Lake watershed in Manitoba.
- Release an improved conservation and management plan for Riding Mountain National Park in Manitoba.
- Establish protected areas in Manitoba Provincial Forests.

- Implement recommended actions in Manitoba with respect to this ecoregion, according to the schedule established in the 1996–98 Action Plan.

Conservation Partners

For contact information, please see appendix G.

- Alberta Wilderness Association
- Canadian Nature Federation
- Canadian Parks and Wilderness Society, Calgary-Banff Chapter
- Canadian Parks and Wilderness Society, Edmonton Chapter
- Ducks Unlimited Canada, Saskatchewan
- Endangered Spaces Campaign, Manitoba
- Endangered Spaces Campaign, Saskatchewan
- Federation of Alberta Naturalists
- Friends of Prince Albert National Park
- Manitoba Future Forest Alliance
- Manitoba Naturalists Society
- The Nature Conservancy, Alberta
- The Nature Conservancy, Manitoba
- Nature Saskatchewan
- Resource Conservation Manitoba
- Watchdogs for Wildlife
- World Wildlife Fund Canada

Relationship to Other Classification Schemes

The Mid-Continental Canadian Forests are located in the Boreal Plains Ecozone. The terrestrial ecoregions that correspond to this region are the Slave River Lowland, the Mid-Boreal Uplands, and the Mid-Boreal Lowland (TEC ecoregions 136, 139–142, 144, 147, 148, and 150–154) (ESWG 1995). In Rowe's (1972) forest classification system, this area corresponds to the following boreal forest sections: Manitoba Lowlands, Mixedwood, Hay River, Lower Foothills and Upper Mackenzie (B.18a, B.18b, B.19a and B.23a).

Prepared by J. Shay, S. Smith, A. G. Appleby, R. Usher, G. Whelan Enns, K. Kavanagh, M. Sims, and G. Mann

Key Number:	**93**
Ecoregion Name:	**Midwestern Canadian Shield Forests**
Major Habitat Type:	**6.1 Boreal Forest/Taiga**
Ecoregion Size:	**538,770 km²**
Biological Distinctiveness:	**Bioregionally Outstanding**
Conservation Status:	**Snapshot—Relatively Stable**
	Final—Vulnerable

Introduction

This ecoregion covers much of northern Saskatchewan, north-central Manitoba (north and east of Lake Winnipeg), and a portion of northwestern Ontario. It is classified as having a subhumid high- to mid-boreal ecoclimate. It is marked by cool summers (except the Lac Seul Upland area, which has warm summers) and very cold winters. The mean annual temperature ranges from –4°C to 0.5°C; the mean summer temperature ranges from 11.5°C to 14°C, and the mean winter temperature ranges from –20.5°C to –14.5°C. In each case, the Lac Seul Upland area represents the warmest temperature in the ecoregion, and the Athabasca Plain, Churchill River Upland, and Hayes River Upland, regions represent the cooler temperatures. Mean annual precipitation ranges from 350 to 700 mm, the wettest areas being in the southeastern portions of the Lac Seul Upland and Hayes River Upland (ESWG 1995).

Permafrost occurs sporadically throughout this ecoregion, except in the area of the Lac Seul Upland. Wetlands are extensive in the regions of the Lac Seul Upland and the Athabasca Plain, and numerous small to large lakes are a prominent feature of the entire ecoregion. Archean rocks form steeply sloping uplands and lowlands in the Churchill River Upland and Hayes River Upland, while the Archean bedrock of the Lac Seul Upland area forms more broadly sloping uplands and lowlands. The ecoregion is covered with undulating to ridged glaciolacustrine or fluvioglacial deposits with occasional hummocky bedrock ridges and knolls (ESWG 1995).

Biological Distinctiveness

A portion of this ecoregion (the Athabasca Plain and the Churchill River Upland) forms part of the continuous coniferous boreal forest that extends from northwestern Ontario to Great Slave Lake in the Northwest Territories.

Forests of this ecoregion are dominated by stands of black spruce (*Picea mariana*) and jack pine (*Pinus banksiana*), with a shrub layer of ericaceous shrubs (Ericaceae) and a ground cover of moss and lichens. Depending on drainage, surficial material, and local climate, trembling aspen (*Populus tremuloides*), white birch (*Betula* spp.), white spruce (*Picea glauca*), balsam poplar (*Populus balsamifera*), and balsam fir (*Abies balsamea*) also occupy significant areas of this ecoregion. Poorly drained areas covered by fens and bogs are dominated by black spruce. Bedrock exposures have few trees and are covered with lichens. Fire is an important disturbance regime in this ecoregion, particularly on conifer-dominated dry sites, such as the Athabasca Plain (ESWG 1995).

Characteristic wildlife include moose (*Alces alces*), black bear (*Ursus americanus*), woodland caribou (*Rangifer tarandus* ssp. *caribou*) (for which there is important winter range in Athabasca Plain), barren-ground caribou (*Rangifer tarandus* ssp. *arcticus*), lynx (*Lynx canadensis*), wolf (*Canis lupus*), beaver (*Castor canadensis*), otter (*Lutra canadensis*), marten (*Martes americana*), ermine (*Mustela erminea*), fisher (*Martes pennanti*), muskrat (*Ondatra zibethica*), snowshoe hare (*Lepus americanus*), red-backed vole (*Clethrionomys gapperi*), red squirrel (*Tamiasciurus hudsonicus*), least chipmunk (*Eutamius minimus*), ducks, geese, pelican (*Pelecanus erythrorhynchos*), sandhill crane (*Grus canadensis*), spruce grouse (*Dendragapus canadensis*), sharp-tailed grouse (*Tympahuchus phasianellus*), willow ptarmigan (*Lagopus lagopus*), common nighthawk (*Chordeiles minor*), red-tailed hawk (*Buteo jamaicensis*), raven (*Corvus corax*), common loon (*Gavia immer*), bald eagle (*Haliaeetus leucocephalus*), gray jay (*Perisoreus canadensis*), hawk owl (*Surnia ulula*), great horned owl (*Bubo virginianus*), herring gull (*Larus argentatus*), double-crested cormorant (*Phalacrocorax auritus*), and several other passerine species (ESWG 1995).

Freshwater lakes make up a significant component of the landscape. The Athabasca plain and basin contain some of the most significant active sand dune systems in the boreal regions of North America.

Conservation Status

Habitat Loss and Degradation

Up to 80 percent of this ecoregion remains as intact habitat, although that number drops to an estimated 65 percent in Manitoba. The principal causes of habitat loss are forestry (rapidly expanding), mining (uranium, nickel, gold, copper), and flooding from hydroelectric development.

Remaining Blocks of Intact Habitat

Most of the ecoregion remains relatively intact.

Degree of Fragmentation

Fragmentation of habitat is increasing. Principal causes are transportation routes, logging activity and roads, and flooding from hydroelectric projects.

Degree of Protection

- Opasquia Provincial Wilderness Park—northwestern Ontario, 4,730 km²
- Woodland Caribou Provincial Wilderness Park—northwestern Ontario, 4,500 km²
- Atakaki Provincial Wilderness Park—eastern Manitoba, 3,930 km²
- Lac La Ronge Provincial Park—central Saskatchewan, 3,011 km²
- Amisk Provincial Park—north-central Ontario, 2,377 km²
- Athabasca Sand Dunes Provincial Wilderness Park—northwestern Saskatchewan, 1,925 km²
- Pipestone River Provincial Waterway Park—northwestern Ontario, 973 km²
- Trout Lake Provincial Park—northwestern Ontario, 78 km²
- Pakwash Provincial Park—northwestern Ontario, 39 km²
- Jan Lake Ecological Reserve—east-central Saskatchewan, 20 km²

Types and Severity of Threats

Major or potential increases in the annual allowable cut in large-scale forestry licenses, new hydroelectric dams and major transmission corridors, and extensive mineral exploration are increasing the level of threat to this ecoregion as a whole.

Suite of Priority Activities to Enhance Biodiversity Conservation

- Establish protected areas in the following locations:
  - area around Cree Lake in Saskatchewan
  - Foster Lake area in Saskatchewan
  - Churchill River corridor in Saskatchewan
  - Reindeer Lake in Saskatchewan
- Make protection-standard upgrades to Tom Lamb Wildlife Management Areas in Manitoba (shared with ecoregion [92]).
- Implement recommended actions in Manitoba with respect to this ecoregion, according to the schedule established in the 1996–98 Action Plan.

Conservation Partners

For contact information, please see appendix G.

- Earth's Environment
- Endangered Spaces Campaign, Manitoba
- Endangered Spaces Campaign, Saskatchewan
- Environment North
- Federation of Ontario Naturalists
- Manitoba Future Forest Alliance
- Manitoba Naturalists Society
- The Nature Conservancy, Manitoba
- Nature Saskatchewan
- Resource Conservation Manitoba
- Saskatchewan Environmental Society
- The Wildlands League
- World Wildlife Fund Canada

Relationship to Other Classification Schemes

The Midwestern Canadian Shield Forests span from northern Alberta east to western Ontario. The Athabasca Plain, Churchill River Upland, Hayes River Upland, and Lac Seul Upland (TEC ecoregions 87–90) are the regions within this ecoregion (ESWG 1995). Rowe's (1972) boreal forest sections include: Central Plateau, Upper and Lower English River, Upper Churchill, Nelson River, Northern Coniferous, Athabasca South, and Northwestern Transition (B.8, B.11, B.14, B.20, B.21, B.22a, B.22b and B.27).

Prepared by J. Shay, A. G. Appleby, G. Whelan Enns, T. Gray, K. Kavanagh, M. Sims, and G. Mann

Key Number:	**94**
Ecoregion Name:	**Central Canadian Shield Forests**
Major Habitat Type:	**6.1 Boreal Forest/Taiga**
Ecoregion Size:	**454,382 km²**
Biological Distinctiveness:	**Bioregionally Outstanding**
Conservation Status:	**Snapshot—Relatively Stable Final—Vulnerable**

Introduction

This ecoregion occupies a U-shaped area stretching from the Ontario-Manitoba border in northwestern Ontario south and eastward to the north shore of Lake Superior and then northeastward into west-central Quebec.

Mean annual temperature ranges from –2°C to 1.5°C, with mean summer temperatures from 12.5°C to 14°C, and mean winter temperatures from –17°C to –12°C. Mean annual precipitation varies across the ecoregion; annual precipitation is as low as 550 mm in the north, increasing to around 700–800 mm near Lake Nipigon, and is as high as 900 mm in the west. This area is described as having a moist and humid mid- to high-boreal ecoclimate (ESWG 1995).

This region is underlain by the acidic Archean bedrock of the Canadian Shield. Ridged bedrock outcrops are covered with calcareous, sandy to loamy till in the north and a thin, acidic, sandy till in the south. The southeastern region is bounded to the north by Paleozoic bedrock of the Hudson Basin, toward which the coverage by wetlands increases. In the western region, sporadic discontinuous to isolated patches of permafrost with low ice content occur. In the east are the Mistassini Hills, with summits more than 1,065 m asl (ESWG 1995).

Biological Distinctiveness

Black spruce is the climatic climax species in the north, as this area is dominated by coniferous forest. Fire occurs frequently, such that forest stands are composed of medium to tall closed stands of black spruce and jack pine (*Pinus banksiana*) with some paper birch (*Betula papyrifera*). South-facing slopes or warmer areas include greater proportions of quaking aspen (*Populus tremuloides*), white birch (*Betula* spp.), white spruce (*Picea glauca*), and balsam fir (*Abies balsamea*). Most of the south is dominated by mixed forest, characterized by white and black spruce (*Picea mariana*), balsam fir, jack pine, trembling aspen, and paper birch. Forest fires are an important natural disturbance in these coniferous forests (ESWG 1995).

Characteristic wildlife include moose (*Alces alces*), caribou (*Rangifer tarandus*), black bear (*Ursus americanus*), lynx (*Lynx canadensis*), snowshoe hare (*Lepus americanus*), wolf (*Canis lupus*), sharp-tailed grouse (*Tympahuchus phasianellus*), ruffed grouse (*Bonasa umbellus*), American black duck (*Anas rubripes*), and wood duck (*Aix sponsa*), as well as Canada goose (*Branta canadensis*) in the northern region and hooded merganser (*Lophodytes cucullatus*) and pileated woodpecker (*Dryocopus pileatus*) in the west (ESWG 1995).

Outstanding features of this ecoregion include areas of rich clay plains, which support some of the most productive boreal forest systems in North America. Major concentrations of disjunct arctic and western plant species are found along the north shore of Lake Superior. This ecoregion marks the most southern distribution of woodland caribou range remaining in North America.

Conservation Status

Habitat Loss

It is estimated that 40 percent or less of this ecoregion remains as intact habitat. Small parts of this ecoregion are heavily altered, with some conversion to pasture, while more extensive areas have been significantly disturbed by large-scale, mechanized logging. More than 50 percent of this ecoregion has been logged, and additional parts of the ecoregion are scheduled for logging in the near future.

Remaining Blocks of Intact Habitat

A large block north of Lake Nipigon in Ontario remains free of logging and major human settlements. In addition, a number of smaller, intact habitat blocks remain in both Ontario and Quebec (e.g., around Lake Mistassini, Quebec),

but most of these are near the northern fringe of the ecoregion.

Degree of Fragmentation

Habitat fragmentation has principally occurred as the result of forestry practices (clear-cuts and logging roads). There is one major transportation corridor through the Ontario portion of the ecoregion. Overall, species response to the habitat fragmentation indicates a relatively low impact but one that is likely to increase as more habitat is altered.

Degree of Protection

- Wabakimi Provincial Park (regulated as of June 1997)—northwestern Ontario, approx. 8,950 km^2
- Pukaskwa National Park—west-central Ontario, northern shore of Lake Superior, 1,877 km^2
- Winisk River Provincial Waterway Park—northern Ontario, 1,525 km^2
- Missinaibi River Provincial Waterway Park—central Ontario, 991 km^2
- Albany River Provincial Waterway Park—northern Ontario, 951 km^2
- Severn River Provincial Waterway Park—northern Ontario, 829 km^2
- Otoskwin-Attawapiskat River Provincial Waterway Park—northeastern Ontario, 825 km^2
- Kesagami Provincial Park—northeastern Ontario, 559 km^2
- Brightsand River Provincial Waterway Park—northwestern Ontario 412 km^2
- Michipicoten Island Provincial Park—island in Lake Superior, 367 km^2

Types and Severity of Threats

Large-scale forest clear-cuts and fire suppression are major threats in this ecoregion. Many forests that have been subjected to mechanized logging are converting to aspen and birch instead of maintaining a dominance of conifer species. This change in ecosystem character is occurring over large areas, particularly in the central portion of this ecoregion. Forestry is rapidly expanding in this ecoregion and presents a significant threat to ecosystem integrity in the next twenty years.

Mining and mineral exploration are also present in both the Ontario and the Quebec portions of the ecoregion, while hydroelectric development is a major threat to large areas of lowland habitat in Quebec. There are additional disturbances to natural habitats from major oil and gas trunk lines and hydro transmission corridors.

Suite of Priority Activities to Enhance Biodiversity Conservation

- Establish protected areas in the following locations:
  - Monts Otish in Quebec
  - Lac Albanel in Quebec
  - Rivière Témiscamie in Quebec
  - Lake Nipigon–Nipigon River corridor in Ontario
- Expand Pukaskwa National Park in Ontario.
- Identify candidate protected areas in the clay-belt section of this ecoregion.

Conservation Partners

For contact information, please see appendix G.

- Conservation de la Nature de l'Environnement du Québec (RNCREQ) Environment North
- Federation of Ontario Naturalists
- Grand Conseil des Cris
- The Nature Conservancy, Quebec
- Northwatch
- Regroupement National des Conseils Régionaux
- Union Québecoise pour la Conservation de la nature (UQCN)
- The Wildlands League
- World Wildlife Fund Canada, Quebec Region

Relationship to Other Classification Schemes

The Central Canadian Shield Forests stretch across northern Ontario and Quebec. This ecoregion includes Lake Nipigon, Big Trout Lake, the Abitibi Plains, and Rivière Rupert Plateau (TEC ecoregions 94–96 and 100) (ESWG 1995). Boreal forest sections (as defined by Rowe 1972) in this region are: Chibougamau-Natashquan, Gouin, Northern Clay, Central Plateau, Superior, Upper English River,

Northern Coniferous, and Forest-Tundra (B.1b, B.3, B.4, B.8, B.9, B.11, B.13a, B.22a, and B.32).

Prepared by B. Meades, A. Perera, L. Gratton, N. Zinger, T. Gray, K. Kavanagh, M. Sims, and G. Mann

Key Number:	**95**
Ecoregion Name:	**Southern Hudson Bay Taiga**
Major Habitat Type:	**6.1 Boreal Forest/Taiga**
Ecoregion Size:	**373,502 km²**
Biological Distinctiveness:	**Bioregionally Outstanding**
Conservation Status:	**Snapshot—Relatively Intact**
	Final—Relatively Intact

Introduction

This ecoregion extends along the lowlands adjacent to Hudson Bay from Manitoba, though Ontario and into a small part of western Quebec next to James Bay. The islands in James Bay (under the jurisdiction of the Northwest Territories) also fall within this ecoregion. Approximately 80 percent of the ecoregion lies within Ontario.

Mean annual temperatures range from –5°C to –2°C, mean summer temperatures from 10.5°C to 11.5°C, and mean winter temperatures from –19°C to –16°C. Mean annual precipitation ranges from 400 mm in the far north to 700 or 800 mm in the far east. Generally, the lowlands receive a mean precipitation of 500–700 mm annually. This ecoregion has a high subarctic ecoclimate in the north, a low subarctic ecoclimate in the lowland, and a high boreal ecoclimate in the James Bay lowland, all characterized by short, cool summers and cold winters (ESWG 1995)

This ecoregion is part of the Hudson Bay Lowland, within the Hudson geologic platform, and is underlain by flat-lying Paleozoic limestone bedrock. Continued uplifting of the land surface due to isostatic rebound creates belts of raised beaches. Along the Hudson Bay coast east of the Nelson River, these well-drained beaches support white spruce, alternating with fens, polygonal peat plateaus, and peat plateaus. North of the Nelson River, beaches are more subdued and the terrain is dominated by fens, polygonal peat plateaus, and peat plateaus. The latter often occur in parallel rows marking the underlying beaches. The coastal areas are dominated by marshes, shallow water, and tidal flats, which are extensive at the junction of the Hudson and James Bays. The Hudson Bay Lowland slopes gently to the northeast and east to Hudson Bay, and its maximum elevation is around 120 m asl in the south. The James Bay Lowland slopes north, toward the bay (ESWG 1995).

Biological Distinctiveness

Stunted black spruce and tamarack dominate the vegetation here. These trees become larger and more dense toward the south. The shrub layer consists of dwarf birch (*Betula* spp.), willow (*Salix* spp.), and northern Labrador tea (*Ledum greenlandicum*). Poorly drained areas support dense sedge (*Carex* spp.), moss, and lichen covers, and the raised beaches present a striking pattern of successive black spruce (*Picea mariana*)–covered ridges alternating with depressional bogs and fens. The area to the south of James Bay acts as a transition between the coniferous mixed forests to the south and the tundra to the north and, as such, has a greater diversity of species, including balsam fir (*Abies balsamea*), white spruce (*Picea glauca*), and black spruce, quaking aspen (*Populus tremuloides*), and paper birch (*Betula papyrifera*). Wetlands make up 50–75 percent of this ecoregion (ESWG 1995).

Characteristic wildlife species include caribou (*Rangifer tarandus*), snow goose (*Chen caerulenscens*), Canada goose (*Branta canadensis*), and snowshoe hare (*Lepus americanus*) throughout most of the ecoregion. Polar bear (*Thalarctos maritimus*), brown lemming (*Lemmus trimucronatus*), arctic fox (*Alopex lagopus*), tundra swan (*Cygnus columbianus*), sea ducks, shorebirds, beluga whale (*Delphinapterus leucas*), and seal (Phocidae) are found along the Hudson Bay coast. Willow ptarmigan (*Lagopus lagopus*) occur south of the Hudson Bay coast, and toward the lower end of James Bay black bear (*Ursus americanus*), wolf (*Canis lupus*), moose (*Alces alces*), lynx (*Lynx canadensis*), ruffed grouse (*Bonasa umbellus*), and American black duck (*Anas rubripes*) are found (ESWG 1995).

This ecoregion contains some of the most extensive wetland complexes in the boreal system in North America. Isostatic rebound continues to occur at high rates in this ecoregion, possibly among the highest in the world. This has created a rapidly emerging shoreline of up to one-meter vertical increase per century. Given the flat topography, this rebound can create expanses of one or two kilometers of newly exposed beach. As a result of this dynamic process and flat landscape, some of the continent's most extensive and productive northern salt marshes in the world occur along the coastline. They support huge breeding colonies of waterfowl (particularly snow geese) and are used by these and additional waterfowl and shorebird species as major staging areas in migration.

This ecoregion supports the continent's (and the world's) southernmost populations of polar bear.

Conservation Status

Habitat Loss and Degradation

It is estimated that 99 percent of this ecoregion remains as intact habitat. Small patches of disturbance occur around scattered communities.

Remaining Blocks of Intact Habitat

Most of the ecoregion remains as an intact habitat block.

Degree of Fragmentation

There is no human-induced fragmentation of any significance. Areas around Churchill, Manitoba, are possibly the most impacted by the presence of a spaceport and railway line servicing a growing tourism trade and active summer seaport.

Degree of Protection

- Polar Bear Provincial Park—northern Ontario, 22,880 km^2
- Wapusk National Park—northeastern Manitoba, 11,475 km^2
- Jog Lake Conservation Reserve—northern Ontario, 480 km^2

Types and Severity of Threats

Most of this ecoregion is too far north for any commercial forest harvest. Hydroelectric developments are possible future threats in the James Bay lowlands of Quebec and Ontario. Fly-in hunt camps need to be carefully monitored to ensure that harvests are maintained at sustainable levels.

Suite of Priority Activities to Enhance Biodiversity Conservation

- Establish a protected area on the Ministikawatin Peninsula in Quebec.
- Expand Polar Bear Provincial Park to better include polar bear denning habitats.
- Raise the protection standards of the Cape Tatnam Provincial Wildlife Management Area, Manitoba, and the Akimiski Island Migratory Bird Sanctuary in the Northwest Territories.
- Raise the protection standards of the Churchill Wildlife Management Area surrounding Wapusk National Park.

Conservation Partners

For contact information, please see appendix G.

- Endangered Spaces Campaign, Manitoba
- Federation of Ontario Naturalists
- Grand Conseil des Cris
- Manitoba Naturalists Society
- The Nature Conservancy, Manitoba
- The Nature Conservancy, Quebec
- Northwatch
- The Wildlands League
- World Wildlife Fund Canada, Quebec Region

Relationship to Other Classification Schemes

The Southern Hudson Bay Taiga stretches along the southern shore of Hudson Bay and James Bay in Manitoba, Ontario, and Quebec and is characterized by the Coastal Hudson Bay Lowland, the Hudson Bay Lowland, and the James Bay Lowland (TEC ecoregions 215–217) (ESWG 1995). Within this region, boreal forest sections include: Hudson Bay Lowlands, East James Bay, Northern Coniferous, and Forest-Tundra (B.5, B.6, B.22a, and B.32) (Rowe 1972).

Prepared by T. Gray, G. Whelan Enns, N. Zinger, K. Kavanagh, and M. Sims

Key Number:	**96**
Ecoregion Name:	**Eastern Canadian Shield Taiga**
Major Habitat Type:	**6.1 Boreal Forest/Taiga**
Ecoregion Size:	**741,510 km²**
Biological Distinctiveness:	**Globally Outstanding**
Conservation Status:	**Snapshot—Relatively Stable**
	Final—Relatively Stable

Introduction

This large ecoregion occupies a significant part of northern Quebec and most of Labrador, stretching from Hudson and James Bays to the west, southern Ungava Bay to the north, and the Atlantic Ocean to the east.

The mean annual temperature in this ecoregion ranges from –6°C to 1°C, with lower temperatures occurring closer to Hudson Bay and warmer temperatures in eastern Labrador. Mean summer temperatures range from 5.5°C to 10°C, and mean winter temperatures range from –18°C to –1°C. Mean annual precipitation varies greatly between 300–400 mm in the area directly south of Ungava Bay and 1,000 mm in the southeastern part of this ecoregion. This large ecoregion is characterized by a range of low, mid, and high subarctic ecoclimates and experiences cool summers and very cold winters, with the exception of the coastal barrens, influenced by the Atlantic Ocean and characterized by short, cool, moist summers and longer winters (ESWG 1995).

The physiography of this ecoregion is rough and undulating and is composed mainly of massive Archean granites, granitic gneiss, and acidic intrusives, with some sedimentary rock found along the coast. Glaciation has given this ecoregion a rolling, morainal plain with numerous small, shallow lakes. Permafrost is found throughout the ecoregion in isolated patches, especially in wetlands (ESWG 1995).

The western portion of this ecoregion is composed of the Larch Plateau and the Richmond Hills and has a hummocky surface with elevations ranging from 150 m asl near the coast of James Bay to 450 m asl farther east, precipitation to flow eastward to Ungava Bay and westward to Hudson Bay. Kaniapiskau Plateau forms the core of Lake Plateau in the central and south-central areas and is composed of massive granulite and charnockite Archean rocks. Portions of the plateau reach elevations of 915 m asl. An escarpment runs from northwest to southeast through the center of the ecoregion, overlooking the Labrador Hills, which are composed of folded Precambrian sedimentary and volcanic rocks. Their surfaces are in the form of sinuous ridges and valleys formed by down-warped, folded, and faulted strata. Summit elevations here range from 360 m asl to 730 m asl. Along the Atlantic Coast, steep-sided, rounded mountains with deeply incised U-shaped valleys and fjords extending inland along the Labrador Sea coast. Discontinuous, sandy, bouldery morainal veneers dominate its surfaces. The southeastern edge of the ecoregion is level to gently undulating peatland, interrupted only by a few conspicuous eskers, exposed bedrock highs, and shallow rivers (ESWG 1995).

Biological Distinctiveness

Vegetation consists of open, stunted stands of black spruce (*Picea mariana*) (climax species) and tamarack, with secondary quantities of white spruce (*P. glauca*) and dwarf birch (*Betula* spp.), willow (*Salix* spp.), ericaceous shrubs (Ericaceae), cottongrass (*Eriophorum* spp.), lichens, and moss. Poorly drained sites support sedge (*Carex* spp.), Labrador tea (*Ledum groenlandicum*), cottongrass, and sphagnum moss (*Sphagnum* spp.). The northwestern boundary of this ecoregion is where the tree limit is reached in Quebec, and trembling aspen reaches its northern limit in the Mecatina River ecoregion in Labrador. Balsam fir (*Abies balsamea*) is restricted to rare sites of medium-textured materials. The Kingarutuk-Fraser River area and Mealy Mountains form extensive tundra barrens with continuous vegetation cover restricted to depressions where snow accumulates and provides moisture throughout the growing season. Throughout the south, open coniferous forest is transitional to the closed coniferous boreal forest to the south, and the tundra and alpine tundra communities in the northern part of this ecoregion. In the south, open stands of lichen and black and white spruce with feathermoss understory dominate. Toward Ungava Bay in the north, vegetative cover becomes more sparse and more open. Along the Atlantic Coast, low, closed to open white spruce forest with a moss understory is found on most slopes; however, coastal heath dominates along headlands and ridges, and cliff summits are mainly exposed

bedrock with mosses and lichens limited to cracks. Salt marshes and plateau bogs are common on large marine terraces near the coast as well. In the eastern tip, extensive string bogs with open water dominate and are surrounded by sedges, brown mosses, and sphagnum mosses (ESWG 1995).

Caribou (*Rangifer* spp.), moose (*Alces alces*), black bear (*Ursus americanus*), wolf (*Canis lupus*), red fox (*Vulpes fulva*), arctic fox (*Alopex lagopus*), wolverine (*Gulo luscus*), coyote (*C. latrans*), snowshoe hare (*Lepus americanus*), grouse (*Dendragapus* spp.), osprey (*Pandion haliaetus*), raven (*Corvus corax*), and waterfowl are species common in this ecoregion. The Atlantic Coast of this ecoregion forms part of the Atlantic migratory flyway and provides important habitat for seabird colonies, as well as seal (Phocidae) whelping areas (ESWG 1995). This ecoregion also provides important habitat for peregrine falcon (*Falco peregrinus*) and the eastern harlequin duck (*Histrionicus histrionicus*), both endangered species in Canada.

Outstanding features of this ecoregion include most of the year-round range of the George River barren ground caribou herd (*Rangifer tarandus* ssp. *arcticus*), the world's largest migrating herd, with an estimated 800,000 animals. String bogs are widespread in this ecoregion and may be considered among the most extensively developed examples in North America. In addition, a small (estimated to be less than 300 animals) and very rare population of land-locked freshwater seals inhabits Lac des Loups Marins, Quebec.

Conservation Status

Habitat Loss and Degradation

Approximately 95 percent of the ecoregion's natural habitat remains intact. About 5 percent of the ecoregion is permanently flooded from hydroelectric projects.

Remaining Blocks of Intact Habitat
The majority of the ecoregion remains intact.

Degree of Fragmentation
Little habitat fragmentation; some fragmentation may result locally from flooding by hydroelectric dams.

Degree of Protection

There are no protected areas in this ecoregion that meet IUCN I–III protection standards.

Types and Severity of Threats

Future hydroelectric projects are proposed for many of the ecoregion's major rivers, particularly in Quebec. Locally, mining activity is significantly increasing in some areas (e.g., Voisey's Bay, Labrador), and a real threat of logging exists in the southeastern portion of the ecoregion in Labrador. Caribou harvest must be closely managed so as not to threaten the George River herd in the future, particularly if a population crash occurs, as has happened in the past.

Suite of Priority Activities to Enhance Biodiversity Conservation

- Protected areas need to be permanently established. The following are some important candidate sites:
  - Mealy Mountains National Park—southeast Labrador
  - Lac Joseph–Atikonak proposed provincial wilderness reserve—southwest Labrador
  - Lac Guillaume-Delisle—west-central Quebec
  - Lac à l'eau claire—west-central Quebec
  - Baie aux feuilles—northeast Quebec
  - Lac Burton-Rivière Roggan et la pointe Louis XIV—west-central Quebec
  - Riviere Koroc—northeast Quebec
  - Monts Pyramides—northeast Quebec
  - Lac Cambrien—northeast Quebec
  - Canyon Eaton—northeast Quebec
  - Collines Ondulées—northeast Quebec
  - Confluence des rivières de la Baleine et Wheeler—northeast Quebec
  - Lac Bienville—central Quebec

Conservation Partners

For contact information, please see appendix G.

- Action: Environment
- Grand Conseil des Cris
- Newfoundland/Labrador Environmental Association
- Protected Areas Association of Newfoundland and Labrador
- The Nature Conservancy, Quebec
- Union Québecoise pour la Conservation de la Nature (UQCN)
- World Wildlife Fund Canada, Quebec Region

Relationship to Other Classification Schemes

This ecoregion covers a vast area across northern Quebec and Labrador. Within the Eastern Canadian Shield Taiga, the following terrestrial ecoregions are found: La Grande Hills, Southern Ungava Peninsula, New Québec Central Plateau, Ungava Bay Basin, George Plateau, Kingurutik-Fraser Rivers, Smallwood Reservoir-Michikamau, Coastal Barrens, Mecatina River, Eagle Plateau, Winokapau Lake North and Goose River West (TEC ecoregions 72–86) (ESWG 1995). The Eastern Canadian Shield Taiga is forested by boreal sections Chibougamau-Natashquan, East James Bay, Hamilton and Eagle Valleys, Northeastern Transition, Newfoundland-Labrador Barrens, and Forest-Tundra (B.1b, B.6, B.12, B.13a, B.13b, B.31, and B.32), as well as Tundra (Rowe 1972).

Prepared by B. Meades, L. Gratton, A. Perera, N. Zinger, K. Kavanagh, M. Sims, and G. Mann

Key Number:	**97**
Ecoregion Name:	**Eastern Canadian Forests**
Major Habitat Type:	**6.1 Boreal Forest/Taiga**
Ecoregion Size:	**480,929 km²**
Biological Distinctiveness:	**Regionally Outstanding**
Conservation Status:	**Snapshot—Vulnerable**
	Final—Endangered

Introduction

This ecoregion is distinguished from the Central Canadian Shield Forests [94] by a more maritime influence and balsam fir as the climatic climax species. The Eastern Canadian Forests characterize forested land in eastern Quebec, much of Newfoundland, and disjunct occurrences in the highlands of New Brunswick and on Cape Breton Island, Nova Scotia.

The ecoclimate of this ecoregion ranges from high- and mid-boreal and perhumid mid-boreal to Oceanic, Atlantic, and maritime mid-boreal. Summers are generally cool, with average temperatures ranging between 8.5°C in the north to 14.5°C in the south. Winter temperatures vary according to proximity to the ocean and continental land mass. Thus, winters tend to be colder in Quebec and Labrador, particularly in the north, where mean temperatures range from −8°C to −13°C. On the Island of Newfoundland, winters are shorter and milder; mean temperatures vary between −5.5°C and −1°C. Precipitation follows a similar pattern, there being less in the western, continental part of the ecoregion (800–1,000 mm) than in the eastern and southern coastal areas and the Island of Newfoundland (1,000–1,200 mm in the north, 1,200–1,600 mm in the south). Coastal areas, especially in southeastern and northern Newfoundland and the Cape Breton Highlands, are particularly prone to heavy fog. Also, sea ice plays a significant role in the adjacent terrestrial climate of this ecoregion around the Strait of Belle Isle (ESWG 1995).

A wide range of physiographic features characterize this ecoregion, most of which are the result of glaciation. In the eastern part of the ecoregion, from Lac St. Jean south to the Gulf of St. Lawrence, and as far west as the Labrador coast, the region is underlain by massive Precambrian and Archean granites and gneisses, and lies between sea level and 600 m asl. There are steep slopes that rise abruptly above the St. Lawrence River, and the interior of this part of the ecoregion is rolling or undulating and glacial drift covered. Also, this portion of the ecoregion is incised by several large, wide river valleys. Isolated pockets of permafrost are found in some parts of this area but become less common to the east and south. As one moves east toward the Atlantic Coast, the surface becomes rougher, and surface deposits become thin and discontinuous, heavily influenced by fluvioglacial processes. The portions of the ecoregion in southeastern Labrador are also characterized by deeply dissected margins. On the Island of Newfoundland, the physiography is also a result of glaciation, but rock outcrops become common amidst hummocky, undulating, and sometimes ridged morainal deposits of varying thicknesses of sand or loam. The southern part of the island is part of the Appalachian peneplain,

composed of a mix of soft, late, mostly unfolded Precambrian sedimentary and volcanic rocks. On the Gaspé Peninsula and in the New Brunswick and Cape Breton Highlands, the Appalachian peneplain is also a factor but is characterized by hummocky to mountainous terrain, underlain by folded Paleozoic sandstones and quartzites. Fluvioglacial deposits occur mostly in the valleys. This is some of the highest terrain on the east coast of Canada, and some peaks of the Appalachian range reach above 1,000 m asl in the Gaspé peninsula. Anticosti Island stands out in this ecoregion, as it is a south-dipping cuesta of Paleozoic carbonate strata, and relief rarely reaches 150 m asl. Along the coastlines of the entire ecoregion differential erosion has played a significant role. Also, especially on the east coast of the Island of Newfoundland, the exposed bedrock terrain can slope up to a 30 percent grade (ESWG 1995).

Biological Distinctiveness

The boreal forest in this bioregion is characterized by a mix of balsam fir (*Abies balsamea*) and black spruce (*Picea mariana*). Balsam fir dominates to the east as a result of the maritime influence of the Atlantic. Paper birch (*Betula papyrifera*), aspen (*Populus tremuloides*), and black spruce are typical of disturbed sites. White spruce (*Picea glauca*) dominates in coastal areas where sea salt spray affects plant distributions. Moss-heath vegetation or barrens are also common in coastal areas affected by high winds. The warmer Lac St. Jean valley is dominated by mixed woods more typical of southern climes (sugar maple [*Acer saccharum*], beech [*Fagus grandifolia*], and yellow birch [*Betula alleghaniensis*] on upland sites, while eastern hemlock [*Tsuga canadensis*], balsam fir, eastern white pine [*Pinus strobus*], and white spruce prevail in valleys) (ESWG 1995).

The entire ecoregion provides prime habitat for many species, including moose (*Alces alces*), black bear (*Ursus americanus*), lynx (*Lynx canadensis*), and red fox (*Vulpes fulva*). Woodland caribou (*Rangifer tarandus* ssp. *caribou*) can also be found through the area, with the exception of the New Brunswick and Cape Breton Highlands. In the central Laurentians, the northeastern portion of the ecoregion, snowshoe hare are common, and the wolf is an important predator. Marten (*Martes americana*), beaver

(*Castor canadensis*), porcupine (*Erethizon dorsatum*), bobcat (*Felis rufus*), and rabbit are common in the Appalachian regions. Goose, ptarmigan (*Lagopus* spp.), and ruffed grouse (*Bonasa umbellus*) are common in the north. Because of the length of coastline associated with this ecoregion, the area also supports a great variety of seabirds like murre (*Uria* spp.), eider (*Somateria* spp.), tern (*Sterna* spp.), and puffin (*Fratercula* spp.). In addition, seasonal bird populations vary significantly, as the eastern portion of the ecoregion is in the path of the Atlantic migratory flyway (ESWG 1995).

This ecoregion exhibits high levels of plant endemism in the Gulf of St. Lawrence, with between 100 and 150 species. Maritime heath vegetation, a continentally unique plant assemblage, occurs in areas on the Island of Newfoundland and the Gulf of St. Lawrence shore in Quebec. Large seabird colonies exist along the shorelines of this ecoregion.

Conservation Status

Habitat Loss and Degradation

Approximately 40 percent of this ecoregion remains as intact habitat. The majority of this occurs along the northern portions of the ecoregion in Quebec. Parts of the ecoregion in the Gaspé, northern New Brunswick, and Newfoundland have been heavily altered by a long history of human settlement. Some areas have been extensively logged and not returned to their original vegetation communities, often remaining as barrens or shrublands. Mining in some localized areas has also resulted in habitat loss (e.g., Matamec, Quebec).

Remaining Blocks of Intact Habitat

Most remaining habitat blocks are in the northern portions of the ecoregion. No major intact habitat blocks remain in Newfoundland or the Gaspé outside of protected areas.

Degree of Fragmentation

Fragmentation of forest habitat is most notable. Little in the way of mature forest habitat remains throughout much of this ecoregion. Road networks through parts of the ecoregion (including logging roads) contribute to habitat fragmentation as well.

Degree of Protection

- Bay du Nord Provincial Wilderness Reserve—south-central Newfoundland, 2,895 km^2
- Monts Valin Provincial Park—north of the Saguenay River, south-central Quebec
- Avalon Provincial Wilderness Reserve—eastern Newfoundland, 1,070 km^2
- Cape Breton Highlands National Park—northern Nova Scotia, 950 km^2
- Parc de la Gaspésie—eastern Quebec, 801 km^2
- Middle Ridge Provincial Wildlife Reserve—south-central Newfoundland, 618 km^2
- Terra Nova National Park—eastern Newfoundland, 405 km^2
- Saguenay Provincial Park—south-central Quebec, 283 km^2
- Polletts Cove, Aspy Fault Protected Area—Cape Breton Island, Nova Scotia, 275 km^2
- Forillon National Park—eastern Quebec, 240 km^2

Types and Severity of Threats

Logging is by far the most extensive threat to this ecoregion. In one logging license area in Quebec, 15,000 square kilometers of forest are scheduled to be logged in the coming twenty-five years. Logging elsewhere, combined with fuelwood harvest by coastal communities, has resulted in very little original forest remaining. Species composition has changed dramatically in historic times since European settlement of this region. Mining and mineral exploration are rapidly expanding in this ecoregion. Locally, peat extraction is of concern to some wetland habitats, particularly in Newfoundland.

Suite of Priority Activities to Enhance Biodiversity Conservation

- Implement the protected-areas strategy that has been launched in Newfoundland and Labrador by the year 2000 and establish appropriate representative protected areas. In the interim, designate the Little Grand Lake proposed ecological reserve.
- Establish more protected areas for the Christmas Mountains and surrounding area in New Brunswick.

- Establish protected areas in Quebec including:
  - Rivière Manitou
  - Monts Groulx
  - Rivière Vaureal watershed
  - Harington Harbour shoreline habitat.

Conservation Partners

For contact information, please see appendix G.

- Action: Environment
- Canadian Parks and Wilderness Society, Nova Scotia Chapter
- Cape Breton Naturalists Society
- Conservation Council of New Brunswick
- Federation of Nova Scotia Naturalists
- Friends of the Christmas Mountains National Park
- Heritage Foundation Terra Nova
- Les Amis de plein air de Cheticamp
- Margarée Environmental Society
- Natural History Society of Newfoundland and Labrador
- The Nature Conservancy, Quebec
- Nature Trust of New Brunswick
- New Brunswick Federation of Naturalists
- New Brunswick Protected Natural Areas Coalition
- Newfoundland/Labrador Environmental Association
- Nova Scotia Nature Trust
- Protected Areas Association of Newfoundland and Labrador
- Sierra Club, Cape Breton Group
- Tuckamore Wilderness Society
- Union Québecoise pour la Conservation de la Nature (UQCN)
- World Wildlife Fund Canada, Quebec Region

Relationship to Other Classification Schemes

This area corresponds to the terrestrial ecoregions of the Central Laurentians (TEC ecoregion 101), the Mecatina Plateau (TEC ecoregion 103), the northern Appalachians (TEC ecoregion 117), and Anticosti Island (TEC ecoregion 102). The Eastern Canadian Forests in Labrador cover the Paradise River and Lake Melville regions (TEC ecoregions 104 and 105).

On the Island of Newfoundland, ecoregions include the Strait of Belle Isle (TEC ecoregion 106), the Northern Peninsula (TEC ecoregion 107), the Maritime Barrens (TEC ecoregion 114), the Avalon Forest (TEC ecoregion 115), and Southwestern, Central, and Northeastern Newfoundland (TEC ecoregions 109, 112, and 113). In New Brunswick, the New Brunswick Highlands (TEC ecoregion 119) are home to Eastern Canadian Forests, and in Nova Scotia, the Cape Breton Highlands (TEC ecoregion 129) are as well (ESWG 1995).

The Eastern Canadian Forests area is also characterized by numerous Rowe (1972) forest regions and sections. Boreal forest sections include: Laurentide-Onatchiway, Chibougamau-Natashquan, Gaspé, Hamilton and Eagle Valleys, Northeastern Transition, Grand Falls, Corner Brook, Anticosti, Northern Peninsula, Avalon, Newfoundland-Labrador Barrens, and Forest-Tundra B.1a, B.1b, B.2, B.12, B.13c, B.28a, B.28b, B.28c, B.29, B.30, B.31, and B.32). In this part of the Acadian forest region are the New Brunswick Uplands and the Cape Breton Plateau (A.1 and A.6). The Saguenay section (GL.7) of the Great Lakes–St. Lawrence forests is also part of the Eastern Canadian Forests.

Prepared by B. Meades, L. Gratton, A. Perera, N. Zinger, L. Jackson, J. Goltz, C. Stewart, K. Kavanagh, M. Sims, and G. Mann

Key Number:	**98**
Ecoregion Name:	**Newfoundland Highland Forests**
Major Habitat Type:	**6.1 Boreal Forests/ Taiga**
Ecoregion Size:	**16,338 km²**
Biological Distinctiveness:	**Nationally Important**
Conservation Status:	**Snapshot—Relatively Stable Final—Relatively Stable**

Introduction

This ecoregion lies entirely within the Island of Newfoundland. It is distinguished from the Eastern Canadian Forests [97] by higher elevational range and rugged topography.

This area has a mean annual temperature of 4°C; mean summer temperature ranges from 11.5°C to 12°C, and the mean winter temperature from −3.5°C to −4°C. The mean annual precipitation varies between 1,000 and 1,400 mm. This region has a maritime high-boreal ecoclimate, with cool summers and cold snowy winters (ESWG 1995).

Rugged, steep slopes are formed of acidic, crystalline Paleozoic and Precambrian rocks and range from sea level to 815 m asl, although most of the region falls in the 300–700 m range. Ridged to hummocky bare rock is common (ESWG 1995).

Biological Distinctiveness

The boreal forest in this ecoregion is characterized by dwarf, open, and sometimes closed cover patches of black spruce (*Picea mariana*) and balsam fir (*Abies balsamea*) alternating with communities of dwarf kalmia (*Kalmia polifolia*) and mosses. Exposed sites support mixed evergreen and deciduous shrubs (ESWG 1995).

This ecoregion contains the northernmost extension of the Appalachian Mountain system in North America. Some arctic plant species occur as disjunct southern populations or extensions of their usual continental range. Similarly, arctic hares (*Lepus arcticus*) are at the southern limits of their continental range in the Newfoundland Highland Forests.

Conservation Status

Habitat Loss

It is estimated that 80–90 percent of the ecoregion remains as intact habitat. Most areas above the tree line have been impacted only as the result of a few transmission corridors, mineral exploration and extraction, and domestic wood harvesting at the middle and lower elevations.

Remaining Blocks of Intact Habitat

The ecoregion is naturally divided into three disjunct habitat blocks. All of these are relatively intact.

Degree of Fragmentation

Human-induced fragmentation is relatively minor.

Degree of Protection

- Gros Morne National Park (upper elevations)—western Newfoundland, 1,942 km²
- Barachois Pond Provincial Park—southwestern Newfoundland, 34 km²
- King George IV Ecological Reserve—western Newfoundland, 19 km²

Types and Severity of Threats

The lower forested slopes of this ecoregion are under threat from increased domestic wood harvest and commercial pulp and paper harvest. Areas at higher elevation (most of the ecoregion) are principally threatened by granite quarries and a high level of mining potential and interest. All-terrain vehicle traffic is extensive in some areas.

Suite of Priority Activities to Enhance Biodiversity Conservation

- Implement the protected-areas strategy that has been launched in Newfoundland and Labrador by the year 2000 and establish appropriate representative protected areas. In the interim, designate the Little Grand Lake proposed ecological reserve.
- Reduce and better regulate domestic wood cutting in Gros Morne National Park.

Conservation Partners

For contact information, please see appendix G.

- Action: Environment
- Natural History Society of Newfoundland and Labrador
- Newfoundland-Labrador Environmental Association
- Protected Areas Association of Newfoundland and Labrador
- Tuckamore Wilderness Society
- World Wildlife Fund Canada

Relationship to Other Classification Schemes

The Newfoundland Highland Forests in eastern Newfoundland are found across the Long Range

Mountains (TEC ecoregions 108, 110, and 111) (ESWG 1995). Within Rowe's (1972) Boreal Forest Region, this area corresponds to Corner Brook and the Newfoundland-Labrador Barrens (B.28b and B.31).

Prepared by B. Meades, L. Gratton, A. Perera, L. Jackson, K. Kavanagh, M. Sims, and G. Mann

Key Number:	**99**
Ecoregion Name:	**South Avalon–Burin Oceanic Barrens**
Major Habitat Type:	**6.1 Boreal Forest/ Taiga**
Ecoregion Size:	**2,030 km²**
Biological Distinctiveness:	**Globally Outstanding**
Conservation Status:	**Snapshot—Relatively Stable Final—Relatively Stable**

Introduction

This small ecoregion comprises the outer headlands of narrow peninsulas in eastern Newfoundland. These headlands are exposed to high winds, cool temperatures, and salt spray for most of the year.

The ecoregion is classified as having an exposed oceanic low-boreal ecoclimate. Mean annual temperature is approximately 5.5°C. Summers are cool, with a mean temperature of 11.5°C, and foggy. Winters tend to be short and relatively mild, with a mean temperature of −1°C. Mean annual precipitation ranges between 1,200 and 1,500 mm (ESWG 1995).

This ecoregion is underlain predominantly by a mixture of late Precambrian sedimentary and volcanic strata, and its elevations rise abruptly from the sea to about 200 m asl. Stream erosion has cut deeply, and the uplands are dissected, rugged, and rocky along the coastline, but inland they present a rolling terrain of low relief. The surface of the uplands is dominated by peat-covered, rolling to hummocky, sandy morainal deposits with slopes that range 5–30 percent. Surficial deposits of glacial till are common, and wetlands cover more than 25 percent of the ecoregion (ESWG 1995). Precambrian fossils have been found at Mistaken Point Ecological Reserve.

Biological Distinctiveness

The moss-heath plant associations of this ecoregion, dominated by blanket bogs and *Racomitrium* heath, are unique to North America. Their closest affinities are in oceanic climates such as Iceland, northern Scotland, and Spitsbergen (ESWG 1995).

The ecoregion supports dense carpets of moss and fruticose lichen, along with closed, low-growing ericaceous shrubs. Dwarf krummholz of balsam fir (*Abies balsamea*) occurs on some upland sites (ESWG 1995).

Characteristic wildlife include caribou (*Rangifer tarandus*), willow ptarmigan (*Lagopus lagopus*), and many seabird species, including one of the world's largest northern gannet (*Morus bassanus*) colonies, situated at Cape St. Mary's (ESWG 1995).

Conservation Status

Habitat Loss and Degradation

It is estimated that 95 percent of this small ecoregion remains intact. Some habitat loss has occurred due to the presence of a few very small coastal communities and associated access roads.

Remaining Blocks of Intact Habitat

The ecoregion is naturally disjunct; each area could be considered as a separate habitat block.

Degree of Fragmentation

Very minor habitat fragmentation exists.

Degree of Protection

- Chance Cove Provincial Park—southeastern Newfoundland, 20 km²
- Cape St. Mary's Ecological Reserve—southeastern Newfoundland, 13 km²
- Mistaken Point Ecological Reserve—southeastern Newfoundland, 3 km²
- Holyrood Pond Provincial Park—southeastern Newfoundland, 2 km²

Types and Severity of Threats

The greatest threat is from all-terrain vehicles (ATVs) in the lowlands. Recreation activities around the seabird colonies will require careful management. Caribou poaching is also a problem in this area.

Suite of Priority Activities to Enhance Biodiversity Conservation

- Establish additional protected areas to capture the landscape-level patterns unique to this area. Priority sites for protected status include the St. Shotts area of the Avalon Peninsula.
- Control ATV traffic into parts of this ecoregion.

Conservation Partners

For contact information, please see appendix G.

- Action: Environment
- Natural History Society of Newfoundland and Labrador
- Newfoundland-Labrador Environmental Association
- Protected Areas Association of Newfoundland and Labrador
- World Wildlife Fund Canada

Relationship to Other Classification Schemes

The South Avalon–Burin Oceanic Barrens occur along the southern tips of the Avalon and Burin Peninsulas in Newfoundland (TEC ecoregion 116) (ESWG 1995). The boreal forest here is also named Avalon (B.30), as described by Rowe (1972).

Prepared by B. Meades, L. Gratton, A. Perera, K. Kavanagh, M. Sims, and G. Mann

Key Number:	**100**
Ecoregion Name:	**Aleutian Islands Tundra**
Major Habitat Type:	**6.2 Tundra**
Ecoregion Size:	**5,479 km²**
Biological Distinctiveness:	**Globally Outstanding**
Conservation Status:	**Snapshot—Relatively Intact**
	Final—Relatively Intact

Introduction

The Aleutian Island chain extends from the Alaska Peninsula almost 1,500 km to the east between the Bering Sea and the Gulf of Alaska. It is composed of a series of sedimentary islands capped by steep volcanoes. Elevations range from sea level to over 1,900 m, with the higher volcanoes glaciated. Vegetation at the higher and more exposed areas consists of dwarf shrub communities codominated by willow (*Salix* spp.) and crowberry (*Empetrum nigrum*). Lower, more protected areas support mesic graminoid herbaceous meadows dominated by bluejoint (*Calamagrostis canadensis*), with a variety of other herbs. Dry graminoid herbaceous communities occur in coastal areas, and bogs support low scrub communities, developing thick peat deposits (Gallant et al. 1995).

The islands experience a maritime climate. Precipitation varies widely, from 530 mm up to 2,080 mm. Generally, larger islands receive more precipitation than smaller ones, and coastal areas more than inland areas. Temperatures range from average lows of –20°C to –7°C in winter to average highs of 10°C to 13°C in summer (Gallant et al. 1995). Most soils form in volcanic ash or cinders over basaltic rock, and dominant soil types are typic haplocryands and typic vitricryands. Higher elevations often are covered in bare rock and basaltic rubble. Only the easternmost part of the archipelago was glaciated during the Pleistocene, but many of the islands show a history of glacial presence.

Biological Distinctiveness

The Aleutian Islands Tundra ecoregion supports many seabird colonies of extraordinary size and global importance. The Pribilof Islands, for example, provide breeding habitat for approximately 3 million seabirds including virtually all of the world's 250,000 red-legged kittiwakes (*Rissa brevirostris*) (Sowls, Hatch, and Lensink 1978). Many of the islands also support endemic species, including the Pribilof Island shrew (*Sorex pribilofensis*) and the Aleutian shield fern (*Polystichum aleuticum*), the only federally listed endangered plant in Alaska.

Conservation Status

Several species introduced for ranching have become feral, including cattle, reindeer, and fox. These have caused habitat alteration through grazing and predation on seabird colonies. Rat introductions also are likely to be impacting the seabird colonies. Pollutants, associated primarily with military development, are locally acute. Radioactivity has persisted from the nuclear testing on Amchitka Island in 1971, according to studies conducted by Greenpeace (1996).

Perhaps the conservation problem of most concern is the decline in almost all species of fish-eating seabirds in the Aleutians. Mortality and population decline are mostly likely a result of trophic changes in the Bering Sea ecosystem due to commercial harvests of fish and whales over the last four decades, according to a study by the National Research Council (1996).

Habitat Loss

The Aleutian Islands are largely intact, with habitat alteration mostly resulting from widely scattered communities, military bases (some abandoned), cattle ranches (both ongoing and in disuse), fox farms (abandoned), and subsistence hunting.

Remaining Blocks of Intact Habitat

The largest islands (Attu, Umnak, Unalaska, Akun) all have blocks of habitat in excess of 1,000 km². The chains of islands are also largely intact.

Degree of Fragmentation

The island chain is highly fragmented naturally, with few islands greater than 2,000–3,000 km², but very little additional anthropogenic fragmentation has occurred.

Degree of Protection

Almost all of the islands are included in the Alaska Maritime National Wildlife Refuge (AMNWR), and many areas are also included in the Aleutian Islands Wilderness. Small areas already developed were excluded from AMNWR or wilderness designation. Designation of protected areas on land, however, does not address the threats from changes in the surrounding marine ecosystems on which the terrestrial systems depend or from the residual effects of radioactivity and other pollutants.

Types and Severity of Threats

Major threats include:

*   continuation of habitat degradation and conversion due to cattle ranching, feral cattle, and introduced reindeer
*   predation of seabirds by feral fox and rats
*   continuation of decline in seabird populations due to commercial fisheries' impacts on the Bering Sea marine ecosystem

Priority Activities to Enhance Biodiversity Conservation

*   Develop a better understanding and management of the marine ecosystem on which the terrestrial systems depend.
*   Continue pressure on the military to clean up and restore bases, particularly as closures and abandonment happen.
*   Eradicate feral cattle and fox from several islands.

Conservation Partners

For contact information, please see appendix G.

*   Alaska Marine Conservation Council
*   Bering Sea Coalition
*   The Nature Conservancy, Alaska

Relationship to Other Classification Schemes

This ecoregion is identical to Gallant et al.'s (1995) ecoregion 114. The ecoregion was delineated at Unimak Pass between the Alaska Peninsula and the "first" of the Aleutian Islands and was primarily based on climate.

Prepared by R. Hagenstein and T. Ricketts

Key Number:	**101**
Ecoregion Name:	**Beringia Lowland Tundra**
Major Habitat Type:	**6.2 Tundra**
Ecoregion Size:	**151,046 km²**
Biological Distinctiveness:	**Regionally Outstanding**
Conservation Status:	**Snapshot—Relatively Intact**
	Final—Relatively Intact

Introduction

The Beringia Lowland Tundra ecoregion is formed by three major disjunct areas along the Bering Sea coast of Alaska from the base of the Alaska Peninsula to Kotzebue Sound, as well as one smaller area on the east side of St. Lawrence Island and St. Matthew Island. The ecoregion is characterized by low, flat, or gently rolling terrain, wet soils, and resulting predominance of wet and mesic graminoid herbaceous vegetation. In better-drained areas, especially in the somewhat more rolling portions of the section surrounding Bristol Bay, dwarf shrub communities occur interspersed with the wet herbaceous tundra, dominated by sedges, including *Eriophorum angustoifolium* and *Carex* spp. Dwarf shrub vegetation is usually dominated by ericaceous species, including crowberry (*Empetrum nigrum*). In some limited areas of favorable soil drainage and microclimate, stands of black and white spruce (*Picea mariana, P. glauca*) occur, with understories of alder (*Alnus* spp.), willow (*Salix* spp.) and dwarf birch (*Betula* spp.). Lakes and ponds cover 15–25 percent of the surface area, and wetlands cover between 55 percent (southern portions) and 78 percent (northern portions) of the region.

The region experiences a climate that is transitional between maritime and continental influences. Average annual precipitation varies widely, ranging from 250 mm in Kotzebue Sound to 860 mm near Bristol Bay. Mean daily minimum temperatures in winter range from −25°C in the northern parts to −10°C along the Alaska Peninsula. Summer high temperatures can reach 18°C. In the Kotzebue Sound and the Yukon-Kuskokwim Delta areas, soils are shallow to permafrost and consist mostly of histic pergelic cryoquepts and pergelic cryofibrists. In the more southerly and better-drained Bristol Bay area, the permafrost table is deep to discontinuous and somewhat more developed soil profiles can occur (Gallant et al. 1995). All but the Yukon-

Kuskokwim Delta was covered in Pleistocene glaciation (Pielou 1994; Gallant et al. 1995).

Biological Distinctiveness

The abundance of surface water makes this ecoregion excellent habitat for waterfowl, and the tidal flats of the coast provide an abundance of habitat for shorebirds. Many species of birds depend on the habitats of this ecoregion for the majority, if not all, of their nesting habitat. These include the arctic loon (*Gavia artica*), Canada goose (*Branta canadensis*), bristle-thighed curlew (*Numenius tahitiensis*), dovekie (*Alle alle*), McKay's bunting (*Plectrophenax hyperboreus*), and white wagtail (*Motacilla alba*). The highest densities of nesting tundra swans, the majority of the world's emperor swans, and 50 percent of the world's black brant (*Branta bernicla*) are supported by the Yukon-Kuskokwim Delta alone (McNab and Avers 1994). This delta is the most expansive area of highly productive waterfowl nesting habitat in Alaska and among the most significant on earth. Izembek Lagoon, on the Alaska Peninsula, is the most important migratory staging area for black brant in Alaska. St. Lawrence and St. Matthew Islands also support large seabird colonies. St. Lawrence Island colonies support approximately 2 million seabirds, including the largest murre (*Uria* spp.) colonies in the eastern Bering Sea (Alaska Fish and Wildlife Research Center 1988).

Mammals include river otters (*Lutra canadensis*), short-tailed and least weasels (*Mustela erminea* and *M. rixosa*), brown bear (*Ursus arctos horribilis*), moose (*Alces alces*), and caribou (*Rangifer* spp.). Populations of all top-level predators are intact, and brown bears reach extraordinary natural densities in the Katmai and Lake Iliamna areas. All five species of North American Pacific salmon are native here, and Bristol Bay supports the largest run of sockeye salmon (*Oncorhynchus nerka*) in the world (McNab and Avers 1994).

Conservation Status

Habitat Loss and Degradation

The Beringia Lowland Tundra is almost entirely intact. Human disturbance consists of small permanent and seasonal settlements, mostly along rivers and coasts. These communities depend on subsistence fishing and on hunting of marine mammals (seals and whales), caribou, moose, and birds. For the most part, the terrestrial impacts of these settlements are local and slight.

Remaining Blocks of Intact Habitat

The largest blocks of intact habitat include the Ahklun Mountains and the Bendeleben Mountains, which contain diverse habitat types.

Degree of Fragmentation

This ecoregion is naturally fragmented into three large mainland areas and numerous islands. The southernmost block of habitat around the Ahklun Mountains is relatively unfragmented. The northernmost block is still mostly intact but is becoming trisected by roads and is increasingly impacted by mining. This area has the potential to be much more fragmented in the future.

Degree of Protection

This ecoregion is represented by several protected areas. Those in the Seward Peninsula–Kotzebue area include:

- Bering Land Bridge National Park—western Alaska
- Selawik National Wildlife Refuge—western Alaska

The Yukon-Kuskokwim Delta is contained completely within the Yukon Delta National Wildlife Refuge. Protected areas of the ecoregion surrounding Bristol Bay and on the Alaska Peninsula include:

- Togiak National Wildlife Refuge—western Alaska
- Egegik, Pilot Point, Cinder River, Point Moller State Critical Habitat Areas—western Alaska
- Izembek National Wildlife Refuge—western Alaska
- Alaska Peninsula National Wildlife Refuge—western Alaska
- Becharof National Wildlife Refuge—western Alaska
- Katmai National Park and Preserve—western Alaska

Finally, a number of islands scattered throughout the ecoregion, especially St. Matthew Island,

are included as units of the Alaska Maritime National Wildlife Refuge.

Types and Severity of Threats

Currently, waterfowl harvests are managed by USFWS, through comanagement agreements with Native communities and through international treaty. Although there is some potential for unsustainable waterfowl harvests from local communities, the harvests are at present well managed and represent less of a threat than habitat destruction in wintering areas for nesting waterfowl.

Commercial fishing is a significant presence and can affect not only the marine ecosystem but the many terrestrial species that rely on the abundant fish runs each year. Salmon fisheries in the area are generally well managed to ensure that sufficient spawning stock returns to the rivers. As with any commercial fishery, there is a concern with overharvesting, particularly of weak stocks in years of generally abundant returns.

Suite of Priority Activities to Enhance Biodiversity Conservation

The biggest gaps in the protected areas of this ecoregion are in the Nushagak River valley and along the west side of the Alaska Peninsula.

Conservation Partners

For contact information, please see appendix G.

- Alaska Marine Conservation Council
- National Audubon Society, Alaska-Hawaii Regional Office
- The Nature Conservancy, Alaska

Relationship to Other Classification Schemes

This ecoregion combines Gallant et al. (1995) ecoregions 109 and 112. Major delineation features are the low topography, poor to moderate drainage, dominant wet tundra and dwarf shrub communities, and coastal position. Combination decisions are based on Gallant et al. (1995) and Küchler (1975).

Prepared by R. Hagenstein and T. Ricketts

Key Number:	**102**
Ecoregion Name:	**Beringia Upland Tundra**
Major Habitat Type:	**6.2 Tundra**
Ecoregion Size:	**97,364 km²**
Biological Distinctiveness:	**Bioregionally Outstanding**
Conservation Status:	**Snapshot—Relatively Stable**
	Final—Relatively Stable

Introduction

The Beringia Upland Tundra consists of three disjunct areas on the Bering Sea coast of Alaska, one comprising the upland and mountainous areas of Seward Peninsula, one corresponding to the hills and mountains of the Ahklun and Kilbuck mountain ranges in southwest Alaska, and one of much smaller extent on the western half of St. Lawrence Island in the northern Bering Sea. These areas are similar in their varied terrain and elevation and corresponding variety of vegetation, habitats, and communities. The ecoregion consists of steep, jagged mountain ranges set among large areas of rolling hills, broad valleys, and lowlands. Elevation ranges from sea level to 500 m in the hilly uplands to over 1,500 m in the tallest ranges. Plant communities respond to these differences in topography and accompanying drainage. Low-lying, poorly drained areas support wet graminoid herbaceous vegetation similar to that of ecoregion [101]. Better-drained areas contain either mesic graminoid herbaceous communities dominated by sedges (*Eriophorum vaginatum, Carex bigelowii*) or low-scrub communities. The scrub communities are dominated by either ericaceous shrubs (e.g., *Arctostaphylos alpina, Vaccinium vitis-idaea, Empetrum nigrum*) or a mix of mountain-avens (*Dryas octopetala*) and dwarf arctic birch (*Betula nana*). Protected, well-drained valley bottoms may contain coniferous forests dominated by white spruce (*Picea glauca*), broadleaf forests of balsam poplar (*Populus balsamifera*), or mixed forests characterized by white spruce and paper birch (*Betula papyrifera*). The steep peaks and ridges of the tallest mountains are almost barren, and some retain cirque glaciers.

Climatic characteristics of the Beringia Upland Tundra ecoregion are also varied, ranging from maritime along the coast to continental in inland areas. An average of 250–1,000 mm of precipitation falls annually, ranging up to 2,000 mm in the higher areas of the Ahklun and Kilbuck Mountains. Average daily winter tem-

peratures range from –24°C to –16°C, and average daily summer temperatures range from 13°C to 19°C. These temperature averages are cooler when taken for the Seward Peninsula portion alone, a trend that reflects the generally colder, harsher climate in the more northern parts of the ecoregion. Permafrost is continuous on the Seward Peninsula, while it is discontinuous to patchy in the Ahklun and Kilbuck Mountains (Gallant et al. 1995). The highlands of the Seward Peninsula were glaciated in the Pleistocene, as were most of the Ahklun and Kilbuck Mountains (Pielou 1994; Gallant et al. 1995).

Wildfire occurrence is common on the Seward Peninsula, where lichens and mosses dry out in summer. Burns are less frequent and smaller in the Ahklun and Kilbuck Mountains (Bailey et al. 1994).

Biological Distinctiveness

A variety of birds use the assorted habitats of this ecoregion to breed, from spectacled eiders (*Somateria fishceri*) and turnstones (*Arenaria* spp.) in the lower portions to bristle-thighed curlews (*Numenius tahitiensis*) along the coast to blackpoll warblers (*Dendroica striata*) in the conifer stands along protected river valleys. The breeding range of the bristle-thighed curlew is restricted completely to this ecoregion. In the Seward Peninsula portion, arctic foxes (*Alopex lagopus*) and tundra hares (*Lepus othos*) are common, and polar bears (*Thalarctos maritimus*) are frequently seen. Caribou (*Rangifer* spp.) and musk oxen (*Ovibos moschatus*) were both introduced there, in the 1890s and 1970, respectively. In the southern portion, beaver (*Castor canadensis*) are numerous, and wood frogs (*Rana sylvatica*) are reported. Across the ecoregion the diversity of lichens and tundra plants is high, including many representatives of the Siberian flora in St. Lawrence Island and the Seward Peninsula. These parts of the ecoregion also support an Asian avifaunal component.

Coastal areas, especially on the Seward Peninsula and St. Lawrence Island, have seabird colonies with abundant populations of cliff-nesting alcids and other species, including common and thick-billed murres (*Uria aalge* and *U. lomvia*) and tufted puffins (*Fratercula cirrhata*). The Walrus Islands in Togiak Bay also support

Alaska's largest walrus (*Odobenus rosmarus*) haul-out site.

Like most ecoregions in Alaska, these areas are large blocks of essentially intact natural habitat with populations of all indigenous plants and animals, including top-level predators, occurring in populations within their range of natural variation.

Conservation Status

Habitat Loss and Degradation

Very scattered pockets of habitat loss are associated with human communities. Human population in this ecoregion is low, consisting mostly of small settlements spread throughout the region. Levels are low, however, and impacts on populations are generally small. Mining activity on the Seward Peninsula is responsible for the majority of human impact; the region of the Bendeleben Mountains, north of Nome, has experienced the most extensive development of any area in the ecoregion.

Remaining Blocks of Intact Habitat

The ecoregion has several widely disjunct blocks, all of them greater than 5,000 km².

Degree of Fragmentation

The mines in this area and the roads between them are increasingly fragmenting and altering the tundra habitat. This development has the potential to interrupt mammal movements and migrations and to result in severe local degradation of the delicate tundra habitats. Recovery from roads, construction sites, and even human trails requires as many as twenty-five or fifty years, longer if areas greater than 1 km² are disturbed (Bliss 1988).

Degree of Protection

Several protected areas exist in the southern (Ahklun and Kilbuck Mountains) section of the ecoregion:

- Wood-Tikchik State Park—southwestern Alaska
- Walrus Islands State Game Sanctuary—western Alaska
- Yukon Delta National Wildlife Refuge—western Alaska

- Togiak National Wildlife Refuge—southwestern Alaska

In the northern section, on the Seward Peninsula, the only protected area is Bering Land Bridge National Park, which includes mostly lowland tundra.

Most of the larger seabird colonies in the ecoregion are also conserved within units of the Alaska Maritime National Wildlife Refuge.

Types and Severity of Threats

The main threat to the ecoregion is the expanding development of mines and mining roads on the Seward Peninsula. Areas in the Ahklun and Kilbuck Mountains show potential for mineral extraction as well, so threats of similar habitat fragmentation and degradation are present there, too.

There have been conservation concerns recently regarding the Kilbuck caribou herd that is at low levels. Additionally, a limited hunt has recently been allowed on walrus in the Walrus Islands.

Suite of Priority Activities to Enhance Biodiversity Conservation

- Create an upland tundra protected area on the Seward Peninsula, which is becoming increasingly roaded and developed. Greater oversight and management of mining activities are also recommended.
- Monitor harvested species, especially caribou and walrus.

Conservation Partners

For contact information, please see appendix G.

- National Audubon Society (for Beringia), Alaska-Hawaii Regional Office
- Sierra Club
- The Wilderness Society

Relationship to Other Classification Schemes

This ecoregion combines Gallant et al.'s (1995) ecoregions 110 and 111. It delineates the upland and mountainous areas between the continuously forested inland areas and the coast. The ecoregion is delineated from the other coastal tundra ecoregions by its higher topography, better drainage, and resulting dominance of more mesic tundra and dwarf shrub communities. Combination decisions were based on Gallant et al. (1995) and Küchler (1975).

Prepared by R. Hagenstein and T. Ricketts

Key Number:	**103**
Ecoregion Name:	**Alaska/St. Elias Range Tundra**
Major Habitat Type:	**6.2 Tundra**
Ecoregion Size:	**169,049 km²**
Biological Distinctiveness:	**Regionally Outstanding**
Conservation Status:	**Snapshot—Relatively Intact**
	Final—Relatively Intact

Introduction

The Alaska/St. Elias Range Tundra is a long belt of high, rugged mountains arcing north from the base of the Alaska Peninsula, east to encompass the Alaska Range, and south to include the Wrangell–St. Elias Range on the Canadian-Alaskan border near Yakutat Bay. The Canadian portion of this ecoregion encompasses the southwestern corner of the Yukon Territory and extreme northwestern British Columbia. Elevations range from sea level at the western end, to 600 meters in the broad, lower valleys, often to over 4,000 meters. Mt. McKinley, the highest point in North America, lies within the ecoregion, with an elevation of over 6,100 meters. The St. Elias Mountains are among the highest in Canada, ranging upward to 6,000 m asl. Peaks stand as isolated blocks separated by broad ice fields. Because the limit of permanent snow is 2,150 m asl, the mountains present great masses of ice and snow, causing great valley glaciers. Permafrost is continuous at high elevation and sporadic and discontinuous at low elevation (ESWG 1995).

The ecoregion, which covers 169,091 km², consists mostly of rocky slopes, ice fields, and glaciers. Where permanent ice and snow fields do not dominate, alpine tundra vegetation is usually dominated by dwarf shrub communities, including mountain-avens (*Dryas octopetala, D.*

integrifolia) and ericaceous species (e.g., *Vaccinium vitis-idaea, Cassiope tetragona*). More protected slopes support low or tall scrub communities, consisting of dwarf birch (*Betula glandulosa* and *B. nana*), willows (*Salix* spp.), alder (*Alnus sinuata* and *A. crispa*), and ericaceous shrubs. Well-drained sites in the lower valleys can support open forests of white and black spruce (*Picea glauca, P. mariana*), or paper birch (*Betula papyrifera*) and quaking aspen (*Populus tremuloides*). Wet sites in this area support cottongrass and sedge.

For much of its distance, this ecoregion is inland of another mountain belt, the Pacific Coastal Mountains [104], resulting in a dominance of continental climate influences. Annual precipitation averages about 400 mm in the lower elevations and is estimated to average over 2,000 mm in the higher elevations. Average daily minimum temperatures in the winter range from –25°C to –34°C, while average daily maximum temperatures in the summer range from 18°C to 22°C. Temperatures are lower throughout the year at higher elevations. Farther south in the Canadian portion of the ecoregion, annual temperature decreases with elevation, but at valley bottoms it is –1.5°C, with a summer mean of 9.5°C and a winter mean of –14°C. Mean annual precipitation increases with elevation, ranging from approximately 300 mm at low elevations, to more than 1,000 mm in the ice fields. Generally, this ecoregion has an alpine and glacierized North Pacific cordilleran ecoclimate (ESWG 1995).

This region is characterized by permanent ice and snow fields and minor areas of rock outcrop, rubbly colluvium, and alpine tundra vegetation. Glaciation was extensive during the Pleistocene, and much of the ecoregion's area remains under permanent ice. Soils are quite poor, shallow over bedrock, and generally do not retain enough moisture to form permafrost (summarized from Gallant et al. 1995).

Biological Distinctiveness

The Alaska/St. Elias Range Tundra ecoregion is an almost entirely intact ecological system, with its full complement of top-level predators within their natural range of variation. There are significant brown bear concentrations in Denali Park, Prairie Creek, and coastal portions in the southwest part of the ecoregion, near Lake Iliamna and Kamishak Bay. Other wildlife includes mountain goat, caribou, moose, Dall sheep, beaver, hare, salmon, and other fish.

Conservation Status

This ecoregion has suffered little habitat loss, degradation, or fragmentation. Minor losses are associated with development at the entrance to Denali National Park, at the north side of Denali Park at Kantishna, surrounding the abandoned Kennicott copper mine in the Wrangell Mountains, and at Nabesna in the Wrangells. Also, coal mining at Healy (northeast of Denali Park) has caused some habitat loss. Several roads cross the ecoregion, but their fragmentation effect is relatively low, and habitat blocks between them are all larger than 10,000km². Human habitation is low, and human use is generally limited to recreation, subsistence, and sport hunting and fishing. Wildfire occurrence is low, with most of the fires occurring in the central part of the ecoregion, and ranging in size from 0.01 km² to 32.9 km² (Gallant et al. 1995).

Degree of Protection

Important protected areas include:

- Lake Clark National Park and Preserve— southern Alaska
- Denali National Park—south-central Alaska
- Denali State Park—south-central Alaska
- Tetlin National Wildlife Refuge—southeastern Alaska
- Wrangell-St. Elias Park and Preserve—Alaska
- Tatshenshini Park—northwestern British Columbia (partially in the ecogregion), 9,580 km²
- Kluane National Park—southwestern Yukon Territory, 22,015 km²
- Kluane Game Sanctuary (low-level protection)—southwestern Yukon Territory

Types and Severity of Threats

No severe threats currently face this ecoregion on a large scale. Some smaller-scale threats include:

- excessive recreational use surrounding national parks and wildlife areas (Impacts could become acute locally.)

- possible overharvest of wildlife, although currently populations are well managed
- Expanded mineral exploration and exploitation (This region contains many extractable mineral resources, including gold, silver, lead, copper, coal, uranium, and molybdenum (Gallant et al. 1995). The world's richest copper deposits were mined at Kennicott in the southern Wrangell Mountains in the early part of the century.)

Suite of Priority Activities to Enhance Biodiversity Conservation

- Manage recreation, tourism, and related facilities in Denali and Wrangell–St. Elias Parks to minimize impact and avoid excessive use of local areas.
- Protect bear populations at Prairie Creek and near Kamishak Bay. Currently, these areas have no formal protection designation. Any protection must happen through cooperative approaches with the Native corporations that own title to the lands in these areas.
- Work with Native corporations that own land within U.S. national parks (especially inholdings in Lake Clark and Wrangell–St. Elias Parks) to ensure compatible land uses.

Conservation Partners

For contact information, please see appendix G.

- Canadian Arctic Resources Committee
- Canadian Parks and Wilderness Association, Yukon Chapter
- Friends of Yukon Rivers
- National Parks and Conservation Association
- The Nature Conservancy, Alaska (regarding Prairie Creek)
- Sierra Club
- The Wilderness Society
- World Wildlife Fund Canada
- Yukon Conservation Society

Relationship to Other Classification Schemes

This ecoregion combines Gallant et al.'s (1995) ecoregions 116 and 118 and the Ecological Stratification Working Group's (1995) ecoregion

173. Major delineation features are high elevation, moderate to very steep terrain, sparse vegetation, and general inland position with accompanying continental climate. Combination decisions were based on Gallant et al. (1995), McNab and Avers (1994), Ecoregions Working Group (1989), Viereck et al. (1992), and Küchler (1975).

On the Canadian side of the border, the St. Elias Mountains (TEC ecoregion 173) cover this ecoregion (ESWG 1995). In addition to Tundra vegetation, this area of southeastern Yukon also contains occurrences of Northern Pacific Coast forests and Boreal forest (Rowe 1972).

Prepared by R. Hagenstein, T. Ricketts, J. Peepre, K. Kavanagh, M. Sims, and G. Mann

Key Number:	**104**
Ecoregion Name:	**Pacific Coastal Mountain Tundra and Ice Fields**
Major Habitat Type:	**6.2 Tundra**
Ecoregion Size:	**137,279 km²**
Biological Distinctiveness:	**Bioregionally Outstanding**
Conservation Status:	**Snapshot—Relatively Intact**
	Final—Relatively Intact

Introduction

The Pacific Coastal Mountain Tundra and Ice Fields ecoregion consists of a steep, very rugged range of mountains that stretches from the Kenai Peninsula along the Gulf of Alaska coast and the Canadian-Alaskan border to the southern end of the Alaska panhandle. In Canada, this ecoregion encompasses the extreme southwestern corner of the Yukon Territory and parts of the coastal mountains in British Columbia south to Portland Inlet. Elevations range from sea level to over 4,500 m, and slopes generally are steeper than 7 degrees, ranging to over 20 degrees. The landscape of this ecoregion is dominated by mountains of great height. Most peaks reach between 2,100 m and 3,050 m, but some are over 5,000 m (Mount Logan is 5,959 m, and King Peak is 5,175 m). In this high, extreme northern part of the ecoregion, the Seward, Hubbard, and Malaspina glaciers are the dominant physiographic influences, and the area is, in fact, part of the largest nonpolar ice field in the

world. In the south, along the Boundary Ranges, glaciation is also a factor, especially from the Grand Pacific and Llewellyn glaciers. The mountains are composed of crystalline gneisses and granite rocks and are cut into several segments by large, steep-sided transverse valleys. Isolated patches of permafrost are found throughout the ecoregion at elevations above 2,500 m.

This ecoregion generally experiences a transitional climate between maritime and continental effects, and climate patterns vary considerably with elevation, latitude, and topography. There are no long-term weather stations in the region, but estimates based on data from low-elevation stations in ecoregion [23] suggest annual precipitation ranging from about 2,000 mm to over 7,000 mm. Soil development is not widespread and consists of poor, gravelly soils that form in till and colluvium. The ecoregion was extensively glaciated in the Pleistocene, and most of it still is (summarized from Gallant et al. 1995). In Canada, this ecoregion is characterized by a combination of glacierized, alpine, subalpine, and maritime North Pacific cordilleran ecoclimate. Because of the great height of the mountains in the region, the mean annual temperature is cold: –0.5°C (with an average summer temperature of 10°C, and –11.5°C in the winter). Also, because of the extreme elevation, and the proximity to the Pacific Coast, the region receives an extraordinary amount of precipitation, from a low of 1,000 mm in the eastern part of the Boundary Ranges, to 2,400 mm in the ice fields of the Fairweather Ranges, to as much as 3,500 mm around Mount Logan, Canada's highest peak (ESWG 1995). In the high alpine, the great majority of this precipitation falls as snow, and the proportion of rain increases in correspondence to lower elevation, which, in rare cases, reaches as low as sea level in southeastern Alaska.

Biological Distinctiveness

Much of the ecoregion lies beneath glaciers and ice fields, and most of the rest is devoid of vegetation. Where the ground is vegetated, communities are dominated by dwarf and low shrub communities, including mountain heath (*Phyllodoce aleutica*) and ericaceous shrubs (e.g., *Cladothamnus pyrolaeflorus*). Mountainous terrain and coastal influence result in three vegetation zones following elevational gradients: alpine tundra vegetation of variable ground cover; sub-alpine forests of alpine fir (*Abies lasiocarpa*), mountain hemlock (*Tsuga mertensiana*), and some Sitka spruce (*Picea sitchensis*) at middle elevations; and closed forests of western hemlock and some Sitka spruce at warmer, more humid, lower elevations. In the north, around Mount Logan, there is no terrestrial vegetation or soil development, due to the extreme elevation. A narrow zone of temperate coniferous rain forest extends between this ecoregion and the coast for much of its length [ecoregion 23]. In several places these forest species extend up mesic washes and river valleys into the Pacific Coastal Mountain Tundra and Ice Fields. These stands are dominated by hemlock (*Tsuga herophylla* or *T. mertensiana*) and subalpine fir or by Sitka spruce. The forests on the Kenai Peninsula represent a transitional forest type between temperate rain forests characteristic of coastal areas and boreal forest and taiga communities characteristic of interior Alaska.

The ecosystems of this ecoregion remain generally intact, with their full range of top predators existing in their natural ranges of variation. Wildlife includes moose (*Alces alces*), mountain goat (*Oreamnos americanus*), grizzly and black bear (*Ursus arctos* and *U. americanus*), wolf (*Canis lupus*), wolverine (*Gulo luscus*), ptarmigan (*Lagopus* spp.), and spruce grouse (*Dendragapus canadensis*). There are also black-tailed deer (*Odocoilus hemionus*) in the lower river valleys. The portion on the Kenai Peninsula holds particular biological interest as a mixing area of populations from the forests of both sides, specifically between the Snow River drainage on the west side to King's Bay in Prince William Sound. Additionally, major rivers that bisect this ecoregion, including the Copper, Alsek, Taku, and Stikine Rivers, provide migratory corridors for waterfowl, passerines, and terrestrial mammals that connect the coastal forests with interior areas. In addition, salmon stocks (*Oncorhynchus* spp.) in this ecoregion are of continental significance.

Conservation Status

This ecoregion remains almost entirely intact, with little habitat loss, degradation, or fragmentation. Habitat loss is restricted to mine sites and the few roads and communities. Other human impacts are mostly limited to recreation and subsistence hunting and fishing. The habitat is essentially intact. Data on wildfire occurrence are few, but occurrence is thought to be very low. In the Chugach Range, in the northern part of the ecoregion, fires are very infrequent and range in size from 0.01 km² to 0.4 km². Large-scale forest disturbance from spruce bark beetle infestations may be a more dominant influence than fires in this area. The overall amount of intact habitat is estimated at 95 percent of the total area of the ecoregion.

Degree of Protection

Important protected areas include:

- Kenai Fjords National Park—southern Alaska
- Chugach National Forest (however, no areas within the Chugach have been designated as wilderness as yet)—southern Alaska
- Wrangell-St. Elias National Park and Preserve—southeastern Alaska
- Glacier Bay National Park—southeastern Alaska
- Tongass National Forest—southeastern Alaska
- Kluane National Park—southwestern Yukon (partially in ecoregion), 22,015 km²
- Tatshenshini-Alsek Provincial Park—northwestern British Columbia (partially in ecoregion), 9,580 km²
- Stikine Provincial Park—northwestern British Columbia, 2,170 km²

And the following wilderness areas:

- Russell Fjord Wilderness—southern Alaska
- Tracy Arm–Fords Terror Wilderness—southern Alaska
- Stikine-LeConte Wilderness—southern Alaska
- Misty Fjords Wilderness—southern Alaska
- Endicott River Wilderness—southern Alaska
- Tatshenshini Wild and Scenic River—southern Alaska

Types and Severity of Threats

Most major threats relate to mining potential and the proposed roads and development associated with mineral exploitation. For example, two proposed roads would cut through the British

Columbia coast range: one at Taku Inlet and one at the Stikine River. The ecoregion contains a variety of extractable minerals such as gold, silver, copper, and zinc, as well as coal, petroleum, and uranium (Gallant et al. 1995).

Much of the area has experienced small-scale gold mining through the present century, and some of this activity continues on a modest scale. There have also been a number of significant mineral deposits located throughout the ecoregion, though few have been fully developed.

The Kenai Peninsula has the highest chance of experiencing significant timber harvest or wildlife exploitation in the future. Planned timber harvests in response to mortality from a spruce bark beetle infestation could cause damage to streams important for salmon spawning and to declining brown bear populations on the Kenai Peninsula.

Suite of Priority Activities to Enhance Biodiversity Conservation

- Monitor mining proposals over time. This ecoregion has experienced a history of small-scale mining and several large-scale mining proposals.
- Address the unsustainable timber harvest practices on the Kenai Peninsula. The best approach could be to build public pressure to use sustainable harvest practices.
- Establish a protected area in the Taku watershed in British Columbia.
- Establish an ecologically representative protected area in the lower Stikine watershed in British Columbia.

Conservation Partners

For contact information, please see appendix G.

- Alaska Center for the Environment
- Alaska Rainforest Campaign
- Canadian Arctic Resources Committee
- Canadian Parks and Wilderness Society, British Columbia Chapter
- Canadian Parks and Wilderness Society, Yukon Chapter
- Friends of the Stikine National Parks and Conservation Association

- Interrain Pacific
- Prince William Sound Science Center
- Southeast Alaska Conservation Council (SEACC)
- Taku Wilderness Association
- World Wildlife Fund Canada
- Yukon Conservation Society

Relationship to Other Classification Schemes

This ecoregion combines Gallant et al.'s (1995) ecoregion 119 with Ecological Stratification Working Group's (1995) ecoregions 184, 185, and 186. Major delineation features include high elevation, moderate to very steep terrain, general lack of vegetation, and coastal position. Combination decisions were based on Gallant et al. (1995), McNab and Avers (1994), Ecoregions Working Group (1989), Viereck et al. (1992), and Küchler (1975).

In Canada, this ecoregion runs along the northeastern border of British Columbia and Alaska, incorporating the Northern Coastal Mountains (TEC ecoregions 185 and 186). The Mount Logan area (TEC ecoregion 184) in the southeast region of Yukon Territory is also included in the Canadian portion of this ecoregion (ESWG 1995). This region is primarily Tundra but incorporates some Northern Pacific Coast vegetation (C.3) (Rowe 1972).

Prepared by R. Hagenstein, T. Ricketts, L. Craighead, J. Peepre, K. Kavanagh, M. Sims, and G. Mann

Key Number:	**105**
Ecoregion Name:	**Interior Yukon/Alaska Alpine Tundra**
Major Habitat Type:	**6.2 Tundra**
Ecoregion Size:	**236,600 km²**
Biological Distinctiveness:	**Regionally Outstanding**
Conservation Status:	**Snapshot—Relatively Stable Final—Relatively Stable**

Introduction

This ecoregion encompasses much of south-central Yukon and a part of east-central Alaska.

There are a number of disjunct outliers of this ecoregion in central Alaska.

The ecoclimate of this ecoregion can be described as a combination of alpine, subalpine, and boreal northern cordilleran. The region experiences short cool summers (mean temperature 10°C to 10.5°C) and long, cold winters (mean temperature –20°C to –23°C), but there is great variation about monthly means. The coldest temperature in North America (–63°C) was officially recorded in this ecoregion on the Kluane Plateau. Mean annual precipitation is lowest in the valleys within the rain shadow of the coastal ranges (less than 300 mm) and increases in the interior ranges and plateaus (up to 600 mm) (ESWG 1995).

This ecoregion contains mountain ranges with many high peaks, and extensive, rolling to undulating hills and plateaus, moderately to deeply incised. These ranges are separated by wide valleys and lowlands whose topography ranges from level to undulating. Almost all of the area was affected by glaciation. Most of the region lies between 900 m and 1,500 m asl, with peaks reaching between 2,100 m and 2,400 m in elevation. Permafrost is common in many parts of the ecoregion, particularly in the north and at higher elevations, but becomes more sporadic and discontinuous to the southwest toward the coast (ESWG 1995) .

Biological Distinctiveness

White and black spruce (*Picea glauca* and *P. mariana*) form the most common forest types, usually within a matrix of aspen (*Populus tremuloides*) or dwarf willow (*Salix* spp.), birch (*Betula* spp.), and ericaceous (Ericaceae) shrubs. These forests are usually open and extensive. Black spruce, scrub willow, and birch are dominant in poorly drained sites, particularly in the lowlands, and black spruce, willow, and paper birch (*Betula papyrifera*) prevail on slopes underlain by permafrost. Balsam poplar (*P. balsamifea*) occur on some floodplains. Lodgepole pine (*Pinus contorta*) and alpine fir (*Abies lasiocarpa*) are also present, depending on time since disturbance and local conditions. At the higher elevations, sparsely vegetated alpine and subalpine communities consist of mountain-avens (*Dryas hookeriana*), dwarf willow, birch, ericaceous shrubs, graminoid species (Gramineae and Cyperaceae), and mosses (ESWG 1995).

Characteristic wildlife include caribou (*Rangifer tarandus*), grizzly and black bear (*Ursus arctos* and *U. americanus*), Dall sheep (*Ovis dalli*), moose (*Alces alces*), beaver (*Castor canadensis*), red fox (*Vulpes fulva*), wolf (*Canis lupus*), hare (*Lepus* spp.), common raven (*Corvus corax*), rock and willow ptarmigan (*Lagopus mutus* and *L. lagopus*), and golden eagle (*Aquila chrysaetos*) (ESWG 1995).

Significant ecological phenomena include some possible Beringia floristic elements on south-facing, low-elevation slopes along the Yukon River valley. Parts of the ecoregion represent unglaciated terrain, while unique periglacial landforms exist in some of the uplands.

Conservation Status

Habitat Loss and Degradation

Approximately 85 percent of this ecoregion was estimated to remain intact. Mining and road development are the primary human factors contributing to habitat loss. In parts of the ecoregion, valley bottoms, which contain the most productive habitats, have been significantly altered by mining activities. In fact, most of the 15 percent habitat loss in this ecoregion represents degradation of valley bottoms.

Remaining Blocks of Intact Habitat

Large blocks of intact habitat remain, although these increasingly have roads dissecting them.

Degree of Fragmentation

Much of the ecoregion (in Yukon) is taiga and boreal forest. Alpine areas are disjunct, although this is the natural condition. Roaded valleys are more problematic, as these create barriers or interfere with seasonal movement of large mammals.

Degree of Protection

There are no protected areas in the Yukon, but the following areas exist in Alaska:

- Yukon-Charley Rivers National Preserve—eastern Alaska
- Streese National Conservation Areas—eastern Alaska
- White Mountains National Recreation Area—eastern Alaska

- Arctic National Wildlife Refuge—small part of the northeastern portion of the ecoregion in Alaska

Note: Many of the upland (alpine) outliers in this ecoregion in western Alaska are not protected and are not part of the lowland preserves surrounding them.

Types and Severity of Threats

Mining continues to expand, and there is already a small amount of logging in the southwestern (Yukon) portion of the ecoregion with possible plans to increase logging of these northern forests.

Suite of Priority Activities to Enhance Biodiversity Conservation

- Recover '40 Mile' and 'Aistiakik' caribou herds.
- Upgrade protection standards of the MacArthur Game Sanctuary (Yukon) to exclude mineral exploration and mining.
- Establish Wellesley Lake candidate protected area.

Conservation Partners

For contact information, please see appendix G.

- Canadian Arctic Resources Committee
- Canadian Nature Federation
- Canadian Parks and Wilderness Society, Yukon Chapter
- Friends of Yukon Rivers
- World Wildlife Fund Canada
- Yukon Conservation Society

Relationship to Other Classification Schemes

The Interior Yukon/Alaska Alpine Tundra includes the Klondike Plateau, the Ruby Ranges, and the Yukon Plateau-North (TEC ecoregions 172, 174, and 176) (ESWG 1995). This region includes Central and Eastern Yukon and Kluane Boreal forest regions (B.26a, B.26b, and B.26c) and Tundra vegetation (Rowe 1972).

Prepared by S. Smith, J. Shay, J. Peepre, K. Kavanagh, M. Sims, and G. Mann

Key Number:	**106**
Ecoregion Name:	**Ogilvie/MacKenzie Alpine Tundra**
Major Habitat Type:	**6.2 Tundra**
Ecoregion Size:	**208,421 km²**
Biological Distinctiveness:	**Regionally Outstanding**
Conservation Status:	**Snapshot—Relatively Stable Final—Relatively Stable**

Introduction

This extremely steep, mountainous ecoregion encompasses the Ogilvie and Wernecke Mountains, the Backbone Ranges, the Canyon Ranges, the Selwyn Mountains, and the eastern and southern Mackenzie Range (these last two are an extension of the Rockies).

Alpine to subalpine northern subarctic cordilleran describes this region's ecoclimate. Weather patterns from the Alaskan and Arctic coasts have a significant influence on this ecoregion. Summers are warm to cool, with mean temperatures ranging from 9°C in the north to 9.5°C in the south. Winters are very long and cold, with very short daylight hours. Mean temperatures range from −19.5°C in the south to −21.5°C in the north, where temperatures of −50°C are not uncommon. Mean annual precipitation is highly variable but generally increases along a gradient from northwest to southeast, the highest amounts (up to 750 mm) falling at high elevation in the Selwyn Mountains. At lower elevations, anywhere from 300 mm (in the north) to 600 mm (in the south) is the average (ESWG 1995).

The bedrock is largely sedimentary in origin, with minor igneous bodies, and much of this is mantled with colluvial debris and frequent bedrock exposures and minor glacial deposits. Barren talus slopes are common. Although parts of the northwest portion of this ecoregion are unglaciated, the majority has been heavily influenced by glaciers. Alpine and valley glaciers are common, especially in the southern and eastern parts of the area where the ecoregion contains broad, northwesterly trending valleys. Valleys tend to be narrower and sharper in the unglaciated northwest. Elevations in the ecoregion also tend to increase as one moves southeast. In the north, in the unglaciated portions of the Ogilvie and Wernecke Mountains, elevations are mostly between 900 m and 1,350 m asl, the highest peaks reaching 1,800 m. In the central part of the ecoregion, elevations can reach above 2,100 m asl, and in the south (Selwyn Mountains) peaks reach as high as 2,950 m. Permafrost is extensive and often continuous throughout the region (ESWG 1995).

Biological Distinctiveness

Subalpine open woodland vegetation is composed of stunted white spruce (*Picea glauca*) and occasional alpine fir (*Abies lasiocarpa*) and lodgepole pine (*Pinus contorta*), in a matrix of willow (*Salix* spp.), dwarf birch (*Betula* spp.), and northern Labrador tea (*Ledum decumbens*). These often occur in discontinuous stands. In the north, paper birch (*B. papyrifera*) can form extensive communities in lower-elevation and mid-slope terrain, but this is less common in the south and east. Alpine tundra at higher elevations consists of lichens, mountain-avens (*Dryas hookeriana*), intermediate to dwarf ericaceous shrubs (Ericaceae), sedge (*Carex* spp.), and cottongrass (*Eriophorum* spp.) in wetter sites (ESWG 1995).

Characteristic wildlife include caribou (*Rangifer tarandus*), grizzly and black bear (*Ursus arctos* and *U. americanus*), Dall sheep (*Ovis dalli*), moose (*Alces alces*), beaver (*Castor canadensis*), red fox (*Vulpes fulva*), wolf (*Canis lupus*), hare (*Lepus* spp.), common raven (*Corvus corax*), rock and willow ptarmigan (*Lagopus mutus* and *L. lagopus*), bald eagle (*Haliaeetus leucocephalus*), and golden eagle (*Aquila chrysaetos*). Gyrfalcon (*Falco rusticolus*) and some waterfowl are also to be found in some parts of the Mackenzie Mountains (ESWG 1995)

Outstanding features of this ecoregion include areas that may have remained ice free during the late Pleistocene—relict species occur as a result. Also, the ecoregion supports a large and intact predator-prey system, one of the most intact of the Rocky Mountain ecosystem. The winter range for the Porcupine caribou herd and full-season range of the Bonnet-Plume woodland caribou herd (5,000 animals) are found in this area. The Fishing Branch Ecological Reserve has

the highest concentration of grizzly bears in North America for this northern latitude.

Conservation Status

Habitat Loss and Degradation

It is estimated that at least 95 percent of the ecoregion is still intact. Mining and mineral, oil, and gas exploration are the principal sources of habitat disturbance and loss.

Remaining Blocks of Intact Habitat
The ecoregion is principally intact.

Degree of Fragmentation
To date, the ecoregion has remained principally intact. Roads are increasingly becoming a concern, as is some of the access associated with mineral exploration.

Degree of Protection

- Tombstone Mountain Territorial Park Reserve—Yukon Territory, 800 km²
- Fishing Branch Ecological Reserve—northwestern western Yukon Territory, 165 km²

Types and Severity of Threats
Most of the threats relate to future access into this northern and fragile ecoregion. Further road development and mineral exploration may result in increased human access. This is already occurring in the western half of the ecoregion.

Suite of Priority Activities to Enhance Biodiversity Conservation

- Enlarge Tombstone Mountain Territorial Park, Yukon Territory
- Establish protected areas in the various mountain ranges that comprise this ecoregion in both Yukon and Northwest Territories.
- Enlarge Fishing Branch Ecological Reserve, Yukon Territory
- Protect the Wind, Snake, and Bonnet-Plume Rivers.
- Develop protected-area proposals for the Keele Peak Area and the Itsi Range, Yukon Territory

Conservation Partners

For contact information, please see appendix G.

- Canadian Arctic Resources Committee
- Canadian Nature Federation
- Canadian Parks and Wilderness Society, Yukon Chapter
- Ecology North
- Friends of Yukon Rivers
- World Wildlife Fund Canada
- Yukon Conservation Society

Relationship to Other Classification Schemes

The North Ogilvie Mountains (TEC ecoregion 168) characterize the northern part of this ecoregion; the Mackenzie Mountains (TEC ecoregion 170) run east-west through Yukon Territory and the Northwest Territories; and the Selwyn Mountains (TEC ecoregion 171) are located in the south section of this ecoregion, which is part of the taiga cordillera ecozone (ESWG 1995). Forest types here are Eastern Yukon Boreal (B.26c), Boreal Alpine Forest-Tundra (B.33), and Tundra (Rowe 1972).

Prepared by S. Smith, J. Peepre, J. Shay, C. O'Brien, K. Kavanagh, M. Sims, and G. Mann

Key Number:	**107**
Ecoregion Name:	**Brooks/British Range Tundra**
Major Habitat Type:	**6.2 Tundra**
Ecoregion Size:	**159,481 km²**
Biological Distinctiveness:	**Bioregionally Outstanding**
Conservation Status:	**Snapshot—Relatively Intact**
	Final—Relatively Intact

Introduction

The Brooks/British Range Tundra ecoregion includes the mountainous belt that extends from almost the Chukchi Sea across northern Alaska and into northern Yukon Territory and extreme northwestern Northwest Territories. This ecoregion consists of three large areas connected along a continuum: the Western Brooks Range,

with relatively low, less rugged mountains and less permanent ice; the Eastern Brooks Range/British Range, with higher, more rugged terrain and more permanent ice; and the lower area near Anaktuvuk Pass that divides the two mountainous areas. Elevations across the ecoregion range from 800 m to 2,400 m, with peaks above 1,800 m retaining the once extensive Pleistocene glaciation (Pielou 1994; Gallant et al. 1995).

Arctic climate prevails, temperatures decreasing with altitude. At Anaktuvuk Pass (the only long-term weather station in the region), the mean daily minimum temperature in winter is –30°C, mean daily maximum temperature in summer is 16°C, and annual precipitation is approximately 280 mm/year (Gallant et al. 1995). Temperatures at higher elevations are probably lower in both seasons. The mean annual temperature for the British-Richardson Mountains is –10°C; mean summer temperature is 6.5°C, and mean winter temperature is –25°C. Summers are short and cool, and winters are extremely cold but more moderate at higher elevations. Major mountain passes are subject to strong outflow winds, resulting in severe wind chill conditions. Mean annual precipitation is approximately 350 mm. The ecoclimate is described as alpine to subalpine northern subarctic cordilleran. Glaciation, frost action, and rapid erosion of steep, unstable slopes have kept soil accumulation low. Where they do accumulate, soils are predominantly pergelic cryaquepts, pegelic cyumbrepts, and lithic cryorthents.

In Canada, permafrost with low ice content is continuous in the southern half of the region, and ice content increases to the north. While the northern portion of the British and Richardson Ranges is unglaciated and peaks can reach an elevation of 1,675 m, the southern parts of the ranges exhibit smooth, rounded profiles composed almost entirely of folded, sedimentary strata of Cambrian and Paleozoic origins. The ecoregion also contains a small portion of unglaciated topography composed of Tertiary sediments and has some excellent examples of periglacial landforms (cryoplanation terraces and summits, in particular) (ESWG 1995).

Biological Distinctiveness

Due to the harsh, mountainous climate and terrain, vegetation cover is sparse and restricted largely to valleys and lower slopes. Subalpine open woodland vegetation is composed of stunted white spruce (*Picea glauca*) and occasional alpine fir (*Abies lasiocarpa*) and lodgepole pine (*Pinus contorta*), in a matrix of willow (*Salix* spp.), dwarf birch (*Betula* spp.), and northern Labrador tea (*Ledum decumbens*). Alpine tundra at higher elevations consists of lichens, mountain-avens (*Dryas hookeriana*), intermediate to dwarf ericaceous shrubs (Ericaceae), sedge (*Carex* spp.), and cottongrass (*Eriophorum* spp.) in wetter sites. The highest latitudinal limit of tree growth (white spruce) in Canada is reached in the British-Richardson Mountains ecoregion (TEC 165). Wet to mesic sites contain mesic graminoid herbaceous communities (including *Carex aquatilis, C. bigelowii, Salix planifolia,* and *S. lanata*), while drier sites support dwarf shrub vegetation (including *Arcostaphylos alpina, Vaccinium* spp., and *Dryas octopetala*) (Gallant et al. 1995). *Dryas*-lichen vegetation is common, with *Dryas integrifolia* or *D. octopetala* constituting 80–90 percent of the vascular plant cover (Bliss 1988).

The Brooks and British Ranges straddle the migration routes of the Porcupine, Central Arctic, and Western Arctic caribou herds (*Rangifer tarandus*). The herds generally migrate from north to south, following river valleys and traditional routes. The Brooks and British Ranges are therefore essential to the annual movements the caribou require, as well as to the free seasonal movement of grizzly bears (*Ursus arctos*) and wolves (*Canis lupus*). Other wildlife includes Smith's longspurs (*Calcarius pictus*), horned larks (*Eremophila alpestris*), Dall sheep (*Ovis dalli*), snowshoe hare (*Lepus americanus*), red fox (*Vulpes fulva*), and arctic ground squirrel (*Citellus parryi*) (McNab and Avers 1994). The eastern Brooks Range supports the northernmost breeding populations of golden eagles (*Aquila chrysaetos*) and gyrfalcons (*Falco rusticolus*) in the United States.

This mountainous region is extremely remote. Although used for subsistence and sport hunting, the ecoregion supports the full range of species characteristic of this area, including top-level predators at viable population levels.

Conservation Status

The Brooks/British Range ecoregion is almost entirely intact. Two major highway corridors cross the ecoregion: the Dalton Highway (and oil pipeline) and the Dempster Highway. These structures can act as a barrier (or filter) to the movement of some species but seldom an impenetrable one. Most historic caribou movements have paralleled the Dalton Highway and pipeline (north-south), but some groups, as well as bears and wolves, are known to move in an east-west direction as well. Both highways provide access to these mountains for hunters, and a buffer of hunting pressure exists bordering the road corridors. Other human disturbance is low and has consisted of some mineral exploration and extraction. Wildfire occurrence is common, and fires average 17.9 km^2 in size (Bailey et al. 1994).

The Red Dog lead-zinc mine, on the western edge of the ecoregion, is the largest development in the ecoregion. The size of reserves has recently doubled through additional exploration work, and the mine has a projected life of forty years. Ore is trucked along a road from the mine site to tidewater on the Chukchi Sea coast.

Degree of Protection

The protected-area system captures a good range of mountain tundra habitat. The most important areas include:

- Gates of the Arctic National Park and Preserve—northern Alaska
- Noatak National Preserve—northern Alaska
- Kobuk Valley National Park—northern Alaska
- Ivvavik National Park—northern Yukon (partially in ecoregion), 10,168 km^2
- Vuntut National Park—northern Yukon (partially in ecoregion), 4,345 km^2

A number of Wild and Scenic Rivers have been designated throughout this ecoregion as well.

Despite this network of protected areas, one of the primary conservation priorities in this ecoregion is maintaining the freedom of movement for caribou and other large mammals. This cannot be accomplished simply by establishing and managing protected areas. The entire ecoregion, as well as those around it, must be managed to minimize barriers to these migrations and to other necessary movements.

Types and Severity of Threats

Major threats include:

- the possibility of mining activity—some mineral exploration and small-scale extraction already have occurred
- an increase in the hunting and recreation impacts on fragile tundra ecosystems and wildlife populations in the vicinity of the two roads, the Dalton and Dempster Highways
- a continued steady increase in recreation activity in the national parks—again, fragile ecosystems with low resilience to human disturbance make this a concern

Suite of Priority Activities to Enhance Biodiversity Conservation

- Manage entire ecoregion (and neighboring ecoregions) to maintain freedom of movement for caribou herds, brown bears, wolves, and other migratory or highly mobile wildlife. Prevent construction of barriers to migration, such as roads or pipelines.
- Monitor development and recreation growth in national parks and surrounding large lakes and rivers to ensure that native fisheries are not being overexploited and habitat damaged through overuse.

Conservation Partners

For contact information, please see appendix G.

- Canadian Arctic Resources Committee
- Canadian Parks and Wilderness Society, Yukon Chapter
- Friends of Yukon Rivers
- Northern Alaska Environmental Center
- Sierra Club, Alaska
- The Wilderness Society
- World Wildlife Fund Canada
- Yukon Conservation Society

Relationship to Other Classification Schemes

This ecoregion combines Gallant et al.'s (1995) ecoregion 102 and Ecological Stratification Working Group's (1995) ecoregion 165. Major delineation features are the steep, high topography and sparse vegetation. Combination deci-

sions were based on Gallant et al. (1995), McNab and Avers (1994), Ecoregions Working Group (1989), and Viereck et al. (1992). The Canadian portion of the Brooks/British Range Tundra is in northern Yukon, extending slightly into the Northwest Territories. The corresponding terrestrial ecoregion is the British-Richardson Mountain region (TEC ecoregion 165) (ESWG 1995). Vegetation in this region is primarily Tundra, with some Boreal Alpine (B.33) (Rowe 1972).

Prepared by R. Hagenstein and T. Ricketts

Key Number:	**108**
Ecoregion Name:	**Arctic Foothills Tundra**
Major Habitat Type:	**6.2 Tundra**
Ecoregion Size:	**123,514 km²**
Biological Distinctiveness:	**Globally Outstanding**
Conservation Status:	**Snapshot—Relatively Stable**
	Final—Relatively Stable

Introduction

The Arctic Foothills Tundra forms a transition between the flat, low-lying Arctic Coastal Tundra to the north and the steep mountainous Brooks/British Range Tundra to the south. It is a series of rounded hills and plateaus that covers 123,512 km² and stretches from the Chukchi Sea east across northern Alaska to the Yukon border. Drainage is better and more defined than in ecoregion [109], the Arctic Coastal Tundra, with less saturated soils and fewer thaw lakes (Gallant et al. 1995). The resulting dominant vegetation consists of mesic graminoid herbaceous (*Eriophorum vaginatum* and *Carex bigelowii*) and dwarf shrub (*Betula nana, Empetrum nigrum, Ledum decumbens, Vaccinium vitis-idaea*) communitites. Open low scrub occurs along some drainages, and a forest community stands along the Noatak River valley, the only forest example in the ecoregion. Wildlife includes caribou, moose, wolf, brown bear, hare, ground squirrel, and waterfowl.

The region experiences an arctic climate, receiving a similar amount of precipitation (140 mm) to that of the Arctic Coastal Plain, but it is somewhat warmer on average than its neighboring ecoregions ([107] or [109]). Thick permafrost underlies the region, and the active layer is generally 1 meter in thickness. Along with the Arctic Coastal Tundra ecoregion to the north, this region was left unglaciated in the Pleistocene and formed part of the extensive Bering Sea Pleistocene refugium (Pielou 1994; summarized from Gallant et al. 1995).

Biological Distinctiveness

The Arctic Foothills Tundra has maintained its full complement of high-level predators and contains important denning sites for brown bear and wolf, which range throughout the ecoregion and neighboring ones. In addition, this ecoregion holds extreme importance for the migration of three distinct caribou herds (Western Arctic, Central Arctic, and Porcupine herds) that cross the region. The Colville River is an extremely important corridor for avian species and moose populations, while the coastal wetlands to the west provide travel routes for many shorebird and waterfowl species. Gyrfalcons, peregrine falcons, and rough-legged hawks breed in very high concentrations along the Colville River bluffs, and several Siberian-Asian bird species commonly breed here.

Conservation Status

This ecoregion is essentially one enormous block of continuous habitat, bisected by one major road (and pipeline) and several more minor tracks. The oil pipeline and accompanying road affect migrations, and local disturbances from mines and settlements are considerable and permanent. Recovery from site disturbances occurs very slowly. Disturbances that destroy the vegetative mat may essentially be permanent. Wildfire disturbance is low, and graminoid and dwarf shrub communities' recovery from fire is generally rapid (Bliss 1988).

Degree of Protection

The only protected area of any size in the ecoregion is the Arctic National Wildlife Refuge, on the far east end. This covers about 15,000 km^2 of the ecoregion. In the west, very small areas are protected in the Alaska Maritime National Wildlife Reserve, the Cape Krusenstern National Monument, and the Noatak National Preserve.

Types and Severity of Threats

The main threats are potential habitat conversion from coal and mineral development in the western half of the ecoregion and continued degradation around existing roads and mines. The current scope of these activities is small, but local impacts can be severe and permanent in the delicate tundra habitats and the vastness of the shallow coal deposits create a significant potential threat. High levels of subsistence hunting and resource use occur, but these are well managed, and wildlife exploitation threats are resultingly low. The trans-Alaska pipeline and Dalton Highway provide easy access to this ecoregion. Once closed to travel by the public, the Dalton Highway now can be traveled without restriction. The increased access to this previously remote area increases the possibilities of overexploitation of wildlife and habitat damage from off-road vehicles.

Suite of Priority Activities to Enhance Biodiversity Conservation

- Protect the Colville River corridor, for avian and moose migration corridors and raptor breeding
- Protect brown bear and wolf denning habitats
- Protect Omalik Lagoon, south of Point Lay, an extremely important location for beluga whales, fish, and subsistence uses
- Prevent the opening of coalfields in the western portion of the ecoregion and protect the habitat values of Omalik Lagoon, the most likely port location for coal shipping.

Conservation Partners

For contact information, please see appendix G.

- Northern Alaska Environmental Center
- The Wilderness Society
- Sierra Club

Relationship to Other Classification Schemes

This ecoregion is identical to Gallant et al.'s (1995) ecoregion 102. Major delineation features are the rolling, hilly topography, better drainage, and dominant mesic herbaceous and dwarf shrub tundra communities.

Prepared by R. Hagenstein and T. Ricketts

Key Number: **109**
Ecoregion Name: **Arctic Coastal Tundra**
Major Habitat Type: **6.2 Tundra**
Ecoregion Size: **103,814 km²**
Biological Distinctiveness: **Globally Outstanding**
Conservation Status: **Snapshot—Relatively Stable**
Final—Relatively Stable

Introduction

The Arctic Coastal Tundra stretches along the northern coasts of Alaska, encompassing the eastern lowlands of Banks Island and Tuktoyaktuk coastal plains in the Yukon Territory, and the lower Anderson and Horton River plains in the Northwest Territories, covering 103,814 km². It is a low, gradually rising plain characterized by poor drainage, wet graminoid herbaceous vegetation communities (*Carex* spp., *Eriophorum* spp.), and many thaw lakes, which cover up to 50 percent of the surface (Gallant et al. 1995), distinguishing this ecoregion from the Low Arctic Tundra ecoregion [110].

The region has arctic climate conditions and is underlain by thick, continuous permafrost. The growing season extends from approximately June 15 to the end of August, although frost can occur in any month. Precipitation is relatively low, ranging from 100 to 300 mm, with occasional higher values in the south, nearer to the taiga plain. Summers are very short and cool, with mean temperatures of 4.5°C on the continental coastal plain to 1°C on western Banks Island. Mean daily minimum temperature in winter is −30°C, and mean daily maximum temperature in summer is 8°C (Gallant et al. 1995). Along with the Arctic Foothills Tundra ecoregion to the south, this region was left unglaciated in the Pleistocene and formed part of the extensive Bering Sea Pleistocene refugium (Pielou 1994; also summarized from Gallant et al. 1995).

All of the ecoregion is low (between 0 and 150 m asl) and wet (25–50 percent wetland).

Also, permafrost is deep and continuous throughout the area, with very high ice content and abundant ice wedges (pingos are also common under the continental coastal plains). On the mainland, glaciation has resulted in level to rolling fluvioglacial, colluvial, and morainal deposits, in addition to marine deposits. Undulating glacial drift and outwash deposits are also common. Banks Island is unglaciated and is characterized by low rolling hills with western-sloping, eroded terraces. It is worth noting that this ecoregion contains the Mackenzie River delta and the distinctive landforms associated with it, which include active alluvial channels, estuarine deposits, and innumerable lakes.

Biological Distinctiveness

Mainland coastal plain areas support a continuous cover of shrubby tundra vegetation, consisting of dwarf birch, willow, northern Labrador tea (*Dryas* spp.), and sedge tussocks. Wetland vegetation is common throughout the ecoregion, including horizontal and low-center lowland polygon fens with small elevated peat mound bogs and marshes along the coast. Sedges and grasses predominate in the wet soils, but in some slightly raised, better-drained areas they give way to dwarf shrub communities. The cover of mosses on the wet soils is nearly continuous and includes aulaomnium, ditrichum, calliefgon, and others (Bliss 1988). Warm sites support tall dwarf birch, willow, and alder, particularly around the Mackenzie delta and the Yukon coastal plain.

The ecoregion supports a wide variety of wildlife, including walrus (*Odobenus resmarus*), beluga whale (*Delphinapterus leucas*), polar bear (*Thalarctos maritimus*), arctic char (*Savelinus alpinus*), caribou (*Rangifer tarandus*), snowshoe hare (*Lepus americanus*), red fox (*Vulpes fulva*), wolf (*Canis lupus*), arctic hare (*Lepus arcticus*), arctic ground squirrel (*Citellus parryi*), seal (Phocidae), seabirds, and waterfowl. Musk ox (*Ovibos moschatus*) are common on Banks Island (McNab and Avers 1994).

The entire Arctic Coastal Plain is an important breeding and calving ground for many species. Three major caribou herds, Western Arctic, Central Arctic, and Porcupine herds, migrate here annually to calve, while many species of waterfowl, passerines, and shorebirds arrive in the summer to breed. The Colville

Delta and Teshekpuk Lake areas are especially important for breeding birds. The coast of the Arctic National Wildlife Refuge (ANWR) in particular is extraordinarily rich and productive, especially considering its relatively constricted area. It represents the center of distribution for musk ox populations in Alaska. Snow geese (*Chen caerulescens*) breed in small areas in this ecoregion, which contains their only breeding sites in Alaska. Kasegaluk Lagoon, in the western section of the ecoregion, is a critical staging area for brant (*Branta bernicla*) and supports 2,500–3,500 beluga whales each summer. The coastal plain, especially in the Colville Delta area, is important for spectacled and king eiders (*Somateria fischeri* and *S. spectabilis*), and yellow-billed loons (*Gavia adamsii*)—all species of concern.

Conservation Status

Over 90 percent of the Arctic Coastal Plain habitat remains intact. The disturbance that has occurred has centered around the community of Barrow and is associated with oil development in the Prudhoe Bay and Kuparuk oil fields. The transportation corridor that includes the Dalton Highway and the Trans-Alaska Pipeline also represents a significant area of disturbance due to direct habitat loss, potential deflection of migratory animals, and increased human access. Oil field development is currently expanding to the west. Arco recently announced a significant discovery in the Colville Delta, and interest in expansion into the National Petroleum Reserve in Alaska is at a high. The oil industry and Alaska's political leadership have a significant interest in opening the coastal plain of the Arctic National Wildlife Refuge to exploration and development. These expansions of industrial activity have potential to interrupt migration routes through an already existing natural "bottleneck" in the coastal plain ecoregion, and to disturb nesting habitats of waterfowl along the Arctic Coastal Plain. Other disturbances are usually local and small in scale, although recovery from vehicle or even human track disturbance is very slow, and constructs such as roads and pipelines can disrupt mammal movements and migrations.

Degree of Protection

- The Arctic National Wildlife Refuge represents the only large protected area in the ecoregion. The portion of the refuge that falls in this ecoregion was designated as a special study area when the refuge was created in 1980 to evaluate the potential for oil and gas development. Although the area is technically within a designated protected area, its future is not secure and may be threatened by potential expansion of oil and gas development.
- Teshekpuk Lake, important for bird migration and breeding, is a Special Management Area within the National Petroleum Reserve in Alaska.
- Ivvavik National Park in northwestern Yukon Territory (only partially in ecoregion) comprises 10,100 km², third largest in Canada. The park is the calving ground for the 160,000 strong Porcupine caribou herd, and its wetlands provide habitat for migrating birds and other wildlife.

Types and Severity of Threats

The main threat is the expansion of the Prudhoe Bay oil field development around the bay and into the Arctic National Wildlife Refuge coastal plain, as well as to the west into the Colville Delta and National Petroleum Reserve in Alaska. This expansion would significantly erode wilderness values and has the potential to disrupt wildlife habitat and use patterns.

Suite of Priority Activities to Enhance Biodiversity Conservation

- Ensure long-term protection for the coastal plain of the Arctic National Wildlife Refuge, to protect caribou migration and calving, bear and wolf movements and foraging, and musk ox activity. These will be increasingly impacted with expansion of oil field development. Manage areas around the reserve to ensure persistence of large-scale phenomena, like migrations.
- Create protected areas in biologically important parts of the National Petroleum Reserve in Alaska, including areas adjacent to the Colville Delta, Teshekpuk Lake, and Kasegaluk Lagoon. Monitor bird populations

in these areas in response to oil development activities.

- Restore caribou calving areas in Prudhoe Bay and Kuparuk oil fields if current indices of decline and stress continue.
- Ensure that the entire coast is managed to retain its ability to support the annual concentrations of migratory waterfowl, which occur throughout the ecoregion.

Conservation Partners

For contact information, please see appendix G.

- Canadian Arctic Resources Committee
- Canadian Parks and Wilderness Society, Yukon Chapter
- Ecology North
- Friends of Yukon Rivers
- Northern Alaska Environmental Center
- Sierra Club
- The Wilderness Society
- World Wildlife Fund Canada

Relationship to Other Classification Schemes

This ecoregion combines Gallant et al.'s (1995) ecoregion 101 with Ecological Stratification Working Group's (1995) ecoregions 14, 32, 33, and 34. Major delineation features are the low topography, poor drainage, dominant wet tundra communities, and coastal position. Combination decisions were based on Gallant et al. (1995), McNab and Avers (1994), Ecoregions Working Group (1989), and Viereck et al. (1992).

Prepared by R. Hagenstein, M. Sims, G. Mann, and T. Ricketts

Key Number:	**110**
Ecoregion Name:	**Low Arctic Tundra**
Major Habitat Type:	**6.2 Tundra**
Ecoregion Size:	**796,674 km²**
Biological Distinctiveness:	**Globally Outstanding**
Conservation Status:	**Snapshot—Relatively Intact**
	Final—Relatively Intact

Introduction

This ecoregion covers a significant east-to-west expanse. Much of it falls within the mainland boundaries of the Northwest Territories, although it also includes part of Quebec. Specifically, it stretches from the Dease Arm Plain in its northwestern reaches, through the plains and lowlands of the central Northwest Territories, to the Ottawa and Belcher Islands in Hudson Bay and the southern and western sections of Southampton Island, to the central Ungava peninsula in northern Quebec.

The great majority of this ecoregion is classified as having a low arctic ecoclimate. The only exceptions are the Dease Arm Plain, stretching from Great Bear Lake to the east side of the Mackenzie River delta, and the Belcher Islands, in the southeastern part of Hudson Bay. The ecoregion experiences short, cool summers, with mean local temperatures ranging from 4°C to 6°C. Winters are long and very cold, with mean local temperatures from –28°C in the northwest to –17.5°C in the southeast. Mean annual precipitation varies widely across the ecoregion, from 200 mm in the northwest around the Amundsen Gulf coast, to 500 mm in northern Quebec. Particular local effects are evident in certain portions of the region, especially in the offshore island systems in the Coronation Gulf and Hudson Bay, where open water in the late summer and early fall moderates local climate and creates drizzly, foggy seasonal weather (ESWG 1995).

The physiography of this ecoregion varies considerably due to its size. The ecoregion is largely underlain by Precambrian granitic bedrock. The terrain consists mostly of broadly rolling uplands and lowlands. Throughout the ecoregion, there are exposures of bedrock. Strung out across the landscape are long, sinuous eskers reaching lengths of up to 100 km in places. A small part of the ecoregion west of the Firth River is unglaciated. The undulating landscape is studded with numerous lakes, ponds, and wetlands. Virtually the entire area is underlain by continuous permafrost, with active layers that are usually moist or wet throughout the summer (ESWG 1995).

Biological Distinctiveness

The ecoregion is characterized by a continuous cover of shrubby tundra vegetation. Permafrost is continuous with low to high ice content, except for discontinuous permafrost in the Ottawa Islands ecoregion and Belcher Island ecoregion. Vegetation in some areas can be described as tundra-subarctic forest transition, including black spruce (*Picea mariana*), tamarack (*Larix laricina*), white spruce (*P. glauca*), dwarf birch (*Betula* spp.), willow (*Salix* spp.), and heath species. In addition, herb and lichen species are very common, in mixture with other vegetation. Some major river valleys support outlying spruce growth, and wetlands are very common in lowlands throughout the ecoregion. Much of the area represents a vegetative transition between the taiga forest to the south and the treeless arctic tundra to the north (ESWG 1995).

There is a great diversity of mammal species that inhabit this ecoregion. Notably, it is the major summer range and calving grounds of some of Canada's largest caribou herds (*Rangifer tarandus*), barren-ground caribou in the west and woodland caribou in the east. Other mammals include grizzly bear (*Ursus arctos*), black bear (*U. americanus*) in northern Quebec, polar bears (*Thalarctos maritimus*) in coastal areas, wolf (*Canis lupus*), moose (*Alces alces*), arctic ground squirrel (*Citellus parryi*), and brown lemming (*Lemmus trimucronatus*). Many migratory birds depend on the ecoregion for primary breeding and nesting grounds. Representative species include the yellow-billed, red-throated, and arctic loons (*Gavia adamsii, G. stellata,* and *G. arctica*), tundra swan (*Cygnus columbianus*), snow goose (*Chen caeruleswcens*), old-squaw (*Clangula hyemalis*), gyrfalcon (*Falco rusticolus*), willow and rock ptarmigan (*Lagopus lagopus* and *L. mutus*), red-necked phalarope (*Phalaropus lobatus*), parasitic jaeger (*Stercorarius parasiticus*), snowy owl (*Nyctea scandiaca*), hoary redpoll (*Carduelis hornemanni*), and snow bunting (*Plectrophenax nivalis*). In the marine environment, typical species include walrus (*Odobenus rosmarus*), seal (Phocidae), beluga whale (*Delphinapterus leucas*), and narwhal (*Monodon monoceros*) (ESWG 1995).

Significant ecological phenomena within this ecoregion include several barren-ground caribou (*Rangifer tarandus* ssp. *arcticus*) herds totaling 1.5 million animals. Globally, the world's total breeding populations of snow and Ross's goose (*Chen rossii*) nesting colonies are found here. In addition, the largest mainland musk ox (*Ovibos moschatus*) population and only "blonde musk oxen" are found in this area.

Conservation Status

Habitat Loss and Degradation

Approximately 95 percent of the natural habitat of this ecoregion is estimated to be intact. There is terrain disturbance in the vicinity of communities, abandoned military sites, and areas with mining exploration and active mines. Some habitats, such as raised beach ridges and areas of glacial outwash, are more threatened, as these gravel deposits are sought after for road, airstrip, and other construction needs.

Remaining Blocks of Intact Habitat

Most of the ecoregion is intact.

Degree of Fragmentation

Caribou seasonal range (behavioral, not habitat) fragmentation is conceivable for the Bathurst herd as a cumulative effect from mining exploration sites and mine construction.

Degree of Protection

- Thelon Wildlife Sanctuary—Northwest Territories, 52,925 km^2
- Tuktut Nogait National Park—Northwest Territories, 22,145 km^2

Conservation sites with lower protection standards include:

- Queen Maud Gulf Migratory Bird Sanctuary (MBS)—Northwest Territories, 62,782 km^2
- Harry Gibbons MBS—Northwest Territories, 1,489 km^2
- McConnell River MBS—Northwest Territories, 329 km^2

Types and Severity of Threats

General threats to arctic regions apply, such as atmospheric fallout resulting in heavy metal and pesticide pollution. There is a risk of site-specific oil and chemical spills and tailing effluent escapes. Mining is a rapidly growing industrial threat in parts of this ecoregion, especially with respect to diamonds and copper. Associated road building is in the planning stages and may be a significant threat for some species and habitats. Ecotourism will need to be carefully managed in order that nesting bird colonies, other sensitive wildlife species, and caribou calving grounds are not disturbed. With increased access, overhunting of caribou is a possibility, and commercial harvesting of both caribou and musk oxen needs to be carefully monitored and controlled. In addition, carnivore deaths resulting from human defense and nuisance kills (especially of wolverine [*Gulo luscus*] and grizzly bears) may further impact predator-prey dynamics.

Suite of Priority Activities to Enhance Biodiversity Conservation

- Upgrade protection standards for Queen Maud Gulf MBS, Northwest Territories, specifically regulations pertaining to mining and mineral exploration.
- Protect two candidate sites (one provincial park and one ecological reserve) in Quebec natural region #25.
- Protect caribou calving grounds by:
  - establishing land-use regulations for caribou migration routes and corridors, especially post-calving aggregations;
  - establishing a buffer zone to connect the Beverly caribou calving grounds to the Thelon National Wildlife Sanctuary and the Queen Maud Gulf MBS;
  - introducing regulations to reduce the removal of gravel-laden landforms—critical habitats for many arctic species during at least a part of their seasonal movements—for road fill and airstrips.
- Adjust wildlife management plans to accommodate losses (of both individuals and important habitat) due to increasing human encroachment, especially with respect to mining and related activities.

Conservation Partners

For contact information, please see appendix G.

- Canadian Arctic Resources Committee
- Canadian Nature Federation
- Ecology North

- Inuvialuit Game Council, Joint Secretariat
- Kativik Aboriginal Government
- Société Makivik
- World Wildlife Fund Canada, Quebec Region

Relationship to Other Classification Schemes

The Low Arctic Tundra extends across northern Northwest Territories and Quebec and along the Ottawa Islands. The terrestrial ecoregions within this area are the following: Dease Arm Plain, Coronation Hills, Bluenose Lake Plain, Bathurst Hills, Queen Maud Gulf Lowland, Chantrey Inlet Lowland, Takijuq Lake Upland, Garry Lake Lowland, Back River Plain, Dubawnt Lake Plain/Upland, Maguse River Upland, Southampton Island Plain, Central Ungava Peninsula, Ottawa Islands, and Belcher Islands (TEC ecoregions 35–49) (ESWG 1995). This expansive region is almost entirely tundra vegetation but includes some transitional subarctic-tundra forest as well (Rowe 1972).

Prepared by A. Gunn, S. Oosenbrug, C. O'Brien, N. Zinger, K. Kavanagh, M. Sims, and G. Mann

Key Number:	**111**
Ecoregion Name:	**Middle Arctic Tundra**
Major Habitat Type:	**6.2 Tundra**
Ecoregion Size:	**1,033,010 km²**
Biological Distinctiveness:	**Bioregionally Outstanding**
Conservation Status:	**Snapshot—Relatively Intact**
	Final—Relatively Intact

Introduction

This ecoregion is very large, stretching from the Banks Island lowlands in the west to the eastern tip of Baffin Island in the east. Principally in the Northwest Territories, it includes a small portion of extreme northern Quebec.

This ecoregion is described as having a predominantly mid-arctic ecoclimate. It incorporates the coldest and driest landscapes in Canada. Summer mean temperatures range from 0.5°C in the north-central part of the ecoregion to 4.5°C in the south-central area.

Winter mean temperatures vary between –30°C in the northern and western part of the ecoregion and –20°C in the northern Ungava Peninsula. Mean annual precipitation is very low, 100–200 mm, with occasional higher values in the southeastern portion of the ecoregion. These measures are the lowest in Canada, and this ecoregion is often referred to as a polar desert. Snow covers the ground for at least ten months of the year (ESWG 1995).

The western portion of the ecoregion is underlain by flat-lying Paleozoic and Mesozoic sedimentary bedrock and consists of lowland plains covered with glacial moraine, marine deposits, and bedrock outcrops. East of Prince of Wales and Somerset Islands, the terrain is composed mainly of Precambrian granitoid bedrock, and tends to consist of plateaus and rock hills. The Arctic Islands circumscribe a variety of oceanic conditions. In the northern half of the ecozone, the waters are ice fast, even through the summer. Toward the south, open waters are more common in the summer, but pack ice usually persists offshore. The permafrost is continuous and may extend to depths of several hundred meters (ESWG 1995).

Biological Distinctiveness

Because of high winds, harsh climate, and shallow, poor soils, vegetation is sparse and dwarfed. Although the area supports some shrub life (arctic willow [*Salix arctica*], for example), plant life is characterized by a continuous or discontinuous cover of herb (saxifrage [*Saxifraga* spp.], *Dryas* spp.) and lichen-dominated vegetation. The size of any shrubs decreases quickly as one moves to the north. In the east, the Meta Incognita Peninsula, Wager Bay Plateau, and Northern Ungava Peninsula are described as having a low arctic ecoclimate where shrub tundra vegetation dominates, while tall dwarf birch (*Betula* spp.), willow (*Salix* spp.), and alder (*Alnus* spp.) occur on warm sites. Vegetative cover tends to be greater on wetter sites confined to coastal lowlands, sheltered valleys, and moist, nutrient-rich corridors along rivers and streams (ESWG 1995).

Characteristic mammalian wildlife include barren-ground caribou (*Rangifer tarandus* ssp. *arcticus*), musk ox (*Ovibos moschatus*), wolf (*Canis lupus*), arctic fox (*Alopex lagopus*), polar bear (*Thalarctos maritimus*), arctic hare (*Lepus*

arcticus), brown lemming (*Lemmus trimucronatus*), and collared lemming (*Dicrostonyx groenlandicus*). Peary caribou (*Rangifer tarandus* ssp. *pearyi*) are found on high Arctic Islands. The ecoregion provides a major breeding ground for many migratory bird species, including snow goose (*Chen caerulescens*), brant (*Branta bernicla*), Canada goose (*Branta canadensis*), eider (*Somateria* spp.) and old-squaw duck (*Clangula hyemalis*). Other birds include red-throated loon (*Gavia stellata*), gyrfalcon (*Falco rusticolus*), willow and rock ptarmigan (*Lagopus lagopus* and *L. mutus*), red phalarope (*Phalaropus fulicaria*), parasitic and long-tailed jaeger (*Stercorarius parasiticus* and *S. longicaudus*), snowy owl (*Nyctea scandiaca*), and snow bunting (*Plectrophenax nivalis*). Marine species, which are found in much higher concentration at the eastern and western coastal margins of the ecoregion, include beluga whale (*Delphinapterus leucas*), walrus (*Odobenus rosmarus*), seal (Phocidae), and narwhal (*Monodon monoceros*) (ESWG 1995).

Significant ecological features of this ecoregion include 80 percent of the world's musk ox population, including the highest density (in Minto Inlet and Thompson valley); global range of Arctic Island caribou (similar to Peary caribou but its taxonomic designation is unresolved); global range (with ecoregion [112]) of high arctic wolf [*Canis lupus arctos*]; calving grounds for six caribou herds on Northwest Territories mainland and others on the islands; and polar bear denning hot spots.

Conservation Status

Habitat Loss and Degradation

Approximately 95 percent of the natural habitat of this ecoregion is estimated to be intact. There is terrain disturbance in the vicinity of communities, abandoned military sites, and mining exploration. Some habitats, such as raised beach ridges and areas of glacial outwash, are more threatened, as these gravel deposits are sought after for road, airstrip, and other construction needs.

Remaining Blocks of Intact Habitat
Most of the ecoregion is intact.

Degree of Fragmentation

This ecoregion is naturally fragmented due to the large number of islands that make up its composition. For much of the year, however, sea ice links the islands, permitting dispersal of large mammals. Caribou seasonal ranges can be interrupted by high-frequency human activity (i.e., behavioral versus habitat fragmentation—this likely would be significant for assessing cumulative effects).

Degree of Protection

- Aulavik National Park—Northwest Territories, 12,200 km^2
- Banks Island No. 1 Migratory Bird Sanctuary—western Northwest Territories, 20,518 km^2
- Dewey Soper MBS—Northwest Territories, 8,159 km^2
- East Bay MBS—Northwest Territories, 1,166 km^2
- Bowman Bay Wildlife Sanctuary—Northwest Territories, 1,079 km^2
- Cape Dorset MBS—Northwest Territories, 259 km^2
- Banks Island No. 2 MBS—western Northwest Territories, 142 km^2 (part of Aulavik National Park)
- Katannilik Territorial Park—Northwest Territories, 15 km^2
- Cape Parry MBS—Northwest Territories, 3 km^2

Types and Severity of Threats

Threats are relatively minor. One threat is atmospheric fallout, resulting in heavy metal and pesticide pollution. There is a risk of oil spills in coastal areas. Ecotourism will need to be carefully managed in order that nesting bird colonies, other sensitive wildlife species, and caribou calving grounds are not disturbed. With increased access, overhunting of caribou is a possibility, and commercial harvesting of both caribou and musk oxen needs to be carefully monitored and controlled.

Suite of Priority Activities to Enhance Biodiversity Conservation

- Complete designation of Wager Bay National Park, Northwest Territories.
- Upgrade protection standards for both migratory bird sanctuaries and territorial parks to permanently exclude mineral exploration and mine development.
- Protect Cap Wolstenholme and Mount de Povungnituk candidate protected areas in Quebec.
- Protect caribou calving grounds from human encroachment and habitat loss. Important calving grounds include:
  - Arctic Island/Peary Type: Banks Island, northwest Victoria Island, Prince of Wales Somerset, south and east Victoria Islands, Boothia Peninsula
  - Pelly Bay, Keith Bay, Wager Bay (barren-ground herds)
  - Lorriland, North and South Melville
  - Baffin and Southhampton
- Protect Minto Inlet, a musk ox refugium and an arctic island wolf denning site. Protection is needed from overharvesting and wetland loss.
- Protect polar bear denning hot spots (including Gateshead Island, north Pelly Bay, and Wager Bay) from human disturbance.

Conservation Partners

For contact information, please see appendix G.

- Canadian Arctic Resources Committee
- Canadian Nature Federation
- Ecology North
- Kativik Aboriginal Government
- Nunavut Wildlife Management Board
- Société Makivik
- World Wildlife Fund Canada, Quebec Region

Relationship to Other Classification Schemes

The Middle Arctic Tundra incorporates many different ecoregions across the Northern Arctic ecozone. These regions include: Banks Island Lowland, Amundsen Gulf Lowlands, Shaler

Terrestrial Ecoregions of North America

Mountains, Victoria Island Lowlands, Prince of Wales Island Lowland, Boothia Peninsula Plateau, Gulf of Boothia Plain (TEC ecoregions 15–21), Melville Peninsula Plateau (TEC ecoregion 23), Foxe Basin Plain, Pangnirtung Upland, Hall Peninsula Upland, Meta Incognita Peninsula, Baffin Upland, Wager Bay Plateau, and Northern Ungava Peninsula (TEC ecoregions 25–31).

Prepared by A. Gunn, S. Oosenbrug, C. O'Brien, N. Zinger, K. Kavanagh, M. Sims, and G. Mann

Key Number:	**112**
Ecoregion Name:	**High Arctic Tundra**
Major Habitat Type:	**6.2 Tundra**
Ecoregion Size:	**463,688 km²**
Biological Distinctiveness:	**Nationally Important**
Conservation Status:	**Snapshot—Relatively Intact**
	Final—Relatively Intact

Introduction

This ecoregion encompasses most of the northern Arctic archipelago, including most of Axel Heiberg and Ellesmere Islands, a significant portion of Baffin Island, the majority of Somerset Island, and all of the remaining Queen Elizabeth Islands.

The entire ecoregion is classified as having a high arctic ecoclimate. The climate is very dry and cold, and mean summer temperatures in the north of the ecoregion, in the Sverdrup and Parry Islands, are as low as –1.5°C (and as high as 2°C in the south, on the Lancaster Plateau in western Baffin Island). Winter mean temperatures range between –32°C in the north and –23°C in the south. Precipitation is also very low, ranging mostly between 100 mm and 200 mm, with a few sites registering greater values (up to 400 mm) in the Baffin Island Uplands. The northern portions of the ecoregion—Ellesmere Island and the Parry and Sverdrup Island groups—generally receive less precipitation than anywhere else in Canada, sometimes as little as 50 mm in a single year (ESWG 1995).

The physiography of this ecoregion varies widely across the landscape. The western portion is underlain by flat-lying Paleozoic and Mesozoic sedimentary bedrock. The landscape

is rugged and mountainous on the north and west coasts of Ellesmere and Axel Heiberg Islands, where ice-covered mountains reach 2,500 m in elevation. To the east and south, the landscape consists mostly of lowland plains covered with glacial moraine, marine deposits, and bedrock outcrops. It is in some cases severely ridged, as on the Parry Island plateau. East of Prince of Wales and Somerset Islands, the terrain is composed mainly of Precambrian granitoid bedrock and tends to consist of plateaus and rocky hills. Most of the Queen Elizabeth Islands lie between 0 m and 1,000 m asl, with greater elevations found in the east (between 300 m and 1,000 m), on the Lancaster and Borden Peninsula Plateaus. The Arctic Islands circumscribe a variety of oceanic conditions. In the northern half of the ecoregion, the waters are ice fast, even through the summer. Toward the south, open waters are more common in the summer, but pack ice usually persists offshore. The permafrost is continuous and may extend to depths of several hundred meters (ESWG 1995).

Biological Distinctiveness

Plant species must be very robust to survive in this ecoregion's climate. In the north, clumps of moss, lichen, and cold-hardy vascular plants such as sedge (*Carex* spp.) and cottongrass (*Eriophorum* spp.) are the dominant vegetation. Arctic willow (*Salix arctica*) and *Dryas* spp. occur infrequently, and mixed, low-growing herbs like purple saxifrage (*Saxifraga oppositifolia*), *Kobresia* spp., and arctic poppy (*Papaver radicadum*) can sometimes be found at lower elevations. In the southeast, the dominant vegetation is generally very similar, but, because of some milder climate, wet areas can develop up to about 60 percent of wood rush (*Luzula* spp.), wire rush (*Juncus* spp.), and saxifrage (*Saxifraga* spp.), with a nearly continuous cover of mosses. In all other instances, the vegetative cover is sparse and discontinuous (ESWG 1995).

Wildlife characteristic of the entire region include musk ox (*Ovibos moschatus*), arctic hare (*Lepus arcticus*), arctic fox (*Alopex lagopus*), and caribou (*Rangifer tarandus*). Polar bears (*Thalarctos maritimus*) are common in coastal areas. Representative birds include king eider (*Somateria spectabilis*), rock ptarmigan (*Lagopus mutus*), northern fulmar (*Fulmarus glacialis*), plover (*Charadrius* spp. and *Pluvialis* spp.), hoary redpoll (*Carduelis hornemanni*), snow bunting (*Plectrophenax nivalis*), and other seabirds. Marine mammals include walrus (*Odobenus rosmarus*), seal (Phocidae), and a variety of whales (Cetacea). In addition, gyrfalcon (*Falco rusticolus*), jaeger (*Stercorarius* spp.), snowy owl (*Nyctea scandiaca*), and narwhal (*Monodon monoceros*) are more common in the south around the Lancaster Plateau and Sound (ESWG 1995).

Significant ecological features of this ecoregion include the global range of the endangered Peary caribou (*Rangifer tarandus* ssp. *pearyi*) and the global distribution of polar desert habitats with globally unique species assemblages of plants, vertebrates, and mammals. This is one of only two ecoregions (the other is [111]) with viable populations of high arctic wolves (*Canis lupus arctos*). The ecoregion also includes nesting sites for Ross's gulls (*Rhodostethis rosea*) and ivory gulls (*Pagophila eburnea*).

Conservation Status

Habitat Loss and Degradation

At least 98 percent of this ecoregion is considered to remain intact. Very small areas of habitat loss are attributed to coastal communities and terrain disturbance in their immediate vicinities. There are also abandoned oil and gas camps and seismic lines.

Remaining Blocks of Intact Habitat
The ecoregion can be considered as intact.

Degree of Fragmentation
This ecoregion is naturally fragmented due to the large number of islands that make up its composition. For much of the year, however, sea ice links the islands, permitting dispersal of large mammals.

Degree of Protection

- Polar Bear Pass National Wildlife Area—northern Northwest Territories, 2,624 km²
- Bylot Island Migratory Bird Sanctuary—northern Northwest Territories, 1,087 km²
- Prince Leopold Island Migratory Bird Sanctuary—northern Northwest Territories, 504 km²

- Seymour Island Migratory Bird Sanctuary—northern Northwest Territories, 8 km²

Types and Severity of Threats

Threats are relatively minor. One threat is atmospheric fallout, resulting in heavy metal and pesticide pollution. There is a risk of oil spills in coastal areas. Ecotourism will need to be carefully managed in order that nesting bird colonies, other sensitive wildlife species, and caribou calving grounds are not disturbed. With increased access, overhunting of caribou is a possibility. This is of particular concern for the population of Peary caribou; in the last few years there has been a precipitous decline in population levels of this endangered animal.

Suite of Priority Activities to Enhance Biodiversity Conservation

- Complete designation of Northern Bathurst Island National Park, ensuring that the area includes Peary caribou calving grounds and that management plans are developed to prevent disturbance on caribou summer areas.
- Protect Bailey Point (Melville Island). This is an important refugium for musk ox and a plant refugium.
- Develop and implement a recovery plan for Peary caribou.

Conservation Partners

For contact information, please see appendix G.

- Canadian Arctic Resources Committee
- Ecology North
- Nunavut Wildlife Management Board
- World Wildlife Fund Canada

Relationship to Other Classification Schemes

The High Arctic Tundra ecoregion includes the Ellesmere Mountains (TEC ecoregions 8 and 10); the Eureka Hills, which are within the Ellesmere Mountains (TEC ecoregion 9); the Sverdrup Islands Lowland to the southwest (TEC ecoregion 11); the Parry Islands Plateau in the south (TEC ecoregion 12) and the

Lancaster (TEC ecoregion 13) and Borden Peninsula Plateaus (TEC ecoregion 22) and Baffin Island Uplands (TEC ecoregion 24) in the east (ESWG 1995).

Prepared by A. Gunn, S. Oosenbrug, C. O'Brien, K. Kavanagh, M. Sims, and G. Mann

Key Number:	**113**
Ecoregion Name:	**Davis Highlands Tundra**
Major Habitat Type:	**6.2 Tundra**
Ecoregion Size:	**87,882 km²**
Biological Distinctiveness:	**Nationally Important**
Conservation Status:	**Snapshot—Relatively Intact, Final—Relatively Intact**

Introduction

Rugged topography distinguishes this ecoregion from the High Arctic Tundra ecoregion [112].

This ecoregion has a high arctic and oceanic high arctic ecoclimate. A humid, extremely cold climate is marked by very short, cold summers. Mean annual temperature is –11.5°C. Mean summer temperature is 1°C, and mean winter temperature is –23°C. Mean annual precipitation is 200–400 mm overall, with 400–600 mm centering around the Cumberland Peninsula (ESWG 1995).

This ecoregion is comprised of the Baffin Mountains, an elevated belt of deeply dissected crystalline rocks that extend along the northeastern flank of Baffin and Bylot Islands. Ice-capped mountains reach 1,525–2,135 m asl. Sloping gently westward, the ecoregion's general aspect is one of a broad, gently warped, old erosion surface etched by erosion along joint systems and zones of weakness. Long arms of the sea penetrate as glacier-filled sounds or fjords; some cut through highlands to Baffin upland to the east. The ecoregion is underlain by deep, continuous permafrost with low ice content. Bare bedrock is common (ESWG 1995).

Biological Distinctiveness

The dominant vegetation is a discontinuous cover of mosses, lichens, and cold-hardy vascular plants such as sedge (*Carex* spp.) and cottongrass (*Eriophorum* spp.).

Characteristic wildlife include arctic hare (*Lepus arcticus*), arctic fox (*Alopex lagopus*), and caribou (*Rangifer tarandus*). Polar bears (*Thalarctos maritimus*) are common in coastal areas. Representative birds include king eider (*Somateria spectabilis*), rock ptarmigan (*Lagopus mutus*), northern fulmar (*Fulmarus glacialis*), plover (*Charadrius* spp. and *Pluvialis* spp.), hoary redpoll (*Carduelis hornemanni*), and snow bunting (*Plectrophenax nivalis*). Marine mammals include walrus (*Odobenus rosmarus*), seal (Phocidae), and a variety of whales (Cetacea) (ESWG 1995).

Among the many ecologically significant features this ecoregion includes are: snow goose (*Chen caerulescens*) nesting colonies (Bylot Island), one of the largest colonies globally; and nesting cliffs at Bylot Island for major colonies of thick-billed murres (*Uria lomvia*) and black-legged kittiwakes (*Riss tridactyla*). A large proportion of the polar bears in the Northwest Territories–Greenland shared population is found in this area in the summer along the coast, and denning and maternal denning sites are found inland. Caribou calving sites (relatively undescribed) are found in higher elevations and seasonal migration to summer feeding in areas in valleys.

Conservation Status

Habitat Loss and Degradation

At least 98 percent of this ecoregion is considered to remain intact. Very small areas of habitat loss are attributed to coastal communities and terrain disturbance in their immediate vicinities.

Remaining Blocks of Intact Habitat

The ecoregion can be considered as intact.

Degree of Fragmentation

The region is not fragmented.

Degree of Protection

- Part of Auyuittuq National Park—Northwest Territories, 21,471 km²
- Bylot Island Migratory Bird Sanctuary (not as highly protected)—Northwest Territories, 259 km²

Types and Severity of Threats

Threats are relatively minor. One is atmospheric fallout, resulting in heavy metal and pesticide pollution. There is a risk of oil spills in coastal areas. Ecotourism will need to be carefully managed in order that nesting bird colonies, other sensitive wildlife species, and caribou calving grounds are not disturbed. With increased access, overhunting of caribou is a possibility.

Suite of Priority Activities to Enhance Biodiversity Conservation

• Complete designation of North Baffin Island National Park.
• Develop management plans and set specific recommendations to protect caribou calving sites.

Conservation Partners

For contact information, please see appendix G.

• Canadian Arctic Resources Committee
• Ecology North
• Nunavut Wildlife Management Board
• World Wildlife Fund Canada

Relationship to Other Classification Schemes

The Baffin Mountains (TEC ecoregion 5) extend along the Davis Highlands Tundra (ESWG 1995).

Prepared by A. Gunn, S. Oosenbrug, C. O'Brien, K. Kavanagh, M. Sims, and G. Mann

Key Number:	**114**
Ecoregion Name:	**Baffin Coastal Tundra**
Major Habitat Type:	**6.2 Tundra**
Ecoregion Size:	**9,103 km²**
Biological Distinctiveness:	**Nationally Important**
Conservation Status:	**Snapshot—Relatively Stable**
	Final—Relatively Stable

Introduction

This ecoregion, characterized primarily as a coastal plain, is situated on the central north coast of Baffin Island and is relatively small in area.

The ecoregion is characterized by a high arctic ecoclimate. The humid, cold arctic climate is marked by short, cold summers (mean temperature 1°C) and long, cold winters (mean temperature −22.5°C). The mean annual temperature is around −11.5°C, and mean annual precipitation ranges between 200 mm and 300 mm (ESWG 1995).

The ecoregion is geologically composed of crystalline Precambrian massive rocks that occur as isolate outliers from peninsulas, and fjords that extend out from the Davis Highlands. It is a gently warped, old erosion surface with permafrost of low ice content (ESWG 1995).

Biological Distinctiveness

The ecoregion has a sparse vegetative cover of mixed low-growing herbs and shrubs, consisting of moss, purple saxifrage (*Saxifraga oppositifolia*), Dryas, arctic willow (*Salix arctica*), *Kobresia*, sedge (*Carex* spp.), and arctic poppy (*Papaver radicadum*). Wet sites can develop up to about 60 percent cover of wood rush (*Luzula* spp.), wire rush (*Juncus* spp.), and saxifrage (*Saxifraga* spp.), along with a nearly continuous cover of mosses.

Characteristic wildlife includes arctic hare (*Lepus arcticus*), arctic fox (*Alopex lagopus*), lemming (Lemminae), and caribou (*Rangifer tarandus*). Polar bears (*Thalarctos maritimus*) are common in the coastal areas; a large proportion of the Northwest Territories–Greenland shared population of polar bears use this coastal ecoregion in summer. Representative birds include king eider (*Somateria spectabilis*), rock ptarmigan (*Lagopus mutus*), northern fulmar (*Fulmarus glacialis*), plover (*Charadrius* spp. and *Pluvialis* spp.), hoary redpoll (*Carduelis hornemanni*), and snow bunting (*Plectrophenax nivalis*) (ESWG 1995).

Conservation Status

Habitat Loss and Degradation

The ecoregion is still intact.

Remaining Blocks of Intact Habitat

Although naturally dissected by fjords, the ecoregion habitat is intact.

Degree of Fragmentation

The highly indented coastline means that the eastern Baffin coast is naturally fragmented. This land pattern could mean that the ecoregion might be vulnerable to anthropogenic fragmentation from any future human land use.

Degree of Protection

There are no protected areas.

Types and Severity of Threats

There are relatively minor threats. Atmospheric fallout resulting in heavy metal and pesticide pollution is among the threats. There is also a risk of oil spills in this coastal area.

Suite of Priority Activities to Enhance Biodiversity Conservation

- Establish a representative protected area.
- Develop a wildlife management plan.

Conservation Partners

For contact information, please see appendix G.

- Canadian Arctic Resources Committee
- Ecology North
- Nunavut Wildlife Management Board
- World Wildlife Fund Canada

Relationship to Other Classification Schemes

The Baffin Island Coastal Lowlands (TEC ecoregion 6) run along the northeast coast of Baffin Island.

Prepared by A. Gunn, C. O'Brien, K. Kavanagh, M. Sims, and G. Mann

Key Number:	**115**
Ecoregion Name:	**Torngat Mountain Tundra**
Major Habitat Type:	**6.2 Tundra**
Ecoregion Size:	**32,288 km²**
Biological Distinctiveness:	**Nationally Important**
Conservation Status:	**Snapshot—Relatively Stable**
	Final—Relatively Stable

Introduction

This ecoregion occupies the northernmost part of Labrador and the western slopes of the Torngat Mountains in Quebec. It is a mountainous region principally covered by arctic vegetation, although these gradually integrate with boreal elements toward the southern end of the region and in valleys and river corridors that penetrate the mountains.

The ecoregion has a moist, low arctic ecoclimate. Its humid, cold climate is characterized by short, cool, moist summers (mean temperature 4°C) and long, cold winters (mean temperature −16.5°C). Mean annual temperature is −6.5°C. Mean annual precipitation is 400–700 mm, with higher values occurring in the high central elevations (ESWG 1995).

The ecoregion is composed of massive Archean granitic rocks that form steep-sided, rounded mountains with deeply incised valleys and glacier-carved, deep, U-shaped valleys and fjords along the Labrador Sea coast. In the west, permafrost is continuous and marked by sporadic ice wedges; it becomes discontinuous though still extensive in the eastern, coastal portion (ESWG 1995).

Biological Distinctiveness

The ecoregion is characterized by a sparse cover of lichen, moss, arctic sedge (*Carex* spp.), grass (Gramineae), and patches of arctic mixed evergreen and deciduous shrubs on sheltered, south-facing valley slopes. Unvegetated rock and tundra (arctic heath made up of lichens, mosses, and sedges) each constitute about 50 percent of the upland surfaces. White birch–willow (*Betula papyrifera–Salix* spp.) thickets growing on less stable scree frequently form a transition zone between the tundra and the very open spruce forests. Arctic black spruce (*Picea mariana*) with mixed evergreen and deciduous shrubs, underlain by moss, is dominant on wetter sites (ESWG 1995).

Wildlife in the ecoregion is characterized mostly by small mammals, although the Torngat Mountains do provide seasonal habitat for polar bear (*Thalarctos maritimus*) and caribou (*Rangifer tarandus*). This ecoregion is home to the only global population of tundra-dwelling black bears (*Ursus americanus*) and completely encompasses the range of the Torngat caribou herd (approximately 10,000 animals). It also provides habitat for the southernmost denning of polar bears on the North American East Coast. In addition, the coastal area of this ecoregion lies along the Atlantic migratory flyway.

Conservation Status

Habitat Loss and Degradation

This ecoregion is still essentially intact.

Remaining Blocks of Intact Habitat

The ecoregion remains as a single block of habitat.

Degree of Fragmentation

Fjords intersect the coastline in a manner that naturally fragments some of the terrestrial habitat. The valley and ridge system in the mountains plays an important role in wildlife movement patterns.

Degree of Protection

There are no protected areas, although negotiations are proceeding to establish a national park covering a significant portion of this ecoregion. A candidate provincial park and a proposal for a provincial ecological reserve overlap much of the Quebec portion of this ecoregion.

Types and Severity of Threats

There is mineral interest throughout this mountainous area, although no major commercial finds have been made. Any development in valleys and lowlands could disrupt seasonal movements of large mammals such as caribou and tundra-dwelling black bears.

This ecoregion is extremely fragile. Many wildlife species are at low population levels due to the low productivity of the land and harsh climatic conditions. These species are particularly vulnerable to human development pressures; any future recreational and tourism activities will need to be strictly managed so as not to interfere with the wildlife of the area.

Suite of Priority Activities to Enhance Biodiversity Conservation

- Establish the Torngat Mountains National Park in northern Labrador.
- Establish the candidate provincial park and ecological reserve on the Quebec portion of the ecoregion.

Conservation Partners

For contact information, please see appendix G.

- Action: Environment
- Canadian Nature Federation
- Labrador Inuit Association
- Natural History Society of Newfoundland and Labrador
- Newfoundland/Labrador Environmental Association
- Protected Areas Association of Newfoundland and Labrador
- World Wildlife Fund Canada, Quebec Region

Relationship to Other Classification Schemes

The Torngat Mountains (TEC ecoregion 7) are in the northernmost region of Labrador. This ecoregion lies over the Quebec-Newfoundland provincial border and is characterized by transitional Boreal subarctic-tundra vegetation (32) and Tundra (ESWG 1995).

Prepared by A. Veitch, K. Kavanagh, M. Sims, and G. Mann

Key Number:	**116**
Ecoregion Name:	**Permanent Ice**
Major Habitat Type:	**6.2 Tundra**
Ecoregion Size:	**112,906 km²**
Biological Distinctiveness:	**Nationally Important**
Conservation Status:	**Not Assessed**

Introduction

This ecoregion occurs on Axel Heiberg, Ellesmere, and Devon Islands. It consists of large polar ice caps that are essentially permanent. The ice-covered Grantland and Axel Heiberg Mountains reach 2,500 m asl. There is little in the way of permanent species occurrences (except transitory).

This ecoregion is classified as having an oceanic, high arctic ecoclimate. Mean annual temperature among the permanent ice is –18.5°C. Because of high elevation and latitude, the area is characterized by very short, very cold summers (mean temperature –2°C) and very long, extremely cold winters (mean temperature –30°C to –35°C). Mean annual precipitation ranges between 200 and 300 mm (ESWG 1995).

A belt of deeply dissected Precambrian crystalline rock extends along the eastern flanks of Devon Island and along Ellesmere Island south of Bache Peninsula. The entire area is underlain by continuous permafrost of low ice content, and ice fields and nunataks are common. To the northwest, the mountains pass abruptly into a narrow, seaward-sloping plateau, and to the east, with decreasing ruggedness into the elevated dissected edge of Eureka upland. The ranges and ridges are transsected by numerous steep-walled valleys and fjords with glaciers (ESWG 1995).

Biological Distinctiveness

Vegetation is very sparse in this ecoregion, as the landscape is dominated by the Ellesmere and Devon Island Ice Caps. Where plant life exists, clumps of moss, lichen, and cold-hardy vascular plants such as sedge (*Carex* spp.) and cottongrass (*Eriophorum* spp.) are the dominant vegetation, and arctic willow (*Salix arctica*) and *Dryas* occur infrequently (ESWG 1995).

The entire ecoregion is characterized by very low species diversity. Represented wildlife include arctic hare (*Lepus arcticus*), arctic fox (*Alopex lagopus*), lemming (Lemminae), musk ox (*Ovibos moschatus*), and caribou (*Rangifer tarandus*). Polar bears (*Thalarctos maritimus*) are common in coastal areas. Characteristic bird species include king eider (*Somateria spectabilis*), rock ptarmigan (*Lagopus mutus*), northern fulmar (*Fulmarus glacialis*), ringed plover (*Charadrius hiaticula*), hoary redpoll (*Carduelis hornemanni*), and snow bunting (*Plectrophenax nivalis*). Marine mammals include walrus (*Odobenus rosmarus*), seal (Phocidae), and whales (Cetacea) (ESWG 1995).

Conservation Status

Habitat Loss and Degradation
The ecoregion remains 100 percent intact.

Remaining Blocks of Intact Habitat
The ecoregion is composed of five disjunct ice caps. All are intact.

Degree of Fragmentation
None.

Degree of Protection
- Northern Ellesmere Island National Park—37,775 km²

Types and Severity of Threats
Settling of air pollutants over the polar areas is of concern to this ecoregion.

Suite of Priority Activities to Enhance Biodiversity Conservation

There are no priority activities. The national park is well representative of the polar ice conditions.

Conservation Partners

For contact information, please see appendix G.

- Canadian Arctic Resources Committee
- Canadian Nature Federation
- Ecology North
- World Wildlife Fund Canada

Relationship to Other Classification Schemes

Canada's permanent ice region covers the Ellesmere and Devon Islands Ice Caps (TEC ecoregions 1–4).

Prepared by K. Kavanagh, M. Sims, and G. Mann

Conservation Partner Contact Information

Action: Environment
PO Box 2549
St. John's, NF A1C 6K1, Canada

Adirondack Council
PO Box D-2
Elizabethtown, NY 12932, USA
518-873-2240, Fax: 518-873-6675

Alabama Natural Heritage Program
Huntington College
Massey Hall
1500 East Fairview Avenue
Montgomery, AL 36106-2148, USA
334-834-4519, Fax: 334-834-5439
alnhp@wsnet.com

Alaska Boreal Forest Council
PO Box 84530
Fairbanks, AK 99708, USA
907-457-8453

Alaska Center for the Environment
519 W. 8th Avenue, Suite 201
Anchorage, AK 99501, USA
907-274-3621

Alaska Marine Conservation Council
Box 101145
Anchorage, AK 99510, USA
907-277-5357, Fax: 907-277-5975

Alaska Rainforest Campaign
1016 W. 6th Avenue, Suite 200
Anchorage, AK 99501, USA
907-274-7246

Alaska Rainforest Campaign
419 Sixth Street, Suite 318
Juneau, AK 99801, USA
907-274-7246

Alberta Wilderness Association
Box 6398, Station D
Calgary, AB T2P 2E1, Canada
403-283-2025, Fax: 403-270-2743

Alliance for the Wild Rockies
PO Box 8371
Missoula, MT 59807, USA
406–721–5420

American Wildlands
40 East Main Street, #2
Bozeman, MT 59715, USA
406-586-8175, Fax: 406-586-4282

Les Amis de plein air de Cheticamp
Box 472
Cheticamp, NS B0E 1H0, Canada

Annapolis Field Naturalists Society
Box 576
Annapolis Royal, NS B0S 1A0, Canada

The Appalachian Mountain Club
5 Joy Street
Boston, MA 02108, USA
617-523-0636

Arizona Sonora Desert Museum
2021 North Kinney Road
Tucson, AZ 85743, USA
520-883-1380

Arkansas Natural Heritage Commission
323 Center Street
Tower Building, Suite 1500
Little Rock, AR 72201, USA
501-324-9150, Fax: 501-324-9618

Association for Biodiversity Information
3421 M Street NW, Box 1737
Washington, DC 20007, USA

Association of Forest Service Employees for
Environmental Ethics
PO Box 11615
Eugene, OR 97440, USA
503-484-2692
afseee@ofseee.org or http://www.afseee.org

Audubon Society
700 Broadway
New York, NY 10003, USA
212-979-3000

Banff/Bow Valley Naturalists
Box 1693
Banff, AB T0L 0C0, Canada
403-762-4160, Fax: 403-762-4160

Bat Conservation International
PO Box 162603
Austin, TX 78716-2603, USA

BC Wild
PO Box 2241, Main Post Office
Vancouver, BC V6B 3W2, Canada
604-669-4802, Fax: 604-669-6833

Bereton Field Naturalists' Club
Box 1084
Barrie, ON L4M 5E1, Canada
705-728-8428

Bering Sea Coalition
730 I Street, Suite 200
Anchorage, AK 99501, USA
907-279-6566

Blomidon Naturalists Society
Box 127
Wolfville, NS B0P 1X0, Canada
902-542-2201

Brandon Naturalists' Society
RR 3, Box 57
Brandon, MB R7A 5Y3, Canada
204-727-2995

Burns Bog Conservation Society
11961 88th Avenue, #202
Delta, BC V4C 3C9, Canada
604-572-0373, Fax: 604-572-0374

California Department of Fish and Game
1416 9th Street
Sacramento, CA 95814, USA
916-653-7664, Fax: 916-653-1856

California Desert Protection League
c/o Elden Hughes
14045 Honeysuckle Lane
Whittier, CA 90604, USA
562-941-5306

The California Native Plant Society
1772 U Street, Suite 17
Sacramento, CA 95815, USA
916-447-2677

California Oak Foundation
1212 Broadway, Suite 810
Oakland, CA 94612, USA
510-763-0282

California Resources Agency
1416 9th Street, Room 1311
Sacramento, CA 95814, USA
916-653-5656

Canadian Arctic Resources Committee
1 Nicholas Street, Suite 412
Ottawa, ON K1N 7B7, Canada
613-241-7379, Fax: 613-241-2244

Canadian Nature Federation
1 Nicholas Street, Suite 606
Ottawa, ON K1N 7B7, Canada
613-562-3447, Fax: 613-562-3371

Canadian Parks and Wilderness Society
British Columbia Chapter
611-207 W. Hastings Street
Vancouver, BC V6B 2H7, Canada
604-685-7445, Fax: 604-685-6449

Canadian Parks and Wilderness Society
Yukon Chapter
30 Dawson Road
Whitehorse, YT Y1A 5T6, Canada
403-668-6321, Fax: 403-668-6321

Canadian Parks and Wilderness Society
Calgary-Banff Chapter
306, 319 10th Avenue SW
Calgary, AB T2R 0A5, Canada
403-281-0522, Fax: 403-260-0333

Canadian Parks and Wilderness Society
Edmonton Chapter
Box 52031, 8210 109th Street
Edmonton, AB T6G 2T5, Canada
403-433-9302, Fax: 403-433-9305

Canadian Parks and Wilderness Society
Ottawa Valley Chapter
Box 3072, Station D
Ottawa, ON K1P 6H6, Canada
613-720-2797, Fax: 613-730-0005

Canadian Parks and Wilderness Society
Nova Scotia Chapter
73 Chadwick Street
Dartmouth, NS B2Y 2M2, Canada
902-466-7168

Canadian Spirit of Ecotourism Society
Box 7
Salmon Arm, BC V1E 4N2, Canada

Cape Breton Naturalists Society
c/o Cape Breton Center for Heritage
    and Science
225 St. George Street
Sydney, NS B1P 1J5, Canada

Carolinian Canada Program
c/o Ontario Ministry of Natural Resources
659 Exeter Road
London, ON N6E 1L3, Canada
519-661-2744

Castle Crown Wilderness Coalition
Box 2621
Pincher Creek, AB T0K 1W0, Canada

Le Centre de Données sur le Patrimoine Naturel
    du Quebec Ministère de l'Environnement
    Direction de la conservation et du patrimoine
    écologique
2360 Ste-Foy Road, 1st Floor
Sainte-Foy, PQ G1V 4H2, Canada
418-644-3358

Chihuahuan Desert Research Institute
PO Box 1334
Alpine, TX 79831, USA
915-837-8370

Clayoquot Biosphere Project
Box 67
Tofino, BC V0R 2Z0, Canada

Clearwater Forest Watch
c/o Inland Empire Public Lands Council
PO Box 174
Spokane, WA 99210, USA
509-327-1699

Le Club de Naturalistes de la Peninsule
    Acadienne
Box 115, RR 2
Tabusintac, NB E0C 2A0, Canada

Collin County Open Space Program
210 S. McDonald Street, Suite 22
McKinney, TX 75069, USA
972-548-4141

Conservation Council of New Brunswick
180 St. John Street
Fredericton, NB E3B 4A9, Canada
506-458-8747, Fax: 506-458-1047

The Conservation Fund
Texas Chapter
PO Box 50271
Austin, TX 78768, USA
505-776-1369

The Conservation Fund
9850 Hiland Road
Eagle River, AK 99577, USA
907-694-9060

Conservation Trust of Puerto Rico
PO Box 4747
San Juan, PR 00902-4747, USA
809-722-5834

Copper Country Alliance
Box 22
Chitina, AK 99566, USA

Copper River Watershed Forum
PO Box 1430
Cordova, AK 99574, USA
907-424-3926

Corporation for the Northern Rockies
306 S. Fifth Street
Livingston, MT 59047, USA
406-222-0730

Craighead Environmental Research Institute
Box 156
Moose, WY 83012, USA
307-733-3387

Craighead Wildlife-Wildlands Institute
5200 Upper Miller Creek Road
Missoula, MT 59803, USA
406-251-5069

Critical Wildlife Habitat Program
c/o Wildlife Branch
Department of Natural Resources
Box 24, 200 Saulteaux Crescent
Winnipeg, MB R3J 3W3, Canada
204-945-7775

Ducks Unlimited Canada
Box 1160, Oak Hammock Marsh
Stonewall, MB R0C 2Z0, Canada
204-467-3000, Fax: 204-467-9436

Durham Region Field Naturalists
499 Stone Street
Oshawa, ON L1J 1A4, Canada
905-428-0584

Earthroots
401 Richmond Street West, Suite 251
Toronto, ON M5V 3A8, Canada
416-599-0152, Fax: 416-340-2429

East Kootenay Environmental Society
Cranbrook-Kimberley Branch
Box 8
Kimberley, BC V1A 2Y5, Canada
604-427-2535, Fax: 604-427-2535

East Kootenay Environmental Society
Creston Valley Branch
Box 1837
Creston, BC V0B 1G0, Canada
604-428-9532, Fax: 604-866-5477

East Kootenay Environmental Society
Elkford Branch
PO Box 855
Elkford, BC V0B 1H0, Canada
604-865-7606

East Kootenay Environmental Society
Golden Branch
PO Box 1946
Golden, BC V0A 1H0, Canada
604-348-2225

East Kootenay Environmental Society
Invermere Branch
PO Box 2741
Invermere, BC V0A 1K0, Canada
604-342-0180, Fax: 604-442-6385

East Kootenay Environmental Society
Sparwood Branch
PO Box 1359
Sparwood, BC V0B 2G0, Canada
604-425-2902

Ecology Action Centre
1553 Granville Street
Halifax, NS B3J 1W7, Canada

The Ecology Center
2530 San Pablo Avenue
Berkeley, CA 94702, USA
510-548-2220

The Ecology Circle
375 Patricia Boulevard
Prince George, BC V5L 3T9, Canada
604-563-5390

Ecology North
4807 49th Street, Suite 8
Yellowknife, NT X1A 3T5, Canada
403-873-6019, Fax: 403-873-3654

Ecotrust
1200 NW Front Avenue, Suite 470
Portland, OR 97209, USA
503-227-6225

The Endangered Habitats League
8424-A Santa Monica Boulevard, #592
Los Angeles, CA 90069-4267, USA
323-654-1456, Fax: 323-654-1931

Endangered Spaces Campaign, Manitoba
63 Albert Street, Suite 411
Winnipeg, MB R3B 1G4, Canada
204-944-9593, Fax: 204-882-2236

Endangered Spaces Campaign, Saskatchewan
3079 Athol Street
Regina, SK S4S 1Y6, Canada
306-586-3863, Fax: 306-586-3863

Environment North
704 Holly Crescent
Thunder Bay, ON P7E 2T2, Canada
807-475-5267, Fax: 807-577-6433

Environmental Coalition of Prince Edward
    Island
RR 1
Cardigan, PEI C0A 1G0, Canada
902-556-4696, Fax: 902-368-7180

Essex County Field Naturalists
Box 23011, Devonshire Mall
Windsor, ON N8X 5B5, Canada
519-252-2473

Essex County Greenbelt Association
82 Eastern Avenue
Essex, MA 01929, USA
508-768-7241

Everglades Coalition
West Palm Beach Office
Contact Person: Ron Tipton, WWF

Federation of Alberta Naturalists
Box 1472
Edmonton, AB T5J 3V9, Canada
403-453-8629, Fax: 403-453-8553

Federation of Nova Scotia Naturalists
c/o Nova Scotia Museum
1747 Summer Street
Halifax, NS B3H 3A6, Canada

Federation of Ontario Naturalists
355 Lesmill Road
Don Mills, ON M3B 2WB, Canada
416-444-8419, Fax: 416-444-9866

Florida Natural Areas Inventory
1018 Thomasville Road, Suite 200-C
Tallahassee, FL 32303, USA
904-224-8207, Fax: 904-681-9364

Ford Alward Naturalist Association
RR 5
Hartland, NB E0J 1N0, Canada
506-246-5572, Fax: 506-246-5572

Friends of the Boundary Waters Wilderness
5th Street SE, Suite 329
Minneapolis, MN 55414, USA
612-379-3835

Friends of Caren
Box 872
Madiera Park, BC V0N 2H0, Canada
604-261-4331

Friends of the Christmas Mountains
Box M-956, Mount Allison University
Sackville, NB E0A 3C0, Canada

Friends of Ecological Reserves
Box 8477, Victoria Central PO
Victoria, BC V8W 3S1, Canada

Friends of Elk Island Society
Site 4, RR 1
Fort Saskatchewan, AB T8L 2N7, Canada
403-492-5494

Friends of McNeil River
PO Box 231091
Anchorage, AK 99523-1091, USA
907-235-6175

Friends of Nature Conservation Society
22 Pleasant Street
Chester, NS B0J 1J0, Canada
902-275-3361

Friends of Okanagan Mountain Park
RR 2, S 26A, C-5
Peachland, BC V0H 1X0, Canada

Friends of Point Pelee
RR 1
Leamington, ON N8H 3V4, Canada
519-326-6173

Friends of Prince Albert National Park
c/o Prince Albert National Park
Box 100
Wakesiu Lake, SK S0J 2Y0, Canada

Friends of the Stikine
1405 Doran Road
North Vancouver, BC V7K 1N1, Canada
604-985-4659

Friends of Strathcona Park
Box 3404, 279 Puntledge Road
Courtenay, BC V9N 5N5, Canada
604-338-1944

Friends of Yoho
Box 100
Field, BC V0A 1G0, Canada
604-343-6393, Fax: 604-343-6330

Friends of Yukon Rivers
21 Klondike Road
Whitehorse, YT Y1A 3L8, Canada
403-668-7370

Fundy Guild
Box 150
Alma, NB E0A 1B0, Canada

Galiano Conservancy Association
RR 1, Porlier Pass Road
Galiano Island, BC V0N 1P0, Canada
604-539-2424, Fax: 604-539-2424

Georgia Natural Heritage Program
Wildlife Resources Division
Georgia Department of Natural Resources
2117 U.S. Highway 278 SE
Social Circle, GA 30279, USA
706-557-3032, Fax: 706-557-3040
email: natural_heritage@mail.dnr.state.ga.us

Gowgaia Institute
Box 638
Queen Charlotte City, BC V0T 1S0, Canada
250-559-8068, Fax: 250-559-8006

Grand Canyon Trust
Route 4, Box 718
Flagstaff, AZ 86001, USA
602-774-7488

Grand Conseil des Cris
277 Duke Street
Montreal, PQ H3C 2M2, Canada
514-861-5837, Fax: 514-861-0760

The Great Bear Foundation
PO Box 1289
Bozeman, MT 59715-1289, USA
406-586-5533, Fax: 406-586-6103

The Great Land Trust
PO Box 101272
Anchorage, AK 99510-1272, USA
907-278-4998

Great Sand Hills Planning District Commission
PO Box 120
Sceptre, SK S0N 2H0, Canada
306-623-4229, Fax: 306-623-4229

Greater Yellowstone Coalition
Meredith Taylor, Wyoming Field Representative
6360 Highway 26
Dubois, WY 82513, USA
307-455-2161, Fax: 307-455-3169

Hamilton Naturalists Club
Box 89052, Westdale PO
Hamilton, ON L8S 4R5, Canada
905-547-5116

Hawaii Natural Heritage Program
The Nature Conservancy of Hawaii
1116 Smith Street, Suite 201
Honolulu, HI 96817, USA
808-537-4508, Fax: 808-545-2019

Heard Museum of Natural History
1 Nature Place
McKinney, TX 75069, USA
972-562-5566

Heritage Foundation Terra Nova National Park
Glovertown, NF A0G 2L0, Canada
709-533-2884, Fax: 709-533-2706

Humboldt State University
Arcata, CA 95521, USA
707-826-3256

Idaho Conservation League
PO Box 884
Boise, ID 83701, USA
208-3445-6933, Fax: 208-344-0344

Illinois Department of Natural Resources
524 South 2nd Street, Room 400 LTP
Springfield, IL 62764, USA
217-782-5597

Indiana Department of Natural Resources
402 West Washington Street, Room W255B
Indianapolis, IN 46204-2748, USA
317-232-4200, Fax: 317-232-8036

Inland Empire Public Lands Council
S. 517 Division
Spokane, WA 99202, USA
509-838-4912

Innuvialuit Game Council, Joint Secretariat
Box 2120 Inuvik, NT X0E 0T0, Canada
403-979-2828

Institute of Agriculture and Natural Resources
University of Nebraska
Lincoln, NE 68583, USA
402-472-7211

Interrain Pacific
119 Seward Street, #19
Juneau, AK 99801, USA
907-586-2301

Island Nature Trust
PO Box 265
Charlottetown, PEI C1A 7K4, Canada
902-892-7513, Fax: 902-628-6331

Islands Trust
1106 Cook Street, 2nd Floor
Victoria, BC V8V 3Z9, Canada

Kachemak Heritage Land Trust
PO Box 2400
Homer, AK 99603, USA
907-235-5263

Kamloops Naturalists
Box 625
Kamloops, BC V2C 5L7, Canada
604-372-5956

Kansas Natural Heritage Inventory
Kansas Biological Survey
2041 Constant Avenue
Lawrence, KA 66047-2906, USA
913-864-7698

Dr. James Karr
School of Fisheries
University of Washington
1013 NE 40th Street
Seattle, WA 98105-6698, USA
206-685-4784

Kentucky Heartwood
660 Mt. Vernon Ridge
Frankfort, KY 40601, USA

Kentucky Natural Heritage Program
Kentucky State Nature Preserves Commission
801 Schenkel Lane
Frankfort, KY 40601, USA
502-573-2886, Fax: 502-573-2355
email: ksnpc@mail.state.ky.us

Kettle Range Conservation Group
PO Box 150
Republic, WA 99166, USA
509-775-3454

Klamath Forest Alliance
PO Box 820
Etna, CA 96027, USA
530-467-5405, Fax: 530-467-3130
email: klamath@sisqtel.net
www.sisqtel.net/users/klamath

Labrador Inuit Association
240 Water Street
St. John's, NF A1C 1B7, Canada
709-722-6160

Laskeek Bay Conservation Society
Box 867
Queen Charlotte City, BC V0T 1S0, Canada
250-559-2345, Fax: 250-559-2345

Louisiana Natural Heritage Program
Department of Wildlife and Fisheries
PO Box 98000
Baton Rouge, LA 70898-9000, USA
504-765-2821, Fax: 504-765-2607

Lower Fort Garry Volunteers
Box 394
Selkirk, MB R1A 2B3, Canada
204-785-8577, Fax: 204-482-5887

Maine Coast Heritage Trust
169 Park Row
Brunswick, ME 04011, USA
207-729-7366, Fax: 207-729-6863

Maine Natural Areas Program
Department of Conservation
93 State House Station
Augusta, ME 04333-0093, USA
207-287-2211

Manitoba Future Forest Alliance
70 Albert Street, Suite 2
Winnipeg, MB R3B 1E7, Canada
204-947-3081, Fax: 204-947-3076

Manitoba Heritage Habitat Corporation
1555 St. James Street, Suite 200
Winnipeg, MB R3H 1B5, Canada
204-784-4362

Manitoba Naturalists Society
63 Albert Street, Suite 401
Winnipeg, MB R3B 1G4, Canada
204-943-9029

Margaree Environmental Protection Association
Box 617
Margaree Forks, NS B0E 2A0, Canada
902-248-2573

Margaree Environmental Society
Box 617
Margaree Forks, NS B0E 2A0, Canada

Maryland Heritage and Biodiversity
  Conservation Programs
Department of Natural Resources
Tawes State Office Building, E-1
Annapolis, MD 21401, USA
410-974-3195, Fax: 410-974-5590

Massachusetts Audubon Society
208 South Great Road
Lincoln, MA 01773, USA
800-AUDUBON [800-283-8266]

McIlwraith Field Naturalists of London
Box 4185
London, ON N5W 5H6, Canada
519-645-2842, Fax: 519-645-0981

Meewasin Valley Authority
402 3rd Avenue South
Saskatoon, SK S7K 3N9, Canada
306-665-6887, Fax: 306-665-6117

Michigan Environmental Council
115 W. Allegan, Suite 10B
Lansing, MI 48933, USA
517-487-9539 FAX: 517-487-9541

Michigan Natural Areas Council
University of Michigan
Botanical Gardens
1800 N. Dixboro Road
Ann Arbor, MI 48105, USA
313-435-2070

Michigan Nature Association
7981 Beard Road, Box 102
Avoca, MI 48006-0102, USA
313-324-2626

Minnesota Department of Natural Resources
500 Lafayette Road
St. Paul, MN 55155-4001, USA
612-296-6157, Fax: 612-297-3618

Mississippi Natural Heritage Program
Museum of Natural Science
111 North Jefferson Street
Jackson, MS 39201-2897, USA
601-354-7303, Fax: 601-354-7227

Missouri Department of Conservation
PO Box 180
Jefferson City, MO 65102-0180, USA
573-751-4115, Fax: 573-751-4467

Missouri Natural Heritage Database
Missouri Department of Conservation
PO Box 180
Jefferson City, MO 65102, USA
573-751-4115, Fax: 573-526-5582

Moncton Naturalists' Club
Box 28036
Highfield Square PO
1100 Main Street
Moncton, NB E1C 9N4, Canada

Montana Wilderness Association
PO Box 635
Helena, MT 59624, USA
406-443-7350

MoRAP
Midwest Science Center
4200 New Haven Road
Columbia, MO 65201, USA

Muskoka Field Naturalists
RR 2, Kingsett Road
Port Carling, ON P0B 1J0, Canada

National Audubon Society
Alaska-Hawaii Regional Office
308 G Street, Suite 217
Anchorage, AK 99501, USA
907-276-7034

National Audubon Society
2631 12th Court SW, #A, PO Box 462
Olympia, WA 98502, USA
206-786-8020

National Biological Service
Florida and Caribbean Science Center
7920 NW 71st Street
Gainesville, FL 32653-3071, USA
352-378-8181

National Cattlemen's Association
Environmental Stewardship Award
PO Box 3469
Englewood, CO 80155, USA
303-694-0305

National Park Service
Great Smoky Mountains National Park
Gatlinburg, TN 37738, USA
615-436-1200

National Park Service
Shenandoah National Park
Route 4, Box 348
Luray, VA 22835, USA
703-999-2243

National Parks and Conservation Association
329 F Street, Suite 208
Anchorage, AK 99501, USA
907-277-6722

National Parks and Conservation Association
Southwest Regional Office
823 Gold Avenue
Albuquerque, NM 87102, USA
505-247-1221

National Wildlife Federation—Western Division
Western Natural Resource Center
921 Southwest Morrison, Suite 512
Portland, OR 97205, USA
503-222-1429, Fax: 503-222-3203

Native Plant Society of Texas
PO Box 891
Georgetown, TX 78627, USA
512-863-9685

Native Prairies Association of Texas
301 Nature Center Drive
Austin, TX 78746, USA
512-327-8181

Natural Areas Preservation Association
4144 Cochran Chapel Road
Dallas, TX 75209, USA
214-352-8370

Natural Heritage Information Centre
PO Box 7000
Peterborough, ON K9J 8M5, Canada
705-745-6767

Natural History Society of Newfoundland
   and Labrador
68 Leslie Street
St. John's, NF A1E 2V8, Canada

Natural History Society of Prince Edward Island
PO Box 2346
Charlottetown, PEI C1A 8C1, Canada

Natural Resources Conservation Service
Department of Agriculture
14th and Independence Avenue, SW
PO Box 2890
Washington, DC 20013, USA
202-720-3210

Nature Action
C.P. 434
Saint-Bruno-de-Montarville, PQ J3V 5G8,
    Canada
514-441-3899

The Nature Conservancy
1815 North Lynn Street
Arlington, VA 22209, USA
703-841-5300, Fax: 703-841-1283
http://www.tnc.org

The Nature Conservancy, Adirondack Chapter
PO Box 65, Route 73
Keene Valley, NY 12943, USA
518-576-2082

The Nature Conservancy
Eastern Regional Office
201 Devonshire Street, Fifth Floor
Boston, MA 02110, USA
617-542-1908

The Nature Conservancy
Midwest Regional Office
1313 5th Street SE, #314
Minneapolis, MN 55414, USA
612-331-0700

The Nature Conservancy
Southeast Regional Office
PO Box 2267
Chapel Hill, NC 27515-2267, USA
919-967-5493

The Nature Conservancy
Western Region
2060 Broadway, Suite 230
Boulder, CO 80302, USA
303-444-1060

The Nature Conservancy of Alabama
2821-C 2nd Avenue
Birmingham, AL 35233, USA
205-251-1155

The Nature Conservancy of Alaska
421 West 1st Avenue, Suite 200
Anchorage, AK 99501, USA
907-276-2584

The Nature Conservancy, Alberta
3400 Western Canadian Place
700 8th Avenue SW
Calgary, AB T2P 1H5, Canada
403-294-7030, Fax: 403-265-8263

The Nature Conservancy of Arizona
300 E. University Boulevard, Suite 230
Tucson, AZ 85705, USA
520-622-3861

The Nature Conservancy, British Columbia
827 West Pender Street, 2nd Floor
Vancouver, BC V6C 3G8, Canada
604-684-1654, Fax: 604-656-7583

The Nature Conservancy of California
California Regional Office
201 Mission Street, 4th Floor
San Francisco, CA 94105, USA
415-777-0487

The Nature Conservancy of Canada
110 Eglinton Avenue West, 4th Floor
Toronto, ON M4R 2G5, Canada
416-932-3202, Fax: 416-932-3208

The Nature Conservancy of Canada,
    Atlantic Canada
Box 8505, 6 Kent Avenue
Halifax, NS B3K 5M2, Canada
902-857-1013

The Nature Conservancy of Canada
Western Canadian Place, 3400
707 8th Avenue SW
Calgary, AB T2P 1H6, Canada
403-294-7064, Fax: 403-265-8263

The Nature Conservancy of Colorado
1244 Pine Street
Boulder, CO 80302, USA
303-444-2985

The Nature Conservancy of Florida
Florida Regional Office
222 S. Westmonte Drive, Suite 300
Altamonte Springs, FL 32714, USA
407-682-3664

The Nature Conservancy of Georgia
1330 W. Peachtree Street, Suite 410
Atlanta, GA 30309-2904, USA
404-873-6946

The Nature Conservancy
Great Lakes Program
8 South Michigan Avenue, Suite 2301
Chicago, IL 60603-3318, USA
312-759-8017

The Nature Conservancy of Hawaii
1116 Smith Street, Suite 201
Honolulu, HI 96817, USA
808-537-4508, Fax: 808-545-2019

The Nature Conservancy of Iowa
431 East Locust, Suite 200
Des Moines, IA 50309-1909, USA
515-244-5044

The Nature Conservancy of Kentucky
642 W. Main Street
Lexington, KY 40508, USA
606-259-9655

The Nature Conservancy of Louisiana
PO Box 4125
Baton Rouge, LA 70821, USA
504-338-1040

The Nature Conservancy of Maine
14 Main Street, Suite 401
Brunswick, ME 04011, USA
207-729-5081

The Nature Conservancy, Manitoba
298 Garry Street
Winnipeg, MB R3C 1H3, Canada
204-942-6156, Fax: 204-947-2591

The Nature Conservancy of Maryland
Chevy Chase Metro Building
2 Wisconsin Circle, Suite 300
Chevy Chase, MD 20815, USA
301-656-8673

The Nature Conservancy of Massachusetts
79 Milk Street, Suite 300
Boston, MA 02109, USA
617-423-2545

The Nature Conservancy of Mississippi
PO Box 1028
Jackson, MS 39215-1028, USA
601-355-5357

The Nature Conservancy of Montana
32 South Ewing
Helena, MT 59601, USA
406-443-0303

The Nature Conservancy of Nebraska
431 North Maple Street
Ainsworth, NE 69210, USA
402-387-1061

The Nature Conservancy of New Hampshire
2 1/2 Beacon Street, Suite 6
Concord, NH 03301, USA
603-224-5853

The Nature Conservancy of New Mexico
212 E. Marcy Street
Santa Fe, NM 87501, USA
505-988-3867

The Nature Conservancy of New York
Central/Western Office
315 Alexander Street, 2nd Floor
Rochester, NY 14604, USA
716-546-8030

The Nature Conservancy of North Carolina
4011 University Drive, Suite 201
Durham, NC 27707, USA
919-403-8558

The Nature Conservancy of Ohio
6375 Riverside Drive, Suite 50
Dublin, OH 43017, USA
614-717-2770

The Nature Conservancy of Pennsylvania
1100 E. Hector Street, Suite 470
Conshohocken, PA 19428, USA
610-834-1323

The Nature Conservancy, Quebec
800 René-Lévesque Boulevard West, Suite 1100
Montreal, PQ H3B 1X9, Canada
514-876-1606, Fax: 514-871-8772

The Nature Conservancy of South Carolina
PO Box 5475
Columbia, SC 29250, USA
803-254-9049

The Nature Conservancy, Tallgrass Prairie
   Office
PO Box 458
Pawhuska, OK 74056, USA
918-287-4440

The Nature Conservancy of Tennessee
50 Vantage Way, Suite 250
Nashville, TN 37228, USA
615-255-0303

The Nature Conservancy of Texas
PO Box 1440
San Antonio, TX 78295-1440, USA
210-224-8774

The Nature Conservancy of Texas
Clymer Meadow Blackland Prairie Preserve
PO Box 26
Celeste, TX 75423, USA
903-568-4139

The Nature Conservancy of Utah
559 E. South Temple
Salt Lake City, UT 84102, USA
801-531-0999

The Nature Conservancy of Vermont
27 State Street
Montpelier, VT 05602-2934, USA
802-229-4425

The Nature Conservancy of Virginia
1233 A Cedars Court
Charlottesville, VA 22903-4800, USA
804-295-6106

The Nature Conservancy of West Virginia
723 Kanawha Boulevard East, Suite 500
Charleston, WV 25301, USA
304-345-4350

Nature Saskatchewan
1860 Lorne Street, Suite 206
Regina, SK S4P 2L7, Canada

Nature Trust of British Columbia
100 Park Royal South, Suite 808
Vancouver, BC V7T 1A2, Canada
604-925-1128, Fax: 604-926-3482

Nature Trust of New Brunswick
Box 603, Station A
Fredericton, NB E3B 5A6, Canada
506-457-2398

Navaho Nation Natural Heritage Program
PO Box 1480
Window Rock, Navaho Nation, AZ 86515, USA
520-871-7603

Nebraska Natural Heritage Program
Game and Parks Commission
2200 North 33rd Street
PO Box 30370
Lincoln, NE 68503, USA
402-471-5469

Nechako Environmental Society
Box 805, Station A
Prince George, BC V2L 4V2, Canada
604-562-6587, Fax: 604-562-4271

New Brunswick Federation of Naturalists
277 Douglas Avenue
Saint John, NB E2K 1E5, Canada
506-532-3482

New Brunswick Protected Natural Areas
   Coalition
180 St. John Street
Fredericton, NB E3B 4A9, Canada
506-452-9902, Fax: 506-458-1047

New Hampshire Natural Heritage Inventory
Department of Resources & Economic
   Development
172 Pembroke Street
PO Box 1856
Concord, NH 03302, USA
603-271-3623

New Jersey Audubon Society
PO Box 125
790 Ewing Avenue
Franklin Lakes, NJ 07417, USA
201-891-1211

New Jersey Conservation Foundation
300 Mendham Road
Morristown, NJ 07960, USA

New York Natural Heritage Program
Department of Environmental Conservation
700 Troy-Schenectady Road
Latham, NY 12110-2400, USA
518-783-3932

Newfoundland/Labrador Environmental
   Association
603 TD Place, 140 Water Street
St. John's, NF A1C 6H6, Canada

Nipissing Naturalists Club
154 Balsam Crescent
North Bay, ON P1B 6M3, Canada

Norfolk Field Naturalists
PO Box 995
Simcoe, ON N3Y 5B3, Canada
519-586-3985

North Carolina Heritage Program
Deparment of Environment, Health & Natural
   Resources
Division of Parks and Recreation
PO Box 27687
Raleigh, NC 27611, USA

North Okanagan Naturalists Club
Box 473
Vernon, BC V1T 6M4, Canada
604-545-2341

Northcoast Environmental Center
879 Ninth Street
Arcata, CA 95521, USA
707-822-6918, Fax: 707-822-0827

Northern Alaska Environmental Center
218 Driveway
Fairbanks, AK 99701, USA
907-452-5021

Northern Appalachian Restoration Project
PO Box 6
Lancaster, NH 03584, USA
603-636-2952

Northwatch
Box 264
North Bay, ON P1B 8H2, Canada
705-497-0373, Fax: 705-476-7060

Northwest Ecosystem Alliance
1421 Cornwall Avenue
Bellingham, WA 98227, USA
360-671-9950
email: nwea@pacificrim.net

Northwest Wildlife Preservation Society
Box 34129, Station D
Vancouver, BC V6J 4N3, Canada
604-736-8750, Fax: 604-736-9615

Nova Scotia Nature Trust
Box 2202
Halifax, NS B3J 3C4, Canada
902-425-7900, Fax: 902-425-7990

Nova Scotia Wild Flora Society
c/o Nova Scotia Museum
1747 Summer Street
Halifax, NS B3H 3A6, Canada

Nunavut Wildlife Management Board
PO Box 1379
Iqaluit, NT X0A 0H0
867-979-6962, Fax: 867-979-7785

Ohio Natural Heritage Data Base
Division of Natural Areas & Preserves
Department of Natural Resources
1889 Fountain Square, Building F-1
Columbus, OH 43224, USA
614-265-6543, Fax: 614-267-3096

Okanagan Silkameen Parks Society
Box 787
Summerland, BC V0H 1Z0, Canada
604-494-8996

Oklahoma Natural Heritage Inventory
Oklahoma Biological Survey
111 East Chesapeake Street
University of Oklahoma
Norman, OK 73019-0575, USA
405-325-1985, Fax: 405-325-7702
http://obssun02.uoknor.edu/biosurvey/onhi/home.

Oregon Natural Resources Council
5825 North Greeley
Portland, OR 97217, USA
503-283-6343

Oregon Wildlife Federation
PO Box 67020
Portland, OR 97267, USA
503-659-9054

Orillia Naturalists' Club
Box 2381
Orillia, ON L3V 6V7, Canada

Dr. Clinton Owensby
Kansas State University
Manhattan, KS 66506, USA

The Ozark Society
PO Box 2914
Little Rock, AR 72203, USA

Pacific Rivers Council
PO Box 10798
Eugene, OR 97440, USA
503-345-0119

Parks and People PEI
Box 1506
Charlottetown, PEI C1A 7N3, Canada

Partners in Flight
Hampton Waterfowl Research Center
Route 1, Box 188-A
Humphrey, AR 72073, USA
501-873-4651

Pender Harbour and District Wildlife Society
Box 220
Madeira Park, BC V0N 2H0, Canada
604-885-9310

Peninsula Field Naturalists
Box 23031, RPO Midtown
St. Catharines, ON L2R 7P6, Canada
905-934-5541

Pennsylvania Natural Diversity Inventory
Central Bureau of Forestry
PO Box 8553
Harrisburg, PA 17105-8552, USA
717-783-0388, Fax: 717-783-5109

Pennsylvania Natural Diversity Inventory
West Natural Areas Program
316 4th Avenue
Pittsburgh, PA 15222, USA
412-288-2777, Fax: 412-281-1792

Pennsylvania Natural Diversity Inventory, East
The Nature Conservancy
34 Airport Drive
Middletown, PA 17057, USA
717-948-3962, Fax: 717-948-3957

PENS—Parc d'environnement naturel de Sutton
PO Box 804
24 N. Main Street
Sutton, PQ J0E 2K0, Canada
514-538-4085

Dr. Dave Perry
Forest Science 310
Oregon State University
Corvallis, OR 97331-4501, USA
541-737-6588

Pickering Naturalists
Box 304
Pickering, ON L1V 2R6, Canada

Prairie Conservation Forum
c/o Bag 3014, YPM Place
530 8th Street South
Lethbridge, AB T1J 4C7, Canada
403-381-5430

Prairie Farm Rehabilitation Administration
603-1800 Hamilton Street
Regina, SK S4P 4L2, Canada
306-780-5070, Fax: 306-780-5018
http://www.agr.ca/pfra/pfintro.htm

Presqu'ile-Brighton Naturalists
Box 192
Brighton, ON K0K 1H0, Canada

Prince Edward Island Salmon Association
PO Box 3315
Charlottetown, PEI C1A 8W5, Canada

Prince Edward Island Wildlife Federation
Central Queen's Branch
14 Ferndale Drive
Charlottetown, PEI C1A 6J3, Canada
902-687-3131, Fax: 902-687-2350

Prince George Naturalists
Box 1092, Station A
Prince George, BC V2L 4V2, Canada
604-564-4643

Prince William Sound Science Center
PO Box 705
Cordova, AK 99574, USA
907-424-5800

Pro Terra-Kootenay Nature Allies
Box 9
Argenta, BC V0G 1B0, Canada
604-366-4387

Protected Areas Association of Newfoundland
    and Labrador
PO Box 1027, Station C
St. John's, NF A1C 5M5, Canada
709-726-2603, Fax: 709-726-2603

Quetico Foundation
48 Yonge Street, Suite 610
Toronto, ON M5T 2C6, Canada
416-941-9388, Fax: 416-941-9236

Quik Iqtani Inuit Association
PO Box 219
Iqaluit, NT X0A 0H0, Canada
867-979-5391

Quinte Field Naturalists
316 Foster Avenue
Belleville, ON K8N 3R5, Canada
613-962-2885

Red Deer River Naturalists
Box 785
Red Deer, AB T4N 5H2, Canada
403-227-2944

Red River Basin Water Management
    Consortium
15 North 23rd Street
PO Box 9018
Grand Forks, ND 58203, USA
701-777-5000

Regroupement National des Conseils Régionaux
    de l'Environnement du Québec (RNCREQ)
31 King Street West
Sherbrooke, PQ J1H 1N5, Canada
819-821-4357

Residents Committee to Protect the
    Adirondacks
Main Street
North Creek, NY 12853, USA
518-251-4257

Resource Conservation Manitoba
70 Albert Street, Suite 2
Winnipeg, MB R3B 1E7, Canada
204-925-3777

Restore the Northwoods
PO Box 440
Concord, MA 01742, USA

REVE (Regroupement écologiste Val d'Or et
    environs)
C.P. 605
Val d'Or, PQ J9P 4P6, Canada
819-738-5261, Fax: 819-825-5361

Rideau Valley Field Naturalists
Box 474
Perth, ON K7H 3G1, Canada
613-273-2758, Fax: 613-273-4255

Rocky Mountain Naturalists
404 Aspen Road
Kimberley, BC V1A 3B5, Canada

Dr. Bill Romme
Biology Department
Ft. Lewis College
Durango, CO 81301, USA
970-247-7322

Saskatchewan Environmental Society
Box 1372
Saskatoon, SK S7K 3N9, Canada
306-665-1915, Fax: 306-665-2128

Saskatchewan Forest Conservation Network
Box 359
Glaslyn, SK S0M 0Y0, Canada
306-342-4689

Saskatchewan Wildlife Federation
Box 788
Moose Jaw, SK S6H 4P5, Canada
306-692-8812

Save Our Shores
Box 3113, Station B
Saint John, NB E2M 4X7, Canada
506-674-1417

Selkirk-Priest Basin Association
509-448-3020

Shuswap Naturalists
Box 1076
Salmon Arm, BC V0E 2T0, Canada
604-675-4565

Sierra Club
730 Polk Street
San Francisco, CA 94109, USA
415-776-2211

Sierra Club
Box 37
Goodlands, MB R0M 0R0, Canada
204-658-3467, Fax: 204-658-3541

Sierra Club
1516 Melrose Avenue
Seattle, WA 98122, USA
206-621-1696

Sierra Club—Alaska Field Office
241 East 5th Avenue, #205
Anchorage, AK 99501, USA
907-276-4048, Fax: 907-258-6807

Sierra Club of British Coumbia
1525 Amelia Street
Victoria, BC V8W 2K1, Canada
604-386-5255, Fax: 604-386-4453

Sierra Club, Cape Breton Group
RR 1, Site 3, Box 7
230 #6 Mines Road
Port Caledonia, NS B1A 5T9, Canada

Sierra Club, Lone Star Chapter
1104 Nueces
Austin, TX 78701, USA
512-477-1729

Sierra Club, Northeast Regional Office
85 Washington Street
Saratoga Springs, NY 12866, USA
518-587-9166, Fax: 518-583-9062

Sierra Club—Northern Plains Field Office
23 North Scott, #27
Sheridan, WY 82801, USA
307-672-0425, Fax: 307-672-6187

Society of Ecological Restoration
1207 Seminole Highway
Madison, WI 53711, USA
608-262-9547

Society of Grassland Naturalists
Box 2491
Medicine Hat, AB T1A 8G8, Canada
403-526-6443

Society for the Protection of New Hampshire
   Forests
54 Portsmouth Street
Concord, NH 03301-5400, USA
603-224-9945, Fax: 603-228-0423

Society for Range Management
1839 York Street
Denver, CO 80206, USA
303-355-7070

The Sonoran Institute
7290 E. Broadway Boulevard, Suite M
Tucson, AZ 85710, USA
520-290-0828, Fax: 520-290-0969

South Carolina Heritage Trust
Wildlife & Marine Resources Department
PO Box 167
Columbia, SC 29202, USA
803-734-3893, Fax: 803-734-6310

South Peel Naturalists
Box 74501, Lorne Park PO
Mississauga, ON L5H 3A5, Canada
905-279-8807

Southeast Alaska Conservation Council
   (SEACC)
419 6th Street, #328
Juneau, AK 99801, USA
907-586-6942, Fax: 907-463-3312

Southern Utah Wilderness Alliance
1471 S. 1100 East
Salt Lake City, UT 84105-2423, USA
801-486-3161

Southwestern Cattle Raisers Association
1301 W. Seventh Street
Fort Worth, TX 76102-2665, USA

Strategies Saint-Laurent
690 Grande-Allée, 4th Floor
Quebec, PQ G1R 2K5, Canada
418-648-8079, Fax: 418-648-0991

Sydenham Field Naturalists
Box 22008
Wallaceburg, ON N8A 5G4, Canada

Taku Wilderness Association
Box 321
Atlin, BC V0W 1A0, Canada
250-651-0047

Tennessee Division of Natural Heritage
Department of Environment & Conservation
410 Church Street
Life and Casualty Tower, 8th Floor
Nashville, TN 37243-0447, USA
615-532-0431, Fax: 615-532-0046

Tetrahedron Alliance
Box 20, Cowley's Site
Sechelt, BC V0N 3A0, Canada

Texas A&M University
Department of Rangeland Ecology and
   Management
College Station, TX 77843-2126, USA
409-845-7332

Texas Center for Policy Studies
PO Box 2618
Austin, TX 78768, USA
512-474-0811

Texas Farm Bureau
PO Box 2689
Waco, TX 76702-2689, USA
254-772-3030

Texas Parks and Wildlife Department
4200 Smith School Road
Austin, TX 78744, USA
512-389-4800 or 800-792-1112
http://www.tpwd.state.tx.us

Texas Parks and Wildlife Department
PO Box 2127
Sulphur Springs, TX 75483, USA
903-945-3132

Tlell Watershed Society
Box 84
Tlell, BC V0T 1Y0, Canada
250-557-4471, Fax: 250-557-4472

Tongass Conservation Society
907-225-5827

TREE (Time to Respect the Earth's
   Ecosystems)
33 Riley Crescent
Winnipeg, MB R3T 0J5, Canada
204-452-9017, Fax: 204-774-4131

Trustees of Reservations
Northeast Regional Office
Crane Memorial Reservation
PO Box 563
Ipswich, MA 01938, USA
508-356-4351

Trustees of Reservations
Southeast Regional Office
Bradley Reservation
2468B Washington Street
Canton, MA 02021, USA
617-821-2977

Tuckamore Wilderness Society
Box 1, Site 12, RR 1
Cornerbrook, NF A2H 2N2, Canada
709-639-1770

Tucson Audubon Society
300 East University
Tucson, AZ 85705, USA
520-629-0510

UQCN—Union Québecoise pour la
   Conservation de la Nature
690 Grande Allée, 4th Floor
Quebec, PQ G1R 2K5, Canada
418-648-2104

U.S. Fish and Wildlife Service
1875 Century Boulevard
Atlanta, GA 30345, USA
404-679-4000

U.S. Fish and Wildlife Service
Department of the Interior
1849 C Street NW
Washington, DC 20240, USA
202-208-5634

U.S. Fish and Wildlife Service
Region 5, NE Regional Office
300 Westgate Center Drive
Hadley, MA 01035, USA
413-253-8300

U.S. Forest Service
Region 3, Great Lakes–Big Rivers Regional
 Office
1 Federal Drive
Federal Building
Fort Snelling, MN 55111, USA
612-725-3563

U.S. Forest Service
Region 6, Mountain-Prairie Regional Office
134 Union Boulevard
PO Box 25486
Denver, CO 80225, USA
303-236-7920

U.S. Forest Service
1720 Peachtree Road, Suite 800
Atlanta, GA 30345, USA
404-347-4177

Utah Wilderness Coalition
PO Box 520974
Salt Lake City, UT 84152-0974, USA
801-486-3161

Vermont Nongame & Natural Heritage Program
Vermont Fish & Wildlife Department
103 S. Main Street, 10 South
Waterbury, VT 05671-0501, USA
802-241-3700

Victoria Natural History Society
Box 5220, Station A
Victoria, BC V8R 6N4, Canada
604-658-8965

Virginia Division of Natural Heritage
Department of Conservation & Recreation
Main Street Station
1500 E. Main Street, Suite 312
Richmond, VA 23219, USA
804-786-7951, Fax: 804-371-2674

Watchdogs for Wildlife
Box 1480
Swan River, MB R0L 1Z0, Canada

Waterton Natural History Association
Box 145
Waterton, AB T0K 2M0, Canada
403-859-2340, Fax: 403-859-2624

West Central Research and Extension Station
University of Nebraska
North Platte, NE 69101-9495, USA
308-532-3611

West Virginia Highlands Conservancy
PO Box 306
Charleston, WV 25321, USA

West Virginia Natural Heritage Program
Department of Natural Resources Operations
 Center
Ward Road, PO Box 67
Elkins, WV 26241, USA
304-637-0245, Fax: 304-637-0250

Western Canada Wilderness Committee
20 Water Street
Vancouver, BC V6B 1A4, Canada
604-683-8220, Fax: 604-683-8229

Western Pennsylvania Conservancy
PNDI-West, 316 Fourth Avenue
Pittsburgh, PA 15222, USA
412-288-2777, Fax: 412-281-1792

White Rock and Surrey Naturalists
Box 75044, White Rock PO
White Rock, BC V4A 9M4, Canada
604-599-6877

Wild Earth
PO Box 455
Richmond, VT 05477, USA
802-434-4077

The Wilderness Society
105 W. Main Street, Suite E
Bozeman, MT 59715, USA
406-587-7331

The Wilderness Society
1424 Fourth Avenue, Suite 816
Seattle, WA 98101-2217, USA
202-624-6430

The Wilderness Society
Southeast Regional Office
1447 Peachtree Street NE, Suite 812
Atlanta, GA 30309, USA
404-872-9453

The Wilderness Society, Alaska Region
430 West 7th Avenue, #210
Anchorage, AK 99501, USA
907-272-9453

The Wildlands League
401 Richmond Street West, Suite 380
Toronto, ON M5V 3A8, Canada
416-971-9453, Fax: 416-979-3155

The Wildlands Project
SouthPAW
PO Box 3141
Asheville, NC 28802, USA

The Wildlands Project
1955 W. Grant Road, Suite 148
Tucson, AZ 85745, USA
520-884-0875

The Wildlife Society
c/o 1129 Queen's Avenue
Brandon, MB R7A 1L9, Canada
204-726-6450, Fax: 204-726-6518

Williams Lake Environmental Society
Box 4222
Williams Lake, BC V2G 2V3, Canada
604-392-2355

Wisconsin Department of Natural Resources
Box 7921
Madison, WI 53707, USA
608-266-2621

World Wildlife Fund Canada
90 Eglinton Avenue E, Suite 504
Toronto, ON M4P 2Z7, Canada
416-489-8800, Fax: 416-489-3611

World Wildlife Fund Canada, Quebec Region
1253 McGill College Avenue, Suite 446
Montreal, PQ H3B 2Y5, Canada
514-866-7800, Fax: 514-866-7808

Wyoming Outdoor Council
201 Main Street
Lander, WY 82520, USA
307-332-7031, Fax: 307-332-6899
email: woc@wyoming.com

Wyoming Wildlife Federation
PO Box 106
Cheyenne, WY 82003, USA
307-637-5433, Fax: 307-637-6629

Yukon Conservation Society
302 Hawkins Street
Whitehorse, YT Y1A 4S2, Canada

GLOSSARY

adaptive radiation   The evolution of a single species into many species that occupy diverse ways of life within the same geographical range (Wilson 1992).

alpha diversity   Species diversity within a single site.

amphibian   A member of the vertebrate class Amphibia (frogs and toads, salamanders, and caecilians).

amphipod   Any of a large group of small crustaceans with a laterally compressed body, belonging to the order Amphipoda.

anadromous   Species that spawn in freshwater and migrate to marine habitats to mature (e.g., salmon).

anthropogenic   Human induced.

aquatic   Growing in, living in, or frequenting water.

aquifer   A formation, group of formations, or part of a formation that contains sufficient saturated permeable material to yield significant quantities of water to wells and springs (Maxwell et al. 1995).

arctic   Referring to all nonforested areas north of the coniferous forests in the Northern Hemisphere (Brown and Gibson 1983).

artesian spring   A geologic formation in which water is under sufficient hydrostatic pressure to be discharged to the surface without pumping.

assemblage   In conservation biology, a predictable and particular collection of species within a biogeographic unit (e.g., ecoregion or habitat).

barrens   A colloquial name given to habitats with sparse vegetation or low agricultural productivity.

basin   See catchment.

beta-diversity   Species diversity between habitats (thus reflecting changes in species assemblages along environmental gradients).

biodiversity   (Also called biotic or biological diversity.) The variety of organisms considered at all levels, from genetic variants belonging to the same species through arrays of species to arrays of genera; families, and still higher taxonomic levels; includes the variety of ecosystems, which comprise both communities of organisms within particular habitats and the physical conditions under which they live (Wilson 1992).

biodiversity conservation   Five classes of biodiversity conservation priorities were determined in this priorities study by integrating biological distinctiveness with conservation status (figure 5.1). The five classes roughly reflect the concern with which we should view the erosion of biodiversity in different ecoregions, and the timing and sequence of response by governments and donors to the loss of biodiversity.

biogeographic unit   A delineated area based on biogeographic parameters.

433

biogeography The study of the geographic distribution of organisms, both past and present (Brown and Gibson 1983).

biological distinctiveness Scale-dependent assessment of the biological importance of an ecoregion based on species richness, endemism, relative scarcity of ecoregion, and rarity of ecological phenomena. Biological distinctiveness classes are globally outstanding, regionally outstanding, bioregionally outstanding, and nationally important.

biome A global classification of natural communities in a particular region based on dominant or major vegetation types and climate.

bioregion A geographically related assemblage of ecoregions that share a similar biogeographic history and thus have strong affinities at higher taxonomic levels (e.g., genera, families). There are six biogeographic divisions of North America, consisting of contiguous ecoregions, designed to better address the biogeographic distinctiveness of ecoregions.

bioregionally outstanding A biological distinctiveness class.

biota The combined flora, fauna, and microorganisms of a given region (Wilson 1992).

biotic Biological, especially referring to the characteristics of faunas, floras, and ecosystems (Wilson 1992).

bog A poorly drained area rich in plant residues, usually surrounded by an area of open water, and having characteristic flora.

boreal forest Type of major habitat occurring in the temperate and subtemperate zones of the Northern Hemisphere that characteristically has coniferous trees with some types of deciduous trees (Brown and Gibson 1983).

canebrake A thicket of cane.

catadromous Diadromous species that spawn in marine habitats and migrate to freshwater to mature (e.g., eels).

catchment All lands enclosed by a continuous hydrologic-surface drainage divide and lying upslope from a specified point on a stream (Maxwell et al. 1995); or, in the case of closed-basin systems, all lands draining to a lake.

centinelan extinction The phenomenon of species going extinct before they have been discovered or described by the scientific community.

chaparral The type of sclerophyllous scrub occurring in the southwestern region of North America with a Mediterranean or xeric climate.

clear-cut A logged area where all or virtually all the forest canopy trees have been eliminated.

community Collection of organisms of different species that co-occur in the same habitat or region and that interact through trophic and spatial relationships (Fiedler and Jain 1992).

conifer A tree or shrub in the phylum Gymnospermae whose seeds are borne in woody cones. There are 500–600 species of living conifers (Norse 1990).

conservation biology Relatively new discipline that treats the content of biodiversity, the natural processes that produce it, and the techniques used to sustain it in the face of human-caused environmental disturbance (Wilson 1992).

conservation status As defined in this book, the assessment of the status of ecological processes and of the viability of species populations in an ecoregion. The different status categories used are extinct, critical, endangered, vulnerable, relatively stable, and relatively intact. The snapshot conservation status is based on an index derived from values of four landscape-level variables. The final conservation status is the snapshot assessment modified by an analysis of threats to the ecoregion over the next twenty years.

conversion Habitat altered by human activities to such an extent that it no longer supports most characteristic native species and ecological processes.

creek The smallest size class of a lotic system, typically associated with headwaters.

critical Conservation status category characterized by low probability of persistence of remaining intact habitat.

deciduous forest Habitat type dominated by trees whose leaves last a year or less; they drop and replace their leaves over periods sufficiently distinct that they are leafless for some portion of the year (Norse 1990).

degradation The loss of native species and processes due to human activities such that only certain components of the original biodiversity persist, often including significantly altered natural communities.

diadromous Species that migrate between freshwater and marine habitats while spawning in one habitat and maturing in another (Nyman 1991).

disturbance Any relatively discrete event in time that disrupts ecosystem, community, or population structure and changes resources, substrate availability, or the physical environment (Fiedler and Jain 1992).

drainage basin See catchment.

ecological processes Complex mix of interactions between animals, plants, and their environment that ensure that an ecosystem's full range of biodiversity is adequately maintained. Examples include population and predator-prey dynamics, pollination and seed dispersal, nutrient cycling, migration, and dispersal.

ecoregion A large area of land or water that contains a geographically distinct assemblage of natural communities that (a) share a large majority of their species and ecological dynamics, (b) share similar environmental conditions, and (c) interact ecologically in ways that are critical for their long-term persistence.

ecoregion-based conservation Conservation strategies and activities whose efficacy are enhanced through close attention to larger (landscape- or aquascape-level) spatial and temporal-scale patterns of biodiversity, ecological dynamics, threats, and strong linkages of these issues to fundamental goals and targets of biodiversity conservation.

ecosystem A system resulting from the integration of all living and nonliving factors of the environment (Tansley 1935).

ecosystem service Service provided free by an ecosystem, or by the environment, such as clean air, clean water, and flood amelioration.

endangered Conservation status category characterized by medium to low probability of persistence of remaining intact habitat.

Endangered Spaces Campaign A science-based campaign launched by World Wildlife Fund Canada in 1989 to ensure that all of Canada's terrestrial regions are represented with protected areas by the year 2000 and Canada's marine regions are represented by 2010. The campaign is guided by the principle that the conservation of Canada's biological diversity requires establishment of a network of protected areas capturing the full range of ecological features characteristic of Canada's natural regions.

endemic A species or race native to a particular place and found only there (Wilson 1992).

endemism Degree to which a geographically circumscribed area, such as an ecoregion or a country, contains species not naturally occurring elsewhere.

endorheic Referring to a closed basin with no natural watercourses leading to the sea.

enduring feature A landform complex or geographic unit within a natural region characterized by relatively uniform origin, texture of surficial material, and topography-relief patterns.

environmental gradients Changes in biophysical parameters such as rainfall, elevation, or soil type over distance.

estuarine Associated with an estuary.

estuary A deepwater tidal habitat and its adjacent tidal wetlands, which are usually semi-enclosed by land but have open, partly obstructed, or sporadic access to the open ocean, and in which ocean water is at least occasionally diluted from freshwater runoff from the land (Maxwell et al. 1995).

evolutionary phenomenon Within the context of WWF regional conservation assessments, refers to patterns of community structure and taxonomic composition that are the result of an extraordinary example of evolutionary processes, such as pronounced adaptive radiations.

evolutionary radiation   See radiation.

exotic species   A species that is not native to an area and has been introduced intentionally or unintentionally by humans; not all exotics become successfully established.

extinct   A species or population (or any lineage) with no surviving individuals.

extinction   The termination of any lineage of organisms, from subspecies to species and higher taxonomic categories from genera to phyla. Extinction can be local, in which one or more populations of a species or other unit vanish but others survive elsewhere, or total (global), in which all the populations vanish (Wilson 1992).

extirpated   Status of a species or population that has completely vanished from a given area but continues to exist in some other location.

extirpation   Process by which an individual, population, or species is destroyed (Fiedler and Jain 1992).

family   In the hierarchical classification of organisms, a group of species of common descent higher than the genus and lower than the order; a related group of genera (Wilson 1992).

fauna   All the animals found in a particular place.

fire regime   The characteristic frequency, intensity, and spatial distribution of natural fire events within a given ecoregion or habitat.

flooded grassland   A grassland habitat that experiences regular inundation by water.

flora   All the plants found in a particular place.

fragmentation   Landscape-level variable measuring the degree to which remaining habitat is separated into smaller discrete blocks; also, the process by which habitats are increasingly subdivided into smaller units (Fiedler and Jain 1992).

freshwater   In the strictest sense, water that has less than 0.5% of salt concentration (Brown and Gibson 1983); in this study, refers to rivers, streams, creeks, springs, and lakes.

genera   The plural of genus.

genus   A group of similar species with common descent, ranked below the family (Wilson 1992).

glade   An open space surrounded by forest.

Global 200   A set of approximately 200 terrestrial, freshwater, and marine ecoregions around the world that support globally outstanding or representative biodiversity as identified through analyses by World Wildlife Fund–United States. One component of the Living Planet Campaign.

globally outstanding   Biological distinctiveness category for units of biodiversity whose biodiversity features are equaled or surpassed in only a few other areas around the world.

grassland   A habitat type with landscapes dominated by grasses and with biodiversity characterized by species with wide distributions, communities being relatively resilient to short-term disturbances but not to prolonged, intensive burning or grazing. In such systems larger vertebrates, birds, and invertebrates display extensive movement to track seasonal or patchy resources.

groundwater   Water in the ground that is in the zone of saturation, from which wells and springs and groundwater runoff are supplied (Maxwell et al. 1995).

guild   Group of organisms, not necessarily taxonomically related, that are ecologically similar in characteristics such as diet, behavior, or microhabitat preference, or with respect to their ecological role in general.

gymnosperm   Any of a class or subdivision of woody vascular seed plants that produce naked seeds, not enclosed in an ovary. Conifers and cycads are examples of gymnosperms.

habitat   An environment of a particular kind, often used to describe the environmental requirements of a certain species or community (Wilson 1992).

habitat blocks   Landscape-level variable that assesses the number and extent of blocks of contiguous habitat, taking into account size requirements for populations and ecosystems to function naturally. It is measured here by a habitat-dependent and ecoregion size–dependent system.

habitat loss Landscape-level variable that refers to percentage of the original land area of the ecoregion that has been lost (converted). It underscores the rapid loss of species and disruption of ecological processes predicted to occur in ecosystems when the total area of remaining habitat declines.

habitat type In this study, a habitat type is defined by the structure and processes associated with one or more natural communities. An ecoregion is classified under one major habitat type but may encompass multiple habitat types.

headwater The source of a stream or river.

herbivore A plant-eating animal, especially ungulates.

herpetofauna All the species of amphibians and reptiles inhabiting a specified region.

hibernacula Microhabitats where organisms hibernate during the winter.

indigenous Native to an area.

intact habitat Relatively undisturbed areas characterized by the maintenance of most original ecological processes and by communities with most of their original native species still present.

introduced species See exotic species.

invasive species Exotic species (i.e., alien or introduced) that rapidly establish themselves and spread through the natural communities into which they are introduced.

invertebrate Any animal lacking a backbone or bony segment that encloses the central nerve cord (Wilson 1992).

isopod A member of the crustacean order Isopoda, a diverse group of flattened and segmented invertebrates. Pillbugs are an example.

karst Applies to areas underlain by gypsum, anhydrite, rock salt, dolomite, quartzite (in tropical moist areas), and limestone, often highly eroded, complex landscapes with high levels of plant endemism.

keystone species Species that are critically important for maintaining ecological processes or the diversity of their ecosystems.

landform The physical shape of the land, reflecting geologic structure and processes of geomorphology that have sculptured the structure.

landscape An aggregate of landforms, together with its biological communities (Lotspeich and Platts 1982).

landscape ecology Branch of ecology concerned with the relationship between landscape-level features, patterns, and processes and the conservation and maintenance of ecological processes and biodiversity in entire ecosystems.

late-successional Species, assemblages, structures, and processes associated with mature natural communities that have not experienced significant disturbance for a long time.

lentic Referring to standing freshwater habitats, such as ponds and lakes (Brown and Gibson 1983).

life cycle The entire lifespan of an organism from the moment it is conceived to the time it reproduces (Wilson 1992).

Living Planet Campaign An ambitious public engagement effort launched by World Wildlife Fund–United States in conjunction with the WWF international network in April 1997 and designed to make the final one thousand days before the year 2000 a turning point in the conservation of some of the earth's most outstanding endangered species and spaces. The campaign aims to engage everyone to take part in leaving our children a living planet—particularly relating to approximately 232 natural habitats that have been identified by WWF as the most outstanding ecoregions on earth. This list is called the Global 200.

lotic Refers to running freshwater habitats, such as springs and streams (Brown and Gibson 1983).

macroinvertebrates Invertebrates large enough to be seen with the naked eye (e.g., most aquatic insects, snails, and amphipods) (Maxwell et al. 1995).

major habitat type   Set of ecoregions that (a) experience comparable climatic regimes; (b) have similar vegetation structure; (c) display similar spatial patterns of biodiversity; and (d) contain flora and fauna with similar guild structures and life histories. Ten major habitat types (MHTs) are defined in this study.

marine   Living in salt water (Brown and Gibson 1983).

mesic   Moist, wet.

mesophytic   Applies to plants that grow under conditions of abundant moisture.

mollusk   An animal belonging to the phylum Mollusca, such as a snail or clam (Wilson 1992).

nationally important   A biological distinctiveness category.

natural disturbance event   Any natural event that significantly alters the structure, composition, or dynamics of a natural community. Floods, fire, and storms are examples.

natural range of variation   A characteristic range of levels, intensities, and periodicities associated with disturbances, population levels, or frequency in undisturbed habitats or communities.

neotropical migrant   Birds, bats, or invertebrates that seasonally migrate between the Nearctic and Neotropics.

non-native species   See exotic species.

nonpoint source   A diffuse form of water quality degradation produced by erosion of land that causes sedimentation of streams, eutrophication from nutrients and pesticides used in agricultural and silvicultural practices, and acid rain resulting from burning fuels that contain sulfur (Lotspeich and Platts 1982).

obligate species   A species that must have access to a particular habitat type to persist.

old-growth forest   A late-successional or climax stage in forest development exhibiting characteristic structural features, species assemblages, and ecological processes.

oribatid   The largest and most abundant group of free-living mites.

phylum   Primary classification of animals that share similar body plans and development patterns.

population   In biology, any group of organisms belonging to the same species at the same time and place (Wilson 1992).

population sink   An area where a species displays negative population growth, often due to insufficient resources and habitat, or high mortality.

prairie   An extensive tract of flat or rolling grassland. A term especially used to refer to the plains of central North America.

predator-prey system   An assemblage of predators and prey species and the ecological interactions and conditions that permit their long-term coexistence.

protected areas   Regions allocated for the protection of natural areas, ecological processes, species, or physical features of the environment, such as nature reserves, national parks, wilderness areas, wildlife sanctuaries, or national monuments and landmarks.

protection   Landscape-level variable that assesses how well humans have conserved large blocks of intact habitat and the biodiversity they contain. It is measured here by the number of protected blocks and their sizes in a habitat-dependent and ecoregion size–dependent system.

pyrogenic   Pertaining to communities or habitats that develop after fire events.

radiation   The diversification of a group of organisms into multiple species, due to intense isolating mechanisms or opportunities to exploit diverse resources.

rarity   Seldom occurring either in absolute number of individuals or in space (Fiedler and Jain 1992).

refugia Habitats that have allowed the persistence of species or communities because of the stability of favorable environmental conditions over time.

regionally outstanding A biological distinctiveness category.

relatively intact Conservation status category indicating the least possible disruption of ecosystem processes. Natural communities are largely intact, with species and ecosystem processes occurring within their natural ranges of variation.

relatively stable Conservation status category between vulnerable and relatively intact, in which extensive areas of intact habitat remain but local species declines and disruptions of ecological processes have occurred.

relictual taxa A species or group of organisms largely characteristic of a past environment or ancient biota.

representation The protection of the full range of biodiversity of a given biogeographic unit within a system of protected areas.

restoration Management of a disturbed and/or degraded habitat that results in recovery of its original state (Wilson 1992).

riparian Referring to the interface between freshwater streams and lakes and the terrestrial landscape.

savanna A habitat largely dominated by grasslands but with woodland and gallery forest elements.

sclerophyll Type of vegetation characterized by hard, leathery, evergreen foliage that is specially adapted to prevent moisture loss, generally characteristic of regions with Mediterranean climates.

sclerophyllous Relating to sclerophyll.

semiaquatic Living partly in or adjacent to water (Brown and Gibson 1983).

seral The stages a natural community experiences after a disturbance event.

shrub steppe Shrub and grass habitats in cooler environments.

shrublands Habitats dominated by various species of shrubs, often with many grass and forb elements.

silviculture The management of forest trees, usually to enhance timber production.

sinkholes Depressions or cavities created by dissolution of limestone bedrock or collapse of caves. Typically found in karst landscapes.

source pool A habitat that provides individuals or propagules that disperse to and colonize adjacent or neighboring habitats.

species The basic unit of biological classification, consisting of a population or series of populations of closely related and similar organisms (Wilson 1992).

species richness A simple measure of species diversity calculated as the total number of species in a habitat or community (Fiedler and Jain 1992).

spring A natural discharge of water as leakage or overflow from an aquifer through a natural opening in the soil or rock onto the land surface or into a body of water (Hobbs 1992).

steppe Arid land with xerophilous vegetation, usually found in regions with extreme temperature ranges and loess soils.

stream A general term for a body of flowing water (Maxwell et al. 1995); often used to describe a mid-sized tributary (as opposed to a river or creek).

subspecies Subdivision of a species. Usually defined as a population or series of populations occupying a discrete range and differing genetically from other geographical races of the same species (Wilson 1992).

subtropical An area in which the mean annual temperature ranges from 13°C to 20°C (Brown and Gibson 1983).

taiga Subarctic habitat type consisting of moist coniferous forest, dominated by spruce and fir, that begins where tundra ends.

tallgrass prairie A type of North American prairie dominated by several species of tall grasses.

taxon (pl. taxa) A general term for any taxonomic category, e.g., a species, genus, family, or order (Brown and Gibson 1983).

temperate An area in which the mean annual temperature ranges from 10°C to 13°C.

terrestrial Living on land.

tributary A stream or river that flows into a larger stream, river, or lake, feeding it water.

umbrella species A species whose effective conservation will benefit many other species and habitats, often due to its large area requirements or sensitivity to disturbance.

ungulate A member of the group of mammals with hoofs, of which most are herbivorous.

vagile Able to be transported or to move actively from one place to another (Brown and Gibson 1983).

vascular plant A plant that possesses a specialized vascular system for supplying its tissues with water and nutrients from the roots and food from the leaves.

vulnerable Conservation status category characterized by good probability of persistence of remaining intact habitat (assuming adequate protection) but also by loss of some sensitive or exploited species.

watershed See catchment.

wetlands Lands transitional between terrestrial and aquatic systems, where the water table is usually at or near the surface or the land is covered by shallow water. These areas are inundated or saturated by surface or groundwater at a frequency and duration sufficient to support a prevalence of vegetation typically adapted for life in saturated soil conditions (Maxwell et al. 1995).

xeric Dryland or desert.

xerophilous Thriving in or tolerant of xeric climates.

zoogeography The study of the distributions of animals (Brown and Gibson 1983).

LITERATURE CITED
AND CONSULTED

Abell, R., D. M. Olson, E. Dinerstein, P. T. Hurley, W. Eichbaum, S. Walters, T. Allnutt, W. Wettengel, and C. J. Loucks. 1999. *A conservation assessment of the freshwater ecoregions of North America.* Washington, DC: Island Press.

Adam, P. 1990. *Saltmarsh ecology.* Cambridge, England: Cambridge University Press.

Agee, J. K. 1993. *Fire ecology of Pacific Northwest forests.* Washington, DC: Island Press.

———. 1994. *Fire and weather disturbances in terrestrial ecosystems of the Eastern Cascades.* PNW-GTR-320. Washington, DC: USDA Forest Service.

Alaback, P. B. 1988. Endless battles, verdant survivors. *Natural History* 8: 45–48.

———. 1991. Comparative ecology of temperate rain forests of the Americas along analogous climatic gradients. *Revista Chilena de Historia: Natural* 64: 399–412.

———. 1993. Biodiversity patterns in relation to climate and a genetic base for the rainforests of the west coast of North America. In R. Lawford, P. Alaback, and E. R. Fuentes (editors), *High latitude rain forests of the west coast of the Americas: Climate, hydrology, ecology, and conservation.* New York: Springer-Verlag.

———. In review. Ecological characteristics of temperate rainforests: Some implications for forest conservation in Prince William Sound.

Alaback, P. B., and G. P. Juday. 1989. Structure and composition of low elevation old growth forests in research natural areas of southeast Alaska. *Natural Areas Journal* 9: 27–39.

Alaska Fish and Wildlife Research Center. 1988. *Populations, productivity, and feeding habits of seabirds on St. Lawrence Island.* Anchorage: Alaska Fish and Wildlife Research Center.

Albert, D. 1994. Michigan's landscape. Pages 5–28 in D. C. Evers (editor), *Endangered and threatened wildlife of Michigan.* Ann Arbor: University of Michigan Press.

Aldon, E. F., C. E. Gonzales Vicente, and W. H. Moir, technical coordinators. 1987. *Strategies for classification and management of native vegetation for food production in arid zones.* General Technical Report RM-150. Fort Collins, CO: USDA Rocky Mountain Forest and Range Experiment Station.

Alexander, B. G., F. Ronco, A. S. White, and J. A. Ludwig. 1984. *Douglas-fir habitat types of northern Arizona.* General Technical Report RM-108. Fort Collins, CO: USDA Rocky Mountain Forest and Range Experiment Station.

Alig, Ralph J., W. G. Hohenstein, B. C. Murray, and R. G. Haight. 1990. *Changes in area of timberland in the United States, 1952–2040, by ownership, forest type, region, and state.* SE-64. Asheville, NC: USDA Southeastern Forest Experiment Station.

Allison, A., S. E. Miller, and G. M. Nishida. 1995. Hawaii biological survey. Page 362 in E. T. LaRoe, G. S. Farris, C. E. Puckett, P. D. Doran, and M. J. Mac (editors), *Our living resources: A report to the nation on the distribution, abundance, and health of U.S. plants, animals, and ecosystems.* Washington, DC: U.S. Department of the Interior, National Biological Service.

Alverson, W. S., D. M. Waller, and S. L. Solheim. 1988. Forests too deer: Edge effects in northern Wisconsin. *Conservation Biology* 2: 348–358.

Ames, C. R. 1977. Wildlife conflicts in riparian management: Grazing. Pages 49–51 in R. R. Johnson and D. A. Jones (technical coordinators), *Importance, preservation, and management of riparian habitat: A symposium.* General Technical Report RM-43. Fort Collins, CO: USDA Rocky Mountain Forest and Range Experiment Station.

Anderson, R. G. 1990. The historic role of fire in the North American grassland. Pages 8–18 in L. C. Scott and L. L. Wallace (editors), *North American tallgrass prairies.* Norman: University of Oklahoma Press.

Andrus, R. E., and E. F. Karlin. 1989. A preliminary report on New Jersey's rare species of sphagnum: A first look at endangered and threatened bryophytes in New Jersey. In E. F. Karlin (editor), *Proceedings of a conference on New Jersey's rare and endangered plants and animals.* Mahwah, NJ: Institute for Environmental Studies.

Angermeier, P. L., and J. R. Karr. 1994. Biological integrity versus biological diversity as policy directives. *BioScience* 44: 690–697.

Armour, C. L., D. A. Duff, and W. Elmore. 1991. The effects of livestock grazing on riparian and stream ecosystems. *Fisheries* 16: 7–11.

Arno, S. F., E. Reinhardt, and J. Scott. 1993. *Forest structure and landscape patterns in the subalpine lodgepole pine type: A procedure for quantifying past and present conditions.* INT-294. Ogden UT: USDA Forest Service, Intermountain Research Station.

Bahre, D. 1995. Human impacts on the grasslands of southern Arizona. Pages 230–264 in M. P. McClaran and T. R. Van Devender (editors), *The desert grasslands.* Tucson: University of Arizona Press.

Bailey, R. G. 1984. Testing an ecosystem regionalization. *Journal of Environmental Management* 19: 239–248.

———. 1987. Suggested hierarchy of criteria for multi-scale ecosystem mapping. *Landscape and Urban Planning* 14: 313–319.

———. 1992. Ecogeographic analysis: A guide to the ecological division of land for resource management. In *Proceedings of the national workshop on taking an ecological approach to management.* WO-WSA-3. Washington, DC: USDA Forest Service, Watershed and Air Management.

———. 1994. *Ecological classification for the United States.* Washington, DC: USDA Forest Service.

———. 1995. *Description of the ecoregions of the United States,* 2nd ed. rev. and expanded. Washington, DC: USDA Forest Service. Includes separate map at 1:7,500,000.

———. 1996. *Ecosystem geography.* New York: Springer-Verlag.

Bailey, R. G., P. E. Avers, T. King, and W. H. McNab (editors). 1994. *Ecoregions and subecoregions of the United States.* Washington, DC: USDA Forest Service.

Bailey, R. G., S. C. Zoltai, and E. B. Wiken. 1985. Ecological regionalization in Canada and the United States. *Geoforum* 16(3): 265–275.

Baker, E. 1984. *An island called California.* Berkeley: University of California Press.

Banfield, A. W. F. 1974. *The mammals of Canada.* Ottawa: National Museums of Canada.

Barbour, M. G. 1988. California upland forests and woodlands. Pages 131–164 in M. G. Barbour and W. D. Billings (editors), *North American terrestrial vegetation.* Cambridge, England: Cambridge University Press.

Barbour, M. G., and W. D. Billings (editors). 1988. *North American terrestrial vegetation.* Cambridge, England: Cambridge University Press.

Barbour, M. G., J. H. Burk, and W. D. Pitts. 1980. *Terrestrial plant ecology.* Menlo Park, CA: Benjamin/Cummings.

Barbour, M. G., and J. Major (editors). 1977. *Terrestrial vegetation of California.* New York: John Wiley.

——— (editors). 1988. Mixed evergreen forest supplement. Page 1010 in *Terrestrial vegetation of California,* new expanded edition. Special Publication No. 9. Sacramento: California Native Plant Society.

——— (editors). 1988. *Terrestrial vegetation of California.* Special Publication Number 9. Sacramento: California Native Plant Society.

Barbour, M. G., B. Pavlik, F. Drysdale, and S. Lindstrom. 1991. California vegetation: Diversity and change. *Fremontia: A Journal of the California Native Plant Society* 19(1): 3–12.

Barbour, M. G., B. Pavlik, F. Drysdale, and S. Lindstrom. 1993. *California's changing landscapes: Diversity and conservation of California vegetation.* Sacramento: California Native Plant Society.

Barnes, B. V. 1991. Deciduous forests of North America. Pages 219–344 in E. Rohrig and B. Ulrich (editors), *Temperate deciduous forests.* Amsterdam: Elsevier.

Barnett, L. 1974. *The American wilderness: The ancient Adirondacks.* Alexandria, VA: Time-Life Books.

Bean, M. J. 1993. Invertebrates and the Endangered Species Act. *Wings: Essays on invertebrate conservation* 17(2): 12–15.

Beauchamp, R. M. 1986. *A flora of San Diego County, California.* National City, CA: Sweetwater River Press.

Beccaloni, G. W., and K. J. Gaston. 1995. Predicting the species richness of neotropical forest butterflies: Ithomiinae (Lepidoptera: Nympalidae) as indicators. *Biological Conservation* 71: 77–86.

Behnke, R. J., and M. Zarn. 1976. *Biology and management of threatened and endangered western trouts.* General Technical Report RM-28. Fort Collins, CO: USDA Forest Service, Rocky Mountain Forest and Range Experiment Station.

Bennett, S. H., and J. B. Nelson 1989. *South Carolina Heritage Trust Program.* Columbia, SC: South Carolina Wildlife and Marine Resources Department.

Benson, L. D. 1957. The floras of North America. Pages 577–640 in L. Benson (editor), *Plant classification.* Boston: D. C. Heath & Co.

———. 1982. *The cacti of the United States and Canada.* Stanford, CA: Stanford University Press.

Berger, J. 1990. Persistence of different-sized populations: An empirical assessment of rapid extinction in bighorn sheep. *Conservation Biology* 4: 91–98.

Berger, J., and J. W. Stinton. 1985. *Water, earth, and fire: Land use and environmental planning in the New Jersey pine barrens.* Baltimore, MD: Johns Hopkins University Press.

Beschta, R. L., C. A. Frissell, R. Gresswell, R. Hauer, J. R. Karr, G. W. Minshall, D. A. Perry, and J. J. Rhodes. 1995. *Wildfire and salvage logging: Recommendations for ecologically sound post-fire salvage logging and other post-fire treatments on federal lands in the west.* Unpublished report. Oregon State University, Corvallis.

Betz, R. F. 1978. The prairies of Indiana. In *Proceedings of the fifth midwest prairie conference.* Ames: Iowa State University.

Bibby, C. J. 1992. *Putting biodiversity on the map: Priority areas for global conservation.* Washington, DC: Integrated Conservation and Development Project (ICDP).

Bingham, B. B. 1992. Distinctive features and definitions of young, mature, and old growth Douglas-fir/hardwood forests. In L. F. Ruggiero, K. B. Aubry, A. B. Cary, and M. H. Huff (editors), *Wildlife and vegetation of unmanaged Douglas-fir forests.* General Technical Report PSW-GRW-285. Portland, OR: USDA Forest Service, Pacific Northwestern Research Station.

Bingham, B. B., and J. O. Sawyer. 1988. Volume and mass of decaying logs in an upland old growth redwood forest. *Canadian Journal of Forest Research* 18: 1649–1651.

Binkley, D., and T. C. Brown. 1993. Forest practices as nonpoint sources of pollution in North America. *Water Resources Bulletin* 29(5): 729–740.

Biota of North America Program (BONAP), North Carolina Botanical Garden. International vascular plant database. University of North Carolina at Chapel Hill.

Birch, T. W., and E. H. Wharton. 1982. *Land use change in Ohio, 1952 to 1979.* NE-70. Radnor, PA: USDA Forest Service, Northeastern Forest Experiment Station.

Bland, W. L., and C. A. Jones. 1993. Agricultural technology makes its mark, 1940 to the present. Pages 177–191 in R. Sharpless and J. C. Yelderman, Jr. (editors), *The Texas Blackland Prairie Land History and Culture. Proceedings of the Sympposium of the Natural Regions of Texas.* Waco, TX: Baylor University Program for Regional Studies.

Bleich, V. C., and S. A. Holl. 1982. Management of chaparral habitat for mule deer and mountain sheep in southern California. Pages 247–254 in *Proceedings of the Symposium on Dynamics and Management of Mediterranean-Type Ecosystems.* General Technical Report PSW-58. Berkeley, CA: USDA Forest Service, Pacific Southwest Forest and Range Experiment Station.

Bliss, L. C. 1988. Arctic tundra and polar desert biome. Pages 2–32 in M. G. Barbour and W. D. Billings (editors), *North American terrestrial vegetation.* Cambridge, England: Cambridge University Press.

Bock, J. H., and C. E. Bock. 1995. The challenges of grassland conservation. Pages 199–222 in A. Joern and K. H. Keeler (editors), *The changing prairie—North American grasslands.* New York: Oxford University Press.

Bolsinger, C. L. 1988. *The hardwoods of California's timberlands, woodlands, and savannas.* Resource Bulletin PNW-RB-148. Portland, OR: USDA Forest Service, Pacific Northwest Research Station.

Bolsinger, C. L., and K. L. Waddell. 1993. *Area of old growth forests in California, Oregon, and Washington.* Resource Bulletin PNW-RB-197. Portland, OR: USDA Forest Service, Pacific Northwest Research Station.

Borland, V., R. Noss, P. Strittholt, P. Frost, C. Carroll, and R. Nawa. 1995. A biodiversity conservation plan for the Klamath/Siskiyou Region. *Wild Earth* 5: 52–59.

Born, J. D., and D. D. Van Hooser. *Intermountain forest survey, remote sensing, and geographic information.* Fort Collins, CO: USDA Forest Service, Intermountain Research Station.

Bowen, E. 1972. *The American wilderness: The high Sierra.* New York: Time-Life Books.

Bowler, P. A. 1990. Coastal sage scrub restoration: I. The challenge of mitigation. *Restoration and Management Notes* 8: 78–82.

Brash, A. R. 1987. The history of avian extinction and forest conversion on Puerto Rico. *Biological Conservation* 39: 97–111.

Braun, E. L. 1950. *Deciduous forests of eastern North America.* New York: Hafner Press/MacMillan.

———. 1989. *The woody plants of Ohio.* Columbus: Ohio State University Press.

Breden, T. F. 1989. A preliminary natural community classification for New Jersey. In E. F. Karlin (editor), *Proceedings of a conference on New Jersey's rare and endangered plants and animals.* Mahwah, NJ: Institute for Environmental Studies.

Briggs, J. C. 1974. *Marine zoogeography.* New York: McGraw-Hill.

Brinson, M. M., B. L. Swift, R. C. Pantico, and J. S. Barclay. 1981. *Riparian ecosystems: Their ecology and status.* FWS/OBS-81/17. Kearneysville, WV: USDA Fish & Wildlife Service.

Brockman, C. F. 1986. *Trees of North America.* New York: Golden Press.

Brode, J. M., and R. B. Bury. 1984. The importance of riparian systems to amphibians and reptiles. Pages 30–36 in R. E. Warner and K. Hendrix (editors), *California riparian systems: Ecology, conservation, and productive management.* Berkeley: University of California Press.

Brooks, D. J. 1993. *U.S. forests in a global context.* RM-228. Fort Collins, CO: USDA Forest Service, Rocky Mountain Forest and Range Experiment Station.

Brooks, R. T., D. B. Kittredge, and C. L Alerich. 1993. *Forest resources of southern New England.* NE-127. Radnor, PA: USDA Forest Service, Northeastern Forest Experiment Station.

Brotherson, J. D., and R. Q. Landers. 1978. Recovery from severe grazing in an Iowa tall-grass prairie. In *Proceedings of the fifth midwest prairie conference.* Ames: Iowa State University Press.

Brown, D. 1972. *The American wilderness: Wild Alaska.* New York: Time-Life Books.

Brown, D. E. (editor). 1994. *Biotic communities: Southwestern United States and northwestern Mexico.* Salt Lake City: University of Utah Press.

Brown, J. H. 1995. *Macroecology.* Chicago: University of Chicago Press.

Brown, J. H., and A. C. Gibson. 1983. *Biogeography.* St. Louis, MO: C.V. Mosby.

Brown, T. C., and D. Binkley. 1994. *Effect of management on water quality in North American forests.* RM-248. Fort Collins, CO: USDA Forest Service, Rocky Mountain Forest and Range Experiment Station.

Buchert, G. P., O. P. Rajora, J. V. Hood, and B. P. Dancik. 1997. Effects of harvesting on genetic diversity in old growth eastern white pine in Ontario, Canada. *Conservation Biology* 11(3): 747–758.

Buckhouse, J. C., and G. F. Gifford. 1976. Water quality implications of cattle grazing on a semiarid watershed in southeastern Utah. *Journal of Range Management* 29: 109–113.

Burgess, T. L. 1995. Desert grassland, mixed shrub savanna, shrub stepp, or semidesert shrub. Pages 31–67 in *The desert grassland.* Tucson: University of Arizona Press.

Burkhead, N. M., and R. E. Jenkins. 1991. Fishes. Pages 321–338 in K. Terwilliger (editor), *Virginia's endangered species.* Blacksburg, VA: McDonald and Woodward.

Burt, H. B., and R. P. Grossenheider. 1980. *A field guide to the mammals: North America north of Mexico.* Boston: Houghton Mifflin.

Byrd, M. A., and D. W. Johnston. 1991. Birds. Pages 477–486 in K. Terwilliger (editor), *Virginia's endangered species.* Blacksburg, VA: McDonald and Woodward.

Caicco, S. L., J. M. Scott, B. Butterfield, and B. Csuti. 1995. A gap analysis of the management status of the vegetation of Idaho. *Conservation Biology* 9: 498–511.

California Native Plant Society. 1984. Serpentine flora, notes on prominent sites in California. *The Lassics Fremontia, a Journal of the California Native Plant Society* 11: 15–17.

———. 1986. Darlingtonia seeps. *The Lassics Fremontia, a Journal of the California Native Plant Society* 14: 18.

Canadian Parks and Wilderness Society, Wildlands League chapter. 1997. *Ontario's forest products industry: A new appetite in the forest.* Fact Sheet #3. Toronto, Ontario: Forest Diversity and Community Survival.

Capp, J., B. Van Zee, P. Alaback, J. Boughton, M. Copenhagen, and J. Martin. 1992. *Ecological definitions for old growth forest types in the Alaska region.* R10-TP-28. Juneau, AK: USDA Forest Service.

Carothers, S. W. 1977. Importance, preservation, and management of riparian habitats: An overview. Pages 2–4 in R. R. Johnson and D. A. Jones (technical coordinators), *Importance, preservation, and management of riparian habitat: A symposium.* General Technical Report RM-43. Fort Collins, CO: USDA Forest Service, Rocky Mountain Forest and Range Experiment Station.

Carothers, S. W., and R. R. Johnson. 1975. Water management practices and their effects on nongame birds in range habitats. Pages 210–212 in D. R. Smith (editor), *Proceedings of the symposium on management of forest and range habitats for nongame birds.* General Technical Report WO-1. Washington, DC: USDA Forest Service.

Carothers, S. W., R. R. Johnson, and S. W. Aitchison. 1974. Population structure and social organization of Southwestern riparian birds. *American Zoologist* 14: 97–108.

Carr, A. 1973. *The American wilderness: The Everglades.* New York: Time-Life Books.

Carter, D. 1995. Utah wilderness: The first decade! *Wild Earth* (Summer 1995): 16–21.

Carter, E. K., and H. T. Odum (editors). 1984. *Cypress swamps.* Gainesville: University of Florida Press.

Chabreck, R. A. 1988. *Coastal marshes: Ecology and wildlife management.* Minneapolis: University of Minnesota Press.

Chadwick, D. H. 1995. What good is a prairie? *Audubon* (Nov.–Dec.).

Chai, D., L. W. Cuddihy, and C. P. Stone. 1989. *An inventory and assessment of anchialine pools in Hawaii Volcanoes National Park from Waha'ula to Ka'aha, Puna and Ka'u, and Hawai'i.*

Chaney, E., W. Elmore, and W. S. Plaits. 1990. *Livestock grazing on western riparian areas.* Denver, CO: U.S. Environmental Protection Agency, Region 8.

Christensen, A. G., L. J. Lyon, and J. W. Unsworth. 1993. *Elk management in the northern region: considerations in forest plan updates or revisions.* INT-303. Fort Collins, CO: USDA Forest Service, Intermountain Research Station.

Christensen, N. L. 1988. Vegetation of the southeastern coastal plain. Pages 317–363 in M. G. Barbour and W.

D. Billings (editors), *North American terrestrial vegetation.* Cambridge, England: Cambridge University Press.

Clark, C. 1974. *The American wilderness: The Badlands.* New York: Time-Life Books.

Clary, W. P., E. D. McArthur, D. Bedunah, and C. L. Wamboldt (compilers). 1992. *Proceedings: Symposium on ecology and management of riparian shrub communities.* INT-289. Fort Collins, CO: USDA Forest Service, Intermountain Research Station.

Cleaves, D. A., and M. Bennett. 1995. Timber harvesting by nonindustrial private forest landowners in western Oregon. *Western Journal of Applied Forestry* 10(2): 66–71.

Clements, F. E. 1920. *Plant indicators.* Publication 290. Washington, DC: Carnegie Institution of Washington.

Cogger, H. G. 1992. *Reptiles and amphibians of Australia.* Ithaca, NY: Cornell University Press.

Colborn, T. E., A. Davidson, S. N. Green, R. A. Hodge, C. I. Jackson, and R. Liroff. 1990. *Great Lakes, great legacy?* Washington, DC: Conservation Foundation; Ottawa, Ontario: Institute for Research on Public Policy.

Collar, N. J., L. P. Gonzaga, N. Krabbe, A. Madrono Nieto, L. G. Naranjo, T. A. Parker III, and D. C. Wege. 1992. *Threatened birds of the Americas: The ICBP/IUCN Red Data Book,* 3rd edition. Cambridge, England: International Council for Bird Preservation.

Collins, B. R., and K. H. Anderson. 1994. *Plant communities of New Jersey: A study in landscape diversity.* New Brunswick, NJ: Rutgers University Press.

Collins, J. T., S. L. Collins, J. Horak, D. Mulhern, W. H. Busby, C. C. Freeman, G. Wallace, and J. E. Hayes, Jr. 1995. *An illustrated guide to endangered or threatened species in Kansas.* Lawrence: University Press of Kansas.

Collins, O. B. 1975. Range vegetation and mima mounds in North Texas. *Journal of Range Management* 28: 209–211.

Conant, R., and J. T. Collins. 1991. *A field guide to reptiles and amphibians: Eastern/central North America.* Boston: Houghton Mifflin.

Conrad, C. E., and W. C. Oechel (editors). 1982. *Dynamics and management of Mediterranean-type ecosystems.* General Technical Report PSW-58. Berkeley, CA: USDA Forest Service, Pacific Southwest Forest and Range Experiment Station.

Constantz, G. 1994. *Hollows, peepers, and highlanders: An Appalachian Mountain ecology.* Missoula, MT: Mountain Press.

Cook, F. R. 1984. *Introduction to Canadian amphibians and reptiles.* Ottawa: National Museum of Natural Sciences, National Museums of Canada.

Cooper, S. V., K. E. Neiman, and D. W. Roberts. 1991. *Forest habitat types of northern Idaho: A second approximation.* INT-236. Fort Collins, CO: USDA Forest Service, Intermountain Research Station.

Cope, E. A. 1982. Noteworthy collections, California: *Abies lasiocarpa. Madroño* 29: 218.

Corn, P. S., and R. B. Bury. 1990. *Sampling methods for terrestrial amphibians and reptiles.* GTR-256. Portland, OR:

USDA Forest Service, Pacific Northwest Research
Station.

Couchiching Conservancy. 1996. *Carden Plain habitat conservation: A report of the Carden Alvar Project.* Couchiching Conservancy.

Council for Agricultural Science and Technology. 1974.
Livestock grazing on federal lands in the 11 western
states. *Journal of Range Management* 27: 174–181.

Covington, W. W. (technical coordinator). 1993. *Sustainable ecological systems: Implementing an ecological approach
to land management.* Fort Collins, CO: USDA Forest
Service, Rocky Mountain Forest and Range Experiment Station.

Covington, W. W., and M. M. Moore. 1992. Postsettle-
ment changes in natural fire regimes: Implications for
restoration of old growth ponderosa pine forests. Pages
87–99 in *Old growth forests in the southwest and Rocky
Mountain regions. Proceedings of a workshop.* Portal, Arizona, March 9–13.

Cowdrey, A. E. 1996. *This land, this South: An environmental history.* Lexington: University Press of Kentucky.

Cox, J., R. Kautz, M. MacLaughlin, and T. Gilbert. 1994.
*Closing the gaps in Florida's wildlife habitat conservation
system.* Tallahassee: Office of Environmental Services,
Florida Fish and Fresh Water Fish Commission.

Crocker-Bedford, C. 1990. Goshawk reproduction and forest management. *Wildlife Society Bulletin* 18: 262–269.

Cronon, W. 1983. *Changes in the Land.* New York: Hill and
Wang.

Cronquist, A. 1997. *Intermountain flora: Vascular plants of
the Intermountain West.* New York: Columbia University
Press.

Crosswhite, F. S. 1980. Dry country plants of the south
Texas plains. *Desert Plants* 2(3): 141–179.

Crowley, J. M. 1967. Biogeography of Canada. *Canadian
Geographer* 11: 312–326.

Crumpacker, D. W. 1984. Regional riparian research and a
multi-university approach to the special problem of
livestock grazing in the Rocky Mountains and Great
Plains. Pages 413–422 in R. E. Warner and K. Hendrix
(editors), *California riparian systems: Ecology, conservation, and productive management.* Berkeley: University of
California Press.

Cubbage, F. W., and C. H. Flather. 1993. Forested wetland area and distribution. *Journal of Forestry* 91(5):
35–40.

Cuddihy, L. W., and C. P. Stone. 1990. Summary of vegetation alteration in the Hawaiian Islands. Pages
467–472 in E. A. Kay (editor), *A natural history of the
Hawaiian Islands.* Honolulu: University of Hawaii
Press.

Culver, D. C., and J. R. Holsinger. 1992. How many
species of troglobites are there? *Bulletin of the Natural
Speleological Society* 54: 79–80.

Currie, D. J. 1991. Energy and large-scale patterns of animal and plant-species richness. *American Naturalist*
137: 27–49.

Curtis, J. T. 1959. *The vegetation of Wisconsin: An ordination
of plant communities.* Madison: University of Wisconsin
Press.

Cushing, D. 1995. *Population production and regulation in the
sea: A fisheries perspective.* Cambridge, England: Cambridge University Press.

(CWWR) Center for Water and Wildland Resources. 1996.
*Sierra Nevada Ecosystem Project Report: Final report to
Congress,* volumes 1–3 & Summary. Davis: University
of California Center for Water and Wildland
Resources.

Dahl, T. E. 1990. *Wetland losses in the United States, 1780's
to 1980's.* Washington, DC: U.S. Department of the
Interior, Fish and Wildlife Service.

Dahl, T. E., and C. E. Johnson. 1991. *Status and trends of
wetlands in the conterminous United States, mid-1970's to
mid-1980's.* Washington, DC: U.S. Department of the
Interior, Fish and Wildlife Service.

Daily, G. C. (editor). 1997. *Nature's services: Societal depen-
dence on natural ecosystems.* Washington, DC: Island
Press.

Dansereau, P. 1966. *Studies on the vegetation of Puerto Rico:
1. Description and integration of the plant communities.*
Special Publication no. 1. Mayaguez: University of
Puerto Rico, Institute of Caribbean Science.

Davies, W. E., J. H. Simpson, G. C. Ohlmacher, W. S.
Kirk, and E. G. Newton. 1984. Engineering aspects of
karst. Map in *National Atlas of the U.S.A., 1:7,500,000.*
Washington, DC: Department of the Interior, U.S.
Geological Survey.

Davis, F. W., J. E. Estes, B. Csuti, J. M. Scott, and D.
Stoms. 1993. *Geographic information systems analysis of
biodiversity in California.* Santa Barbara: University of
California, Department of Geography.

Davis, M. B. 1993. *Old growth in the East.* Richmond, VT:
Wild Earth Publications.

———. 1995. Threatened eastern old growth. *Wild Earth*
(Fall 1995): 54–56.

———. (editor). 1996. *Eastern old-growth forests: Prospects
for rediscovery and regeneration.* Washington, DC: Island
Press.

Davis, S. D., S. J. M. Droop, P. Gregerson, L. Henson, C.
J. Leon, J. Lamein Villa Lobos, H. Synge, and J. Zan-
tovska. 1986. *Plants in danger: What do we know?*
Gland, Switzerland: IUCN.

Davis, S. M. 1990. Sawgrass and cattail production in relation to nutrient supply in the Everglades. Pages
325–341 in R. R. Sharitz and J. W. Gibbons (editors),
Freshwater wetlands and wildlife. Ninth annual symposium, Savannah River Ecology Laboratory. Charleston,
SC.

Davis, S. M., and J. C. Ogden (editors). 1994. *Everglades:
The ecosystem and its restoration.* Delray Beach, FL: St.
Lucie Press.

Davis, W. S., and T. P. Simon (editors). 1995. *Biological
assessment and criteria: Tools for water resource planning
and decision making.* Sydney: CRC Press.

Deane, J. G., and D. Holing. 1985. Ravage the rivers, ban-
ish the birds. *Defenders* 60: 20–33.

Debevoise, N., and C. Rawlins (editors). 1996. *The red
desert blues: The industrialization of southwest Wyoming.*
Lincoln: Wyoming Outdoor Council.

DeGraff, R. M. 1991. Breeding bird assemblages in managed northern hardwood forests in New England. Pages 154–171 in J. G. Rodiek and E. G. Bolen (editors), *Wildlife and habitats in managed landscapes.* Washington, DC: Island Press.

DeGraff, R. M., V. E. Scott, R. H. Hamre, L. Ernst, and S. H. Anderson. 1991. *Forest and rangeland birds of the United States: Natural history and habitat use.* Agricultural Handbook 688. Washington, DC: USDA Forest Service.

DellaSala, D. A., K. A. Engel, D. P. Volsen, R. L. Fairbanks, J. C. Hagar, W. C. McComb, and K. J. Raedeke. 1994. *Effectiveness of silvicultural modifications of young-growth forests as enhancement for wildlife habitat on the Tongass National Forest, southeast Alaska.* Juneau, AK: USDA Forest Service. Unpublished report.

DellaSala, D. A., and D. M. Olson. 1996. Seeing the forest for more than the trees. *Wildlife Society Bulletin* 24(4): 770–776.

DellaSala, D. A., D. M. Olson, S. E. Barth, S. L. Crane, and S. A. Primm. 1995. Forest health: Moving beyond rhetoric to restore healthy landscapes in the inland Northwest. *Wildlife Society Bulletin* 23: 346–356.

DellaSala, D. A., D. M. Olson, and S. L. Crane. 1995. Ecosystem management and biodiversity conservation: Applications to inland Pacific Northwest forests. Pages 139–160 in D. M. Baugartner and R. L. Everett (editors), *Proceedings of ecosystem management in western interior forests.* Pullman: Washington State University.

DellaSala, D. A., J. C. Hagar, K. E. Engel, W. C. McComb, R. L. Fairbanks, and E. G. Campbell. 1996a. Effects of silvicultural modifications of temperate rainforest on breeding and wintering bird communities, Prince of Wales Island, southeast Alaska. *Condor* 98: 706–721.

DellaSala, D. A., J. R. Strittholt, R. F. Noss, and D. M. Olson. 1996b. A critical role for core reserves in managing inland northwest landscapes for natural resources and biodiversity. *Wildlife Society Bulletin* 24: 209–221.

DellaSala, D. A., W. Wettengel, A. Hackman, K. Kavanagh, D. Olson, and E. Dinerstein. 1997. *Protection and independent certification: A shared vision for North America's diverse forests.* Washington DC: WWF. (map poster)

DeMeo, T., J. Martin, and R. A. West. 1993. *Forest plant association management guide Ketchikan Area, Tongass National Forest.* R10-MB-210. Ketchikan, AK: USDA Forest Service.

de Steiguer J. E., J. M. Pye, and C. S. Love. 1990. Air pollution damage to U.S. forests. *Journal of Forestry* 88(8): 17–22.

Deyrup, M., and T. Eisner. 1993. Last stand in the sand. *Natural History* 12: 42–47.

Diamond, D. D. 1993. *Plant communities of Texas.* Austin: Texas Natural Heritage Program, Texas Parks and Wildlife Department.

Diamond, D. D., and T. E. Fullbright. 1990. Contemporary plant communities of upland grasslands of the Coastal Sand Plain, Texas. *Southwestern Nature* 35: 385–392.

Diamond, D. D., D. H. Riskind, and S. L. Orzell. 1987. A framework for plant community classification and conservation in Texas. *Texas Journal of Science* 39: 203–221.

Diamond, D. D., and F. E. Smeins. 1984. Remnant grassland vegetation and ecological affinities of the upper coastal prairie of Texas. *Southwestern Nature* 29: 321–334.

Diamond, D. D., and F. E. Smeins. 1985. Composition, classification and species response patterns of remnant tallgrass prairies in Texas. *American Midland Naturalist* 29: 321–334

Diamond, D. D., and F. E. Smeins. 1993. The native plant communities of the Blackland Prairie. Pages 66–81 in R. Sharpless and J. C. Yelderman, Jr. (editors), *The Texas Blackland Prairie Land History and Culture.* Symposium of the Natural Regions of Texas, Baylor University Program for Regional Studies. Baylor University, Waco, Texas.

Dick-Peddie, W. A. 1993. *New Mexico vegetation: Past, present, and future.* Albuquerque: University of New Mexico Press.

Dickson, J. G., F. R. Thompson III, R. N. Conner, and K. E. Franzreb. 1993. Effects of silviculture on neotropical migratory birds in central and southeastern oak pine forests. In D. M. Finch and P. W. Stangel (editors), *Status and management of neotropical migratory birds, 1992 September 21–25.* General Technical Report RM-229. Fort Collins, CO: USDA Forest Service, Rocky Mountain Forest and Range Experiment Station.

Diesch, S. L. 1970. Disease transmission of waterborne organisms of animal origins. Pages 265–285 in T. L. Willrich and G. E. Smith (editors), *Agricultural practices and water quality.* Ames: Iowa State University Press.

Dinerstein, E., D. M. Olson, D. J. Graham, A. L. Webster, S. A. Primm, M. P. Bookbinder, and G. Ledec. 1995. *A conservation assessment of the terrestrial ecoregions of Latin America and the Caribbean.* Washington, DC: World Bank.

Dobbyn, J. 1994. *Atlas of the mammals of Ontario.* Toronto: Federation of Ontario Naturalists.

Dobson, A. P., J. P. Rodriguez, W. M. Roberts, and D. S. Wilcove. 1997. Geographic distribution of endangered species in the United States. *Science* 275: 550–553.

Donham, M. 1992. Songbird data for forest defenders. *Wild Earth* (Spring): 60–61.

Doolittle, J. 1974. *The American wilderness: Canyons and mesas.* New York: Time-Life Books.

Doppelt, B., M. Scurlock, Chris Frissell, and James Karr. 1993. *Entering the watershed: A new approach to save America's river ecosystems.* Eugene, OR: Pacific Rivers Council; Washington, DC: Island Press.

Douglas, M. S. 1947. *The Everglades: River of grass.* Miami, FL: Hurricane House.

Drost, C. A., and G. M. Fellers. 1996. Collapse of a regional frog fauna in the Yosemite area of the California Sierra Nevada, USA. *Conservation Biology* 2: 414–425.

Duffy, D. C., and A. J. Meier. 1992. Do Appalachian herbaceous understories ever recover from clearcutting. *Conservation Biology* 2: 196–201.

Eastman, J. R. 1992. *IDRISI Version 4.0: User's Documentation.* Worcester, MA: Clark University Graduate School of Geography.

Echeverria, J. D., P. Barrow, and R. Roos-Collins. 1989. *Rivers at risk: The concerned citizen's guide to hydropower.* Washington, DC: Island Press.

Ecoregions Working Group, Canada Committee on Ecological Land Classification. 1989. *Ecoclimatic regions of Canada: First approximation.* Ecological Land Classification Series, no. 23. Ottawa: Sustainable Development Branch, Canadian Wildlife Service, Conservation and Protection Environment.

Ecotrust, Pacific GIS, and Conservation International. 1995. *The rain forests of home: An atlas of people and place. Part 1: Natural forests and native languages of the coastal temperate rain forest.* Portland, OR: Authors.

Edey, M. A. 1972. *The American wilderness: The northeast coast.* New York: Time-Life Books.

Eggen-McIntosh, S., K. B. Lannom, and D. M. Jacobs. 1994. Mapping forest distributions of Central America and Mexico. Pages 273–281 in *GIS/LIS Proceedings.* Asheville, NC: USDA Forest Service, Southern Forest Experiment Station.

Ehrlich, P. R., D. S. Dobkin, and D. Wheye. 1992. *The imperiled and extinct birds of the United States and Canada including Hawaii and Puerto Rico.* Stanford, CA: Stanford University Press.

Ekman, S. 1953. *Zoogeography of the sea.* London: Sidgewick and Jackson.

Elias, T. S. (editor). 1987. The problem of the Salmon Mountains. Pages 155–158 in *Conservation and management of rare and endangered plants.* Sacramento: California Native Plant Society.

Environment Canada. 1993. Canada, terrestrial ecoregions. Map (1:7,500,000). *The National Atlas of Canada,* 5th edition. Hull, Quebec: Environment Canada.

Epperly, S. P., J. Braun, and A. Veishlow. 1995. Sea turtles in North Carolina waters. *Conservation Biology* 2: 384–394.

Ernst, C. H., J. E. Lovich, and R. W. Barbour. 1994. *Turtles of the United States and Canada.* Washington, DC: Smithsonian Institution Press.

EROS (Earth Resource and Observation System). 1996. *North America landcover characteristics: USGS land use and landcover classification.* Washington, DC: National Mapping Division, U.S. Geological Survey.

ESWG (Ecological Stratification Working Group). 1995. *A national ecological framework for Canada.* Ottawa: Agriculture and Agri food Canada, Research Branch, Centre for Land and Biological Resources Research; and Environment Canada, State of the Environment Directorate, Ecozone Analysis Branch. (map)

Evans, D. L. 1994. *Forest cover from landsat thematic mapper data for use in the Catahoula ranger district geographic information system.* SO-99. Asheville, NC: USDA Forest Service, Southern Forest Experiment Station.

Everett, R., P. Hessburg, J. Lehmkuhl, M. Jensen, and P. Bourgeron. 1994. Old forests in dynamic landscapes. *Journal of Forestry* 92: 224–225.

Everett, R. L., and D. M. Baumgartner. 1994. *Symposium proceedings: Ecosystem management in Western Interior Forests.* Fort Collins, CO: USDA Forest Service.

Ewell, J. J., and J. L. Whitmore. 1973. *The ecological life zones of Puerto Rico and the Virgin Islands.* Forest Service Research Paper No. ITF 18. Washington, DC: USDA Forest Service.

Eyre, F. H. (editor). 1980. *Forest cover types of the United States and Canada.* Washington, DC: Society of American Foresters.

FAO. 1990. *Forests resources assessment 1990: Global synthesis.* Forest Research Paper 124. New York: FAO.

Farb, P. 1963. *Face of North America: The natural history of a continent.* New York: Harper & Row.

Farrar, J. L. 1995. *Trees in Canada.* Ottawa: Fitzhenry and Whiteside and the Canadian Forest Service.

Federation of Ontario Naturalists. 1997. *End of the road. Ontario's roadless wilderness.* Toronto: Author.

Feibleman, P. S. 1973. *The American wilderness: The bayous.* New York: Time-Life Books.

Felger, R. S., and M. B. Johnson. 1995. Trees of the northern Sierra Madre Occidental and Sky Islands of southwestern North America. Pages 71–83 in L. F. DeBano, P. F. Folliott, A. Ortega Rubio, G. J. Gottfried, R. H. Hamre, and C. B. Carleton (technical coordinators), *Biodiversity and management of the Madrean Archipelago: The Sky Islands of southwestern United States and northwestern Mexico.* General Technical Report RM-GTR-264. Golden, CO: USDA Forest Service.

Felger, R. S., and M. F. Wilson. 1995. Northern Sierra Madre Occidental and its Apachian outliers: A neglected center of biodiversity. Pages 36–51 in L. F. DeBano, P. F. Folliott, A. Ortega Rubio, G. J. Gottfried, R. H. Hamre, and C. B. Carleton (technical coordinators) *Biodiversity and management of the Madrean Archipelago: The Sky Islands of Southwestern United States and Northwestern Mexico.* General Technical Report RM-GTR-264. Golden, CO: USDA Forest Service.

Fenneman, N. M. 1938. *Physiography of eastern United States.* New York: McGraw-Hill.

Fergus, C. 1993. Scrub: Learning to love it. *Audubon Magazine* 95(3): 100–104.

Fiedler, P. L., and S. K. Jain (editors). 1992. *Conservation biology: The theory and practice of nature conservation, preservation, and management.* New York: Chapman and Hall.

Fincher, J., and M. Smith. 1994. *A discriminant-function approach to ecological site classification in northern New England.* NE-686. Radnor, PA: USDA Forest Service, Northeastern Forest Experiment Station.

Fisher, R. N., and H. B. Shaffer. 1996. The decline of amphibians in California's great central valley. *Conservation Biology* 10: 1387–1397.

Flader, S. L. (editor). 1983. *The Great Lakes forest: An environmental and social history.* Minneapolis: University of Minnesota Press.

Flannery, T. 1994. *The future eaters.* New York: George Braziller.

Flather, C. H., S. J. Brady, and D. B. Inkley. 1992. Regional habitat appraisals of wildlife communities: A landscape-level evaluation of a resource planning model using avian distribution data. *Landscape Ecology* 7(2): 137–147.

Flather, C. H., L. A. Joyce, and C. A. Bloomgarden. 1994. *Species endangerment patterns in the United States.* RM-241. Fort Collins, CO: USDA Forest Service, Rocky Mountain Forest and Range Experiment Station.

Fleischner, T. L. 1994. Ecological costs of livestock grazing in western North America. *Conservation Biology* 8: 629–644.

Fleischner, T. L., D. E. Brown, A. Y. Cooperrider, W. B. Kessler, and E. L. Painter. 1994. Society for Conservation Biology position statement: Livestock grazing on public lands in the United States of America. *Society for Conservation Biology Newsletter* 1(4): 2–3.

Fleishman, E., and D. D. Murphy. 1993. *A review of the biology of the coastal sage scrub.* Stanford, CA: Center for Conservation Biology, Department of Biological Sciences, Stanford University. Unpublished report, May 10 update.

Flores, M. G., L. J. Jimenez, S. X. Madrigal, T. F. Takaki, X. E. Hernandez, and R. J. Rzedowski. 1971. *Mapa de tipos de vegetacion de la Republica Mexicana.* Map at a scale of 1:2,000,000. Mexico City: Secretaria de Recursos Hidraulicos.

Florida Department of Natural Resources, Division of Recreation and Parks, Bureau of Plans, Programs, and Services. 1975. Map no. 21, General distribution of sand pine scrub. Page 105 in *Florida environmentally endangered lands plan.* Tallahassee.

Flowers, R. W. 1993–94. Endangered invertebrates and how to worry about them. *Wild Earth* (Winter): 25–31.

Folkerts, G. W. 1982. The Gulf Coast pitcher plant bogs. *American Scientist* 70: 260–267.

Folliott, P. F., G. J. Gottfried, D. A. Bennett, C. Hernandez, M. Victor, A. Ortega Rubio, and R. H. Hamre (technical coordinators). 1992. *Ecology and management of oak and associated woodlands: Perspectives in the southwestern United States and northern Mexico.* RM-218. Fort Collins, CO: USDA Forest Service, Rocky Mountain Forest and Range Experiment Station.

Ford. R. G. In preparation. *A conservation assessment of the marine ecoregions of the United States.* Washington, DC: World Wildlife Fund.

Foreman, D. 1993. Eastern forest recovery. *Wild Earth* (Summer): 25–28.

Foreman, D., and H. Wolke. 1992. *The big outside: A descriptive inventory of the big wilderness areas of the United States.* New York: Harmony Books.

Foreman, D., H. Wolke, and B. Koehler. 1991. The Earth First! wilderness preserve system. *Wild Earth* (Spring): 33–38.

Forest Ecosystem Management Assessment Team. 1993. *Forest ecosystem management: An ecological, economic, and social assessment.* Washington, DC: USDA Forest Service, National Oceanic and Atmospheric Administration, Bureau of Land Management, Fish and Wildlife Service, National Park Service, U.S. Environmental Protection Agency.

Forsyth, A. 1985. *Mammals of the Canadian wild.* Camden East, Ontario: Camden House.

Foti, T. L., and S. M. Glenn. 1991. The Ouachita Mountain landscape at the time of settlement. Pages 49–66 in D. Henderson and L. D. Hedrick (editors), *Restoration of old growth forests in the interior highlands of Arkansas and Oklahoma.* Morillton, AR: Ouachita National Forest and Winrock International Institute for Agricultural Development.

Franklin, J. F. 1988. Pacific Northwest forests. Pages 103–130 in M. G. Barbour and W. D. Billings (editors), *North American terrestrial vegetation.* Cambridge, England: Cambridge University Press.

Franklin, J. F., and C. T. Dyrness. 1973. *Natural vegetation of Oregon and Washington.* Corvallis: Oregon State University Press.

Franklin, T., and D. DellaSala. 1996. Conservation strategies for the inland Pacific Northwest. *Wildlife Society Bulletin* 24(2): 178–179.

Franzreb, K. E. 1987. Perspectives on managing riparian ecosystems for endangered bird species. *Western Birds* 18: 3–9.

Frayer, W. E., D. D. Peters, and H. R. Pywell. 1989. *Wetlands of the California Central Valley status and trends: 1939 to mid-1980's.* Portland, OR: U.S. Fish and Wildlife Service, Region 1.

Frest, T. J., and E. J. Johannes. 1991. *Present and potential candidate molluscs occurring within the range of the northern spotted owl.* Report prepared for the Northen Spotted Owl Recovery Team, Other Species and Ecosystems Committee, Seattle, WA.

Frier-Murza, J., and J. C. Sciascia. 1989. Legal protection for New Jersey's endangered and nongame wildlife species. In E. F. Karlin, (editor), *Proceedings of a conference on New Jersey's rare and endangered plants and animals.* Mahwah, NJ: Institute for Environmental Studies.

Fryar, R. D. 1991. Old growth stands of the Ouachita National Forest. Pages 105–114 in D. Henderson and L. D. Hedrick (editors), *Restoration of old growth forests in the interior highlands of Arkansas and Oklahoma.* Morillton, AR: Ouachita National Forest and Winrock International Institute for Agricultural Development.

Gagné, W. C. 1988. Conservation priorities in Hawaiian natural systems. *BioScience* (38)4: 264–271.

Gagné, W. C., and C. C. Christensen. 1985. Conservation status of native terrestrial invertebrates in Hawai'i. Pages 105–126 in C. P. Stone and J. M. Scott (editors), *Hawaii's terrestrial ecosystems: Preservation and management.* Proceedings of a symposium held June 5–6, 1984, at Hawaii Volcanoes National Park. Honolulu: University of Hawaii.

Gagné, W. C., and L. W. Cuddihy. 1990. Vegetation. Pages 4–114 in W. L. Wagner, D. R. Herbst, and S. H. Sohmer (editors), *Manual of the flowering plants of Hawaii.* Honolulu: University of Hawaii Press.

Gaines, D. 1977. The valley riparian forests of California: Their importance to bird populations. Pages 57–85 in A. Sands (editor), *Riparian forests in California: Their ecology and conservation.* Institute of Ecology Publication No. 15. Davis: Institute of Ecology, University of California.

Gallant, A. L., E. F. Binnian, J. M. Omernik, and M. B. Shasby. 1995. *Ecoregions of Alaska.* U.S. Geological Survey Professional Paper 1567. Washington, DC: U.S. Government Printing Office.

Ganey, J. L., R. B. Duncan, and W. M. Block. 1992. *Use of oak and associated woodlands by Mexican spotted owls in Arizona. Ecology and management of oak and associated woodlands: Perspectives in the southwestern United States and northern Mexico.* RM-218, pages 125–128. Fort Collins, CO: USDA Forest Service, Rocky Mountain Forest and Range Experiment Station.

Garrison, G. A., A. J. Bjugstad, D. A. Duncan, M. E. Lesis, and D. R. Smith. 1977. *Vegetation and environmental features of forest and range ecosystems.* Washington, DC: USDA Forest Service Agricultural Handbook No. 475.

Gedney, D. R., and B. A. Hiserote. 1989. *Changes in land use in western Oregon between 1971–74 and 1982.* PNW-RB-165. Portland, OR: USDA Forest Service, Pacific Northwest Research Station.

Gehlbach, F. R. 1993. *Mountain islands and desert seas: A natural history of the U.S.-Mexican borderlands.* College Station: Texas A&M University Press.

Gilbert, C. R. (editor). 1992. *Rare and endangered biota of Florida: Volume 2. Fishes.* Gainesville: University Press of Florida.

Gill, F. B. (editor). 1978. *Zoogeography in the Caribbean.* Special Publication No. 13. Philadelphia: Academy of Natural Sciences of Philadelphia.

Gillen, R. L., W. C. Krueger, and R. F. Miller. 1984. Cattle distribution on mountain rangeland in northeastern Oregon. *Journal of Range Management* 37: 549–553.

Gon, S. 1996. Unpublished map of Hawaiian ecoregions. The Nature Conservancy of Hawaii.

Good, R. E., N. F. Good, and J. W. Andresen. 1979. The Pine Barren plains. Pages 283–295 in T. Forman (editor), *Pine Barrens: Ecosystem and landscape.* New York: Academic Press.

Goodrich, S. 1992. Summary flora of riparian shrub communities of the intermountain region with emphasis on willows. In *Proceedings: Symposium on Ecology and Management of Riparian Shrub Communities.* Fort Collins, CO: USDA Forest Service, Intermountain Research Station.

Gottfried, G. J. 1992. *Ecology and management of the southwestern pinyon-juniper woodlands. Ecology and management of oak and associated woodlands: Perspectives in the southwestern United States and northern Mexico.* RM-218. Fort Collins, CO: USDA Forest Service, Rocky Mountain Forest and Range Experiment Station.

Gould, F. W. 1962. *Texas plants: A checklist and ecological summary.* College Station: Texas A&M University, Texas Agricultural Expansion Station.

Gould, F. W., G. O. Hoffman, and C. A. Rechenthin. 1960. *Vegetational areas of Texas.* Texas Agriculture Experimental Station Leaflet 492. College Station: Texas A&M University.

Government of Canada. 1991. *The state of Canada's environment.* Ottawa: Environment Canada.

Greenland, D., and L. W. Swift, Jr. (editors). 1990. *Climate variability and ecosystem response.* SE-65. Asheville, NC: USDA Southeastern Forest Experiment Station.

Greenpeace. 1996. *Nuclear flashback: The return to Amchitka.* Washington, DC: Greenpeace.

Greller, A. M. 1988. Deciduous forest. Pages 287–316 in M. G. Barbour and W. D. Billings (editors), *North American terrestrial vegetation.* Cambridge, England: Cambridge University Press.

Greswell, R. E., A. Barton, and J. L. Kershner (editors). 1989. *Practical approaches to riparian resource management: An educational workshop.* Billings, MT: Bureau of Land Management.

Griffith, G. E., J. M. Omernik, T. F. Wilton, and S. M. Pierson. 1994. Ecoregions and subregions of Iowa: A framework for water quality assessment and management. *Journal of the Iowa Academy of Sciences* 101(1): 5–13.

Groom, M. J., and N. Shumaker. 1993. Evaluating landscape change: Patterns of worldwide deforestation and local fragmentation. Pages 22–44 in P. M. Kareiva, J. M. Kingsolver, and R. B. Huey (editors), *Biotic interactions and global change.* Sunderland, MA: Sinauer Associates.

Groombridge, B. 1982. *The IUCN amphibia reptilia red data book.* Gland, Switzerland: IUCN.

Grubb, T. G., J. L. Ganey, and S. R. Masek. 1997. Canopy closure around nest sites of Mexican spotted owls in north-central Arizona. *Journal of Wildlife Management* 61(2): 336–342.

Gruell, G. E. 1983. *Fire and vegetative trends in the northern Rockies: Interpretations from 1971–1982 photographs.* INT-158. Fort Collins, CO: USDA Forest Service, Intermountain Forest and Range Experiment Station.

Gunderson, L. H., and W. F. Loftus 1993. The Everglades. Pages 199–255 in W. H. Martin, S. G. Boyce, and A. C. Echternacht (editors), *Biodiversity of the southeastern United States: Lowland terrestrial communities.* New York: John Wiley.

Habeck, J. R. 1990. Old-growth ponderosa pine-western larch forests in western Montana: Ecology and management. *Northwest Environmental Journal* 6: 271–292.

Habeck, J. R., and R. W. Mutch. 1973. Fire-dependent forests in the eastern Rocky Mountains. *Journal of Quaternary Research* 3: 408–424.

Hackney, C. T., S. M. Adams, and W. H. Martin (editors). 1992. *Biodiversity of the southeastern United States: Aquatic communities.* New York: John Wiley.

Hall, E. R. 1946. *Mammals of Nevada.* Davis: University of California Press.

Hamrick, J. L., A. F. Schnabel, and P. V. Wells. 1994. Distribution of genetic diversity within and among populations of Great Basin conifers. Pages 147–161 in K. T. Harper, L. L. St. Clair, K. H. Thorne, and W. W. Hess (editors), *Natural history of the Colorado Plateau and Great Basin.* Niwot: University Press of Colorado.

Handley, C. O. Jr. 1991. Mammals. Pages 539–543 in K. Terwilliger (editor), *Virginia's endangered species.* Blacksburg, VA: McDonald and Woodward.

Hanes, T. L. 1995. California chaparral. Pages 417–469 in M. G. Barbour and J. Major (editors), *Terrestrial vegetation of California.* Special Publication No. 9. Davis: California Native Plant Society.

Hanski, I., and M. Gilpin (editors). 1991. *Metapopulation biology: Ecology, genetics, and evolution.* San Diego, CA: Academic Press.

Hansson, L. 1992. Landscape ecology of boreal forests. *Tree* 7(9): 299–302.

Harding, L. E., and E. McCullum (editors). 1994. *Biodiversity in British Columbia. Our changing environment.* Ottawa: Environment Canada, Canadian Wildlife Service, Ministry of Supply and Services.

Harper, K. T., L. L. St. Clair, K. H. Thorne, and W. W. Hess (editors). 1994. *Natural history of the Colorado Plateau and Great Basin.* Niwot: University Press of Colorado.

Harris, R. R., D. C. Erman, and H. M. Kerner (coordinators). 1992. *Syposium on biodiversity of northwestern California.* October 28, 1991, Santa Rosa, California. Wildland Resources Center Report No. 29, Division of Agriculture and Natural Resources, University of California, Berkeley.

Hawaii Natural Heritage Program. 1987. *Biological overview of Hawaii's natural area reserves system.* Honolulu: Prepared for Hawaii State Department of Land and Natural Resources.

Hawaii Natural Heritage Program. 1997. *Summary listing of Hawaiian natural community types.* Honolulu: Author (current database).

Hawaiian endangered species task force defines its priorities. 1992. *Herbarium Pacificum News* 9: 10.

Hayden, B. P., G. C. Ray, and R. Dolan. 1984. Classification of coastal and marine environments. *Environmental Conservation* 11(3): 199–207

Hayward, G. D., and J. Verner (technical editors). 1994. *Flammulated, boreal, and great gray owls in the United States: A technical conservation assessment.* Washington, DC: USDA Forest Service, General Technical Department.

Heady, H. F. 1995. Valley grassland. Pages 491–514 in M. G. Barbour and J. Major (editors), *Terrestrial vegetation of California.* Special Publication No. 9. Davis: California Native Plant Society.

Hedges, B. S. 1996. Historical biogeography of West Indian vertebrates. Pages 163–196 in D. G. Fautin, D. J. Futuyma, and F.C. James (editors), *Annual review of ecology and systematics.* Palo Alto, CA: Random House.

Hefner, J. M., and J. D. Brown. 1984. Wetland trends in the southeastern United States. *Wetlands* 4: 1–11.

Hejl, S. J. 1992. The importance of landscape patterns to bird diversity: A perspective from the northern Rocky Mountains. *Northwest Environment Journal* 8: 119–137.

Henderson, D., and L. D. Hedrick (editors). *Restoration of old growth forests in the interior highlands of Arkansas and Oklahoma: Proceedings of the conference.* Morillton, AR: Ouachita National Forest and Winrock International Institute for Agricultural Development.

Henderson, R. A., and E. J. Epstein. 1995. Oak savannas in Wisconsin. Pages 230–232 in E. T. LaRoe, G. S. Farris, C. E. Puckett, P. D. Doran, and M. J. Mac (editors), *Our living resources.* Washington, DC: U.S. Department of the Interior, National Biological Service.

Henjum, M. G., J. R. Karr, D. L. Bottom, D. A. Perry, J. C. Bednarz, S. G. Wright, S. A. Beckwitt, and E. Beckwitt. 1994. *Interim protection for late-successional forests, fisheries, and watersheds: National forests east of the Cascade crest, Oregon and Washington.* Bethesda, MD: Wildlands Society.

Hernández, H. M., and R. T. Bárcenas. 1995. Endangered cacti in the Chihuahuan desert: I. Distribution patterns. *Conservation Biology* 9(5): 1176–1188.

Heusser, C. J. 1989. North Pacific coastal refugia: The Queen Charlotte Islands in perspective. Pages 91–106 in G. G. E. Scudder and N. Gessler (editors), *Proceedings of the Queen Charlotte Islands First International Symposium.* Vancouver: University of British Columbia.

Hilyard, D., and M. Black. 1987. Coastal sage scrub restoration (California). *Restoration and Management Notes* 5: 96.

HINHP (Hawaii Natural Heritage Program). 1996. *Hawaiian Natural Community Database.* Honolulu: The Nature Conservancy of Hawaii.

Hinkle, C. R., W. C. McComb, J. M. Safley Jr., and P. A. Schmalzer. 1993. Mixed mesophytic forests. Pages 203–254 in W. H. Martin, S. G. Boyce, and A. C. Echternacht (editors), *Biodiversity of the southeastern United States: Upland terrestrial communities.* New York: John Wiley.

Hobbs, H. H., III. 1992. Caves and springs. Pages 59–131 in C. T. Hackney, S. M. Adams, and W. H. Martin (editors), *Biodiversity of the southeastern United States: Aquatic communities.* New York: John Wiley.

Hoekstra, J. M., R. T. Bell, A. E. Launer, and D. D. Murphy. 1995. Soil arthropod abundance in coast redwood forest: Effect of selective timber harvest. *Environmental Entomology* 24: 246–252.

Holing, D. 1987. Hawaii: The eden of endemism. *The Nature Conservancy News* 37(1): 7–13.

Holland, R. F. 1978. *The geographic and edaphic distribution of vernal pools in the great central valley, California.* Berkeley: California Native Plant Society.

Holland, R. F., and S. Jain. 1995. Vernal pools. Pages 515–553 in M. G. Barbour and J. Major (editors), *Terrestrial vegetation of California.* Special Publication No. 9. Davis: California Native Plant Society.

Holm, L. G. 1997. *The world's worst weeds.* New York: John Wiley.

Holm, L. G., D. L. Plucknett, J. V. Pancho, and J. P. Herberger. 1977. *The world's worst weeds: Distribution and biology.* Honolulu: University Press of Hawaii.

Holthausen, R. S., M. J. Wisdom, J. Pierce, D. K. Edwards, and M. M. Rowland. 1994. *Using expert opinion to evaluate a habitat effectiveness model for elk in western Oregon and Washington.* PNW-RP-479. USDA Forest Service, Pacific Northwest Research Station.

Hornbeck, J. W., and W. B. Leak. 1992. *Ecology and management of northern hardwood forests in New England.* NE-159. Radnor, PA: USDA Forest Service, Northeastern Forest Experiment Station.

Howarth, F. G., and W. P. Mull. 1992. *Hawaiian insects and their kin.* Honolulu: University of Hawaii Press.

Howarth, F. G., G. Nishida, and A. Asquith. 1995. Insects of Hawaii: Status and trends. In E. T. LaRoe, G. S. Farris, C. E. Puckett, P. D. Doran, and M. J. Mac (editors), *Our living resources: A report to the nation on the distribution, abundance, and health of U.S. plants, animals, and ecosytems.* Washington, DC: U.S. Department of Interior, National Biological Service.

Hubricht, L. 1985. The distributions of the native land mollusks of the eastern United States. *Fieldiana: Zoology* 24: 1–191.

Hummel, M. 1989. *Endangered spaces: The future for Canada's wilderness.* Toronto: Key Porter Books.

Ingersoll, C. A., and M. V. Wilson. Restoration of a western Oregon remnant prairie. *Restoration and Management Notes* 9(2): 110–111.

IUCN. 1992. *Protected areas of the world: A review of national systems: Volume 4. Nearctic and Neotropical.* Gland, Switzerland: IUCN.

IUCN. 1994. *Guidelines for protected areas management categories: Part II. The management categories.* Gland, Switzerland: IUCN.

Jackson, D. D. 1975. *The American wilderness: Sagebrush country.* New York: Time-Life Books.

Jackson, D. D., and P. Wood. 1975. *The American wilderness: The Sierra Madre.* Alexandria, VA: Time-Life Books.

Jacobi, J. 1985. *Vegetation mapping of the Hawaiian Islands.* Overlays of maps at 1:24,000, Mauna Loa Research Station, HI: U.S. Fish and Wildlife Service.

Jain, S. 1976. *Vernal pools: Their ecology and conservation.* Publication no. 9. Davis: University of California, Institute of Ecology.

James, H. F., and S. L. Olson. 1991. Descriptions of thirty-two new species of birds from the Hawaiian islands: Part II. Passeriformes. *Ornithological Monographs* 46: 1–88.

Jeffries, D. L., and J. M. Klopatek. 1987. Effects of grazing on the vegetation of the blackbrush association. *Journal of Range Management* 40: 390–392.

Jenkins, R. E. 1985. Information methods: Why the heritage programs work. *Nature Conservancy News* 35(6): 21–23.

———. 1988. Information management for the conservation of biodiversity. Pages 231–239 in E. O. Wilson (editor), *Biodiversity.* Washington, DC: National Academy Press.

Jenkins, R. E., and N. M. Burkhead. 1994. *Freshwater fishes of Virginia.* Bethesda, MD: American Fisheries Society.

Jensen, D. B., M. S. Torn, and J. Harte. 1993. *In our own hands: A strategy for conserving California's biological diversity.* Berkeley: University of California Press.

Jezerinac, R., G. W. Stocker, and D. C. Tarter. 1995. *The crayfishes (Decapoda: cambaridae) of West Virginia.* Columbus: Ohio Biological Survey, College of Biological Sciences, Ohio State University.

Johnson, A. S. 1989. The thin, green line: Riparian corridors and endangered species in Arizona and New Mexico. Pages 35–46 in G. Mackintosh (editor), *In defense of wildlife: Preserving communities and corridors.* Washington, DC: Defenders of Wildlife.

Johnson, C. G. Jr. 1994. *Forest health in the Blue Mountains: A plant ecologist's perspective on ecosystem processes and biological diversity.* PNW-GTR-339. Portland, OR: USDA Forest Service, Pacific Northwest Research Station.

Johnson, R. R., and L. T. Haight. 1985. Avian use of xeroriparian ecosystems in the North American warm deserts. Pages 156–160 in R. R. Johnson, C. D. Ziebell, D. R. Patton, P. F. Folliott, and F. H. Hamre (technical coordinators), *Riparian ecosystems and their management: Reconciling conflicting uses.* General Technical Report RM-120. Fort Collins, CO: USDA Forest Service, Rocky Mountain Forest and Range Experiment Station.

Johnson, R. R., L. T. Haight, and J. M. Simpson. 1977. Endangered species vs. endangered habitats: A concept. Pages 68–79 in R. R. Johnson and D. A. Jones (technical coordinators), *Importance, preservation, and management of riparian habitat: A symposium.* General Technical Report RM-43. Fort Collins, CO: USDA Forest Service, Rocky Mountain Forest and Range Experiment Station.

Johnson, T. H. 1988. *Biodiversity and conservation in the Caribbean: Profiles of selected islands.* Cambridge, England: International Council for Bird Preservation.

Johnston, M. C. 1963. Past and present grasslands of southern Texas and northeastern Mexico. *Ecology* 44: 456–466.

Johnston, V. R. 1994. *California forests and woodlands: A natural history.* Berkeley: University of California Press.

Kantrud, H. A., G. L. Krapu, and G. A. Swanson. 1989. *Prairie basin wetlands of the Dakotas: A community profile.* Biological Report 85 (7.28). Washington, DC: U.S. Department of the Interior, Fish and Wildlife Service.

Kareiva, P., and U. Wennergren. 1995. Connecting landscape patterns to ecosystem and population processes. *Nature* 237: 299–302.

Kartesz, J. T. 1994. *A synonymized checklist of the vascular flora of the United States, Canada, and Greenland,* 2 volumes. Portland, OR: Timber Press.

Kartesz, J. 1997. Personal communication. Director, Biota of North America Program, University of North Carolina.

Kauffman, J. B. 1988. The status of riparian habitats in Pacific Northwest forests. Pages 45–55 in K. Raedeke (editor), *Streamside management: Riparian wildlife and forestry interactions.* Contribution no. 59. Seattle: Institute of Forest Resources, University of Washington.

Kauffman, J. B., and W. C. Krueger. 1984. Livestock impacts on riparian ecosystems and streamside management implications: A review. *Journal of Range Management* 37: 430–437.

Kaufmann, M. R., R. T. Graham, D. A. Boyce Jr., W. H. Moir, L. Perry, R. T. Reynolds, R. L. Bassett, P. Mohlhop, C. B. Edminster, W. M. Block, and P. S. Corn. 1994. *An ecological basis for ecosystem management.* RM-246. Fort Collins, CO: USDA Forest Service, Rocky Mountain Forest and Range Experiment Station.

Kaufmann, M. R., W. H. Moir, and W. W. Covington. 1992. Old-growth forests: What do we know about their ecology and management in the southwest and Rocky Mountain regions? Pages 1–11 in *Old-growth forests in the southwest and Rocky Mountain regions: Proceedings of a workshop.* Portal, Arizona, March 9–13.

Kavanagh, K., and T. Iacobelli. 1995. *A protected areas gap analysis methodology: Planning for the conservation of biodiversity.* Discussion Paper. Toronto: WWF Canada.

Kavanaugh, D. H. 1989. The ground-beetle (Coleoptera: Carabidae) fauna of the Queen Charlotte Islands: Its composition, affinities, and origins. Pages 131–146 in G. G. E. Scudder and N. Gessler (editors), *Proceedings of the Queen Charlotte Islands First International Symposium.* Vancouver: University British of Columbia.

Kay, E. A., 1994. *A natural history of the Hawaiian Islands: Selected readings II.* Honolulu: University of Hawaii Press.

Keeler-Wolf, T. 1995. *A manual of California vegetation.* Sacramento: California Native Plant Society Press.

Keeley, J. E. 1990. The California valley grassland. Pages 2–23 in A. A. Schoenherr (editor), *Endangered plant communities of Southern California: Proceedings of the 15th Annual Symposium.* Special Publication No. 3. Claremont: Southern California Botanists.

Keeley, J. E., and S. C. Keeley. 1988. Chaparral. Pages 165–207 in M. G. Barbour and W. D. Billings (editors), *North American terrestrial vegetation.* Cambridge, England: Cambridge University Press.

Keener, C. S. 1983. Distribution and biohistory of the endemic flora of the mid Appalachian shale barrens. *Botanical Review* 49: 65–115.

Kellogg, E. (editor). 1992. *Coastal temperate rain forests: Ecological characteristics, status, and distribution worldwide.* Portland, OR: Ecotrust; and Washington, DC: Conservation International.

Kendeigh, S. C., H. I. Baldwin, V. H. Cahalane, C. H. D. Clarke, C. Cottam, W. P. Cottam, I. McT. Cowan, P. Dansereau, J. H. Davis Jr., F. W. Emerson, I. T. Haig, A. Hayden, C. L. Hayward, J. M. Linsdale, J. A. Macnab, and J. E. Potzger. 1950–51. Nature sanctuaries in the United States and Canada: A preliminary inventory. *Living Wilderness* 15(35): 2–43.

Kennedy, C. E. 1977. Wildlife conflicts in riparian management: Water. Pages 52–58 in R. R. Johnson and D. A. Jones (technical coordinators), *Importance, preservation, and management of riparian habitat: A symposium.* General Technical Report RM-43. Fort Collins, CO: USDA Forest Service, Rocky Mountain Forest and Range Experiment Station.

Kirk, R. (editor). 1996. *The enduring forests: Northern California, Oregon, Washington, British Columbia, and southeast Alaska.* Seattle, WA: Mountaineers Press.

Klopatek, E. V., R. J. Olson, C. J. Emerson, and J. L. Jones. 1979. Land use conflicts with natural vegetation. *Environmental Conservation* 6: 191–200.

Knauth, P. 1972. *The American wilderness: The north woods.* New York: Time-Life Books.

Knight, D. H. 1994. *Mountains and plains: The ecology of Wyoming landscapes.* New Haven, CT: Yale University Press.

Knopf, F. L. 1994. Avian assemblages on altered grasslands. *Studies in Avian Biology* 15: 247–257.

———. 1995. Declining grassland birds. Pages 296–298 in E. T. LaRoe, G. S. Farris, C. E. Puckett, P. D. Doran, and M. J. Mac (editors), *Our living resources.* Washington, DC: U.S. Department of the Interior, National Biological Service.

Korte, P. A., and L. H. Fredrickson. 1977. Loss of Missouri's lowland hardwood ecosystem. In K. Sabol (editor), *Transactions of the Forty-Second North American Wildlife and Natural Resources Conference.* Washington, DC: Wildlife Management Institute.

Krebs, C. J. 1989. *Ecological methodology.* New York: Harper & Row.

Krever, V., E. Dinerstein, D. Olson, and L. Williams (editors). 1994. *Conserving Russia's biological diversity: An analytical framework and initial investment portfolio.* Washington, DC: World Wildlife Fund.

Kricher, J. C., and G. Morrison. 1993. *A field guide to the ecology of western forests.* Boston: Houghton Mifflin.

Kruckeberg, A. R., and D. Rabinowitz. 1985. Biological aspects of endemism in higher plants. *Annual Review of Ecologist Systemics* 16: 447–479.

Küchler, A. W. 1964. *Potential natural vegetation of the conterminous United States* Map and illustrated manual. New York: American Geographical Society.

———. 1975. *Vegetation maps of North America.* Lawrence: University of Kansas Libraries.

———. 1985. Potential natural vegetation of Alaska and Hawaii. Map in *National Atlas of the USA.* Reston, VA: Department of the Interior, U.S. Geological Survey.

Kuusela, J. 1990. *The dynamics of boreal coniferous forests.* Report No. 112. Helsinki: Finnish National Fund for Research and Development.

Landers, L., and D. Wade. 1994. Disturbance, persistence and diversity of the longleaf pine–bunchgrass ecosystem. In *Proceedings of the 1993 Society of American Foresters National Convention.* Bethesda, MD: Society of American Foresters.

Langner, L. L., and C. H. Flather. 1994. *Biological diversity: Status and trends in the United States.* RM-244. Fort Collins, CO: USDA Forest Service, Rocky Mountain Forest and Range Experiment Station.

LaRoe, E. T., G. S. Farris, C. E. Puckett, P. D. Doran, and M. J. Mac (editors). *Our living resources: A report to the nation on the distribution, abundance, and health of U.S. plants, animals, and ecosystems.* Washington, DC: U.S. Department of the Interior, National Biological Service.

Lattin, J. D., and A. R. Moldenke. 1992. *Ecologically sensitive taxa of Pacific Northwest old growth conifer forests.* Report prepared for the Northern Spotted Owl Recovery Team Other Species and Ecosystems Committee, Oregon State University. Corvallis, OR.

Laurance, W. F. 1991. Ecological correlates of extinction proneness in Australian tropical rain forest mammals. *Conservation Biology* 5: 79–89.

Lauriault, J. 1989. *Identification guide to the trees in Canada.* Ottawa: National Museum of Natural Sciences.

Laymon, S. A. 1984. Riparian bird community structure and dynamics: Dog Island, Red Bluff, California. Pages 587–597 in R. E. Warner and K. Hendrix (editors), *California riparian systems: Ecology, conservation, and productive management.* Berkeley: University of California Press.

Lee, D. S., C. R. Gilbert, C. H. Hocutt, R. E. Jenkins, D. E. McAllister, J. R. Stauffer Jr. 1980. *Atlas of North American freshwater fishes.* Raleigh: North Carolina State Museum of Natural History. North Carolina Biological Survey.

Lehmkuhl, J. F., P. F. Hessburg, R. L. Everett, M. H. Huff, and R. D. Ottmar. 1994. *Historical and current forest landscapes of eastern Oregon and Washington: Part I. Vegetation pattern and insect and disease hazards.* PNW-GTR-328. Portland, OR: USDA Forest Service, Pacific Northwest Research Station.

Leverett, R. 1993. The southern Appalachians: Plant paradise . . . imperiled. *Wild Earth* 3(2): 34–38.

Lewis R. R., III, and R. Summer. 1992. Coastal ecosystems. *Restoration and Management Notes* 10 (Columbian Quincentennial Issue): 1.

Licht, D. S. 1994. The great plains: America's best chance for ecosystem restoration, part 2. *Wild Earth* (Fall): 31–36.

Lindsey, C. C. 1989. Part 2: Biotic characteristics of the Queen Charlotte Islands. Pages 107–108 in G. G. E. Scudder and N. Gessler (editors), *Proceedings of the Queen Charlotte Islands First International Symposium.* Vancouver: University of British Columbia.

Liogier, H. A., and L. F. Martorell. 1982. *Flora of Puerto Rico and adjacent islands: A systematic synopsis.* San Juan: Editorial de la Universidad de Puerto Rico.

Little, E. L. Jr. 1976a. Conifers and important hardwoods. In *Atlas of United States trees,* volume 1. Washington, DC: USDA Forest Service.

———. 1976b. Minor western hardwoods. In *Atlas of United States trees,* volume 3. Washington, DC: USDA Forest Service. Washington, DC.

———. 1977. Minor eastern hardwoods. In *Atlas of United States trees,* volume 4. Washington, DC: USDA Forest Service.

Little, E. L. Jr., and F. H. Wadsworth. 1964. *Common trees of Puerto Rico and the Virgin Islands.* Agriculture Handbook No. 429. Washington, DC: USDA Forest Service.

Lodge, D. J., and W. H. McDowell. 1991. Summary of ecosystem level effects of Caribbean hurricanes. *Biotropica* 23: 373–378.

Longhurst, J. W. S. (editor). 1991. *Acid deposition: Origins, impacts and abatement strategies.* New York: Springer-Verlag.

Longhurst, W. M., R. E. Hafenfeld, and G. E. Connolly. 1982. Deer-livestock relationships in the western states. Pages 409–420 in L. Nelson, J. M. Peek, and P. D. Dalke (editors), *Proceedings of the wildlife-livestock relationships symposium.* Moscow: University of Idaho, Forest, Wildlife, and Range Experiment Station.

Loope, L. L., O. Hamann, and C. P. Stone. 1988. Comparative conservation biology of oceanic archipelagos. *BioScience* 38(4): 272–282.

Lotspeich, F. B., and W. S. Platts. 1982. An integrated land-aquatic classification system. *North American Journal of Fisheries Management* 2: 138–149.

Lovejoy, T. E. 1980. Discontinuous wilderness: Minimum areas for conservation. *Parks* 5: 13–15.

Lovingood, P. E., and R. E. Reiman. 1985. *Emerging patterns in the southern highlands.* A reference atlas produced by the Appalachian State University, University of South Carolina, and Tennessee Valley Authority, Economic Development and Analysis Branch. Volume 2, agriculture. Boone, NC: Appalachian Consortium.

Lowe, C. H. 1985. Amphibians and reptiles in southwest riparian systems. Pages 339–341 in R. R. Johnson, C. D. Ziebell, D. R. Patton, P. F. Folliott, and F. H. Hamre (technical coordinators), *Riparian ecosystems and their management: Reconciling conflicting uses.* General Technical Report RM-120. Fort Collins, CO: USDA Forest Service, Rocky Mountain Forest and Range Experiment Station.

Lowe, C. H., and D. E. Brown. 1994. Introduction. Pages 8–16 in D. E. Brown (editor), *Biotic communities of the southwestern United States and northwestern Mexico.* Salt Lake City: University of Utah Press.

Lugo, A. E. 1992. Preservation of primary forests in the Luquillo Mountains, Puerto Rico. *Conservation Biology* 8: 1122–1131.

MacLean, C. D. 1990. *Changes in area and ownership of timberland in western Oregon: 1961–86.* PNW-RB-170. Portland, OR: USDA Forest Service, Pacific Northwest Research Station.

MacMahon, J. A. 1988. Warm deserts. Pages 231–264 in M. G. Barbour and W. D. Billings (editors), *North American terrestrial vegetation.* Cambridge, England: Cambridge University Press.

Madson, J. 1993. *Tall grass prairie.* Helena, MT: Falcon Press.

Mahr, M. 1996. A natural diversity "hot spot" in Yellowstone country. *Wild Earth* 6(3): 33–36.

Majumdar, S. K., R. P. Brooks, F. J. Brenner, and R. W. Tiner Jr. (editors). 1989. *Wetlands ecology and conservation: Emphasis in Pennsylvania.* Easton: Pennsylvania Academy of Science.

Mann, C. C., and M. L. Plummer. 1995. California vs. gnatcatcher. *Audubon* (Jan–Feb): 39–49.

Marcot, B. G. 1997. Biodiversity of old forests of the west: A lesson from our elders. Pages 87–106 in K. A. Kohm and J. F. Franklin (editors), *Creating a forestry for the 21st century: The science of ecosystem management.* Washington, DC: Island Press.

Marcus, M. D., M. K. Young, L. E. Noel, and B. A. Mullan. 1990. *Salmonid-habitat relationships in the western United States: A review and indexed bibliography.* RM-188. Fort Collins, CO: USDA Forest Service, Rocky Mountain Forest and Range Experiment Station.

Mardin, P. G., and A. M. Schwartz. 1981. Comparative regional issues: Land use and environmental planning in the Adirondacks and Appalachians. Pages 89–98 in W. Summerville (editor), *Appalachia/America.* Boone, NC: Appalachian Consortium.

Marquis, D. A. 1975. *The Allegheny hardwood forests of Pennsylvania.* General Technical Report NE-15. Washington, DC: USDA Forest Service.

Martin, W. H., S. G. Boyce, and A. C. Echternacht (editors). 1993. *Biodiversity of the southeastern United States: Upland terrestrial communities.* New York: John Wiley.

Master, L. 1991. Aquatic animals: Endangerment alert. *The Nature Conservancy News* (March–April): 26–27.

Mauk, R. L., and J. A. Henderson. 1984. *Coniferous forest habitat types of northern Utah.* Ogden, UT: USDA Forest Service, Intermountain Station.

Maxwell, J. R., C. J. Edwards, M. E. Jensen, S. J. Paustian, H. Parrot, and D. M. Hill. 1995. *A hierarchical frame-*

work of aquatic ecological units in North America (Nearctic Zone). General Technical Report NC-176. St. Paul, MN: USDA Forest Service.

McClaran, M. P. 1995. Desert grasslands and grasses. Pages 1–30 in M. P. McClaran and T. R. Van Devender (editors), *The desert grasslands.* Tucson: University of Arizona Press.

McClellan, R. 1992. Southern Rockies ecosystem project. *Wild Earth* 2(1): 77–79.

McFarlane, R. W. 1992. *A stillness in the pines: The ecology of the red-cockaded woodpecker.* New York: W. W. Norton.

McGhie, R. G. 1996. *Creation of a comprehensive managed areas spatial database for the conterminous United States.* Summary project technical report NASA-NAGW-1743. Santa Barbara: Remote Sensing Research Unit, University of California.

McGregor, R. L., and T. M. Barkley (editors). 1977. *Atlas of the flora of the Great Plains.* Ames: The Great Plains Flora Association and Iowa State University Press.

McLaughlin, S. P. 1995. An overview of the flora of the sky islands, southeastern Arizona: Diversity, affinities, and insularity. Pages 60–70 in L. F. DeBano, P. F. Folliott, A. Ortega Rubio, G. J. Gottfried, R. H. Hamre, and C. B. Carleton (technical coordinators), *Biodiversity and management of the Madrean Archipelago: The Sky Islands of southwestern United States and northwestern Mexico.* Fort Collins, CO: Rocky Mountain Forest and Range Experiment Station; Tucson: School of Renewable Natural Resources, University of Arizona.

McNab, W. H., and P. E. Avers (compilers). 1994. *Ecological subregions of the United States: Section descriptions.* WO-WSA-5. Washington, DC: USDA Forest Service, Ecosystem Management.

McNab, W. H., and R. G. Bailey. 1994. *Map unit descriptions of subregions (sections) of the United States.* U.S. Geological Survey, National Atlas Series.

McPherson, G. R. 1992. Ecology of oak woodlands in Arizona. In *Ecology and management of oak and associated woodlands: Perspectives in the southwestern United States and northern Mexico.* RM-218. Fort Collins, CO: USDA Forest Service, Rocky Mountain Forest and Range Experiment Station.

Means, D. B., and G. Grow. 1985. The endangered longleaf pine community. *ENFO Report* 85(4): 1–12.

Medin, D. E. 1990. *Birds of an upper sagebrush-grass zone habitat in east-central Nevada.* INT-433. Ogden, UT: USDA Forest Service, Intermountain Research Station and Range Experiment Station.

Meffe, G. K., and C. R. Carroll. 1994. *Principles of conservation biology.* Sunderland, MA: Sinauer Associates.

Mehl, M. S. 1992. Old-growth descriptions for the major forest cover types in the Rocky Mountain region. Pages 106–120 in *Old-growth forests in the southwest and Rocky Mountain regions: Proceedings of a Workshop.* Portal, Arizona, March 9–13.

Meisner, M. 1993–94. Key words of conservation and environmental discourse. *Wild Earth* (Winter): 75–80.

Meldahl, R. S., J. S. Kush, D. J. Shaw, and W. D. Boyer. 1994. Restoration and dynamics of a virgin, old growth longleaf pine stand. Pages 532–533 in *Proceedings of the 1993 Society of American Foresters National Convention.* Bethesda, MD: Society of American Foresters.

Merrill, T., R. G. Wright, and J. M. Scott. 1995. Using ecological criteria to evaluate wilderness planning options in Idaho. *Environmental Management* 19(6): 815–825.

Miller, D. H., and F. E. Smeins. 1988. Vegetation pattern within a remnant San Antonio prairie as influenced by soil and microrelief variation. Pages 62–67 in *The prairie: Roots of our culture; foundation of our economy.* Symposium of the Tenth North American Prairie Conference, Denton, Texas.

Miller, R. R., J. D. Williams, and J. E. Williams. 1989. Extinctions of North American fishes during the past century. *Fisheries* 14(6): 22–38.

Milton, S. J., W. Richard, J. Dean, M. A. duPlessis, and W. R. Siegfried. 1994. A conceptual model of arid rangeland degradation. The escalating cost of declining productivity. *BioScience* 44(2): 70–76.

Mitchell, J. C. 1991. Amphibians and reptiles. Pages 411–422 in K. Terwilliger (editor), *Virginia's endangered species.* Blacksburg, VA: McDonald and Woodward.

Mitchell, J. E. 1993. The rangelands of Colorado. *Rangelands* 15(5): 213–220.

Mladendoff, D. J., T. A. Sickley, R. G. Haight, and A. Wydeven. 1995. A regional landscape analysis and prediction of favorable gray wolf habitat in the Northern Great Lakes Region. *Conservation Biology* 9(2): 279–294.

Moeur, M. 1992. *Baseline demographics of late successional western hemlock/western red cedar stands in northern Idaho research natural areas.* INT-456. USDA Forest Service, Intermountain Research Station.

Moldenke, A. 1990. One hundred twenty thousand little legs. *Wings: Essays on Invertebrate Conservation* 15(2): 11–14.

Mondt, R. 1995–96. Real work and wild vision. *Wild Earth* (Winter).

Monsen, S. B., and N. Shaw (editors). *Managing intermountain rangelands: Improvement of range and wildlife habitats.* General Technical Report INT-157. Ogden, UT: USDA Forest Service, Intermountain Forest and Range Experiment Station,

Monsen, S. B., and S. G. Kitchen. 1994. *Proceedings: Ecology and management of annual rangelands.* INT-GTR-313. Ogden, UT: USDA Forest Service, Intermountain Research Station.

Montgomery, J. A. 1993. The nature and origin of the Blackland Prairies of Texas. Pages 24–40 in R. Sharpless and J. C. Yelderman Jr. (editors), *The Texas Blackland Prairie land history and culture.* Proceedings of the Symposium of the Natural Regions of Texas, Baylor University Program for Regional Studies, Baylor University, Waco, Texas.

Montgomery, J. D. 1989. Rare and endangered pteridophytes in New Jersey. In E. F. Karlin (editor), *Proceedings of a conference on New Jersey's rare and endangered plants and animals.* Mahwah, NJ: Institute for Environmental Studies.

Mooney, H. A. 1995. Southern coastal scrub. Pages 471–489 in M. G. Barbour and J. Major (editors), *Terrestrial vegetation of California.* Special Publication No. 9. Davis: California Native Plant Society.

Moore, Janet R. 1996. *Issues related to a long-term management strategy for Manitoba's tall-grass prairie preserve.* Winnipeg, Manitoba: Critical Wildlife Habitat Program, Wildlife Branch, Department of Natural Resources.

Moreno, J. A., and L. Lusquis. 1985. La proliferaciùn de gorriones exùticos en Puerto Rico. *Revista Sociedad Ornithologaea de Puerto Rico* 1: 8–10.

Morgan, B. J. 1992–93. Indigo blues: The destruction of Gulf Hammock. *Wild Earth* (Winter): 33–38.

Morris, T. H., and M. A. Stubben. 1994. Geologic contrasts of the Great Basin and Colorado Plateau. Pages 9–26 in K. T. Harper, L. L. St. Clair, K. H. Thorne, and W. M. Hess (editors), *Natural history of the Colorado Plateau and Great Basin.* Boulder: University Press of Colorado.

Mosconi, S. L., and R. L. Hutto. 1982. The effect of grazing on the land birds of a western Montana riparian habitat. Pages 221–233 in L. Nelson, J. M. Peek, and P. D. Dalke (editors), *Proceedings of the wildlife-livestock relationships symposium.* Moscow: Forest, Wildlife, and Range Experiment Station, University of Idaho.

Moser, D. 1974. *The American wilderness: The Snake River country.* New York: Time-Life Books.

Mowat, F., and E. May. 1991. James Bay or the largest hydro development in North America. *Wild Earth* (Fall): 20–25.

Mueggler, W. F. 1988. *Aspen community types of the intermountain region.* INT-250. Ogden, UT: USDA Forest Service, Intermountain Forest and Range Experiment Station.

———. 1992. *Cliff Lake bench research natural area: Problems encountered in monitoring vegetation change on mountain grasslands.* INT-454. Ogden, UT: USDA Forest Service, Intermountain Research Station.

———. 1994. *Sixty years of change in tree numbers and basal area in central Utah aspen stands.* INT-RP-478. Ogden, UT: USDA Forest Service, Intermountain Research Station.

Mueller, R. F. 1992. Central Appalachian wilderness in perspective: The Monongahela National Forest. *Wild Earth* 2(2): 56–60.

Mueller-Dombois, D., and H. Ellenberg. 1974. *Aims and methods of vegetation ecology.* New York: John Wiley.

Muldavin, E., F. Ronco Jr., and E. F. Aldon. 1990. *Consolidated stand tables and biodiversity data base for southwestern forest habitat types.* RM-190. Fort Collins, CO: USDA Forest Service, Rocky Mountain Forest and Range Experiment Station.

Murphy, D. 1993. California's vanishing butterflies. *Defenders of Wildlife* (Fall): 17–21.

Musselman, R. C. (technical coordinator). 1994. *The glacier lakes ecosystem experiments site.* RM-249. Fort Collins, CO: USDA Forest Service, Rocky Mountain Forest and Range Experiment Station.

Mutch, R. W., S. F. Arno, J. K. Brown, C. E. Carlson, R. D. Ottmar, and J. L. Peterson. 1993. *Forest health in the Blue Mountains: A management strategy for fire adapted ecosystems.* PNW GTR 310. Portland, OR: USDA Forest Service.

Myers, J. M. 1995. Puerto Rican parrots. Pages 83–86 in E. T. LaRoe, G. S. Farris, C. E. Puckett, P. D. Doran, and M. J. Mac (editors), *Our living resources: A report to the nation on the distribution, abundance, and health of U.S. plants, animals, and ecosystems.* Washington, DC: U.S. Department of the Interior, National Biological Service.

Myers, R. L. 1985. Fire and the dynamic relationship between Florida sandhill and sand pine scrub vegetation. *Bulletin of the Torrey Ecological Society of America* 64: 62.

Myers, R. L., and J. J. Ewel. 1990a. *Ecosystems of Florida.* Orlando: University of Central Florida Press.

Myers, R. L., and J. J. Ewel. 1990b. Problems, prospects, and strategies for conservation. Pages 619–632 in R. L. Myers and J. J. Ewel (editors), *Ecosystems of Florida.* Orlando: University of Central Florida Press.

Naiman, R. J., J. Magnuson, D. M. McKnight, and J. A. Stanford (editors). 1995. *The freshwater imperative: A research agenda.* Washington, DC: Island Press.

National Research Council. 1994. *Rangeland health: New methods to classify, inventory, and monitor rangelands.* Washington, DC: National Academy Press.

National Research Council. 1996. *The Bering Sea ecosystem.* Washington, DC: National Academy Press.

The Nature Conservancy. 1996. *Priorities for conservation: 1996 annual report card for U.S. plant and animal species.* Arlington, VA: Author.

———. 1997. *Designing a geography of hope: Guidelines for ecoregion-based conservation in The Nature Conservancy.* Arlington, VA: Author.

Neelands, R. W. 1968. *Important forest trees of the eastern United States.* Atlanta: USDA Forest Service, Southern Region.

Nelson, C. R. 1994. Insects of the Great Basin and Colorado Plateau. Pages 211–254 in K. T. Harper, L. L. St. Clair, K. H. Thorne, and W. M. Hess (editors), *Natural history of the Colorado Plateau and Great Basin.* Boulder: University Press of Colorado.

Neves, R. J. 1991. Mollusks. Pages 251–263 in K. Terwilliger (editor), *Virginia's endangered species.* Blacksburg, VA: McDonald and Woodward.

Newmark, W. D. 1991. Tropical forest fragmentation and the local extinction of understory birds in the eastern Usambara Mountains, Tanzania. *Conservation Biology* 5: 67–78.

Niering, W. A. 1992. The New England forests. *Restoration and Management Notes* 10(1): 24–28.

Nikiforuk, A., and E. Struzik. 1989. The great forest sell-off. *Toronto Globe and Mail Business Magazine* (November).

Niles, L. J., K. E. Clark, and S. Paturzo. 1989. Status, protection and future needs of endangered and threatened birds of New Jersey. In E. F. Karlin (editor), *Proceedings of a conference on New Jersey's rare and endangered plants and animals.* Mahwah, NJ: Institute for Environmental Studies.

Norse, E. 1990. *Ancient forests of the Pacific Northwest.* Washington, DC: Island Press.

Norton, L. J. 1972. *The American wilderness: Atlantic beaches.* New York: Time-Life Books.

Noss, R. F. 1987. From plant communities to landscapes in conservation inventories: A look at The Nature Conservancy (USA). *Bioregional Conservation* 41: 11–37.

———. 1989. Longleaf pine and wiregrass: Keystone components of an endangered ecosystem. *Natural Areas Journal* 9(4): 211–213.

———. 1992. The Wildlands Project: Land conservation strategy. *Wild Earth* (Special Issue: The Wildlands Project): 10–25.

———. 1993. A conservation plan for the Oregon coast range: Some preliminary suggestions. *Natural Areas Journal* 13: 276–290.

———. 1996. Protected areas: How much is enough? Pages 91–120 in R. G. Wright (editor), *National parks and protected areas.* Cambridge, MA: Blackwell.

Noss, R. F., and A. Cooperrider. 1994. *Saving nature's legacy: Protecting and restoring biodiversity.* Washington, DC: Defenders of Wildlife and Island Press.

Noss, R. F., E. T. LaRoe, and J. M. Scott. 1995. *Endangered ecosystems of the United States: A preliminary assessment of loss and degradation.* Biological Report 28. Washington, DC: USDA National Biological Service.

Noss, R. F., and R. L. Peters. 1995. *Endangered ecosystems of the United States: A status report and plan for action.* Washington, DC: Defenders of Wildlife.

Nuzzo, V. A. 1986. Extent and status of Midwest oak savanna: Presettlement and 1985. *Natural Areas Journal* 6(2): 6–36.

Nyman, L. 1991. *Conservation of freshwater fish: Protection of biodiversity and genetic variability in aquatic ecosystems.* Göteborg, Sweden: SWEDMAR.

Office of the Forest Service, Area of Resources Administration. 1978. *The public forests of Puerto Rico.*

Ohmart, R. D. 1995. Ecological condition of the east fork of the Gila River and selected tributaries: Gila National Forest, New Mexico. Pages 312–317 in D. W. Shaw and D. M. Finch (technical editors), *Desired future conditions for southwest riparian ecosystems: Bringing interests and concerns together.* General Technical Report RM-GTR-272. Washington, DC: USDA.

———. 1996. Historical and present impacts: livestock grazing on fish and wildlife resources on western riparian habitats. Pages 245–279 in P. Krausman (editor), *Rangeland wildlife.* Denver, CO: Society for Range Management.

Old-Growth Policy Advisory Council. 1993. *Interim report on conserving old growth red and white pine.* Ottawa: Ontario Ministry of Natural Resources.

O'Leary, J. R. 1989. California coastal sage scrub: General characteristics and future prospects. *Crossosoma* 15: 4–5.

———. 1990. Californian coastal sage scrub: General characteristic and considerations for biological conservation. Pages 24–41 in A. A. Schoenherr (editor), *Endangered plant communities of Southern California.* Special Publication No. 3. Proceedings of the 15th Annual Symposium, Southern California Botanists, Claremont, California.

Ollinger, S. V., J. D. Aber, C. A. Federer, G. M. Lovett, and J. M. Ellis. 1995. *Modeling the physical and chemical climate of the northeastern United States for a geographic information system.* NE-191. Radnor, PA: USDA Forest Service, Northeastern Forest Experiment Station.

Olson, D. M. 1992. *The northern spotted owl conservation strategy: Implications for Pacific Northwest forest invertebrates and associated ecosystem processes.* Report submitted to the Northern Spotted Owl EIS Team, USDA Forest Service (November). Portland, OR: The Xerces Society.

Olson, D. M., B. Chernoff, G. Burgess, I. Davidson, P. Canevari, E. Dinerstein, G. Castro, V. Morisset, R. Abell, and E. Toledo (editors). 1997. *Freshwater biodiversity of Latin America and the Caribbean: A conservation assessment.* Proceedings of a workshop. Washington, DC: World Wildlife Fund.

Olson, D. M., and E. Dinerstein. 1998. The Global 200: A representation approach to conserving the earth's most biologically valuable ecoregions. *Conservation Biology* 3: 502–515.

Olson, D. M., E. Dinerstein, G. Cintrón, and P. Iolster (editors). 1996. *A conservation assessment of mangrove ecosystems of Latin America and the Caribbean.* Report from a workshop. Washington, DC: World Wildlife Fund.

Omernik, J. M. 1995a. Ecoregions: A framework for managing ecosystems. *George Wright Forum* 12(1): 35–51.

———. 1995b. *Level III ecoregions of the continent.* Washington, DC: National Health and Environment Effects Research Laboratory, U.S. Environmental Protection Agency. Maps at 1:7,500,000 scale.

Oregon-Washington Interagency Wildlife Committee. 1979. *Managing riparian ecosystems for fish and wildlife in eastern Oregon and eastern Washington.* Oregon-Washington Interagency Wildlife Committee. (Available from Washington State Library, Olympia, Washington.)

Orians, G. H. 1993. Endangered at what level? *Ecological Applications* 3: 206–208.

Ortiz, P. R. 1989. A summary of conservation trends in Puerto Rico. Pages 851–854 in C. A. Wood (editor), *Biogeography of the West Indies: Past, present, future.* Gainesville, FL: Sandhill Crane Press.

Pacific Rivers Council. 1995. *Coastal salmon and communities at risk: Briefing book.* Eugene, OR: Pacific Rivers Council.

Page, L. M., and B. M. Burr. 1991. *A field guide to freshwater fishes: North America north of Mexico.* Boston: Houghton Mifflin Company.

Parvin, R. W. 1989. Reclaiming a big thicket gem. *Nature Conservancy Magazine* 39(3): 22–26.

Pase, C. P. 1994. 111.5 Alpine tundra. Pages 27–33 in *Biotic communities of the American Southwest.* Albuquerque, NM: USDA Forest Service.

Pase, C. P., and D. E. Brown. 1994. Rocky Mountain (Petran) and Madrean montane conifer forests. Pages 43–48 in D. E. Brown (editor), *Biotic communities of the southwestern United States and northwestern Mexico.* Salt Lake City: University of Utah Press.

Patrick, R. 1994. *Rivers of the United States: Volume 1. Estuaries.* New York: John Wiley.

Paysen, T. E. 1982. Vegetation classification: California. Pages 75–80 in *Proceedings of the Symposium on Dynamics and Management of Mediterranean-Type Ecosystems.* General Technical Report PSW-58. Berkeley, CA: USDA Forest Service, Pacific Southwest Forest and Range Experiment Station.

Peck, S. B. 1981. Zoogeography of invertebrate cave faunas in southwestern Puerto Rico. *National Speleological Society Bulletin* 43: 70–79.

———. 1997. Origin and diversity of the North American cave fauna. Pages 60–66 in I. D. Sasowsky, D. W. Fong, and E. L. White (editors), *Conservation and protection of the biota of karst.* Special Publication No. 3. Charles Town, WV: Karst Waters Institute.

Peet, R. K. 1988. Forests of the Rocky Mountains. In M. G. Barbour and W. D. Billings (editors), *North American terrestrial vegetation.* Cambridge, England: Cambridge University Press.

Peterle, T. J. (editor). 1992. *2020 vision: Meeting the fish & wildlife conservation challenges of the 21st century.* West Lafayette, IN: Wildlife Society.

Petrides, A. P. 1988. *A field guide to eastern trees: Eastern United States and Canada.* Boston: Houghton Mifflin.

———. 1992. *A field guide to western trees: Western United States and Canada.* Boston: Houghton Mifflin.

Philibosian, R., and J. A. Yntema. 1977. *Annotated checklist of the birds, mammals, reptiles, and amphibians of the Virgin Islands & Puerto Rico.* Frederiksted, St. Croix, U.S. Virgin Islands: Information Services.

Pielou, E. C. 1994. *A naturalist's guide to the Arctic.* Chicago, IL: University of Chicago Press.

Pierce, R. S., J. W. Hornbeck, C. W. Martin, L. M. Tritton, C. T. Smith, C. A. Federer, and H. W. Yawney. 1993. *Whole-tree clearcutting in New England: Manager's guide to impacts on soils, streams, and regeneration.* NE-172. Radnor, PA: USDA Forest Service, Northeastern Forest Experiment Station.

Platts, W. S. 1979. Livestock grazing and riparian/stream ecosystems: An overview. Pages 39–45 in O. B. Cope (editor), *Proceedings of the forum: Grazing and riparian/stream ecosystems.* Denver, CO: Trout Unlimited.

———. 1981. *Influence of forest and rangeland management on anadromous fish habitat in western North America: No. 7, Effects of livestock grazing.* General Technical Report PNW-124. Portland, OR: USDA Forest Service, Pacific Northwest Forest and Range Experiment Station.

———. 1983. Vegetation requirements for fisheries habitats. Pages 184–188 in S. B. Monsen and W. Shaw (compilers), *Managing inter-mountain rangelands: Improvements of range and wildlife habitats.* General Technical Report INT-157. Washington, DC: USDA Forest Service.

Platts, W. S., and R. L. Nelson. 1989. Characteristics of riparian plant communities and streambanks with respect to grazing in northeast Utah. Pages 73–81 in R. E. Greswell, B. A. Barton, and J. L. Kerschner (editors), *Practical approaches to riparian resource manage-*

ment: An educational workshop. Billings, MT: Bureau of Land Management.

Povilitis, T. 1995. The Gila River–Sky Island region: A call for bold conservation action. *Wild Earth* (Fall): 73–77.

Powell, D. S., J. L. Faulkner, D. R. Darr, Z. Zhu, and D. W. MacCleery. 1993. *Forest resources of the United States, 1992.* RM-234. Fort Collins, CO: USDA Forest Service, Rocky Mountain Forest and Range Experiment Station.

Primack, R. B. 1993. *Essentials of conservation biology.* Sunderland, MA: Sinauer Associates.

Puerto Rico Department of Natural Resources. 1989. *Natural areas of conservation priority in Puerto Rico based on the Puerto Rico Natural Heritage Division data bank.* San Juan, Puerto Rico: Planning Area, PRDNR.

Raffaele, H. A. 1989a. The ecology of native and introduced granivorous birds in Puerto Rico. Pages 541–566 in C. A. Wood (editor), *Biogeography of the West Indies: Past, present, future.* Gainesville, FL: Sandhill Crane Press.

———. 1989b. *A guide to the birds of Puerto Rico and the Virgin Islands.* Princeton, NJ: Princeton University Press.

Randall, K. E. 1995. The Arizona riparian area advisory committee: An experience in defining desired conditions. Pages 216–226 in D. W. Shaw and D. M. Finch (technical editors), *Desired future conditions for southwest riparian ecosystems: Bringing interests and concerns together.* General Technical Report RM-GTR-272. Lakewood, CO: USDA Forest Sevice.

Ray, G. C., and B. P. Hayden. 1992. *Large marine ecosystems.* Washington, DC: American Association for the Advancement of Science.

Reed, P., G. Haas, F. Beum, and L. Sherrick. 1989. Nonrecreational uses of the National Wilderness Preservation System: A 1988 telephone survey. Pages 220–228 in H. Freilich (compiler), *Wilderness benchmark 1988: Proceedings of the National Wilderness Colloquium.* General Technical Report SE-51. Asheville, NC: USDA Forest Service, Southeastern Forest Experiment Station.

Rehfeldt, G. E. 1994. Genetic structure of western red cedar populations in the interior West. *Canadian Journal of Forest Research* 24: 670–680.

Reichman, O. J. 1987. *Konza prairie: A tallgrass natural history.* Lawrence: University Press of Kansas.

Reid, N., J. Marroquín, and P. Beyer-Münzel. 1990. Utilization of shrubs and trees for browse, fuelwood, and timber in the Tamaulipan thornscrub, northeastern Mexico. *Forest Ecology and Management* 36: 61–79.

Reid, T. S., and D. D. Murphy. 1995. Providing a regional context for local conservation action: A natural community conservation plan for the Southern California coastal sage scrub. *Science & Biodiversity Policy* (BioScience Supplement): 84–90.

RESTORE: The North Woods. 1994. *Maine woods: Proposed national park & preserve: A vision of what could be.* Concord, MA: Author.

Rhodes, R. 1974. *The American wilderness: The Ozarks.* New York: Time-Life Books.

Ribe, T. 1993–94. Human fear diminishes biological diversity in Rocky Mountain forests. *Wild Earth* (Winter): 39–43.

Richardson, C. J. 1983. Pocosins: Vanishing wastelands or valuable wetlands? *Bioscience* 33: 626–633.

Richardson, C. J., and J. W. Gibbons. 1993. Pocosins, Carolina bays and mountain bogs. Pages 257–310 in W. H. Martin, S. G. Boyce, and A. C. Echternacht (editors), *Biodiversity of the southeastern United States: Lowland terrestrial communities.* New York: John Wiley.

Richerson, P. J., and K. Lum. 1980. Patterns of plant species diversity in California: Relation to weather and topography. *American Naturalist* 116: 504–536.

Ricketts, T., E. Dinerstein, D. M. Olson, and C. Loucks. 1999. Who's where in North America: Patterns of species richness and the utility of indicator taxa for conservation. *BioScience* 49(5).

Riemann, H. R., and W. A. Befort. 1995. *Aerial photo guide to New England forest cover types.* NE-195. Radnor, PA: USDA Forest Service, Northeastern Forest Experiment Station.

Rietveld, W. J. (techical coordinator). 1994. *Agroforestry and sustainable systems: Symposium proceedings.* RM-GTR-261. Fort Collins, CO: USDA Forest Service, Rocky Mountain Forest and Range Experiment Station.

Riskind, D. H., R. George, G. Waggerman, and T. Hayes. 1987. Restoration in the subtropical United States. *Restoration and Management Notes* 5(2): 80–82.

Risser, P. G., E. C. Birney, H. D. Blocker, S. W. May, W. J. Parton, and J. A. Wiens. 1981. *The true prairie ecosystem.* Stroudsburg, PA: Hutchinson Ross Publishing.

Roath, L. R., and W. C. Krueger. 1982. Cattle grazing influence on a mountain riparian zone. *Journal of Range Management* 35: 100–103.

Robbins, W. G., and D. W. Wolf. 1994. *Landscape and the intermontane Northwest: An environmental history.* PNW-GTR-319. Portland, OR: USDA Forest Service, Pacific Northwest Research Station.

Robichaud, C., and K. H. Anderson. 1994. *Plant communities of New Jersey: A study in landscape diversity.* New Brunswick, NJ: Rutgers University Press.

Robinson, S. K., J. A. Grzybowski, S. I. Rothstein, M. C. Brittingham, L. J. Petit, and F. R. Thompson. 1993. Management implications of cowbird parasitism on neotropical migrant songbirds. In D. M. Finch and P. W. Stangel (editors), *Status and management of neotropical migratory birds, 1992 September 21–25.* General Technical Report RM-229. Fort Collins, CO: USDA Forest Service, Rocky Mountain Forest and Range Experiment Station.

Robison, H. W., and R. T. Allen. 1995. *Only in Arkansas: A study of the endemic plants and animals of the state.* Fayetteville, AR: University of Arkansas Press.

Romme, W. H., D. W. Jamieson, J. S. Redders, G. Bigsby, J. P. Lindsey, D. Kendall, R. Cowen, T. Kreykes, A. W. Spencer, and J. C. Ortega. 1992. Old-growth forests of the San Juan National Forest in southwestern Colorado. Pages 154–165 in *Old-growth forests in the southwest and Rocky Mountain regions.* Proceedings of a workshop, Portal, Arizona. March 9–13.

Roundy, B. A., E. D. McArthur, J. S. Haley, and D. K. Mann (compilers). 1995. *Proceedings: Wildland shrub and aridland restoration symposium.* INT-GTR-315. Ogden, UT: USDA Forest Service, Intermountain Research Station.

Rowe, J. S. 1972. *Forest regions of Canada.* Ottawa: Canadian Forestry Service, Department of Fisheries and the Environment. Text and national map at 1:6,700,000 scale.

Rudis, V. A., and J. B. Tansey. 1995. Regional assessment of remote forests and black bear habitat from forest resource surveys. *Journal of Wildlife Management* 59(1): 170–180.

Ruggiero, L. F., K. B. Aubry, S. W. Buskirk, L. J. Lyon, and W. J. Zielinski (technical editors). 1994. *American marten, fisher, lynx, and wolverine.* General Technical Report RM-254. Fort Collins, CO: USDA Forest Service.

Russell, F. 1973. *The American wilderness: The Okefenokee swamp.* New York: Time-Life Books.

Rzedowski, J. 1994. *Vegetación de México.* México: Editorial Limusa.

Sabadell, J. E. 1982. *Desertification of the United States: Status and issues.* Washington, DC: Bureau of Land Management.

Saenz, L. 1986. Grasslands as compared to adjacent *Quercus garryana* understories under two different grazing regimes. *Madroño* 33: 40–47.

The Salmon Mountains refugium, Klamath Region, California. 1969. Abstracts of the papers presented at the XI International Botanical Congress, Seattle, WA.

Samson, F. B., P. Alaback, J. Christner, T. DeMeo, A. Doyle, J. Martin, J. McKibben, M. Orme, L. Suring, K. Thompson, B. G. Wilson, D. A. Anderson, R. W. Flynn, J. W. Schoen, L. G. Shea, and J. L. Franklin. 1989. Conservation of rainforests in southeast Alaska: Report of a working group. Pages 122–133 in *Transactions of the 54th North American Wildlands and Natural Resources Conference.* Washington, DC: Wildlife Management Institute.

Saunders, D. A., R. J. Hobbs, and C. R. Margules. 1991. Biological consequences of ecosystem fragmentation: A review. *Conservation Biology* 5: 18–32.

Sawyer, J. O. 1996. Northern California. In R. Kirk (editor), *The enduring forests: Northern California, Oregon, Washington, British Columbia, and southwest Alaska.* Seattle, WA: Mountaineers Press.

Sawyer, J. O., and D. A. Thornburgh. 1988. Montane and subalpine vegetation of the Klamath Mountains. Pages 699–732 in M. G. Barbour and J. Major (editors), *Terrestrial vegetation of California.* Special Publication No. 9. Davis: California Native Plant Society.

Sawyer, J. O., D. A. Thornburgh, and J. R. Griffin. 1995. Mixed evergreen forest. Pages 359–381 in M. G. Barbour and J. Major (editors), *Terrestrial vegetation of California.* Special Publication No. 9. Davis: California Native Plant Society.

Schmidly, D. J., D. L. Scarbrough, and M. A. Horner. 1993. Wildlife diversity in the Blackland Prairies. Pages 82–95 in R. Sharpless and J. C. Yelderman Jr. (editors), *The Texas Blackland Prairie land history and cul-*

ture. Symposium of the Natural Regions of Texas, Baylor University Program for Regional Studies, Baylor University, Waco, Texas.

Schmidt, K. 1996. Rare habitats vie for protection. *Science* 274: 916–918.

Schmidt, W. C., and F. Holtmeier (compilers). 1994. *Proceedings: International workshop on subalpine stone pines and their environment: The status of our knowledge.* INT-GTR-309. Ogden, UT: USDA Forest Service, Intermountain Research Station.

Schoenherr, A. A. 1992. *A natural history of California.* Berkeley: University of California Press.

———. (editor). 1990. *Endangered plant communities of Southern California: Proceedings of the 15th Annual Symposium.* Special Publication No. 3. Claremont, CA: Southern California Botanists.

Schroeder, W. A. 1978. Mapping the pre-settlement prairies of Missouri. In *Fifth Midwest Prairie Conference proceedings.* Ames: Iowa State University.

———. 1981. *Presettlement prairie of Missouri.* Natural History Series No. 2. Jefferson City: Missouri Department of Conservation.

Schweitzer, D. F. 1989. A progress report on the identification and prioritization of New Jersey's rare lepidoptera: 1981–1987. In E. F. Karlin (editor), *Proceedings of a conference on New Jersey's rare and endangered plants and animals.* Mahwah, NJ: Institute for Environmental Studies.

Scott, J. A. 1986. *The butterflies of North America: A natural history and field guide.* Stanford, CA: Stanford University Press.

Scott, J. M., F. Davis, B. Csuti, R. Noss, B. Butterfield, C. Groves, J. Anderson, S. Caicco, F. D'Erchia, T. C. Edwards, J. Ulliman, and R. G. Wright. 1993. Gap analysis: A geographical approach to protection of biological diversity. *Wildlife Monographs* 123: 1–41.

Scott, J. M., T. H. Tear, and F. W. Davis. 1996. *Gap analysis: A landscape approach to biodiversity planning.* Bethesda, MD: American Society for Photogrammetry and Remote Sensing.

Scott, S. L. 1995. *Field guide to the birds of North America.* Washington, DC: National Geographic Society.

Sellards, E. H. 1914. *Florida state geological survey: Sixth annual report.* Tallahassee, FL: State Geological Survey.

Shafer, C. L. 1995. Value and shortcomings of small reserves. *BioScience* 45: 80–88.

Shaffer, T. L., and W. E. Newton. 1995. Duck nest success in the prairie potholes. Pages 300–302 in E. T. LaRoe, G. S. Farris, C. E. Puckett, P. D. Doran, and M. J. Mac (editors), *Our living resources: A report to the nation on the distribution, abundance, and health of U.S. plants, animals, and ecosystems.* Washington, DC: U.S. Department of the Interior, National Biological Service.

Sharitz, R. R., and W. J. Mitsch. 1993. Southern floodplain forests. Pages 311–372 in W. H. Martin, S. G. Boyce, and A. C. Echternacht (editors), *Biodiversity of the southeastern United States: Lowland terrestrial communities.* New York: John Wiley.

Shelford, V. E. 1963. *The ecology of North America.* Chicago: University of Illinois Press.

———. 1933. Ecological Society of America: A nature sanctuary plan unanimously adopted by the Society, December 28, 1932. *Ecology* 14: 240–245.

———. (editor). 1926. *Naturalist's guide to the Americas.* Baltimore, MD: Williams and Wilkins.

Shinn, C. W. 1993. *British Columbia log export policy: Historical review and analysis.* PNW-RP-457. Portland, OR: USDA Forest Service, Pacific Northwest Research Station.

Shlisky, A. J. 1994. Pages 201–206 in *Proceedings of Society of American Foresters, Bethesda, Maryland, 1993.* Bethesda, MD: Society of American Foresters.

Shreve, F., and I. L. Wiggins. 1964. *Vegetation and flora of the Sonoran desert,* volume 1. Stanford, CA: Stanford University Press.

Sigler, J. W., and W. F. Sigler. 1994. Fishes of the Great Basin and the Colorado Plateau: Past and present forms. Pages 163–210 in K. T. Harper, L. L. St. Clair, K. H. Thorne, and W. M. Hess (editors), *Natural history of the Colorado Plateau and Great Basin.* Boulder: University Press of Colorado.

Sims, P. L. 1988. Grasslands. Pages 265–285 in M. G. Barbour and W. D. Billings (editors), *North American terrestrial vegetation.* Cambridge, England: Cambridge University Press.

Skeen, J. N., P. D. Doerr, and D. H. Van Lear. 1993. Oak-hickory-pine forests. Pages 1–33 in W. H. Martin, S. G. Boyce, and A. C. Echternacht (editors), *Biodiversity of the southeastern United States: Upland terrestrial communities.* New York: John Wiley.

Skinner, M. W., and B. M. Pavlik. 1994. *Inventory of rare and endangered vascular plants of California.* Sacramento: California Native Plant Society.

Skole, D. L., and C. Tucker. 1993. Tropical deforestation and habitat fragmentation in the Amazon: Satellite data from 1978 to 1988. *Science* 260: 1905–1910.

Smeins, F. E., and D. D. Diamond. 1983. Remnant grasslands of the Fayette Prairie, Texas. *American Midland Naturalist* 110: 1–13.

Smeins, F. E., D. D. Diamond, and C. W. Hanselka. 1991. Coastal prairie. Pages 269–290 in R. T. Coupland (editor), *Ecosystems of the world: Natural grasslands—introduction and western hemisphere.* New York: Elsevier.

Smith, J. P. 1973. The Klamath region. *California Native Plant Society Newsletter* 8: 3–6.

———. 1988. Endemic vascular plants of northwestern California and southwestern Oregon. *Madroño* 35: 54–69.

Smith, N. S., and R. G. Anthony. 1992. *Ecology and management of oak and associated woodlands: Perspectives in the southwestern United States and northern Mexico.* RM-218. Fort Collins, CO: USDA Forest Service, Rocky Mountain Forest and Range Experiment Station.

Smith, O. D., Jr. 1991. Older stand of the Ozark–St. Francis National Forests: Type, extent, and location. Pages 115–137 in D. Henderson and L. D. Hedrick (editors), *Restoration of old growth forests in the interior highlands of Arkansas and Oklahoma.* Morillton, AR: Ouachita National Forest and Winrock International Institute for Agricultural Development.

Smith, W. H. 1991. Air pollution and forest damage. *Chemical and Engineering News* 69(45): 30–43.

Smith, W. P. (editor). 1993. *Fourth meeting of the southeast management working group: Abstracts.* General Technical Report SO-95. Asheville, NC: USDA Forest Service, Southern Forest Experiment Station.

Snyder, D. B. 1989. On the edge of extirpation: New Jersey's most critically imperiled flora. In E. F. Karlin (editor), *Proceedings of a conference on New Jersey's rare and endangered plants and animals.* Mahwah, NJ: Institute for Environmental Studies.

Snyder, J. R., A. Herndon, and W. B. Robertson. 1990. South Florida rockland. Pages 230–277 in R. L. Myers and J. J. Ewel (editors), *Ecosystems of Florida.* Orlando: University of Central Florida Press.

Society of American Foresters. 1980. Douglas-fir–tanoak–madrone. Pages 111–112 in F. H. Eyre (editor), *Forest cover types of the United States and Canada.* Washington, DC: Society of American Foresters.

Sohmer, S. H., and S. Gon. 1996. Hawaiian region: CPD Site PO6 Hawaiian Islands, Hawaii, USA. Pages 549–555 in WWF and IUCN, *Centres of Plant Diversity,* volume 2. Cambridge, England: IUCN.

Sohmer, S. H., and R. Gustafson. 1987. *Plants and flowers of Hawaii.* Honolulu: University of Hawaii Press.

Solomon, D. S., and W. B. Leak. 1994. *Migration of tree species in New England based on elevational and regional analyses.* NE-688. Radnor, PA: USDA Forest Service, Northeastern Forest Experiment Station.

Southeast Alaska Conservation Council. 1993. *Alaska rainforest atlas.* Juneau, AK: Author.

Sowls, A. L., S. A. Hatch, and C. J. Lensink. 1978. *Catalog of Alaskan seabird colonies.* Washington, DC: U.S. Department of Interior, Fish and Wildlife Service.

Specht, R. L. 1982. General characteristics of Mediterranean-type ecosystems. Pages 13–19 in *Proceedings of the symposium on dynamics and management of Mediterranean-type ecosystems.* General Technical Report PSW-58. Berkeley, CA: USDA Forest Service, Pacific Southwest Forest and Range Experiment Station.

Spellerberg, I. F. 1992. *Evaluation and assessment for conservation.* New York: Chapman & Hall.

Stebbins, G. L., and J. Major. 1965. Endemism and speciation in the California flora. *Ecological Monographs* 35: 1–35.

Stebbins, R. C. 1985. *A field guide to western reptiles and amphibians.* Boston: Houghton Mifflin.

Steinhart, P. 1990. *California's wild heritage: Threatened and endangered animals in the golden state.* Sacramento: California Department of Fish and Game.

Stephenson, S. L., A. N. Ash, and D. F. Stauffer. 1993. Appalachian oak forests. Pages 255–304 in W. H. Martin, S. G. Boyce, and A. C. Echternacht (editors), *Biodiversity of the southeastern United States: Upland terrestrial communities.* New York: John Wiley.

Stevens, L. E., B. T. Brown, J. M. Simpson, and R. R. Johnson. 1977. The importance of riparian habitat to migrating birds. Pages 156–164 in R. R. Johnson, and D. A. Jones (technical coordinators), *Importance, preservation, and management of riparian habitat: A symposium.* General Technical Report RM-43. Fort Collins, CO: USDA Forest Service, Rocky Mountain Forest and Range Experiment Station.

Stevens, W. K. 1996. Salvation at hand for a California landscape. *New York Times,* February 27, B5.

Stranahan, S. Q. 1993. *Susquehanna, river of dreams.* Baltimore, MD: Johns Hopkins University Press.

Strickland, S. S., and J. W. Fox. 1993. Prehistoric environmental adaptations in the Blackland Prairie. Pages 97–121 in R. Sharpless and J. C. Yelderman Jr. (editors), *The Texas Blackland Prairie land history and culture.* Symposium of the Natural Regions of Texas, Baylor University Program for Regional Studies, Baylor University, Waco, Texas.

Strittholt, J. R., and R. E. J. Boerner. 1995. Applying biodiversity gap analysis in a regional nature reserve design for the edge of Appalachia, Ohio. *Conservation Biology* 9: 1492–1505.

Suring, L. H., D. C. Crocker-Bedford, R. W. Flynn, C. S. Hale, G. C. Iverson, M. D. Kirchhoff, T. E. Schenck, L. C. Shea, and K. Titus. 1993. *A proposed strategy for maintaining well-distributed, viable populations of wildlife associated with old growth forests in southeast Alaska.* Anchorage, AK: USDA Forest Service.

Swift, B. L. 1984. Status of riparian ecosystems in the United States. *Water Resources Bulletin* 20(2): 223–228.

Szaro, R. C. 1989. Riparian forest and scrubland community types of Arizona and New Mexico. *Desert Plants* 9(3-4): 69–138.

Tanner, O. 1974. *The American wilderness: New England wilds.* New York: Time-Life Books.

Tansley, A. G. 1935. *Introduction to plant ecology: A guide for beginners in the study of plant communities.* London: George Allen & Unwin.

Taylor, R. J., and R. W. Valum. 1974. *Wildflowers 2: Sagebrush country.* Beaverton, IL: Touchstone Press.

Tester, J. R. 1995. *Minnesota's natural heritage: An ecological perspective.* Minneapolis: University of Minnesota Press.

Thomas, J. W. (editor). 1979. *Wildlife habitat in managed forests: The Blue Mountains of Oregon and Washington.* Agriculture Handbook No. 553. Washington, DC: USDA Forest Service.

Thomas, J. W., C. Maser, and J. E. Rodiek. 1979. Riparian zones in managed rangelands: Their importance to wildlife. Pages 21–31 in O. B. Cope (editor), *Proceedings of the Forum: Grazing and Riparian/Stream Ecosystems.* Denver, CO: Trout Unlimited.

Thompson F. R., III. 1993. Simulated responses of a forest-interior bird population to forest management options in central hardwood forests of the United States. *Conservation Biology* 7(2): 325–333.

Thornburgh, D. A. 1977. Montane and subalpine vegetation of the Klamath Mountains. Pages 699–732 in M. G. Barbour and J. Major (editors), *Terrestrial vegetation of California.* Special Publication No. 9. Davis: California Native Plant Society.

Thornburgh, D. A., and W. F Bowman. 1970. Extension of the range of *Abies lasiocarpa* into California. *Madroño* 20: 413–415.

Thorne, R. F. 1995. Montane and subalpine forests of the transverse and peninsular ranges. Pages 537–557 in M.

G. Barbour and J. Major (editors), *Terrestrial vegetation of California.* Special Publication No. 9. Davis: California Native Plant Society.

Tiner R. W., Jr. 1984. *Wetlands of the United States: Current status and recent trends.* Washington, DC: U.S. Department of the Interior, Fish and Wildlife Service.

Tiner, R. W., and D. G. Burke. 1995. *Wetlands of Maryland.* Hadley, MA: U.S. Fish and Wildlife Service, Ecological Services, Region 5; Annapolis, MD: Maryland Department of Natural Resources.

Tipton, R. 1996. Director, U.S. ecoregional conservation, World Wildlife Fund, Everglades restoration, personal communication.

Tisdale, E. W. 1961. Ecologic changes in the Palouse. *Northwest Science* 35(4): 134–138.

Tukey, J. W. 1977. *Exploratory data analysis.* Reading, MA: Addison-Wesley.

Turner, J. T. 1994a. Great Basin desertscrub. Pages 145–155 in D. E. Brown (editor), *Biotic communities in southwestern United States and northwestern Mexico.* Salt Lake City: University of Utah Press.

———. 1994b. Mojave desertscrub. Pages 157–168 in D. E. Brown (editor), *Biotic communities in southwestern United States and northwestern Mexico.* Salt Lake City: University of Utah Press.

Turner, R. M., J. E. Bowers, and T. L. Burgess. 1995. *Sonoran Desert plants: An ecological atlas.* Tucson: University of Arizona Press.

Tweit, S. J. 1995. *Barren, wild, and worthless: Living in the Chihuahuan Desert.* Albuquerque: University of New Mexico Press.

USDA, 1989. *An analysis of the wildlife and fish situation in the United States: 1989–2040.* A technical document supporting the 1989 USDA Forest Service RPA assessment. Fort Collins, CO: USDA Forest Service, Rocky Mountain Forest and Range Experiment Station.

———. 1991. *Questions and answers on a conservation strategy for the northern spotted owl.* Miscellaneous Publication. Portland, OR: USDA Forest Service, Pacific Northwest Research Station.

———. 1995. *Interior West global change workshop.* RM-GTR-262. Fort Collins, CO: USDA Forest Service, Rocky Mountain Forest and Range Experiment Station.

USDA Forest Service. 1991. *Tongass Land Management Plan: Supplement to the draft environmental impact statement.* R10-MB-149. Juneau, AK: USDA Forest Service Tongass National Forest.

———. 1994a. *Ecological subregions of the United States: Section descriptions.* Washington, DC: Forest Service.

———. 1994b. *Final supplemental environmental impact statement on management of habitat for late-successional and old growth forest related species within the range of the northern spotted owl,* volume 1. Washington, DC: Forest Service.

———. 1994c. *Karst vulnerability assessment report.* Juneau, AK: USDA Forest Service, Tongass National Forest Ketchikan Area Thorne Bay.

USDA Forest Service and USDI Bureau of Land Management. 1996. *Status of the interior Columbia Basin: Sum-mary of Scientific Findings.* General Technical Report PNW-GTR-385. Portland, OR: Forest Service, Pacific Northwest Region; Washington, DC: Bureau of Land Management.

———. 1997. *Eastside draft environmental impact statement,* vol. 1. Portland, OR: Forest Service, Pacific Northwest Region; Washington, DC: Bureau of Land Management.

U.S. Department of Commerce. 1980. *Census of population, 1980.* Washington, DC: Census Bureau.

U.S. Fish and Wildlife Service. 1992. *Hawaii's extinction crisis, a call to action: A report on the status of Hawaii's natural heritage.* Washington, DC: Hawaii State Department of Land and Natural Resources, U.S. Fish and Wildlife Service, The Nature Conservancy of Hawaii.

———. 1993. *Lower Rio Grande National Wildlife Refuge: Final environmental assessment.* Albuquerque, NM: U.S. Fish and Wildlife Service.

U.S. General Accounting Office. 1988. *Rangeland management: More emphasis needed on declining and overstocked grazing allotments.* GAO/RCED-88-80. Washington, DC: U.S. General Accounting Office.

Van Der Valk, A. (editor). 1989. *Northern prairie wetlands.* Ames: Iowa State University Press.

Van Velson, R. 1979. Effects of livestock grazing upon rainbow trout in Otter Creek, Nebraska. Pages 53–55 in O. B. Cope (editor), *Proceedings of the Forum on Grazing and Riparian/Stream Ecosystems.* Denver, CO: Trout Unlimited.

Van Vuren, D. 1982. Comparative ecology of bison and cattle in the Henry Mountains, Utah. Pages 449–457 in L. Nelson, J. M. Peek, and P. D. Dalke (editors), *Proceedings of the wildlife-livestock relationships symposium.* Moscow: USDA Forest Service, Wildlife, and Range Experiment Station, University of Idaho.

Vance-Borland, K., R. Noss, J. Strittholt, P. Frost, C. Carroll, and R. Nawa. 1995. A biodiversity conservation plan for the Klamath/Siskiyou region. *Wild Earth* 5(4): 52–59.

Vasek, F. C., and R. F. Thorne. 1988. Transmontane coniferous vegetation. Pages 797–834 in M. G. Barbour and J. Major (editors), *Terrestrial vegetation of California.* Special Publication No. 9. Davis: California Native Plant Society.

Viereck, L. A., C. T. Dyrness, A. R. Batten, and K. J. Wenzlick. 1992. *The Alaska vegetation classification.* General Technical Report PNW-GTR 286. Portland, OR: USDA Forest Service, Pacific Northwest Research Station.

Viers, S. D., Jr. 1982. Coast redwood forest: Stand dynamics, successional status, and the role of fire. Pages 119–141 in J. E. Means (editor), *Forest succession and stand development research in the northwest.* Corvallis: Oregon State University Forest Research Laboratory.

Vogl, R. J. 1982. Chaparral succession. Pages 81–85 in *Proceedings of the symposium on dynamics and management of Mediterranean-type ecosystems.* General Technical Report PSW-58. Berkeley, CA: USDA Forest Service, Pacific Southwest Forest and Range Experiment Station.

Waddell, K. L., D. D. Oswald, and D. S. Powell. 1989. *Forest statistics of the United States, 1987*. PNW-RB-168. Portland, OR: USDA Forest Service, Pacific Northwest Research Station.

Wagner, F. H. 1978. Livestock grazing and the livestock industry. Pages 121–145 in H. P. Brokaw (editor), *Wildlife and America*. Washington, DC: Council on Environmental Quality.

Wagner, W. L., and V. A. Funk. 1995. *Hawaiian biogeography: Evolution on a hot spot archipelago*. Smithsonian series in comparative evolutionary biology. Washington, DC: Smithsonian Institution Press.

Wagner, W. L., D. R. Herbst, and S. H. Sohmer. 1990. *Manual of the Flowering Plants of Hawaii*. Honolulu: University of Hawaii Press, Bishop Museum.

Walker, B. S. 1973. *The American wilderness: The Great Divide*. New York: Time-Life Books.

Walker, L. C. 1991. *The southern forest: A chronicle*. Austin: University of Texas Press.

Wallace, D. R. 1983. *The Klamath knot*. San Francisco: Sierra Club Books.

Wallace, D. R. 1992. The Klamath surprise: Forestry meets biodiversity on the West Coast. *Wilderness* 56: 10–33.

Wallace, R. 1972. *The American wilderness: The Grand Canyon*. New York: Time-Life Books.

———. 1973. *The American wilderness: Hawaii*. New York: Time-Life Books.

Ware, S., C. Frost, and P. D. Doerr. 1993. Southern mixed hardwood forest: The former longleaf pine forest. Pages 447–493 in W. H. Martin, S. G. Boyce, and A. C. Echternacht (editors), *Biodiversity of the southeastern United States: Lowland terrestrial communities*. New York: John Wiley.

Waring, R. H., and J. F. Franklin. 1979. Evergreen coniferous forests of the Pacific Northwest. *Science* 204: 1380–1386.

Warner, R. E., and K. M. Hendrix. 1984. *California riparian systems: Ecology, conservation, and productivity*. Berkeley: University of California Press.

Warshall, P. 1995. Southwestern Sky Islands ecosystems. Pages 318–322 in E. T. Laroe, G. S. Farris, C. E. Puckett, P. D. Dornan, and M. J. Mac (editors), *Our living resources: A report to the nation on the distribution, abundance, and health of U.S. plants, animals, and ecosystems*. Washington, DC: U.S. Department of the Interior, National Biological Service.

Watkins, T. H. 1989. *Time's island: The California desert*. Sacramento, CA: Gibbs-Smith.

Wauer, R. H. 1992. *A naturalist's Mexico*. College Station: Texas A&M University Press.

Weaver, J., R. Escano, T. Puchlerz, and D. Despain. 1985. A cumulative effects model for grizzly bear management in the Yellowstone ecosystem. In *Proceedings: Grizzly Bear Habitat Symposium*. Missoula, Montana.

Weaver, J. E., and F. E. Clements. 1938. *Plant ecology*, 2nd edition. New York: McGraw-Hill.

Weaver, J. E. 1954. *North American prairie*. Lincoln, NE: Johnsen Publishing.

Wells, W. G., II. 1982. Hydrology of Mediterranean-type ecosystems: A summary and synthesis. Pages 426–430 in *Proceedings of the symposium on dynamics and management of Mediterranean-type ecosystems*. General Technical Report PSW-58. Berkeley, CA: USDA Forest Service, Pacific Southwest Forest and Range Experiment Station.

West, N. E. 1988. Intermountain deserts, shrub steppes, and woodlands. Chapter 7 in M. G. Barbour and W. D. Billings (editors), *North American terrestrial vegetation*. Cambridge, England: Cambridge University Press.

Westman, W. E. 1982. Coastal sage scrub succession. Pages 91–99 in *Proceedings of the symposium on dynamics and management of Mediterranean-type ecosystems*. General Technical Report PSW-58. Berkeley, CA: USDA Forest Service, Pacific Southwest Forest and Range Experiment Station.

———. 1983. Xeric Mediterranean-type shrubland associations of Alta and Baja California and the community/continuum debate. *Vegetatio* 52: 3–19.

Westman, W. E., and J. F. O'Leary. 1986. Measure of resilience: The response of coastal sage scrub to fire. *Vegetatio* 65: 179–189.

Wharton, C. H., W. M. Kitchens, E. C. Pedleton, and T. W. Snipe. 1982. *The ecology of bottomland hardwood swamps in the Southeast: A community profile*. FWS/OBS 81/37. Washington, DC: U.S. Fish and Wildlife Service.

Whelan, R. J. 1995. *The ecology of fire*. Cambridge, England: Cambridge University Press.

Wheller, M. W. 1994. *Freshwater marshes: Ecology and wildlife management*, 3rd edition. Minneapolis: University of Minnesota Press.

White, P. S., E. Buckner, J. D. Pittillo, and C. V. Cogbill. 1993. High elevation forests: Spruce fir forests, northern hardwoods forests, and associated communities. Pages 305–338 in W. H. Martin, S. G. Boyce, and A. C. Echternacht (editors), *Biodiversity of the southeastern United States: Upland terrestrial communities*. New York: John Wiley.

Whitney, C. G. 1984. Fifty years of change in arboreal vegetation of Heart's Content, and old growth hemlock–white pine–northern hardwood stand. *Ecology* 65: 403–408.

———. 1990. Multiple pattern analysis of an old growth hemlock–white pine–northern hardwood stand. *Bulletin of the Torrey Botanical Club* 117: 39–47.

———. 1994. *From coastal wilderness to fruited plain: A history of environmental change in temperate North America, 1500 to the present*. Cambridge, England: Cambridge University Press.

Whitney, S. 1985. *Western forests*. New York: Alfred A. Knopf.

Whittaker, R. H. 1960. Vegetation of the Siskiyou Mountains, Oregon and California. *Ecological Monographs* 30: 279–338.

———. 1961. Vegetation of the Pacific Coast states and the central significance of the Klamath region. *Madroño* 16: 5-23.

Wiken, E. B. (compiler). 1986. *Terrestrial ecoregions of Canada.* Ecological Land Classification Series No. 19. Hull, Quebec: Environment Canada.

Wiken, E. B., and K. Lawton. 1995. North American protected areas: An ecological approach to reporting and analysis. *George Wright FORUM* 12(1): 25–34.

Wiken, E. B., D. M. Welch, G. R. Ironside, and D. G. Taylor. 1981. *The northern Yukon: An ecological land survey: Ecological land classification series.* Lands Directorate, Vancouver, British Columbia: Environment Canada.

Wikramanayake, E., E. Dinerstein, P. Hedao, D. M. Olson, L. Horowitz, and P. Hurley. In press. *A conservation assessment of the terrestrial ecoregions of the Indo-Pacific region.* Washington, DC: World Wildlife Fund.

Wikramanayake, E., E. Dinerstein, J. Robinson, U. Karanth, A. Rabinowitz, D. M. Olson, T. Mathew, P. Hedao, M. Connor, G. Hemley, and D. Bolze. 1997. *A framework for identifying high priority areas and actions for the conservation of tigers in the wild.* Washington, DC: World Wildlife Fund and Wildlife Conservation Society.

Wilburn, J. 1985. Redwood forest. *Outdoor California* (Jan.–Feb.): 13–16.

Wilcove, D. S., M. J. Bean, R. Bonnie, and M. McMillan. 1996. *Rebuilding the ark: Toward a more effective Endangered Species Act for private land.* Washington, DC: Environmental Defense Fund.

Wilcove, D. S., C. H. McLellan, and A. P. Dobson. 1986. Habitat fragmentation in the temperate zone. Pages 237–256 in M. E. Soulé (editor), *Conservation biology: The science of scarcity and diversity.* Sunderland, MA: Sinauer Associates.

Wild Earth. 1991. New forestry threatens Kalmiopsis. *Wild Earth* (Fall): 14–15.

———. 1993. Desert tortoise vs. nuclear dump. *Wild Earth* (Summer): 22–23.

The Wildlife Society. 1996. The Wildlife Society position statement on livestock grazing on federal rangelands in the western United States. *The Wildlifer,* issue no. 274 (Jan.–Feb.): 10–13.

Williams, J. E., J. E. Johnson, D. A. Hendrickson, S. Contreras-Balderas, J. D. Williams, M. Navarro-Mendoza, D. E. McAllister, and J. E. Deacon. 1989. Fishes of North America endangered, threatened, or of special concern: 1989. *Fisheries* 14(6): 2–20.

Williams, M. 1989. *Americans and their forests.* Cambridge, England: Cambridge University Press.

Williams, R. L. 1973. *The American wilderness: The northwest coast.* New York: Time-Life Books.

———. 1974. *The American wilderness: The Cascades.* New York: Time-Life Books.

Willson, G. D. 1995. The Great Plains. Pages 295–296 In E. T. LaRoe, G. S. Farris, C. E. Puckett, P.D. Doran, and M. J. Mac (editors), *Our living resources: A report to the nation on the distribution, abundance, and health of U.S. plants, animals, and ecosystems.* Washington, D.C.: U.S. Department of Interior, National Biological Service.

Wilson, E. O. 1985. The biological diversity crisis: A challenge to science. *Issues in Science and Technology* 2: 20–29.

———. (editor) 1988. *Biodiversity.* Washington, DC: National Academy Press.

———. 1992. *The diversity of life.* Cambridge, MA: Belknap Press.

Winchester, N. N., and R. A. Ring. 1997. *Conservation of arthropod biodiversity: Reevaluation of the centinelan extinction concept.* A paper presented at the Annual Meeting of the Society for Conservation Biology, Victoria, British Columbia, June 6–9.

Wirtz, W. O., II. 1982. Postfire community structure of birds and rodents in Southern California chaparral. Pages 241–246 in *Proceedings of the symposium on dynamics and management of Mediterranean-type ecosystems.* General Technical Report PSW-58. Berkeley, CA: USDA Forest Service, Pacific Southwest Forest and Range Experiment Station.

Wissmar, R. C., J. E. Smith, B. A. McIntosh, H. W. Li, G. H. Reeves, and J. R. Sedell. 1994. *Ecological health of river basins in forested regions of eastern Washington and Oregon.* PNW-GTR-326. Portland, OR: USDA Forest Service, Pacific Northwest Research Station.

Wittbecker, A. 1995. Saving common places: The palouse. *Wild Earth* (Spring): 54–58.

Wolke, H. 1991. *Wilderness on the rocks.* Tucson, AZ: A Ned Ludd Book.

Woodward, S. L., and R. L. Hoffman. The nature of Virginia. Pages 23–48 in K. Terwilliger (editor), *Virginia's endangered species.* Blacksburg, VA: McDonald and Woodward.

World Wide Fund for Nature and International Union for Conservation of Nature. 1995. Hawaiian Islands. Pages 549–555 in *Centres of plant diversity*, volume 2. Oxford, England: Information Press.

Wright, H. A., and A. W. Bailey. 1980. *Fire ecology and prescribed burning in the Great Plains: A research review.* General Technical Report INT-77. Ogden, UT: USDA Forest Service Intermountain Forest and Range Experiment Station.

———. 1982. *Fire ecology: United States and southern Canada.* New York: John Wiley.

Wright, J. W., and L. J. Vitt (editors). 1993. *Biology of whiptail lizards (genus* Cnemidophorus*).* Norman, OK: Oklahoma Museum of Natural History.

Wuerthner, G. 1986. *Idaho mountain ranges.* Helena, MT: American Geographic Publishing.

Yahner, R. H. 1995. *Eastern deciduous forest: Ecology and wildlife conservation.* Minneapolis: University of Minnesota Press.

Yorks, T. P., N. E. West, and K. M. Capels. 1992. Vegetation differences in desert shrublands of western Utah's Pine Valley between 1933 and 1989. *Journal of Range Management* 45: 569–578.

Youngblood, A. P., and R. L. Mauk. 1985. *Coniferous forest habitat types of central and southern Utah.* INT-187. Ogden, UT: USDA Forest Service, Intermountain Research Station.

Zhu, Z. 1994. *Forest density mapping in the lower 48 states: A regression procedure.* SO-280. Asheville, NC: USDA Forest Service, Southern Forest Experiment Station.

Zinke, P. J. 1988. The redwood forest and associated north coast forests. Pages 679–698 in M. G. Barbour and J. Major (editors), *Terrestrial vegetation of California.* Special Publication No. 9. Davis: California Native Plant Society.

AUTHORS

Taylor H. Ricketts
Conservation Analyst
Conservation Science Program
World Wildlife Fund–United States
and
Staff Scientist
Center for Conservation Biology
Stanford University

Eric Dinerstein, Ph.D.
Chief Scientist and Director
Conservation Science Program
World Wildlife Fund–United States

David M. Olson, Ph.D.
Senior Scientist
Conservation Science Program
World Wildlife Fund–United States

Colby J. Loucks, M.E.M.
Conservation Analyst/GIS Specialist
Conservation Science Program
World Wildlife Fund–United States

William Eichbaum, L.L.B.
Vice President
U.S. Program
World Wildlife Fund–United States

Dominick DellaSala, Ph.D.
U.S. Program
World Wildlife Fund–United States

Kevin Kavanagh, M.Sc.
Director
Endangered Spaces Campaign
World Wildlife Fund Canada

Prashant Hedao, M.L.A.
GIS Specialist
Conservation Science Program
World Wildlife Fund–United States

Patrick Hurley
Research Assistant
Conservation Science Program
World Wildlife Fund–United States

Karen Carney
Research Assistant
Conservation Science Program
World Wildlife Fund–United States

Robin Abell, M.S.
Conservation Analyst
Conservation Science Program
World Wildlife Fund–United States

Steven Walters, M.S.
GIS Specialist
Conservation Science Program
World Wildlife Fund–United States

CONTRIBUTORS

Jonathan Adams
Pangolin Words, Inc.
10102 Haywood Circle
Silver Spring, MD 20902, USA

Will Allen
Conservation Fund
PO Box 374
Chapel Hill, NC 27514, USA

Sandy Andelman
The Nature Conservancy of Washington
217 Pine Street, Suite 1100
Seattle, WA 98101, USA

Alan G. Appleby
Endangered Spaces Campaign
World Wildlife Fund Canada
c/o 3079 Athol Street
Regina, Saskatchewan S4S 1Y6, Canada

James F. Bergan
The Nature Conservancy of Texas
PO Box 163
Collegeport, TX 77428-0163, USA

John Broadhead
Gowgaia Institute
PO Box 638
Queen Charlotte City, BC V0T 1S0, Canada

Dirk Bryant
World Resources Institute
1709 New York Avenue
Washington, DC 20006, USA

Steve Buttrick
The Nature Conservancy
Northeast Regional Office
201 Devonshire Street, 5th Floor
Boston, MA 02110, USA

Steve Chaplin
The Nature Conservancy
Midwest Regional Office
1313 5th Street Southeast
Minneapolis, MN 55414, USA

Roberta Clowater
New Brunswick Protected Natural Areas Coalition
180 St. John Street
Fredericton, NB E3B 4A9, Canada

Terry Cook
The Nature Conservancy of Texas
PO Box 5190
Fort Hood, TX 76544-0190, USA

Jim Cooperman
Shuswap Environmental Action Society
c/o RR #1, S10, C2
Chase, BC V0E 1M0, Canada

Robin Cox
The Nature Conservancy of California
California Regional Office
201 Mission Street, 4th Floor
San Francisco, CA 94105, USA

Lance Craighead
c/o American Wildlands
40 East Main, Suite 2
Bozeman, MT 59715, USA

David Culver
American University
Biology 8007, Hurst 6
4400 Massachusetts Avenue NW
Washington, DC 20016, USA

Mary Davis
PO Box 131
Georgetown, KY 40324, USA

Dennis Demarchi
Ministry of Environment, Lands and Parks
Wildlife Branch
780 Blanchard Street
Victoria, BC V8V 1X4, Canada

Jim Eidson
The Nature Conservancy of Texas
State Headquarters
PO Box 1440
San Antonio, TX 78295-8774, USA

Amy Farstad
Biota of North America Program
Coker Hall, CB #3280
The University of North Carolina at Chapel Hill
Chapel Hill, NC 27599-3280, USA

Thomas Fleischner
Prescott College
Environmental Studies
220 Grove Avenue
Prescott, AZ 86301, USA

R. Glenn Ford
Ecological Consulting, Inc.
2735 N.E. Weidler Street
Portland, OR 97232, USA

Dave Foreman
The Wildlands Project
PO Box 13768
Albuquerque, NM 87192, USA

Steve Gatewood
The Wildlands Project
1955 West Grant Road, #148
Tucson, AZ 85745-1147, USA

Jim Goltz
New Brunswick Federation of Naturalists
126 Wilsey Road, Suite 17
Fredericton, NB E3B 5J1, Canada

Sam Gon
The Nature Conservancy of Hawaii
Pacific Regional Office
1116 Smith Street, Suite 201
Honolulu, HI 96817, USA

Louise Gratton
591 Rosenberry
Sutton, PQ J0E 2K0, Canada

Tim Gray
The Wildlands League
401 Richmond Street West, Suite 380
Toronto, ON M5V 3A8, Canada

Anne Gunn
Department of Resources, Wildlife and Economics
    Development
Box 2668, Bretzlaff Drive
Yellowknife, NT X1A 2P9, Canada

Arlin Hackman
World Wildlife Fund Canada
90 Eglinton Avenue East, Suite 504
Toronto, ON M4P 2Z7, Canada

Randy Hagenstein
The Nature Conservancy of Alaska
421 West First Avenue, Suite 200
Anchorage, AK 99501, USA

Ron Heyer
National Museum of Natural History
Room W201, MRC162
10th and Constitution Avenue NW
Washington, DC 20560, USA

Bob Holland
3371 Ayres Holmes Road
Auburn, CA 95603, USA

Tony Iacobelli
World Wildlife Fund Canada
90 Eglinton Avenue E
Suite 504
Toronto, ON MYP 2Z7, Canada

Laura Jackson
Protected Areas Association of Newfoundland and
    Labrador
PO Box 1027, Station C
St. John's, NF A1C 5M5, Canada

John Kartesz
Biota of North America Program
Coker Hall, CB #3280
The University of North Carolina at Chapel Hill
Chapel Hill, NC 27599-3280, USA

James MacMahon
College of Science
Utah State University
Logan, UT 84322-4400, USA

Kate MacQuarrie
Island Nature Trust
PO Box 265
Charlottetown, PEI C1A 7K4, Canada

Geoff Mann
c/o World Wildlife Fund Canada
90 Eglinton Avenue East, Suite 504
Toronto, ON M4P 2Z7, Canada

Bill Meades
Canadian Forest Service
Great Lakes Forestry Centre
1219 Queen Street East
PO Box 490
Sault Ste. Marie, ON P6A 5M7, Canada

Rod Mondt
1955 West Grant Road, #148
Tucson, AZ 85745, USA

Janet Moore
Critical Wildlife Habitat Program
PO Box 24
200 Saulteaux Cres
Winnipeg, MB R3J 3W3, Canada

David Neaves
PO Box 3550
Amahim Lake, BC V0L 1C0, Canada

Jim Nelson
851 Wollaston Street, Suite 3
Victoria, BC V9A 5A9, Canada

Reed Noss
Department of Fisheries and Wildlife
104 Nash Hall
Oregon State University
Corvallis, OR 97331, USA

Chris O'Brien
c/o Ecology North
4807 49th Street, Suite 8
Yellowknife, NT X1A 3T5, Canada

Sebastian Oosenbrug
Department of Resources, Wildlife and Economic
    Development
Wildlife and Fisheries Division
600, 5102 50th Avenue
Yellowknife, NT X1A 3S8, Canada

Gordon Orians
University of Washington
Department of Zoology
University of Washington
Box 351800
Seattle, WA 98195-1800, USA

Juri Peepre
Canadian Parks and Wilderness Society
Yukon Chapter
30 Dawson Road
Whitehorse, YT Y1A 5T6, Canada

Ajith Perera
Ontario Forest Research Institute
1235 Queen Street East
Sault Ste. Marie, ON P6A 5N5, Canada

Robert Peters
Defenders of Wildlife
1101 14th Street NW, Suite 1400
Washington, DC 20005-5605, USA

Steve Primm
World Wildlife Fund–United States
1250 Twenty-fourth Street NW
Washington, DC 20037-1175, USA

Scott Robinson
Illinois Natural History Survey
607 East Peabody Drive
Champagne, IL 61820, USA

Jon-Paul Rodriquez
Department of Ecology and Evolutionary Biology
Princeton University
Eno Hall
Washington Road
Princeton, NJ 08544-1003, USA

John Sawyer
Department of Biological Sciences
Humboldt State University
Arcata, CA 95521, USA

Rick Schneider
Nebraska Natural Heritage Program
Game and Parks Commission
2200 North 33rd Street, PO Box 30370
Lincoln, NE 68503, USA

Jennifer Shay
University of Manitoba
Botany Department
Room 505, Buller Building
Winnipeg, MB R3T 2N2, Canada

Marni Sims
World Wildlife Fund Canada
90 Eglinton Avenue East, Suite 504
Toronto, ON M4P 2Z7, Canada

Phillip Sims
USDA Southern Plains Research Station
2000 18th Street
Woodward, OK 73801, USA

Fred E. Smeins
Texas A&M University
Department of Rangeland Ecology and Management
College Station, TX 77428, USA

George Smith
Canadian Parks and Wilderness Society, BC Chapter
c/o RR #4, S19, C43
Gibsons, BC V0N 1V0, Canada

Scott Smith
Yukon Land Resource Unit
Agriculture Canada
PO Box 2703
Whitehorse, YT Y1A 2C6, Canada

Randy Snodgrass
World Wildlife Fund–United States
1250 Twenty-fourth Street NW
Washington, DC 20037-1175, USA

Colin Stewart
Federation of Nova Scotia Naturalists
73 Chadwick Street
Dartmouth, NS B2Y 2M2, Canada

Jim Strittholt
Earth Design Consultants
800 Northwest Starker, Suite 31
Corvallis, OR 97330, USA

Emma Underwood
World Wildlife Fund–United States
1250 Twenty-fourth Street NW
Washington, DC 20037-1175, USA

Robyn Usher
Gaia Consultants
1035, 510–5th Street Southwest
Calgary, AB T2P 3S2, Canada

Alasdair Veitch
Government of Northwest Territories
PO Box 130

Norman Wells, NT X0E 0V0, Canada
Alan Weakley
The Nature Conservancy
Southeast Regional Office
101 Conner Drive, Suite 302
Chapel Hill, NC 27514, USA

Wesley W. Wettengel
World Wildlife Fund–United States
1250 Twenty-fourth Street NW
Washington, DC 20037-1175, USA

Gaile Whelan-Enns
Endangered Spaces Campaign
World Wildlife Fund Canada
c/o 63 Albert Street, Suite 411
Winnipeg, MB R3B 1G4, Canada

Chris Williams
World Wildlife Fund–United States
1250 Twenty-fourth Street NW
Washington, DC 20037-1175, USA

Don Wilson
Smithsonian Institution
NMNH, BDP, MRC 180
10th and Constitution NW
Washington, DC 20560, USA

Kim Wolfe
World Wildlife Fund–United States
1250 Twenty-fourth Street NW
Washington, DC 20037-1175, USA

Nathalie Zinger
World Wildlife Fund Canada
1253 McGill College Avenue, Suite 446
Montreal, PQ H3B 2Y5, Canada

INDEX

7877